WORLD RAILWAY ACCIDENTS 1901-1915

Francis Voisey

ISBN: 978-1-9162011-4-9

This book is dedicated to Signalman Brian Stone,
who first awakened my lifelong interest in railways way back in 1966.

ACKNOWLEDGMENTS

Thanks to the staff of the Exeter Reference Library and the University of Exeter Library for their assistance while I was researching this book, to my friend June Nightingale for encouraging me to go ahead and get it finished when my enthusiasm was waning a little, and to my publishing team, without whose guidance and assistance it is doubtful if I would have been able to navigate the mysteries of getting my first book into print.

CONTENTS

INTRODUCTION

This book is the third of what it is hoped will be a series detailing major railway accidents worldwide from the early days to the present. A good many books have been written over the years about railway accidents, some concentrating on single disasters, others covering accidents in a given country or a stated geographical area. Accidents in countries such as Great Britain, Germany, the Netherlands, the United States, Australia, New Zealand and soon have been documented comprehensively in one or more books, but no attempt has been made previously to detail every significant accident worldwide. This is clearly a very ambitious project, but two volumes dealing with the nineteenth century having been completed, this book will attempt to detail the major accidents which occurred during the first fifteen years of the twentieth century, between 1901 and 1915, with further volumes hopefully to follow dealing with the remainder of the twentieth century. As before an attempt will be made to include all accidents involving five or more fatalities to train passengers. A number of less serious accidents which have been considered of interest have also been included. The accidents have been grouped into chapters according to their primary cause and are dealt with as far as possible chronologically within each chapter.

The gathering of information about early twentieth century accidents in distant parts of the world has not been easy, though it is much easier nowadays than when this project began in the days before the internet. Outside of Great Britain and Ireland few countries seems to have had a formal body charged with investigating all serious railway accidents before late in the twentieth century. Reliance has largely had to be placed on contemporary newspaper reports and previously published works on the subject, though the author has tried as much as possible to keep to primary sources. If the book appears to be biased towards accidents which occurred in Great Britain and the United States this is probably partly because these are the countries about which information on accidents has been easiest to obtain, though these were also in any case the countries where many of the most serious accidents did occur.

Contemporary spellings of place names have been used where appropriate and for the many American accidents, terms such as railroad, engineer, caboose and so on have been used instead of railway, driver, brake van etc.

By 1901 railways were to be found almost everywhere, though some countries only had very small systems or perhaps just one line.

The names and the boundaries of countries have undergone many changes over the years. German and Italian unification had been completed by 1881, but Eastern European countries for example such as the Czech Republic, Slovenia, Slovakia, Estonia and so on did not exist before 1919 or later. Australia became a single country in 1901 but South Africa not until 1910, whilst India up until 1948 included the present countries of Pakistan and Bangladesh. Newfoundland did not become part of the confederation of Canada until 1949. The names of some other countries and of numerous places have changed. The names of locations which were in use at the time an accident occurred have been given, together with present names where appropriate, as for example Ürsküb (now Skopje).

Since the writing of the first two volumes of this series information on a number of additional accidents has come to light, mostly ones in the United States before and during the American Civil War plus various mishaps in India from 1870 onwards. For the benefit of readers of the earlier two volumes and to provide a more complete record these accidents are briefly detailed below. They have been listed in chronological order, regardless of type.

2 August 1853 On the Belvidere Delaware Railroad train of ten platform cars, carrying about 200 labourers who had been working on the railroad and on repairs to the Delaware Division of the Pennsylvania Canal, was being propelled by a locomotive at eight or nine miles an hour near Bulls Island station, New Jersey, United States, when a cow jumped onto the track in front of it. The leading car was thrown from the track and five more cars followed before the engineer could stop his train. Some of the cars were overturned and others thrown upon one end. Ten labourers were killed outright, and an eleventh man died of his injuries.[1]

5 September 1856 During the evening a cattle train with a passenger car attached arrived at Port Byron, New York, United States, on the New York Central Railroad, and came to a stand on the main track. The switch tender then fell asleep, having omitted to change the points. As a result the following *Cincinnati Express* was not diverted into the side track when it arrived at 10 p.m. and instead collided directly with the aforementioned passenger car. The express was only travelling at six miles an hour but the smoke-pipe and safety-valve of its engine, the *J.H. Chedell*, were broken off, searing steam filled the car and the six drovers occupying it were scalded to death. The coroner's jury duly found Switch Tender Thomas Fogarty responsible for the accident. He presumably subsequently stood trial for manslaughter, but it has not proved possible to find any record of this.

7 May 1857 The westbound night express of the Michigan Southern Railroad ran into a cow near Riga, Michigan, United States. The tender of the engine was derailed, the first passenger car was thrown crossways over the track and the second passenger car rammed the rear of the first and pushed it into the leading baggage car. Five persons in the first passenger car were killed instantly and several were seriously injured.[2]

31 May 1863 On this date an extra train of box cars, carrying between 500 and 600 negroes 'belonging' to Confederate General Wade Hampton and his brother was en route

[1] *Weekly National Intelligencer*, Washington, D.C., United States, 6th August 1853

[2] *Augusta Weekly Chronicle and Sentinel, Augusta*, Georgia, United States, 20th May 1857 (taken from Chicago Journal)

from Mississippi to Columbia, South Carolina via the 'Augusta Railroad.' Near Bamberg, South Carolina, United States, several of the box cars became detached from the train and came to a stand on the open line. It was half an hour after midnight and the occupants of the box cars were presumably asleep and did not know what had happened. When the engineer of a following passenger train became aware of the obstruction he did everything possible to pull up, but a collision was unavoidable. Two box cars were wrecked and eleven of the slaves were killed and three others probably fatally injured. The only casualty on the passenger train was the coloured fireman, who slightly sprained his leg in jumping from the engine.[3]

6 June 1863 The boiler of a locomotive attached to a passenger train of the Kentucky Central Railroad, which was due to leave Nicholasville, Kentucky, United States, for Covington, exploded, killing and injuring a number of soldiers of a Massachusetts regiment. Three men were killed outright and two of the twelve or so injured had reportedly died by the time of the initial report. When the explosion occurred Engineer Frank Lockwood was away at dinner and Fireman George Staller and Conductor James Taylor were also absent from the engine. Equally lucky was the telegraph operator, who was not in his office, which was badly damaged. A second report stated that four soldiers had lost their lives, together with a coloured man employed about the depot and a teamster. There was some talk among the surviving soldiers of lynching the engineer.[4]

14 September 1863 Very brief reports appeared in Southern newspapers of an accident on the 'State Road' of Georgia (the Georgia Railroad and Banking Company) near Ackworth, Georgia, United States. In what was probably a 'cornfield meet' an up troop train and a down freight train were in collision. A second report on the accident stated that 12 people had been killed and about 60 wounded. Both trains were 'running out of schedule time.'[5]

25 September 1863 At 6.30 p.m. a westbound passenger train consisting of a locomotive and twenty coaches left Kalamazoo, Michigan. Between Mattawan and Lawton, Michigan, United States, some 'drunken rowdies,' as they were described, cut the bell rope, and pulled out the link between the cars, dividing the train. The line hereabouts being on a falling grade the engineer did not discover what had happened until he stopped at Lawton. Two minutes later he saw the rear part of the train approaching and applied steam to get the front of the train out of the way. This sudden application of power caused the train to again become divided behind the engine and a collision occurred between the two sets of cars. Five people were killed and 22 injured, some so seriously that they were not expected to recover.[6]

11 October 1863 On the morning of the day of the accident a northbound Confederate troop train left Atlanta, Georgia, United States, carrying men of the 1st Tennessee Battalion and 15th Tennessee Regiment. A few miles beyond Marietta it ran off the track. No one was hurt but it took most of the day to get it back on the road, when it set off again, much behind time. In the opposite direction a southbound train of about 15 cars,

---

[3] *Atlanta Southern Confederacy*, Atlanta, Georgia, United States, 2nd June 1863 (taken from Charleston Courier, 1st June)

[4] Cincinnati Daily Commercial, Cincinnati, Ohio, United States, 8th June 1863

[5] *Tri Weekly Courier*, Rome, Georgia, United States, 15th September 1863

[6] *Sandusky Daily Commercial Register*, Sandusky, Ohio, United States, 28th September 1863

carrying only a handful of sick soldiers, was heading for Atlanta. The Adairsville operator telegraphed to Marietta to find if the track was clear and the Marietta operator, unaware of the delay to the northbound train, replied that it was. The Adairsville operator duly permitted the southbound train to proceed and the two trains met on a curve near Etowah, Georgia, both going at full speed. Eighteen soldiers died in the wreck and 67 were injured.[7]

5 March 1864 The gross carelessness of an engineer was stated to have caused a rear collision at Anderson, Tennessee, United States. The first of two northbound freight trains, each of which included one passenger car, so they were really mixed trains, stopped for water at Anderson. The second train appeared around a curve travelling at full speed and despite frantic signalling from the crew of the first train made little attempt to pull up before it had run into the caboose in front. Smashed lamps and an overturned stove then set the wreckage ablaze. Mrs Keys, Mrs Vogel, Mrs Jordan and George Thriatt of the 4th Minnesota Regiment died in the burning car, a Miss Pichett soon succumbed to her injuries and others were supposedly fatally hurt. The dead women were all from Huntsville. The engineer responsible fled the scene.[8]

1867 A number of Australian newspapers in May and June reported upon an accident near Nikitino station, Russia, 'on the Moscow line.' A passenger train reportedly ran off the rails on a curve and eight carriages fell down a 50ft high embankment. Thirty persons were killed and over 100 seriously injured. One report added that peasantry stripped the dead of any valuables.[9] Unfortunately none of the reports provides a date for the accident. Given that there was no telegraphic communication with Australia at this date it must at any rate have occurred some weeks earlier and could be the same one as that which happened between Koslow and Riäsan on 20th February.

20 October 1870 A bridge over the Cheyair River, India, 'on the road to Cuddapah,' on the Madras Railway was washed away, and a train was precipitated into the water. Five railwaymen and three passengers lost their lives, the passengers being named as the Rev and Mrs Seymour and a Mr Charry. The foundations of the bridge, 'which rested on treacherous ground, had been undermined by scour.' It was not clear if the structure had collapsed before the train reached it or while the train was crossing it. A commission of experienced officers was appointed to investigate the condition of all similar bridges along the line.[10]

14 August 1875 The official report on the Jubbulpore accident stated that it occurred on a branch of the East Indian Railway, not the Great Indian Peninsula, that it happened three miles from Katnee, and was supposed to have been caused by the engine running over its own cowcatcher, which had worked loose and dropped onto the track. This report added that the train had been travelling at high speed, the carriages had overridden one another and been wrecked, and 9 passengers had lost their lives and 13 been injured.[11]

25 December 1876 The mail train and a special goods train should have crossed one another at 2 a.m. at Amboor station, India, on the Madras Railway. Forgetting about the

---

[7] *New Albany Daily Ledger*, New Albany, Indiana, United States, 14th October 1863

[8] *The Age*, Philadelphia, Pennsylvania, United States, 11th March 1864

[9] *The Perth Gazette and Western Australian Times*, 24th May 1867

[10] Report to the Secretary of State for India in Council on Railways in India for the year 1870-71

[11] Ibid 1875-76

crossing and receiving no signal to stop from the station staff the driver of the mail train continued and the two trains collided on the single line two miles further on. One composite and two third-class carriages were smashed, together with three waggons. Ten passengers were killed on the spot and 12 lives were lost in all. The mail train driver and the Amboor stationmaster and assistant stationmaster were punished by fines and imprisonment.[12]

25 September 1877 An Illinois Central Railroad passenger train was thrown from the track at 3 a.m. at Sanduval, Illinois, United States, killing seven passengers and injuring a number of others. Masked men were seen from the carriage windows in the bright moonlight, but they then disappeared. It was supposed that the derailment had resulted from an act of sabotage, the saboteurs intending to rob the train passengers, but that seeing the magnitude of the derailment they had become frightened and had made their getaway.[13]

July 1879 The flooding of a small stream carried away the embankment of the Rajputana Railway near Bandiqui station, India, at an unspecified date in July. A passenger train fell into the gap and eight 'native' passengers and the guard were killed.[14]

31 October 1879 The official report stated that 22 passengers died and 46 were injured in the accident near Arconum. Many of the passengers were, it stated, members of a detachment of the Queen's Own Sappers and Miners, en route to Burma.[15]

26 May 1881 No. 16 up local train on the Sind, Punjab and Delhi Railway, travelling from Jullundur Cantonment to Lahore, was derailed 4 miles from Amritsar, India, due to the breakage of the rear axle of waggon No. 1520. Several vehicles towards the rear of the train were derailed but the dead and injured passengers were all riding in third-class carriage No. 253, which was overturned and smashed to pieces. Four passengers were killed outright and 19 injured, two of them it was feared fatally. It was decided upon inspection that the flaw in the axle which had failed could not have been detected before the derailment and that therefore no one was to blame.[16]

17 January 1882 The up mail train, running under a caution message, overtook a goods train just over a mile from Shikohabad station, India, on the East Indian Railway. Thick fog prevented the driver of the mail train from seeing the train in front. Thirteen 'native' passengers lost their lives and 15 were injured. The European guard of the goods train jumped from his van just before the collision and was seriously injured. The mail train driver subsequently stood trial for causing the accident but in the circumstances was acquitted by the High Court in Allahabad and the railway company belatedly recognised the need for a modification of the rules permitting trains to follow one another in foggy weather.[17]

16 December 1884 A horse and buggy carrying members of the Rangiora Presbyterian Church choir was struck on a level crossing by a train from Christchurch, New Zealand, near that place. David Kerr, the son of a builder, was killed outright and two of his sons,

---

[12] Ibid 1876 77

[13] *Richmond Daily Despatch*, Richmond, Virginia, United States, 27th September 1877

[14] Report to the Secretary of State for India in Council on Railways in India for the year 1879-80

[15] Ibid

[16] Administration Report on the Railways in India for the year 1881-82

[17] Report to the Secretary of State for India in Council on Railways in India for the year 1882-83

Maggie Marrall and Donald M'Kay were fatally injured. Seven others were badly hurt.[18]

15 April 1889 An up mixed train on the Eastern Bengal State Railway was caught by a tornado at bridge 23, just north of Madhnagar station, India. The train consisted of an engine and tender, 11 coaches, 19 waggons and a van. Of these all except the engine and tender, a third-class carriage and the rear van were derailed, ten vehicles falling into a stream and 19 remaining on the embankment. Seven passengers died, 80 were hurt and there was much damage to rolling stock, though the track and bridge escaped unscathed.[19]

27 October 1889 On 27th, not 28th October as previously detailed, a down mail train leaving Hathras Road station on the East Indian Railway was wrongly turned into a dead-end siding, rammed the buffers and fell down a 4ft high embankment. The women's carriage behind the leading brake van was broken up, 17 passengers died and 29 were injured. The pointsman responsible was sentenced to a year's rigorous imprisonment.[20]

5 December 1891 The accident occurred between Okara and Satghara stations, India, and involved No. 1 up and No. 8 down mail trains. The assistant stationmaster at Satghara permitted train No. 1 to come from Okara after the line clear which had been previously been received from Okara had already been delivered to the driver and guard of the down train. At the moment that he was telegraphing to Okara to accept the up mail the down mail left his station. According to the official report 41 passengers and two railway servants lost their lives and 28 were injured. The Montgomery District Magistrate sentenced the assistant stationmaster to a months' rigorous imprisonment and a fine of 100 rupees.[21]

12 May 1892 The engine of a ballast train on the Gonda-Bahramghat branch of the metre gauge Bengal and North Western Railway ran into a bullock at mile 8 between Maijapur and Gonda Kacheri, India. Sixteen open trucks were thrown off the track and down an embankment and eight coolies were killed and seven injured.[22]

4 April 1895 On the Oudh and Rohilkhand Railway No. 151 up mela passenger train collided with No. 46 down goods train on the single line between Jaunpur City and Mihrawan stations. Seven passengers died, nine were injured, and there was considerable damage to rolling stock. The accident was found to have been the fault of the Mihrawan assistant stationmaster, who was prosecuted, convicted, and sentenced to two years' rigorous imprisonment. (A mela is a Hindu festival).[23]

14 March 1896 An unusual accident occurred on the Bahramghat branch of the Oudh and Rohilkhand Railway. Through carelessness on the part of the line jamadar No. 23 up standard gauge mixed train was diverted onto the metre gauge line. The engine tender and six waggons were derailed, four overturned, the derailed vehicles were more or less damaged, and the branch was blocked for some time. It does not seem that anyone was hurt though, and the two years' rigorous imprisonment imposed on the man at fault seems

---

[18] *Newcastle Morning Herald and Miner's Advocate*, Newcastle, New South Wales, 18th December 1884

[19] Administration Report on the Railways in India for the year 1889-90

[20] Ibid

[21] Ibid 1891-92

[22] Ibid 1892-93

[23] Ibid 1895-96

a harsh punishment.[24]

28 June 1897 Due to a defect in a slip coupling a slip coach collided with the main part of the train at Agarpara, India, on the Eastern Bengal State Railway. Nine passengers were slightly injured, and a slip carriage was damaged.[25] This is the only slip coach accident that the author has come across outside of England.

19 September 1897 A report from Vienna stated that a passenger express and a freight train carrying troops had collided near Kafosvara, Austria. Four train hands and six soldiers were killed, and another 30 soldiers were believed to have been fatally injured.[26]

23 September 1897 The railway company is stated in the official report to have been the Southern Mahratta rather than the Mysore State as stated in the newspaper report. The location of the accident is given as having been between Channapatna and Mudgeri stations and the death toll given in *The Times* was apparently much exaggerated. The official report states that 16 passengers and four railwaymen died, and the same number of passengers and two railway servants were hurt.[27]

[24] Ibid 1896-97

[25] Ibid 1897-98

[26] *Roanoke Times*, Roanoke, Virginia, United States, 22nd September 1897

[27] Administration Report on the Railways in India for the year 1897-98

CHAPTER 1

DERAILMENTS: TRACK DEFECTS

The tracks over which trains run are obviously a vital part of the railway and any problems with the track are likely to lead to the train being derailed. By 1901 the technology involved in laying railway track had much advanced since the early days, but train speeds and weights had also much increased, while accidents occurred from time to time due to failures to keep the track in a good state of maintenance, to insufficient allowance being made for hot weather expanding the rails, to damage caused to the track from floodwaters and so on. Also included in this chapter, though they were not strictly due to defects in the track, are accidents due to trains having become derailed on open points. Broken rails were another hazard which caused mishaps and worse from time to time.

Nine soldiers, eight of them new recruits on their way to join the Tenth Infantry regiment in San Francisco and bound from there to the Philippines, where the United States was at war with Filipino nationalists, were among those involved in the derailment of Erie Railroad passenger train No. 5, the *New York – Chicago Limited*, just on the Amasa side of Greenville, Pennsylvania, United States, at 7.10 a.m. on Thursday 7th February 1901. At least three of the soldiers were killed, including Sergeant Harry A. Hart, who was in charge of the squad. Another of the soldiers was badly hurt, the other five escaped. Initially it was believed that 15 people had lost their lives in all, but the final number of dead was probably only five, with about 14 others injured. The train had been in charge of Conductor Sam Randall, of Meadville, Pennsylvania, with Engineer Brett Luce and Fireman George Eckert. It had been about two hours late and was running at high speed when it left the track on a sharp curve, with the Shenango River 40 feet below on one side of the track and a steep bluff rising on the other. The engine was the first to leave the track and after running a short distance it overturned and ended up half buried in the hillside. Engineer Luce and Fireman Eckert jumped and escaped with painful bruises. The train was stated to have consisted entirely of vestibuled Pullman cars, a mail car, a combination baggage and smoking car, a day coach and three sleepers, drawn by a new Atlantic class locomotive. It was in the combination car that all the deaths and injuries occurred when it was almost completely telescoped by the steel mail car ahead of it, which 'went through it as if it were paper.' The derailment was attributed to either a broken rail or a broken wheel flange.[28]

As soon as news of the accident reached Meadville a wrecking train of two coaches and a baggage car, equipped with 15 cots and a quantity of blankets, was despatched to the scene. Four surgeons accompanied the train, together with Superintendent Belnap and Trainmaster Allen. It took several hours to release all the injured from the wreckage, they were then taken on the train to the Spencer hospital in Meadville.[29]

The initial reports of the derailment of a passenger train near Lofa (or Lo Fa) in China on Sunday 28th April 1901 attributed it to the collapse of a culvert on the Peking-Tientsin railway 'on the section of the line repaired by the Germans.' It was stated that 11 Chinese had been killed and 40 other passengers had been injured, including two Americans, one of whom could not recover.[30]

A follow-up report blamed the accident on an accumulation of sand on the track 30 feet from a bridge, which had caused the engine and four coaches to capsize. The number of dead was now stated to be 19, all of them Chinese, with another 23 injured, while an American private called Kennedy had been slightly hurt. It was expected that traffic over the line would be resumed late on the 29th.[31]

The washing away of a culvert by floodwaters following an exceptionally severe local storm led to a major disaster near Cass, Indiana, United States, half an hour after midnight on Wednesday 26th June 1901. Involved in the accident about five miles east of Logansport was Wabash Railroad passenger train No. 3, a westbound limited vestibule express which was stated to have been one of the fastest and handsomest trains on the system. When it left Peru, Indiana, a few miles east of Cass, bound for St Louis, it consisted of two portions, one from Toledo, Ohio, and the other from Buffalo, New York. The train was formed of engine No. 182, a baggage and express car, an emigrant car, a combination baggage and smoking car, three chair cars, three sleeping cars and the private car of Superintendent W. M. Cotter of the St Louis Iron Mountain and Southern Railway, and was in charge of Conductor Brownlee, with Engineer John Butler and Fireman John Adams.

When the express left Peru it was running more than an hour late and as it approached Cass it was running at about 50 miles an hour. Engineer Butler had just eased off a little prior to negotiating a curve at Cass when without warning there was a jolt and his big engine shot off the line into a ditch, taking the following cars with it. The locomotive buried itself in the soft soil and the remainder of the train, except the rear Pullman car and the private car at its rear, came down on top of it. For a moment after the wreck there was silence. 'Not a shriek was heard from the muddle of broken cars and maimed people at the bottom of the ravine,' a newspaper reporter wrote, and he continued, 'Then out of the

[28] *Hamilton Daily Republican*, Hamilton, Ohio, United States, 7th February 1901

[29] *New York Times*, New York, United States, 8th February 1901

[30] *The Times*, London, England, c. 29th April 1901

[31] Ibid, c. 30th April 1901

blur of red lights, smoke and steam came the first groans of the injured.'

Some of the less badly injured and uninjured passengers made their way across fields to alert nearby farmers. 'Soon lanterns were seen dotting the darkness in all directions as the first of the rescuers hurried to the wreck.' By the light available from the torches and lanterns some of the injured were removed from the wreckage and laid on the side of the embankment. Meanwhile the Cass operator was busy telegraphing all the neighbouring towns for assistance. Within a short time, every physician in Peru had reported to the railroad station there, ready to board a relief train to the scene, while further help came from Lafayette and Danville, Illinois.[32]

When dawn broke it could be seen that the first eight vehicles of the train had piled up on the engine. The rear Pullman car was left straddling the culvert, with one truck on each abutment, while Superintendent Cotter's car remained on the rails. Engineer Butler and Fireman Adams survived the immediate wreck, though the fireman was severely injured and the engineer less badly hurt. Due to the speed of the train their engine had shot across the gap where the culvert had given way to bury itself in the soft earth on the far side. Its momentum exhausted, it had then fallen back into the gap. At the bottom of the pile of wrecked coaches was an emigrant car in which had been riding a group of Italian immigrants. This car was ground into the soft earth by the following cars falling on top of it and was the place where all the fatalities occurred. Twelve of the Italian immigrants were killed outright and two or three others it was feared fatally injured. A thirteenth death was that of the two-month-old daughter of Frank Parks, a Finn.[33] About 30 other passengers were injured. The Italians had all been on their way to Colorado. As their interpreter was one of the dead their names could not be immediately obtained though the names of all, but one woman, were determined subsequently. Conductor Brownlee was fortunate enough to have just left the emigrant car when the wreck happened. He escaped unhurt.

A newspaper reporter interviewed one of the Pullman porters, named Henry Boyd, who had been riding in the rear sleeping car. Boyd related, "The cries of the injured were frightful. I climbed down among the broken cars, and began to pull the passengers out, fearing the wreck would catch fire. I called on uninjured passengers; but they were dazed and gave little assistance. P. W. White, porter for Carter's special car, assisted me heroically. He risked his life in creeping under the debris and we carried out the dead and the dying. Many begged us to knock them in the head and end their sufferings. Some of the helpless people were pinned fast under timbers and we helped them out. The most of the dead were in the combination baggage and smoker car and in the chair car behind, occupied by Italian emigrants. Many children and women were crushed into shapeless form and others were frightfully mangled. We carried out twelve dead. We waited for over an hour for assistance from Peru and Logansport. It was dark and nobody knew where to find farm houses for help. We could do nothing to relieve the sufferings of the victims. It was frightful to hear the shrieks of agony and their piteous pleadings to be killed. My car stood directly over the washout, one end on either bank, and how it escaped going down is more than I can explain." Porter Boyd added that the train was so late because of a near escape from an accident several hundred miles to the east. A large tree had fallen across the track and but for a warning from a railroad employee the

[32] *The Fort Wayne Journal-Gazette*, Fort Wayne, Indiana, United States, 27th June 1901
[33] *The Logansport Pharos*, Logansport, Indiana, United States, 27th June 1901

express would have run into it. It was just stopped in time but then had to wait while the obstruction was removed.

Another survivor, passenger D.E. Collins of Toledo, Ohio, walked to Logansport with news of the disaster. He stated, "In the chair car I saw two women and a boy. They were not injured. Seven or eight people were lying on the bank beside the train motionless as I left the scene of the disaster. I suppose they were dead. The smoker was telescoped, and two sleepers and the dining-car were off the track. I heard cries and groans from every part of the wreck."

As had happened before and was to happen again not everyone was a hero; some decided to use the disaster to their advantage. Thus, local farm boys David Tucker and Ernest Zimmerman were arrested at the scene charged with robbing the dead and injured passengers of money and valuables. 'Two policemen were at the wreck. But for their determined attitude the prisoners would have been roughly handled by an indignant crowd.'

Wabash Railroad Superintendent Gould stated, "I learn from a person living near Cass that two black clouds gathered and came in contact about over the place where the express was wrecked. The culvert which washed out was a stone box three feet square, and under a twenty-five foot fill. The water was about five feet above the opening of the culvert and about 200 feet wide, causing the earth above to slide. We have never observed high water at this point before." [34]

Because one or two of the victims of the accident had died in Peru hospital, in Miami county, whereas the disaster had occurred in Cass county, two separate inquests were held. That held in Peru county concluded first, with Coroner Yarling coming to the conclusion that the victims had come to their deaths by reason of a cloudburst having washed out the track, and that the Wabash company had used ordinary care with regard to the culvert and track and was not to be held responsible for the accident.[35]

The Cass county inquest, held by Coroner Charles Smith, came to the directly opposite conclusion. He ruled that 'the deceased came to their deaths by reason of the wreck of a passenger train on the Wabash railroad, caused by the giving way of the track on the night of June 25, 1901.

'I further find that at the point where the wreck occurred there was and had been for many years prior to said wreck, a stone box or culvert, three feet by three feet five inches in capacity, being used by said railway company for the purpose of carrying the water which accumulated to the north of said railway track, under the railroad right-of-way. That during ordinary rainfalls, large quantities of water came down from the hills and surrounding country to the north, and that said stone box or culvert was insufficient in capacity to carry the water which naturally flowed in that direction during heavy rainfalls, and that the railroad company might have known of the insufficiency of said culvert had they used due care in its inspection.

'I find also that the giving way of the railroad embankment under the tracks was caused by reason of the water backing up to the north of the grade, occasioned by the insufficient capacity of the culvert above mentioned, and that the negligence of said

[34] *The Indianapolis News*, Indianapolis, Indiana, United States, 26th June 1901

[35] *Logansport Daily Journal*, Logansport, Indiana, United States, 20th July 1901

railway company in failing to provide a culvert of sufficient capacity and endurance, was the direct cause of said wreck, and of the persons killed thereto.' [36]

The Wabash management were doubtless less than happy with this coroner.

At some point, probably soon after the disaster, Cass station seems to have been renamed Danes, presumably after W. S. Danes, the president of the Wabash Railroad at the time.

As was often the case the first reports of the casualties in an accident on the Canadian Pacific Railway east of Ingolf, Ontario, Canada, on 14th August 1901, were grossly exaggerated, with rumours that 150 men had died, and many others had been injured. In truth only two men lost their lives and four more, one of them the fireman of the train, were injured. The train involved was a westbound extra, the first section of two or more trains carrying men from the Maritime Provinces (the eastern provinces of New Brunswick, Nova Scotia and Prince Edward Island) to Manitoba to help with the harvesting of grain, which was a very labour intensive job in the years before the process was mechanised; the trains which carried the men being known as harvester's specials. The engine and five coaches of the special involved in the accident left the track due to a broken rail. Those who lost their lives were Dan White, of Kent City, New Brunswick; and Donald McKegan, of Cape Breton, Nova Scotia. The injured fireman was named Thomas Corwin.

Tourist passenger train No. 9 on the Galveston Harrisburg and San Antonio Railroad was wrecked at 2.35 a.m. on the morning of Friday 7th March 1902, on a curve 2½ miles from Maxon station, Texas, United States. The train had left San Antonio at 9.15 p.m. the previous evening, bound for California. It included of one tourist car each from Cincinnati, Ohio, Washington D.C. and St Paul, Minnesota; a Pullman car from New Orleans and two ordinary day coaches. The train was stated to have been running at its normal speed when it was completely derailed except for one private car, and the wreckage immediately caught fire. The blaze was seen by the station agent at Tabor, seven miles distant, and he immediately set off for the scene as fast as he could go on a hand car. 'On reaching the scene a wild scene greeted him. Passengers who were able to get out were running hither and thither, frantic in their efforts to rescue those who were in the burning cars. It is thought that few passengers in the two day coaches were saved as both cars were piled up on top of the baggage and express cars and were the first to catch fire. The Pullman and tourist sleeping car passengers were rescued from the flames and if reports are true, none were killed in those cars, although several were injured.' [37]

By the following day it was stated at Houston, Texas, that there had been 12 deaths, including Engineer W. Price and Fireman H. Bertsch. Another report from San Francisco, California, announced that 'Two day and mail cars, one coach, one chair car, three tourist sleepers and one first-class sleepers were completely destroyed by fire' and listed the names of the engineer and fireman, a news agent and 11 passengers as 'missing', implying that their bodies had been consumed in the fire, which was probably the result

[36] *The Logansport Times*, Logansport, Indiana, United States, 26th July 1901

[37] *The Daily Californian*, Bakersfield, California, United States, 7th March 1902

of the wooden coaches being set alight by burning engine coals. About 25 other passengers were injured. A private car owned by Thomas Ryan of New York and carrying him, and his family, was marshalled at the rear of the train. It could be pulled clear before the fire reached it and none of its occupants were hurt.[38]

Conductor Stockwell was certain that the train had only been running at its ordinary speed and was of the opinion that an obstruction on the line, rather than a broken rail, had caused the derailment.[39] Some of the passengers thought that the train had been travelling too fast.[40]

Four people were killed outright and 23 were injured, at least one of whom it was feared would die, according to an early report of the derailment of the *California Limited*, the most important passenger train of the Atchison Topeka and Santa Fe Railroad, near Cama switch, five miles west of Medill, Missouri, United States, on 29th April 1902. Other reports the next day spoke of five deaths. Those who were killed outright were Isaac Weirthmer, of San Diego, California; the young son of Henry C. Gates, of Australia; and Rufus Buster and C.C. Fairbanks, both of Chicago, Illinois, and both waiters in the dining car. The train was one of the fastest in the country. It had left Fort Madison, Iowa, an hour late and was travelling at 65 miles an hour as it passed through Wyaconda, just before Cama switch. Beyond Wyaconda there was a falling gradient and experienced railwaymen put the train's speed at 75 miles an hour at Cama. The line had a succession of curves hereabouts, and a mile or so on the Medill side of Cama, on one of the curves, the train left the track. The engine kept to the rails but the coaches behind were scattered about the track. Behind the engine were a mail car, a combined baggage and smoking car, a dining car and seven Pullman cars. The baggage and dining cars were the most badly damaged and it was in these that most of the casualties occurred. Thanks to a quick brake application by Engineer Marvin M. Ball the rear Pullman cars, although derailed, did not go down the embankment. Neither the engineer nor his fireman, Dan Morgan, was hurt. Conductor Charles Sargent was very badly bruised.

One of the brakemen of the train walked to Medill while a second made his way to the Hayes farmhouse, from where he was able to telephone the first news of the accident to Fort Madison. Another person who showed his devotion to duty was Mail Clerk Rankin, who was badly bruised but refused to go to the company hospital at Fort Madison until someone had relieved him to take charge of the registered pouches in his van. A relief train from Fort Madison was followed by a wrecking train and physicians from Wyaconda, Medill, Kahoka and other points attended the injured.

Both Isaac Weirthmer and the Gates family had been on their way from Australia; Weirthmer to visit relatives in Germany and the Gates family to educate their children in England – involving long and one would have thought expensive journeys across the Pacific Ocean, the United States (probably San Francisco to New York) and then the Atlantic Ocean. Mrs Gates was prostrated by the loss of her son. Other passengers on the train included Charles Markell, the Australian representative of the great Parke, Davis & Co drug company, S. Tenner French, long-time manager of the White Star Line shipping

[38] *The Tri-County Gazette*, Mentone, Indiana, 13th March 1902

[39] *Santa Fe New Mexican*, Santa Fe, New Mexico, United States, 8th March 1902

[40] *Oakland Tribune*, Oakland, California, United States, 8th March 1902

company and celebrated baritone singer Max Heinrich of the *Gran Opera* company.[41]

On 3rd May a coroner's jury at Kahoka acquitted the trainmen of all blame for the derailment, attributing it to a defective switch and defective track. Excessive speed on the curve does not seem to have been considered as a cause.[42]

A train of Belgian pilgrims, who were on a pilgrimage to Lourdes via Paris, was involved in an accident at Moyenneville, France, on the Chemins de fer du Nord line between Amiens and Compiègne, on 6th May 1902. The train was known as the 'brown train' because each colony of pilgrims was distinguished by the colours of certain badges worn by the accompanying nuns and priests. The 350 passengers, who were mostly women and children, had been singing hymns when the leading carriage became derailed, followed by several others. The engine kept the track and unaware of the derailment the driver kept steaming for some distance. Because the accident happened in open country it was some time before help arrived. Six women, a man and a 12-year-old girl lost their lives and about 40 passengers were injured. Due to a lack of medical facilities many of the wounds were dressed with straw. Heavy rain had caused the track to sink and the rails to spread.[43 44]

Track defects were possibly the cause of a derailment of a Southern Railroad excursion train on a high embankment near Berry, Alabama, United States at 9.30 a.m. on the morning of 1st September 1902, though here again the cause of the derailment had not been definitely established by the time of the last newspaper report on the disaster. The train was running at 30 miles an hour and had just entered a curve on a 60-foot-high embankment when the tender of the engine left the rails without warning, dragging the engine and the leading four cars of the ten-car train after it. A coupling snapped behind the fourth car and the remaining six cars remained on the rails, but the engine and leading cars fell down the embankment. 'The cars, which were packed with passengers, turned completely over several times and were crushed like egg shells, killing and crippling the inhabitants. Persons who have returned from the scene of the wreck say it is indescribable.'

As might be expected the death toll was high. A report from Birmingham, Alabama, spoke of 30 persons being killed outright and 81 injured, physicians considering that no less than 29 of those injured could not survive. The names of those killed who could be identified were listed in the paper; the whites being listed first and the blacks afterwards. Only three whites lost their lives, Engineer J. W. Crook, Trainmaster H. M. Dudley and a Roscoe Shelby, of Columbus, Mississippi, possibly the conductor of the train or else a passenger. The identifiable black passengers who died were named as Jim Smith, of Isola, Mississippi; Ezel Patterson, of West Point, Mississippi; and Lillie Martin and child, Charly Carp, Ed Clark, Nute Green and Willie Thompson, all of Columbus, Mississippi. Another 19 bodies could not be given names. John McRevey, the negro fireman of the

[41] *The Evening Gazette*, Burlington, Iowa, United States, 30th April 1902
[42] *The Durango Democrat*, Durango, Colorado, United States, 4th May 1902
[43] *The Times*, London, England, 7th and 8th May 1902
[44] *Lloyds Weekly Newspaper*, London, England, 11th May 1902

train, was one of those thought to be fatally injured.[45]

In 1901, at the request of local businessmen, the French Chemins de fer du Nord company inaugurated a new fast morning train from Lille to Laon to provide a better service for the eastern districts of the country. On about 17th September 1902 this train, express No. 2828, had a narrow escape at Douai station, when it collided with some waggons. Ten days later, on 27th September 1902, it suffered a major disaster at Arleux, France. On the day of the accident the train consisted of a Nord company locomotive and 12 Chemin de fer de l'Est company vehicles; two vans and ten first, second and third class carriages, occupied by 95 passengers. It had left Lille at 8.11 a.m. and consisted of two portions which were to be separated at Busigny; one portion going forward to Reims via Laon and the other for Nancy and Dijon via Charleville-Mezieres.

The railway from Douai to Cambrai, through Arleux, was single track with passing places and there were three different sets of points giving access to different sidings on the approach to Arleux station. With a view to improving the loading of beet traffic the goods facilities at this station were being reorganised and in connection with this work the points had been disconnected from the signal box and were being worked locally by levers on the spot. At the same time the Bourret locking between the points and the signals had been disconnected so that the signal protecting the points could be cleared regardless of their position. Porter Moreau, the railwayman stationed at the facing points concerned in the accident, declared afterwards that he had left them set for the main line ten minutes before the arrival of the well-filled business train, but when the latter train reached the connection, running at 60 or 70 miles an hour, its driver found that he was being diverted into the siding. He immediately applied the steam brakes but the strain of the heavy locomotive negotiating the turnout at high speed proved too much for the track, which gave way, and the engine overturned. The leading van of the train climbed up onto the tender of the locomotive and the following carriages piled up behind the van. Three of the carriages were completely crushed by those following them and only the last vehicle remained on the line undamaged. The wooden bodies of the coaches, even those which had not been crushed by those behind, did not stand up very well to the shock of the derailment, and most of those on the train were either killed or hurt. The final number of dead was stated to be 24, with 67 injured.

A reporter described the scene after the accident. 'The sight was inexpressibly pitiful. A military officer, with the ribbon of the Legion of Honour in his buttonhole, lay alongside a sister-of-mercy {nun}, and next to her a pretty little fair-haired girl hardly three years old, whose sweet, childlike expression made it impossible to believe that she was not simply sleeping. Others, torn, bleeding and dismembered, with legs and arms missing, and still others almost decapitated, completed the ghastly row of victims.

'M Louis Balle, a dealer in chemical products, who resides in the Rue Solferino, Lille, had a miraculous escape from death, for he lay for more than three hours crushed beneath the roofs of the carriages, and when finally rescued his body was literally black and he could hardly breathe. I saw him later and heard from his own lips the story of his wonderful experience.

'"I was in a second-class compartment in the middle of the train," he said, "with six

[45] *The Harvard Herald*, Harvard, Illinois, United States, 4th September 1902

other passengers. Three or four minutes after we passed Arleux we were violently shaken, and before we had time to look out of the window there was a tremendous concussion, and our carriage was immediately reduced to splinters. When I recovered from my bewilderment, I found myself buried up to the waist in ruins of all sorts. Piercing cries were rising all round me, and just above my head was the next carriage, which had risen on the top of the wreckage of our carriage. When the railway servants came to me they tried to drag me out, but found it impossible, and for over three hours I suffered excruciating mental and physical agony and watched my fellow passengers in the wagon above my head being carried off by the rescuers. What tortured me more than anything was the fear that before my turn came the carriage overhead would collapse on top of me and crush me to death. My legs were benumbed, and my breathing was becoming exceedingly painful, when at last, after what seemed an eternity of suffering, the rescuers were able to extricate me, and I was placed on a cushion in a field. I returned to Lille in a wagon with two other injured passengers, one of whom was a Customs officer, who had both legs terribly crushed. On the way death put an end to his agony.'" [46]

The initial assistance to the victims was provided by local railwaymen and by bargemen who had witnessed the accident from their vessels on an adjoining canal, and they were soon joined by local townspeople. A medical officer who had been a passenger on the train also gave first aid, although he himself was injured. Doctors later arrived by motor car from Lille and Douai and on two relief trains, which conveyed the injured to those two places at 11 a.m. and 12 noon. The bodies of the dead were laid out in a room at the station. Emile Maruéjouls, the Minister of Public Works, dispatched his son Pierre to the scene.

Regarding the cause of the accident, in view of the porter's categorical assertion that he had left the points set for the main line, the possibility of malicious interference with them was suggested, but this was ruled out in view of the fact that the points had been in full view of the station staff the whole time. More plausible was the possibility of the points having shifted due to the vibration of the train, especially as they had been previously damaged, and a makeshift repair made to them.

As usual there were two separate investigations into the disaster, a judicial one conducted by the Douai Public Prosecutor and one conducted by State investigators. Porter Moreau and Arleux stationmaster Crespin were the only ones initially charged with responsibility for the accident and the two railwaymen were arrested but allowed bail. Criticisms having been raised by the State investigators that only subordinates were being pursued, four higher Nord company officials – a section manager, a district manager and two inspectors – were subsequently also charged in connection with the disaster, it being alleged that they had been negligent in the organisation and supervision of the remodelling work. The verdict of the Douai Criminal Court in December 1902 was that only Porter Moreau, Stationmaster Crespin and the Nord district manager were guilty. Porter Moreau was sentenced to four months' imprisonment and a 50-franc fine, Stationmaster Crespin to six months' gaol and a 200-franc fine and the district manager was fined 1,000 francs. On appeal the porter's conviction was upheld but the stationmaster had his prison sentence dropped and his fine reduced to 50 francs.[47] [48]

---

[46] *The Daily Mail*, London, England, 29th September 1902.

[47] Wikipedia – Accident ferroviaire d'Arleux

Four hoboes died and 47 box cars were in the ditch following the derailment of a Great Northern Railroad freight train just on the Butte side of Midvale, Montana, United States. What was described as a defective – presumably meaning broken – rail caused the accident.

At 2.45 p.m. on the afternoon of Friday 24th July 1903 two trains were due to cross one another at Buchholz-Königstraße station, in Saxony, Germany, and the passenger train from Weipert was already waiting in the station. As the opposing train, also a passenger working, ran into the station, its last three carriages were derailed on the points because the signalman had reversed them prematurely. Five passengers lost their lives and an unstated number were injured. On 16th September the signalman is recorded as having found himself in court at Chemnitz charged with having caused the accident through his negligence, but the outcome of his trial is unknown.[49]

Very brief reports in several American and Australian newspapers spoke of the derailment of a train in Spain on 12th December 1903 due to the collapse of an embankment near Cordova.[50] [51] Some of the reports spoke of the train falling from a bridge and 12 or 15 deaths, with only two carriages escaping destruction, but the true figure was six dead with another 19 injured, while no bridge was involved. The train involved in the accident was Madrid Malaga mail train No. 101 and the accident occurred between Baena and Alcaudete stations. Heavy rain led to the waterlogging and subsidence of an embankment some 15 metres high, though a passenger also stated that the train, being late, was running at high speed. Those killed included the fireman and a brakeman.[52]

The failure of the engine of a freight train was the initial factor which led to a bad accident at Godfrey, Kansas, United States, on Monday 21st December 1903. Following the freight train was the *Meteor*, the St Louis-San Francisco Railroad's fast train. The circumstances of the accident were simple enough. Because of the problems with its engine the crew of the freight train were ordered to hold the main track at Godfrey while the passenger express was diverted onto the siding. The brakeman of the freight train, J. A. Bartley, should have gone back and flagged the *Meteor* to warn its engineer to reduce speed but he failed to do so. The passenger train was behind time and was running at full speed as it approached Godfrey, attempting to regain a few minutes. Unaware that his train would not be holding the main track as usual and not having been cautioned, Engineer B. A. Dewees did not reduce speed and his engine and the whole of the train left the track at the switch and all except the rear sleeper overturned. Such was the speed of the train at the moment of derailment that the engine and leading baggage car came to a stand almost 60 feet from the track. The baggage car was completely wrecked and the

[48] *The Standard*, London, England, 29th September 1902
[49] *Eisenbahnunfälle in Sachsen* by Claus Schlegel, pg. 95 (in German)
[50] *The Cedar Rapids Weekly Gazette*, Cedar Rapids, Iowa, 16th December 1903
[51] *The Brisbane Courier*, Brisbane, Queensland, Australia, 16th December 1903
[52] *La Vanguardia*, Barcelona, Spain, 14th December 1903

smoker badly damaged. Engineer Dewees was killed, as were Fireman Theodore Bishard, Conductor George Hoyt, news agent John Bluebacker, James H. Wyman, Asa Moreland, James Kirkpatrick, and brothers Lon and Joseph Corbin, the latter five all presumably passengers. In addition express messenger John Bell, mail clerk H. B. Darlington and passengers Sheridan Knable and E. B. Garroway were, it was thought, fatally injured. Fourteen other passengers were seriously injured, including another brother of the Corbins. The brothers had been on their way home to Oklahoma to spend Christmas there. Brakeman Bartley made himself scarce after the accident.[53]

Another report on the accident added that fire broke out in the chair cars soon after the derailment and contributed to some of the injuries. The accident occurred in the early hours of the morning and in the darkness and confusion it was some time before the uninjured train crew and passengers could help the victims, while it was several hours before a relief train with physicians left Fort Scott, Kansas, for the scene, and 11 a.m. before it arrived back there with the dead and injured.[54]

A broken rail caused the derailment of a northbound Ferrocarril Central Mexicano (Mexican Central Railway) passenger train running at high speed a few miles south of Zacatacas, Mexico, on Monday 18th April 1904. The engine, baggage car and first, second and third-class coaches went down an embankment. Two Pullman cars at the rear of the train stayed on the track and their occupants escaped with a severe shaking. The initial report spoke of eight dead and 35 injured, the dead including the American engineer and express messenger and the Mexican fireman.[55]

A second report stated that Engineer Lee was not killed outright but was in hospital, mortally wounded, while Conductor W. B. Moore was in a grave condition, and that both were under arrest(!) It added that Express Messenger W. F. Bostwick, of El Paso, had been killed instantly and Messenger Ramora Medina of El Paso and two helpers were seriously injured. The dead also included a Pullman car passenger who had been in the second-class coach at the moment of the derailment. This report spoke of 11 dead and 49 injured, 16 seriously.[56]

A third report, three days later, in the same newspaper, gave the account of H. G. Caldwell, 'travelling freight and passenger agent for the Santa Fe railroad,' who had been a passenger on the train and who maintained that the accident had been much more serious than officially stated. Mr Caldwell alleged that no less than 69 persons had died in the accident and many more had been seriously injured. Describing the accident, Caldwell told the reporter, '"Darkness was all that was lacking to complete the horror of the situation. It happened just after noon. We were about four miles out from Zacatecas, and the after-dinner cigarettes were scarcely finished when the wreck came. Many died while we were working with them; others died on the way back to Zacatecas, and still others died after reaching the city. For all I know still more may have died since I left."' [57]

---

[53] *The Iowa Recorder*, Greene, Iowa, United States, 23rd December 1903
[54] *The Lincoln Daily Star*, Lincoln, Nebraska, United States, 21st December 1903
[55] *Dubuque Telegraph Herald*, Dubuque, Iowa, 21st April 1904
[56] *The Sun*, Baltimore, Maryland, United States, 22nd April 1904
[57] Ibid, 25th April 1904

An accident similar to that at Godfrey, Kansas, noted above, occurred to a St Louis Iron Mountain and Southern Railroad train at Wickes Siding, near Kimmswick, Missouri, United States, on 30th April 1904, but this time the fault lay with the engineer of the derailed train rather than with a brakeman. It was stated after the accident that Engineer John M. Bailey, who died in the accident, had received definite orders to approach the siding at Kimmswick at moderate speed, and the said orders were found in the pocket of his overalls afterwards. Conductor J. C. Austin, who survived, also swore that he had read out the orders to the dead engineer.[58]

As at Godfrey the events leading up to the accident began with a problem with a freight train. In this case a box car attached to a southbound fast freight train suffered a broken journal. The disabled car had to be left on the main track at Wickes Siding and the remainder of the freight train proceeded without it. The main track at the siding thus being blocked, the dispatcher issued orders for all northbound and southbound trains to pass through Wickes via the siding. Five trains did so safely. Conductor Austin of train No. 18 read the order regarding the use of the siding at Wickes to Engineer Bailey at DeSoto, a few miles below Kimmswick. Despite this the train passed through Kimmswick at a high rate of speed and reached the switch at Wickes, a short distance beyond, travelling at almost 45 miles an hour, the engineer having either forgotten the order or mistaken where he was. The speed was too great for the engine to take the switch and it overturned, burying Engineer Bailey beneath tons of steel. Master mechanic A. E. Tarbor, who was also riding in the engine cab, was likewise killed instantly. Fireman Gumbert jumped but was fatally injured anyway.

Behind the engine the leading baggage car ended up in a ditch 100 feet away. In it was express messenger B. C. DeGroat, who died soon after being removed from the wreckage. The smoker in the rear of the baggage car and the coach behind the smoker overturned and were demolished. Five passengers riding in the smoker lost their lives and many of those in the following coach were badly hurt. One of the four sleeping cars also overturned but the remaining three stayed upright and those in them were unhurt.

After the accident news of it was telegraphed to DeSoto and St Louis and relief trains with physicians soon arrived. The wrecking train had already been ordered out to remove the disabled box car and so was on hand to deal with the more serious problem, but the line was blocked for several hours.[59]

The train was a special proceeding to St Louis for the St Louis World's Fair (officially the Louisiana Purchase Exposition) which opened on the day of the accident and ran until the beginning of December.[60]

At 4.25 a.m. on Monday 1st August 1904, train No. 5 steamed out of Midland Junction, Western Australia, for a short run to Pickering Brook, 15 miles away. It was made up of engine No. 134, nine empty waggons and a guards van and was in charge of Guard Adam Anderson, with Driver George Geer and Fireman William Lyons. Forty-five

---

[58] *The Daily Free Press*, Carbondale, Illinois, United States, 2nd May 1904

[59] *Cedar Falls Gazette*, Cedar Falls, Iowa, United States, 10th May 1904

[60] Wikipedia – Louisiana Purchase Exposition

minutes later Guard Anderson limped onto the veranda of the stationmaster's house at Midland Junction and rousing the night stationmaster told him that his train had been wrecked by a washout at Poison Gully and that both his mates were dead. He explained that he had been busy with his duties in his van when he had been suddenly hurled to its far end, along with an 8 cwt iron safe. When he had struggled out of his van and gone to the front of the train he had found the engine pointing skywards on the far side of a raging torrent. Hesitating long enough only to confirm that the driver and fireman were beyond human aid he had struggled back to Midland Junction.

Night Stationmaster G. F. Veitch at once arranged for an engine to be readied and sent a messenger to Dr Fergusson Stewart, and before long the stationmaster, the doctor and half a dozen men were on their way to the scene. When they arrived, they found that the guard's report had been all too accurate. Driver Geer was lying on the left bank of the stream, fearfully mangled, while Fireman Lyons was buried beneath the whole weight of the engine tender. There was a twenty-foot gap in the line where it had been washed away.[61]

It seemed, at first sight, like a tragic accident, very sad but one which could hardly have been avoided. By the conclusion of the inquest, though, it was clear that it had resulted from a number of failings and oversights and could easily have been prevented, leading a Perth newspaper to write that, 'In truth, reading a report of the case, it is hard to find any excuse or mitigating circumstances for the dreadful blundering, laxity and mental fog which preceded and led up to the tragedy.' [62]

On the first day of the inquest proper it emerged that Ganger Patrick Wall, upon whose section of track the washout had occurred, had fallen sick on Friday, 29th July and had not gone back to work until 4th August. Ganger Wall said that on Saturday morning, 30th July, he had told one of his men, named Anderson, that he was sick and had asked Anderson to perform some repairs. The ganger said he had told Inspector Mattim that he was ill on Monday morning. He could not say if Anderson had patrolled the line on Saturday and said, confusingly, first that he would have patrolled the line on Sunday had he felt well enough to do so and then that he never patrolled it on Sundays, although he knew that the first train ran on Mondays before he patrolled the line at about 7.30 a.m.

Repairer Charles Anderson admitted that Ganger Wall had appointed him as acting ganger on Saturday because Wall was sick. It had not occurred to him to patrol the length on Sunday, 31st, even though it had been stormy. They received no pay for work done on Sundays. He had seen very rough weather on Sundays before when the line had not been patrolled.

Number Taker Harold Masters stated that he was also the telephone operator at Midland Junction. He had been on duty from 4 p.m. until 6 a.m. the next morning. He did not recollect receiving any telephone messages between 4 p.m. and midnight. He had not been in the room all the time and was not there from 5 p.m. until after 6. Booking clerks, guards and porters did not usually answer the phone.

Ganger B. Fermhough stated that he was in charge of No. 14 gang on No. 1 section of the Eastern line. He had patrolled the line on Sunday 31st, had noticed water over the line at 22 miles and 76 chains, and had reported the fact to the night stationmaster. He had

---

[61] *The West Australian*, Perth, Western Australia, 2nd August 1904
[62] *The Sunday Times*, Perth, Western Australia, 14th August 1904

then gone to Ganger Wall's house, told him of the state of the line and asked Wall if the latter intended to patrol his own length and whether Wall intended to send up to Statham's siding for his men, but Wall, who had been in bed, had replied that he was not well and had no one to send. Ganger Fermhough said that he had then gone to look after his own length and because it was flooding he had been unable to assist Ganger Wall. He recalled a washout at Poison Gully about ten years previously but then the culvert had not been carried away. [63]

Frederick Lewis, a quarryman, said that he had found the washout at Poison Gully at about 5 p.m. on Sunday afternoon. The rails, with the sleepers attached to them, were suspended over a distance of about half a chain (11 yards). The gully resembled a small river. He had told Mrs Bacon, the manager's wife at the station siding, what he had seen, and she had telephoned 55, Midland Junction, and repeated what he had told her. She had been asked whether there was any water on the rails, and he had said no. She had told him that the ganger had been or was along the line. The call was made at a quarter to 6. He had not actually said that the line was dangerous for traffic but thought that the telephone call was sufficient to warrant its inspection.

August Jensch, another quarryman at Statham's quarry, stated that he had arrived at Poison Gully some time after 6. There the earth had given way under his feet and striking a match he had found the sleepers hanging in the air. He had reported the washaway at the camp but finding that Mr Lewis had already reported it he had been satisfied.

Mrs Eliza Bacon testified that she was the wife of the manager of Statham's Quarry. Between 5.30 and 5.40 Mr Lewis had come to her door, reported the washaway, and said that he thought it would be best if she advised the station at Midland Junction by telephone. She had done so, and someone had replied, "The ganger has just come in and reports six inches of water on the rails." She had asked Lewis if that was the spot he meant and he had replied, "No, there is no water on the rails at the spot I mean. It's a washaway."

She had then told the person at Midland Junction, "The man says there is no water on the rails at the spot he means."

The reply was, "Right."

She had said, "I thought it best to ring you," and the railwayman at Midland Junction had rung off.

Traffic Foreman George Arnold testified that he had taken over from Stationmaster Veitch at midnight that Sunday night, when the latter had told him that the line was under water at Mileage 22 and 76 chains, and that he had cancelled trains Nos. 1, 2, 11 and 12; also that Ganger McGee was out examining the line and would advise him whether it would be possible to run No. 5 goods. Ganger McGee had duly come to him at 1.55 a.m. and told him that the water had receded and that it was safe to run train No. 5, provided that drivers were cautioned to run slowly at that point. He (Arnold) had duly warned Driver Geer immediately before he started. The traffic foreman added that he had received no reports of problems at any other point and considered therefore that the whole line was safe.

Ganger McGee stated that his section of line ran from 21 miles 22 chains to 22 miles

[63] *The West Australian*, Perth, Western Australia, Australia, 11th August 1904

49 chains on the Eastern line and did not extend to the point where the Canning line branched off (or thus to Poison Gully, this being in Ganger Wall's length). At 10 p.m. on Sunday, when he had been off duty, he had received a verbal message from the night stationmaster (Mr Veitch) to the effect that the line was under water at 22 miles 76 chains and asking him to investigate. He had done so and found the water running over the line to a depth of 2 in. to 5 in. He had then gone to Ganger Wall's house, where Mrs Wall had told him that her husband was sick and in bed. He did not see or leave a message for Ganger Wall and saw none of Wall's men. Returning to the reported problem spot he had found that the water had receded somewhat. He had then gone as far as the Helena Bridge, checking that that was safe, had not thought it necessary to go any further and returning a third time to the place giving trouble had found that the ballast was washed away. Going back to Midland Junction he had told the stationmaster that the line was unsafe at about 11.30 p.m. He had then gone back and effected repairs sufficient to make the line passable at reduced speed and had so advised the stationmaster (actually the night foreman of course) at about 2 a.m.[64]

At the resumed inquest Driver William Watts stated that at about 7 p.m. on the Sunday evening Night Stationmaster Veitch had told him that the Canning line was in a very bad state, that he had had a message from Statham's and that Poison Gully was in a very bad state. The stationmaster had asked his opinion and he had advised against sending out any trains that night.

Acting Fireman Albert Brown stated that he had heard someone, he did not know who, tell Driver Watts that there had been a message from Statham's to the effect that the line was partly under water, and was very bad at Poison Gully. Between 9 and 10 Watts had mentioned to him the possibility of their having to go out on a light engine to examine the line.

Acting Bookkeeper Percy Miller denied receiving any telephone message regarding a problem at Statham's siding that evening.

Mrs Bacon was recalled and said that she had not used the words 'Poison Gully' when she telephoned Midland Junction and had not even been aware that there was any point along the line with that name.

Fireman Cecil Marsden said that he had spoken to Stationmaster Veitch at 7.25 p.m. that Sunday evening and the night stationmaster had told him that one of the gangs had reported that there was water over the line at about 23 miles, near Cobweb Siding. The stationmaster had then mentioned Poison Gully; he thought that it was the stationmaster who had mentioned that name first, but he would not swear it. He (Marsden) had never had any fear of an accident at the Poison Gully culvert.

Night Stationmaster Veitch denied having told Driver Watts of having had any message from Statham's. He said that it was Fireman Marsden who had mentioned Poison Gully to him as a place that might be a possible danger spot and he had then asked the fireman how Poison Gully had acquired its name, and had been told that it was because of poisonous weeds which grew there. Later he had seen Driver Watts and had told him that he had been advised that the Canning branch was not good at 22 miles 76 chains and had mentioned Poison Gully, having heard the name from Marsden. He could not suggest any reason why Watts, Brown and Marsden should perjure themselves as

[64] *The Swan Express*, Midland Junction, Western Australia, Australia, 13th August 1904

regards to himself unless they were biased.[65]

The inquest ended with a verdict from the jury censuring Night Stationmaster Veitch and Acting Ganger Anderson. The jury found that, 'We are of opinion that Engine Driver George Geer and Fireman William Lyons came to their deaths on Monday, August 1, in a railway accident, owing to the collapse of a culvert at Poison Gully, caused by flooding after an excessive rainfall. Your jury regret to record the disregard displayed to Rule 291 of the W. A. Government Railways Regulations by the officer and ganger in charge of the section during such climatic influences. We also beg to recognise the willingness and assistance rendered by Gangers McGee and Fermhough - rendered outside their own responsibilities.'

Mr N. K. Ewing, instructed by the W. A. Amalgamated Society of Railway Employees, asked if the jury would care to name the officers referred to in their rider. The foreman of the jury stated that the rider applied to Night Stationmaster Veitch and Acting Ganger Anderson. Mr Ewing replied that he thought it only fair to point out that Rule No. 291 referred only to permanent way men, and at no time during the inquest had it been suggested that Mr Veitch was in charge of the section.[66]

Like two of the accidents previously mentioned the next to be described was not strictly due to track defects but was a derailment on a set of facing points in the United States, or on an open switch as they were termed in that country. Its circumstances are very well described by Shaw in his classic work, *A History of Railroad Accidents, Safety Precautions and Operating Practices*. The location of the disaster was the small station of Mentor, Ohio, and the date was 21st June 1905.

Mentor was a station on the Lake Shore and Michigan Southern Railroad main line between Buffalo and Chicago, which formed, with the New York Central and Hudson River Railroad main line to Buffalo, part of what was advertised as the Water Level Route between New York and Chicago. The two companies were in fierce competition for the lucrative New York-Chicago traffic with the Pennsylvania Railroad, and in 1902 had inaugurated the *Twentieth Century Limited* express to cater for upper class and business passengers between the two cities. At the beginning of 1905 this express was scheduled to cover the approximately 900-mile journey in 20 hours, but the rival Pennsylvania company cut the schedule of its trains, and so from Sunday 18th June the schedule of the *Twentieth Century Limited* was cut to 18 hours, with talk of reducing it even further to 17 hours.[67] The high speeds involved in the fast schedules were, however, met with some misgivings by members of the travelling public. As far as the railroad company was concerned special precautions were taken to ensure its safety. Orders were given that the lines were to be kept clear of other trains, reportedly an hour in advance of its booked time, the best engineers, firemen and conductors were chosen to work it and the train equipment was inspected at every division point. The expresses were in fact treated in much the same way as British Royal trains.

At 9.20 p.m. on the evening of Thursday 21st June the eastbound express, train No. 26, was travelling at 60 or 70 miles an hour as it approached Mentor. Engineer Allen Tyler

---

[65] *The West Australian*, Perth, Western Australia, Australia, 17th August 1904
[66] Ibid, 18th August 1904
[67] *The Sun*, Baltimore, Maryland, United States, 22nd June 1905

and Fireman Aaron Gorham were keeping a good lookout, and Fireman Gorham confirmed after the accident, from his bed in Cleveland hospital, that the signal light at the facing points near the station had been showing a white, 'All clear' light. The next moment the locomotive was being hurled into the ditch. The combination car piled up on the top of the engine, and fire broke out in the wreckage to add to the horror, while the Chicago sleeper, marshalled behind the combination car, was swung sideways and struck and demolished the freight depot. Then a moment after the crash the boiler of the locomotive exploded 'with terrific force', scattering steam and fire.

Nineteen people died in the Mentor disaster including Engineer Tyler and three other railroad men: Head Brakeman Brant, Baggageman N. B. Walters and Porter W. D. Mickey. The passengers who died included New York attorney John R. Bennett and T. R. Morgan, second vice president of the Wellman-Seavers-Morgan company of Cleveland, Ohio.

A reporter described the scene. 'The train was crowded, practically all the accommodation being taken when it left the city. It was behind time and the greatest of speed was being made to make up the lost time. The crash came without the slightest warning. The scenes following the wreck were appalling. The night was dark save for the light from the blazing coach that was crushed and splintered on top of the engine. Men swarmed about it, combatting the flames with the means at hand, groping their way through the blinding, scalding steam that rose in clouds, hunting for the injured, whose piteous cries were such as chilled the hearts of those who heard them. The water supply was small, and the means at hand for fighting the flames were pitifully inadequate, but the zeal of the rescuers wrought great things for the first five minutes.' The rescuers were led by Conductor Alexander Hammond, who had escaped injury in the crash.[68]

Conductor Hammond rushed to the switch immediately after the derailment. He found the points locked open for the siding. The switch light was out. No train had passed over the connection since train No. 10, the eastbound Chicago and Boston service 45 minutes earlier. Lake Shore manager W. H. Marshal expressed the opinion that the mechanism must have been tampered with by an unknown person as the points were working perfectly well after the accident, and he determined to make a thorough investigation into what had happened. Lake Shore Assistant General Superintendent Moon thought likewise that the switch had been opened by some unknown malicious person, probably a crank. Meanwhile President William Newman of the New York Central stressed that he did not regard the speeds necessary to maintain the 18-hour schedule as being unsafe.[69]

The hunt for the mysterious saboteur proved fruitless and after a while suspicion fell instead on the Mentor night operator, Walter F. Minor. It was his duty to check that the points were correct before the passing of the *Twentieth Century Limited*, and it seemed likely that having failed to check them earlier Minor had impulsively operated the points right in front of the express, though he steadfastly denied having done so. At the inquest into the deaths of those killed in the accident, held in Cleveland on 17th August, Coroner Siegelstein stated in his verdict that, "I find that the switch was opened by Walter F. Minor, the night telegraph operator in charge of the night station. I reach this finding by deduction – i.e. after having read over the notes on the testimony taken at the Painesville

---

[68] Ibid, 23rd June 1905

[69] *The Boston Daily Globe*, Boston, Massachusetts, United States, 22nd June 1905

inquest, which I attended, and the testimony taken at my own inquest. I find that the time (according to observations taken at this office) from which the headlight of the Twentieth Century Limited is seen approaching from the depot platform at Mentor until it reaches the switch is a maximum of 31 seconds. During this brief period of time, if Mr Minor did not throw the switch on the night of June, and another person did, then that person must have appeared at the switch from some unknown place, unlocked the lock with a key, removed the lock, taken the coupling pin out, turned the switch, put the coupling pin in place again, placed the lock in the hole of the coupling pin, locked the lock, and disappeared – a physical impossibility in my opinion. In the 500 folios of typewritten testimony taken at my inquest there appears not one word tending to show the presence of anyone but Minor, who was in charge of the station that night, in the immediate vicinity of the depot or the switch immediately preceding the wreck or afterwards. Hence, I can only come to the conclusion, which is reinforced by other points of the evidence, that Walter F. Minor opened the switch, but whether he did so of his own volition or by telegraphic orders of a superior officer, I am not prepared to say." [70]

After the accident there were calls for the schedule of the *Twentieth Century Limited* to be relaxed, and its timing was eventually increased to 20 hours, but not until 1912.[71]

On 1st September 1905 disaster struck what the London *Times* referred to as the *Cromer Express*, the 9.27 a.m. Great Eastern Railway passenger train from London Liverpool Street to the English seaside resort via Ipswich and Norwich. The express consisted of an engine and tender and 14 vehicles: a brake van, a third-class carriage, a first-class carriage, three third-class carriages, a composite carriage, a first-class carriage, a third-class, another brake van, two third-class carriages, a composite and a third-class carriage at the end of the train. All went well until the train arrived at Witham station at about 10.30 a.m., but there most of its vehicles left the track on the approach to the station platforms. The engine, tender and three leading vehicles broke loose from the remainder of the train and came to a stand a little beyond the platforms, little damaged. The next five vehicles in contrast suffered severe damage. The fifth and sixth vehicles of the train ended up on the down island platform; the former demolished the porters' and ticket collector's rooms and killed a porter, and eight passengers in the latter carriage were killed instantly. A ninth passenger in this carriage, a child, later succumbed to their injuries and another passenger, whose position in the train could not be determined, also died. The rear six vehicles were only slightly derailed. In addition to the ten passengers and one railwayman who lost their lives, 66 passengers and five railway servants were injured.

Some of the passengers related their stories to a *Times* reporter. 'A young man named Arthur Powell, of Walthamstow, who was travelling in the middle of the train at the time of the accident, said: - The train was just getting up speed after leaving Chelmsford, and was swinging along at a good pace when the accident occurred. The first I felt of it was a sudden shaking, as though the train had run off the metals and was going over sleepers. I do not know what occurred next, for I knew no more until I found myself lying on the line, having been pulled out from among the wreckage. My cap was afterwards taken out

[70] *The Evening Independent*, Massillon, Ohio, United States, 17th August 1905

[71] Wikipedia – Twentieth Century Limited

from the wrecked train, and while I was resting, I saw a gentleman taken dead out of the compartment where I had been sitting. The other gentleman with whom I had travelled was also injured, and he complained of severe pain in his spine. I was going to Cromer.

'Mr W. E. Sell, of Lower Edmonton who was travelling in a coach which was overturned, said: -There were six of us in our compartment, and the first indication we had of anything amiss was a violent bumping, after which we were thrown across the compartment. The train seemed to travel about 50 yards after this, and the carriage in which we were was overturned. I succeeded in getting clear of the wreckage by climbing out of the door onto the roof.' [72]

An up train was approaching the station at the moment of the derailment but Signalman Fred Parish in Witham Station Junction box, seeing that something was wrong, threw his signals back to danger, sent the 'Obstruction Danger' signal to Witham East box and the up train was stopped short of the wreckage.

Lt-Col P. G. von Donop investigated the accident on behalf of the Board of Trade. He noted in the conclusion to his long report that the express that the train had departed from its last stop at Chelmsford at 10.17 a.m. and the accident had occurred at almost exactly 10.29 a.m., giving an average speed over the nine miles from Chelmsford to Witham of 45 miles an hour. The speed through Witham station itself was probably somewhat higher on the falling gradients, perhaps as high as 70 miles an hour, but there was no speed restriction at that point.

The most important evidence as to the cause of the derailment, the inspecting officer stated, was that of Head Guard John Smith and Shunter Frank Hume. Head Guard Smith testified that he had been riding in the leading brake van of the train, next to the engine. Just before reaching the public overbridge south of Witham station his van had received a severe jerk as if one of its near side wheels had struck something very hard. The van had risen into the air on that side and when it came down the wheels did not regain the rails. The derailed van then jumped about continuously until the train came to a standstill. Guard Smith was confident that the initial jerk he had felt had been at the crossover just short of the public overbridge.

Shunter Hume had been standing on the up side opposite the same crossover. He saw the engine and first two vehicles pass over the connection safely, but the third vehicle dropped off the line and the track was at once ploughed up.

Driver John Hills and Fireman George Ward on the engine of the express agreed that neither their locomotive nor its tender had left the rails at all, but that just before reaching the station platforms they had felt a jerk as if the guard had applied his brake. Driver Hills had then applied the engine brake and looking back they had seen that the train was off the road.

Standing by the crossover when the express passed them had been Foreman Platelayer Robert Pryke and Platelayers Arthur Newman and Morris Pavelin. All said that they had no idea how the derailment which had happened right in front of them had occurred, and it seemed extraordinary, Lt-Col von Donop declared, that they could give no assistance in clearing up the matter.

Marks on the permanent way, the inspecting officer continued, taken in conjunction

---

[72] *The Times*, London, England, 2nd September 1905

with the statements of the guard, engine crew and shunter, left no doubt that the derailment had commenced immediately ahead of the V-crossing of the crossover. The sequence of events following the derailment was easy to determine, the problem, as in most cases of derailment, was to distinguish cause from effect, and to find whether it was a problem with the rolling stock or with the track which had caused it. As regards the rolling stock the vehicles of the train had been examined the night before the accident, at Liverpool Street the same morning and again, less thoroughly, during the train's journey at Chelmsford. No problems had been found. Examination of the rolling stock after the derailed showed only damage likely to have been caused by it. Several items had also been found alongside the track but these, it seemed probable, were debris from the damaged carriages which had fallen from them as they were being moved out of the way after the accident.

Regarding the permanent way, it was not of the highest standard; the 85 lbs to the yard rails had been reduced to 82 lbs by wear and the ballast, while ample in quantity and well drained, was of gravel and not entirely satisfactory. The company were in the process of relaying with 95 lb rails and substituting slag ballast. There were, however, no special weaknesses in the track, while any obstruction near the crossover would have been noticed by the platelayers. There was, however, the fact that immediately before the derailment the platelayers had been at work at the knuckle timber, which was the very point where it had begun, and the inference that the accident was in some way connected with their works could not be ignored. Foreman Pryke stated initially that all he had done was loosen the ballast a little on each side of the knuckle timber and that nothing had been done to make the line unsafe to run over at high speed.

A month after the inquiry had ended further evidence came to light. Shunter Hume stated that he had been standing near the crossover waiting to cross the line and had noticed as the train approached that one of the keys at the wing rail of the V-crossing was out. As the train passed the rail immediately jumped up and although the engine and leading two vehicles negotiated the obstruction safely the third vehicle did not. Hume admitted that he had thought the missing key had caused the accident and excused his keeping quiet about the matter for a week by saying that he had injured himself in the rescue work and had been confined to bed for several days. He had then told a train examiner named Kisby, who had not informed his superior officers until 14th October because he had, he said, thought it in the interest of the Company to say nothing.

Supporting Shunter Hume's allegation was the evidence of Driver Hills and Fireman Ward that they had become concerned because the three platelayers had been unusually slow in moving away from the track, appearing to wish to get some job done before the train arrived, and that the eyes of all three of the trackmen had appeared focused on one particular spot. Predictably Foreman Pryke and his men absolutely denied the truth of the statements of the shunter and the engine crew. Lt-Col von Donop though, believed them, and found that it was impossible to resist the conclusion that the accident had been caused by a fatal weakening of the track at the knuckle timber, possibly because the men had exceeded their foreman's instructions, and that they had desperately been trying to put matters right when the express arrived. He concluded that the accident pointed to the necessity of great care being taken by platelayers when they were carrying out repairs to the track. He also drew attention to the fire which had broken out and destroyed one carriage due to the rupture of a gas cylinder. In the present case this had not resulted in any further deaths or injuries, but it showed the need for a safer means of illumination than gas, probably electricity.[73]

Spreading track or a broken rail were given as the cause for the derailment of Atchison Topeka and Santa Fe Railroad passenger train No. 1, known as the *California Limited.* The train had left Chicago at 10 p.m. the night before and the accident happened a mile east of Sheffield, Missouri, United States, at 10 a.m. on 30th October 1905. The tender of the engine was the first to leave the track and two following baggage and mail cars, a smoker and a chair car then piled up in a heap of wreckage 30 ft high. Varying reports stated that the derailment had occurred in a 50 ft deep granite cutting and that the derailed train went down an embankment! Reportedly ten or more Pullman cars, tourist sleeper and dining cars remained on the track. Twelve people died in the accident, all men. The victims included Baggageman J. F. Capps and Porter William Harrison. Two mail clerks were also among those killed. Thirty or more others were injured. A trainman ran to Sheffield to give the alarm while uninjured passengers began the rescue work. Two wrecking trains, carrying surgeons, were despatched to the scene of the accident.[74] The passengers on the train were mostly 'colonization tourists'.[75]

Westbound Great Northern Railroad passenger train No. 3 bound for Spokane was derailed 1½ miles east of Camden, Washington, United States, at 4.18 p.m. on the afternoon of Monday 23rd July 1906. The train was reportedly travelling at high speed as it neared a bend and its passengers were said to have had a premonition of disaster. Emerging from the west portal of a tunnel the engine left the track and plunged 50 feet down the hillside into the Diamond Lake, followed by the mail coach, baggage car, smoker and day cars. As the coaches struck the water there was a blinding flash from the explosion of acetylene gas tanks and fire added to the horror of the scene. Some of the passengers nevertheless succeeded in smashing coach windows and fought their way out before the smoking car was filled with water. Uninjured train crew and passengers extinguished the fire in the first-class coach, but not before some of the survivors had been badly burned. At the time of the first reports on the accident the bodies of two of those killed, passengers George H. Curtice and T. J. Dolbrow, had been recovered, while seven or possibly more bodies lay at the bottom of the lake, waiting for the arrival of divers from Seattle for their recovery.[76]

One theory as to the cause of the accident was that Engineer Edward Munson had lost control of his engine, or else had lost his head, for the train was going faster than was safe around the numerous curves. The bodies of the engineer and Fireman George Bell were among those still in the lake, as were the bodies of Baggageman George R. Strickland and mail clerk Charles Danner. Ten passengers were injured and about the same number were shaken up. The engine, tender, mail and baggage cars were completely submerged, and the next three vehicles were thrown on their sides among the rocks and boulders at the water's edge. Only the dining car and the *Chelan* palace car remained on the track.

A Camden merchant named C. L. Maynard had a different theory as to the reason for

[73] Board of Trade report
[74] *The Sedalia Evening Sentinel*, Sedalia, Missouri, United States, 30th October 1905
[75] *Burlington Evening Gazette*, Burlington, Iowa, United States, 30th October 1905
[76] *The Weekly Sentinel*, Fort Wayne, Indiana, United States, 25th July 1906

the derailment. In a lengthy statement he attributed it to the condition of the track. "I do not think that there is the slightest doubt that the Great Northern's deadly wreck at Camden was due entirely to the rotten state of the roadbed. It has been the common talk of the district that it has been absolutely rotten. Even to the lay eye the state of the track speaks for itself. Not one spike in a hundred touches the rail it is supposed to hold in place. And almost at any point one can step down and pick these spikes from the ties. Fully 90 per cent of the ties are so rotten that they can be picked to pieces with the fingers, and in hundreds of places between the scene of the accident and the station at Camden there are spaces of from four to six feet where the track is absolutely unsupported by the ties. This is due to the fact that many of the ties have been burnt away by the fire from the engines until there is a clear space of two inches or so between the top of the tie and foot of the rail. Often in such places the ties have been broken by the weight of passing trains, and thus the rail is absolutely unsupported for a considerable space. I am aware that a tie often appears rotten when in reality it has a solid core and all due allowance has been made for that. A number of ties are rotten only on the surface, but even then, the wood that has rotted away has loosened the creepers used to keep the rails from sliding. This is so that in dozens of places the rails have slid for spaces of from six inches to two or three feet. In sliding thus the spikes are naturally dragged out of place, and no attempt seems to have been made to have them put right. In many places the rails have slid to such an extent that the heads of the ties have come together to form an angle. The braces are all twisted and out of place. Dozens of them can be picked up from the track by the fingers and quite a few are lying clear of the rails altogether. There is not one of them in 50 which is doing the work for which it was placed there. The sleepers are all broken almost without exception and can be picked up from under the rails by the hundreds. In places the rails are lying altogether out of the perpendicular, and if a line were dropped from the outside edge of the rail to the ties on which it is placed it would be found to be out of the perpendicular by quite an inch. Just a little more than this would mean a serious accident, for this occurs in places quite as dangerous as that in which the accident took place. In many places the rails have been worn to such an extent that the part on which the flange of the wheels touches has been worn flush with the fish plates at the points of the rails, the fish plate showing where the wheels of a passing train have struck it. If these rails come much more worn nothing is more certain than that sooner or later a contact with a fish plate will throw a train off the track. In my own mind I am absolutely convinced that the accident was caused purely and simply by a spreading rail. There are many other possible explanations, but a cursory examination of the roadbed will show that this is the most probable. At the actual scene of the accident the ties appear to be thoroughly rotten. It may be possible that they have a solid core, but they certainly do not look it. The people of Camden and neighbourhood have long expected an accident in the vicinity. All who were seen agreed that the two section foremen in charge of the sections there are conscientious men, who examine and report on the track regularly. It is said that these reports are taken no notice of at all at headquarters." [77]

Quite a monologue and it is hoped that the author will be forgiven for reproducing it in full.

[77] *The Columbian*, Columbia Falls, Montana, United States, 28th July 1906

In October 1906 what was then the British colony of Ceylon (now Sri Lanka) suffered disastrous flooding in both the low country around Colombo and in the hills, with a large area of land under water and many fatalities. On Saturday 27th October subsidence of the track due to the heavy rainfall led to the derailment of a mixed train in the hills near Hatton. The *Times* stated that 'Ten natives, including the engine driver and the fireman, were killed, and many others were injured. There was one European on the train, and he escaped.' Advices from the R.M.S. (Royal mail ship) *Modavia*, published in several Australian newspapers, spoke of 15 deaths in the accident.[78] [79]

Heavy rains were responsible for another disaster on 12th March 1907, this time at Alkmaar in the Transvaal, southern Africa on the Central South African Railways. A culvert collapsed beneath a train on the Delagoa Bay line, wrecking the engine and three carriages of a passenger train and killing 11 people. Nine other passengers, Driver Rance and Fireman Alford, were injured. A Johannesburg correspondent to *The Times* in London noted that the accident had caused deep gloom throughout the Transvaal, as one of those who had lost his life had been Dr Adam Jameson, the former Commissioner of Lands, who had been proceeding home after retiring from his office only the previous week. The correspondent added that no negligence was attributable to the authorities regarding the disaster, despite a momentary outburst of indignation in Pretoria that morning because of a rumour of recent reductions in staff patrolling the line. It was officially stated that the rumour was totally unfounded, and that there had in fact been an increase in the number of 'night watchmen'.[80]

For the next accident we return once more to the United States and to another due to an open switch. This time the accident occurred in the far west, at Colton, California, on the Southern Pacific Railroad. Late on the afternoon of Thursday 28th March 1907 train No. 9, the westbound *Sunset Express*, running very late, was derailed on a set of facing points at the entrance to the yard at Colton, which had been left set for the diverging route, while travelling at 40 miles an hour. The results were very serious. Ten of the 14 coaches of the train were derailed, four of them being smashed completely. A party of Italians, numbering almost a hundred, were occupying two immigrant cars, which 'were literally smashed into kindling wood', and most of those killed were Italians from New York and New Orleans. In contrast the passengers in the Pullman cars escaped injury. By the end of the day 23 bodies had been recovered, eight very seriously injured victims were lying in Colton hospital and another 50 less seriously injured persons had been forwarded to Los Angeles on a special train. With the exception of mail clerk Albert Shie all those who lost their lives were passengers, including Southern Pacific fireman H. I. Walters of Sacramento, who was travelling as a passenger. All but three members of the *Florence Roberts* theatrical company escaped injury.

Mail clerk Shie had promised his wife only a few days earlier that he would give up his job riding on the trains after she had pleaded with him for years to do so because of the ever present dangers of the job. He had written a letter to his father-in-law in

---

[78] *Sydney Evening News*, Sydney, Australia, 15th November 1906
[79] *The Times*, London, England, 30th October 1906
[80] Ibid, 14th and 19th March 1907

Standsville, Ohio, apprising the latter of his intention to retire from the mail service and adding that he and his wife would come and visit her father before he (Shie) got another job. The unposted letter was found on the body of the dead mail clerk.

Another of the dead passengers was a George Sharpe of Muncie, Indiana. Sharpe used to live in Los Angeles but had moved to Muncie three years earlier after separating from his wife. He had told a friend that he was returning to Los Angeles in the hope of a reconciliation with her. He was unaware that his wife had divorced him a year earlier and had remarried and was travelling westwards happy at the thought of seeing her again. Friends of the couple had gone to the depot at Los Angeles to break the bad news to him, only to find that he had lost his life en route.

Already at 1 a.m. the next morning Coroner Van Wye was taking preliminary testimony from the Colton switching crew to determine if they should be held responsible for the wreck. Switch Foreman L. A. Alvord explained that when he had come on duty that morning train No. 9 had been chalked up the board as running eight hours late. It had not appeared by the time he went for lunch and when he returned the board was silent about it. As the eight hours had then elapsed, he assumed that it must have gone by in his absence, and he only realised that it had not when he heard the engineer of his switching engine whistling frantically to the express to pull up. It had then been too late to get to and alter the points as the express had been halfway across the Santa Ana bridge. Supported by his subordinates, Switch Foreman Alvord blamed the Colton operator for not keeping him up to date with train movements.

Speaking later Coroner Van Wye stated that he was of the opinion that according to the company rules Alvord had a good defence. The information displayed in the morning was that the *Sunset Express* was running eight hours late, and should have therefore arrived at 1.45 p.m., whereas in fact it was actually 10 hours late. The operator had failed to notify the switching crew, who were remote from any telegraphic communication, of the extra delay to the express. Furthermore, the rules of the company were that trains running more than ten minutes late must enter yards limits running under complete control. The yards limits at Colton extended over a mile eastward of the points where the derailment occurred.[81]

The coroner's jury, however, found Switch Foreman Alvord and his men guilty of criminal negligence for leaving the switch open, the engineer of the express likewise guilty of criminal negligence for running too fast, and censured the Southern Pacific company for operating trains using fragile cars.[82]

Engineer Clarence E. Warmington, who had been in charge of the locomotive of the express, died from his injuries on Sunday 7th April. It was reported that he had received orders to make up 15 minutes' time over the 30 miles east of Colton. The last six miles into Colton had been covered in five minutes and his train had regained the whole 15 minutes when it reached the switch. "Hot Stuff" Warmington, as he was known, was stated to have been 'one of the ablest railroad engineers in the country', [83] though his apparently reckless running on the day of the accident in an attempt to regain a few minutes would seem to indicate otherwise.

---

[81] *The Ogden Standard*, Ogden City, Utah, United States, 29th March 1907

[82] *The Logansport Pharos*, Logansport, Indiana, 4th April 1907

[83] *The Daily Californian*, Bakersfield, California, United States, 9th April 1907

On Thursday 11th April Coroner Van Wye swore out warrants on charges of manslaughter against Switch Foreman Alvord, Switchman J. G. Crusemeyer and (presumably unaware of his death several days earlier) Engineer Warmington. [84] No report of any trial has come to light.

In mid-May the Interstate Commerce Commission announced that it intended to make a thorough investigation of the disaster and commenting on a testimony forwarded to him by Coroner Wye, I.C.C. Secretary Edward A. Mosley remarked that, 'We note it is charged that the railroad company supplied to the Italian passengers cars that were unfit for service. We wish to have all possible light on this subject, also to know all the circumstances surrounding the wreck.' [85]

Westbound Canadian Pacific Railway train No. 1, known as the Transcontinental Express, which had left Quebec, Montreal, on the morning of Tuesday 9th, was derailed by a broken rail near Pardee, Ontario, Canada, 23 miles east of Chapleau, on Wednesday 10th April 1907. Five cars went down an embankment into a ditch and they immediately caught fire, either from overturned stoves, broken oil lamps or a cooking range, all of which were given as possible causes. Trapped in the cars, a number of the passengers burnt to death. The railway company stated the following day that 15 lives had been lost, and 30 or 40 others were injured, mostly slightly. Most of the victims were immigrants from England, who had come across to Canada on the SS *Kensington*, their journeys having been made under the auspices of the Salvation Army. Those killed were named as William Day, of London, England; Winifred and Susan Horton, of Charing Heath, Kent, England; C. H. Golding, his brother, sister-in-law and two children, of Plaistow, Essex, England; Mrs August D. Jorklund and two children, of Worcester, Massachusetts, United States; Mrs R. B. Champion and baby; and Mrs R. Davidson and baby.[86]

By Friday 12th the *Winnipeg Tribune* newspaper had much more on the disaster, including the statements of the husbands and fathers of four of those who died. Mr R. Davidson said, "What makes me feel worst of all is the statement in the papers that I was with my family at the time of the accident and made no attempt to save them. I had left my dear ones to go and make tea at the end of the car, and after the crash came I cut my hands in many places breaking the glass windows to get out of the car and go to them, but when I got where they were I was too late, before I could find them the heat was so intense that it drove me back."

Mr Champion had a similar story. "The papers said I was with my dear wife and baby at the time of the accident, would to God I had been. I would have saved her life. I was in the lavatory two cars distant when the shock came, the door of which jammed; after great difficulty I got out through the window, but by the time I reached the car in which were my wife and family, it was almost consumed by the flames." [87]

The same newspaper was full of the stories of other survivors of the wreck, of which one must suffice, that of James and S. A. Knight, of Cambridge, England, bound for Brandon, Ontario and R. Johannesen, of Aberdeen, Scotland, heading for Winnipeg,

[84] *The Emery County Progress*, Castle Dale, Utah, United States, 13th April 1907
[85] *Santa Fe New Mexican*, Santa Fe, New Mexico, United States, 14th May 1907
[86] *The Winnipeg Tribune*, Winnipeg, Manitoba, Canada, 11th April 1907
[87] Ibid, 12th April 1907

Manitoba. '"It was about 11.45 in the morning when the accident occurred," they said, "and we were at a point about 1½ miles from Pardee, a little station west of Chapleau. We would judge that the train was running at the rate of about 35 miles an hour. There were nine coaches on the train at the time. Including two baggage cars, three colonist cars, a dining car, a sleeper and two first-class cars. We were in the colonist car next to the second baggage car, in which coach the majority of the passengers were men. It was almost dinner time and a great many people were cooking and otherwise preparing meals. The first we knew of the accident was when we felt the cars bumping over the sleepers. Immediately every man was struck dumb and for the moment the silence was deathly. Every man looked horror stricken, with open mouths and staring eyes, knowing instinctively that something was going to happen yet not having the power to move. Before we had time to think we were all dashed to the ground by the baggage falling on us and the coach overturning, and the scene was one of the wildest that could be imagined. People were screaming and moaning, window panes were being smashed open, and the injured, with blood streaming from their faces, were fighting desperately to crawl through the upper windows. We saw one man with his ear torn completely off and his eye knocked out.

'"When we got into the open, we found that we were standing on ice and we afterwards discovered that the wreck occurred along the beach at Huron Bay, on Lake Superior. The engine and first baggage car never left the track, but the second baggage car was overturned, also two tourist cars, another one of these being off the track somewhat, but not overturned. It was most fortunate that the accident did not occur a little later in the season, as in that case the casualties would have been very great, as several would have drowned. The bank where we went over was about ten feet in height, and our car must have slid yards on the ice, which was otherwise most fortunate, as otherwise it might have telescoped.

'"The wreck occurred on a sharp curve and was evidently due to the outside rail spreading. There was no broken rail and the accident was probably caused by a fish plate coming loose. Allowing the rail to become free after the engine had passed over it.

'"Fanned by the brisk wind from the lake the fire which broke out almost before the cars stopped rolling burned intensely. The scene that met our gaze is too terrible for even the strongest imagination to grasp. Amid the crackling of the fire, with its flying smoke and cinders, came the wailing and screaming of the victims, the moaning of the dying the terrified cries of infants and above all the cry that struck home to the heart of every man present, freezing the blood in the veins. Help! Help! It came. My God, help! Fire! The cries were frenzied and terrible to hear. Occasionally one of the acetylene gas cylinders underneath cars exploded with a crash. From the burning baggage car came a long drawn out and melancholy howl of a bloodhound, accompanied by the frightened barks of four other dogs as they scratched desperately at their boxes in an effort to free themselves.

'"Before the fire became too intense many men made heroic attempts to save life, and among other things we noticed was a man in a soldier's greatcoat, who climbed back through a window and brought out a woman. This sight was thrilling in itself but we were to be eye-witnesses of one of the greatest tragedies possible and to witness the scene of a mother sacrificing her life for her child. The woman had barely been taken from the cars before she remembered her child, and forgetting everything else, she swiftly climbed back through the window despite the efforts of the soldier to detain her, and she was never seen alive again. It was a terrible affair.'" [88]

The soldier was identified in another newspaper as Mr James Gibson, late 12th Lancers, who was going west to settle.[89]

The inquest into the disaster opened at Chapeau on 18th April. Roadmaster Fought, in charge of the section of line between Chapeau and White River, confirmed that the derailment had been caused by a broken rail. He said that breaks were more common at the time of year when the accident occurred due to frost getting in the rail at night and thawing occurring during the day. There was also a possibility that the track had spread on the curve due to the action of the frost. The rail involved weighed 75 lbs/yard. Under the instructions of the general superintendent it had been sent to North Bay to enable the carbon content of its steel to be determined. Continuing the evidence, an unnamed employee stated that the acetylene gas which had made the fire so intense had been stored in tanks placed beneath the cars from which it could not escape if its supply was turned off by a main lever in a closet at the end of each car. However, brakemen often forgot to turn off this lever, and he thought that must have been the case in the tourist car, permitting the gas to escape and feed the fire. No special instructions had been issued as to how to turn off the gas and even old experienced employees sometimes forgot to. The inquest was then adjourned to hear the evidence of the Conductor Joseph LaRiviere and Engineer Herbert Gay of train No. 1, also of Section Foreman Antotone Bouillion and labourers Charles Leon and Alf Devars. The newspaper noted that the last three were all French-Canadian and could only speak a few words of English.[90]

At the close of the inquest on 19th April the jury returned a verdict of 'Accidental death', adding a strong recommendation that a safer form of train lighting than acetylene gas should be employed.[91]

Eydtkuhnen – Posen – Berlin fast train D 52 was derailed between Tremessen and Talsee stations, Germany, at about 1 a.m. on Wednesday 7th August 1907. Fortunately the train was booked to stop at Tremessen and hence was not travelling at full speed. The two engines overturned and the leading three vehicles, a luggage van and third- and second-class carriages, were 'smashed to pieces'. The engine crews managed to jump clear, but eight passengers were killed and another 25 were more or less seriously hurt.[92]

Work was going on with the track at the point of derailment. Preceding train D 52 had been a stopping passenger train, an express which did not stop at Tremessen, and two goods trains. The driver of the second goods train, the one which had immediately preceded train D 52, had felt that there was something wrong with the track at the spot where the derailment occurred but instead of stopping at Talsee he had not reported the matter until he reached the next station at Gnesen. It was then too late to warn train D 52 at Tremessen, but a warning was sent to a crossing keeper near where the track repairs were in process. This keeper then attempted to warn the crew of the train with a lamp but was too late.

---

[88] Ibid.

[89] *The Times*, London, England, 13th April 1907

[90] *The Winnipeg Tribune*, Winnipeg, Manitoba, Canada, 19th April 1907

[91] Ibid, 20th April 1907

[92] *The Times*, London, England, 8th August 1907

The accident was one of a number of similar ones at this time which arose from the view that it was not necessary to impose a speed restriction over a section of line on which routine permanent way repairs were being undertaken. It was only later fully understood that if the track had been disturbed too much it was necessary to restrict the speed of trains over it until the track had settled down again.[93]

Train No. 6 of the Chesapeake and Ohio Railway was derailed a mile below Kanawha Falls, West Virginia, United States, at 8 p.m. on Monday 2nd September 1907. Described as a local, the train also included a combination mail and express car, and it was this car, together with the baggage car and just the leading passenger coach of the train, which left the track. The vehicles ran along the ties (sleepers) for perhaps a hundred yards before plunging down a steep embankment to the edge of the Kanawha River. A broken coupling ensured that the remaining two passenger coaches remained on the track and their occupants escaped injury. Five or possibly seven people died in the accident and between six and 20 were injured. Opinions varied amongst the railroad staff as to the reason for the accident but spreading rails were thought to be the most likely cause.[94] [95]

A Sète to Bordeaux train met with an accident on the line between Toulouse and Montauben near Grisolles station, France, on Tuesday 8th November 1908. A broken rail caused the train to leave the track and a coach was smashed up against the parapet of a bridge, while the following carriages were also derailed. Ten passengers died in the accident and about 60 were injured, including nine postal workers. The dead were mostly reservists returning from a stretch of military service. M. Barthou, the Minister of Public Works, went to the scene of the accident.[96] [97]

The small South American country of Ecuador had one of its few bad railway accidents on 24th February 1909. A northbound passenger train on the main line left the track 'due to a displaced rail' and was hurtled over a cliff 100 ft high near Riobamba. Twenty-five passengers were killed and another 40 injured.[98]

No. 2 up Calcutta-Madras mail train, which was fortunately only carrying a few passengers, was running between Minjur and Ennore, India, when it was derailed near the latter place on 19th June 1909 by a 'defective rail'. The coaches of the train piled up on the derailed locomotive and 15 passengers were killed and 15 injured. 'Rescue parties extricated the bodies of some of the dead from the wreckage and all of the injured were taken out and rushed to Madras Hospitals.' [99] [100] The derailment occurred on the 5 ft 6 in

[93] *Schatten der Eisenbahngeschichte* by Hans Joachim Ritzau pg. 213 (in German)
[94] *The Punxutawney Spirit*, Punxutawney, Pennsylvania, United States, 3rd September 1907
[95] *Rushville Daily Republican*, Rushville, Indiana, United States, 3rd September 1907
[96] *The Times*, London, England, 9th November 1907
[97] Wikipedia – Liste des accidents ferroviaires en France au XX siècle
[98] *The Times*, London, England, 25th February 1909
[99] Ibid, 21st June 1909
[100] *The Evening Observer*, Dunkirk, New York, United States, 22nd June 1909

gauge Madras and Mahratta Railway; the official report attributed it to 'kinks in the line' at mile 13-4.[101]

Wabash Railroad fast passenger train No. 4, in charge of Conductor William Feye, with Engineer W. A. Flowers and Fireman Lee Bond, met with an accident at 10.10 p.m. on Saturday night, 24th July 1909, near Hull's Point, Missouri, United States. At the point of the accident problems had been foreseen with damage to the railroad from the adjoining Missouri River, and a barrier of 75 feet of earth had been provided between the railroad and the river. A watchman had seen no sign of any danger a couple of hours before the accident. He had been stationed at the spot for a week or so since excessive rainfall had caused the river to run very high, but no damage to the track had occurred in that time and no slow order had been issued to train crews. It was theorised that the strong current of the river had undermined the sandy soil, unnoticed and undiscovered. The undermining had extended over the 75 feet between the river and the track and for an equal distance on the far side of the line. When the train arrived, the track, if it had not subsided before, gave way beneath the weight of the engine and vehicles and engine, baggage car, mail car, a coach and two Pullman cars were precipitated into the water. A fortunate feature of the derailment was that in falling the mail car was split open, permitting the mail clerks to escape. The rear vehicles of the train initially remained on the track, and another fortunate circumstance was that they were drawn back when they were by the relief train which took the injured to Kansas City. Twenty minutes after the coaches had been removed from the scene another 100 yards of track where the coaches had been standing fell into the river as well.

Engineer Flowers, Fireman Bond, Baggageman Harry Echer and passengers William Turner, Jess Rohannan, Charles Anthony and Donald King (a two-year-old child) lost their lives. It was believed that three other passengers had been fatally hurt.[102] A woman doctor who had been a passenger on the train was said to have worked like a Trojan to care for the injured, assisted by other women passengers.[103]

At 6.30 on the morning of Wednesday 15th December 1909 Southern Railway passenger train No. 11 was derailed two miles from Reidsville, at the Reedy Fork Trestle, North Carolina, United States. The train in question ran locally between Richmond, North Carolina, and Atlanta, Georgia with sleeping cars from Richmond to Charlotte, North Carolina and from Norfolk, Virginia, to Charlotte. The locomotive, mail and three baggage cars passed over the trestle safely, but the rear two coaches and the two sleeping cars were derailed and plunged down the 50 ft embankment at the far end of the trestle and into the small creek. The Richmond Pullman car, *Ephesus*, was totally demolished, and its wreckage ended up in the creek. [104] Practically all the passengers in the derailed rear coaches were either killed or injured. Pullman Conductor D. C. Nolen, Pullman Porter Isaac Donlin, Flagman Charles Bagby and 11 passengers were killed, and 30 other passengers were injured. Luckier than most were New York millionaire George Gould

[101] Administration Report on the Railways in India for the year 1909

[102] *Moberly Weekly Monitor*, Moberly, Missouri, United States, 27th July 1909

[103] *Atchison Daily Globe*, Atchison, Kansas, United States, 26th July 1909

[104] *The Enterprise*, Williamston, North Carolina, United States, 24th December 1909

and his son, travelling in one of the Pullmans from Norfolk to Thomasville, North Carolina, who were only slightly hurt.[105] Conductor G. H. Goble also had a very narrow escape from drowning.[106]

The cause of the derailment was later identified as a broken rail 200 feet from the trestle.[107]

What might be regarded as a major catastrophe caused principally by track defects occurred on the Chicago Great Western Railroad between Green Mountain and Gladbrook, Iowa, United States, on 21st March 1910. The events leading up to the disaster began with the washing away of a bridge on the Chicago Rock Island and Pacific line at Shellsburg, near Cedar Rapids, Iowa. As a result the Rock Island's Twin City Express was diverted over the Chicago Great Western line. The Twin City Express was a combination of two trains, one of which, train No. 19, had left Chicago, Illinois, at 4.15 p.m. the previous day and the other train, No. 21, which had left St Louis, Missouri, two hours earlier. [108] Double-headed, the combined train, which because of the earlier accident had been diverted over the Chicago and North Western and Chicago Great Western lines between Cedar Rapids and Waterloo via Marshalltown, was supposedly travelling at 25 miles an hour when it was derailed. The accident occurred at about 8.15 a.m.

The ten-coach train was being operated with the engines running tender first at the moment of the accident. At the front of the train, behind the two locomotives and forming part of train No. 21, were the *Colonia* Pullman sleeping car, a smoker car and a ladies' day coach, Chicago Burlington and Quincy Railroad coach No. 5616. The first engine, No. 820, left the track in a deep cutting, taking the second engine, No. 1009, with it. Both buried themselves in a bank of thick, wet, yellow clay, stopping the train so suddenly that the Pullman car was catapulted into the smoker, and that car into the ladies' day coach, which was telescoped for half its length. [109] It was in these two coaches that most of the deaths and injuries occurred. The accident occurred about midway between Green Mountain and Gladbrook stations. Early reports put the number of dead at 42, and the number eventually rose to 55. Those who lost their lives included half a dozen railwaymen; Engineers Henry Mott and R. S. Robinson, Conductor Jacob Neuholz (or Nausolz) of train No. 21, Brakemen C. Harter and Ross Charter and Porter Archie Price.

News of the disaster was telephoned from a nearby farmhouse and doctors arrived by motor car, as had happened as early as 1902 following the Arleux accident in France. The physicians came from Gladbrook, Green Mountain and Marshalltown, the bad condition of the roads on the low black soil making their journeys difficult. Meanwhile the uninjured passengers did what they could, women tearing up their skirts to make bandages.[110] Train No. 419 was a short distance behind the wrecked train and after the disaster was brought into use as a relief train and brought the injured and the bodies of

[105] *New Holland Clarion*, New Holland, Pennsylvania, United States, 18th December 1909
[106] *The Gettysburg Times*, Gettysburg, Pennsylvania, United States, 16th December 1909
[107] *The Citizen*, Berea, Kentucky, United States, 23rd December 1909
[108] *Las Vegas Optic*, Las Vegas, New Mexico, United States, 21st March 1910
[109] *The Vindicator and Republican.*, Estherville, Iowa, United States, 23rd March 1910
[110] *Las Vegas Optic*, Las Vegas, New Mexico, United States, 21st March 1910

those killed to Marshalltown, while between 1 and 2 p.m. a Chicago Great Western breakdown train arrived from Oelwin to clear the line.

A reporter for the *Waterloo Evening Courier* managed to interview one of the passengers after the accident: 'E. J. Kline, living at 1508 West Fourth Street, was a passenger on the ill-fated train, and escaped most miraculously. He arrived in Waterloo this afternoon at 3.20 and was the first eye witness to tell the tragic story of the disaster. "I am very lucky to be alive," said Mr Kline this afternoon to a Courier reporter. "I was sitting in the front of the smoking car with two gentlemen, strangers to me, and desiring to get a drink, I remarked, 'Excuse me gentlemen, I am going to the water tank.' Before I reached the tank there came an awful shock and crash. When I came to I saw the bodies of three women hanging through the floor of a car above me. They were moaning but their moans became more and more feeble until they ceased altogether. As soon as I could stir myself about, I glanced at the other end of the coach and found my seatmates both dead, horribly crushed. When I crawled out of the coach, I was surprised to see several of the unhurt passengers running like scared rabbits. They seemed to have completely lost their heads. I called to the conductor and asked him about rescuing the injured and removing the dead. I grabbed the axe that the coaches usually carry and used it to cut away the roof. I was anxious to get to the women whose dangling bodies had attracted my startled gaze when I first recovered consciousness. But cut as hard as I could, I could not penetrate the tin roof of the car. The axe always rebounded. Later we got a rope from a farmer and did better work then."' [111]

On 30th March the Iowa state railroad commissioners spent all day examining the causes of the derailment without coming to a definite conclusion. It emerged that it had been John White, the Chicago Great Western engineer who was acting as pilot, who had made the decision that the train should be operated over the line with the engines running tender first, because he considered that the locomotives were too heavy to be turned on the wye at Marshalltown. Dispatcher F. C. Balkie at Des Moines thought that the engines should have been turned, and several dispatchers said that equally large engines had been turned successfully at Marshalltown before, but Balkie did not order White to turn the engines, leaving the decision up to him. As pilot, White had absolute authority over the Rock Island enginemen and conductors but in any case none of them had raised any objections or appeared concerned about the engines running tender first, though there seemed to be a general opinion that the speed of the train had caused the light tender of the leading engine to pitch and roll and led to the accident. [112] Initially it had been stated that the derailment had resulted from a broken rail.

By 8th April three people had been arrested for robbing the victims of the accident.[113]

The railroad commissioners issued their report on the accident at the beginning of June. They stated that the cause of the derailment could not be known with certainty, but that certain suggestive facts could be stated:

1. The track had been in a dangerous condition. It lay upon a bed of clay which was wet and springy due to improper drainage. The track had been unable to withstand the weight of two 67-ton engines and their tenders. That a fallen brake beam had

---

[111] *Waterloo Evening Courier*, Waterloo, Iowa, United States, 21st March 1910
[112] *Waterloo Semi-Weekly Courier*, Waterloo, Iowa, United States, 1st April 1910
[113] Ibid, 8th April 1910

contributed to the derailment by damaging the ties and rails was possible but unlikely. Observations made by the commission over the previous three years had shown that sufficient labour and money was not being spent on the maintenance of track on many roads. The first requirement for safety was a good track and roadbed. Too much public attention had been focused on passenger and freight rates and not enough on safety. The commission was aware that at the time of year when the derailment occurred there were likely to be soft and dangerous spots on any road, but with good foundations the danger could be minimised. The soft track and poor drainage appeared to be the primary cause of the derailment.

2. Although all the employees considered it safe to do so, the commission strongly considered that there was a great danger in running engines tender first, and this should be only done when absolutely necessary, over short distances, and then at a much lower speed than 25 miles an hour. A tender leading was top heavy and likely to be thrown out of balance, quite apart from the fact that it restricted the view of the engineer.

3. It was perfectly clear that it was extremely dangerous to run so heavy a train over a line in such a condition as that between Marshalltown and Waterloo. It could not be said with certainty, but it was very likely that if trains 19 and 21 had been run separately the accident would have been avoided. The operation of a heavy train, consisting of two engines and ten cars, most of them sleepers and of steel construction, over anything but a first-class track must be condemned.

4. No one riding in the rear seven cars of the train had been hurt. It was the placing of the light wooden day coaches at the front of the train between heavy steel coaches which had caused such an appalling loss of life. The lighter coaches should have been marshalled at the rear of the train.

In conclusion the commission stated that they felt that the time had come for the railroads to pay greater attention to public safety, and to be compelled to use the most modern scientific appliances and methods to that end. For example modern block systems to prevent collisions, and to scrap all the old wooden cars and replace them with steel ones. Questions of safety were of infinitely greater importance than those of speed, passenger and freight rates and other matters, and should receive much greater attention in the future.[114]

On Saturday 29th August 1911 a schoolteacher's excursion train from Utica, New York, to Atlantic City, New Jersey, met with an accident on the Pennsylvania Railroad at Martin's Creek, New Jersey, United States. The train had begun its journey on the Delaware Lackawanna and Western Railroad, transferring to the Pennsylvania at a point close to the accident. The train was running as the second section of train No. 573 and consisted of an engine, a combination car, three coaches and a dining car, all the vehicles having wooden bodies and steel platforms. Some of the coaches were lit by acetylene, others by Pintsch gas (a mixture of hydrocarbons and hydrogen). Slightly less than a mile south of Martins Creek station, while travelling at 50 miles an hour, the train left the track and went down an embankment. In the derailment the acetylene gas tank of coach No. 100 was punctured and fire broke out almost immediately. The train also reportedly sideswiped an oil tank as it left the track and oil ran out over the wreckage. The blaze

114 *Waterloo Evening Courier*, Waterloo, Iowa, United States, 6th June 1910

broke out at both ends, started by engine coals at the front and an upset stove at the rear. Engineer Walter Vancy, Conductor Charles Person, Baggageman Harry Wilmer and nine passengers died in the accident and 101 people, including Fireman George Parsons, were injured, many seriously. All the coaches were destroyed by fire and the heat caused the acetylene tank to explode some hours later.[115]

Some of the victims were cremated in the fire to the point that little or nothing remained of them. A number of Uticans went to the scene of the scene of the accident to try and find out what had become of friends or relatives. One of them was a Thomas Walker, the father of Besse Walker, one of those killed. An elderly man, he was heartbroken at what had happened. He could find nothing of his daughter, not even a scrap of dress or a shoe. More fortunate was Miss E. M. Cahill, of New York City, who heard of the accident and knew that her sister, Miss Daldee Cahill, an Utica teacher, had been on the train. "I have been through two awful experiences of this kind," she said, "and I can't stand another. Had my sister been killed I am sure I too should have died. I had another sister killed and a brother killed in a railroad wreck." She discovered that her sister had registered in at a Martins Creek hotel on the Saturday night, leaving for home on the Sunday morning, and had not been badly hurt. Upon learning this Miss Cahill publicly offered up a fervent prayer of thanksgiving and then collapsed.[116]

At the point of the accident the section foreman was engaged, with a gang of 17 men, in throwing, realigning and surfacing the curve. During the morning a flagman was used in both directions to protect the work and between noon and 1 p.m. a flagman remained to protect against trains from the north, the direction from which train No. 573 came. This flagman was also withdrawn at about 1 p.m., an hour before the accident. In the intervening period three southbound trains had passed the point safely, but all had been travelling at a lower speed than the excursion.

This accident was the very first for which the Interstate Commerce Commission published a report. They concluded that there could be no doubt that the derailment had been caused by uneven, irregular and insecure track which was not safe for a train to pass over at a speed of 50 miles an hour. In the opinion of the commission the work should have been protected by a flagman until it had been completed. The rules of the Pennsylvania company, as interpreted by an official, did not require flag protection for such work, and the commission considered that the rules should be amended so that they did. In addition they pointed out that it was the combination of wooden cars and gas lighting which had led to the high number of casualties and that had steel coaches and electric lighting been employed the accident would probably have been much less serious.[117]

A derailment and collision just outside of the French capital on the late evening of Saturday 18th May 1912 resulted from a faulty set of points. The 9.30 p.m. departure from Paris Nord to Creil travelled as far as an overbridge carrying the Rue de Doudeauville when it met an incoming suburban train from Montsoult. The incoming train had passed the first 15 coaches of the outgoing one when its driver suddenly found his engine being

[115] Interstate Commerce Commission report
[116] *Lock Haven Express*, Lock Haven, Pennsylvania, United States, 1st May 1911
[117] Interstate Commerce Commission report

diverted into the path of latter and his engine cut through the its three coaches, sparing only the brake van at the rear.[118] (The newspaper reports regarding this accident vary considerably in their descriptions of it, another newspaper stating that it was the outgoing train that was diverted into the path of the incoming one, and furthermore that the incoming train was not a suburban working from Montsoult, but an express from Chantilly, a little further out from the capital). Regardless of which train ran into which, the collision killed nine people on the spot, including according to the London *Times* the driver of the incoming train. Another four male passengers died later of their injuries. The guard of the outgoing train escaped and was able to stop several other approaching trains. After the collision the wreckage was speedily removed, and normal working was resumed.

A reporter for the *Standard* newspaper managed to find a London commercial traveller who had been a passenger in the outgoing train, which he had just managed to jump into as it was pulling out of Paris Nord station. '"There were four others in my compartment," he said, and we had just got going when there was a startling shock accompanied by a stupendous noise. The far side of the compartment had been torn off like so much cardboard. The passengers in my compartment were thrown together, and I saw that they were injured in the head. I myself received a terrific blow on the foot, but when I had taken things in I clambered out and looked around. Next to me was what looked like a bare platform. It was the framework and floor of a carriage, the body of which had been simply cut away from it by the colliding train. On and under the wheels were mutilated bodies and human remains, and at the end I saw the body of a man which had been terribly crushed between two coaches. I shall never forget the cries of the injured nor the actions of those who were almost mad with panic. Some of them shouted, 'Fire!' But I shouted to them that there was no danger of fire. A few yards away a soldier with both legs crushed was making a piteous appeal for someone to put an end to his misery. Two of the women passengers who had been in the next compartment were completely dismembered. Further on lay a young woman to whom I had meant to bring help, but I found that her head had been entirely severed from the body."' [119]

The Chemins de fer du Nord announced that the accident had resulted from the fracture of some rodding between the points and the signal box. A similar accident a few hundred metres away, near the Pont Marcadet, the previous December, had resulted in four deaths and a dozen injured, while the Calais express had been derailed only a few weeks earlier due to defective points.[120]

On Sunday 13th April 1913 the Central Limited real estate interests (corresponding to British estate agents) chartered a train to carry 700 prospective buyers of land free of charge the short distance from Montreal to Chambly, Quebec, Canada, on the Central Vermont Railway branch. The train was making its return journey to Montreal and travelling at between 20 and 30 miles an hour with the engine running tender first when it left the track a few hundred yards from East Greenfield station. The tender jumped to the right of the track and the engine to the left. Both engineer and fireman were thrown from

---

[118] *The Times*, London, England, 20th May 1912

[119] *The Standard*, London, England, 20th May 1912

[120] *The Times*, London, England, 20th May 1912

the cab and the fireman received injuries which proved fatal within a short time while the engineer escaped with painful bruises. The three leading coaches were derailed but remained upright, however there was telescoping of the ends of the coaches and it was the passengers in the first two who suffered the most. The train was crowded to capacity, with passengers standing on the platforms of the coaches, and it was they who fared the worst. Blinding steam from the boiler of the engine also penetrated the coaches, impeding the efforts of the surviving passengers to escape and causing painful scalding.

The passengers from the four rear carriages, which remained on the rails, went forward and attempted to assist those in the front of the train, but 'Their efforts were fruitless in the case of some of the victims whose bodies were so firmly wedged in the shattered remains of the coaches that they were only released after hours of work by the wrecking crew. Between the first and second coaches two bodies were crushed, one being that of little Maggie Deer, aged 10, and the other that of a yet unidentified man. They were still unreleased at midnight. A heartrending incident was the presence of the child's mother, herself badly shaken by the accident, who begged the rescuers to release her child. Another man, Rochon by name, was pinned under a truck, and although almost disembowelled, lived for several hours in an unconscious condition while efforts were made to effect his release.'

The accident happened at about 5.20 p.m. and it was an hour before the wrecking crew arrived with the necessary equipment and midnight before the relief train reached Montreal. A strict censorship of all information by the officials made identification of the victims difficult. The cause of the derailment was not entirely clear, but it was attributed to spreading rails. In addition to the fireman six passengers died in the accident and 15 were injured.[121]

Just before the First World War, at the height of an immigration boom, large numbers of poor people from Great Britain and elsewhere in Europe moved to Canada to start a new life in the Prairies, lured by government advertisements promising cheap land and a clean and healthy life in western farming communities. Thus a party of immigrants, mostly from Ireland and Scotland but others from Scandinavia and as far away as Russia, left Glasgow, Scotland, on 15th June 1913 and arrived at Montreal, Quebec, Canada after a voyage across the Atlantic on the SS *Pretorian* steamship of the Allen line. They spent the night there before boarding a Canadian Pacific Railway train to continue their journey westwards to Winnipeg, Manitoba, riding in spartan colonist cars with hard benches and little padding, or if they could afford it slightly less spartan tourist cars. The train was in charge of Conductor Daniel Cameron, with Engineer Ben Chapman, and left Ottawa Central station at 1.30 p.m. on 25th June 1913. It consisted, behind the engine, of baggage and mail cars, three colonist (third-class) cars, a first-class coach, two tourist (second-class) cars, a dining car and a Pullman sleeping car in that order. The passengers had only travelled a short distance further, and their train was skirting the Ottawa River at McKellar, Ontario,[122] when it left the rails. The locomotive, baggage and mail cars and leading colonist car remained on the track, but the next two colonist cars were derailed and toppled towards the river, one car turning over twice and the other just falling onto its

[121] *The New Westminster News*, New Westminster, British Columbia, Canada, 14th April 1913
[122] Webpage – Today in Ottawa's History: The McKellar Train Disaster

side. The coupling broke behind the third colonist car and the first-class coach following ended up with its nose pointing towards the river and its rear still on the track-bed. The tourist cars coming next were derailed and tilted over on the opposite side, away from the river, the dining car was derailed but remained more or less in a straight line and only the front trucks of the Pullman car at the rear of the train left the track.

The accident was witnessed from the lineside by Mr J. B. Carron. '"It was not running very fast," said Mr Carron. "I don't think it was going more than 25 miles an hour. Some of the children waved their handkerchiefs through the windows. Then I noticed the immigrant and colonist cars begin to wobble and there was a crash as they left the steel. The engine and the first two coaches went ahead and stopped about 100 yards west of the wreck, but the colonist and immigrant coaches toppled over on the brink of the bank and then rolled over. I rushed up the bank side of the track, but the car was then on its side in the river."'

By the next day it was known that eight people had died in the accident, another two were thought to be fatally hurt and over 50 were injured, all of them immigrants from the two colonist cars, except for Conductor Cameron, who had been punching tickets in one of the cars and who managed to escape from the river with a dislocated shoulder. The dead who had been identified were Patrick Mulvenna, aged 25, from County Antrim, Ireland; John Hogg, aged 30, from County Derry, Ireland; John Moodie, aged 17, of Sanday, Orkney, Scotland; Mrs Jane McNealy, aged 40, of Glasgow, Scotland and John Peace, aged 21, also of Glasgow, Scotland. Moodie and Hogg were bound for Calgary, Alberta; Mrs McNealy and Peace for Edmonton, Alberta and Mulvenna for Winnipeg, Manitoba. Another occupant of the colonist cars was a lucky survivor, 'Thomas Wallace, aged 45, a Scottish immigrant who was going to Edmonton was among those who had a miraculous escape.

'Wallace, together with two of his friends, David Wilkie, of Glasgow, and James Clements, of Londonderry, was in the immigrant coach which plunged into the river. He found himself struggling in the water but was fortunately close to a window. He broke the glass with his fist and dived down, coming up on the edge of the bank. To his swimming strength Wallace attributes his escape. He said he pulled one little boy out of the water and that another sank as he was doing so. The child who succumbed was afterwards found. The name could not be ascertained.' The newspaper report added that, 'There was remarkably little panic among the passengers. The Russians took things stolidly, while the British immigrants showed their breed.' On the other hand the newspaper reported that there were pathetic scenes as injured victims were removed, and some of the injured were taken away to hospital so quickly that their relatives did not know where they had gone. Most of the deaths were caused by passengers being hurled against the sides of the cars as they toppled over, though a handful of passengers drowned, including a woman and a boy who had been looking out of a window and who were trapped beneath the overturned cars. The water was only about three feet deep.

After the accident assistance came from people nearby, who were soon joined by doctors and nurses from Ottawa, summoned by telephone and arriving in automobiles. The injured were taken to two Ottawa hospitals. Wrecking trains from Ottawa and Smith's Falls soon arrived to clear the wreckage.

A brief preliminary inquiry into the accident was held by Chief Engineer Mountain, Inspector Lalonde and Chief Operating Officer A. J. Nixon of the Canadian Pacific. They theorised that the most likely cause of the derailment was that the track had been 'kinked' by the sun, and after withstanding the passage of the locomotive and leading vehicles had given way. A sketch was made of the scene and matters deferred for a day or so pending a more thorough investigation and the coroner's inquest.[123]

No progress had been made in determining exactly how the accident had occurred by the conclusion of the inquest in mid-July. The jury was told that the derailment of the train could have been caused by a "sun kink", some other defect in the track or a defect in the train itself. They could only return a verdict that the death of passenger John Peace had resulted from 'a cause unknown to them'.[124] The report of the commission has not been found but it seems probable that it was not able to come to any more concrete conclusions.

Abnormally heavy rain caused the Salko River to rise to an unprecedented extent and it washed away the approach embankment just beyond the south abutment of a bridge near Panchra station, India, on the Ondal – Sainthia branch of the East Indian Railway, early on the morning of Saturday 28th June 1913. The passenger train which left Panchra at 6 a.m. fell into the breach in the embankment, 10 of the 18 vehicles of the train being completely destroyed and the engine and remaining vehicles badly damaged. The bridge itself had not been directly injured by the flood but the impact upon it of the derailed vehicles at the rear of the train caused it to collapse as well. Twenty-three people lost their lives in the disaster and about the same number were injured.

The official report on the accident continued, 'A careful examination of this part of the line had been made only days before the date of the accident, and the occurrence can only be attributed to causes which could not have been foreseen; the flood level is reported to have been eight feet above any previously recorded. On the same occasion a light engine ran into a breach caused by a washaway in the railway embankment near the Bukleswar bridge; the engine crew, with the exception of the fireman, who was killed, fortunately escaped.

'In these cases it was not possible for the drivers to have seen the breaches in time to have pulled up, as it was raining at the time when the disasters occurred and the breaches would in any event not have been easy to notice at a distance owing to the dark masses of the bridge, etc., in the background.' [125]

A month and a day after the McKellar accident, on Saturday 26th July 1913, another train provided mainly for the benefit of Europeans hoping for a better life in the New World featured in a major disaster, this time in Denmark. Scandinavians wishing to emigrate to the United States or Canada had to make their trans-Atlantic journeys from a British port, first crossing the North Sea. Those from Denmark mostly used the DRDS steamers from the Danish port of Esbjerg to Harwich in England, continuing across England by train to an Atlantic port such as Liverpool. To cater for them the Danish State Railway set up an express train from København to Esbjerg, train No. 1029, commonly

[123] *The New Westminster News*, New Westminster, British Columbia, Canada, 26th June 1913
[124] *Winnipeg Free Press*, Winnipeg, Manitoba, Canada, 16th July 1913
[125] Railways in India: Administration Report for the Year 1913-14

known as *Emigranten*.

The day of the accident was a very warm one. Being a summer Saturday it was also a busy one, so train No. 1029 was strengthened (from Fredericia) to ten coaches and provided with a second locomotive so as to be able to keep time, engine No. K505 being used as the pilot engine and No. A 154 as train engine, with Driver C. F. Pedersen and Fireman R. H. Rasmussen (were they short of drivers?) in charge of the respective locomotives. The train should have departed from Fredericia at 2.50 p.m., but it had to wait for train No. 934 from Århus, four carriages of which formed part of its forward makeup, and it did not therefore get away until 3.06 p.m. There was thus an incentive to run fast to regain time, and although two minutes more had been lost leaving Kolding due to the number of passengers, three minutes were regained by Lunderskov. After the accident various witnesses, both railwaymen and passengers, were questioned as to the speed of the train between Lundeskov and Bramminge, the location of the accident. Most thought that the train was going faster than usual. Foreman N. A. Jensen at Brørup estimated speed of the train as 120 km/hr, but his assistant J. A. Weilling doubted it was going that fast, though he agreed that the express seemed to be running faster than usual.[126]

New 45 kg/m track had been laid in the Bramminge area the year before. On the morning of the accident work was in progress raising a stretch of track which was too low on the slight curve west of the station. Three rail lengths had been dealt with before dinner. After the meal break the ganger and one of his men went away on an errand while the remaining three platelayers removed the gravel ballast from the ends of the sleepers over 11 rail lengths. When the ganger and his subordinate returned at 3 p.m. the work of lifting the track was resumed. At this point two lengths of rail began to bend slightly; they were straightened using crowbars and ballast was shovelled up around the ends of the sleepers. The rail lifting was then halted until train No. 1029 had passed. The train soon appeared and as it was passing them there was a terrific bang and a cloud of dust, which when it settled revealed that the train had left the track.

Driver Pederson on the pilot engine noticed a bump as the train reached the derailment point, while his fireman, A. P. Sloth, observed a bend in the track; the crew on the train engine also noticed a bump and the next moment their engine was in the ditch; most of the train carriages followed the engine into the ditch on the north side of the line. The pilot engine was not derailed but broke loose and came to a stand 200 metres down the line. The three wooden carriages in the middle of the train suffered the most, being crushed between the rearmost three and the front four. In all 15 passengers died in the accident and 54 were injured, all but one of the deaths occurring in carriage AA 20. The more modern carriages had withstood the shock of the derailment better.

Among the passengers on the train was Goods Traffic Assistant Frans Buch of Århus, who noted that the derailment had happened at about 4.11 p.m. Unhurt, he had climbed out of a window and seeing the magnitude of it he also learnt that no one had yet gone for help. The derailed train had brought down the telegraph wires, so he made his way to an inn on the western edge of Bramminge from where he was able to telephone Esbjerg for a relief train and doctors. Meanwhile another railwayman on the train, named Jensen, had made his way to Bramminge station and informed the official there, A. L. Hansen, who

[126] Webpage Jernbanen.dk: Brammingeulykken

contacted two local doctors and also informed Esbjerg. Both doctors chanced to be out on their rounds and only reached the accident scene some time later but a nurse from København who chanced to be on holiday in Bramminge reached the accident site earlier, and a quickly organised relief train from Esbjerg arrived at 4.58 p.m. with the first physicians. A second relief train arrived from Fredericia at 6.10 p.m. The uninjured passengers were forwarded to Esbjerg by motor car. The steamer to Harwich was delayed for five hours to accommodate them.[127]

The accident attracted much coverage in the press, probably not only because of its severity but because one of those killed was what the London *Times* referred to as a 'Socialist member of Parliament', Peter Sabroes. The MP's mother was awarded 100 kroner/year in compensation, his wife 2,400 kroner annually and each of his children 500 kroner annually until they reached the age of 18. Eight of the others who were killed were Germans, including the then renowned opera singer Borre and his wife. King Christian X visited the accident site the next day.

Investigations into the accident were undertaken both by the railway authorities and, because there had been fatalities, by the police. However, the cause of the derailment remained unclear. There were suggestions that the track had been distorted by the heat and that the excessive speed of the train had been responsible. The engine crews denied that they had been going faster than permitted and pointed out that in any case a recent reduction in the maximum permitted speed had been imposed solely to save coal, and not for safety reasons. A calculation made on the basis of recorded train times gave an average speed between Gørding and Bramminge of 120 km/hr, but the report noted that this was not a reliable indication, given that a slight variation in times would give a considerable different figure. The speedometer of engine A 154 was faulty, but its driver alleged that he had not exceeded 90 km/hr. A test train was run but the results of the trial were not very helpful. In the end the police investigation blamed the two old four-wheeled waggons and the old bogie waggon (AA 20) plus the condition of the track. The railway company investigation concluded that the speed of the train had been too high, that the use of the three old carriages had no significance in causing the derailment and that the principal cause of the accident had been the kinking of the track. Finally an independent assessor, Professor Lütken, of the Polytechnic Teaching Institute, gave his opinion. He estimated the speed of the train at the moment of derailment as 105 km/hr and attributed the derailment to the excessive speed of the train, exacerbated by the condition of the track. To the author, who is by no means an expert on the subject, it would appear that the disturbance of the track over a considerable length caused by the removal of the ballast beneath the sleepers on a hot day led to the beginnings of the buckling observed by Fireman Sloth on the train engine, which then developed into a kink sufficient to cause derailment beneath the train itself, the speed of the train having little to do with the matter.[128]

In December 1913 the Ministry of Justice announced that it was not believed that anyone could be held responsible for the accident and so no charges would be brought against anyone. By 1915 compensation of 715,000 kroner had been paid to victims. Another consequence of the accident was that the marshalling of old carriages between newer ones was prohibited.

[127] *The Times*, London, England, 28th July 1913

[128] Webpage Jernbanen.dk: Brammingeulykken

Japan was affected by a severe typhoon over a period of several days around 28th August 1913. Miles of railway line on the island of Hokkaido were affected by flooding and many bridges washed away. At one point a passenger train was hurled into a river, seven of its passengers being killed and 40 injured.[129]

Twelve people were initially reported to have lost their lives and more than a hundred injured, some fatally, when passenger train No. 18 of the Central of Georgia Railway, with Conductor Goodroe, Engineer Cox and Fireman Richardson left the rails and plunged down a steep embankment 17 miles south of Eufaula and near Clayton, Alabama, United States on Thursday 13th November 1913. The train of five cars was crowded with excursionists en route from Ozark to Eufaula for a fair. The derailment was blamed on a broken rail. As the train rounded a curve the rear three of the five cars left the track, broke free of the front two, went down the embankment and were demolished. The lucky passengers in the leading two cars went to the assistance of their less fortunate fellow passengers while a message was quickly sent to Clayton, three miles away, and relief trains, with doctors and nurses despatched from both Eufaula and Ozark. The injured were taken by train to Clayton.[130] Different newspapers spoke of a final total of 15 or even 22 dead in the accident but the Interstate Commerce Commission (ICC) stated that only nine died, with no less than 389 passengers and one railwayman being injured.

According to the ICC report the rail which broke had weighed 56 lbs to the yard and had been 40 years old. The track was laid on untreated pine and cypress ties and without either tie plates or ballast, the track being laid directly on the sandy soil. So severely were the sleepers decayed that many of the spikes could be pulled out by hand. In the nine-mile section of track within which the accident had occurred 14 broken rails had been found in the previous month, but only one replaced. Section Foreman Beasley, who was in charge of this length, stated that he had insufficient men to maintain the track properly, while Track Supervisor Minor was of opinion that the track was reasonably safe and not particularly in need of attention, he being content to allow four or six rotten sleepers beneath a rail provided they were not together. The line normally only saw two trains a day. After the accident inspection disclosed 16 broken rails within a mile of track west of the derailment. Many of the breaks had been repaired by joining the rails with fishplates; in a few a hole had simply been drilled in the web at the end of one rail and the two rails joined with straps without bolts. Such poor track is reminiscent of course of that which allegedly led to the 1906 Camden, Washington derailment.

In addition the highest permitted speed between stations according to the time card was 33 miles an hour, though Section Foreman Beasley said that the train passed him a mile from the scene of the accident running at 40 to 45 miles an hour. Engineer Cox said that he had reduced speed to 5 miles an hour at a trestle east of Clayton and had not reapplied steam. Fireman Richardson estimated the speed of the train as 20 to 25 miles an hour at the point of derailment. Three passengers on the train did not notice any reduction of speed and had remarked that the train was going fast. Conductor Goodroe, preoccupied with collecting tickets, did not notice the train's speed, and said none of the passengers

[129] *The Sun*, Sydney, New South Wales, Australia, 1st September 1913
[130] *The Washington Post*, Washington, D.C., United States, 14th November 1913

had mentioned it to him.

The ICC considered that the track had been in a deplorable condition. With the condition of the track and the lack of sufficient staff to properly patrol it and carry out repairs disasters such as that which had taken place could be expected to occur. The derailment had been directly caused by a broken rail, poor track and too high a speed but with the condition of the track it was doubtful even if the comparatively low speed required by the schedule was safe. Furthermore there was nothing to prevent the engineer running at 60 miles an hour at certain points, provided that he did not go below the minimum time between stations. A definite maximum speed should be imposed, consistent with running trains relatively safely over the patched rails and rotten sleepers.[131]

Our next accident took place on Christmas Eve, Wednesday 24th December 1913, in the Netherlands. To cope with the anticipated extra Christmas traffic train No. 146 had been strengthened by the addition of a baggage van and six extra carriages so that engine No. 76 ended up pulling a train of 15 vehicles. The train left Groningen eight minutes late and after delays crossing two trains left Assen 22 minutes late. After Assen the line crossed a drawbridge over the Oranjekanaal, and immediately beyond the bridge was a loop siding, used for the loading of silt from Ems-Dollard estuary. The two sets of facing points leading from the single line into the siding were worked by individual hand levers at the spot; they normally stood closed and locked for the main line and were found to be properly locked after the accident. As the train was passing over the facing points leading to the siding its last two vehicles left the track. They were dragged along the ground for some distance, then the train became divided into three parts. The penultimate vehicle of the train, carriage AB 2088, tipped over and collided with a stack of sleepers which had been stacked in the six feet between the main line and the siding. The side of the carriage was thereby ripped out and five passengers in it were killed, one was seriously injured, and another suffered a broken arm. Those elsewhere in the train escaped injury.

The bridge tender saw the accident and immediately summoned assistance from nearby Beilen. An hour and a half after the derailment the train, minus its rear two vehicles, continued to Zwolle with all but the seriously injured passenger, who was freed from the wreckage later in the evening and taken back to Groningen. The next day the wreckage was cleared, the damaged track repaired, and normal working resumed with a speed restriction at the accident spot.

The accident was only the second serious one to occur in the Netherlands, the other being the Capelle collision 14 years earlier, furthermore one of its victims was the son of the Prime Minister, Pieter Cort van Linder, and it caused something of a sensation in the country with calls for the results of the investigation into it to be made public. In the end there were in fact two investigations, one undertaken by the Maatschappij tot Exploitatie van Staatsspoorwegen (the railway company – usually abbreviated to SS) and the other by the Raad van Toezicht op de Spoorwegdiensten (the official accident investigation board). The investigations came to very different conclusions, the railway company finding that the points had been defective and the accident investigation board that carriage AB 2088 had lurched slightly as it left the drawbridge, sufficiently to throw its

[131] Interstate Commerce Commission report

front axle off at the points. The points were replaced with a new set on 29th December.[132]

Torrential rain, lasting according to a newspaper correspondent almost 36 hours, undermined the track near Lebanon, Missouri, United States, and caused the derailment of the St Louis-San Francisco Railroad's passenger train No. 5, the southbound *Texas Limited*, in charge of Conductor Laughlin, at 2.57 a.m. on Tuesday 15th September 1914. The coaches plunged into a water-filled ditch 12 ft deep. It was of course dark when the accident occurred, and Engineer O'Brien had little warning of any problem. He survived the derailment and afterwards telegraphed his company headquarters, as follows: 'I took charge of the engine at Newburgh, Mo. I was running very slowly after we left Lebanon. About two miles west of Lebanon I saw a quantity of water rushing down a ravine which we entered. I stopped the engine but was not able to stop quick enough to avoid the water. The engine, mail car, baggage car and smoker were overturned and buried in water.'

By the following day 26 bodies had been recovered from the submerged coaches. Many of them were scantily clad and identification was proving difficult. Telephone and telegraph communication in the area around Lebanon had been interrupted by the floods while several relief trains sent to the scene had been held up by washouts elsewhere. By the time of the newspaper report the next day only six of those who lost their lives had been identified; Mrs Elizabeth Rostetler was from Alliance, Ohio while the others were all from Missouri. They were: Henry Waggoner, of St Louis; W. W. Childers, of Billings and John Meyers, his wife and daughter from Thayer. It was believed that many of those who died had, yet again, been immigrants and that the dead would thus largely be foreigners. 'Those who escaped from the wreck did so by breaking the windows and swimming to the edge of the gulley. The fate of the mail clerks and baggagemen is not yet known.' [133]

Heroine of the wreck was Nurse Mona Campbell of St Louis. Having escaped by smashing a window she climbed out onto the side of the car and standing in the water felt about it until she touched the hand of a drowning passenger. As soon as she found one, she pulled them out with all her strength and in this way was able to save the lives of five people. On a less happy note a man who managed to escape found that his wife and child were pinned under the water in the chair car; he reportedly became a raving maniac. Although his driver swam to safety the fireman of the train was one of the victims.[134] The final number of dead was 27, all passengers except the fireman and one mail clerk.

At the point of derailment a creek had formerly been crossed at two different points by trestles, but to avoid the danger of their catching fire these had been filled in in 1904, with the stream being diverted to form a waterway running along the south side of the track between the two points where it had formerly crossed the railroad. As a result the stream met the fill almost at right angles and 180 feet of rip-rap had been laid to protect the embankment. The creek was an arroyo, dry for part of the year and often carrying little water. The local section foreman said that because of the heavy rain he had set off from his section house, 5½ miles west of Lebanon, at 1 a.m. to inspect the track. He had

[132] *Spoorwegongevallen in Nederland* by R. T. Jongerius, pgs. 91-8 (in Dutch)

[133] *The Oelwein Daily Register*, Oelwein, Iowa, United States, 16th September 1914

[134] *Bureau County Tribune*, Princeton, Illinois, United States, 18th September 1913

never known of a problem in the vicinity of the derailment and hence had gone along the line in the opposite direction and had not gone to the scene until after the accident. Older employees could remember no problems there since at least 1876.

The Interstate Commerce Commission, in a short report, found no one to blame for the accident and concluded simply that it been due to the track having been undermined by high water following heavy rains.[135]

A mixed train was derailed between Monyó and Kölnök, Hungary, on Tuesday 20th April 1915 due to the subsidence of an embankment. Five passengers died and 17 were seriously injured.[136]

On Sunday 22nd August 1915 a passenger train was derailed between Eschenau and Forth, Germany, on the Nürnberg – Gräfenberg line, due to a broken rail. Five people died and others were injured.[137]

In what was almost a repeat of the Lebanon disaster a passenger train was derailed on Saturday 16th October 1915 near Randolph, Kansas, United States, on the Union Pacific Railroad's Manhattan branch. Unusually for the time the train involved did not consist of a steam locomotive and coaches but was a gasoline (petrol) powered McKeen motorcar (rail motor), a concept pioneered by the Union Pacific as early as 1905.[138] Westbound train No. 579, that involved in the accident, consisted of motor car No. 13 and trailer No. 4, the latter a combination mail and baggage car. The train was filled with teachers going to the county convention. In charge of it was Conductor Knittle, with Motorman O. M. Warren. As at Lebanon there had been very heavy rain in the time leading up to the derailment, with an estimated five inches of precipitation in some places in the previous 18 hours in the catchment area of Fancy Creek, which ran alongside of the railroad for a short distance before it was crossed by a steel truss bridge.

The train left Manhattan, Kansas, at 7 a.m. and the derailment occurred at about 7.30. Motorman Warren (according to the Interstate Commerce Commission report, Warner in the newspaper) was thrown through the window of the car as it left the track and was able to wade ashore. He described his experiences. "'I couldn't see anything the matter with the track ahead of me," said Warner tonight. "I supposed though the dirt had been washed from under the track at the end of the bridge. There was enough clinging to ties to deceive the eye. The first hint I had that anything was wrong was when I was thrown through the glass in front of the car into about four feet of water. I was unhurt and got to my feet in time to see the front end of the car hanging straight down over the abutment. It seemed to hang for a half minute, then slowly settled down in the water, standing almost perpendicular and all the time I could hear shouts of men and screams of women imprisoned in the car as it settled. I ran to get help and met men running from the direction of Randolph. By the time we got back there were many women in the water,

135 Interstate Commerce Commission report
136 *Eisenbahnunfälle Bind 1* by Ludwig Stockert, pg. 80 (in German)
137 *Eisenbahn-Katastrophen in Deutschland* by Hans Joachim Ritzau, pg. 32 (in German)
138 Webpage – Utah Rail.com: Union Pacific motor cars

and we helped them out.'" [139] All of those were in the smoking compartment at the front end of the car, which was totally submerged, lost their lives. In all 11 passengers died in the accident and 41 passengers and three railwaymen were injured. The trailer did not leave the rails.

The Interstate Commerce Commission (ICC) investigation into the accident found that the swollen waters of Fancy Creek had washed away and undermined the embankment on the approach to the bridge, leaving rails, ties and some of the fill suspended in the air so that the track gave way under the weight of the train. Neither the motorman nor the conductor had anticipated any danger.

Section Foreman Well said that the water in the creek had been low the night before. He too had not expected any problems there and in the morning had instead gone to the western end of his section, where he considered there was more risk. He considered the Fancy Creek bridge to be the safest of the 10 bridges in his section. He could remember no problem there in the previous 11 years.

Chief Engineer R. L. Huntley stated that the bridge had been there 30 years. Floods in 1903 had washed away its eastern approaches and there had been periods of high water since, but inspections had failed to reveal any danger. He considered that very heavy rains in the territory drained by Fancy Creek had caused a rush of water which had struck and washed away the bank, then formed an eddy which had undermined the embankment. The ICC concluded that the accident had resulted from the washing away of the bridge approaches due to conditions which could not have been anticipated.[140]

[139] *The Telegraph-Herald*, Dubuque, Iowa, United States, 17th October 1915
[140] Interstate Commerce Commission report

CHAPTER 2

DERAILMENTS - BROKEN TYRES, BROKEN AXLES AND OTHER VEHICLE DEFECTS

One of the frequent causes of derailments of trains other of problems with the track was defects in the trains themselves, such as broken tyres, wheels and axles. Such accidents were not as frequent by the 1880s as in the early days of railways but they still happened quite often and were to a large degree unavoidable. This chapter will look at some of them.

At 16:54 on Thursday 31st July 1902, No. 19 up mixed train was entering Khatauli station, India, on the North Western State Railway, when it became derailed on the set of facing points leading to the platform loop line. (The Parliamentary report uses the 24-hour clock even at this early date). The engine of the train buried itself up to the buffers in a bank and the leading brake van, a second-class carriage, a bogie third-class carriage and the following vehicle were completely wrecked. Eighteen passengers were killed outright or died soon afterwards and 27 were injured, 10 of whom subsequently died. A report in the London *Times* stated that 16 'natives' had died and 21 'natives' and nine Europeans had been seriously injured. Some of the Europeans were possibly among those who later died of their injuries. The derailment was blamed on the flange of the right leading wheel of the engine having worn too sharp, plus the train having been running too fast.[141] [142]

A bad accident in Canada on Thursday 7th May 1903 involved a tie gang who were riding in a boarding car of a westbound train after having distributed ties (sleepers) along the line. The men were asleep when the accident happened about a mile and a half west of Dexter, Ontario, on the Canadian Pacific Railway. A broken wheel on the boarding car caused it and four other cars to leave the track. The car was marshalled between a dining

---

[141] *The Times*, London, England, 2nd August 1902

[142] Administration Report on the Railways in India for the Calendar Year 1902

car in front and a flat car and caboose behind. When the derailment occurred the sleeping (boarding) car was telescoped by the following flat car and the wreckage of the wooden coach was immediately set alight by a log fire which had been burning in the car to provide warmth. The fire spread very rapidly, and 12 of the 20 men died, and their bodies were largely consumed by the fire. Of the eight who escaped one managed to run through the end door of the car but was badly burned, and the others escaped after one of them fell through a window and the others followed. Those who lost their lives were F. Bystrom, E. Erikson, F Hickman, A. Johnson, A. Kunkall and L. Larson, who were all Swedes, and G. Dangluk, Fred Fluk, Ivon Hudena, Margk Kurlee, G. Lundmark and A. Watin, all from Galicia in Eastern Europe. Those who escaped were mostly badly hurt and it was feared initially that a number of them would die but all appear to have survived. The foreman, a timekeeper and a workman in the car ahead escaped with slight injuries and one passenger, Jas Reid of Port Arthur, who had been riding in the second car, somehow managed to sustain a broken arm and dislocated shoulder, possibly while helping in the rescue work.

As soon as news of the accident was received in Fort William, Ontario, a wrecking train was sent out from there, accompanied by the Canadian Pacific surgeon and other doctors from the town but before Dexter was reached word was received that the services of the other doctors were not required, and they returned to Fort William on No. 2 passenger train. Dr Hamilton went on to the wreck, bandaged up the injured men and they were placed in a caboose and reached Fort William at about 10 a.m. The men, probably all recent immigrants, had been engaged at Winnipeg.[143]

At the beginning of this chapter we come for the second time in this volume to an accident in the British Isles. The accident next to be considered was a derailment at Waterloo, England; not the famous London terminus but the station on the outskirts of Liverpool on the Southport branch. Involved in the accident on Wednesday 15th July 1903 was the 4.30 p.m. Lancashire and Yorkshire Railway express passenger from Liverpool Exchange to Southport. The train was made up of engine No. 670, a four wheels coupled tank locomotive and six bogie vehicles: three third-class carriages, two composite carriages and a brake composite at its rear. In charge of the train was Guard George Shepherd, and on the engine were Driver William Lloyd and Fireman Edward Rigby. The train was well filled with passengers, mostly businessmen on their way home.

Guard Shepherd told the subsequent Board of Trade inquiry into the accident that his train had left Liverpool Exchange on time and had been checked by the distant signals at Exchange Junction and Sandhills No. 1 cabins, but without loss of time. The train had run smoothly through Seaforth station, he had put his hand on the hand brake as was customary to keep the couplings tight as the express was going down the bank, and he had then felt a jolt. This had been about the nearest bridge to Waterloo station. He had put on the vacuum brake and the train had quickly come to a stand. He then found that the leading bogie of his van was derailed. The express had been running at its normal speed, which was about 50 miles an hour through Waterloo. It had to run at up to 60 miles an hour to cover the 17¼ miles from Liverpool to Birkdale in the scheduled 20 minutes.

---

[143] *Manitoba Free Press*, Winnipeg, Manitoba, Canada, 8th and 12th May 1903

Driver Lloyd agreed that the usual speed through Waterloo station was 45 to 50 miles an hour. He had not shut off steam going down the bank from Seaforth Junction but had had the engine in midgear. As he was approaching the footbridge at Waterloo his engine had given a sudden leap and left the rails. There had been no warning, such as a grinding noise. He could remember nothing more. The locomotive had been running smoothly up to that point.[144]

On leaving the track the engine mounted the island platform of the station, knocked down the supports of a footbridge and overturned, badly damaged. The two leading carriages were wrecked. Fireman Rigby and four passengers were killed outright, a fifth passenger, Miss Edith Morgan, of Southport, died of her injuries in Bootle Hospital early the next morning, and Mr G. Newstead, a cashier at Lloyds Bank, died a month later.[145]

Initially the derailment was blamed on the renewal of the rails and sleepers in connection with the electrification of the branch, which was then in progress, coupled with the high speed of the train, but the Board of Trade investigation, conducted by Major E. Druitt, reached a different conclusion. The inspecting officer noted that the engine had derailed to the right, or inside of the curve. Driver Lloyd, who was seriously scalded, had stated that the engine had jumped up bodily without any warning. There was no sign of any obstruction on the line and the permanent way appeared to be in good order. No work in connection with the coming electrification had yet been done at the accident site. However, examination of the line had led to the discovery of an axle box pin lying close to the inside rail of the down (to Southport) line 138 yards in the rear of the station footbridge. Then 65 yards short of the footbridge a spring bridge of one of the coupled wheels of the engine was found in the six foot (the space between the down and up lines). Finally 16 yards from the footbridge the whole spring of one of the coupled wheels was found almost buried in the ballast. Ten yards beyond the buried spring severe damage to the track began.

Major Druitt pointed out that had excessive speed alone been the cause of the derailment the train would have been derailed to the left or outside of the curve, whereas the opposite was the case. He considered that the most probable cause of the accident was the dropping of the spring of the right trailing coupled wheel of the engine. His only recommendation was that the speed limit through the curve be reduced from 50 to 35 miles an hour.[146]

The inquest on six of the victims at the Town Hall, Waterloo, was postponed until 4th August pending the recovery of Driver Lloyd from his injuries. It returned the predictable verdict of 'Accidental death', the jury adding that from the evidence they were unable to arrive at the precise cause of the derailment.[147]

A broken wheel was believed to be the cause of the derailment of a Wabash Railroad passenger train near Pendleton, Missouri, United States, on Tuesday 6th September 1904. Eight passengers died in the accident and over 30 were injured. The southbound train had left Des Moines, Iowa, for St Louis, Missouri, at 6.40 a.m., and consisted of a

---

[144] Board of Trade report
[145] *The Times,* 16th and 17th July and 17th August 1903
[146] Board of Trade report
[147] *The Standard*, London, England, 5th August 1903

locomotive, baggage car, smoking car, day coach, dining car and Pullman sleeping cars. The day coach left the track and breaking loose from the smoker plunged down an embankment, dragging the dining car with it. The heavy dining car fell onto the day coach, killing and injuring the passengers in it. There was only one employee in the dining car, who was injured.

Passenger O. E. Sailor told a newspaper reporter, "The train was an hour late and was running about seventy miles an hour. Two miles west of Warrenton, while turning a curve, the rear coach was thrown from the track and turned over twice, going down an embankment thirty feet high. There were about fifty passengers in the coach and every one of them was badly injured. Four were killed outright, and two more died while we were there. The doctor said eighteen of them were fatally injured and could not live till morning. We could not learn their names but were told they were told from La Plata and Macon, and some from Kirksville, Mo., en route to the fair.

"One man, his wife and little daughter were killed outright, one lady was caught under a pair of flying trucks and had to be dug out. One gentleman jumped off the hind end of the coach as it was turning over, and his head was crushed to a jelly. In the next coach from the rear end, which was a dining car, only six passengers were in it, and all of them were injured. Both coaches were torn to pieces."

Another passenger, Mr J. M. Davis, said, "The wreck, as I understand it, was due to some track repairs. The section men were placing a rail in the main track and the flagman was sent out, but the engineer failed to heed his signals to stop, and when the train came around the curve the rail, which was not sufficiently spiked, turned over and let the rear cars off after the front part of the train had passed over the loose rail. The dining car ran on the end of the ties before it turned over. The dead and dying passengers were taken to Warrenton and a relief train to the injured back to Moberly. The only one of the train crew that was injured was a brakeman, and he was dug out of a cinder pile with a broken leg.

"It was a horrible sight to see the little children that were bleeding and crying. Broken arms and legs were numerous. No one was hurt except those in the two rear coaches." [148]

A Louisville and Nashville Railroad passenger train which had left Knoxville, Tennessee, the night before was on time and running into Louisville Union station, Kentucky, United States, at 8 a.m. on Monday 28th May 1906 when it became derailed. It had passed Magnolia Street and was moving at moderate speed at the time. The engine and leading combination baggage and mail car of the train had passed safely over a set of facing points when a flange on a wheel of the following smoking car broke and threw the switch open. Derailed, the smoking car was dragged along the ties for 200 feet and then collided with the box cars of a freight train standing on a siding and its roof was torn off. The front end of the day coach which came next was broken in, seats were ripped up and all its windows broken. A hot water heater which had been at the forward end of the day coach was hurled into the middle of it, killing passenger Howard Coleman, of Stanford, Kentucky. In all eight passengers in the smoker and day coach lost their lives and were 22 injured, one it was thought fatally. Another newspaper account spoke of nine deaths. Sleeping cars from Knoxville and Nortonville at the rear of the train were also derailed

[148] *Sedalia Weekly Sentinel*, Sedalia, Missouri, United States, 9th September 1904

but remained intact and their occupants were unhurt.[149] [150]

A special train carrying Shriners (members of a fraternal organisation known in full as the Ancient Arabic Order of the Nobles of the Mystic Shrine) returning from the annual meeting of members at Los Angeles to New York and Pennsylvania, had travelled no further than Honda siding, California, United States, when it met with disaster on Saturday 11th May 1907. The siding was just south of Surf station, on the Southern Pacific Railroad coast line, where the railroad ran along the shore of the Pacific Ocean for 100 miles north of Santa Barbara. Carrying 145 Shriners and their friends from Ismalia Temple, Buffalo, New York; Rajah Temple, Reading, Pennsylvania and neighbouring places, the train was running at 50 miles an hour as it reached the siding. The front of the engine passed safely over the facing points, but its tender was derailed dragging the engine down the embankment onto the beach, followed by the leading baggage car and the dining car, which fell onto the wrecked engine. Both the baggage car and the dining car were badly damaged. When the derailment occurred at 2.35 p.m. there were 32 passengers in the dining car, most of whom were killed, either by being crushed or by exposure to scalding steam from the locomotive. Almost all the diners in the car at the time were from the Rajah Temple, Reading, these having been the last to go forward to the dining car. The next three coaches were also derailed but the last two remained on the track. Fire broke out in the wreckage but was extinguished by uninjured passengers from the rear of the train. Engineer Frank Champlain was thrown out of his locomotive in the crash. He picked himself up and ran a mile for help before realising that he had broken his arm and suffered severe scalds. The day after the accident it was known that 31 people had died and a score were injured, some of whom it was feared would die. The final death toll was 33. At least 15 of those who lost their lives came from Reading, Pennsylvania.

A newspaper reporter described the aftermath of the accident. 'Mrs John W. Cutler, of Binghampton, N.Y., was in the baggage car at the time of the crash to rearrange her trunk. Her body was driven through the floor, and the car had to be jacked up before the body could be released. Mrs Fred Greenwood, of Binghampton, was with Mrs Cutler, and also went down under the tons of baggage and broken timbers, but she was covered by an arch of trunks. When the rescuers burrowed their way to where the two women lay Mrs Greenwood reached out and grasped the foot of one of the men and shouted, "I'll not let go until you get me out." Then scalding steam enveloped her, and she was horribly burned. She was rescued alive, however, and was among those taken to San Luis Obispo. When Miss Cora Young, of Cleveland, was taken from the shattered dining car she was still alive, but frightfully injured. Her entire body had been showered with boiling water, and when women friends removed her corset the imprint of the stays was left deep in the scalded flesh. George Hagerman, of Reading, Pa, refused the aid of his brother nobles after they had dragged him, fatally hurt, from the wreck. "I am dying," he said, "go help the women."'

A relief train left Santa Barbara, 61 miles to the south, at 4 p.m. and took 90 minutes to reach Honda. Simultaneously with that from Santa Barbara a second relief train set off

[149] *Harrisonburg Daily Times*, Harrisonburg, Virginia, United States, 29th May 1906
[150] *The Warren Mail*, Warren, Pennsylvania, United States, 31st May 1906

from San Luis Obispo, about the same distance away to the north, with doctors and nurses, and reached Honda just before the Santa Barbara one. Meanwhile the bodies of 25 of the dead had been laid out in the sand alongside the track, while the injured, some unconscious and dying, rested on bedding and plush seats taken from the Pullman cars. When the two relief trains arrived it was decided to convey the injured in one train to San Luis Obispo and the bodies of those killed in the other train to Santa Barbara. The line must have been quickly cleared, for the report mentioned the Santa Barbara train having to wait two hours in the sandy desert to cross a stream of further northbound trains carrying Shriners from Los Angeles to their homes in the east. In all with various waits to meet opposing trains the return journey to Santa Barbara took 6½ hours.

One local inhabitant was determined to help with the rescue and would not be dissuaded, 'When the relief train arrived at Sutton {this is probably a misprint for Sudden, a place a few miles south of Honda} on its way to the wreck Miss Maria Fuller, 19 years old, of Los Angeles, jumped on board. "I want to go with you to the wreck," she said to Dr W. A. Taylor, Southern Pacific division surgeon. "You cannot go," replied the surgeon. Ten minutes later, when the train was running 50 miles an hour, a wisp of dark hair blowing above the car platform drew the attention of Dr Taylor and the conductor. A moment more and they had dragged the girl from the brake beam of the tender. Brought into the car, Miss Fuller smiled away the doctor's glowering looks. "I really had to go," she said, "I am a member of the Southgate Eastern Star {another fraternal organisation, open to both men and women} and I am pledged to give aid whenever I can." On the way back she did her best to comfort Mrs W. A. Wasson, whose husband was killed.'[151]

Initially the accident was blamed on a faulty switch, despite the fact that the Islam Temple special had passed the siding travelling at 60 miles an hour only ten minutes earlier. Inquests held on the victims shortly afterwards at both Santa Barbara and San Luis Obispo, where some of those injured had died, failed to cast much light on the matter; the former jury simply admitted that they could not say how the derailment had happened and that at San Luis Obispo blamed the accident on 'defective equipment' without specifying what equipment they had in mind.[152]

Subsequently the Southern Pacific conducted its own investigation into the disaster. This found that the line at Honda was straight, with 75 lb per yard rails, stone ballast and a roadbed of sand. The train speed had been high, but not in excess of that permitted. There was no evidence that the switch had been tampered with. Automatic signals were installed; should the point blade have failed to come within a quarter inch of the stock rail circuit breakers would have thrown their stop and distant arms to danger. A signal maintainer had passed the spot only a few hours before the accident and had found nothing wrong. After the accident one point rail was discovered to have had a small piece broken off, as if by a heavy blow. An extended investigation had failed to disclose the cause of the accident. The only tenable theory was that some obstruction, such as a brake shoe, had fallen against the points between the front of the locomotive and the first derailed wheel. That wheel had then struck the obstruction and forced it against the points such that they were forced open, derailing the train. The obstruction had then been pushed out of the points, they had returned to their normal position and the rear of the train had continued straight ahead. Examination of the engine failed to show any missing

[151] *The Sun*, Baltimore, Maryland, United States, 13th May 1907
[152] *The Galveston Daily News*, Galveston, Texas, United States, 14th May 1907.

parts likely to cause the derailment, and the tender was too badly damaged to draw any firm conclusions, but it seemed likely that something which fell from the tender had made the its rear truck and the forward truck of the baggage car take the wrong side of the points.[153]

One of the most popular expresses of the London Brighton and South Coast Railway was involved in an accident at Stoat's Nest, England, on Saturday 29th January 1910. The 3.40 p.m. departure from Brighton to London was a very popular train, as it enabled an arrival at the capital in time for dinner, and on Saturdays it was usually run in two portions, one to London Victoria and the other to London Bridge. It was the first, Victoria portion which was involved in the accident. On the day of the accident the train consisted of an Atlantic class tender engine, a leading brake van, Pullman cars *Albert Victor* and *Prince*, a third-class brake composite, a third-class carriage, a composite carriage, Pullman car *Princess Patricia*, two composite carriages and a third-class brake composite, in that order. In charge of the train was Head Guard James Wood and on the engine were Driver Thomsett and his fireman. The train left Brighton on time.

On duty in Stoat's Nest South box that afternoon was Signalman Albert Knight. He accepted the first portion of the 3.40p.m. train from Cane Hill box on the recently opened Quarry avoiding line at 4.24 p.m., got it accepted from Stoats Nest Central box and cleared his signals for it at 4.25. The train passed at 4.28 p.m. at its usual speed. As it passed, he noticed sparks coming from one of the carriages in the centre of the train and then observed that a vehicle had left the track. There were a set of facing points 50 yards south of his box, but he had not replaced his signals at this point and so thanks to the interlocking could not therefore have shifted the points, which were not damaged in the accident and which worked normally afterwards. He had been about to send the "Stop and Examine" bell signal to the Central box when the train left the road. The sparks looked like those coming from a blacksmith's anvil. He had no idea as to the cause of the derailment.

Signalman Frank Bowers in Stoat's Nest Central box had had the express accepted from the North box and cleared his signals for it at 4.26 p.m. He saw the engine and front vehicles of the train pass his box at a low speed a couple of minutes later and was surprised it was such a short train, noticed that it had no tail lamp and then looked back and saw that the rear of the train had been wrecked in the station.

Stoat's Nest Stationmaster Alfred Chalker was near the porter's room on the up local platform. Hearing a noise, he had turned around to see a carriage sliding up the ramp of the down local through platform, broadside on. The carriage demolished a water column and then stopped.

Driver Thomsett said that all had gone normally until he felt the Westinghouse brakes come on near the Central box and he had assumed that a brake pipe had burst. He had first tried to maintain enough pressure to get to East Croydon, but found that he could not, and sensing that something else must be wrong he had looked back and found that his train had parted, whereupon he had brought the front portion of his train to a stand near the North box, and found that he only had four vehicles attached to his engine.

---

[153] *Reno Evening Gazette*, Reno, Nevada, United States, 2nd June 1907

Head Guard Wood, riding in the front van, could offer little useful information. He put the speed of the train at 35 to 40 miles an hour approaching Stoat's Nest. He had noticed his van shaking about the South box, looked out and saw that his train had parted and tried to attract his driver's attention. Just then the Westinghouse brakes came on and the train stopped. He had not noticed anything unusual with the running of the train and none of the passengers had complained to him of any discomfort.

The six rear vehicles of the train had become derailed as it approached Stoat's Nest station and the leading one of them, the fifth vehicle in the train, which was a bogie lavatory third, No. 1325, had swung broadside across the line, mounted the ramp of the station platform and been totally wrecked. Five passengers died in this carriage and two men named William Rose and Beaumont Willard, who had been standing on the platform, were also killed. Sixty-five other passengers were injured, though only eight seriously. The passengers who lost their lives were Mrs Henry Beere, aged 55, of London W.C.; Mrs Emma Side, of London W.; Mrs Elizabeth Cate, aged 51, of Brighton; Mrs Mary Spencer, aged 51, of Fulham; and Miss Charlotte Carter, aged 25, of Brighton.

As the accident occurred in a populated area help was soon at hand. Doctors and nurses from the nearby Cane Hill Asylum were early on the scene, and a young doctor and his wife chanced to be on the train, returning from their honeymoon. They escaped injury and the doctor assisted his colleagues from the asylum. A party of Boy Scouts also gave valuable aid. Traffic was not seriously interfered with as trains could pass the accident site on other lines at reduced speed.

Mr L. Irvine, a passenger who had been riding in a second-class compartment of the carriage which was eighth in the train described his experience to a reporter. 'There was another gentleman in the compartment with me. The train was going at a considerable speed, when it suddenly began to rock from side to side, but not in a particularly alarming way. Then it came to a standstill, and I remarked to my companion, "There must be something wrong, and we must be off the line." We both got out of the carriage and I saw that the wheels of the carriage were completely embedded in some clay which had been cut up by the train. I could not see the engine, and it looked as if the first part of the train had gone on. There was a third-class carriage on the far-side of the rails lying on its side. I saw two ladies taken out of the upturned third-class carriage, and they were evidently dead, their bodies being dreadfully mangled. What struck me at the time as being most remarkable was that I heard no screaming. I next saw another lady carried out of the overturned carriage. She was moaning and crying out, "Mind my arm, it's dreadful." She seemed to be in a terrible state, and I should think she died on the way to hospital. I next saw two men apparently dead taken from behind the Pullman car. A passenger, who, I should think, was a doctor, quickly made some splints by breaking up portions of the woodwork of the carriage. I left the scene in less than half an hour by a train for Victoria, and in the same carriage were two ladies who had been travelling in the third-class carriage which I had seen lying on its side. One of them was a young lady. She said she had not been injured beyond a severe shaking. Both of them were able to get out of the carriage.' [154]

Lt-Col P. G. von Donop investigated the accident on behalf of the Board of Trade. The inspecting officer noted that none of the railwaymen on the train were able to throw

[154] *The Times*, London, England, 31st January 1910

any light on the derailment. He detailed the positions of the derailed vehicles after the derailment and then remarked that an examination of the line on the Brighton side of Stoat's Nest had shown unusual rubbing marks on the inside cheek of the left-hand rail. The marks had appeared intermittently from about Star Lane box, two miles from Stoat's Nest, as if the flange of a wheel had been unduly pressing against the rail. The first mark of actual derailment was six feet north of the facing points near the South box. From there the marks continued to a check rail which it was evident that a wheel had mounted. It was clear, the inspecting officer continued, that the derailment had occurred with carriage 1325, the one which was totally wrecked. There was no sign of the axle box of carriage 1325 having overheated, and the sparks seen by Signalman Knight had evidently been caused by the wheel having been grinding against the flanges of the bogie frame of the carriage, the wheel having shifted on its axle. The shifting of the wheel had made the wheel pair wide to gauge, which accounted for the unusual marks back along the line. In the station the displaced wheel had mounted a check rail and complete derailment had occurred at a diamond crossing, where the pressure of the following vehicles had forced the fifth coach broadside across the up main and down local lines.

In conclusion Lt-Col von Donop stated that it was impossible to say at exactly what point the final shifting of the wheel on the axle took place. There was nothing wrong with the permanent way and hence the derailment must be entirely attributed to the looseness of the wheel on its axle. Derailments due to such a cause were very rare, though two had occurred on the Great Western Railway in 1895 and 1898 at Maidenhead and Exeter respectively. Since those two derailments the Great Western required the pressure required to drive all wheels onto their axles to be automatically recorded, and the Brighton company now intended to do the same. The inspecting officer recommended that the measure should be generally adopted.[155]

Mr George Spencer, the husband of Mrs Mary Spencer, one of the passengers who lost their lives, was himself hurt in the accident and subsequently suffered from depression, doubtless grieving over the loss of his wife. At the beginning of April it all proved too much, and he committed suicide.[156]

Stoats Nest station was renamed Coulsdon and Smitham Downs 1911 and Coulsdon North 1923.

Following the disasters at Villepreux-les-Clayes and Saujon the Chemins de fer de l'État suffered a third accident on Saturday 10th September 1910 on the Paris St Lazare – Cherbourg main line at Bernay, France.

Express train No. 322 left Cherbourg for Paris at 1.10 p.m. on the afternoon of the accident. At Caen four more coaches were added and the engines were changed; from there the now 22 vehicles of the train were hauled by a Pacific class engine with Driver Pierre Mallard and Fireman Gustave Bertaud. The journey onwards to Paris went without incident until the train reached Bernay. Immediately after passing through the station there the locomotive left the rails as it was crossing a bridge over the Charentonne River, overturned, and after ploughing along the ground for 150 metres came to a stand at the Boucheville level crossing. The tender shot past the engine and the leading van of the

155 Board of Trade report
156 *The Times* 4th and 5th July and April 1910

train knocked down a wall of the crossing keeper's house. The six following vehicles were telescoped and broken up and most of the remainder of the train was reportedly also telescoped and overturned, except for the last five, which were derailed but remained upright.

As the accident had happened close to Bernay station and in the centre of town help was immediately forthcoming and proceeded under the supervision of Sub-Prefect Louis Bigot. Driver Mallard, Fireman Bertaud and three passengers were killed outright, and 50 people were injured, 15 seriously, and were taken to the hospital, hotels and even private houses according to their condition. Four of them died from injuries and scalds within the next two days and three more later, bringing the number of dead up to 12.

The accident caused a sensation and was discussed at length in parliament, both as regarding its causes and as regarding the responsibilities of the state railway company.

Examination of marks on the track showed that the derailment had originated at a set of points, but its exact course was unclear. The rear of the engine, rather than the front, had been the first to leave the track, excluding the points moving under the train, but there was disagreement between the investigators as to whether the engine or the leading van had been the first to become derailed. All agreed, however, that the derailment had resulted from a combination of high speed, poor track and unstable rolling stock.

As to the speed of the train expresses were authorised to run at 100 kilometres an hour over the Cherbourg line and so-called *rapides* at 120. Surviving passengers were unanimous that the train had been going too fast, and that they had expected a derailment, but their memories were perhaps biased by the trauma of their ordeals. They put the speed at 110 km/h at the time of the derailment. Conductor Alfred Joly, on the other hand, maintained that his driver, because of the heterogenous make-up of his train, had been travelling at no more than 80 or 90 km/h through Bernay. The locomotive had been fitted with a tachograph, but this was too badly damaged in the accident to yield any information, despite the best efforts of experts at the Sotteville workshops. That the speed of the train had not been excessive was indicated by the fact that over the last few kilometres from Saint-Mards-de-Fresne, where it had already been running four minutes late, it had lost two more minutes, rather than regaining any time.

The state of the track was another possible factor in the derailment. Most lines with equivalent traffic were provided with Vignoles rails 18 or 24 metres long and weighing over 40 kg/m. The Cherbourg line still had 8-metre lengths of double-headed rail weighing only 38¾ kg/m. Criticisms in the press of the speeds being run over inferior track on the Cherbourg line were more or less admitted by railway company managers, while a group of parliamentary deputies who visited the scene several months after the accident reportedly found worn rails, loose fishplates, rotten sleepers and so on.

According to the regulations express and *rapide* trains travelling at over 100 k/hr must be composed entirely of vehicles marked with a star to indicate that they were safe to travel at such a speed. The drivers of trains not composed entirely of such vehicles were to be given a written notice at their departure station ordering them not to exceed 90 k/hr. Train No. 322 had been made up of a mixture vehicles from the État, Nord and Ouest systems, with varying buffer heights, some as light as 7 or 8 tonnes, others weighing 20 tonnes, some very old and so on. It was determined that Driver Mallard had been issued a speed restriction order before he left Cherbourg. If the derailment began with the locomotive this mixture of vehicles could not in itself have caused the derailment, though it doubtless made its effects worse and the rough riding probably influenced the passengers' perception of the speed at which the train was running.

The Pacific locomotive pulling the train was one of a batch of 50 ordered in 1910. It had only been brought into use on 1st July and had run no more than 9,200 kilometres. As it weighed 123 tonnes it was clearly not suitable for the light and poorly maintained track of the Cherbourg line. Its use had been condemned by the railway unions and was also subsequently criticised by parliament. To avoid criticism of the État management in their choice of locomotive the accident was blamed, however, on insufficient play in the leading bogie of the locomotive, presented as a design fault affecting the entire class. This judgement was confirmed by a subsequent derailment of another Pacific locomotive at Rambouillet, though again this had occurred on very poor track. The untimely zeal of an État engineer was blamed and there were lengthy debates in parliament regarding locomotive design and train speeds. The Pacific locomotives were relegated to working freight trains pending replacement of their bogies, though they would have been safe enough on passenger trains on better track.

The accident also re-ignited the controversy over the taking over of private railway companies by the state.[157]

No. 5 up mixed train was derailed between Lakhminia and Lakho stations, India, on the Bengal and North Western Railway on 4th May 1911. Six passengers were killed outright and 14 injured, two of whom subsequently died. The derailment was due to the breakage of the buffer of a covered goods waggon.[158]

Six boys (as they were described) of New Cambria and Bevier, Missouri, who were stealing a ride on a Chicago Burlington and Quincy Railroad freight train, were thought to have lost their lives when it was derailed half a mile east of Lingo, Macon county, Missouri, United States, on Sunday 8th October 1911. One of those who died was named as Floyd Summers of New Cambria. The derailment was attributed to a broken drawbar. Nine cars of cattle and horses which formed part of the train all went into the ditch. It was expected to take six to eight hours to clear the wreckage and determine the number of dead.[159]

Over the years the English London and North Western Railway's Irish Mail train was involved in several bad accidents, for example the Abergele disaster in August 1868. On Saturday 14th August 1915 the second portion of the train was in another bad accident near Weedon, England.

Signalman Walter Cater in Heyford North box had been offered and accepted the 8.45 a.m. up passenger train from Birmingham to London Euston by his colleague at Weedon No. 2 box at 9.47 a.m. and received the 'Train entering Section' signal for it at 9.53. Meanwhile in the opposite direction Heyford South box had offered him the second

---

157 Wikipedia – Accident ferroviaire de Bernay

158 Administration Report on the Railways in India for the year 1911

159 *The Salt Lake Herald Republican*, Salt Lake City, Utah, United States, 9th October 1911

portion of the Irish Mail, the 8.30 a.m. from London Euston to Holyhead, at 9.51 a.m., sent it on at 9.52 and it had passed him at 9.53. Express trains normally only took two or three minutes between Heyford and Weedon, but he saw nothing of the Birmingham train in the next ten minutes, nor had he received the 'Train out of Section' for the Irish Mail. Finally at 10.07 a.m. the fireman of the Birmingham train arrived at his box and told him that their engine had broken down and their train had come to a stand in Stowe Hill Tunnel.

Signalman Walter Norton at Weedon No. 2 was likewise wondering why he had seen nothing of the Irish Mail and why Heyford North had not cleared out the Birmingham train. Just before 10 he rang Signalman Cater regarding the fact that the latter had not cleared the Birmingham train and Signalman Cater enquired about the Irish Mail. Realising that something must be wrong, Signalman Norton then informed his stationmaster, and also called Hillmorton (Rugby) and Bletchley and told them that there was a problem and trains had best be diverted via the Roade loop through Northampton. The two signalmen exchanged 'Obstruction Danger' signals at 10.08.

The second part of the down Irish Mail was double-headed. In charge of the pilot engine, *Precedent* class No. 1189, was Driver Thomas Hadfield. His train had left London Euston on time at 8.32 a.m. and all had gone all right until they emerged from Stowe Hill Tunnel, travelling at about 60 miles an hour. An up train passed them at that point, and he noticed ballast being thrown up, some of it striking his train. His engine then gave a lurch and ran off the line, at the same time breaking away from the train engine. It came to a stand about 200 yards down the line. Seeing that the up line was obstructed he went forward to protect it, leaving his fireman with the engine but accompanied by the driver of the train engine.[160]

Behind the engines the coaches of the Irish Mail had been badly wrecked. The leading brake van and following Post Office sorting and parcels vans had broken loose from the remainder of the train, gone down an embankment about 12 feet high on the down side into a turnip field, turned completely over and been smashed. The next five coaches crossed the line onto the up side and came to rest at different points and with varying degrees of damage. The sixth coach suffered most, being telescoped by the seventh, while the eighth was battered by the leading dining car. The two dining cars remained upright, and the rear coaches were not derailed. After the accident the uninjured passengers quickly began the rescue work while doctors arrived from Northampton, Rugby and Weedon; the Northampton branch of the St John's Ambulance Association provided several motor ambulances and several local military establishments also assisted. Nine passengers were killed outright, a tenth died later and 21 were more or less seriously injured.

Private James Bocherty, D.C.M., an Irish soldier who had fought at Ypres, was going home, presumably on leave, to Waterford, Ireland. He related, 'I was asleep when the train went off the line. I was awakened by being flung from one end of the carriage to the other. The coach dashed on, and finally came to a standstill resting on its side. The other occupants were in an amazing heap. The carriage partitions and doors were showers of broken glass. With other soldiers on the train I set to work to aid the injured, whose cries were most distressing. Scrambling along the coach we found four women passengers

[160] Board of Trade report

practically buried in the debris. Two were already dead, and the other two were badly injured and unconscious. In one compartment I came across a young lady who was very excited but unhurt. She had £50 in gold, which had fallen out of her bag and was scattered all over her carriage. I helped her to collect it.' [161]

By Monday 16th it was apparent that the derailment of the Irish Mail had been caused by a problem with the Birmingham train. What exactly went wrong emerged at the Board of Trade inquiry. There Driver F. Brightland related that he had left Birmingham New Street at 8.45 a.m. with *George the Fifth* class engine No. 1489. They had stopped at Stechford and Coventry and on arrival at Rugby he had gone around to oil his engine, when he had noticed that the pin was missing from the right driving pin washer. A fitter had been called, and in about five minutes the pin had been replaced with another one. They started again from Rugby, next stop Willesden, and all went well until they were passing the Weedon No. 1 down distant signal, when his fireman had suddenly shouted, "Whoa!" and applied the brake. He then felt a lurch and saw ballast flying up, whereupon he brought his train to a stand as quickly as he could, by then his engine had entered Stowe Hill Tunnel. He and his fireman then examined their engine and found that the right-hand coupling road had become detached from the crank pin at one end and was badly bent. They found that the engine was too badly damaged to take the train forward and he sent his mate, Fireman F. G. Williams, forward to Heyford North box for assistance. The down Irish Mail had passed them just as they were stopping but they had no idea that it was in any danger.

Fitter Edward Oldham at Rugby said that an engineman had come over and had asked him if he could put a pin in the crank pin washer of the engine of the 9.35 train to London. He had replied that he did not have one, whereupon it was suggested that he could get one from shunting engine No. 665, which was working at the rear of the train. He had agreed and borrowing a hammer had gone to that engine and knocked the pin out, then taken the pin, along with the hammer and a spanner, to engine No. 1489. There he had put the pin in and driven it in tight. Having no chisel on him he had used the jaw of the five-eighths spanner to open the split part of the pin. The washer had been in its place and had not worked out at all. He could not account for the pin having come out only a few miles after leaving Rugby.

The inspecting officer, Lt-Col E. Druitt, remarked that from the evidence there could be little doubt that the Irish Mail had been derailed by the detached coupling rod of the Birmingham train pushing the down road out of alignment. Signalman Norton had found the washer at Weedon station and a member of the breakdown gang had found the split pin near Welton station, 8 miles from Rugby and nearly 5 miles short of Weedon. The derailment had occurred slightly under a mile south of Weedon. Brake and Carriage Examiner Frederick Elson had watched Fitter Oldham at work and both swore that the fitter had opened its split end, but the inspecting officer stated that in that case it was very difficult to see how the pin could have worked out, and he could only conclude that in fact Fitter Oldham, in his haste, had not opened the split pin as he said. Lt-Col Druitt concluded that the cause of the accident was most unusual one, and he believed it was the only one of its kind on record.[162]

---

161 *The Times*, London, England, 16th August 1915

162 Board of Trade report

Adjourned inquests at Northampton on 6th and Weedon on 9th September returned identical verdicts of 'Accidental death' and identical recommendations that spare crank pins could be kept on locomotives and that pins should never be swopped from one engine to another.[163]

[163] *The Standard*, London, England, 7th and 10th September 1915

CHAPTER 3

OTHER DERAILMENTS

Other chapters in this book cover derailments caused by track defects, defects with the trains themselves and those caused deliberately by acts of vandalism or sabotage. This chapter will deal with derailments due to other causes such as excessive speed, collisions with animals and other accidental obstructions, traction shocks and so on, together with those for which no cause was apparent.

On the evening of Friday 15th February 1901 the 5.55 p.m. passenger service from Sydney, Australia, to Hurstville left the Redfern station at the usual time in charge of Guard Alfred Bull. It consisted of a small suburban tank engine and seven suburban corridor coaches, and was crowded with passengers going home from work, who appeared to be 'mostly of the artisan and labouring classes'. All went well until the train was about a quarter mile beyond Sydenham station, when the engine suddenly left the rails, rolled over, then righted itself, continued along the permanent way for 30 or 40 yards, demolished the gates and gateposts at the Bridge street level crossing, then plunged down an embankment and through a fence, ending up facing the way that it had come. Near the level crossing the engine had become detached from its train, and the leading carriage, a second-class smoker filled with passengers, ran on a short distance, then turned over and was 'smashed into splinters', while the second carriage stopped at right angles across the track, also badly damaged. The remaining carriages were little damaged, and their occupants escaped almost unhurt.

After the accident there was, according to witnesses, a period of total silence, before uninjured passengers and local residents began the work of assisting those in the leading two carriages, many of whom were shockingly mangled. Axes were procured to chop open the bodies of the wooden carriages and their dark recesses were explored by rescuers by the light of candles for further victims. The injured were laid on cushions procured from the other carriages before being taken to Marrickville Cottage Hospital and when that was full to the Prince Alfred Hospital in Sydney. The bodies of those who had died were taken to the South Sydney Morgue. It was reported the next day that eight people had died in the accident and 24 had been injured. Fireman Sydney King on the engine was among those killed. One more passenger died of his injuries later.[164]

The inquest upon Herbert Darnley, one of the passengers killed, was an extended affair, with much evidence tendered regarding the speed of the train, the condition of the track and the stability or otherwise of the engine pulling the train.[165] It was not finally concluded until Monday 6th May, when Driver Matthew Doyle gave evidence. The latter stated that on the day of the accident he had been in charge of F class engine No. 363 of the Hurstville train. He had been in good health. Four or five days earlier he had been asked to provide an explanation for the late running of his train and he considered that he had provided a satisfactory explanation. He denied saying to anyone before he left Redfern that he would take care that day that he was not late. Nothing unusual had happened when he felt the engine leave the rails. He estimated that his train had been at 25 to 30 miles an hour at the time and could not account for the derailment.

The jury returned a verdict of 'Accidental death', and that the derailment had been caused by the high speed at which the engine was driven and by the defective characteristics of the locomotive. They added a rider that F class tank engines should not be used on passenger trains.[166]

Subsequently the accident was investigated by a board of inquiry which included experts from South Australia, Victoria and Queensland. Its observations were made public in mid-June. The report of the board stated that, 'We are unanimously of opinion that the accident was due to a combination of causes – namely the excessive speed of an engine which was unsuitable for high speed on a line which at or near the point of derailment was not in good order.' The *Sydney Morning Herald* noted that the board's report squared with that of the artisans of the coroner's jury except that the jury had not commented on the state of the road.

Commenting upon the report, the New South Wales railway commissioners stated that there was no necessity for high speed and that the F class engines were a legacy from the previous administration. The engines had hither run fairly safely but a new type of locomotive had since been introduced for the suburban service. As regards to the track the commissioners said that though the line at the point of derailment was not equal in condition to adjacent sections which had recently been re-laid with 80lb rails it was still equal to traffic at a proper speed.

The newspaper editorial considered that the jury's verdict and board's report were together a damaging criticism of the railway service. The commissioners must act. The permanent way must be upgraded to an improved condition. The F class engines must be withdrawn from hauling passenger trains as soon as possible and the frequently excessive speed of the trains, due either to negligent drivers or to the unreasonable timings of the trains, must be curtailed. The *Morning Herald* concluded that, 'The Commissioners now have both the popular and the expert verdict on the Sydenham accident. It is their business to prevent a recurrence.'[167]

---

164 *The Evening News*, Sydney, New South Wales, Australia, 16th and 18th February 1901

165 *The Sydney Morning Herald*, Sydney, New South Wales, Australia, 18th, 27th and 28th February, and 1st, 2nd, 5th, 6th, 8th, 9th, 13th, 14th, 15th, and 16th March 1901

166 *The National Advocate*, Bathurst, New South Wales, Australia, 7th May 1901

167 *The Sydney Morning Herald*, Sydney, New South Wales, Australia, 15th June 1901

Armoured train No. 15, which left Pretoria at 7 p.m. on Monday 5th May 1902 for Pietersburg, was derailed on a sharp curve at Daspoort Cutting, Transvaal, and fell down and embankment. Killed outright were Lieut. Campbell, of the Cape Garrison Artillery and 10 soldiers, while 11 men were injured. The fireman of the train, whom the report on the accident stated 'knew the line well' was also killed. Mention in the brief report that the fireman was very familiar with the line would seem to imply that his driver was not, and that the derailment had been due to excessive speed.[168] Four of the injured soldiers may have died later as another source states that in all 15 died.[169]

A cablegram from Lisbon spoke of the derailment of the Madrid express near Guarda, Portugal, from an unknown cause, on 7th July 1902. Nearly all the carriages were wrecked. Six persons were killed and 27 injured.[170]

The derailment of a light engine from an unstated cause led to a more serious accident at about 6 p.m. on Saturday 21st November 1903. Shortly after it had gone off the line the engine was run into by a Berne-Lausanne express about 300 metres from Palézieux station, Switzerland, and the tender, leading luggage van and front passenger carriage of the express were smashed, the carriage being telescoped into the luggage van. Six people in the leading coach died and seven were injured. Those who lost their lives were two children named Gratchoff and their governess, Miss Ricket, of St Petersburg; and three Swiss passengers – Mlle Josephine Bterki, of Ste Ursanne; M. Grunewald, sonin-law of M. Monier, a Federal Judge; and Mlle Berkschi, of Vevey. M. Boecken, of St Petersburg, and three Swiss officers were among those injured.[171] [172]

'The most terrible accident, it is said, that ever occurred on the Baltimore and Ohio railroad left its harvest of death and ruin at Laurel Run, two miles west of Dawson station, this evening...' Thus began a newspaper report of an accident in Pennsylvania, United States, which occurred on Monday 23rd December 1903. Eastbound train No. 12, the *Duquesne Limited*, the pride of the company, was running at probably 60 miles an hour or more as it neared Laurel Run, with Engineer William Thornley and Fireman Thomas Cook on the engine. It consisted behind the locomotive of a baggage car, a smoker, a day coach, two Pullman cars and a Pullman dining car.

A few minutes earlier a freight train, which included a flat car loaded with heavy timber, had passed in the opposite direction on the adjoining track. As it went by Laurel Run the stays on one of the cars holding the timber gave way and the car's load fell onto the running line on which the passenger train was approaching. Engineer Thornley apparently never saw the obstruction, for the engine brakes were never applied, the train simply ploughed into the mass blocking its path. The timbers were more than a match for the train and the engine was thrown off the line and into the side of the cutting. Behind

---

168 *The Times*, London, England, 9th and 10th May 1902
169 Webpage – The South African Military History Society, *Military History Journal* Vol 11 No. 5 The Anglo-Boer War
170 *The Weekly Tribune*, South Haven, Michigan, United States, 18th July 1902
171 *The Times*, London, England, 23rd November 1903
172 *The Standard*, London, England, 23rd November 1903

the engine the tender was thrown into the air and ended up some way down the line, the baggage car went down the embankment to the right and the smoker mounted the boiler of the engine and knocked off the steam dome. The day coach was considerably damaged, the two Pullman cars not very much. In the Pullman dining car the sudden stop resulted in all the tables and chairs in it sliding to the front end where they ended up in a heap, but its few occupants escaped with bruises. Many passengers in the day coach were injured, but it was the passengers in the smoker, who in the words of the newspaper reporter were 'mostly foreigners and negroes', that suffered most, scalded by searing steam, and almost everyone in it was killed outright or fatally injured. Engineer Thornley and Fireman Cook also lost their lives.

On the opposite bank of the Youghiogheny River the disaster had been witnessed by the operator in the Virgin Run office of the Pittsburgh McKeesport and Youghiogheny Railroad, whose line paralleled the Baltimore and Ohio one on the other side of the river, and he immediately telegraphed his company's station at Dickerson Run, from where a message was telephoned to the Baltimore and Ohio's Dawson station. Meanwhile Baggagemaster Thomas Dom had struggled out of the wreckage of his car. He was injured, but he knew that westbound fast train No. 49 was almost due. Both tracks were blocked, it was dark, and he had somehow to warn the train. Taking a box of matches from his pocket he lit one after another and whirled them around and around. Fortunately Engineer Moses Johnson saw the signal, realised there might be danger, applied his brakes and stopped his train in time, reportedly only three feet short of the wreckage.

Doctors Bell and McDonald of Dawson, on hearing of the wreck, hurried to the scene in a horse and buggy, while Dr Smith Buttermore, who presumably did not have a horse and buggy to hand, simply ran to the scene, arriving not too long after the other two. They were assisted by Miss Wright, a nurse from the Allegheny General Hospital who chanced to be in Dawson. By now fire had also broken out in the wreckage, though it does not seem to have contributed much to the deaths or injuries. The bodies of 38 of the victims were laid out in a long row along the embankment. Before they had lain there very long thieves had apparently begun to rob the bodies of jewellery and money and police had to be drafted in to stop them.[173]

By the close of the inquest on the victims of the accident on 6th January 1904 the number of dead had climbed from an initially estimated 40 or 50 to 65 with the death in hospital of a number of scalded patients. At the inquest Conductor James Garlitz of the Connellsville to Pittsburgh freight stated that his train had had 57 cars: 20 loaded and 37 empty. There had been only one waggon loaded with timber. At Broad Ford, the station before Dawson, his train had got time on trains 46 and 93. His train had been travelling at about 20 miles an hour, with its next stop at Vista, about 10 miles beyond Laurel Run. He had been in his caboose writing out reports between Broad Ford and Vista. At the latter place he had been asked if his train had lost any timber and had said that he did not think it had, and only had discovered the opposite when he had then inspected the car. Evidence was offered by a number of other witnesses and at the conclusion of the proceedings the jury returned a verdict that the accident had been unavoidable and that no blame should be attached to the Baltimore and Ohio company.[174]

---

[173] *The Sun*, Baltimore, Maryland, United States, 24th December 1903

[174] *The Courier*, Connellsville, Pennsylvania, United States, 7th January 1904

The Pennsylvania Railroad suffered what might be considered an unlucky disaster at about 1.40 a.m. on Thursday 11th May 1905 at Lochiel Siding, Pennsylvania, United States, near Harrisburg. The Steelton switching engine was working on the eastbound track and its flagman signalled a following eastbound freight train to stop. The freight train was only travelling at six miles an hour when Engineer John Bless applied his brakes in response to the flagman's signals. Unfortunately the rear cars of his train lacked air brakes, and their continued momentum on the slightly falling gradient caused the middle of the train to buckle. The thirty-fifth car, loaded with steel billets, was derailed and fouled the opposite running line just as the second section of train No. 19, the westbound *Cincinnati Express*, was approaching. The express struck the thirty-fifth car and was brought to a stand just down the line. It was not greatly damaged in the initial collision, but the thirty-sixth car of the freight train was loaded with dynamite, assigned to H.S. Kerbaugh and Co, contractors at Columbia, Pennsylvania. The *Cincinnati Express* came to a stand with its third sleeping car directly opposite the thirty-sixth car of the freight train. The collision with the derailed thirty-fifth car had damaged the dynamite car behind it and spilt some of its load. In this complex accident the boiler of the passenger engine reportedly exploded, coals from the locomotive firebox ignited gas escaping from ruptured tanks of the passenger cars and started a fire, there were a couple of small explosions, probably caused by spilt sticks of dynamite, before the remainder of the 20,000 pounds of dynamite (probably more accurately described as Rockchief blasting powder) in the thirty-six car erupted in a massive explosion. The fire spread, and firemen who had arrived at the scene and who had at first feared to approach closer than 300 yards for fear of further explosions, then found that the intense heat interfered with their rescue work.

Survivors told of their various ordeals. 'Henry Silverman, a New York jewellery salesman, had a satchel in his berth which contained $17,000 worth of fine jewellery samples for the Christmas holidays. Mr Silverman was compelled to leave everything in the passenger coach to escape before {being} burned to death. He could not find his bag in the darkness and had to see his jewels perish in the flames.'

John B Reynolds, a newspaper man from Pittsburgh, who had been returning home from New York and escaped with minor injuries, related that '"The first intimation I had of the wreck was when I heard an awful crash and was thrown out into the aisle of the car. I was dazed for a little time and only realized my position and what had happened when I felt a woman grabbing me and screaming, 'For God's sake help me.' I pushed her out of the window and a fellow passenger handed out a child which belonged to her. He then left the train and called for me to jump through the window. Just then there came a terrific explosion. As I dropped to the ground a missile struck me down. I don't know how long I lay there, but when I recovered my senses, I crawled across the tracks under a freight train and rolled down the embankment on the other side. I was in my night clothes and all my other clothing and belongings were lost. I never want to witness such as sight as that which followed the collision. Women were screaming, children were crying, and strong men were wandering about dazed and helpless. The tracks were strewn in all directions with half-naked men and women, some of them were dead and others seriously injured.''' [175]

At the inquest on 22 victims of the disaster a week after it had occurred the coroner's jury returned a verdict exonerating the Pennsylvania company and its employees, finding that all the rules of the company had been strictly observed, there had been no negligence and the railwaymen were to be commended for the way in which they had cared for the wounded and distressed after the accident.[176] The final number of dead was probably 23, including Engineer H. K Thomas of the express and two other railwaymen, while over 100 were injured.

Twenty-two soldiers and a guard died in an unspecified accident, presumably a derailment, on the Trans-Siberian railway at Yourta station, Russia, on Thursday 24th August 1905. An officer and 47 privates were injured.[177]

On Monday 11th September 1905 it was not a steam railroad but a train on the then newly electrified elevated railways which suffered a bad derailment in New York city in the United States. At 53rd Street on the Ninth Avenue line there was a junction with the Sixth Avenue line, which diverged on a sharp curve over which there was a eight miles an hour speed restriction. Ninth Avenue trains, which continued straight ahead from the junction, faced no restriction of speed. The junction points were controlled by an adjacent signal tower.

On the morning of the accident a Ninth Avenue train approached the junction. It was travelling at 30 miles an hour, but Towerman Cornelius A. Jackson had just previously sent a Sixth Avenue train down the diverging route and he had not reset the points for the main line. The leading car swung around the curve but the strain on the couplings between it and the second car proved too great and they gave way, whereupon the latter car flew off the rails and was pitched from the elevated structure to the street below. Pedestrians on the sidewalk (pavement) below heard a loud rumbling, and upon looking up saw a shower of sparks, followed by splinters and the sound of splitting timbers. The outer guard rail of the elevated structure and the car fell to the street with a deafening crash as passengers jumped from its platforms and windows. The car ended up in a diagonal position with its front end on the ground and the rear end up in the air. Those passengers who had not jumped were thrown into a pile at the end of the car. The third car almost fell upon them as it jumped partly off the elevated structure and came to a stand wedged between a building at the south-east corner of 9th Street and 53rd Street. The remaining cars of the train were derailed but did not fall into the street. Happening within a big city, huge crowds soon flocked to the scene so the first task of the police was to clear them away to enable the rescuers to work to free the passengers trapped in the wreckage. Almost every ambulance in Manhattan went to the scene to remove the injured and dead. There were also three alarms of fire and numerous fire engines went to the site. The firemen easily dealt with a small blaze caused by burning insulation, then set about the work of freeing the victims. The bodies of eight dead were removed from the wreckage and by the next day four more had died in hospital. The final number of dead

[175] *The Racine Daily Journal*, Racine, Wisconsin, United States, 11th May 1905

[176] *The News*, Frederick, Maryland, United States, 18th May 1905

[177] *Marion Weekly Star*, Marion, Ohio, United States, 26th August 1905

was 14, at least one of whom was not a passenger but a pedestrian who had been crushed by the falling train. Over 40 others were injured, many severely.

Soon after the accident Coroner Scholer ordered the arrest of all the railwaymen immediately involved and Towerman Jackson, Conductor J. W. Johnson and Guards Timothy Higginson, J. McDavitt, W. L. Berry and B. Clark were all taken into custody. Motorman Paul Kelly could not be found by the following day. The arrested men were all granted bail.

One theory as to how the accident had occurred was that Towerman Jackson, when he had seen that it was not a Sixth Avenue but a Ninth Avenue train, had reversed the points beneath the wheels of the train. This Towerman Jackson denied, saying that the train had been carrying markers indicating that it was a Sixth Avenue train, and that he had expected it to slow down to take the junction, but it had not done so.

Conductor Johnson made a statement after his arrest. "I was on the platform of the second car. We had just left Fifty-Ninth street and were going at a fair rate of speed as it is a long run to Fiftieth street. The car was pretty well crowded. There were a number of men on the platform and in the aisle, there were a number of shop girls, who were laughing and joking. When we reached the curve the first car went on, and the second and third cars went together like a jackknife, sticking right up in the air and practically closing against each other. In consequence, from where I was on the front platform of the second car at the start, I was dropped to the rear platform of the first car. Then the second car dropped to the street. I held on to the iron rail like death and was straight up in the air. I saw girls and women climbing over the seats of the car that they might escape. Being an old railroader I pulled out my watch and looked at it. It was exactly 7 a.m. The signals were set wrong. That was the trouble. They were set green and yellow, which is for a Sixth avenue train."

Coroner Scholer pre-empted an inquiry into the cause of the accident by stating at the start, "I am satisfied that the whole accident was due to the criminal neglect of the switchman. In the first place he made a mistake in turning the Ninth avenue train into Fifty-Third street and then added to his blunder by turning the switch back again in an attempt to divert the train into Ninth avenue again. The last mistake caused the terrible accident and loss of life."

However, E. P. Bryan, vice president of the Interborough Rapid Transit Company, the owners of the line, gave a statement on behalf of Towerman Jackson. "The Ninth avenue train which met with the accident displayed Sixth avenue signals. On the assumption the signals were correct I gave the train the Sixth avenue track." General Manager Frank Hedley stated that after the accident the switch had been set for a Sixth Avenue train.[178]

The inquiry into the accident began on Wednesday 13th. Superintendent S. D. Smith said that 102 trains an hour passed through the junction during the rush hour. General Manager Hedley stated that Ninth Avenue trains were limited to 25 miles an hour through the junction, Sixth Avenue trains to eight miles an hour. The wrecked train had been displaying the correct markers for the Ninth Avenue route but the two fixed signals protecting the junction, situated 500 feet and 50 feet from the switch, had been cleared for the Sixth Avenue branch. Towerman Jackson declined to make a statement.[179] Also

---

Marion Weekly Star, Marion, Ohio, United States, 26th August 1905

[179] Ibid, 14th September 1905

probably on the 13th George Hughes, a fellow motorman to whom Motorman Kelly had gone after the accident before disappearing, was examined by Coroner Scholer. Hughes said that Kelly had told him that before he had started his train had carried two white disks showing it to be a Ninth Avenue train, and that the signals at the junction had both been green, that is clear for straight ahead.[180] This last assertion contradicted, of course, the observations of General Manager Hedley made after the accident and the interlocking would in any case have prevented the clearing of the signals for the main line with the junction points set for the branch, so it would seem that Motorman Hughes was mistaken. Robert B. Shaw in his book has suggested, doubtless correctly, that repeated trips down the Ninth Avenue line had dulled the motorman's responsiveness to the signals.[181]

The inquiry found both Towerman Jackson and Motorman Kelly responsible for the accident. In March 1907 Towerman Jackson was sentenced to five years in Sing Sing gaol after having been found guilty of manslaughter.[182] Motorman Kelly was finally apprehended in San Francisco, California, in July 1907. In January 1908 he was found guilty of second degree manslaughter and sentenced to a prison term of between one year and six months and two years and six months.[183] Towerman Jackson seems to have had a rough deal, and even the newspaper commented that he had received a severe sentence.

A mail train bound for Vladikavkas left the metals near Rostoff-on-the-Don, Russia, on the evening of Friday 6th October 1905. Twenty-seven people lost their lives and another 35 were injured.[184]

On Saturday 16th June 1906 a passenger train was derailed at Pogranichnaia station, Manchuria, on the Chinese Eastern Railway. The initial report spoke of about a hundred persons dead or injured. A follow up report, however, stated that the initial figures had been exaggerated, and in fact only six people had died and ten had been injured.[185]

In the early years of the twentieth century there was competition in England between the Great Western and London and South Western Railway companies for the boat traffic from Plymouth to London, though until 1906 the Great Western company was hampered by having a considerably longer route via Bristol, one of several circuitous routes from which it derived the nickname 'The Great Way Round.' On Sunday 1st July 1906 it opened its direct route via Westbury, shortening the distance that its trains had to cover by 20 miles, and on the very same day the London and South Western suffered a major disaster to one of its boat trains at Salisbury.

The train involved in the Salisbury accident was carrying a mere 43 first-class passengers who had arrived in Plymouth on the American Line steamship SS *New York,*

---

[180] *The Nutley Sun*, Nutley, New Jersey, United States, 16th September 1905
[181] *A History of Railroad Accidents, Safety Precautions and Operating Practices* by Robert B. Shaw, pg. 387.
[182] *New York Times*, New York, New York, United States, 22nd March 1907
[183] *The Sun*, Baltimore, Maryland, United States, 1st February 1908
[184] *The Times*, London, England, 9th October 1905
[185] Ibid, 18th and 21st June 1906

together with its driver, fireman, guard, ticket collector and two waiters employed by Messrs Spiers and Pond. It consisted behind the engine of a van, three first-class carriages and a kitchen van, and was in charge of Guard William Harrison. The train departed from Stonehouse Pool, Plymouth at 11.1 p.m. on 30th June. Reversal was necessary at Devonport station, from where the train had a clear run to London Waterloo apart from a momentary stop at the Great Western station at Exeter St Davids (imposed by the Great Western company on all South Western trains to show its rival who was in charge) and a stop at Templecombe to change engines (made necessary because the South Western had no water troughs and the locomotives had insufficient water for the whole journey from Plymouth to London). The engine crew was also changed at Templecombe; from there the train was driver by Driver W. J. Robins, with Fireman Arthur Gadd, with engine No. 421.

Guard Harrison told the subsequent Board of Trade inquiry into the accident that he had looked at his watch at Salisbury West box and the train had then been a minute late. In February the company had issued a revised timing for trains between Salisbury East and West boxes and since then drivers had usually braked earlier, running through the station at 30 miles an hour, but on the night in question as they were passing the West box he saw that the needle of the vacuum brake did not move, indicating that the brake had not been applied from the engine. He had therefore just gently applied the hand brake to alert his driver. He then released the brake, the application of which he did not think would have reduced the speed of the train. The driver still failed to apply his brakes, and he (the guard) caught hold of the brake handle again, when he was suddenly thrown across the van as the train left the track. Guard Harrison said he was aware that the reverse curves at the east end of Salisbury station were very severe. Driver Robins was a stranger to him. He did not think that the passengers were in the habit of giving any gratuities to the engine crew after arrival at Waterloo and they would have had no opportunity to do so previously. The speed of the train at the moment of the derailment he put at not less than 50 miles an hour. He did not hear any whistling between Templecombe and Salisbury.

Signalman Herbert Munday was on the night shift at Salisbury West box. He said that he had received the 'Train entering Section' signal from Wilton box at 1.54½ and the train had passed him at 1.56½. Up passenger trains, including boat trains, usually slackened speed at his up main home signal, 200 yards in the rear of his box, and passed him at about 35 miles an hour. On the night of the accident the boat train had passed him at 60 miles an hour and without it brakes having been applied. He had no doubt that the train had been running at an unusually high speed. The train had whistled at his distant signal. He considered that the times in his train register were fairly accurate, and they booked to half minutes.[186]

Having passed through the station at much too high a speed the boat train was wrecked on the sharp curves at its eastern end. The derailed train struck a down milk train of 21 milk and brake vans which chanced to be passing and also a light engine standing in the down bay platform road. The engine and tender of the boat train overturned, its three leading vehicles were completely destroyed and the fourth was very badly damaged. The kitchen van suffered less. Slightly over half of the passengers on the train (24) were killed outright or fatally hurt and seven more were seriously injured. Driver

[186] Board of Trade report

Robins, Fireman Gadd, Guard Chenneours of the milk train and Fireman Charles Chick of the light engine were also killed, the ticket collector and two waiters on the boat train were injured and the driver of the light engine was badly scalded.

Mr G. A. Strachan, of Royleston Street, Boston, Massachusetts, was travelling with his aunt, Mrs Witton, also of Boston. They were pinned down by debris in the accident but not much hurt. Mr Strachan spoke to a reporter for the London *Times*. 'The oscillation of the train was something terrible. I have been accustomed to fast travelling in the States, on the broad gauge with heavy coaches, but I am of opinion that this train was running faster than anything I have previously known. The coaches swayed from side to side and I really feared an accident.' Later he added, 'I had a presentiment that the boat express would never reach London. I have journeyed across the American continent, and knowing what American travelling is, I thought I was going to have a slow time after leaving the ship's tender at Plymouth. I was very much mistaken. My aunt, Mrs Witton, and I had some conversation as to whether I should pull the communication cord, but it seemed to be thought that if I did such a thing I should be laughed at.'

Mr Walter Bellatti, of Jacksonville, Illinois, was also unhurt, and stated, 'I was on the first passenger car, but in the latter part of it. I had just finished dinner, when I heard a loud crash and found myself all of a heap. The first part of the car was completely smashed and fell over on its side. The part I was in ran along the rails until past the engine, which had struck another engine on another line. The first section of the car, as I have explained, was smashed. That in which I was seated only turned over after it had passed the engine. I stopped to look after some of the passengers who were dead or dying. I am now going to the Hôtel Métropole and will not be seen by anybody. Fortunately I escaped with nothing more than a shaking.' [187]

A fortunate aspect of the accident was that it occurred in the county town of Wiltshire, and so plenty of help was immediately on hand. Staff from the local infirmary were soon on the scene, together with ten doctors, some of whom came into the city by motor car. A police sergeant on night duty half a mile away heard the crash and leaving his beat hurried to the station. He removed four of the dead unassisted and was still working at the scene at midday. Breakdown trains with steam cranes were sent from Exmouth Junction and Nine Elms and a salvage corps travelled down from Waterloo.

The Board of Trade report on the accident was lengthy. The investigation was conducted by Major J. W. Pringle, who faced the problem that the person who could best account for the accident was no longer able to do so. As usual the inspecting officer began his report by detailing the circumstances of the disaster and describing the location, the makeup of the train and so on. The evidence of a large number of witnesses was reproduced and his conclusions were then given. Major Pringle began by noting that prior to the commencement of the boat train service in April 1904 every booked train stopped at Salisbury (a handful of specials did not). As soon as the new train was put on the danger of running through Salisbury too fast was recognised, and a speed restriction through the station of 30 miles an hour was imposed. In February 1906 a timing of one minute was brought in between the East and West boxes, 755 yards apart, which implied an average speed of only 25 miles an hour. More recently two notices had been issued to enginemen regarded running too fast in connection with boat trains doing so in the

[187] *The Times*, London, England, 2nd and 3rd July 1906

Basingstoke area. Turning to the timing of the train the signal box times showed that it had lost four minutes in the 20 miles between Templecombe and Dinton, then regained two minutes in the eight miles between Dinton and Salisbury West box, travelling at an average speed of 70 miles an hour. The evidence of Guard Harrison and Signalman Munday agreed that the brakes had not been applied at that point. Inspector John Spicer, Yard Foreman Phillip Rook and Carriage Examiner William Abbot, who had all been on the station, were sure that the train had run through at high speed. Signalman Edward Haines in Salisbury East box put the time of receiving the 'Train entering Section' signal from the West box at something over 1.56½ and booked it as 1.57, thus confirming Signalman Munday's time. All the evidence showed, therefore, that the express had passed through the station at a very high speed, possibly as much as 70 miles an hour.

It was the first time that Driver Robins had worked the boat train. He was a fully qualified driver, a total abstainer, and 'bore a fair character'. Shunter Mullett at Templecombe had stated that he had remarked to Robins that the boat train was running well to time, to which the driver had replied, "Yes, I shan't get into Waterloo before time, else I shall have to go up to see the governor." The driver had also signed the several notices issued reminding him of the need to go slowly through the station. His recklessness was therefore inexplicable. That he had whistled at the Salisbury West distant signal showed that he had been on the alert. It was possible that not having driven through Salisbury before he had not recognised the danger of the short curve, but there was also a facing crossover from the up local to the up through line to be negotiated, with reverse curves with radii of 7½ chains, which any driver must recognise as dangerous except at low speeds. No evidence had been found that passengers were giving drivers tips for running at high speed and there was nothing wrong with the track or the engine. The overturning speed of the engine on the curve he calculated at 63 to 65 miles an hour.

Guard Harrison had been re-examined and had stated that when the driver had failed to apply the vacuum brake even after his slight application of the hand brake, he (the guard) had been afraid to apply the vacuum brake from his van for fear of breaking the couplings or else damaging the train. Guard Harrison had also made differing statements at the Coroner's inquest to those he made at the Board of Trade inquiry. Major Pringle said that he realised that the guard's position was a difficult one, in that he was reluctant to take over control of the train from his driver, but he could have possibly prevented the accident, and had shown himself to be little fitted for responsibility.

In conclusion Major Pringle considered the subject of speed restrictions and advised imposing one of 15 miles an hour at the east end of Salisbury station.[188]

At the inquest on the victims on 16th July the jury returned a verdict that the victims had met their deaths 'by the derailing of an express train on the morning of July 1st, 1906, and that such derailment was caused by the excessive speed at which the train was travelling contrary to the company's regulations.' The jury added a rider that 'We are also of opinion that the driver of every train not stopping at Salisbury should have his attention especially called before starting to the regulations affecting his journey, which we consider was not done in this case.'[189]

A recent author who has made a special study of the accident suggests the possibility

---

188 Board of Trade report

189 *The Times*, London, England, 3rd, 4th, 5th, 6th, 7th, 10th, 11th, 17th and 21st July 1906

that Driver Robins, who had probably not been able to enjoy a great deal of rest in the day or so before the accident, had suffered a microsleep as his train was entering Salisbury station,[190] which seems a very reasonable possibility and might also explain previous excessive speed derailments which had occurred during the night hours at Preston in 1896 and at Aylesbury in 1904. As will be seen there were to be two more disasters from the same cause in the next year or so, these also taking place during the early hours.

On Friday 21st September 1906 the *Times* newspaper in London carried the headline 'The wreck of a Scottish Express'. An accident similar in several respects to the Salisbury disaster had occurred on the East Coast main line at the north end of the Great Northern Railway station at Grantham, England. Again it was an accident which happened in the early hours, again it was a derailment due to excessive speed around a curve and again the driver and fireman did not survive to give witness to how it had occurred. There were also differences; at Grantham the train was booked to stop at the station which it instead ran through at speed and the derailment happened not on a curve in the main line itself but because the train was diverted onto the Nottingham branch, which went off sharply to the left from the main line.

The train involved in the Grantham disaster of Wednesday 19th September 1906 was the 8.45 p.m. down express from London King's Cross to Edinburgh, which consisted of an Atlantic class engine No. 276, a brake van, a mail van, two brake vans, a third-class carriage, a composite carriage, two sleeping cars, two composite carriages and two brake vans, in charge of Conductor Thomas Rowlett, with Guard William Knighton, Driver Frederick Fleetwood and Fireman Ralph Talbot.

The Board of Trade inquiry into the Grantham accident was a very thorough one and the report into it was long and derailed one, like that into the Salisbury disaster. This time the investigation was carried out by Lt-Col P. G. von Donop. The first person whose evidence appeared in the report was Guard Knighton, who said that he had been riding in the Glasgow brake van, the fourth vehicle in the train. The journey had been normal as far as Peterborough, where the engines had been changed and an additional vehicle added to the rear of the train. There was 20 inches of vacuum showing on the gauge in his van at Peterborough, from where the train on time departed at 10.20 p.m. Steam was usually turned off at Saltersford box, but he could not say if it had been on this occasion. As they were passing Grantham South box, he had realised that they were still travelling at a terrific speed and had flown to the vacuum brake, only to find that there was no vacuum, indicating that the brakes had been applied. He estimated the speed of the train through the station at 40 miles an hour, though it was slackening. He had looked out of the van window to see how the facing points at the junction with the Nottingham branch were lying, knowing that they would be all right if they were set for the main line. With that his van was diverted onto the branch, and seemed to take the curve without becoming derailed, but the next thing he knew was he was lying at the foot of the embankment.

Signalman Alfred Day was on duty in Grantham South box, assisted by Telegraph Boy Thomas Sellars. Signalman Day said that he had received the 'Train entering Section' signal for the 8.45 p.m. train from Saltersford at 11.01, and it had passed his box

[190] *Salisbury 1906: An Answer to the Enigma?* by Norman Pattenden, pgs. 69-72

at 11.02. His signals had all been clear for it, but it had appeared to be going too fast to stop at the station. He could not say if steam had been shut off but was sure that the brake had not been applied as there were no sparks from the wheels, nor was there the sound of a braking train. He could see the driver and fireman, they appeared to be looking out of their respective glasses and not doing anything. He remarked to Sellars as to the train's excessive speed and the latter went to the box window. Telegraph Boy Sellars agreed with his signalman that the train brakes had not been applied.

Grantham South box was slightly over a quarter mile south of the station, with Grantham Yard box at the south (Peterborough) end of the up platform and Grantham North box just beyond the station at the junction with the Nottingham branch. Signalman Isaac Crowson was in Grantham Yard box. He told the inquiry that the train had passed him at 11.03, having been accepted by the North box. His signals were off. The train, he thought, had passed him at a terrific rate, perhaps 50 miles an hour. The speed had been about that of a through non-stop train. The north box distant signal was at danger. He did not see the driver and fireman.

Various railwaymen were standing on or near the down platform waiting for the train to arrive.

Inspector George Pile saw the train coming in much faster than he expected, he estimated between 40 and 45 miles an hour. The North box signals were at danger. He thought that steam was turned off but that the brakes had not been applied. He heard no whistling. Turning around to watch the train he almost immediately heard the noise of an explosion and then flames shot up into the air and he had rushed down to the scene.

Travelling in the rear of the train was Conductor Rowlett. Like Guard Knighton he said that he had realised that something was wrong as his train was passing Grantham South box. He had put the window down and saw that they were travelling too fast. He went for the brake but before he could reach it was thrown head over heels in his van and could not get up. His van did not go down the embankment but came to a stand partly hanging over a bridge. He was sure that the brakes had been applied and the wheels were skidding at least from the South box.

On duty in Grantham North box was Signalman Richard Scoffin, who testified that he had accepted the 8.54 train from the Yard box at 10.57. At that time an up Leicester goods train had just come off the Nottingham branch and he had left the double line junction set for the branch in accordance with normal practice, as the 8.45 was due to stop five minutes in the station, during which time he might have had another train coming off the branch. His acceptance of the passenger train was of course perfectly in accordance with the block regulations as with the points set for the branch, he had the requisite clearing point. He received the 'Train entering Section' signal for the train at 11.04 and had turned around to see it arriving in the station. Instead he saw and heard it approaching his box. His main line signals were necessarily all at danger, locked by the points. He did not see the driver or fireman. He thought the train had been going at 40 miles an hour as it passed him. The brakes seemed to be hard on, with the wheels skidding. There was no whistling. He heard, but did not see, the derailment. After the accident he had blocked the line in all directions. He had seen flames arising from different parts of the train almost at once.

Close to the junction with the branch some vehicle or vehicles became derailed to the left and collided with the parapet of a road underbridge. Sixty-five feet of the parapet wall was demolished and almost the whole train was wrecked at this point. The engine overturned, the tender went down the embankment, as did five of the following vehicles of the train. Two more vehicles went halfway down the embankment and the remaining three were derailed but stayed on the embankment. Fire broke out almost at once both on the vehicles which had gone down the embankment and those which had stayed at the top. Driver Fleetwood, Fireman Talbot, nine passengers and a Post Office mail attendant were killed outright, and two seriously injured passengers died later.

The Board of Trade inspecting officer noted that the train had been due to arrive at Grantham at 11 p.m. exactly and had actually passed the Yard box at 11.03, having lost three minutes in the 29¼ miles from Peterborough. It had thus taken 43 minutes to cover the distance at an average speed of a little less than 41 miles an hour and its speed over that section could not be said to have been excessive. A more detailed examination of the signal box passing times showed an average speed of 39 miles an hour over the 20¾ miles between Peterborough and Corby, 35 miles an hour over the 3 miles of rising gradient between Corby and Stoke and just under 60 miles an hour between Stoke and Grantham South box, or 52¾ to the Yard box. With the signal box clocks varying and the boxes being close together absolute reliance could not be placed on these times, but it was clear that the train had lost time between Peterborough and Corby but had averaged between 50 and 60 miles an hour between Stoke and Grantham.

There could be no doubt, Lt-Col von Donop stated, as to the indications of the Grantham signals. The South and Yard box signals had been clear for the train to enter the station, the North box signals had been at danger. Given the closeness of the boxes he did not consider that the Yard box distant should have been lowered before the platform starter was cleared. Despite the adverse signals and the fact that the train was booked to stop at the station it had not done so. The only two men who could have given direct evidence as to what happened had been killed, however the two guards had survived. Lt-Col von Donop reviewed the statements of Conductor Rowlett and Guard Knighton and just like his fellow inspecting officer, Major Pringle, in his investigation into the Salisbury disaster, Lt-Col von Donop was very critical of the two guards for their not having realised sooner what was happening and applied the brakes to avert disaster, and considered that they had not been equal to their responsibilities.[191]

Taking the evidence of the various witnesses it could be concluded that the train had passed through the station at 50 miles an hour or possibly more, that steam had only been turned off just before the accident and that the brakes were probably not applied until the train was passing through the station. The reason for the failure of the train to stop could only lie in a failure of the regulator or vacuum brake, or of a failure of the driver. Examination of the locomotive after the accident had shown nothing wrong with the regulator, nor was there any evidence of a problem with the brakes. The cause must therefore lie in the actions of Driver Fleetwood and to a lesser extent Fireman Talbot. The latter was regarded as an exceptionally reliable driver and from the evidence of W. H. Brooks, his regular fireman, a very cautious one.

Fireman Talbot, who was on the engine that night only because Fireman Brooks was off sick with a bad cold, was not an ordinary fireman, but a premium apprentice, gaining

[191] Board of Trade report

some experience firing engines before going into a management job, perhaps to become a locomotive superintendent or even a locomotive engineer. In contrast to working class Fleetwood and Brooks he was the son of a City of London merchant who had been educated at Abingdon Grammar School. Ordinary railwaymen were not very happy with the use of premium apprentices as firemen, and on the evening of Driver Fleetwood's funeral the Doncaster branch of the Amalgamated Society of Railway Servants unanimously passed a resolution 'That having regard to the terrible accident at Grantham, this meeting of railwaymen feels it its duty to point out that in their opinion the cause of the catastrophe was closely connected with the presence on the engine of a premium apprentice acting as a fireman, whose want of proper experience would not allow him to check momentary forgetfulness on the part of the driver, or to take charge of the engine in the case of a sudden seizure or death.' [192]

The question of Driver Fleetwood having been drunk was raised at the inquiry. Evidence was given that he was not a teetotaller but his mother-in-law, a Mrs Dunkerley, who lived at his house and had cared for one of his children after his wife had died, stated that she had never seen him drunk. Fireman Brooks and other witnesses agreed that they had never seen Fleetwood 'worse for liquor'. Lt-Col von Donop considered various theories to account for the accident. That they had simply forgotten to stop at the station seemed unlikely, for they had worked the same train the night before, in addition to which the North box distant signal had been at danger. A second possibility was that they had failed to realise how close they were to the station, but the approach, with Saltersgate High Bridge and Spittlegate Bridge, together with various signals, was distinctive.

A third possibility was that Driver Fleetwood had been taken ill on the footplate. He had been off work due to illness three times in the previous year, due, according to Mrs Dunkerley, to sciatica. He had also been taken ill once in the cab, but Fireman Brooks said that he had still been able to drive the train. Unless an employee chose to withdraw money from the sick fund he did not need to be examined by a doctor when he went off sick and Fleetwood had not been so examined; the inspecting officer thought that this should be changed. That Driver Fleetwood had suffered a sudden illness was, Lt-Col von Donop concluded, the most likely explanation for the accident, but its exact cause must remain a mystery.

Finally with regard to the suddenness with which the post-accident fire broke out the inspecting officer remarked that it was difficult to resist the conclusion that it was due to the ignition of escaping gas, and that the accident pointed, like previous ones, to the unsuitability of gas as an illuminant for railway vehicles.[193]

A whole book has been written on the Grantham accident, in which the author mentions another possible cause for the disaster. He points out that Driver Fleetwood had mentioned to another driver at Peterborough before setting out on his fateful journey that the previous night with Talbot as fireman eight minutes had been lost between Peterborough and Grantham, and that he did not intend to have it happen again. The extra vehicle added at Peterborough – a brake van carrying a coffin – had added 13 tons to the weight of the train and probably contributed to the loss of three minutes between Peterborough and Corby, perhaps not helped by the less than expert firing of Talbot, and

[192] *The Grantham Rail Crash of 1906* by Harold Bonnett, pg.14
[193] Board of Trade report

it was possible that Driver Fleetwood, determined, as he had said to his fellow driver, not to lose time again, had run the train extra hard from Stoke summit downhill to Grantham and had deliberately delayed the braking, but had then misjudged matters and been unable to pull up in time.[194]

At the inquest on 12 of the victims of the accident the jury returned a verdict that 'The 12 deceased persons were accidentally killed through the express train running off the rails; that the driver and fireman were both competent men to work the train, but the evidence was not conclusive to show whether the brakes were applied in sufficient time to effect a stop at Grantham.' The jury also added a rider that it would be safer if a wider curve were added at the Nottingham branch line and if the points on the main line at the junction in question were set clear as soon as a train had passed.[195]

Signalman Scoffin died in February 1914 aged only 44. Despite being found blameless at the Board of Trade inquiry, if perhaps not by the coroner's jury, he had blamed himself ever since for not having reset the road for the main line.[196]

On the evening of Saturday 6th October 1906 the 20:40 express from Milan to Rome was derailed at the bridge over the River Po just short of Piacenza station, Italy. One of the carriages of the train left the rails, dragging a second with it, and they were completely wrecked. By the following day it was known that five passengers had been killed and about 50 first- and second-class passengers injured. Among those killed were Professor Lino Ferriani,[197] a noted criminologist, and members of his family. The derailment was initially attributed to a broken axle. The Rome correspondent of the London *Times* wrote, 'Hitherto, in spite of their disorganization and mismanagement, the Italian railways have enjoyed extraordinary immunity from disasters of this kind. It is to be hoped that the condition of the rolling stock will not add danger to the existing discomforts of railway travelling in Italy.' [198]

That a broken axle had caused the accident later seemed less likely, as the first carriage to become derailed had been as new one. It was suggested instead that a set of points must have been to blame. The Italian government commissioned an inquiry into the cause.[199]

Electrification of the New York Central and Hudson River Railroads New York suburban lines was first proposed towards the end of the 19th century as a solution to the problems of congestion on the main line into Manhattan. The scheme was given a considerable stimulus due to a rear collision in Park Avenue Tunnel on 8th January 1902 (see Tunnel Accidents chapter) blamed partly on steam and smoke obscured signals, which was followed by a city ordinance requiring the railroad company to discontinue

---

194 *The Grantham Rail Crash of 1906* by Harold Bonnett, pgs. 30-6
195 *The Times*, London, England, 21st, 22nd, 24th, 25th, 27th and 28th September and 9th, 10th and 19th October 1906
196 *The Grantham Rail Crash* of 1906 by Harold Bonnett, pg. 38
197 *The Express and Telegraph*, Adelaide, South Australia, 9th October 1906
198 *The Times*, London, England, 8th October 1906
199 Ibid, 9th October 1906

using steam locomotives within the city limits from 1st July 1908.[200] The new electrified system was hailed as a great achievement when it came into full operation in February 1907. It therefore came as a considerable shock that only three days after the inauguration of the new trains one of them suffered a disastrous derailment.

The train involved in the accident was a White Plains and Brewster express, train No. 25, which had left New York Grand Central terminus at 6.13 p.m., filled with homebound matinee crowds and perhaps some commuters, though the day on which the accident occurred, 16th February 1907, was a Saturday. The train consisted of two electric locomotives, a combination baggage and smoking car and four passenger coaches, all the vehicles being of wooden construction. Commanding the train was Conductor L. W. Hopkins, and it was being driven by Motorman Edward R. Rogers and his assistant, George A. Jacobs. The first stop of the train was at White Plains. It was running at high speed when it reached 205 Street, near the Woodlawn Road area of the Bronx, where there was a sharp curve. The two locomotives and leading car negotiated the curve safely, but the following cars left the track and were dragged along on their sides for a hundred yards before the couplings finally gave way and they ended up in a heap at the foot of a low embankment. A 'sheet of electric flame' enveloped the rear car and for a moment it was feared that fire would add to the disaster, burning those trapped in the wreckage, but it did not spread far before it was extinguished by firemen. The live third rail was broken by the falling cars, cutting the current. Most of the deaths and injuries in the accident occurred through passengers being thrown through or against the car windows and some were described as having been 'ground to pieces'. A feature of the aftermath of the accident was the appearance on the scene of what a newspaper described as 'ghouls', who went through the clothing of victims and searched pocketbooks for valuables until police arrived at the scene and drove them away, which reportedly took them an hour. Italian workmen living in the vicinity were blamed, rightly or wrongly, by the newspaper. By midnight the number of dead had risen to 20 and the final figure was 23. Some, though, had lucky escapes, 'In one instance two policemen and a citizen worked almost an hour to extricate a woman and her child from underneath a seat in the last car. She was Mrs Charles Gilchrist, and her daughter was Hannah, four years old. Mrs Gilchrist escaped with only a long rip across her forehead. The baby, whom she had protected with her body, was absolutely unharmed.[201]

Initially the accident was blamed on a defective rail, which was later found wound around the wheel of a car with one end protruding through the floor of the car. There also suggestions that a flange of one of the locomotive wheels had broken or that the track had spread.[202] Soon there were questions regarding the speed of the train, with Motorman Rogers, in charge of the leading engine of the train, telling Coroner Schwannecke that he had been running on schedule time but apparently admitting that he had been running at 70 miles an hour. Speed around the curve was restricted to 57 miles an hour. The electric locomotives were not equipped with speedometers and it was possible that Motorman Rogers, accustomed to driving steam locomotives, had underestimated his speed with the quieter running of the electric engine.

The coroner's jury returned a verdict on the victims killed in the accident on 4th

[200] Webpage – American Rails.com: New York Central's Electrification.

[201] *Cranbury Press*, Cranbury, New Jersey, United States, 22nd February 1907

[202] *Baltimore Sun*, Baltimore, Maryland, United States, 17th February 1907

March. It took them only 40 minutes to decide that the accident had been caused by 'the train running at a speed in excess of that which has been proved to be safe for such a train on the track laid to the existing curvature at that point,' and they concluded that, 'The New York Central and Hudson River Railroad Company did not take all the necessary, possible and proper precautions to safeguard its passengers at this point and, consequently, was culpably negligent, and that the responsibility for the existing condition seems to be divided between the construction and operating departments.'

The jury recommended that the New York Central and other railroad companies in the state be compelled to reinforce the fastenings of the outer rails on all curves such as that involved in the accident by additional spikes and rail braces and increased superelevation, and that until such strengthening had been completed the companies should be compelled to reduce the speed of trains through such curves to so-called equilibrium speed, or one that was absolutely safe, to protect the lives of passengers and employees. In addition, the jury recommended that the New York Central Company should be compelled to install recording instruments or similar so that motormen could know with reasonable exactness the speed at which their trains were travelling. 'This information should cover speeds at the three different positions of the controlling lever handle and be properly tabulated and placed in a prominent position in the cab of each electric locomotive.'

Over the protests of A. V. Harris, the general counsel for the railroad company, that his action was not justified by the jury's verdict, Coroner Schwanneke committed the President of the New York Central, William H. Newman and the Board of Directors of the company to the Grand Jury with a view to their possibly standing trial on manslaughter charges.[203]

In the end Alfred H. Smith, Vice President and General Manager of the New York Central and Hudson River Railroad, was the sole person chosen to stand trial. The matter dragged on until the end of the year, but on 18th December at his trial Supreme Court Justice Kellogg directed the jury to acquit him, and the matter was closed.[204]

Train No. 1 of the Great Northern Railroad, the *Oriental Limited*, was derailed at 1.23 a.m. on the morning of Monday 15th April 1907 three-quarters of a mile from Bartlett, North Dakota, United States. Fire broke out immediately after the derailment and the whole train of eight coaches was consumed except the rear Pullman and observation cars. It was at first feared that 11 passengers had died in the accident, including nine Greek or Italian labourers, eight of whom were initially unaccounted for, but the only victim positively identified at the time of the first report was Mail Clerk William Jones. Engineer Peter Fergueson, Fireman Roy Curtis, Mail Clerk C. H. Fodness and about eight passengers were more or less seriously injured.

A relief train, 'well supplied with physicians', was made up at Devil's Lake and soon arrived at the scene of the accident. The injured were taken to the Deaconness hospital in Grand Forks and it was reported from there later in the day that all were expected to recover. A wrecking train left Grand Forks for the scene after some considerable delay, having been employed clearing up the derailment of an ore train near Ojata, which had happened on the Sunday morning.

---

[203] *New York Times*, New York, New York, United States, 5th March 1907
[204] Ibid, 19th December 1907

The cause of the derailment was unclear. Engineer Fergueson was of the opinion that the track had been tampered with, having noticed his engine in a very unusual manner a few moments before it left the rails and went down a steep embankment. [205] Not long afterwards, however, blame for the accident was placed on the new General Superintendent Bryan of the Central Division Great Northern, who, it was said, had recently issued orders for trains to make speed regardless of the condition of the track, and had announced that engineers who had been cautious and had failed to make up time would be liable to dismissal. It was also alleged that at the time of the derailment the train had been travelling, not, as previously stated, at 35 miles an hour but at 60, and that the roadbed at the point of derailment had been in a terrible shape.[206]

The initial death toll was later corrected to five; Mail Clerk Jones and four unidentified foreign labourers.[207]

The engine and two coaches of a 'composite' (mixed?) train left the metals between Sowkino and Dobromino stations, Russia, on Saturday 20th April 1907. Five persons were killed and six injured.[208]

A passenger train was derailed on the Algarve railway near Sibola, Portugal, on Saturday 25th May 1907. Seven passengers died and 21 were injured.[209]

The Canadian National Exhibition had been held annually in Toronto, Ontario, since 1878 in August and the beginning of September.[210] On Tuesday 3rd September 1907 the Canadian Pacific Railway had arranged a special train to run from Markdale to Toronto to convey visitors to the exhibition, leaving Markdale at 6.30 a.m. to arrive in Toronto at 10.20. From Markdale the train consisted of an engine and four coaches with Conductor Matthew Grimes and Engineer George Hodge. It was already well filled when it left there and by the time that it departed from Orangeville, about 45 miles from Markdale, at 9.30 a.m. it was 'crowded to the utmost capacity,' though three more coaches were there attached to it. It was also an hour late. Nine miles further on, on the Horseshoe Curve beyond Caledon station, where the line descended the Caledon mountain with steep gradients and severe curvature, the train, running much too fast, left the rails. Engineer Hodge and his fireman jumped and survived, though they were hurt. Two coaches turned upside down and were demolished and the next two were badly smashed. Seven passengers lost their lives and about 20 were injured, some seriously.[211]

A few days after the accident the *Winnipeg Tribune* carried detailed accounts from some of the survivors. 'Mr and Mrs R. A. Jelly, of Selburne, and baby Irene, came into Toronto Junction on the special last night, after being cared for by relatives at Caledon East. The

---

205 *The Fargo Forum and Daily Republican*, Fargo, North Dakota, United States, 15th April 1907
206 Ibid, 17th April 1907
207 *The Big Timber Pioneer*, Big Timber, Montana, United States, 25th April 1907
208 *The Times*, London, England, 22nd April 1907
209 *The Marshfield Times*, Marshfield, Wisconsin, United States, 29th May 1907
210 Wikipedia – Canadian National Exhibition
211 *Winnipeg Tribune*, Winnipeg, Manitoba, Canada, 4th September 1907

little girl was quite unhurt, and by almost a miracle her parents escaped with only a severe shaking up. They were sitting in the fifth coach of the train, which was completely overturned and carried several lengths from the track. "The man Robinson, who was killed, sat just facing us," said Mr Kelly. "The three of us were thrown right into the hat rack when the car turned over. There are four dead from that car alone. Not one seat but was smashed to pieces. I feel sure there are bodies under that wreck yet," he added.

'One of the saddest incidents of the wreck was the serious injury of Mr and Mrs Conn, of Heathcote, who were married only on Monday. They were in the coach behind the baggage car, and Mrs Conn was so pinned down by wreckage that some difficulty was experienced in releasing her. "I was talking to a man across the aisle on a matter of business," Mr Conn relates, "and I really was not paying any attention to the speed of the train. The car gave a sudden sway and never came back. I tore away what debris I could with my hands," he said, describing his wife's rescue, "and kept calling for an axe, which came at last, and we got her out. Armstrong, who was killed, was in the same car, and I wanted to go to his assistance, but I thought it my duty to see to my wife first." Mrs Conn is very seriously injured indeed and at one time yesterday afternoon her pulse stopped. Her recovery is now confidently expected. Her husband is badly bruised and cut.'

The same newspaper reported that after the accident Engineer Hodge had sat on the upturned cab of his engine and wept, sobbing like a child.[212]

The inquest on those who died in the accident opened at Caledon East on 11th September, when various witnesses testified as to the excessive speed of the train. Passenger Joseph Stephenson, of Flesherton, related that the train had gone through Melville Junction 'like a streak of lightning', while Joseph Ferguson, who lived near the scene of the accident, stated, "I was standing in the yard loading a gun when I saw the train coming. The rate was so fast that I did not need a second glance to know that it would not make the curve. I have seen freights running away going around the curve, but none so fast as that passenger train. Before you could say anything, it was into the ditch. As the train went around the curve it swept away two gates and two gate-posts as clean as if they had not been there. I ran over and helped those who were pinned down by the wreck."

Road Superintendent A. L. Smith, asked if the outer rail on the curve had been lowered several inches when the track was renewed, stated that it had not been altered. Questioned as to a report that Engineer Hodge had been drinking, the superintendent replied that there had not been a mark against the engineer in the previous five years and he did not believe a rumour of him having been drinking early on the morning of the accident.[213]

The inquest ended on 18th September with the jury finding Engineer Hodge and Conductor Grimes responsible and they were arrested on a charge of manslaughter. The jury also blamed the Canadian Pacific company for putting inexperienced men in charge of passenger trains.[214]

After a week-long trial Engineer Hodge and Conductor Grimes were acquitted on Saturday 16th November. 'When the jury's verdict was announced the court room at

[212] Ibid, 7th September 1907
[213] Ibid, 13th September 1907
[214] *The Evening Telegram*, Salt Lake City, Utah, United States, 18th September 1907

Brampton, where the trial took place, was the scene of great rejoicing and cheering by the friends of the prisoners, and the officials were unable to restore order for several minutes.' [215]

It seems likely that the excessive speed of the train which led to the derailment was due to Engineer Hodge attempting to make up some lost time, and that perhaps like Driver Fleetwood at Grantham he made an error of judgement in running too fast too far. Why the train was an hour late does not seem to have been recorded.

Running too fast in an effort to make up time was apparently also the cause of a derailment at Norris, Iowa, United States, only three days after the Caledon accident, on 6th September 1907. The northbound Chicago Rock Island and Pacific Railroad express was ten minutes late as it approached Norris, where a southbound freight train was stood in the siding awaiting it. 'Thundering along at a terrific speed' the express just reached the siding when it left the track. The express engine crashed violently into that of the freight engine, wrecking both, the baggage and mail cars of the express were telescoped into one another and its smoking car, in which a number of men were riding, was demolished. It was in the smoking car that all the deaths and injuries occurred, the passengers in the two following day cars escaping with a severe shaking. The only person to escape unhurt from the smoker was a news dealer who was thrown through its broken roof and landed by the side of the track unhurt. Engineer Al Mason and Fireman Henry McCann stuck to their posts and were thrown clear of the wreckage, while Engineer Kinch of the freight train also escaped. The only railwayman injured was Baggageman Myers.[216] The injured were treated by a trained nurse who chanced to be on the train and brought in a special train to the Presbyterian Hospital in Waterloo, Iowa. Eleven passengers were killed outright and 12 were injured.[217] Three of those injured died in hospital the next day, bringing the number of dead up to 14.[218] The accident occurred at about 10.15 a.m.

Rather strangely it seems to have been quickly decided that no one was to blame for the accident, and it was announced that only a formal inquest into it would be held, in which witnesses would testify to completely exonerate all concerned. [219] Railroad Commissioner N. S. Ketchum spoke out however recommending that smoking cars should be abolished in Iowa and smoking compartments at the ends of each coach substituted, with the coaches built of steel, as Pullman cars already were. The new cars would also have side entrances in contrast to the entrances at the ends which were a feature of all current cars. He noted that smoking cars were inevitably old vehicles relegated to use as smoking cars by the company and invariably of the lightest construction, without reinforced ends. As a result they were easily telescoped in collisions, as had happened at Norris Siding. The heavier coach would pass through the lighter one and mow down its occupants or push them to the end to be crushed to death.[220]

---

[215] Winnipeg Tribune, Winnipeg, Manitoba, Canada, 18th November 1907

[216] *The Telegraph-Herald*, Dubuque, Iowa, United States, 7th September 1907

[217] *Brainerd Daily Despatch*, Brainerd, Minnesota, United States, 7th September 1907

[218] *The Journal*, Winchester, Indiana, United States, 11th September 1907

[219] *The Des Moines Daily News*, Des Moines, Iowa, United States, 7th September 1907

[220] *The Evening Gazette*, Burlington, Iowa, United States, 10th October 1907.

By 1907 Japan had had an interest in the Korean peninsula for several decades. In 1905 Korea became a Japanese protectorate and in 1910 it was to become a fully-fledged Japanese colony.[221] It is not surprising therefore that when there was a serious train accident near Seoul, Korea, on Tuesday 1st October 1907, many of the casualties were Japanese soldiers. Early on the morning of the day a southbound train from Seoul was derailed and 42 people, including 30 soldiers, were killed and injured. Without stating the cause of the derailment, if known, the report added that the responsibility for it had not been fixed.[222]

Early on the morning of Tuesday 15th October 1907 the drama which had been played out the previous year first at Salisbury and then at Grantham was to be repeated again, this time at the Great Western and London and North Western Joint station at Shrewsbury, England. Again signalmen witnessed a passenger train rushing past their boxes much faster than it should have done, again the train was derailed on a sharp curve with disastrous results and again the driver and fireman responsible failed to survive the accident to explain what had gone wrong.

Signalman John Beddleston was on duty that day in Shrewsbury No. 2, Crewe Bank, box, having come on at 10 p.m. the night before to work until 8 a.m. that morning. He was one of the first witnesses at the Board of Trade inquiry into the accident, where he related that he had accepted the 1.20 a.m. passenger train from Crewe from Harlescott Crossing box at 2.03 a.m. He had received the 'Train entering section' signal for it at 2.06 a.m. and it had reached him at 2.07 a.m. His box was rather less than two miles from Harlescott Crossing (it was noted in the report that the boxes was actually one mile and 1090 yards apart – if the distance between the two boxes had actually been covered in one minute exactly that would have given an average speed between them of 97 miles an hour but of course box times were only recorded to the nearest minute or sometimes half minute, and had the train taken two minutes the average speed would have come down to slightly under 49 miles an hour). When the train passed his box his signals had all been at danger, as the train had been accepted by the box ahead at Crewe Junction under the 'Warning Arrangement', whereby it had to be stopped at Crewe Bank box and its driver instructed that the line was only clear to the Crewe Junction home signal. As the train neared his box, he could see that it was coming at great speed. He had rushed to the window and waved a red hand lamp towards it, but its driver had appeared to take no notice of it. He had then gone to the block shelf and sent the 4-5-5 'Train Running Away on Right Line' emergency signal to the junction box, after which he had gone to the telephone to ring Crewe Junction, but by then the accident had occurred as the boxes were only 579 yards apart. He did not see the driver and could not be sure if steam had been applied or whether the brake had been on, but he had not seen any fire flying from the wheels. He had worked the box for 30 years and had never seen a train pass travelling so fast before. He had known full well that an accident would happen because there was a speed limit of 10 miles an hour around the curve at the junction. The train had not whistled, and he had not seen the guard though he had only got a glance at the train from

[221] Wikipedia – Korea under Japanese rule.

[222] *Burlington Evening Gazette*, Burlington, Iowa, United States, 1st October 1907

underneath the block shelf. Crewe Junction had acknowledged the 4-5-5.

Signalman Edwin Ward said that he had had no time to answer the 4-5-5 before the crash came. Practically as soon as he had received it the train had been, in his words, 'all across the junction.' It had run past all his signals at danger and had been going at a terrific speed, he estimated 60 miles an hour. He had acknowledged the 4-5-5 from Crewe Bank and then set the 'Obstruction Danger' signal in all directions. He had accepted the 1.20 train under the 'Warning Arrangement' because the Stockport train was already standing at the platform on the line which the 1.20 train was to run to and fish trucks had first to be detached from the rear of the Stockport train via the scissors crossing worked from Shrewsbury Central box. That operation had to be done first before the 1.20 train, the Bristol train, could be admitted to the station. The two trains were re-marshalled in the station before one departed for Bristol and the other for Cardiff. Not all trains were accepted under the warning but four or five out of six were. He did not hear any whistling, see any steam coming out of the funnel or any sign that the brakes were applied. He had never seen a train approach his box at anything like that speed before.

Guard Henry Birch said that he had been in charge of the Bristol and West of England train. They had left Crewe eight minutes late, with a non-stop run to Shrewsbury. His train had consisted of 15 vehicles; a London and North Western (LNW) brake van, three Great Western (GW) composite carriages, two LNW brake vans, an LNW Post Office van, an LNW Post Office sorting carriage, an LNW parcel van, an LNW composite carriage, an LNW brake van, a West Coast Joint Stock composite carriage, a West Coast Joint Stock brake van, a Caledonian covered truck and a West Coast Joint Stock fish van. The engine of his train had been No. 2052 (a practically new 4-6-0 of the *Experiment* class), with Driver Samuel Martin and Fireman Frederick Fletcher. He knew Driver Martin, but not the fireman. The driver had shown no signs of having had too much to drink or of being unwell. He had begun to think that the train was going too fast before they reached the Crewe Bank distant. Seeing the Crewe Bank signals at danger he had gone to apply the continuous brake but when he got to the handle, he found that the driver had already applied it with full force. He estimated that the speed had been reduced to 40 miles an hour before the crash. He had never known a train approach Shrewsbury going so fast before; drivers usually began to brake before the Crewe Bank distant. It was common for trains to come in under the warning. He could not explain the accident.

The results of the derailment were disastrous. The engine and tender were overturned on their right side, the leading two vehicles of the train were entirely demolished, the third vehicle ended up on the Great Western line to Chester, the next six vehicles were piled up in a heap to a considerable height, the tenth vehicle was piled up against the heap of wreckage and damaged beyond repair, the next two vehicles were derailed and thrown on their sides, the next two derailed and damaged and the rear fish van remained on the rails. Of the 70 passengers and 10 Post Office sorters on the train 11 passengers and three Post Office servants were killed, together with two guards, Driver Martin and Fireman Fletcher. Twenty-three passengers, the other seven Post Office servants and a guard were injured, some seriously and most of the remaining passengers were slightly hurt.[223]

A newspaper reporter found a couple of passengers to interview. 'Mrs Maybee, of Exmouth, who was travelling in a carriage with her little girl and a number of ladies from

[223] Board of Trade report

Hereford and Mid-Wales, said: - Suddenly the carriage seemed to open, and the next minute she was involved in wreckage. She kept her presence of mind, and naturally her first care was for her little girl. Happily the child was comparatively uninjured, and she set to work to help the passengers in a worse plight. Many involved in the debris were extricated, and agonizing screams were heard in every direction. She assisted five or six of the injured to the ladies' waiting room, and in the absence of doctors and dressings she tore off her petticoat for bandages. The waiting room presented a horrible sight as the dead and wounded were brought in, many of the latter suffering from ghastly wounds. One man was bleeding from arteries with no one to staunch the flow. But doctors quickly arrived and, with an ample supply of dressing and restoratives, at once set to work to relieve those who were not yet beyond help.

'Mr Atherton, of Llandrindod Wells, who was one of the passengers travelling with a brother, said: - I was sitting in one of the carriages with my brother beside me, and two ladies were the other occupants of the compartment. Suddenly the lights vanished, and I was "fired out" on to the track. I afterwards ascertained I had sustained a scalp wound. Looking round after I had been thrown out of the compartment, I saw a man pinned under a great deal of ironwork wreckage. Nearby was a lady who was crying out, "Jack! Jack!" I went and tugged at the wreckage until I got the man free, and the instant he was liberated he fell back dead. I assisted to get the body from the debris to the platform, but I saw nothing more of the lady, though I have since ascertained that she was unhurt.'[224]

With this being the third major disaster in England due to an excessive speed derailment in just over a year the Board of Trade report upon it by Lt-Col H. A. Yorke was comprehensive. He described in great detail the circumstances of the accident itself, the layout at the Crewe end of Shrewsbury station, the locomotive involved, the signalling and applicable regulations and so on. Finally he came to his conclusions. With regard to the running of the train the inspecting officer noted that it had gained five minutes over the 18¾ miles between Whitchurch and Shrewsbury, covering the distance in 19 minutes at an average speed of approximately 60 miles an hour. Guard Birch stated that the train's brakes had been applied between the Crewe Bank distant and home signals and its speed reduced to 40 miles an hour before the derailment. The problem was that the guard's assertion that the brakes had been applied that early did not agree with the statements of Signalmen Beddleston and Ward or those of other witnesses, while the most eloquent witness of all was the degree of destruction caused by the accident, which plainly indicated that the speed of the train must have been at least 60 miles an hour at the moment of derailment, if not greater. Given its excessive speed, the only question had been whether the train would become derailed first or run far enough to collide with the rear of the train standing at the platform.

There could be no question of any problem with the permanent way, which was of a high class.

Suggestions had been made in the press that the brakes of the train had failed, and the coroner's jury had added to their verdict of 'Accidental Death' a rider to the effect that the brake power of the train had been insufficient. The jury had refused to hear evidence regarding the brake power of the train and their rider, Lt-Col Yorke implied, had been meaningless, unless they meant to imply that the brakes had been out of order.

[224] *The Times*, London, England, 16th October 1907

Allegations of brake failures, the inspecting officer continued, were easy to make and very difficult to disprove. There was however no evidence of brake failure on the present occasion and it must be taken that the brakes had been working normally.

Another suggestion was that the regulator had stuck open, but no problem had been found with it after the accident, other drivers could not remember any problems with that type of regulator, and it had been found closed after the accident.

The gauge glass had been found broken after the accident but had probably broken as a result of it and in any case, protectors fitted to the glasses made the possibility of scalding from steam or hot water from a broken one unlikely.

Having dismissed as extremely improbable the accident being the result of any defect with the track, brakes or engine one was left with the human element. Driver Martin had supposedly been in good health, he was not colour blind, his eyesight was good and although not a teetotaller he was described as a sober man. He was an experienced driver. He did however have 13 entries against him for different offences such as failing to stop at stations and being absent without leave, including two for overrunning signals. Fireman Fletcher was said to be a strict teetotaller, steady and thrifty. He had never before fired for Driver Martin.

On the Sunday night Martin and Fletcher had worked trains from Crewe to Manchester and back, thence to Holyhead, arriving there at 5.19 a.m. on Monday morning. They had slept in the company lodging house at Holyhead on Monday, worked a train from Holyhead to Crewe on Monday evening, done some shunting at Crewe, thence taken over the 1.20 a.m. train. Martin was well acquainted with the line to Shrewsbury, Fletcher less well so. Nevertheless Martin had failed to apply the brakes at the normal point and in fact had shot past the Crewe Bank signals and box at high speed. The inspecting officer did not believe Guard Birch's statement that he had gone to apply the vacuum brake before passing Crewe Bank box. This did not agree with the evidence of other witnesses and was merely what he should have done. There was evidence from witnesses such as a police constable that Martin had whistled near the junction and it seemed likely that he had realised his danger at the last moment, when it was too late. The question was why he had not done so earlier.

The possibility of Martin having been taken suddenly ill had been raised, and his body had been subjected to a post mortem, which had revealed nothing of significance.

The unpleasant possibility that Martin had been drunk must also be considered. There was no evidence that Martin had ever been intoxicated before. A bottle which had been Martin's property and which he had carried with him on the engine was found in the wreckage. It had been kept under lock and key at Shrewsbury and was produced at the inquiry, but at some point after it had been found it had been thoroughly washed out, destroying any evidence of what it had contained. The poor driver's stomach contents were also examined but contained only food and a brown liquid which did not smell of alcohol (and was doubtless tea).

Lt-Col Yorke then turned to what he declared was the simplest explanation for the disaster, that Driver Martin and possibly also Fireman Fletcher had fallen asleep. The accident had occurred during the early hours when human vitality was at its lowest and the desire to sleep was at its greatest. It could be theorised that Martin had been awake at Hadnall (the station before Harlescott Crossing) but had then fallen asleep. The speed of the train had risen on the falling gradient (the line fell at 1 in 117 and 1 in 124 coming into Shrewsbury) and by the time that Martin had woken up or been roused by the fireman they were past Crewe Bank box. Fireman Fletcher had either fallen asleep as well or more likely had been preoccupied with getting the engine fire in order and filling the boiler and had not realised that his driver was asleep. The inspecting officer said that he had received many letters from retired enginemen relating how they too had fallen asleep on their engines. He pointed out that soldiers in an enemy's country, where their lives depended on vigilance, had been known to fall asleep on horse-back and even on camel-back and accidents from enginemen falling asleep were well recognised in America but had rarely, if ever, previously been noted as a risk factor for an accident in England.

Various suggestions, Lt-Col Yorke stated, had been made with a view to preventing future accidents of the same type. Some, such as eliminating all dangerous curves, were clearly impossible. Another was the idea of having a third man on the footplate. Such a man would have little to do, would be in the way, might well distract the driver and fireman by conversation or argument, or could simply fall asleep as well. The provision of speed indicators, by this date compulsory on French railways, was another idea, but clearly of no use were the driver asleep.

The idea of cab signals on the engine was being looked at by several companies (notably the Great Western beginning the previous year) but such a device would need to be wholly reliable and have distinct audible signals for both "danger" and "safety". Special equipment would need to be fitted both on the track and on every locomotive. Automatic train stops were a similar idea and there had been numerous electrical and mechanical inventions, all of which had proved to be too complicated or unsuitable for other reasons, at least on main line railways. Automatic train stops were in use on some underground electric railways in London. The question of cab signals and train stops was still unsettled, though it was being looked into in the United States by the Interstate Commerce Commission. In the meantime the extended use of detonator placers in signal boxes, for use in clear weather as well as fogs, should be considered. Such detonator placers could be provided in Crewe Bank box, which might also be repositioned so that it was further away from the Crewe Junction home signal.

Finally the inspecting officer recommended that the testing of train brakes before a train began its journey should be witnessed by a platform inspector or stationmaster, and that the 'Red Books' which drivers signed to indicate that they were familiar with a stretch of line should in future be witnessed by a responsible officer rather than by a young cleaner, as was currently the practice.[225]

Mentioned in the Board of Trade report but not commented on, doubtless because the condition was not recognised at the time, was that Driver Martin was 52 years old and weighed 14 stone. As an overweight male over the age of 40 he therefore had three risk factors for obstructive sleep apnoea, which could have been another factor tending to cause him to doze off.

On Thursday 31st October 1907 evening train No. 10 on the Kempenich-Brohl metre gauge line (the Brohltalbahn) became derailed on a 50-metre radius curve where the line

[225] Board of Trade report

fell at 1 in 40 at Oberzissen, Germany. The derailment occurred at the point of transition between the rack railway and the adhesion worked section and was due to the driver's failure to brake early enough. Two ballast waggons (?) were overturned, the remainder of the goods waggons were smashed, and the passenger carriage was upset. Five died and seven were injured.[226]

A brief report from St Petersburg stated that a telegram from Samara had stated that a mail train coming from Tashkent had gone off the line on Friday 6th March 1908 between Tschergotur and Kurandy stations, Russia. The mail van had fallen down an embankment; 12 persons had been killed and 43 injured, 36 of them seriously.[227] Several American newspapers, who had probably got their facts wrong, stated that a head-on collision had occurred at Samara station itself on the same date and with exactly the same number of dead and injured.[228]

A westbound Grand Trunk Railway train was wrecked at Dustin, probably a place in Ontario, Canada, on Saturday 16th May 1908. Nine persons were known to be dead and there were thought to be more in the wreckage. More than 25 others were injured, some of them seriously. Despite the apparent magnitude of the accident no report on it has been found in a Canadian newspaper and only brief details in some American ones.[229]

A brief report from St Petersburg spoke of the derailment of a passenger train near Penza, Central Russia on Saturday 23rd May 1908. Some of its carriages were telescoped and 12 persons were killed and 30 injured.[230]

No. 239 up passenger train was derailed at the site of a new diversion at mile 140-19 between Zeyawadi and Nyaungchidauk stations, Burma, on the Burma Railways, on 24th March 1909 due to its attempting to negotiate the diversion at unreduced speed. A few minutes after the accident fire was noticed in one of the bogie third-class carriages and eventually the whole train was burnt out except the two rear vehicles. Seven passengers and two railwaymen died, and 12 passengers and two railwaymen were injured. The diversion should have been marked by a warning light but owing to the negligence of its attendant this light had been allowed to go out.[231]

A train crowded with holidaymakers was derailed at midnight at Gaikas Loop, near Cathcart, Cape Province, South Africa, at about 1 a.m. on Thursday 5th January 1911.[232] The coaches all left the rails and plunged down an embankment, the engine and

---

226 *Eisenbahn-Katastrophen in Deutschland* by Hans Joachim Ritzau, pg. 77 (in German)
227 *The Times*, London, England, 7th March 1908
228 *The Newark Advocate*, Newark, Ohio, United States, 7th March 1908
229 *The Advertiser-Courier*, Hermann, Missouri, United States, 20th May 1908
230 *The Washington Post*, Washington, D.C., United States, 24th May 1908
231 Administration Report on the Railways in India for the year 1909
232 *The South African Railway Magazine*, Vol 5 No 3, March 1911

presumably leading guard's van alone remaining on the track. Some of the coaches overturned and in the pitch darkness there was reportedly the wildest confusion, while a number of passengers were pinned under the wreckage for four hours. Six coaches were wrecked. A doctor travelling as passenger who suffered a broken rib was the first to treat the injured, assisted by unharmed passengers. The initial report spoke of 17 dead – 16 Europeans and one 'native'. A second report spoke of 22 dead, but a third report corrected the number to 18.[233] [234]

Two bad accidents occurred in the United States on consecutive days towards the middle of July 1911, on opposite sides of the country. The first accident, on Monday 10th July 1911, occurred near Dike (alternatively spelt Dyke), Oregon, on the Oregon Trunk Railroad. Southbound train was derailed about 1½ miles north of Dike on "Shoo Fly" curve. The engine, No. 709, turned over to the left and three cars were also derailed. The baggage and smoking cars ran past the engine; the coach, which was the third vehicle in the train, came to a stand alongside of the engine. It was in this coach that all the fatalities were found. Louis Rising, of the Warm Springs Indian School, died a few minutes after the accident; Mrs J. W. Rasmus, of Ellsworth, Wisconsin and her six-month-old baby; Mrs C.H. Baker, of Sheridan, Oregon; Mrs L. J. Rising, of Warm Springs, Oregon (presumably the wife of Louis Rising); and S. L. Arthur, of Seattle, Washington, died the next day and a seventh passenger died later. Several other passengers and the fireman were seriously injured. The deaths were due to steam entering the coach from the damaged boiler of the engine. The seventh passenger who died was probably either Morris Baker, aged 12, or his brother Hester, aged six, whose mother had already died. Their father it was thought would recover. The Bakers had been on the way to Redmond, Oregon, where Mr Baker had recently purchased a new home.

Conductor Cass and Engineer Myles of train No. 102 had been issued a notice reading: 'Approach third crossing under control. Run slow over bridges at second and third crossings and Wapinta Creeks. Reduce to 10 miles around 'Shoo Fly,' two miles north of Dike.' Engineer Myles told an inquest jury the day after the accident that he had been running at 15 miles an hour when it happened, but train passengers and Roadmaster A. S. McCurdy, who was riding on the train, disagreed. The roadmaster stated, "I was standing at the rear door of the car and when the car started to turn, I got out. I examined the track after the accident and found it in good condition. I was unable to find any fault with the track. There was a slow order on that part of the track which was issued by the dispatcher to all trains." He said that having noticed that the train appeared to be speeding he had checked its rate by counting the rail clicks, by which he had deduced a speed of 53 miles an hour. Asked by a juryman what he believed had caused the derailment he stated, unequivocally, "excessive speed."[235] The Interstate Commerce Commission came to the same conclusion, though they only put the speed of the train at 25 miles an hour. They found that contributing to the derailment were the irregular superelevation of the track on the curve and the fact that the track was wide to gauge, and that without these defects the speed of the train alone would not have led to the derailment.

[233] *The Times*, London, England, 6th, 9th and 10th January 1911

[234] *The Standard*, London, England, 6th January 1911

[235] *The Morning Oregonian*, Portland, Oregon, United States, 12th July 1911

Engineer Myles alleged that the accident was due to a 'sun kink'. His superiors agreed with him, the General Superintendent of the company writing a report in which he stated, 'It was unfortunate that Roadmaster McCurdy gave the 53 mile an hour speed evidence before the Coroner's Jury. He could have, as well as not, say that he could not say what did cause the trouble, and that, with all due regard for the truth, as he nor no other human can say what the derailing cause was. A 'sun kink' would have better fitted surrounding conditions than any other probable cause, and have materially eased the tension, from a railroad point of view. To catch the speed, at the limit sat by McCurdy, by counting the rail joints as the train passed over them, by visual observation, is, with anything like a reasonable degree of accuracy, an impossibility. The truck wheel of the engine locking so far back on the frame would have strongly supported the theory that the track was excessively out of line, similar to sun kink action.' [236]

If there was any doubt as to the true cause of the Dike derailment there was none as to the reason for the accident the following day on the East Coast of the United States near Bridgeport, Connecticut, on the New York New Haven and Hartford Railroad. Train No. 72, the eastbound *Federal Express*, consisted on the early morning of Tuesday 11th July 1911 of engine No. 813, a Department of Commerce & Labor Fish Commission car, a Pennsylvania Railroad baggage car, a day coach and six Pullman cars. The train had started out from the Harlem River terminus at 1.52 a.m., 57 minutes late. It was approaching Bridgeport on track No. 2, but the fish car, which contained tanks of fish for the Connecticut lakes, had to be left at Bridgeport, and the applicable siding could only be reached from track 4. Some 15 minutes before the arrival of train No. 72 the dispatcher therefore directed Towerman Hemingway at Burr Road tower, 1½ miles west of Bridgeport, to divert train No. 72 through a facing crossover from track 2 to track 4, and the towerman set up the route accordingly. Due to the limitations of the layout, such as the proximity of a through girder underbridge over Fairfield Avenue, it was not practicable for anything other than a low-speed No. 8 double slip crossover to be installed at Burr Road, the speed through which was restricted to 15 miles an hour. The crossover was however fully protected by signals, with a distant signal kept at caution, a home signal at danger and a dwarf signal for the crossing movement. There was no approach locking installed, so it would have been possible for the towerman to reverse the road in front of the approaching train, but this was not the cause of the accident.

On the morning of the disaster train No. 72 approached Burr Road tower at full speed, notwithstanding the adverse signals, and became derailed on the crossover while still running at an estimated 55 or 60 miles an hour, at 3.32 a.m. Following the derailment the engine and cars of the train demolished the 85 ft south girder of the Fairfield Avenue bridge, ran along the ties and ground, overturned and fell down the embankment to the street below. Engineer Curtis, Fireman Ryan, eight passengers and an off duty railwayman lost their lives.[237] Fire broke out in the wreckage but was quickly extinguished by the Bridgeport fire department, who then set about with the rescue work.[238]

The whole train apart from the rear two Pullman cars left the rails. A newspaper

---

[236] Interstate Commerce Commission report

[237] Ibid.

[238] *The Newark Advocate*, Newark, Ohio, United States, 13th July 1911

reported on the fortunate escape of the occupants of these two vehicles, 'The two Pullmans at the rear end were occupied by St Louis league baseball teams, twenty-two young athletes under the command of their manager, Roger Bresnahan. Bresnahan and his Cardinals boarded the train at Philadelphia and complained sorely because the trainmen tagged their sleepers to the rear end of the *Federal.* They protested because they thought it would be hot so far back, but young Mr Bresnahan, as he stood in pajamas {sic} that had been pink, but had gotten inky black from dirt and sweat, gave thanks fervently that the railroad people did not do what he had asked.' [239]

One theory briefly raised after the accident was that the switches had opened after part of the train had passed over them but the Interstate Commerce Commission (ICC) stated that the distance that the engine had travelled after leaving the track coupled with the direct testimony of all the witnesses made it certain that the train had been travelling at high speed. Quite apart from the signals having been at danger Engineer Curtis knew that the fish van had to be left at Bridgeport, that he would have to be on track 4 to deposit the van, and that the crossover at Burr Road was the last point at which his train could be crossed to that line. The ICC noted that Curtis was an experienced engineer who had worked fast freight and passenger trains and was familiar with the rules, and the tracks and signals at the location of the derailment. There was no explanation as to the engineer's mistake and any theory would be mere conjecture.[240] One idea noted in a newspaper report was that the engineer had been overcome by the heat, another strong possibility, given the hour at which the derailment occurred, would seem to be that Engineer Curtis had simply dozed off, and Fireman Ryan had either failed to notice or had also fallen asleep.

The ICC noted that the best means of preventing such accidents would be some form of automatic train control, such that the speed of the train was reduced to that which was safe to negotiate the crossover, but that such a system was still in the process of development. Failing that, on high speed lines the signals should be maintained at danger for all which it was wished to divert through a low-speed crossover and the switches not set for such a crossing movement until the train had been safely brought under control. Approach locking should also be provided at all interlocking plants working facing crossovers on high-speed routes to prevent the road being reset for a diverging route in the face of a train approaching with the signals clear for the through route.[241]

Another accident due to excessive speed occurred at Müllheim, Baden, Germany, on Monday 17th July 1911, to an international fast train from Basel, Switzerland, to Frankfurt and Berlin. At the time of the accident improvements were being carried out in the Müllheim station area, including the construction of a new pedestrian subway, and in connection with these the Basel – Freiburg line was blocked, and trains were being temporarily diverted to another track via a crossover through which speed was limited to 20 kilometres an hour. The connection was protected by a fixed signal 326 metres in the rear of the crossover, with its associated distant signal, and notices had been issued informing enginemen of the diversion. The international train nevertheless approached

---

[239] *The Fairfield Daily Journal*, Fairfield, Iowa, United States, 12th July 1911

[240] Interstate Commerce Commission report

[241] Ibid.

the crossover at full speed and was predictably derailed. The engine tender left the track, the leading luggage van overturned, and the following passenger coach was thrown diagonally across the line, whereby the next two carriages struck it and were telescoped. The remainder of the train suffered less badly, and a mail van and the restaurant car escaped altogether. Fourteen people died in the accident and 32 were injured. Following the accident all the motor cars in town were requisitioned to remove the injured, and the fire brigade were assisted by a detachment of soldiers. Most of the victims were from Basel and Wiesenthal. 'The train usually carries foreign passengers from France and Switzerland, who join it at Basel.'[242]

A report in the London *Times*, based on information from *Reuters* news agency, surmised that the train had suffered a brake failure, but the truth was rather different. The day before the accident the driver of the express had consumed a considerable quantity of alcohol and had had only three hours' sleep. Then on the day of the accident instead of drinking coffee, which he said he did not like, had taken more alcoholic refreshment, so that in total he had consumed a large amount of alcohol in the 18 hours before the accident, plus had had little sleep. Halfway between his starting point and Müllheim the driver was apparently sat dozing. At Schliengen he was definitely asleep and at Istein he failed to reduce speed at an adverse distant signal, but the main signal then cleared anyway. A curve at Rheinweiler was taken at 100km/h and approaching Müllheim the speed had risen to 115 km/h. The distant and stop signals were at danger. The fireman, who had been busy with the fire, now tried to wake the driver to get him to apply the brakes and himself closed the regulator. He dared not do any more. He probably was not aware of the speed restriction and the danger they were in, and firemen did not take the driving of locomotives out of the hands of their drivers. The train was still travelling at 107 km/h when it left the rails. What was less explicable than the fireman's reluctance to take charge of the engine was the lack of action by the guard, who must have been aware of the speed restriction and have realised that the train was going too fast. He could have prevented the accident by applying the brake in his van, but he failed to do so. It has been a feature of others though, both before and since, that guards are very reluctant to take the control of trains away from their drivers, as noted above in the Salisbury, Grantham and Shrewsbury accidents.

It emerged at the subsequent court case stemming from the disaster that the driver had a history of previous offences, including an incident back in 1903 when in a drunken state after a night out he had contrived to run an engine into a turntable pit, after which he was taken off driving for three years. He also reportedly suffered from mild cerebrovascular disease. Following the court case at Freiburg the driver was sentenced to two years and four months' imprisonment and the guard to six months in gaol.[243]

On Monday 9th October 1911 an express train from São Paulo to Rio de Janeiro was derailed near Piraly, Brazil. A freight train then collided with the wreckage whereupon 'a panic ensued'. Six persons were killed and 20 were injured.[244]

[242] *The Times*, London, England, 18th July 1911

[243] *Eisenbahn-Katastrophen in Deutschland* by Hans Joachim Ritzau, pgs. 60-5 (in German)

[244] *The Galveston Daily News*, Galveston, Texas, United States, 10th October 1911

Ditton Junction, England, the point where the main line from Liverpool to Weaver Junction and the south diverged from that to Warrington and Manchester, was the location of a major disaster on Tuesday 17th September 1912. Only one train was involved in the accident, a passenger working to Liverpool, consisting of a *Precedent* class engine and tender, two Great Western Railway horse boxes, two third-class coaches, a composite coach, a third-class brake composite, two composite carriages and a third-class brake composite. The remaining seven vehicles of the train were all the property of the London and North Western Railway, on which company's line the accident occurred, and all but two of them were gas lit, while the two horse boxes were lit by oil lamps. The train was in charge of Guard George Boardman, with Driver Robert Hughes and Fireman A. Lunn.

Part of the train consisted of a portion from the Cambrian Railways at Afonwen, the remainder possibly came from Holyhead. Guard Boardman joined it at Bangor and a Manchester portion was detached at Llandudno Junction, beyond which the train called at Chester, where the horse boxes were added, and from where it formed the 5.30 p.m. to Liverpool. Stops were made at Helsby, Frodsham and Runcorn, the latter place only about 2½ miles from Ditton Junction. From Runcorn Bridge over the River Mersey to Ditton Junction the line fell at 1 in 114 to Ditton Junction. There were multiple tracks at Ditton Junction, including facing connections between the down Crewe line and the goods line and between the down Crewe line and the down slow line, the Crewe lines themselves continuing to and from Liverpool as the down and up fast lines. Speed through the down main (Crewe line) to slow crossover was limited to 15 miles an hour.

Signalman William Highett told the Board of Trade inquiry into the accident that he had accepted the Chester train from Ditton Bank box at 5.55 p.m. He had then set the road from the down fast to down slow and cleared his home (No. 4) and distant (No. 3) signals. Ditton Bank had given him the 'Train Entering Section' signal at 6.03 p.m. and as far as he could remember the train had passed him at 6.06 p.m. He had been surprised to see that it was travelling at about 45 miles an hour. Trains usually negotiated the crossover at 15 to 25 miles an hour and he had never had to report a driver for going too fast. He saw the engine jump the tracks at the diamond crossing and realised that the train was derailed. After the accident he had sent the 'Obstruction Danger' signal, had telephoned various junctions to arrange diversions, and the stationmaster had arrived at his box five or six minutes later. The time of the sending of the 'Obstruction Danger' signal had been entered as 6.06 by his booking boy at 6.20, as the boy had been out of the box earlier. The Chester train was booked to take the slow line from his box in order to keep the fast line clear for the 2.40 p.m. from London Euston but if the Chester train was running late and the London train had gone there was no need to divert it. The train did not whistle. He saw its driver standing on the proper side of the footplate with his hand on the regulator.

Following the derailment, the engine overturned and ploughed along the ground, tearing up the permanent way until it collided with the pier of a road overbridge. The collision with the bridge pier was so violent that the engine 'broke its back', the barrel of the boiler being torn away from the firebox. The tender was wedged up against the firebox, mortally injuring the driver and fireman. The first six vehicles of the train continued past the engine and tender into the station and were all more or less shattered. The first horse box was broken up but the animal in it survived; the second horse box was split into two by striking the bridge pier and the horse in it was killed. The first passenger carriage ended up near the leading horse box, smashed to pieces. The next two coaches were jammed against the wall supporting the station booking office and the sixth came to a stand under the bridge. The last three vehicles of the train were also derailed and stopped on the ballast short of the overbridge. Then a few minutes after the derailment Stationmaster James Derby and others noticed that fire had broken out in the wreckage resting against the wall of the booking office. He at once directed Porters Peter Kavanagh and Gallagher and Lampman John Ryan to get out the fire hose. This was soon done, and water was directed at the fire from free jets, while those passengers who were able to do so made their way out of the train. Assistance was provided from the adjoining sleeper and creosoting depot, and Porter Kavanagh distributed tools and "extincteurs" from the rear van of the train to willing helpers, included many local inhabitants who had witnessed the accident and fetched ambulance appliances from the porter's room. Despite all these efforts the fire increased in volume and intensity until the coaches closest to and under the bridge were blazing furiously.[245] According to a report in the London *Times* some of the passengers were burnt to death before they could be released,[246] a suggestion rejected by the Board of Trade inspecting officer, LtCol H. A. Yorke, who stated that the blackening of the bodies by dirt, water, ashes and burnt wood had simply made them appear to have been burnt. The Widnes fire brigade arrived at 7.30 but by then the fire was more or less under control. The bodies of 12 dead passengers were removed from the train; 39 passengers and the guard were injured, eight passengers seriously. A 13th passenger died in Liverpool Infirmary on 25th September.

A newspaper reporter wrote, 'Mr Charles Birch, Clubmoor, Liverpool, who was travelling in the end compartment of the second carriage of the train, telling of his experiences, said that he noticed that immediately on crossing the points approaching Ditton Station the carriages began to sway. "We were off the line. I shouted to my fellow passenger as I sprang up, "Jump on the seat." The woodwork of the preceding compartment was forced halfway into ours, and I could see the end of the front carriage coming into ours as we came violently to a standstill. We got out of the carriages by the window. A large number of people were upon the spot, and women brought water, pillows and sheets. The fortitude of some of the injured was remarkable. I offered to help one man away, but he said, "Go and look after someone else. There's worse than me." From the wrecked engine steam was escaping, and afterwards flames burst out suddenly."'

As was often the case some passengers were lucky. 'Mr and Mrs Frank E. Brown, Sefton Park, Liverpool, and their 15-months-old child, had a remarkable escape. Mr Brown was standing up to get some article from the rack, when he was flung against the door. When he came to his senses, he found himself in an overturned carriage, but was able to make his escape by the window. His wife he found pinned under a carriage with two dead people lying across her. With the help of a navvy he got her out in about half an hour. A few minutes later a woman found the child amongst the wreckage. The three other passengers in Mr Brown's compartment were killed.

In other incidents, 'One man leapt from the burning train, and while the coaches were ablaze and imprisoned passengers were calling for assistance, ran to and fro on the permanent way demanding that the train should be sent on without delay as he wished to

[245] Board of Trade report
[246] *The Times*, London, England, 18th September 1912

reach Liverpool quickly. In another instance a lady passenger who had been dazed, but not otherwise injured, made her way across fields to a cottage nearby two miles distant, and, knocking on the door, asked the inmates what had happened. A nurse whose foot had been severed, on being carried on to the station platform, proceeded to take strips from her clothing to bandage her injury. A baby girl was found alone in one coach, and she has been removed to a cottage near the station until claimed. It is believed that the mother has died and that the father is in the Royal Infirmary, Liverpool. On the body of one man taken from the train a gold watch was found, melted by the heat. Yesterday the repairing gang found money and jewellery to the value of about £400 among the *débris*.' [247]

Lt-Col Yorke began the summary of his conclusions as to the cause of the disaster by noting that it was the worst accident to have occurred in England since that at Shrewsbury in October 1907, to which it bore some similarities. The signals cleared by Signalman Highett should have indicated to Driver Hughes that he must immediately reduce his speed to 15 miles an hour, but the signalman, Stationmaster Derby and Ganger William Sefton, who had witnessed the train's approach, had all been surprised at its high speed. Ganger Sefton estimated the speed of the train as having been 60 miles an hour, which from the amount of destruction was probably not an exaggeration. Driver Hughes had begun as a cleaner in June 1891 and been promoted to extra driver in June 1904. He had signed the book as knowing the road between Chester and Liverpool in July 1908. According to the Company record he had made the journey 144 times as fireman and ten times as driver, the train had been diverted from fast to slow six times when he was a fireman but never before when he was driving. His numerous trips over the route as fireman were probably not that relevant as the signals were on the driver's side. Fireman Lunn was only a passed cleaner, had only been over the route once before and had practically no knowledge of it, plainly he could not be expected to pick up Diver Hughes' error.

Guard Boardman, the inspecting officer continued, was better acquainted with the line. The latter admitted not having seen any of the signals at either Ditton Bank or Ditton Junction and made the usual excuse as to having been preoccupied dealing with a large amount of luggage. The inspecting officer conceded that Rule 170b, which stated that guards must give careful attention to luggage, parcels, despatches and other packages entrusted to them, contradicted Rule 177a which stated that the guards of passenger trains must carefully watch the running of trains when approaching important junctions, termini and stations where the train was booked to stop and apply the continuous brake in emergency to stop the train. On busy lines the guard could hardly do both and in any case the construction of modern rolling stock made it difficult for the guard to see the signals without looking out of the window.

Turning to the signals Lt-Col Yorke noted that the down main (fast) and down main (fast) to slow signals were the same height and each had its own distant signal, both distants also having the same height. Signalman Highett said that Driver Hughes had been standing in his normal position and it was clear that up to the last moment he had been completely unaware of the danger he was running into and had imagined that he was continuing to Liverpool on the down fast line. The accident therefore clearly arose from Driver Hughes' lack of familiarity with the line and he should have asked for a pilotman. But the matter did not end there, for the system for ascertaining and testing the

[247] Ibid, 19th September 1912

driver's knowledge of the road was clearly inadequate. Lt-Col Yorke noted that drivers were merely asked at six month intervals by the assistant foreman if they had knowledge of the road and signals, and if they said 'Yes' they were passed as competent to work over that line. The inspecting officer remarked that 'If all examinations were conducted on such lines the results would be startling, and incompetence and ignorance would reign supreme.' Instead a responsible officer from headquarters should be appointed as inspector to examine and test drivers on their knowledge in the same way as signalmen were routinely examined on the block rules and regulations.

Since the accident the Company had altered the arrangement of the signals, putting those applying to the diverging route at a lower height. This the inspecting officer thought was an undoubted improvement, but better would be to dispense with the diverging distant signal altogether and maintain the remaining signal at danger if the road was set for trains to cross over to the slow line. Were the Company to persist in retaining the separate distant a board, illuminated at night, should be fixed to it indicating the maximum authorised speed over the diversion. It was illogical to lower a distant signal, indicating to a driver that he could proceed without a check, when in fact he had to reduce speed to 15 miles an hour for a crossover. In the same paragraph Lt-Col Yorke went on to criticise the Company for their failure to employ speed restriction boards, despite the fact that there were over 1,000 speed restrictions on the system, varying from 5 to 50 miles an hour, and he asked if it was reasonable to ask the drivers to carry them all in their heads.

Regarding the fire which had broken out soon after the derailment the inspecting officer pointed out that it had raised the question once more of the gas lighting of carriages. The gas cylinders had probably only been half full at the time of the derailment and those of the third, fourth and fifth coaches had only been dented, but their connections had been broken, allowing the gas to escape. A very small spark would be enough to ignite the gas, such as from a red-hot cinder or a spark from a wheel, and the immediate appearance of the fire, the speed at which it had spread, the fierceness with which it had burnt and the fact that it had taken two hours to extinguish all pointed to gas as being its cause. Major Pringle had recommended the substitution of electricity after the 1910 Hawes Junction disaster and it was to be regretted that no unanimous action had been taken by the railway companies in this direction, mainly it appeared because of the cost. In fairness to the London and North Western Company it had been using gas lighting for 30 years and had carried 2,937,000,000 passengers without any previous fires from that cause. An increasing number of its carriages were being lit using electricity, but electricity had not yet been adopted as standard.

As far as the provision of tools for emergency use, also recommended by Major Pringle, it was satisfactory to note that the Company had decided to equip all its brake vans with axes, saws, crow bars and fire extincteurs, and 1,000 had already been so equipped.

Finally Ganger Sefton and Signalmen Highett agreed that trains usually ran through the fast to slow crossover at speeds greater than the authorised 15 miles an hour, the ganger stating 35 miles an hour, the signalman up to 25 miles an hour. There was no reason to doubt their statements, but no drivers had been reported. Instructions as to who should report drivers for such breaches, who they should be reported to and whether the reports should be verbal or written should be added to the rule book.[248]

The inquest on all but the last person to die in the accident was held in Widnes Town Hall on Friday 27th September. At its conclusion the jury returned a verdict that 'The deceased accidentally lost their lives through the derailment of the train they were travelling in at the V crossing near to the No 1 signal box, the cause of the derailment being excessive speed.' They apparently did not make any recommendations.[249]

Animals on the line caused a bad accident near Torreón, Mexico, on the evening of Wednesday 18th September 1912. The driver of a military train on the Ferrocarril Central Mexicano (Mexican Central Railway) saw three mules on the track and whistled repeatedly to try and get them to move. They stubbornly refused to budge, the train struck them and was derailed. Eleven soldiers died in the wreck and another 47 were injured, ten it was feared fatally.[250]

American passengers who used the New York New Haven and Hartford Railroad must have been less than happy when less than 15 months after the Bridgeport derailment, on Thursday 3rd October 1912, there was another accident of the same type at Westport, Connecticut, United States. The train involved in this accident was second No. 53, en route from Boston, Massachusetts to New York, New York. It consisted of engine No. 1014, a mail car, a baggage car, four parlor cars, three coaches and a smoking car and was in charge of Conductor Jenkins with Engineer George Clark and Fireman J. J. Moker. The train left Bridgeport 12 minutes late. Both sections of train No. 53 had left Bridgeport, 11 miles east of Westport, on track No. 1 because track No. 3 was occupied by work trains between the two places. There was a crossover at Westport, controlled from a 40-lever mechanical tower, where the trains could be switched back to track No. 3, and the dispatcher instructed the Westport towerman to so divert both trains back to track No. 1. Speed through the No. 10 crossover was limited to 15 miles an hour and, as at Bridgeport, it was protected by a home signal 240 feet in its rear and a distant signal 4,075 feet in its rear. Intermediate between these signals and 2,017 feet from the distant signal was a stop signal protecting a drawbridge over the Saugatuck River. Finally there was a dwarf signal alongside the home signal protecting the crossover. A distant signal at caution could therefore indicate either that the drawbridge was open, that the train was to cross to the other track, or both.

Train first No. 53 reached Westport at 4.53 p.m. and passed through the crossover safely at 15 miles an hour. Towerman Michael Coyle left the crossover in the same position for the second section of the train but had to replace the dwarf signal to danger before he could accept the second section from the next tower in the rear at Greens Farms. Having done so he then cleared the dwarf signal again. Train second No. 53, however, approached the crossover at high speed. The locomotive left the track 87 feet west of the crossover, ran about 400 feet and overturned. The tender went down the 15-foot-high embankment, the mail car also went down the embankment and the baggage car ran past the engine. The leading parlor car was forced against the engine firebox and the

[248] Board of Trade report

[249] *The Times*, London, England, 28th September 1912

[250] *Indiana County Gazette*, Indiana, Pennsylvania, United States, 25th September 1912

second parlor car was forced on top of the first. The next three vehicles were derailed, the last three remained on the track. A bridge girder tore a hole in the engine boiler, allowing steam and hot water to escape and giving the appearance of a boiler explosion. All the vehicles in the train were of wooden construction and fire broke out in the wreckage, probably from contact with the engine firebox, and three of the parlor cars were consumed.[251]

Given the extent of the destruction it was perhaps surprising that only Engineer Clark, Fireman Moker and five passengers lost their lives. Four of the dead passengers were Miss Mary Hamilton, Mrs James Cox Brady and Mrs C. Southerland Ransom, who were sisters; and Mrs E. Palmer Gavitt, all of Albany, New York; the sisters had been returning from a funeral at Hartford, Connecticut.[252] The fifth passenger must have died later, and their name does not seem to have been recorded. One newspaper report spoke mysteriously of eight victims, so possibly another passenger died later.[253]

The Interstate Commerce Commission (ICC) made a full investigation of the accident. Its report noted that it was a duplicate of that at Bridgeport the previous year. Engineer Clark had been 55 years of age and had entered the service as a fireman in 1883. He was promoted to engineman in 1888 but had been dismissed in December of the same year for having been responsible for a collision. He had been taken on again as a fireman in 1893 and been promoted to engineer in 1894. Since then he had been disciplined for running past home signals at danger at New Rochelle in 1898 and Naugatuck Junction in 1905.

The ICC concluded that it was clear that the cause of the accident was the disobedience of the signals by Engineer Clark and thus his failure to reduce speed for the crossover. However, the commission was less than happy at the failure of the New Haven Company to take any action in response to its recommendations following the Bridgeport disaster other than, as the vice president of the Company had put it, "stiffening up the discipline." No new devices had been installed because, "We have not been able to find any that we are satisfied would help." Regarding a lengthening of the crossovers so that they could be negotiated at a higher speed the vice president stated that it had not been done because, "I would think in the end it would make a worse situation on a railroad than a better one," and, "It is simply putting a premium on the violation of a rule." The commission noted that it was hard to understand how such a conclusion could be reached.

As to the recommendation made after Bridgeport that all signals be kept at danger until the train came to a stand before a diverging route was set up, the vice president stated that he doubted if it would increase safety as an engineer overrunning signals would simply go into the back of a train on the straight track instead. The commission noted that in the case of the present accident Engineer Clark had applied the brakes in emergency at the stop signal and had the route still been set for straight ahead his train would have been brought safely to a stand two train lengths beyond the crossover.

The vice president had stated that he knew of no form of automatic train control which would meet the conditions required. The commission referred him to their own report which stated that they had concluded that there were several types of automatic train control which were sufficiently well developed to warrant their installation, even if the

251 Interstate Commerce Commission report

252 *Norwich Bulletin*, Norwich, Connecticut, United States, 4th and 7th October 1912

253 Ibid, 10th October 1912

safety device was still at an experimental stage. If the railroads, the commission felt, had put the same effort into the development of automatic train control as they had to interlocking and block signalling much more progress would have been made but the railroads had been decidedly lax in developing the system. The commission did not feel that legislation to force the issue would be wise, but the railroads should be urged and expected to work on the development of such a system.[254]

The ICC noted that the track conditions at Westport were generally good, despite some of the ties being badly worn, and stated that the track had been safe (at the inquest a former New Haven employee named George Everett had declared that the derailment had been due to rotten ties and had told a newspaper that he had been intimidated by other employees for stating such in his evidence before the coroner).[255]

The commission noted that the notional 30 miles an hour speed limit over the Saugatuck River appeared to be habitually disregarded. Towerman Coyle had testified that trains usually crossed the drawbridge at 45 to 55 miles an hour. He had never reported this and had never been told to do so. Vice President Horn stated that he had not been personally aware of this, but he considered a speed of 50 miles an hour over the drawbridge to be safe and that with regards to exceeding the speed limits over the drawbridge and through the crossover, "One is safe and the other is suicide." Trains were expected to make up time if they were running late and a bulletin had been issued stating that train crews should make every effort to keep trains on time, though engineers were not expected to run at unsafe speeds or disregard slow orders. The ICC report also included a copy of what in Great Britain would be called a 'Please explain' or a 'Form 1'; that is a letter to an engineer asking him to account for a delay to a train. The delay caused to the train in the copy reproduced in the ICC report was as little as one minute. The engineer involved in the reproduced letter told investigators that he did not regard it as a criticism but merely an effort by the officials to reduce delays, but the commission was of the opinion that such notices were likely to put undue pressure on engineers to keep time, even to the point of taking chances to do so.

The ICC echoed a comment made by Conductor Jenkins at the inquest that had the train had steel rather than wooden cars lives would have been saved. The accident demonstrated the inherent weakness of wooden cars and their unsuitability for high-speed trains and the company was urged to comply with the commission's recommendation in its last annual report that steel cars should be substituted.[256]

The inquest into the accident, held by Coroner Jon Phelan, found that Engineer Clark was entirely to blame for it.

Brief reports appeared in several Australian newspapers of the derailment of a passenger train near Cairo, Egypt, on the evening of Monday 9th December 1912. Three first-class carriages were telescoped, and four people were killed, and many injured according to the initial reports.[257] The number of dead seems to have later climbed to six.

---

[254] Interstate Commerce Commission report

[255] *Joplin Morning Tribune*, Joplin, Missouri, United States, 23rd October 1912

[256] Interstate Commerce Commission report

[257] *The Bathurst Times*, Bathurst, New South Wales, Australia, 11th December 1912.

Six men died in a derailment near Garyville, Louisiana, United States, on Saturday 25th October 1913. Because it happened on a logging railway and probably also because those killed were all negro labourers the accident was barely mentioned in the press and the Interstate Commerce Commission did not bother to investigate it. A cow stepped in front of the train and a flat car toppled over into a water-filled ditch, pinning the victims beneath it.[258] Another report wrote, probably erroneously, of the train having collided with a car rather than with a cow and added that there had been 150 men on the train, half of whom had been white (the colour of the people involved was clearly regarded as important) but that all the whites had escaped injury. This report stated that 15 of the labourers had been injured.[259]

A branch from the Uganda Railway to Lake Magadi, Kenya, was not completely opened until 1915; the lake being an important source of the mineral trona, used for glass production, steel making and so on. While the line was still under construction there it suffered a serious accident, presumably a derailment, near Lake Magadi on about Tuesday 24th February 1914. A telegram from Nairobi to the *African World*, reported in a London newspaper, stated that 18 people had died in the accident and 30 were injured, but that none had been Europeans.[260]

On Sunday 9th August 1914 an express train became derailed on the Galatz – Bucharest main line in Romania, between Independentza and Piscu, as it attempted to negotiate a sharp curve. The pilot engine of the double-headed train buried itself in the ground and four carriages ended up piled up upon one another and were completely broken up. Six people died and 19 were seriously injured. The derailment was due to excessive speed, but a contributory factor might have been that the track was waterlogged following heavy rain.[261]

The First World War had been progress rather more than a month when a troop train carrying 500 men of the 5th Infantry (Kaffrarian Rifles) on their way to embark for the Western Front was derailed in the Hex River Pass, Cape Province, South Africa, at about dusk on Thursday 10th September 1914. According to newspaper reports the train was crawling as usual down the steep winding pass when its engine left the rails on one of the numerous curves, followed by its 10 coaches, which toppled about 20ft down an embankment. Nine soldiers were killed and 94 injured.[262] Two hospital trains and a breakdown train were sent to the scene from Cape Town and Worcester and another breakdown train was despatched from Touws River, to where the injured soldiers were conveyed. Exceptional efforts were made, it was reported, to clear the single track as there were 10 other urgent trains bound for Cape Town which were being held up, these

[258] *The Constitution*, Atlanta, Georgia, United States, 26th October 1913
[259] *Canton Daily News*, Canton, Ohio, United States, 26th October 1913
[260] *The Standard*, London, England, 28th February 1914
[261] *Eisenbahnunfälle Bind 1* by Ludwig Stockert pg. 95 (in German)
[262] Ibid, 29th September 1914

being no doubt more troop trains.[263]

Announcing the accident to the House of Assembly Mr Burton, the Minister of Railways, stated that it had probably been caused by the recent heavy rains.[264]

An account published on line a few years ago by a South African newspaper alleges that the accident was due either to a reckless engine driver or a brake failure and that the train was travelling at high speed when it left the track, but this account, which also states that the accident occurred at 2 a.m. rather than at dusk, does not seem very credible, given that it contradicts contemporary accounts in several respects.[265]

Advice was received in London, England by the Exchange Telegraph Company from Vise, Belgium by way of Amsterdam regarding the derailment of a train of 40 carriages carrying wounded German soldiers between Gemmenich, Belgium and Aachen, Germany on Wednesday 9th December 1914. According to the report some of the carriages were destroyed and the rest badly damaged. Fifty-two soldiers were killed, and a large number injured.[266] There had been 1,500 wounded on the train.[267]

Strict censorship meant that news of a major catastrophe in Mexico on Monday 21st June 1915 did not reach the United States until brought there by an American traveller some days later. A train carrying at least six hundred Carranzo soldiers, their wives and children left the track and plunged into a small canyon near Montemorelos, between Tampico and Monterrey. The train had been heavily loaded and was travelling at full, possibly excessive, speed. The train was made up of box cars with men, women and children crowded inside, riding on top and 'as is a Mexican revolutionist custom', riding underneath the cars. There were reportedly three hundred deaths and hundreds of others injured. Many of the passengers were ground under the wreckage.[268] [269]

A leave train coming from Berlin and carrying soldiers returning from their Christmas furloughs was derailed at Bentschen station, Germany, on the Frankfurt-Posen line at 3.30 a.m. on Tuesday 28th December 1915 as the result of its driver attempting to negotiate either a curve or a crossover at excessive speed. The initial report from Posen spoke of the death of 18 soldiers and injuries to 47.[270] Other reports spoke of 19 or 23 dead and put the number of injured as high as 79.[271] [272] Given the time at which the derailment occurred it is tempting to suppose that the driver might have fallen asleep.

---

263 *The Times*, London, England, 12th September 1914
264 *The Evening News*, London, England, 11th September 1914
265 Webpage: TIMES LIVE A Blast from the Past (Published 23rd October 2011)
266 *The Oelwein Daily Register*, Oelwein, Iowa, United States, 11th December 1914
267 *The Mercury*, Hobert, Tasmania, Australia, 11th December 1914
268 *The New York Times*, New York, New York, United States, 2nd July 1915
269 *The King City Chronicle*, King City, Missouri, United States, 9th July 1915
270 *The Standard*, London, England, 29th December 1915
271 *Eisenbahn-Katastrophen in Deutschland* by Hans Joachim Ritzau, pg. 134 (in German)
272 *Eisenbahnunfälle Bind 1* by Ludwig Stockert pg. 96 (in German)

CHAPTER 4

REAR COLLISIONS

The risks associated with one train running into the back of another had been low in the formative years of the railways when trains were few and train speeds low, but as the nineteenth century progressed train speeds and traffic intensity grew faster than advances in brake power, rolling stock construction and signalling and as a result by the 1880s serious rear collisions had become common. This chapter is therefore a long one, even though a great number of less serious accidents have not been included.

On the morning of Friday 7th June 1901, a British troop train from Pretoria to Pietersburg was run in two sections owing to the large amount of traffic. The first section stopped en route at the West York Camp at Daspoort, Transvaal, to pick up an escort. The men comprising the escort were being entrained in an armoured truck at the rear of the train when the second section suddenly came around a sharp curve and ran into it. A large number of trucks were derailed, nine men were killed, and five others were injured. Colonel Wilson, of Kitchener's Scouts, was travelling by the train but escaped with a slight contusion.[273]

On Friday 20th December 1901 in Germany express passenger train D 31 from Paderborn passed Neuenbeken station, three minutes late, passed Blockstelle Keimberg and entered the Keimberg – Schiernberg block section. Between the two places the express stopped out of course when it ran into a stray horse which had wandered onto the line. After 15 minutes the Keimberg signalman had passenger train P 399 to hand and his colleague at Schiernberg had still not cleared back train D 31. Thinking that he must have forgotten to do so the Keimberg signalman called his colleague via the telegraph to remind him, whereupon the Schiernberg signalman, either forgetting about train D 31 or becoming confused, replaced the signals which he had pulled off for the first train, cleared it back and accepted the second. The driver of train P 399 must have only caught sight of the train in front at the last moment, if at all, because of thick fog and his engine

[273] *The Echo*, London, England, 8th June 1901

pushed the banking engine at the rear of the first train right through the last third-class carriage. This was lit by gas, but for once there was no fire. Even so 12 people died and nine were injured.[274]

A brief report in a French newspaper spoke of what appears to have been a rear collision near Gatchina, Russia, not far from St Petersburg, on or about Thursday 3rd July 1902,. An express from St Petersburg collided with an excursion train and several carriages were wrecked. A driver, fireman and 10 passengers were killed and almost 110 passengers were more or less seriously injured.[275]

Five men died in a rear collision at Malden, Illinois, United States, at 3.30 a.m. on the morning of Thursday 25th September 1902 on the Chicago Burlington and Quincy Railroad. The men were all riding in the way car of a stock train which was run into by a following passenger train. When the stock train stopped a brakeman was sent back to protect the line against passenger train No. 4, which was shortly due, but he did not have time to go back very far before the passenger train approached. Conductor Robert Swain saw the danger and shouted to the occupants of the way car to get out. Seven of them and a boy succeeded in doing so, five of the six who did not were killed.

Those who lost their lives were J. H. Carruthers, residence unknow; W. L. Collins, of Batdol, Illinois; Hans Hasmussen, of Ipava, Illinois; A. H. Wegner, of Prairie City, Illinois, and Leroy Woolsey, of Knoxville, Illinois.[276]

The inquest on A. H. Wegner, representing the victims of the accident, took place a few days later. Engineer W. B. Walters, of freight train No. 1707 east, said that his train had had 44 cars plus the way car. As his train had arrived at Malden he had whistled for the operator to clear the train order board for him to proceed but the operator had come out of his office violently waving a red and white light for him to stop, and he had pulled up. After he had been stopped a couple of minutes he had been signalled to draw forward, then Brakeman J. D. Hainline had come forward and told him that passenger train No. 4 had run into them.

Conductor Swain stated that as soon as the train had stopped Rear Brakeman W. H. Lewis had gone back to flag the passenger train, which was due at Malden at 3.25 a.m. Like his engineer he stated that the line was worked on the block system, but it was only two minutes after they had stopped that the passenger train ran into them. It was apparently true that the line was worked on a block system of sorts, but it was not an absolute block system, engineers simply being cautioned if a previous train was still in the block section. It had been raining at the time of the collision, but earlier fog had cleared. They had taken all the precautions that they usually took but had been running on the time of the passenger train for the last ten minutes.

Rear Brakeman Lewis said that he had seen the electric headlight of the approaching passenger train. He had got off as soon as his train had stopped and had got back 15 or 16 lengths when the passenger train passed him. He had waved his red and white lights at it,

---

274 *Eisenbahn-Unfälle in Europa* by Erich Preuß pg. 11(in German)
275 *La Petite Presse*, Paris, France, 5th July 1902
276 *The Republic*, St Louis, Missouri, United States, 26th September 1902

flagging the passenger as hard as he could, but had had no time to put down a torpedo. The rear lights of his train had been burning as usual.

The evidence of Brakeman Hainline did not really add anything to the inquiry.

Fireman P. C. Pierson of the passenger train stated that the only light he had seen was a white light which he took to be the light of the station semaphore. He had not seen the red tail light of the freight train.

The last witness was Engineer C. A. Healy of the passenger train who stated that his train had left the previous station at Princeton, six miles from Malden, about 14 minutes late, having been given a green board there signifying that the block was not clear. According to him there was a slight, low fog at Malden. He had applied the brakes, expecting to find the semaphore at Malden up, and had reduced the speed of his train to about 12 miles an hour, then had released the brakes when he had seen the semaphore at clear. When the rear lights of the freight train had appeared out of the fog it had been about five car lengths away. An emergency brake application had brought the speed of his train down to about six miles an hour at the moment of the collision.

Without bothering to call the unnamed Malden operator to give evidence the inquest jury simply returned a verdict that A. H. Wegner had met his death as a result of injuries received in a rear collision at Malden. A local newspaper remarked that its readers might wonder what the inquest had been held for.[277]

A goods train ran into the back of another goods train in Bloemfontein station, Orange River Colony, on Saturday 4th October 1902. The rear truck of the front train was filled with 'natives', six of whom were killed outright and ten injured, some very seriously. The accident occurred at exactly the same place as a train had collided with a light engine on 24th August, which was either a coincidence or perhaps more likely there was some problem with the signalling. The driver and fireman of the colliding train were arrested, and an inquiry was to be held.[278]

The Southern Pacific Railroad suffered a major disaster at Byron, California, United States on the evening of Saturday 20th December 1902. Two passenger trains were involved. The first, the *Owl*, had stopped out of course and was run into by the *Stockton Flyer*. The initial investigation into the accident placed the blame on Engineer Milton McGuire of the *Flyer*. The *Owl,* which had left Oakland at 5.5 p.m., had come to a stand at a switch half a mile beyond Byron station on a stretch of track which was dead straight for seven miles in the direction from which it had come. The collision occurred at 7.12 p.m., by which time it was pitch dark, but although there was no moon it was not foggy, and the stars were shining. The front train carried red and green lights at its rear. Rear Brakeman Cole had also gone back 1,000 feet to protect his train.

The *Owl* had been running on time when near Byron steam had reportedly escaped from a flue, and extinguished the fire in the firebox. Arrangements were being made for the engine of a freight train to take the failed train either to Byron or beyond, when the

[277] *Bureau County Tribune*, Princeton, Illinois, United States, 3rd October 1902
[278] *The Times*, London, England, 6th October 1902

collision occurred. The *Flyer* had left Oakland half an hour after the *Owl*. Engineer McGuire said that he had seen the tail lamps of the stopped train before he had seen the flagman's signals. He had applied his brakes and done his best to pull up, but the momentum of his train had pushed it on to destruction.

The *Owl* was in charge of Conductor William Dolan with Engineer Louis C. Kerr and Fireman Walter J. Sewell. It had the usual composition; the rear vehicle behind the diner was a day coach for the use of Fresno passengers. It was this coach in which most of the casualties were found, nearly every one of its occupants being killed, mangled or scalded when the engine of the *Flyer* ploughed right through it 'as if it were made of pasteboard', stopping only when it reached the dining car. The express was allegedly running at 60 miles an hour at the moment of the collision. The only fortunate feature of the crash was that there was no fire. Twenty-four passengers were killed outright and 20 or more injured. Engineer McGuire and his fireman, Edward J. Joyce, remained on their engine and were among the injured.

Dr Davidson, of Fresno, chanced to be a passenger on one of the trains, was the first physician on the scene. He was joined by Dr Bird of Byron Springs, who brought with him several nurses, while Drs Theodore Olmstead and J. P. H. Dunn accompanied the wrecking train sent from Oakland, The injured were taken to the hotel in Byron Springs and to the Southern Pacific hospital in San Francisco.[279]

A local paper carried a piece about one of the passengers who lost his life. 'Charles A. Sessions of 1258 Eighth street of this city {Oakland}, who was killed instantly in the wreck at Byron, was on his way to Fresno to meet his son Nathan, who was coming to Oakland from Arizona to spend Christmas with his parents. Sessions had planned to reach Fresno before the train on which his son was speeding northward would arrive at that point. The meeting was to be a surprise to the young man and the two would travel together back to Oakland. The train on which his son was riding passed slowly by the wreck and continued on its way to Oakland. The young man passed within fifteen feet of his father's mangled body, yet all the while he was ignorant of his parent's death. It was only when he reached Oakland that he was informed of it. Charles Sessions was the head sales agent of the James P. Taylor Coal Company and was fifty years of age. He was one of the oldest and most prominent Elks on this side of the bay. The funeral will be conducted from his late residence by the Oakland Lodge of Elks, No 171.'[280]

The reporter of another newspaper interviewed two passengers who survived the collision. 'Thomas Harney and Gus Tower, well known mining men, formerly of this city {Bakersfield} but recently of Tonopah, Nev, were on the wrecked train coming to Bakersfield. They were in the Bakersfield sleeper, which is placed some distance ahead. They arrived here safely on Sunday morning and are now at the Arlington. "We were sitting in the car and had been taking a nap when the crash came," said Mr Harney last night. "Our engine had been doing badly all the way and at this blind siding near Byron we met a freight and stopped to exchange engines. The Stockton train was just behind us and there was a straight track for ten miles, so that there was no possible excuse for the accident. Just a minute before the crash I heard the whistle of the Stockton train, then I felt the jolt and heard the crash of the breaking up of the cars. When I went out it was

[279] *Oakland Tribune*, Oakland, California, United States, 22nd December 1902
[280] Ibid.

something terrible. The air was full of screams and shrieks of agony and terror from the people imprisoned in the wrecked car and the cries of hysterical women. Everything was in confusion at the time...The accident was utterly inexcusable. I talked to the fireman of the Stockton train, Ed Joyce, on the train on the way to Fresno. He was bruised and scalded somewhat, and I helped him some of the way and when I asked him if they saw a flag out and he said, "No, we didn't see any flag, but we saw the rear of the Owl. I don't exactly know why we didn't stop but I guess the engineer thought the train was moving and miscalculated his distance, as there was no schedule stop at that point."' [281]

Conductor Dolan was the first to give evidence at one of the three inquests on the victims of the disaster. He said that they had lost 20 minutes by Byron. There his engineer had come to him and said, "My engine is dead. See if you can't get that freight engine to haul you over to Tracy." There had been markers at the rear end of his train, and he had sent his rear-end and front-end (brakemen) back with lights. Torpedoes or fusees were only used in foggy or cloudy weather. He had been in the depot when the agent's wife had cried out that his train had been run into. He thought that the engineer of the *Flyer* had expected to find a freight train 1,000 feet further on than his train was.[282]

A number of the uninjured passengers died in the days after the disaster and the final number of dead was probably 27. As usual by the time that the inquests had run their courses the newspapers had lost interest in the matter, especially perhaps as a number of other accidents happened around that date, including a railroad disaster in Canada on Boxing Day. It would seem, however, from the brief reports which can be found, that the blame for the accident, initially placed on Engineer McGuire, was later shifted to the crew of the *Owl* for not having protected their train as soon as they should have done, although they knew that the *Flyer* must be close behind. The San Francisco jury passed a verdict that 'the railway employees did not take full advantage of all precautions which are provided for in the railway regulations' – a somewhat vague verdict. They also recommended the installation of a block system along the line.[283]

Twelve labourers died and about the same number were injured in a rear collision on the Great Northern Railroad at the ironically named Happy Hollow, three miles west of Chiwakum, Washington, United States, at 4.40 a.m. on Tuesday 20th January 1903. They were riding in an extra eastbound freight train from Shikomish, in charge of Conductor Shortley, which consisted of an engine, the three outfit (box) cars in which they were sleeping in bunks, and some cars loaded with lumber. When their train ran into the back of a rotary snowplough the box cars were telescoped by the lumber cars and only two of the labourers escaped death or injury. The more fortunate two were thrown bodily into the air, landing 30 feet away on an embankment. Those who survived the accident escaped with little clothing and almost died from exposure before they could be rescued. Nine cars were piled up in a heap and nine lengths of track were torn up.[284] The snowplough, which was in charge of Conductor Franks with Engineer Glauser, was little damaged. Engineer L. E. Adams and Fireman Lewis Breker of the freight train had lucky

---

[281] *The Daily Californian*, Bakersfield, California, United States, 22nd December 1902
[282] Ibid, 26th December 1902
[283] *The Morning Oregonian*, Portland, Oregon, United States, 31st December 1902
[284] *The Daily Huronite*, Huron, South Dakota, United States, 22nd January 1903

escapes.[285]

As soon as news of the accident was received a special train with doctors and assistants was sent from Leavenworth and reached the accident site at 8 a.m. Four injured men went sent back to Leavenworth at once with Dr Hoxy and five others were sent there afterwards as soon as they could be released from the wreckage. Five other men, more or less injured and almost naked, had walked several miles to Merritt.[286]

'APPALLING DISASTER!' screamed the headline in a Plainfield, New Jersey newspaper in late January 1903. 'Plainfield plunged into deepest gloom,' it continued. The shock headline was fully justified as far as the town was concerned for 20 or so of its citizens had been killed in a railroad collision at 6.45 p.m. the evening before, Tuesday 27th January 1903.

Central of New Jersey Railroad train No. 13 was booked to leave New York City at 5.45 p.m. It consisted of seven coaches and a combination baggage and smoking car, and the rear car was also used as a smoker. It ran fast to Elizabeth, New Jersey and then called at Westfield, Netherwood, Plainfield and Somerville. Running on track 2 of the quadruple-track line the train was normally switched to track 4 at Cranford to allow the 6 p.m. departure from New York, the Pennsylvania express, to overtake it, but on the day of the accident it was ordered to proceed to Dunellen before switching to track 4. Then shortly after leaving Westfield station its engineer discovered that he had an overheated journal, the packing of which had caught fire. The train was stopped to let the journal cool; the flagman went back to protect his train and some of the passengers got off to see what was wrong. The problem was fixed, the flagman was called in, the passengers had just resumed their seats and the train was setting off again when the express appeared behind them, running at full speed. The flagman vainly waved his lamp towards the onrushing express and then jumped aside as it ploughed into the rear coach of the forward train.

Quite apart from the flagman the line was protected by automatic electro-pneumatic block signals which witnesses testified had been working normally and been at danger when the express passed them. Gateman William Vernon at the Broad Street crossing in Westfield saw the express overrunning signals and waved a red lantern at it, exclaiming, "My God, they'll all be killed. He hasn't seen it."

The Philadelphia express consisted of a combination baggage and passenger coach, a passenger coach and two Pullman cars. It was in charge of Conductor George McErven, with Engineer James H. Davis and Fireman Samuel McCarthy. It was running at 40 miles an hour when the brakes were applied a few moments before the collision. The engine, No. 27, of the express reduced the rear coach of the local train to kindling wood, throwing its passengers into the air. The next car was demolished, the third, containing many women passengers, was split in two; the fourth had its rear end stove in and even the fifth was partially derailed. The front three cars escaped and were subsequently used as a hospital train to convey the injured to Plainfield. None of the vehicles of the express was derailed, though its windows were broken and its passengers badly shaken. Debris covering engine No. 27 and its tender immediately caught fire. Eastbound train No. 126,

[285] *The Evening Telegraph*, St John's, Newfoundland, 27th January 1903
[286] *Chehalis Bee Nugget*, Chehalis, Washington, United States, 23rd January 1903

a local passenger working from Somerville, was also passing at the time of the collision. Its side walls were damaged, its windows smashed, and its passengers shaken up. At the time of the newspaper report, written the next day, the bodies of 13 victims had been identified, seven remained in the morgue unidentified and four people were said to be 'missing'. There were 35 injured, including Engineer Davis and Fireman McCarthy.[287] Engineer Davis subsequently died of his injuries.

At the opening of the inquest on those who lost their lives in the collision on 3rd February an antemortem statement made by Engineer Davis was read. He stated that on the outward journey he had left Philadelphia with engine No. 323, but it had developed a 'hot-box' (overheated axle box) and he had had to leave it at Trenton Junction and continue with engine No. 27. This locomotive had had a crack in its steam chest on the left side which had prevented him seeing every signal. He had had to shut the engine off to see the signals. Those at Cranford and Roselle had been clear. Then he had to fiddle with the injector and was aware that he had missed one signal. The first he knew of the accident was seeing the local train 60 feet ahead when he had applied the brakes. Conductor John Day of the Easton local also testified on the first day of the inquest. He said he had received the order at Elizabeth to stay ahead of the express on track 2 to Easton. He insisted that his engineer was justified in stopping his train because of the 'hot-box'. Two of the old rear cars of his train had Pintsch gas lighting.[288]

At the resumed inquest Policeman Knapp testified that he had found Engineer Davis (who by this time had died of his injuries) lying on the ground after the accident and had asked him if he had seen the red light. Davis had replied, "I saw the red light. I expected it to turn white." There was evidence that the view of the engineer had been partially obscured by leaking steam and after the collision he had told his conductor, "I could not see them for steam."

Train dispatcher Michael Meaney, who was questioned at length, stated that he had decided to keep the Easton local train on track 2 as far as Dunellen because track 4 was blockaded by a freight train between Westfield and Garwood. As an extra precaution he had sent an order to Cranford to be given to the express that it was to overtake the local train at Dunellen, but the express had passed there without taking the order. He could have arranged for the express to receive the order before it had left Jersey City, but the regulations were that orders must be delivered to the point nearest to where they came into effect.[289]

Later in the inquest Fireman McCarthy gave evidence from his hospital bed. He confirmed that the engine had been leaking steam for the whole journey, but he could not give any useful evidence as to the visibility of the signals through the steam as he had been preoccupied with his firing.

Back in the council chambers Brakeman Harry O'Neil testified to going back 600 or 700 yards when the Easton local train had stopped, being recalled, then running back as the Philadelphia express had come rushing towards him. He had also tried to flag the eastbound train but there was no time.

Engineer Bernard Munn stated that the fact that Engineer Davis might have been

---

[287] *The Daily Press*, Plainfield, New Jersey, United States, 28th January 1903
[288] Ibid, 4th February 1903
[289] *New York Times*, New York, New York, United States, 6th February 1903

working on a leaking injector should not have prevented him looking out for signals.[290]

The inquest jury found a very comprehensive verdict on 13th February. They found that the accident had been caused by the failure of Philadelphia and Reading Railroad engineer Davis (the line was owned by the Central of New Jersey, but the Reading had trackage rights). He had run past the red light placed on the ground at Cranford to indicate there were orders for him, the distant signal preceding Westfield station, the lantern shown by the Broad Street crossing flagman, the red lantern shown by the brakeman of the Easton local and the block signal protecting the local train. There were, however, mitigating circumstances. Engine No. 27 had been leaking an unusual amount of steam, he had missed seeing the Westfield distant signal because at that moment he had been adjusting the injector, having failed to receive the order at Cranford he was unaware that the local was on his track and the local had stopped out of course with a 'hot-box'.

The jury acquitted the Central of New Jersey and the Philadelphia and Reading of any criminal responsibility for the collision. They made a number of recommendations and suggestions:

1. Stoves for heating should be prohibited (two stoves in the coaches had probably contributed to the fire).
2. Electric headlamps should be substituted for oil headlamps.
3. Should a third man be placed in the cab whose sole duty was to look out for signals?
4. Stricter rules regarding the observation of signals should be enforced.
5. Some form of automatic train control should be provided.
6. A State Railroad Commission should be appointed to investigate the above questions with the power to force railroad companies to adopt devices which would improve safety.[291]

The final death toll in the accident was 23, comprising 22 passengers and Engineer Davis.

A disaster at Durand, Michigan, United States, on the Grand Trunk Western Railroad, involved two sections of a circus train. The first section of a Benjamin E. Wallace circus train arrived at Durand at 3.30 a.m. on the morning of Friday 7th August 1903 and came to a stand on the main line. It was showing a red light at its rear. Fifteen minutes later it was run into by the second section, the engine of which was driven by Engineer Charles M. Probst. Probst jumped before the collision, along with Fireman P. E. Colter and Head Brakeman W. F. Benedict, all three of them escaping with slight injuries.

Seventeen persons were killed outright in the collision and four more had died by the end of the day. Grand Trunk Trainmaster James McCarthy and Chief Special Officer A. W. Large were among those killed, the other dead included circus labourers. The circus performers had been travelling in the rear vehicles of the second section and had thus escaped injury. All of Durand's physicians were involved in caring for the injured,

[290] *The Daily Press*, Plainfield, New Jersey, United States, 7th February 1903
[291] Ibid, 14th February 1903

together with doctors from neighbouring towns, while surgeons from Detroit set off for the scene in the early morning, among them Dr Ben Brodie, chief surgeon of the railroad company. The Richelieu Hotel in Durand was used as a temporary hospital. An elephant, two camels and a bloodhound valued at $1,000 were also killed and two camels were badly injured. One report stated that the trumpeting of the elephants and the noises of the other animals had added to the horror of the scene, though another stated that the elephants had remained surprisingly calm and had been led out of the wreckage without trouble. The train had been on its way from Charlotte to Lapeer.[292]

Three cars of the first section were telescoped in the collision and the engine and five cars of the second section were demolished, even though the second section had only been travelling at 15 miles an hour when the impact occurred. Grand Trunk trainmen had been asleep in the caboose at the rear of the first section and the circus labourers had been sleeping in the other two cars; five elephants and several camels had been in the wrecked cars at the front of the second section. It was feared at first that some of the animals had escaped and the fire alarm was sounded, rousing the whole town. A fortunate circumstance was that there was a wrecking train in the Durand yards anyway which was on the scene with tools and equipment in a few minutes. The other wrecked cars of the second train contained only canvas and waggons. After the accident the circus people pitched their tents and camped near the scene of the wreck. The Wallace Bros believed that their financial loss would be heavy, and this was their second wreck within a month.[293]

By 14th August the number of those killed in the collision stood at 23. Engineer Probst claimed that the air brakes had failed to work, and his fireman and the brakeman confirmed this. However, Superintendent Brownlee stated that the brakes had been tested after the accident and found to work perfectly. A company official theorised that the engineer had exhausted the air in the reservoir of his air brakes by several applications between Lansing and Durand and had failed to recharge it. There was a strong possibility that Probst had fallen asleep. He was supposed to approach yard limits such as Durand with his train under control, but he had clearly not, he had also failed to react to the warning fusee laid by a flagman who had gone back three-quarters of a mile and had not even shut off steam until aroused by his fireman. It emerged that he had taken no rest before taking out the circus train and the Interstate Commerce Commission considered that there was 'a strong ground for the inference that he was asleep at his post'.[294]

The jury at the inquest on 27th August on those killed in the collision blamed the accident on a failure of the air brakes and added that it would not have occurred had Engineer Probst checked the air gauge after Charlotte. They also criticised the lack of hand brakes on the train cars. They stopped short of committing Engineer Probst on a charge of manslaughter.[295]

The next accident to be considered is not mentioned in Robert Shaw's classic book on

---

[292] *The Evening Record*, Traverse City, Michigan, United States, 7th August 1903

[293] *New York Times*, New York, New York, United States, 8th August 1903

[294] *A History of Railroad Accidents, Safety Precautions and Operating Practices* by Robert B. Shaw, pgs. 227-8

[295] *The Evening Record*, Traverse City, Michigan, United States, 28th August 1903

American railroad accidents but it was a major disaster, nonetheless. It occurred on Saturday 17th October 1903, on the Pennsylvania Railroad close to Washington Crossing, New Jersey, United States. The victims were mostly Italian labourers, together with a few negroes, and they were on their way to repair washouts along the road. The work train on which they were riding consisted of two coaches, in which they were seated, followed by two flat cars. The train had stopped near the crossing to receive orders respecting the movement of passenger trains when it was run into by a following gravel train, which pushed the flat cars into the coaches in front of them. There were about 180 men in the first train and at the time the initial report on the accident was made the bodies of 14 of them had been removed from the wreck and one more was thought to be still buried in the debris. About 40 others were injured.

After the collision the remaining labourers reportedly became frantic and attempted to physically assault the crew of the colliding train. Police were called from nearby Trenton but before they had arrived the men were calmed down by their foreman. A special train with physicians was sent from Trenton as soon as news of the accident reached there and the dead and injured were taken to that place. The railroad authorities were reluctant to discuss the matter but the reported attributed the collision to either the flagman of the first train failing to go back far enough or to the engineer of the second train failing to see him.[296]

Trainmaster James Gordon told the inquest on the labourers that the work train No. 1229 had stopped near Washington Crossing to pick up 12 cars of gravel at 6.13 a.m. Noticing that one of the cars was very unevenly loaded with gravel he had ordered for it to be cut out, an operation which took some time. He had sent Flagman Jacob Saums back to protect his train. He heard the second train coming and not hearing Engineer Robert Reed acknowledge the warning of Flagman Saums he had sent Frank Parker (or Parks) back as a second flagman. The second train then whistled in acknowledgement of one of the flagman, he thought of Saums, followed by a series of danger toots. The collision had then occurred at 6.30. Gordon said he thought it would take 600 yards for the second train to stop as it had 20 cars, six equipped with air brakes, and a conductor, two brakemen and a flagmen to operate the hand brakes. Both trains had started from Coalport; he assumed that the second would not have started out before the usual interval of five minutes. The job of taking on the cars had been done as quickly as possible and Flagman Saums had gone back at once with red and white lanterns. The atmosphere had been variable, clear at one moment, fog preventing objects from being seen from more than a few feet the next. According to the regulations when the flagman was recalled he should have left two torpedoes on the rails, but he had not been called in before the collision. Reed's train had only been going at about seven miles an hour and the amount of damage done was surprising.

Flagman Saums said that he had got back 500 yards when the second train appeared. He had waved his red lamp at it, but the engineer had not whistled, unless he had done so before. Reed's train had been going at about 20 miles an hour. Flagman Saums said that he had only had two lanterns when he went back, no flags or torpedoes. He did not think that there were any torpedoes carried on the first train. Red lights had been burning at rear of the tool car of the work train. There was a lumber gondola behind the tool car, but

[296] *Newport Daily News*, Newport, Rhode Island, United States 17th October 1903.

the red lights should have been visible several feet above the gondola.[297]

At the resumed inquest the next day Engineer Reed testified that he had only been running at six miles an hour when he saw Flagman Saums' light only 50 yards in the rear of the first train. He had applied the brakes, put down sand, whistled twice and then jumped on the canal side of the track He was sure that it had been Flagman Saums he had seen and not Parker.

Fireman James Watson of the second train put its speed at ten miles an hour. He did not see Saums until they were close to the first train. He had jumped when they were four car lengths short of the work train.

Conductor Johnson Dicks said that they had got a clear signal at Wilburtha (about three miles from Washington Crossing), then at Somerset (about a mile further on) he had got a danger signal from Engineer Reed and he had immediately applied the brakes. Conductor Augustus Conover of the work train had told him at Coalport that the front train would be stopping at Washington Crossing.[298]

Engineer Reed and Fireman Watson were recalled when the inquest resumed on Saturday 24th and Engineer Fenton of the work train corroborated Engineer Reed's testimony as to the short distance the flagman's lantern could be seen in the fog.

Foreman William Houghton, who had been riding on the gondola car, agreed that the fog had been thick and said he had not seen the colliding train before it had whistled, when he had shouted to his men to jump.

Labourer Frank Stout said that hearing the whistle he had paused at the door only long enough to shout to his fellow workmen to jump for their lives. None had taken any notice or had had time to follow him, the collision occurring as he reached the ground. Excused from the witness box after giving his evidence Stout 'amazed Coroner Disbrow and the jurymen' by asking the jury to be lenient with the railroad men and saying that 'they had acted like gentlemen'.[299] Stout's outburst was doubtless especially amazing because he was one of the negro labourers.

The inquest jury finally reached a verdict on 4th November, finding Flagman Saums mainly responsible. Their verdict read in part, 'We find that the fault of the collision was due to Flagman Jacob Saums, who did not go back a sufficient distance to protect his train with the weather conditions that prevailed the morning of October 17th, 1903. We also find that Johnson Dicks, conductor of train 1229, did not use due precautions in informing Engineer Robert Reed that the preceding train was to make a stop at or near Washington Crossing, both Conductors Augustus Conover and Johnson Dicks having dispatcher's orders for a point further on. We also find that the Pennsylvania Railroad Company should maintain a block signal system on the Belvidere division to the end that such accidents may not be due entirely to the fault of their hands.' Prosecutor Crossley stated that the matter would probably be brought before the grand jury.[300]

The final number of fatalities in the collision has been given as 17.

---

297 *Trenton Times*, Trenton, New Jersey, United States, 23rd October 1903
298 Ibid, 24th October 1903
299 Ibid, 26th October 1903
300 Ibid, 4th November 1903

The disastrous rear collisions in the United States continued with one at Kentwood, Louisiana, on the Illinois Central Railroad on Saturday 14th November 1903. According to the first report on the accident in a Sunday newspaper the McComb accommodation train which had left New Orleans at 6.30 p.m. should have been side-tracked at Kentwood out of the way of the Northern express bound for Chicago. The former train was running behind time and was caught up by the express. A large number of passengers in the rear coach of the accommodation lost their lives; almost all of them were blacks as this was the "Jim Crow" coach, reserved for negro passengers, (such coaches had been in use in Louisiana since 1890 when the state's Separate Car Law was passed). One early report spoke of 40 dead and 23 injured, of whom all but one of the dead and 20 of the injured were blacks. The bodies of the victims were jammed between the engine of the express and the passenger coach and many of them badly burnt, the wreckage having caught fire after the collision. Physicians from McComb and Kentwood went to the scene to attend to the injured. Many of the victims were reportedly section hands who had only just boarded the train.[301]

A more detailed explanation of how the accident had occurred was given in another newspaper on the Monday, two days after it had occurred. Train No. 32, the McComb accommodation, was running 45 minutes late, but had orders to proceed to Chatawa, Mississippi, a few miles north of Kentwood, in front of train No. 6, the limited express, which was itself five minutes late passing Tangipahoa, just south of Kentwood. Being 45 minutes late the accommodation was running on the time of the limited and was thus just five minutes ahead of it at this point. The station agent at Tangipahoa reportedly had an order to hold the limited train until the interval between it and the accommodation had increased to 10 minutes and had exhibited the red semaphore to stop it, but Engineer Jones of the limited had ignored the signal. Meanwhile at 7.32 p.m. train No. 32 drew into Kentwood. The "Jim Crow" coach at its rear was filled with negro section hands and labourers from up and down the line who, according to the custom of the line, were making a free journey home to spend the Sunday with their families in Mississippi (the Louisiana – Mississippi state line lies about 5 miles north of Kentwood). There were 45 or 50 of them altogether, of whom it was known by Monday that 26 had died and 16 were more or less seriously hurt. Ahead of the "Jim Crow" coach was the ladies' coach, which had several occupants as far as Kentwood, but most of them had alighted there.

The accommodation was still stopped at Kentwood when the limited ran into it, completely telescoping the "Jim Crow" coach. Then fire broke out in the wreckage. 'Lampton {the Kentwood station agent} and his clerk, Newland, saw a flame begin to grow in the wreckage and connected the hose in their own premises to play upon the ruins seventy-five yards away. Meanwhile the terror of the scene grew. The flames from the burning coaches began to roar with a fury that outstripped the noise of the escaping steam. The shrieks and screams wailed well and far into the night. The mass of the crowd, now increased to over 1,000 persons, stood far back from the scene of torment, and shuddering moans and bursts of prayer answered the shrieks and tortured agonies of the victims in the burning wreck.

'At last the tide turned. The citizens and trainmen had turned three streams of water on the fire and began to gain upon the blazing hell. It was thought that all had escaped from the white coach. Andrew Kizar was dying from the mangling received between the

[301] *The Fort Wayne Journal-Gazette*, Fort Wayne, Indiana, United States, 15th November 1903

negro coach and the white one.

'At last the flames were out. Then the work of taking out the dead began, and that of caring for the wounded. In one mass of wreckage, jammed ten feet above the ground on the front of No. 6 engine, fifteen bodies were found in a space not six feet square.'

The disposal of the bodies of most of those killed was insensitive, if not downright callous, 'Twenty-three of the negroes, including the woman, were thrown into twenty-one pine boxes and buried in a trench just outside the negro burial ground at Kent's mills. No service of any kind was held. The first internment, in the early hours, was so shallow the health authorities objected, and had the entire task done over. The body of Jerry Austin was shipped to his friends in Canton, while that of Wm Lawson was sent to Crystal Springs. Of the injured all except Bud Wilson, McMullen and an unknown negro were sent to the New Orleans charity hospital. Andrew Kizar was sent to Cristal {sic} Springs.' [302] (Kizar, whose Christian name was later given as Edward, was the only white male passenger to be killed outright, he was from McComb City. One white woman, still unidentified several days later, also lost her life.)[303]

'Today people have had time to remember the incidents that were seared from their minds yesterday by the fearful horror of the holocaust. Today people have been asking the causes which resulted in the wreck and have been endeavouring in vain to locate the blame or find the fault.'

Larry Hawthorne, the one passenger who escaped unhurt from the "Jim Crow" coach had a mental breakdown as a result of what he had gone through, the same newspaper report writing of 'one demented darky roaming the streets of Kentwood and muttering and moaning of his fight with fire and steam'.[304]

The final number of people who died in the accident remains unclear. A report from New Orleans on 15 spoke of 28 negro men dead, one negro woman and one white man, Andrew C. Rose, foreman of carpenters, who died of his injuries.[305] This does not therefore include Andrew (or Edward) Kizar or the unidentified white woman, whose deaths would increase the total to 32, which is the figure quoted by Shaw. Some newspapers, however, continued to speak of 40 dead.

Responsibility for the accident was disputed, Engineer Jones of the express claiming that he had understood that he had a clear road to Chatawa.[306] It has not been possible to trace any newspaper reports of an inquest.

An excursion trip from New York to the resort of Greenwood Lake, organised by the German-American First Plattdeutscher club of Hoboken, ended in tragedy on Sunday 10th July 1904. The club had chartered a special train of 12 cars to convey them to the resort, which was at the end of a branch of the nominally independent New York and Greenwood Lake Railway, which was worked by the Erie Railroad. The special was only about ten miles short of its destination when it stopped for water at a tower a quarter of a

---

[302] *The Biloxi Daily Herald*, Biloxi, Mississippi, United States, 16th November 1903

[303] Waterloo Press, Waterloo, Indiana, United States, 19th November 1903

[304] *The Biloxi Daily Herald*, Biloxi, Mississippi, United States, 16th November 1903

[305] The Gleaner, Kingston, Jamaica, 26th November 1903

[306] *The Sun*, Baltimore, Maryland, United States, 16th November 1903.

mile beyond Midvale station. The train was packed with over 800 excursionists, two or three people sharing a seat and men and women standing in the aisles. The picnickers were in a happy mood, with much singing and shouting, perhaps stimulated by more than a little alcohol, though this is not mentioned in the newspaper reports. Suddenly there was a crash and the men and the women in the front coaches were thrown to the ground, while those at the rear of the special fared much worse.

Following the special was the regular branch train, the engine of which came around the corner and crashed into the rear coach, which was lifted up and telescoped into the coach in front. 'As the trains crashed part of the floor of the rear car gave way, and as it telescoped into the rear car in front of it heads were mowed down as if by a scythe, bodies were mangled beyond recognition and men and women were pinned in the wreck so tightly that it required several hours work to release them. One of the saddest features of the accident was that most of the killed and injured had piled into the last two cars in order to be near one another during the trip. Most of the dead also lived within five blocks of one another in Hoboken, and the scene in and about the undertaker's establishment after the wreck resembled on a smaller scale the tragedies about the morgue after the Slocum disaster.' (The Slocum disaster was a fire on the steamship General Slocum less than a month earlier on New York's East River. Over 1,000 of its mostly German-American passengers had died either from the fire or from drowning). Sixteen passengers died.

The engineer and fireman of the regular train disappeared after the accident and the railroad officials refused to disclose their whereabouts.[307]

The engine crew of the regular train probably feared being assaulted or even lynched but a subsequent investigation determined that they were not at fault. A couple of days after the accident Erie Railroad General Superintendent Van Keuren stated that the Midvale operator and the conductor and flagman of the excursion train had been dismissed for negligence. It had been shown that Operator William Richards had failed to replace his station signal to danger after the excursion train had passed, thus misleading Engineer John Landmasser of the regular train into believing that he had a clear road ahead of him. Equally Flagman Ernest Hillier of the excursion had only gone back very few car lengths when he had had time to go back a much greater distance, and Conductor Vanderhoff of the excursion had failed to see that the flagman had properly performed his duty.[308]

The report of the inquest on Henry Kanzer, one of those killed, noted that the verdict of the coroner's jury was almost a foregone conclusion. The jury found Operator Richards and Flagman Hillier jointly responsible for the wreck. The flagman, by his own admission, had only gone back 300 or 400 feet at any time and had partially retraced his steps when the regular train appeared. He said that he had heard the engine of the excursion puff as if it were setting off and had gone back for fear of being left behind, as commonly happened. Witnesses also testified that his flag had been rolled up. Driver Landwasser testified that the flagman's flag had been furled and he had seen the flagman when the latter was 50 feet from his (Landwasser's) train and 250 feet from the excursion. He had made an emergency brake application and put down sand. He could do

[307] *The New Brunswick Daily Times*, New Brunswick, New Jersey, United States, 11th July 1904
[308] *Middletown Daily Argus*, Middletown, New York, United States, 14th July 1904

no more.

The operator admitted that his semaphore signal had not been working properly that morning. He said that it would not drop properly but stuck at an angle of 45 degrees. Henry Meade, an engineer who had been stood on the platform at Midvale, said that he had seen the excursion stop at the water tower and had called to Richards that his signal was still showing clear, but it was not changed.[309] It seemed that Operator Richards had either not been informed of the running of the excursion or had forgotten about it and had assumed that it was the regular train. Not expecting, therefore, another train behind it, he had cleared it back to the preceding station but not bothered to replace his signal. The operator's story of the signal malfunctioning was likely an afterthought. A writer to the New York *Times* a few days later noted that if the Midvale signal had actually been out of order that morning the operator should, according to the regulations, have stopped the regular train with a hand signal, instructed him that the signal was out of order and to proceed with caution. The same writer went on to take issue with a claim that "rotten" coaches had been used on the excursion train, noting that even a Pullman car would not withstand the impact of a locomotive travelling at say 50 miles an hour.[310]

Operator Richards and Flagman Hellier stood trial for manslaughter towards the end of November but the jury were unable to agree on a verdict.[311] No report of a retrial has been found.

A dense fog was a factor in a rear collision on the Chemins de fer du Nord at La Chapelle station, France, only two kilometres outside of Paris Nord station, at 11.10 p.m. on Friday 23rd December 1904. The trains involved in the accident were two expresses, one from Lille and the other the boat train from Boulogne. The Lille express had been stopped by signals outside of the Nord station to allow another train to precede it into the terminus. It was just restarting when it was run into by the Boulogne train, which had overrun a carré (stop signal). The luggage van (another account says two vans) at the rear of the Lille express and the second-class carriage preceding it were smashed to pieces and 12 passengers were killed outright and two more died of their injuries later. Another 20 passengers were injured. A correspondent for a London newspaper pointed out though that the accident would probably have been much worse had the trains not been composed of new rolling stock, with corridor coaches which offered a great resistance to impacts, while 'As there were no third-class carriages, the victims of the accident were all persons of the better class. One of those who was killed on the spot was M. Marcelin Berthelot, the grandson of the illustrious chemist and Academician of the same name, who is described as heart-broken. The young gentleman was on leave of absence from Lille, where he was performing his Military service in the 43rd Regiment of the Line.' [312] The collision, in the darkness and fog, combined to cause something of a panic among the surviving passengers. The injured were conveyed to the nearby Lariboisière Hospital and the injured to the morgue. The injured were visited in the hospital by a representative of the French President and by the Prefect of the Seine Department. All the deaths and injuries were to French passengers in the Lille train, Captain Churchward, the Paris agent

[309] *New York Times*, New York, New York, United States, 19th July 1904
[310] Ibid, 23rd July 1904
[311] *Orange County Times*, Middletown, New York, 25th November 1904
[312] *The Standard*, London, England, 26th December 1904

of the English South Eastern and Chatham Railway, assuring a newspaper correspondent that none of the passengers on the Boulogne express, containing passengers from London, had been injured, though three of them had been severely shaken. They had, however, gone away without making any complaint. Their nationalities were not known.

The correspondent for the London *Times* managed to get the details of an interview with a passenger from the colliding train. 'Mr Douglas Story, who was in the Boulogne express at the time of the accident, has given an account of his personal experience to the *New York Herald* (Paris edition). At 10.45 they passed St Denis and people began to collect their luggage in preparation for reaching Paris. The train was cautiously feeling its way through a white fog at a speed of not more than 15 miles an hour. Suddenly there was a jar. The passengers were thrown into each other's arms, and hand boxes and travelling bags came tumbling on their heads. They fancied something had gone wrong with the air-brake. Soon after there came down the train the rumour of a serious collision. Mr Story hurried forward. There was, he says, no panic and no unseemly excitement. Gendarmes and passengers, railway porters, and soldiers on furlough were engaged in extricating the injured. Later on the doctors and the ambulance corps came to their assistance, and finally "the dead were dug out from under the engine-gruesome figures in the light of the porter's lanterns."' [313]

At the time of the accident the working of the traffic was disorganised due to a fog which made it impossible to see six paces ahead. The driver of the Boulogne express stated that he only saw the carré at danger at the very last moment through the fog, when he had applied the brakes, but the rear of the Lille express had then been no more than 20 yards ahead of him. There was, apparently, also a failure of the signalling apparatus, as with the signal at danger a detonator should have been automatically placed on the rail, but the apparatus was not working properly so the driver received no warning explosion.

The accident led to questions in the French parliament regarding measures which were being taken to increase railway safety.[314]

A month after the French accident at La Chapelle fog played a role in an accident across the English Channel early on the morning of Thursday 19th January 1905. The location was near Cudworth, England, on the Midland Railway main line between Leeds and Sheffield. Six signal boxes were directly concerned in the events leading up to the accident; from the north (Leeds) end these were Carlton Main Colliery Sidings, Cudworth North Junction, Cudworth South Junction, Cudworth Station North, Cudworth Station South and Storrs Mill. The four Cudworth boxes were close together and their distant signals were interlocked so that it was impossible for the signalman in one of them to lower his up fast line distant signal unless the distant signals of the boxes on the south (Sheffield) side of him had already been lowered. In addition, the Carlton Main signalman had instructions not to lower his up fast line distant signal unless his indicator showed that the Cudworth North Junction distant signal was off. A clear Carlton Main distant signal should therefore mean that all the up fast signals up to the Storr's Mill distant were clear. There were four passenger lines as far as the south end of Cudworth station, beyond which the slow lines became goods lines.

[313] *The Times*, London, England, 26th December 1904
[314] *The Standard*, London, England, 26th December 1904

The trains involved in the accident were the 2.25 a.m. up mail train from Leeds to Sheffield and the 3.05 a.m. up express from Leeds to London St Pancras, which ran into the mail train at high speed opposite Storr's Mill box. A down express had been close at hand and had collided with the wreckage and a fire had broken out afterwards. Four passengers, a guard and a fireman were killed outright, and 13 passengers and five railwaymen were injured. One of the injured passengers later died. The accident occurred at 3.37 a.m.. It had begun to get foggy at 1.15 a.m. and the density of the fog had gradually increased.

The Board of Trade inquiry into the accident into the accident took evidence from a large number of witnesses, which was reproduced in full in the report, beginning with that of the various signalmen.

Signalman Thomas Brown testified that he had been on duty at Carlton Main Colliery Sidings. He had accepted the up mail train from Monckton Main Colliery at 3.18 a.m. and had it accepted by Cudworth North Junction in turn but the distant signal for the latter cabin had not been lowered and so in accordance with his instructions he had not lowered his own up fast distant signal either. The mail had passed him at 3.20 a.m. and been cleared back from the North Junction at 3.22. The Scotch express had been offered to him at 3.28, was accepted by the North Junction at the same time, passed him at 3.32 and was cleared back at 3.33. The mail train had only been travelling at about 10 miles an hour, the Scotch express was going considerably faster. He had heard that the fogsignalmen had been called out at 1.30 a.m. and fogsignalman Blackburn, who fogged his up fast distant, had signed on at 3 a.m. and set off for the distant at 3.10. He did not think Blackburn would have reached the signal, 1,080 yards away, before the express passed. Unhappy at the speed at which the express had been running he had telephoned the man at Cudworth South Junction and told him that he was not sure that the express would stop at the North Junction signals. The South Junction man told him that the mail train had left the slow line and gone on towards Storr's Mill.

Signalman Alfred Shirley at Cudworth North Junction said that he had booked the mail train as passing him at 3.25. He had accepted the express at 3.31, cleared his signals other than the distant for it and it had passed him at 3.37. The only thing that struck him as unusual about the express was that it was running fast. It was foggy, and his fogsignalman, William Crossley, had arrived at 3.45. The mail had almost stopped at his box, the express was going at he thought between 50 and 60 miles an hour, too fast to see the signals. He did not consider he would have been justified in sending the 'Train Running Away' signal but he tried phoning Cudworth Station South box. By the time that he had got through the express had passed there.

Signalman John Whyman at Cudworth South Junction said that he had set the road for the mail train to pass from the up fast to the up slow and had tried to lower his distant signal, but it had been frozen on. The mail train had passed him at 3.28 on the up fast and was cleared back from the Station North cabin at 3.29. He had accepted the express from the North Junction at 3.31 and had it accepted from the Station North cabin on the up fast but as it was foggy had not cleared any of his signals for it until he received the 'Train Entering Section' signal from the North Junction. Signalman Brown at Carlton Main had then phoned him and told him to look out for the express, which was running fast. As he was putting the telephone down again the express had rushed past him before he received the 'Train Entering Section' signal from the North Junction. He forwarded the 'Train

Entering Section' to the Station North cabin at 3.31. He did not send the 'Train Running Away' signal as the Station North cabin had accepted the train. He tried to phone the Station South cabin but could not get through as Signalman Shirley was on the line.

Signalman Frederick Southam at Cudworth Station North said that the mail train was booked to stop at Cudworth, it had passed him slowly. When the express was offered to him, he had offered it forward to the Station South cabin, but it had not been accepted for two minutes. After it had been accepted, he had lowered his home signal and the express had passed him at once at a highest rate of speed, and too fast given the fog and the fact that his distant signal had been at danger, but he had had no time to do anything. He did not think the use of the 'Train Running Away' signal would have been justified, and the section was too short anyway.

Signalman Henry Taylor at Cudworth Station South cabin said that he had accepted the mail train at 3.16 and it had been accepted by Storr's Mill at 3.19. The mail had stopped at the station at 3.27 and left at 3.33. The express had been offered to him at 3.30 but he had refused it till the mail had cleared the junction at 3.33. He had replaced his starting signal behind the mail train, after which all his up signals had been at danger. He had received the 'Train Entering Section' signal for the express as soon as he had accepted it and about a minute and a half later, he had heard a train approaching, running at high speed, which he knew must be the express. He had picked up a hand lamp and run to the south end of the cabin, but the train was past him before he could show a red light to the drivers. He had then sent the 'Train or Vehicles Running Away on Right Line' signal to Storr's Mill and it was acknowledged. He thought that his signals had not been visible in the fog from more than 20 yards. The express had passed him at what he estimated as 60 miles an hour as if all the signals had been off. He had not placed fog signals on the line in accordance with Rule 85 as he had expected Storr's Mill to have cleared the mail before the express passed him. If he had had a fog-signalling machine worked from his box he would have used it. He had written instructions to allow the up mail to proceed from Cudworth if the express was three or four minutes behind it.

Signalman John Gambles at Storr's Mill related that had accepted the mail train at 3.20 after crossing a goods train, received the 'Train Entering Section' signal for it at 3.32 and had it accepted by Darfield box. At 3.34 he had accepted the 12.01 a.m. down express from London St Pancras from Darfield, and it had been sent on at 3.36. Also at 3.36 he had received the 'Train Running Away' signal for the up express. He had checked that the mail train had passed his up distant signal before replacing it, then forwarded the running away signal to Darfield in the hope that the latter signalman might be able to stop the down express but getting on the phone to Darfield he learnt that the down express had passed there. He had heard the sound of the express running into the mail as he was sending the running away signal to Darfield. Finding that it was too late to stop the down express at Darfield he had shouted from his cabin for someone to warn it, and an engine whistle had been blown. He had asked the Cudworth station staff to call his fogsignalmen at 1.15; they had arrived just after the collision.[315]

In charge of the pilot engine of the Scotch express, No. 154, was Driver Oliver Mugg. Driver Mugg stated that he had had no trouble in seeing the signals as far as Royston cabin, which was out of switch at night, but he had failed to see the Carlton Main up

[315] Board of Trade report

distant signal, which was a lower arm under the Royston starter. He did not explode any detonators there. He had looked at his fireman, named Robert Pattison (the one who lost his life in the accident) and the latter had lifted his hand, from which he (Driver Mugg) had understood that the fireman had seen that the distant was off. He was aware that this signal was the key one which indicated if the line was clear through Cudworth station and had he not seen the fireman apparently indicate that it was off he would have shut off steam and applied the brakes. He admitted that he had also failed to see any of the signals going through Cudworth junctions and station from his position on the right-hand side of the engine but he said that the fireman had been looking out on the left-hand side and he had assumed that his mate must have seen the signals at clear as he had not said anything. He had also thought that all must be all right as no detonators had been exploded. He did not see the Storr's Mill up distant signal either, and he was unaware that there was anything in front before the collision. He had been relying on his fireman and anxious to keep time.

Driver James Buggins was in charge of the train engine of the Scotch express, No. 833. He stated that there had been a little fog already at Waterloo Colliery, six miles out of Leeds, but he had seen all the signals up to and including the home at Royston Junction (two cabins before Royston Station). He had seen no signals at Hodroyd and Monckton Main box, nor the Royston Station box starter with Carlton Main box's distant below it. Like Driver Mugg he was aware that if this signal was off it indicated a clear road as far as Storr's Mill. He had relied on Driver Mugg having seen it. After that he had seen no signals or signal boxes except the up directing signal at Cudworth North Junction until the Storr's Mill up distant, which he was sure had been off. He had heard no detonators go off. He repeated that he had left the working of the train to the driver of the pilot engine. Driver Buggins' fireman, J. R. Whitham, had little to add to his mate's evidence.

Guard John Frater was in charge of the train. He said that he had been assisted from Carlisle by Guard James Weston, of Edinburgh, who was killed, and from Leeds by Guard Alexander, of St Pancras. Because of the fog he had seen none of the signals. After the accident he found that some goods men were protecting the down line, and Guard Alexander had gone back to protect the up line, so he had turned his attention to the passengers. For five minutes after the collision there was no sign of fire, then it broke out at the leading end of the sleeping car sixth from the engine. All the train except the Stranraer portion was eventually more or less destroyed by fire.

Guard Frederick Alexander agreed with the statement of his colleague and his evidence was mainly concerned with details of the rescue work after the accident had occurred.

Driver Elijah Spurr of the mail train stated that it had been very foggy, but he had seen all of the signals. A van had been detached at Cudworth and he had left there eight minutes late because of the fog. His train had been bumped in the vicinity of Storr's Mill box, throwing him and his fireman backwards on the footplate. He had seen that vehicles of his train were fouling the down line, then the midnight express from London approaching, and had opened his engine whistle and kept it open till the express engine had passed. It had been more or less foggy from Woodlesford and had become so dense that at Royston and Cudworth he had had to actually stop his train to see the signals. Fireman Simeon Pearson could add nothing to Driver Spurr's evidence.

Goods Guard Frank Woods said that he had seen that the down line was fouled, and he managed to put down one detonator in front of the down express.

Driver James Pearson of the down express said that there had been variable fog all the way from Leicester. The Storr's Mill distant had been at danger and as he had approached the overbridge carrying a Great Central Railway colliery branch over the line, he had heard an alarm whistle. He had made an emergency brake application, applied steam and reversed his engine, which then collided with the fish truck. He did not hear the detonator laid by Guard Woods.

Porter George Atyeo of Cudworth explained that Foreman John Jerram had instructed him to call out the fogsignalmen at 1.20 and he had gone to the house of call-man Percy Hill, whose duty was to call the other fogsignalmen, but had been advised by Hill's father that Hill was sick. He had returned to the station and the foreman had told him to call Ganger George Moore instead, which he had done.

Ganger Moore said that he did not usually have anything to do with the fogsignalling and did not know the addresses of the other fogsignalmen. He had not been aware that Percy Hill was unwell. He had got up and set off at 2 a.m. to the house of Platelayer Kettlewell, through fog and over slippery roads, arriving there at 2.20 a.m.

Platelayer William Kettlewell said that he in turn had got up and set off to call out Platelayers Blackburn, Woolstone and Potter. He knew where the last two lived and which street Blackburne lived in, but not the number, so that he had to knock on the doors of three houses (whose occupants were probably not best pleased at being woken up) before he found Blackburne. He had then called Woolstone and Potter.

Platelayer Alfred Blackburne, who fogged the crucial Carlton Main up distant signals, agreed that Kettlewell had roused him at 2.35. He said that he had left his house about ten minutes later and had arrived at the signal box at 3.8 a.m. He had collected his lamp, flags and detonators, lit the lamp and set off for the distant signals. He had been just at the signal post when the express had passed. It had been so slippery that he had ended up on his back at that moment. A fogsignalman's job was certainly no sinecure.

Major J. W. Pringle investigated the accident. He noted that the only signalling irregularity was the acceptance of the up express by Signalman Taylor at Cudworth Station South before Storr's Mill had cleared out the mail in contravention of Regulation 4(e). This regulation decreed that in foggy weather and in the absence of fogsignalmen a train should not be accepted from box in the rear until the previous train had passed the box ahead – so-called 'double block working'. Whilst generally adopted by British railway companies this regulation was not used by the Midland Company, and Signalman Taylor was thus justified in acting as he had.

It must be accepted, Major Pringle stated, that the up express had passed eight distant signals and five stop signals at danger. Driver Mugg had said that he had relied on his fireman. His statement as to the fireman's indicating that the Carlton Main distant was clear could not now be corroborated as the fireman was dead. But the responsibility of viewing signals did not rest equally with the driver and fireman but mainly with the driver. Nor was it in accordance with general practice, even in clear weather, for the driver to rely on his fireman, while in foggy weather Rules 143 and 144 laid down that drivers must run cautiously and stop if they could not see fixed signals or a fogsignalman's green hand signal. That it was possible to observe the signals at low speed was shown by the evidence of Driver Spurr, who had driven with a full sense of his responsibility. Drivers Mullins, of the down goods train, and Pearson, of the down express, had also been able to make out the signals.

Driver Buggins, who had more experience, was much less responsible for the accident, but should have taken action when he realised that Driver Mugg was driving at much too high a speed for the conditions.

As regards the fogsignalmen the inspecting officer considered that the only person who might be criticised was call-man Hill, who should have reported his illness to the stationmaster. The arrangements for calling out the fogsignalmen could not, however, be described as complete and satisfactory. The fogsignalmen should be provided with houses close to where they were needed, either railway properties or houses rented by the Company. Two call-men should be provided, who knew where all the fogsignalmen lived, and it should not be the job of one fogsignalman, such as Woolstone, to call out other men before proceeding to his post. It should also be possible for fogsignalmen to collect detonators at some point on their way to the distant signals without having to go to the signal boxes first.

It was unfortunate that Signalman Brown at Carlton Main had not rung Cudworth Station South, instead of Cudworth South Junction and that Signalman Whyman did not send the 'Train Running Away' signal, while the collision showed the necessity for the signalmen at least in busy boxes being able to place detonators on the line from their boxes, as those of some companies already could.

Various inventions had been proposed to do away with the necessity for manual fog-signalling altogether, but such an invention must give a positive as well as a negative indication and be entirely reliable. No such system had yet come to fruition. In the meantime, the inspecting officer made five suggestions by which the present system could be improved:

(a) Signalmen, at least at important traffic centres, should be provided with mechanical devices to allow them to comply with Rule 85 (placing detonators on the line at their box in foggy weather in the absence of fogsignalmen).

(b) Fogsignalmen at key signals should be required to live as close as possible to the signals they fogged.

(c) Where possible electric communication should be proved between the signal box or station and the fogsignalman's residence.

(d) The arrangements for calling out fogsignalmen at night should be revised and elaborated to prevent delay.

(e) The supply of detonators to fogsignalmen at the most important signals called for consideration.

Regarding the Midland Railway's reluctance to enforce block regulation 4(e) (double block working) because they said it would cause serious difficulty in working the traffic, Major Pringle said that if they did not do so it was all the more reason for them to improve their fogsignalling arrangements.

There was no clear evidence as to the cause of the fire as the coaches were lit by electricity with only a small gas reservoir for heating water in the sleeping car. The fire could have originated from the engine firebox, from electrical fusing or from gas.

Major Pringle ended his report with words of praise for their conduct Drivers Pearson and Spurr, Signalman Gambles, Guard Alexander and Goods Guard Woods.[316]

The passengers who died immediately in the accident were two brothers from Perth, Scotland, Adrian and Alistair Kinloch, on their way to Bradfield College, Berkshire; Harry Bradley, from Redcar, 'believed connected with the turf' and Joseph Gray, a wire-rope splicer, employed by Haggie Brothers hemp works of Gateshead and on his way to Rotherham.

Driver Mugg was perhaps lucky, for the jury at their inquest returned a verdict that there was not sufficient evidence to convict any one criminally. The jury commended Signalman Gambles, Driver Spurr and Goods Guard Woods for their efforts to prevent a second collision.[317] [318] The fifth passenger to die was a Robert Brough, an artist, who died in Sheffield Royal Hospital on Saturday 21st. At his separate inquest the jury simply returned a verdict that he had died from injuries sustained in the accident.[319]

A rear collision at Clifton, Pennsylvania, United States, on the Pittsburgh Fort Wayne and Chicago Railroad a few miles from Pittsburgh, on Friday 3rd March 1905, involved trains carrying members of the Ohio National Guard and of the Tippecanoe Club of Cleveland, a Republican Party political club. The first train carried Companies A, B, C and D of the Engineer corps of the National Guard under the command of Major J. R. McQuigg. The second train made up of six Pullman cars, a day coach and a baggage car, was on its way to Washington, D.C. with 200 club members. The first train consisted of day coaches. There had been a problem for several hours with a 'hot box' on one of the cars of the first train, and a stop had been made at Rochester, about 15 miles short of Clifton to attend top it. It was hoped that following this stop the train would be able to struggle as far as Pittsburgh, but beyond Clifton the 'hot box' began to play up again and the train was brought to a stand. A flagman went back to provide rear protection but before he could get far enough the Tippecanoe train came rushing along. Engineer Andrew Long saw the rear lights of the first train, whistled, applied the brakes and reversed his engine. The rails were slippery, and the train could not be stopped in time. With locked wheels the locomotive ploughed right through the rear car of the first train and part way through the next one. The collision exploded the gas tank of the Pintsch train lighting, leaving the cars momentarily in darkness, though only for a moment. Then the wreckage caught fire, lighting up the whole valley and alerting residents of Clifton and Emsworth who had not heard the collision. Members of the train crew hurried forward to begin the rescue work and were joined by Tippecanoe Club members who had escaped with a shaking, though some of the occupants of the special train had suffered broken limbs in the collision. Doctors from Emsworth and area treated the injured until the arrival of a relief train an hour after the accident.

According to a railroad employee the first train only a few yards in advance of a red signal and thus had minimal protection from the fixed signals. There was also talk that the flagman had not done his job properly. Frank H. Delhi, leader of the drum corps of

---

[316] Ibid.
[317] *The Times*, London, England, 20th and 27th January 1905
[318] *The Standard*, London, England, 21st January 1905
[319] *The Times*, London, England, 24th January and 2nd February 1905

the Tippecanoe Club, related that he had been shaken up but had gone forward immediately after the collision. He said that he had seen a trainman standing at the rear of the first train with a lantern in his hand and had heard him say, "It's my fault, I should have gone back."

The day after the accident it was stated that eight persons had died and nearly 40 were injured. Those killed included Pullman Conductor C. T. Scott and Pullman Porter George Kay. The other dead were probably all members of the National Guard.[320]

Confronted with Frank Delhi's allegation, Flagman Craven of the first train totally denied having made any admission after the wreck and stated that he had not been responsible for the collision.[321] An inquiry into the accident, concluded a few days later, agreed, as it concluded that it had been the fault of Engineer Long, of Allegheny, for running past signals, and he was dismissed. The engineer then allowed himself to be interviewed by a newspaper reporter. "I will never run again on a railroad, as I am past the age when railroads give men charge of engines, so no other railroad is open to me. I have served 33 years and last Friday night I passed a signal light. No matter what the circumstances were, according to railroad rules when an engineer passes a signal light he is discharged from the service. I am very sorry that the accident occurred. I'm sorry for the people who lost their lives and their relatives. I'm also sorry for myself." [322]

The coroner's inquest on those killed the accident was held on 20th March, the number of dead having been corrected to seven. Its jury recommended that Engineer Long be referred to the grand jury on a charge of manslaughter and censured Master Mechanic W. C. A. Henry, Car Inspector Lawrence Riester and Stationmaster John Heagen for not carrying out more carefully the orders for car equipment, especially fire extinguishers. They also recommended that 'better discipline be employed in regard to the block system'.[323]

A disastrous collision occurred on Sunday night, 26th November 1905, at Baker Bridge station, Massachusetts, United States, 18 miles outside of Boston, on the Fitchburg division of the Boston and Maine Railroad. A local train left Boston as usual at 7.16 p.m. and was scheduled to run on the main line as far as South Acton, where it diverged onto the branch. Thirty minutes later it was followed by the Montreal Express. The local train stopped at various intermediate points while the express ran fast to Concord, the next station beyond Baker Bridge, and so normally gradually overhauled the accommodation train, but the trains were still timetabled to be five minutes apart at Lincoln (the station before Baker Bridge), Concord and Concord Junction (between Concord and South Acton). In what might be regarded as typical American practice even at this relatively late date the line was still worked on the time interval system. Staff at intermediate stations and other manned locations were required to display a green flag or light to a train encroaching upon the minimum interval of five minutes for a passenger train or ten minutes for a freight train, while the crew of a train which was running late and encroaching upon the time of a following one were ordered to frequently throw off

---

[320] *Steubenville Herald-Star*, Steubenville, Ohio, 4th March 1905
[321] *The Evening Independent*, Massillon, Ohio, 6th March 1905
[322] *The Altoona Mirror*, Altoona, Pennsylvania, United States, 9th March 1905
[323] *The Evening Independent*, Massillon, Ohio, United States, 21st March 1905

red fusees, the lights of which could be seen from up to a mile away. The engineer of a train seeing a green light exhibited at a station or other point or a burning fusee on the track was obliged to slow up or stop his train so as to avoid overtaking the train in front. These obligations were all set out in company Rule 26.

On the evening in question the Marlborough train was running a little late, but there was evidence that its rear brakeman had dropped off a red fusee near Silver Hill station, a green light had been displayed at Lincoln Great Road Crossing, a second fusee was dropped between that point and Lincoln station, a green light was shown at Lincoln station, and a third fusee was dropped off approaching Baker Bridge station. Notwithstanding these five warnings the express ran into the accommodation train just beyond Baker Bridge, telescoping the rear cars of the local train, which then caught fire, and killing 15 passengers, Fireman Charles Barnard of the express and Brakeman Donat Gauthier, who had thrown down three fusees in vain. Forty or so other passengers were injured, some seriously.[324]

Boston and Maine Railroad President Lucius Tuttle told a newspaper reporter the day after the accident that he would be surprised if Engineer Horace Lyons were not brought before a grand jury to face a manslaughter charge, for by the engineer's own admission he was criminally responsible for the disaster. Questioned by Superintendent A. S. Cheever and Assistant Superintendent E. A. Smith after the collision the engineer reportedly said, "I saw the green light which warned me of the train ahead. I cannot explain why I did not obey the rules." [325] The engineer was said to have suffered a complete mental collapse after the accident.[326]

The inquest on the victims of the accident, at Concord on 29th November, was held by Judge Keyes in private despite the protests of both the railroad company and newspapermen. Thirty-one witnesses appeared including Engineer Charles Perritt, Fireman Earl Boardman, Assistant Conductor James Smith and Brakeman William Jones of the Marlborough local train; Engineer Fred Wade on the train engine and Conductor William Fletcher of the Montreal express and Stationmaster Thomas McHugh at Lincoln (Baker Bridge station was an unstaffed flag station). At the conclusion of the inquest Engineer Lyons was, as expected, committed to appear before the grand jury on a charge of manslaughter and being unable to pay the $10,000 bail was lodged in gaol.[327]

The accident was also investigated by the Massachusetts Railroad Commissioners in December and it emerged early on that Engineer Lyons had worked long hours driving the engines of local freight trains on the days preceding the that of the accident before being put in charge of the express as a spare driver.[328] Lyons did not appear at the inquiry at that point, but he testified on its last day, when he stated that most of his experience as an engineer was on shifting engines working in yards, but he had done five days as a freight train engineer and one day working a passenger train. He had previously worked two and a half years as a fireman. He saw the fusee and light at Lincoln and shut off steam and did not afterwards reapply it. He put his speed at 25 miles an hour there. Between Lincoln and Baker Bridge he had used the air brakes and had reduced the speed

[324] *Boston Daily Globe*, Boston, Massachusetts, United States, 27th November 1905
[325] *Boston Post*, Boston, Massachusetts, United States, 28th November 1905
[326] *Boston Daily Globe*, Boston, Massachusetts, United States, 28th November 1905
[327] Ibid, 30th November 1905
[328] *Boston Post*, Boston, Massachusetts, United States, 12th December 1905

of his train to 20 miles an hour. He had not seen the local train in front until his engine was within 100 feet of it, when he had applied the brakes in emergency and sat in his seat the impact. He had had a headache that afternoon (a migraine perhaps) but had not mentioned it to anyone; he had been confident of his ability to drive the express.[329]

No report of a trial of Engineer Lyons has been found and it seems likely that the manslaughter case against him was not pursued in view of his mental condition following the disaster and the view of the inquest jury that part of the blame for the accident lay with the company for putting such an inexperienced engineer in charge of an express, with an even less experienced fireman, as well as for not having installed a more sophisticated signalling system on a main line.

The Boston and Maine Railroad suffered another rear collision on Thursday 4th October 1906, at Lansingburg, New York, United States, and the accident was again accompanied by fatalities, though not as many as at Baker Bridge the previous year. This time the accident involved a passenger train and a troop train, which the newspapers described as a military special. The passenger train, which was due in Troy at 3.18 p.m., consisted of a baggage car, a smoker, a day coach and two Pullman parlor cars. As it was running an hour late it was stopped at Lansingburg depot waiting to cross the 4.20 p.m. departure from Troy when the following military train ran into it. The latter train was carrying 350 men and 400 horses of the 14th United States Cavalry from Fort Ethan Allen to Newport News, Virginia, from where they were to sail to Cuba, where the United States was about to intervene in a disputed presidential election in that country to protect American interests.[330]

Lansingburg station stood on a heavy grade and sharp curve. The troop train, which included 11 horse cars and seven Pullmans, reportedly appeared suddenly without warning and striking the rear of the regular train smashed the two Pullman cars 'like eggshells', and they were swept from the track and rolled down an embankment. The engine of the special kept the track for a dozen yards and then turned turtle, while the leading vehicle of the troop train was telescoped into its tender. The horse cars were marshalled at the front of the troop train and the soldiers were unhurt and soon assisting with the rescue operation. Many of the passengers on the regular train had left the train before the collision and were pacing up and down the track, doubtless annoyed at the delay, but five were killed outright and about 20 injured. Among those killed was half of a newly-wed couple. 'One man's grief was pitiable. His name was J. W. Dacey and he ran up and down the track crying for his wife. In a few minutes her lifeless form was identified. They had been married Wednesday night at Arlington, Mass., and were on their honeymoon, having planned an extensive trip.'

After the collision Engineer Thomas Holloran surrendered himself at the Lansingburg police station, was locked up and would make no statement. His fireman, supposedly one F. A. Brown, had disappeared. Meanwhile 'Lieutenant Colonel Hardie, in command of the cavalrymen, had established martial law in the vicinity of the wreck and the troopers aided local police in keeping order. Picket lines were established and maintained during the night.' Two injured horses from the telescoped leading horse box were shot. The

---

329 *Boston Daily Globe*, Boston, Massachusetts, United States, 28th December 1905
330 Wikipedia – Second Occupation of Cuba

injured passengers were taken to the nearby Leonard hospital.[331]

The accident was inquired into on Saturday 6th by the Massachusetts Railroad Commissioners. Boston and Mayne Assistant Superintendent M. B. Snyder blamed the accident which had killed six passengers on disobedience of orders. He stated that an order had been issued for the soldier's train to precede the express (No. 5, the train which it ran into) but for two excursion trains to precede the troop train from Melrose to Troy. The excursion trains had duly overtaken the troop train at Melrose and Engineer Holloran had been about to follow from the siding when train No. 5 had thundered past at 60 miles an hour.

Conductor Charles Cluoser of train No. 5 testified that the troop train should have preceded his train to Troy and Conductor Jeremiah Collins of the troop train agreed that his orders gave his train precedence over train No. 5 from Melrose to Troy. Brakeman Stewart of train No. 5 testified that he had gone back 1,400 feet to flag the troop train and that Engineer Halloran could have seen his light at that point from another 400 feet back. The engineer had whistled to acknowledge his signal.

Meanwhile a seventh passenger, Herbert Mitchell, town clerk and treasurer of Falmouth, Massachusetts, died of his injuries on the Saturday morning. His wife had been one of those killed outright.[332]

The railroad commissioners issued their report on the collision a week or so later. It read, 'We are of the opinion that the responsible officials i.e. the conductors and enginemen of trains No. 5 and No. 1,060 contributed to the cause of the accident by not complying fully with the orders they had received. We are also of the opinion from the testimony given that the rear brakeman of train No. 5 failed in the performance of his plain duty, in that he did not go back a sufficient distance from the rear of his train.

'The place of collision is located where the road curves sharply to the right, in a direction for the trains approaching Lancingburg, around the point of a hill, that would prevent the engineman of train No. 1,060 from seeing the train ahead until within about 700 feet in its rear. The grade is descending, in the direction the trains were running, about 40 feet per mile, and is practically the same for several miles beyond the curve.

'Our conclusion is that train No 1,060, consisting of an engine and twenty-seven cars, was running at a speed of not less than thirty miles an hour and probably considerably in excess of this, and we believe that it was a physical impossibility to stop the train in time to prevent the collision after engineman Holloran saw the flagman and answered his signal.

'The main line of the Fitchburg division of the Boston and Maine Railroad is double tracked. The Troy branch on which the accident occurred is a single track line. This portion of the road is at all times more or less congested owing to the number of train movements. In addition to the regular train movements of the afternoon of Oct 4 over this division there were two special excursion trains and troop train No. 1,060. The Fitchburg division is now operated under what is known as the double order system.

'We recommend that the Boston & Maine Company install a first class automatic block signal system, also that the Troy branch be double tracked. Conditions imperatively

---

331 *The News*, Frederick, Maryland, United States, 5th October 1906

332 *Daily Kennebec Journal*, Augusta, Maine, United States, 8th October 1905

demand that this work commence and be carried to a conclusion as speedily as possible.'[333]

Southern Railway President Samuel Spencer, riding in his private car with a number of guests on a hunting trip, was killed along with six others in a rear collision a mile north of Lawyers, Virginia, United States, early on the morning of Thursday 29th November 1906 (Thanksgiving Day). The party had been en route to Friendship, North Carolina. Three of the others who died were Spencer's fellow guests, Philip Schuyler, a retired New York capitalist; Francis Redwood, of Baltimore and Charles Fisher of Baltimore. Of President Spencer's guests only E. A. Merrill, his private secretary, survived. The others who died were D. W. Davis, of Alexandria, Virginia, his private dispatcher; porter J. W. Shaw and an unidentified body, presumed to be that of another porter. About eight people were injured, including E. A. Merrill.

President Spencer's private car was attached to the rear of the *Florida Limited*, train No. 33, at Washington D.C. the train then consisting, from the engine, of a postal car, express car, baggage and second-class composite car, a first-class car, two sleeping cars and the President's private car, No. 100.[334] The journey proceeded normally almost as far as Lawyers, where the train became divided due to a defective coupling, the brakes came on automatically, and it came to a stand at 6.10 a.m. The line was protected by a manual block system and though a flagman apparently did go back to protect the stopped train he only went a short distance.

The pressure of holiday traffic meant that train No. 33 was late leaving Washington, and the *Southwest Limited*, train No. 37, was thus close behind it. Train first No. 37 should have been held at Rangoon, the previous telegraph office, but its engineer received clear signals there. By the time that he saw the flagman and the green flags at the rear of train No. 33 it was far too late to stop. President Spencer's car was broken in two and immediately caught fire. The accident would have been ever worse had it occurred a couple of miles further on, for it happened at the crest of a hill where the train's speed was reduced to 30 miles an hour, a little further on it would have risen to over 60 miles an hour.

As is often the case in disasters there were incidents of tragedy, heroism and at the same time inhumanity. Thus, 'Dispatcher Davis was alive but fatally injured when he was taken from under the wreckage. He asked that word be sent to his wife and child. The death of Mr Davis was touching. He stated to his rescuer that he knew that he was dying and knew that the end could not be far off. "Place your finger on my mouth," he said, "it feels so cool and good." He pleaded with his rescuer, who was also a passenger on the train, not to leave him, and for ten minutes this man stayed with him until he saw that nothing could be done.'

Others' actions were less worthy. 'One of the worst phases of the accident was the heartless manner in which a few passengers and some of the attaches, especially some of the porters of the Atlanta train, ransacked cars for plunder. F. M. Curtis, of Jamestown, saw one negro porter go through a lady's grip. He saw him throw away such things as were of no value to him and appropriate those things that he wanted. Mr Curtis declared

[333] *The Daily Independent and Enterprise*, Bath, Maine, United States, 17th October 1905
[334] *The Evening News*, Roanoke, Virginia, United States, 18th December 1906

that he would have certainly killed the porter if he had had something to do it with. Mr Curtis, who was on his way to High Point, N.C. on a business mission, was the hero of the hour. It was claimed by some of his fellow passengers that to his work and generalship belongs the credit of the rescue of a dozen persons who would have perished but for his efforts.' [335]

Responsibility for causing the collision clearly rested between the Operator Jacobs at Lawyers and G. D. Mattox, the operator at Rangoon. By Monday 3rd December the Southern Railway officials were of the opinion that the blame lay with Operator Mattox at Rangoon, but he denied it. According to him Lawyers had accepted train No. 33 at 6 a.m., at a moment when both his fellow operator, named Clemmer, and a student operator were in the office. The train had passed him at about 6.06 and he had reported its passing to Lawyers. He could not recall if Operator Jacobs at Lawyers had acknowledged his report, but it was his duty to do so. Train first No. 37 was then reported as passing Wilmers, the telegraph office in the rear of Rangoon, and he began calling Lawyers. As train first No. 37 came into view he got through to Lawyers and the operator there accepted the second train at 6.12. As train first No. 37 was passing him at 6.14 he was talking to the Lawyers operator who wanted to know what time No. 33 had passed him. He told the man at Lawyers the time written on his (the Rangoon) block sheet and then asked Lawyers, "What time you?"

Lawyers had simply replied, "By," but did not give a time for the passing of train No. 33 and had then opened his key so he could not be contacted. Train first No. 37 was by then on its way to Lawyers. He had Clemmer and the student operator as witnesses.[336]

The accident was therefore a classic example of the sort of situation which resulted in a manual block failure, as illustrated by numerous examples over the years in Great Britain, where manual block working was much more widespread. In the usual scenario a train stopped out of course in mid-section for some reason; perhaps the engine broke down, or a passenger pulled the communication cord, or the train ran into some animals on the line, or whatever. Then, when the usual time for the train to pass through the section had elapsed either the signalman at the box in advance assumed that it had passed unnoticed and cleared it out, or, less commonly, the signalman at the box in the rear wrongly assumed that it had been cleared out by the man in advance, and in either case a following train was allowed to enter the occupied block section.

The investigation into the accident was held in Washington and concluded on 17th December, when a statement was issued that its decision was that Operator Mattox was indeed to blame and had violated three separate rules. Rule 118 stated that when a train entered a block section the signalman must send 'OS {on section} train No ...' and the time to the office in advance, while when a train had passed 200 feet beyond a home block signal he must give 'BC {block clear} train...' and the time to the signalman in the rear. Operator Jacobs at Lawyers emphatically denied that he had given the BC message for train No. 33 as Mattox stated or in fact that he had even received the OS signal from Mattox for it. Mattox admitted that he had not received a time from Lawyers for the BC message, but said that Lawyers had informed him that train No. 33 was by and then opened his key so that Mattox could not communicate with him. In these circumstances,

335 *The Morning Herald*, Hagerstown, Maryland, 1st December 1906
336 *The Washington Post*, Washington, D.C., United States, 3rd December 1906

under Rule 134, relating to a failure of communication, Mattox should have stopped train No. 37 and obtained permission from the despatcher before allowing it to proceed. Finally according to Rule 116 Mattox should have examined his block record and if the block was clear given 'BW {block wanted} for train…' His block sheet did not give this information as there was no time for train No. 33 passing Lawyers entered, as Mattox admitted.[337]

Abnormally severe weather conditions and possibly driver intoxication were the causes of a disaster at Elliot Junction, Scotland, on the Dundee and Arbroath Joint line, which occurred on Friday 28th December 1906; coincidentally the 27th anniversary of the fall of the Tay Bridge, not many miles away.

Heavy snow had fallen in eastern Scotland on the two days prior to the 28th and a blizzard was raging at the time of the accident. The severe weather had seriously interfered with the working of the joint line, which was the property of the Caledonian and North British Railway Companies and formed an important link between Aberdeen, Dundee and the south, carrying a heavy passenger and goods traffic. By 7 p.m. on December 27th heavy snow falling on the telegraph wires had brought down many telegraph poles between Arbroath and Elliot Junction stations and caused a breakdown of all communication between the two places. Later that night the same thing happened between Elliot Junction and Easthaven, the next station to the south. Overnight three North British passenger trains were snowed up between Arbroath and Montrose and early on the morning of 28th the Caledonian line north of Arbroath was also blocked by snow drifts, but down passenger and goods trains continued to be despatched from Dundee to Arbroath in the hope that the Caledonian line could be forced open and North British trains reach Aberdeen via Guthrie and Kinnaber Junction.

Matters were worsened between 7 and 8 a.m. on 28th when a southbound double-headed 41-waggon North British goods train which had reached Arbroath from Aberdeen via the Caledonian route ran into a snow drift at Dowrie Siding, a mile south of Elliot Junction on the up line to Dundee. The train became divided into three portions and the front of the train continued some way towards Easthaven before it was discovered what had happened. Driver James Boyd on the leading engine then proceeded to Easthaven with the front of the train and there he and Signalman James Strachan decided that the best plan of action was for Boyd to return northwards over the down line to Elliot Junction, cross over there to the up line and come back towards Easthaven behind the detached waggons, propelling them to Easthaven through the snow drifts. It was found upon arrival at Elliot Junction that the mains crossover there was unusable due to frozen points and the engine had to go on to Arbroath to cross over. Returning to the breakaway waggons, the engine began to propel them through the drifts but rather predictably some of them soon became derailed. Guard John Morris of the goods train then struggled back to Elliot Junction to get permission for the un-derailed waggons to be drawn back to there. This was obtained and the engine began to draw them back when it ran into another snow drift and its tender was derailed in turn. The up line was now completely blocked. The time was just after 10 a.m. Efforts were then made to re-rail the engine with ramps, but it was found that lifting gear was required. It was at this point that it was decided to

[337] *The Sun*, Baltimore, Maryland United States, 18th December 1906

open up single line working over the down line. Traffic Inspector Charles Souter walked to Easthaven, arriving there at 1.30 p.m., appointed himself pilotman, got Signalman Strachan to sign the necessary forms and left Easthaven for Elliot Junction at 1.40. Back at Elliot Junction box at 1.58 the single line working forms were completed at that end. An up passenger train from Arbroath which had been waiting three hours at Elliot Junction was despatched over the down line to Easthaven at 2.19 p.m., followed by a light engine for Dundee whose crew had been on duty since 4 a.m. with Traffic Inspector Souter on board. This engine reached Easthaven at 2.35. There was now a gap in the traffic as a down passenger train became stuck in the snow at Easthaven's down distant at 3.01 p.m. and could not move until an engine had come from Carnoustie to assist it. The train eventually reached Easthaven at 4.8 p.m. and Traffic Inspector Souter boarded it to pilot it to Elliot Junction.

Meanwhile back at Arbroath at 3 p.m. standing partly alongside the up platform was a local train in charge of Guard James Anderson with Driver John Ogilvie and Fireman Harry Henson. The engine and four leading vehicles were off the south (Elliot Junction) end of the platform and the next vehicle was under the Keptie Street bridge, where the platform was very narrow. Behind this train was one which had left Edinburgh at 7.35 a.m. for Aberdeen but because of the blockage had managed to get no further than Arbroath. The train was in charge of Guard James Kinnear with Driver George Gourlay and Fireman Robert Irvine. It had arrived at Arbroath at 10.41.

The local train left Arbroath for Dundee (East) at 3.13 p.m. with about 50 passengers, most of whom had got into the four rear coaches, the only ones easily accessible. It was stopped at Arbroath South box, where Signalman James Beattie cautioned Driver Ogilvie to proceed with caution as there was a total block failure between his box and Elliot Junction. Driver Ogilvie testified at the subsequent Board of Trade report that there had been a violent gale blowing and snow was both falling and drifting, but both he and Guard Anderson saw the Elliot Junction distant and home signals at danger. Permanent Way Inspector Alexander Hackney, who was riding on the engine, instructed Driver Ogilvie to pass the home signal at danger and draw cautiously into the station, which the driver did, arriving at the platform at Elliot Junction at 3.19 p.m. Because of the block failure Signalman William Haggart in Elliot Junction did not know that the local train was coming; it had not whistled, and he could not see it until it reached the platform. The train could not proceed over the single line of course as Traffic Inspector Souter, the pilotman, was still at Easthaven waiting for the arrival of the down passenger train stuck at the Easthaven down distant signal. After about ten minutes John Carnegie, the Elliot Junction stationmaster, decided to get the passengers out of the local train and have it drawn forward in case another train should arrive from Arbroath. He had just given instructions for this to be done when the accident happened.

As it was obvious that there was no chance of the North British train being able to proceed to Aberdeen Stationmaster John Grant at Arbroath determined to send its passengers back to Dundee and Edinburgh and it was allowed to start for Dundee (Tay Bridge) at 3.28 p.m., 15 minutes after the local train. The stationmaster told Driver Gourlay to stop at all stations and to run cautiously and the driver had replied, "All right."

Signalman Beattie at Arbroath South said that he had kept his signals at danger, flagged the Edinburgh train past his up home, stropped it dead at his box and told Driver Gourlay that there was no communication to Elliot Junction and to, "Look after himself," in the same manner as he had cautioned the local train. Driver Gourlay asked how long the local train had gone and was told ten minutes. He then authorised the driver to pass the starting signal at danger. He had not been aware that there was single line working beyond Elliot Junction and so could not tell the driver of it. Driver Gourlay had appeared to be perfectly fit for duty.

Driver Gourlay told the inquiry that he had not known that they were to return to Edinburgh until just before 3.26 when he had run around his train to bring his engine to the Dundee end of the train. They had started away at (according to him) 3.26. Guard Kinnear had told him that his first stop for passengers was Dundee and the agent (stationmaster) had told him that he must stop at Elliot Junction as there was single-line working beyond there. He had almost stopped at Arbroath South box; he could not remember if the signalman had come out of his box (Signalman Bettie said that he had gone to the foot of the box steps for both trains). He could not hear what the signalman said above the noise of his engine. Driver Gourlay said that as he left Arbroath there had been a perfect blizzard with heavy drifting of snow; he had been running tender first with a big load of coal. It had been still daylight, though with dusk falling, and he had seen the arm of the Elliot Junction up distant signal hanging off. The home signal had also been hanging off, likewise the up starter. He had been running at 12 or 14 miles an hour. There were no fogmen out, and he had not exploded any detonators and he therefore considered that he was justified in continuing, intending to stop at the Easthaven end of the station. With the blizzard and the snow piled up on the tender he had seen nothing of the local train in front. Someone had given him 'a good drink' on the platform at Elliot Junction after the collision.[338]

Driver Gourlay's train collided violently with the rear of the local train, the force of the collision completely destroying its last three carriages. Twenty-one passengers were killed, six of them off-duty railwaymen travelling home. Among those killed was A. W. Black, M.P. for Banffshire.[339] Driver Gourlay's engine (No. 324) and tender were overturned, and Fireman Robert Irvine was trapped beneath it. He could only be released after many hours and died shortly afterwards. Eight passengers were seriously injured; 13 passengers, Driver Gourlay and two guards of his train were slightly hurt.

Major J. W. Pringle inquired into the accident on behalf of the Board of Trade. He noted that Driver Gourlay had been instructed by Stationmaster Grant at Arbroath to stop at all stations to Dundee (Tay Bridge) and to go cautiously. The same instructions had been given to Guard Kinnear, who had repeated them to his driver, as vouched for by Assistant Guard James Hardie and several other witnesses. General Inspector William McLellan, who travelled on the train, twice warned Gourlay to be careful and to keep a good look-out for the train in front, and Gourlay was cautioned yet again at Arbroath South cabin.

Guard Kinnear said that the snow on the windows of his van made it impossible to see any signals between Arbroath and Elliot Junction, though he and Assistant Guard Hardie appeared to have been mainly occupied with parcels and mail anyway.

Inspector McLellan, travelling on the Edinburgh train as a passenger, had been looking out for the signals but could see one of them. He considered the speed over the

---

338 Board of Trade report

339 *The Times*, London, England, 31st December 1906

latter half of the journey to have been too high for the conditions, perhaps 30 miles an hour.

A Mr George Harcourt, who had seen Driver Gourlay's train approaching Elliot Junction from an adjoining field, considered that it had been travelling faster than other trains had during the storm, but no faster than trains did under normal conditions.

Major Pringle stated that it was clear that the Elliot Junction signals had not been cleared by the signalman and that although their arms were drooping a little below the horizontal due to the weight of the frozen snow, perhaps by 10 degrees, their indications were certainly not such as should have been taken as clear, especially as Rule 74 stated that imperfectly displayed signals must be treated as danger signals.

Various explanations, the inspecting officer continued, had been offered to account for the collision. One was that Driver Gourlay had misjudged where he was, but the features of the route, with three overbridges, a cutting giving way to an embankment, a mains crossover and so on made this unlikely, as did the driver's own statement.

Another point raised was that Gourlay's engine had been running tender first. While agreeing that this put enginemen under a serious disadvantage because there was no protection from the weather, the driver's view was obstructed by coal on the tender, dust flying up from the track and the coal prevented such a good look-out and so on, the inspecting officer did not consider that tender running greatly obstructed the view of tall semaphore signals. He did however consider that tender running was resorted to far too frequently on the Joint line. There was a turntable at St Vigean's Junction, just north of Arbroath station, which was large enough to take Driver Gourlay's engine, which had been unusable the previous evening because of the snow, and Driver Gourlay said that he had not asked to try and use it because of this and the fact that the down line was blocked by trains. Major Pringle thought that Gourlay's engine should have been run to St Vigeans Junction over the up line in the wrong direction and an attempt made to turn it. In any case several other engines had run between Arbroath and Elliot Junction tender first without a problem.

The third explanation was the condition of Driver Gourlay. After the accident he had been given three-quarters of a glass of neat brandy as a restorative. A little later a passenger had suggested to Police Sergeant Alexander that Gourlay was drunk. He was then examined by three doctors, Richard Kelly, James Gilruth and John Duncan, all of whom formed the opinion that he was drunk, that his symptoms were not the result of shock or injury and that the brandy that he had been given after the accident would not fully account for his condition. Other witnesses, such as Driver Ogilvie, were emphatic that Gourlay had not been drunk but suffering from shock. There was also evidence that Gourlay had accepted a drink from a member of the public at about noon, between 12 and 12.30 he had bought threepence worth of whisky at the Victoria Bar, adjoining Arbroath station, and at 1.20 p.m. he had been treated to the same amount. At later moments several witnesses stated that he appeared to have been drinking, but Inspector McLellan considered him to be perfectly steady at 3 p.m. The evidence was therefore very conflicting but Major Pringle stated that he was forced to accept that Driver Gourlay's lack of caution was due to alcohol, possibly accentuated by the severe weather. Breathalysers and blood alcohol measurements were of course not available in 1906.

Turning to other considerations Major Pringle considered that in such conditions all the semaphore signals should have been maintained at danger and handsignalling substituted. Where signals were too far away to be seen from the box in these conditions flagmen should have been placed at them with hand signals and detonators. Again according to Rule 78 fogsignalmen should have been appointed by the stationmasters, but they were apparently not used on the Joint line because fogs were rare on the east coast of Forfarshire. Driver Gourlay asserted that not having exploded any detonators he was justified in assuming that the signal was clear but according to Rules 76 (c) and (d) this was only after he had seen the fogsignalman's hand signal or had heard his voice. Guard Anderson, on the other hand, was not obliged to go back and lay detonators behind the local train because it was under the protection of fixed signals.

The inspecting officer was critical of the apparent general disorganisation and lack of communication on the Joint line. No notice was sent from Elliot Junction to Arbroath that the up line was blocked or that single line working had begun. An up passenger train was despatched from Arbroath at 10.45 a.m. although the line had been blocked since 8 a.m. and single-line working was not begun until 2 p.m. Goods trains were sent from Dundee for Montrose and Aberdeen when the line was blocked beyond Arbroath. After the accident the North British headquarters in Edinburgh did not learn of it until 11 p.m. and the steam crane from St Margarets (Edinburgh) did not in consequence reach Elliot Junction until 2 a.m. the next morning. Despite the most severe snowstorm for 15 years not a single extra man had been employed. Snow ploughs had been employed during the night before the accident but with the turntable at St Vigeans Junction out of use had had to return to Dundee with the plough trailing and thus had not properly cleared the up line. Headquarters had not been informed of the derailment at Dowrie Siding and Driver Boyd had not requested any help from the surfacemen (platelayers) to extricate the waggons left behind; while his plan to propel 30 waggons through snow drifts aided only by his fireman and guard was doomed to failure. The delay in opening up single-line working would not have occurred had Inspector Souter had a light engine at his disposal.

Major Pringle noted that in general the system of control and working of the Joint line did not appear to be satisfactory. No lifting gear or snow ploughs were directly available to Joint line staff, but an application had to be made to either the Caledonian or North British Companies, resulting in delays. Again there were no steam cranes or heavy breakdown equipment closer than Edinburgh (North British) or Glasgow (Caledonian). There was little discipline at Arbroath. Various unauthorised persons had been allowed on the footplate of Gourlay's engine and the proximity of the Victoria Bar was undesirable.

In conclusion Major Pringle stated that Driver Gourlay had received repeated and very explicit instructions at Arbroath to drive cautiously and stop at every station. Nevertheless he had disregarded the distant and home signals at Elliot Junction and the speed at which the collision had occurred as shown by the amount of damage caused, together with the position of the regulator and brake handles on his engine, showed that he could not have stopped at the proper place nor had he been prepared to do so. The coal on the tender must have considerably interfered with his view but should not have prevented him from seeing the signals. Driver Gourlay must therefore be considered primarily responsible for the collision. The fact that Gourlay had left his engine several times at Arbroath, had allowed unauthorised persons onto his engine and the evidence at two inquiries led to the reluctant conclusion that Gourlay's conduct must have been at least in part due to alcohol. Precautions taken by the traffic department of the Joint line might, however, have prevented the accident. Firstly a fogsignalman should, given the weather conditions, have been stationed at the Elliot Junction up distant signal in accordance with Rules 78 to 85. The explosion of his detonator would probably have alerted Gourlay. Secondly in view of the breakdown in block working and the snow affecting the signals the home signal should have been fastened at danger and a groundman stationed there with hand signals and detonators.

The inspecting officer went on to consider contributory factors to the collision. These he stated were:

(a) The general lack of initiative and intelligence displayed by the Joint line staff in exceptional circumstances.

(b) The grave error of Stationmaster Grant in allowing the Edinburgh train to leave so closely behind the local.

(c) The failure of Inspector Hackney to bring in extra permanent way staff.

(d) The delay in single-line working due to the lack of a pilot engine for Inspector Souter and the want of lifting gear, jacks and so on to re-rail the goods engine.

(e) The well-meant but ill-advised action of members of the general public in treating railwaymen to intoxicants.

Finally Major Pringle went on to consider several proposals made to prevent another accident:

(1) Speed indicators. These he said were in general use on the Continent and combined with automatic recorders could show the speed of a train at any moment, the length of stoppages, whether drivers had conformed to speed restrictions and so on. There were, he said, obvious advantages in the use of such instruments, but he could not see how such an instrument would have prevented the present collision, given that there were no permanent speed restrictions between Arbroath and Elliot Junction and it was unlikely that the driver would have consulted the recorder to see how far he had run so early in the journey.

(2) Another suggestion was to have a third man on the footplate just to observe speed restrictions and signals, thus entirely relieving the guard of that duty. The inspecting officer noted that in the United States and Canada a third man was employed on the largest locomotives where the work of firing was too much for one man. In Great Britain a travelling inspector might travel on the footplate for disciplinary or other purposes or a pilotman if the driver did not know the road, but these were special circumstances. To employ a third man all the time posed many difficulties, for example it would be expensive, there would be little room for a third person in the cab and unless he had complete control of the regulator and brakes, he could not be held responsible for safety. If he had then the driver's job would become practically a sinecure.

(3) Automatic devices to control trains as used already on the London underground and tube lines were likely to be much more useful than a third man in the cab. Fogging machines at signals which placed detonator on the rail when the arm was at danger were another idea but gave no safety indication. Another arrangement whereby a bell in the cab indicated a clear signal and a whistle a danger signal (clearly a reference to the Great Western Railway's pioneering automatic warning system, development of which had

begun the previous year) had not stood the test of time long enough to be regarded as wholly reliable.

(4) Larger telegraph poles had been suggested to carry the telegraph wires, but these would present a greater surface area to the wind and in any case much of the problem had been the heavy load of frozen snow on the wires rather than the failure of the poles themselves. Immunity could only be guaranteed by putting the wires under cover or underground but then there would be problems with locating faults unless cabling were employed, and he considered that the cost of this could only be justified in exceptional circumstances.[340]

At the conclusion of the public inquiry into the accident on 18th January 1907 the jury found Driver Gourlay principally to blame. Echoing the findings of Major Pringle they also found that given the weather conditions the Elliot Junction stationmaster should have had fogsignalmen out, that the Arbroath stationmaster should not have despatched the Edinburgh train so soon after the local, and that the lack of appliances to keep the line clear and the want of proper supervision to see that the rules were obeyed had contributed to the accident.[341] Driver Gourlay had already been arrested on a charge of having caused the accident on 1st January.

Driver Gourlay stood trial at Edinburgh on a charge of culpable negligence in March 1907 and the jury found him guilty after 50 minutes by a majority of ten to five, recommending him to the utmost leniency on the grounds of his previously unblemished character, the exceptional weather conditions and the lax state of affairs at Elliot Junction. Lod Justice Clark sentenced him to five months' imprisonment.[342] A petition containing reportedly 70,000 signatures was subsequently sent to the Secretary of Scotland calling for Gourley's release, and though he was not freed his sentence was in response reduced to one of three months in gaol.[343]

Three days after the Elliot Junction disaster, on Sunday 30th December 1906, the United States suffered a worse accident in its capital city at Terra Cotta, D.C. A Baltimore and Ohio Railroad local passenger working, train No. 66 from Frederick, Maryland, to Washington Union, D.C., consisting of an engine and three coaches, had just started from Terra Cotta station when it was run into by an engine and six empty coaches. The three cars of the local train were crowded, with passengers standing in the aisles, and the empty stock train telescoped the two rear cars. It was initially reported that 38 passengers had been killed and about 80 injured. The collision occurred at about 6.35 p.m.

The first people on the scene after the collision were residents of Terra Cotta, described at the time as a sparsely settled village. Physicians arrived after three quarters of an hour and wrecking trains were sent from Washington, and from Baltimore, Maryland, together with police and the fire department. 'Many deeds of heroism were

[340] Board of Trade report
[341] *The Times* 1st, 2nd, 3rd, 4th, 5th, 7th, 11th, 12th, 14th, 15th, 16th, 17th, 18th and 19th January 1907
[342] Ibid, 21st February and 12th and 13th March 1907
[343] Ibid, 25th March and 1st April 1907

performed as the injured lay moaning from pain. Rev Father A.M. Mark, of St Anthony's Church, hurried to the scene, and was in time to administer the last rites to many of the dying. He worked heroically and his words of comfort cheered many poor souls.'

Despite the destruction the passengers at the front of the train escaped. 'Benjamin T. Maynard, of Fostoria, Alexandria County, Virginia, was one of those who escaped without injury. Until Kensington was reached, Mr Maynard was seated in the second car talking with his cousin, Mrs Catherine Hughes, of 1438 Concoran street, northwest. When the train pulled out of Kensington Mr Maynard left the second car and went ahead to the smoker. "Gross carelessness not only on the part of the freight train which ran into us, but on the part of the entire company, was the cause of this terrible disaster," said Mr Maynard. "I asked the engineer of the freight train why he did not pay attention to the signals, and the man answered, 'I did not see any; besides, I thought your train had reached Washington over an hour ago.' Of course it was gross carelessness on the part of the company to allow two trains in the same block. Accompanied by my cousin I boarded the train at Frederick," he said. "Everything went well until we reached Kensington, where the danger signals were displayed. At Rockville the same signals were in view and the same at Takoma. After we left Takoma it became very misty. Our train went at a pretty lively clip to Terra Cotta. The train following us was well up, and we were just pulling out of the station, when all at once the rear train hit the last car with a terrific smash. The car was knocked to splinters, and the middle coach was telescoped. The smoker being the only car that escaped destruction. There were about fifteen persons in our car, and every one of them escaped uninjured, with the exception of a few cuts inflicted by flying glass. I immediately left the car and went in quest of my cousin, whom I found lying on the bank a few yards distant. She was cut about the head and her right shoulder was dislocated. I put her into the coach and went around trying to help those who were suffering. I was standing on the bank, about three feet from the train, when I saw a woman toss a baby from a window in the second coach. With another man, who was near me, I hurried to the spot and picked the infant up to find that it had not received a scratch. We took the baby to Mrs Fisher's house, in Brookland, and left it. The whole place around the wreck was one of awful disorder. I saw a woman lying under the forward trucks of the rear car with both legs and head cut off, and three men in pretty nearly the same condition."' [344]

Immediately after the accident all the crew of the colliding train were arrested: Engineer Harry Hillebrand, Fireman Ira McClelland, Conductor Frank Hoffmeier, Brakeman Ralph Rutter and Baggagemaster W. A. Norris were all taken to the tenth precinct police station to be held until after the inquest.

The line was worked on the manual block system, with block posts at Silver Spring, Takoma Park and University Place, Terra Cotta station being between the latter two points. It was alleged that Operator Wilton Dutrow had given Engineer Hillebrand a double green caution signal, while Operator M. W. Phillips at Takoma Park stated that his signal had been at red as train No. 66 had not yet been cleared back by University Place. Engineer B. L. Vermillion of train No. 66 stated that he heard the empty stock approaching and realizing that it was not going to stop had opened the throttle of his engine and started to pull away when the collision took place. Conductor George Neagle of the accommodation train, clearly suffering severely from shock, said that his train had

---

[344] *The Washington Post*, Washington, D.C., United States, 31st December 1906

just been going into Terra Cotta station; it had in fact already stopped there before the accident and picked up four or five passengers. Engineer Hildebrand was driving engine No. 2120, 'one of the largest on the B & O road'. He said that it had been foggy, and he had seen no signal at Takoma Park.[345]

A key point in explaining how the accident had happened was the fact that Takoma Park was only open as a signal station from 6.30 a.m. to 6.30 p.m. At the latter hour its signals were extinguished, the block section then extending from Silver Spring to University Place. The collision had occurred two minutes before it was due to close. There was no doubt that Takoma Park was still open and its signal red but the engineer, not seeing the signal in a thick fog had assumed that the signal station had closed and there was no signal to be seen.

At the inquest into the accident on 3rd January there were questions raised as to whether the late running local should not have been protected by torpedoes, but Trainmaster J. W. Kelly stressed that the block system was absolute.[346]

The next day (4th January) Chief Train Dispatcher Thomas Dent at Camden station, Baltimore, told the inquest that the operator at Takoma Park had advised him at 6.40 p.m. that the empty train had run past the red signal there. There being no operator at Terra Cotta he had then contacted University Place, who said that the accommodation had not reached there and there was no sign of it. Dent was questioned by the Commissioner Clements of the Interstate Commerce Commission as to whether he had not anticipated danger from the faster empty stock catching up the accommodation and whether the empty stock should not have been allowed to run in front of the local, but he persisted that this was not done and that there had been no danger had not the engineer overrun a red signal. Asked if there was any rule for the engineer to look out for a signal at Takoma Park, Dent said there was not, but it was nevertheless the duty of the engineer to do so.

Engineer Vermillion, Conductor Neagle and Fireman J. Crawford of train No. 66 were then called, followed by Trainmaster Kelly again, who stated that a double green at Silver Spring would indicate that there was a train in the block, with permission to use the crossover at Takoma Park, protected by a flagman. A double green at Silver Spring for the empty train would have been an incorrect signal. If no signal were displayed at Takoma Park the engineer should have been able to see if the signal station was open or closed by the position of the block target.

Engineer Hildebrand was then called. He had received order No. 100 at Washington Junction directing engine 2120 to run extra to Locust Point via "Y" Washington, and that extra 1865, the Frederick local and extra 2120 were to follow one another carefully under green signals. He had seen a double green at Silver Spring at close range. It had been 6.31 by his watch there and Takoma Park, he had assumed, would be closed, as it shut at 6.30. He saw no signal at Takoma Park but whistled for the station and a crossing. The local train he had seen nothing of before he had run into it at 20 or 25 miles an hour. There should have been a flagman posted at Stotts, half a mile short of Takoma Park. It emerged that he had only had eight hours sleep in the 48 hours previous to the accident, but he said that he had not been sleepy. Questioned further, Hildebrand admitted that his service record was not that good; he had been suspended three times, for allowing his

---

345 Ibid.
346 Ibid, 4th January 1907

boiler to run short of water, for failing to stop after exploding a tornado and for running through a set of points. He also admitted to sometimes taking a drink while on duty but denied having done so before the collision.

Hildebrand's statement that he had received a double green signal at Silver Spring was disputed by Operator Dutrow who was positive that a white signal had been displayed. He said that the empty stock had passed him at 6.28, not 6.31 as asserted by Hildebrand. The latter stated that his speed through Takoma Park had been 25 miles an hour, Operator Phillips there and other witnesses put it at 50 miles an hour. Operator Phillips' statement that his red target was displayed at Takoma Park was verified by several other witnesses, though it was only visible at short range through the fog. Phillips testified that when the empty stock had run past his signal, he had informed University Place, '2120 by my red target, going like hell.' He had expected the empty train to stop and had not put down torpedoes, even though there was a thick fog. Because of the down grade trains ran past his signal several times a week, but they always came back. Phillips' testimony was corroborated by Henry Page, a former telegraph operator who had been in the office that night. It was also explained that the normal position of the block signals was red, unless they were held at white by the operator working a lever inside the station.

Samuel Queen, a teamster, a policeman, J.T. Kennedy, and Takoma Park gateman Frederick Lempke all testified that the Takoma Park signal had been red when the empty stock passed it. Calvin Ramsberg, operator at University Place, corroborated the messages Phillips said he had sent and J. G. Trumble confirmed that the Silver Spring had been white, as stated by Operator Dutrow, and not double green.[347]

When the inquest was resumed on Tuesday 8th January evidence was given of a narrowly avoided second rear collision later on the night of the accident due to a second train being admitted to the Kensington-Silver Spring section before the first had cleared. Operator Dutrow said that he had not cleared back the first train.

Operator W. F. Matthews at Randolph, Maryland, testified that he had closed his office 18 minutes early at 6.12 that Sunday night as he had not been expecting another train (probably so he could go home a bit earlier after a long day, or, as he told the inquiry, because he was anxious to go to church). Another train did appear, though, the empty stock. According to the rules with no signal displayed at Randolph, Engineer Hildebrand should have stopped there to investigate, but he did not.

Samuel Styles said that he had been at Stotts station and had flagged train No. 66 so as to put on a little girl. The accommodation train had left at 6.30 by his watch and the equipment train had gone by only 30 seconds later (this would seem to be inaccurate as the local had had time to run to the short distance to Terra Cotta, stop, pick up several passengers and just restart from there before the collision). He and fellow passengers estimated the speed of the equipment train as 50 miles an hour and he considered himself a good judge of speed. Various other witnesses also gave evidence.[348]

On 9th January the coroner's jury found Operator Dutrow rather than Engineer Hildebrand directly responsible for the collision, and Engineers Hildebrand and Vermillion, Conductors Hoffmier and Neagle, Assistant Dispatcher W. E. McCauley, Chief Dispatcher Dent and Trainmaster Kelley indirectly responsible.

347 Ibid, 5th January 1907

348 Ibid, 9th January 1907

The jury found that the Baltimore and Ohio's manual block system did not afford sufficient protection. They recommended that operators and signalmen should be paid higher wages to ensure the employment of competent men, that auxiliary stations (presumably meaning ones not open continuously) should be discontinued, that operators should not have other duties, that timepieces should be more rigidly inspected and there should be a compulsory rule for the comparison of enginemen's watches.[349]

The day after the report of the jury's verdict there was a piece published regarding several important points brought up by the *Railroad Gazette* but not apparently considered by the jury. One obvious one, to those familiar with British practice, was the use of stop signals without associated distant signals. Another was the lack of supervision of the long hours worked by the engineers, as mentioned earlier Hildebrand had worked 48 hours with only eight hours' rest (two four-hour intervals) before the collision. Another was the extinguishing of signal lights when the block stations were closed. Some railroads such as the Chicago and Alton, the Chicago Milwaukee and St Paul and the Seaboard Air Line, kept them lit, in accordance with the usual British practice.[350]

In the end, however, it was Engineer Hildebrand, Fireman McClelland, Conductor Hoffmeier and Brakeman Rutter, that is the crew of the equipment train, who stood trial for manslaughter. On 23rd December 1907 the jury, faced with evidence as conflicting as at the inquest, acquitted them all after three hours' deliberation.[351]

The Interstate Commerce Commission report on the accident was issued in January 1908. It found that 'This wreck was caused by disregard of rules and signals upon the part of the men in charge of the extra train.' And stated that, 'Many references were made to incidents or instances of more or less important infections of rules by employees and of absence of rigorous discipline by the company. This wreck was cause by disregard of rules and signals upon {the} part of the men in charge of the extra train. Inquiry did not disclose act or omission on part of other employees or defect in the operating rules of the company which contributed to this awful occurrence.'[352]

A mistake by a signalman was the apparent cause of an accident at Acqua Bella, near Milan, Italy, on Monday 20th January 1908. An omnibus train from Milan to Treviglio and Bergamo (another report stated for Pavia) was stopped by signals not long after starting its journey. It waited at the signal for 20 minutes before the latter was cleared and was just on the point of restarting when it was run into by the 9 p.m. express from Milan to Rome, running at full speed under clear signals. The rear carriages of the local train were smashed, and the locomotive of the express and the rear carriages of the front train were thrown foul of the opposite running line. Shortly afterwards a Genoa-Milan arrived at the and ran into the wreckage fouling its path. Seven passengers were killed outright and 24 injured, most of them from the Bergamo train, and many who were run down by the Milan train after alighting from their own following the first collision. Travelling in the Rome express were Signor Carcano, Minister of the Treasury, and a number of Senators returning to Rome for re-opening of the Nasi trial (Munzio Nasi, a former

[349] Ibid, 10th January 1907
[350] Ibid, 11th January 1907
[351] *The News*, Frederick, Maryland, United States, 24th December 1907
[352] *The Washington Post*, Washington, D.C., United States, 7th January 1908

minister, had been found guilty of embezzlement the previous year); none of them was hurt. The Senators 'remained a long at the scene of the disaster, assisting the authorities in organizing the work of rescue and clearing the lines'. In addition to those killed and injured one of the engine drivers was reportedly missing.

Two priests and some pupils from a nearby religious institution arrived on the scene soon after the accident and reportedly rendered great service in rescuing many of the injured from the wreck and allaying panic among the survivors.

There was a dense fog at the time of the accident and it would seem that the signalman had forgotten all about the local train, which he could not see because of the fog, and the driver of the train had failed to remind the signalman that he was there. After the accident the signalman had fled, and a warrant was issued for his arrest.[353]

Following the accident Signor Bianchi, general manager of the State Railways, held a conference with Signor Bertolini, and they agreed to open an immediate inquiry into the cause of the accident. Signor Alzona, assistant general manager of the State Railways, thereby left Rome for Milan to open the inquiry.[354]

A later report spoke of the removal of nine bodies from the wreckage.[355]

On the evening of Easter Monday, 20th April 1908, telephone messages began being received at Melbourne, Victoria, Australia, of an appalling railway accident at Sunshine (until shortly before known as Braybrook Junction). A train from Ballarat, due in Melbourne Spencer-Street at 10.27 p.m., had been standing at the platform at Sunshine when it was run into by one from Bendigo, due in Melbourne at 10.55. As soon as news of the accident reached Melbourne a relief train with Drs Buchanan and Pearle, 20 ambulancemen and all the necessary appliances was set up and started off at about midnight. More men were picked up at Footscray and the train reached Sunshine at 12.50 a.m., two hours after the accident had occurred.

'The sight there (at Braybrook Junction, as the newspaper still referred to the location) was something pitiable. The Ballarat train, crowded with passengers, stood on the station, but the back end of it was completely smashed to pieces. It appears that owing to some reason yet to be explained, the train from Bendigo, drawn by two engines, smashed into the Ballarat train while at Braybrook Junction station. Two first-class carriages and a guard's van were smashed into a conglomerate, the wreckage being strewn all over the line. From a hasty glance it appears that at least the bodies of two carriages, very much smashed up, are overlying the undercarriages of two others, the sides of which are burst all over the line in the form of wreckage. The impact was so great that it shifted the roofs off two of the carriages, and these roofs form an obstruction across both lines and overlap onto the platform opposite. The sight on the platform is appalling, people lying all over the platform with hideous wounds on them.'

A passenger on the Ballarat train told his story. 'Norman Hodges, a young man, about twenty years of age, living at Yarra-street, South Yarra, said that he went up to Ballarat on Friday, and was returning yesterday. He was in a carriage a short distance from the

353 *The Times*, London, England, 22nd January 1908

354 *The Standard*, London, England, 22nd January 1908

355 *The Monitor and New Era*, London, England, 25th January 1908

end of the train, and luckily escaped injury. "I was asleep at the time of the collision," he said, "and was awakened by the violent impact. It was some time before I could realise what had happened. The carriage in which I was seated was filled with people, and the utmost consternation prevailed. A number of the women in the screamed and there was a general rush to see what was the matter. Looking around, I saw a terrible scene. I can hardly describe it. The train was in was just moving out of the railway station, and the Bendigo train had crashed into us. Two carriages appeared to be telescoped. They were hopelessly smashed to pieces, and the unfortunate passengers were struggling amongst the ruins. Some were screaming, others were crying out for help, and a few appeared to be helpless, and were groaning. Added to this was the excitement of people rushing to give assistance whilst others searched the carriages for their friends. The most terrible thing, however, was the blaze of light as the debris caught fire. It was some time before anything like proper help could be obtained, and it was difficult to know whom to assist first. People-some of them seriously hurt, others slightly wounded, and others merely suffering from shock-were lying all across the platform. Some of them were reclining on the seats. Those more unfortunate were still being rescued from the train and given to the care of their friends. All the railway officials gave what attention they could. The fire bell was rung, and the members of the fire brigade were soon in attendance. Then those who were not hurt were got together, and we came on with the train.'[356] The fire which broke out in the wreckage was blamed on the ignition of escaping gas.

Initially it was reported that 20 passengers had died in the accident but by the time that inquest opened on 4th May the number of dead had risen to 44. The final death toll was probably 45.

The inquest proper, as opposed to two preliminary hearings, began on 18th May with the evidence of Guard John Fraser of the Ballarat train. He testified that his train had left Deer-park, the previous station, with the staff (the Ballarat branch was single track) at 10.41 and had arrived at Sunshine at 10.47p.m. The train of eight carriages had been too long for the platform there and having stopped to let the passengers in the front carriages disembark, it had to be drawn forward to enable those in the rear coaches to alight and to unload the luggage van. Guard Fraser related that he had seen the Bendigo train about an engine's length away and had waved his lamp for his driver to pull away but there was no chance of avoiding the collision, which happened at 10.49 by his watch. There had been a red tail light and two red side lights on the van of his train. His train had lost 24 minutes pulling up twice at stations with short platforms and ten minutes at Bacchus Marsh taking on extra carriages.

Guard Fraser was followed by Driver Abraham Monk of the pilot engine and Driver Nathaniel Edgar of the train engine of the 7.15 Ballarat train. Then came Porter Edgar Wallace, of Deer-park, who gave the Ballarat train the electric train staff, followed by one of the key witnesses, Sunshine Stationmaster George Kendall. Stationmaster Kendall testified that he had accepted the Ballarat train from Deer-park at 10.19 and the Bendigo train from Sydenham at 10.31. He had intended to accept the Bendigo train under the "1-5" warning signal but could not recollect if he had actually done so. He had in turn had the Ballarat train accepted from West Footscray, the next station towards Melbourne, and had cleared his home, starting and distant signals for it. As the Ballarat train was being drawn forward, he saw the Bendigo train just inside his home signal, clearly not going to

[356] *The Age*, Melbourne, Victoria, Australia, 21st April 1908

pull up. He had waved a hand signal at the engine of the Bendigo train from the platform, then stood back waiting for the collision, expecting any moment to be dashed to pieces. Afterwards he asked the driver of the Bendigo why he had not stopped and was told that the brakes had failed.

Stationmaster Kendall admitted that he should have accepted the Bendigo train under the "1-5" warning signal instead of under full "Line clear". He said that he must have mechanically accepted the train as he had. He was permitted to accept passenger trains under the warning arrangement, except expresses, and to allow trains from both Ballarat and Bendigo to approach his station at the same time, providing one was cautioned. The trains were usually half an hour apart. He had a staff of two porters and a lad but had to work the signal box at times as well as performing his other duties. He also had to supervise Deer-park station and the Federal Manure Company's siding.[357]

At the resumption of the inquest the next day the service record of Driver Leonard Milburn, in charge of the pilot engine of the Bendigo train, was gone over. It was not a good one, with about 16 cautions, fines and reprimands for offences such as allowing his boiler to run low of water, causing a collision, breaking waggon drawbars and so on. This was followed by the evidence of passengers on the Bendigo train.[358]

The lengthy inquest continued over a number of days with expert evidence as to the brakes of the Bendigo train working normally, the possibility of interference with the Sunshine signals and so on. By 17th June Guard William Darcy of the 6.50 Bendigo train was giving evidence. He said that he thought Driver Milburn had had his train under control and that the home signal was often cleared after the train had passed the distant at danger. Driver Milburn had told him after the collision that he had reversed his engine and that the brakes would not work.

Driver Gilbert Dolman, in charge of the train engine of the Bendigo train, then testified. He said that the train had been travelling at 50 miles an hour at the Sunshine distant, where Driver Milburn had applied the brakes. The speed of the train had been reduced to 40 miles an hour. Then, 200 or 300 yards from the home signal the train had seemed to get away as if the brakes had been released. When he saw the home signal at danger, he attempted to apply the brake but there was no air. His fireman applied the hand brake as did Driver Milburn's fireman.

Driver Dolman's fireman, Barkly Deveney, agreed that the train had seemed to get away after passing the distant signal. He wondered if Driver Milburn had seen the home signal cleared but then saw Driver Dolman reversing his engine and heard Driver Milburn give three short, sharp whistles. He (Fireman Deveney) had applied the hand brake and looking out had seen three red lights and had shouted to Driver Dolman, "There's a train standing at the platform." [359]

Driver Dolman and his fireman gave a little more evidence at the resumed inquest the next day and were followed by Driver Milburn's fireman, William Tomlinson, who confirmed that his driver had whistled and applied the brake at the distant signal. Halfway between the distant and home signals Diver Milburn made another brake application. He (the fireman) realised that the brake application had not worked, and he

[357] *The Argus*, Melbourne, Victoria, Australia, 19th May 1908
[358] Ibid, 20th May 1908
[359] Ibid, 18th June 1908

had shouted, "Look out." He had applied the hand brake and his driver had made an emergency brake application and reversed the engine. Seeing the train in front he had cried, "For God's sake, can't we do anything else?" but Driver Milburn had done all he could. Fireman Tomlinson did not agree with his guard that they had been going as fast as 50 miles an hour at the distant signal.

Driver Milburn then gave evidence. He said that he had had one glass of ale with his dinner at Bendigo, that was all. Concerning the journey, he said that he had whistled 200 yards from the Sunshine distant for the crossing and applied the brake. He had subsequently made a further application but 200 yards from the home signal the train had appeared to run into the engine. He had then made an emergency brake application, but it did not have the desired effect. He had then opened the whistle and reversed the engine. He denied saying, "There is no one to blame but myself," after the accident. Driver Milburn broke down into tears twice while giving his evidence.[360]

Driver Milburn had to testify again at the inquest the following day, when his disciplinary record was gone through with him in detail. Driver Dolman and Fireman Deveney were recalled and there was testimony from experts as to the action of the Westinghouse brake.[361]

The inquest jury finally reached their verdict on 9th July. They found Driver Milburn, Driver Dolman and Stationmaster Kendall all guilty of manslaughter. The case against the stationmaster was later dropped and on 9th October the jury took 4½ hours to find Drivers Dolman and Milburn not guilty.[362]

The irregular acceptance of both trains by Stationmaster Kendall was obviously a factor in the collision but it still would not have happened had the Bendigo train not overrun the signals. Other factors behind the accident were probably that the Bendigo train, like the Ballarat one, was running late, being 20 minutes behind time, and that Driver Milburn had been on duty for over 12 hours. It seems likely that he was running fast to try and make up time and because of his tired condition he had mishandled the Westinghouse brakes when approaching Sunshine. Possibly, also, being unaware that the Ballarat train was also running late, he had been expecting a clear run through the station.

On Thursday 1st October 1908 an excursion train on the Toledo and Ohio Central Railroad, returning from a fair at Bowling Green to Toledo, was stopped at Sugar Ridge, Ohio, United States, about ten miles short of its destination, because of a 'hot-box'. Before it could be got moving again it was run into by a following freight train. Its rear car was telescoped, and two others jammed together. Six passengers lost their lives and others were injured. Those who lost their lives were Richard Radecut and Joe Gasler, of Sugar Ridge; J. F. McCroy, of Toledo; J. G. Boyd, of East Toledo; a person named Stonebridge, of Bunbridge and a sixth unidentified person. A relief train was sent to the scene from Toledo. It was unclear if a flagman had been sent back to protect the rear of the passenger train.[363]

---

[360] Ibid, 19th June 1908
[361] Ibid, 20th June 1908
[362] *The Colac Herald*, Colac, Victoria, Australia, 12th October 1908
[363] *The Evening Tribune*, Albert Lea, Minnesota, United States, 2nd October 1908

One report blamed the collision on the carelessness of the crew of the freight train.[364]

Between Slidell and New Orleans, Louisiana, trains of the New Orleans Great Northern Railroad had trackage rights over the line of the New Orleans and Northeastern Railroad. New Orleans and Northeastern Railroad local train No. 9 from Hattiesburg and other places in Mississippi was due to arrive in New Orleans 20 minutes before New Orleans Great Northern fast train No. 61 from Covington, Louisiana. On the morning of Wednesday 11th November 1908, the Northeastern train was running 20 minutes late.

Engineer Wayne Blackman of the Great Northern train stated later that when his train joined the Northeastern tracks at Slidell he had been given the usual clear signals and he had continued towards New Orleans at 45 miles an hour. As his train approached Little Woods station the rear of the Northeastern local suddenly loomed up out of the fog. He had jammed on the brakes, but his engine had ploughed into the rear car of the local, tearing through the last few cars and causing death and destruction. No one on the Great Northern train was seriously hurt. Engineer Blackman stuck to his post and was unhurt; Fireman McCarty jumped and suffered possibly serious internal injuries. There were no telegraph facilities at Little Woods, there was serious delay in getting a message to New Orleans, and it was two hours before a relief train arrived, though New Orleans was only 12 miles distant. The collision occurred just after 8 a.m.

The wreckage caught fire after the collision, and the initial efforts of the uninjured passengers and the few fishermen and hunters from the camp on Lake Pontchartrain which adjoined the station had to be directed to trying to extinguish the blaze. In this they were partly successful, but at the expense of having little time to devote to the badly injured passengers, several of whom died in the smouldering debris. Rude bandages wound around wounds were however later credited by surgeons with saving the lives of more than one victim. As was often the case the women passengers were especially active, tearing up their underskirts to make bandages. Meanwhile the men had torn cushions from the seats of the remaining carriages and the injured were laid on these cushions and tended by the women until the arrival of the relief train. Only one doctor was initially present, Dr Henry Tarleton, of Covington, who had doubtless been a passenger on the express. He worked heroically in tending the most seriously injured and directing the efforts of others.

'Among the sights that greeted the rescuers was that of a small boy apparently unable to extricate himself from the wreckage, which was fast settling down upon him. A score or more of men made a heroic effort to get the child, but the wreckage and escaping steam from the damaged locomotive cut off their way, and they were forced to stand idly by while the child cried futilely for help until he died. This child proved to be Willie Attaway, three-year-old son of Mrs Alphonse Attaway, of Slidell, La. His little body was taken out of the wreck several hours later, horribly mutilated.'

Who was to blame for the collision, which killed eight passengers and injured many others, some fatally, was disputed by the two railroad companies. Northeastern officials claimed that the express should have been proceeding cautiously after having been warned at Slidell that the local was late. The Great Northern trainmen, on the other hand, claimed that they had received no warning at Slidell and that even if they had done the

[364] *Elyria Evening Telegraph*, Elyria, Ohio, 2nd October 1908

local should have been protected by a flagman while it was stopped at Little Woods station.[365]

A report in another newspaper placed the blame squarely on the crew of the Northeastern local train for failing to protect it despite the fact that they knew the express was due and the foggy conditions. This newspaper carried the stories of several passengers involved in the wreck. 'Jake Salman was seated, engaged in conversation with C. B. Lowry, who was killed, in the coach ahead of the rear coach, when the crash came. Mr Lowry, it seems, was crushed between the engine and the inner side of the coach, but Mr Salman, though struck by the engine and partially crushed by the broken timber, had space enough to permit his not being crushed to death. However, Mr Salman received such severe internal injuries that it is hardly expected that he will live very long. Besides receiving a fracture of several ribs and sustaining other internal injuries, one of his ears was mashed almost completely off. He also sustained a broken collar bone. Much valuable and timely assistance was rendered by Inspector O'Connor, and his valuable assistants set about in seeing to the welfare of the injured, and the manner of procedure in extricating the dead bodies from under the pile of debris.

'A. Lannes, a passenger in the Northeastern train when the crash came, gave the following account of the accident on his arrival at the Terminal station. "Our train was proceeding at a moderate rate of speed," he said. "It could not have been going more than eight or ten miles an hour. Suddenly there was a crash and a roar that sounded like a terrific explosion. The New Orleans and Great Northern had crashed into us from the rear. The next moment the air resounded with the shrieks of the injured. Women were begging to be put out of their misery and panic-stricken men, many of them covered with blood, were pouring out of the train. I hope never to witness such a sight again. It will linger with me for many years, I fear. It was about 8 o'clock or 8.05 when the collision occurred. To my mind the engineer of the New Orleans and Great Northern was in a large measure responsible for the catastrophe, and this seemed to be the opinion of a number of other passengers too. The terrific rate of speed at which that train was running absolutely precluded the possibility of avoiding an accident when the advance train was discovered. I saw seven bodies on the side of the track after the collision and was informed that four more had been killed. As to the number of injured, I cannot possibly say. There seemed to me to be dozens of them. The spot resembled a perfect shambles. Jake Salman, I was informed, had both legs broken and could not possibly recover. He was taken to Slidell."'

Despite the collision allegedly having been due to a failure to protect train No. 9 it was Engineer Blackburn and his conductor, A.B. Keyes, who were arrested at the scene,[366] and at the subsequent inquest the jury likewise found them to have been at fault.[367]

On 28th December 1908 southern Italy and the island of Sicily were devastated by the Messina earthquake which cost between 75,000 and 82,000 lives.[368] In the relief

---

[365] *The Sun*, Baltimore, Maryland, United States, 12th November 1908

[366] *The Biloxi Daily Herald*, Biloxi, Mississippi, United States, 12th November 1908

[367] *Appomattox and Buckingham Times*, Appomattox, Virginia, United States, 18th November 1908

[368] Wikipedia – 1908 Messina earthquake

operation following the catastrophe, on Monday 4th January 1909, a freight train ran into and telescoped a relief train crowded with injured persons, killing five of them, at Nizza di Sicilia on Sicily.[369]

On Monday 22nd February 1909 the United States Navy battle fleet popularly known as the Great White Fleet docked at Hampton Roads, Virginia, after a round-the-world tour begun in December 1907 by order of President Theodore Roosevelt, with the stated aim of making courteous visits to numerous countries, and doubtless to display US naval power.[370] In order to cater for the numerous people wishing to welcome the fleet home the Pennsylvania Railroad southbound *New York and Norfolk Express*, train No. 49, 11.22 p.m. from Philadelphia, Pennsylvania, to Norfolk, Virginia, was run in two sections, the railroad company having also offered tickets for the train at reduced rates. It was the first section, consisting behind the locomotive of an express car, a combined baggage and mail car, three day coaches and six Pullman sleeping cars, which was involved in a bad accident. Both sections of the train were timetabled to change engines at Delmar, Delaware, from where two New York Philadelphia and Norfolk locomotives were waiting to take them forward to Norfolk, Virginia, en route to Old Port Comfort (Hampton Roads). Difficulties reportedly arose because another special train was standing in Delmar station, so that the two locomotives were unable to reach a siding and had instead to stop on the main line behind the special. It was, according to the initial investigation into the accident, a failure to flag the engines which caused it.

Engineer Benjamin Ewing of the first section had already reduced speed, he said, when peering through the fog he suddenly saw the rear lights of the two engines on the main line. He applied the brakes in emergency and he and his fireman, T. Henry Esham, stayed at their posts. All three engines were thrown from the track in the collision and the express and combined mail and baggage cars were telescoped by the heavier day coaches and Pullman cars behind them and then set on fire by coals spilt from engine No. 5074. The old mail car had a double crew of four clerks because of the quantity of mail for the officers and men of the fleet. The occupants of the mail and baggage cars were trapped in the wreckage and burnt to death. In all seven people lost their lives. They were: mail clerks John Wood, of Wilmington, Delaware; William Cochran, of Philadelphia, Pennsylvania; and R. M. Davis, of Marydel, Delaware; mail weigher Harvey Wilhelm, of New Castle, Delaware; express messenger W. Oliver Perry, of Cambridge, Maryland; Engineer George Davis, on his way to Delmar to pick up a freight train to work back to Perryville, Maryland; and Baggagemaster John McCready, of Wilmington, Delaware. An eighth corpse was also found but turned out to be a body being shipped to Norfolk, Virginia. The collision occurred at 2.50 a.m. and the fire was not put out almost 12 noon by firemen from Salisbury and Delmar. Engineer Ewing sustained head and facial injuries and lost several teeth. Fireman Esham was buried by the coal in the tender but crawled out unhurt and began searching for his driver.

Another fatality in the accident was *Princess Trixie*, an educated horse. She had been exhibited at Keith's Theatre, Philadelphia, for the previous two weeks and was on her way to Norfolk. Twenty years old, the horse had been shown all over the world by her

---

369 *Roswell Daily Record*, Roswell, New Mexico, United States, 4th January 1909
370 Wikipedia – Great White Fleet

owner, W. Harrison Barnes, was a favourite with vaudeville audiences and had appeared three times before King Edward VII of England. The horse was insured for $5,000. It was in charge of Louis Brockaway, who was one of only two people injured.

Superintendent A. G Whitney stated, "The wreck was due to the failure of a flagman to warn the approaching express. Ira Simpson, of Delmar, was the engineer of locomotive No. 4, the latter of the two engines standing on the main track. It should have been the place of his fireman to go back and warn the oncoming train with a danger signal. Engineer Ewing, of the fated train, informs me that this was not done. Ewing says it was foggy and that, although the track was perfectly straight, he did not see the first of the two idle engines until it was about 40 yards away. He was running, he said, at about 10 miles an hour, but was unable to stop. His regular stop. His regular stop would have been about 200 yards farther on. It may be that the story of the fireman of Engine No. 4 will show that he made some attempt to warn the first section of No. 49. The investigation is now under way and is being conducted rigidly." [371]

A conflicting report in another newspaper stated that the engines had not been on the main track at all but had been in a siding just slightly foul of the main line. They had been struck by the baggage car so hard the latter vehicle was overturned and escaping gas had been set alight and started the fire. This report also pointed out that the express was within yard limits and should thus not have been travelling at more than six miles an hour.[372] It has not been possible to further clarify the exact circumstances of the accident.

Chicago, Illinois, United States, was the scene of a rear collision early on Tuesday 28th September 1909. The accident happened at 12th Street, on the line of the Pittsburgh Cincinnati Chicago and St Louis Railroad, usually known as the Panhandle, but one of the trains involved was a cattle train belonging to the Chicago Milwaukee and St Paul Railroad. The Panhandle train was a passenger working bound from Chicago to Cincinnati, in charge of Conductor John McAuliff with Engineer Mark Wallace. The Cincinnati Flyer, known as train No. 12, had left Chicago Union station at 12.01 a.m.[373] and consisted of a baggage car, a day coach, a dining car and four sleeper cars. There were only a few men in the day coach, who were not hurt.

The passenger train was running at good speed when it came around a corner and Engineer Wallace saw the red lights of the stock train a short distance ahead. Both he and his fireman, Ed Clark, jumped before the collision. Neither of them was hurt and they later worked the train forward. The cattle train was reportedly running at only 10 miles an hour, the passenger train at 25 and only a minute behind it just before the impact. In the collision the caboose of the stock train was telescoped into the two cattle cars in front of it and the wreckage then caught fire. Riding in the caboose were Conductor Charles Bond, who was killed instantly and a number of stockmen, who were killed or injured. Firemen were summoned at once, engine company No. 66, in charge of Captain Murphy, arriving first. They hacked away at the burning wreckage with axes and first released those of the stockmen who were still alive before turning to the bodies of the dead. Six

---

[371] *The Sun*, Baltimore, Maryland, United States, 23rd February 1909.

[372] *The Philadelphia Inquirer*, Philadelphia, Pennsylvania, United States, 23rd February 1909

[373] *The Hamilton Telegraph*, Hamilton, Ohio, United States, 30th September 1909

bodies were reportedly recovered and four more were in sight of the rescuers.[374] Later it was stated that there had been only six deaths in all.

Another newspaper explained how it thought the accident had happened. 'The cause of the wreck was simply failure on the part of the C M & St P trainmen to flag properly resulting in the Panhandle train coming on them and being unable to stop. The Panhandle tracks in the city are used by the St Paul stock trains as far as Brighton Park, where they go to the stock yards but the crews of these trains running peril of the Panhandle freight trains are usually very careful about flagging. At one time an order was issued on the terminal division forbidding the use of the Panhandle tracks by foreign trains between the hours of 7 p.m. and 4 a.m., and had this order been in force the accident would not have occurred. The wreck occurred about three blocks south of 12th street and the stock train had passed the crossing there ten minutes ahead of the passenger train. The engineer on the St Paul train claimed that their engine was moving slowly when struck but the Panhandle men are of the opinion that it was standing. The question of responsibility will be settled in the investigation which will be held in a few days, but the local men can hardly be censured for the accident.'

This newspaper added that after the collision while the uninjured passengers were being helped away 25 steers broke loose and caused another panic as they plunged madly about the tracks in the darkness. Finally they were corralled by the crew of a passing train and driven into a lineside enclosure. Neither the six dead nor any of the injured listed by this paper included Conductor Bond of the stock train.[375]

At the inquest upon the victims of the collision in late October the jury found the crews of both trains to have been at fault. They found that Rear Brakeman W. W. McDonald of the stock train had failed to protect the rear of his train when it was coming to a stand by flagging it or by throwing down a fusee. He did not bear the whole responsibility for the accident though, they considered, as they were of the opinion that Engineer Wallace had exceeded the speed limit of 12 miles an hour at that point laid down by the Panhandle trainmaster. They noted that there was no block system in force on the line and that the tracks were in the process of being elevated over a considerable distance with changes of gradient and sharp curves. Chicago Milwaukee and St Paul trains bound for the stock yards were admitted to the Panhandle tracks at Western Avenue by a switch tender who was a joint employee of the Panhandle and St Paul companies. The admittance of the stock trains was not reported to the trainmaster or Panhandle train dispatcher by the switch tender, nor was it required to be. Up to 40 trains ran nightly, many of them extras with no scheduled times, without being controlled by the trainmaster or dispatcher. The jury considered that the dispatching and running of night time passenger trains over a line congested with "foreign" and extra trains, without notifying the passenger engineer of such trains, at a time when the elevation of tracks necessitated the removal of known signals, was fraught with great danger and reckless if not criminal.

In view of the conflicting evidence the jury found it impossible to indict and name any individual, but they recommended that the coroner submit the evidence given at the inquest to the state attorney and the grand jury.

---

[374] *The Goshen Democrat*, Goshen, Indiana, 1st October 1909

[375] *Logansport Daily Tribune*, Logansport, Indiana, United States, 30th September 1909

The jury recommended the installation of the block system between Western Avenue and the Union stockyards as soon as possible and pending that the use of extra signalmen; that the trainmaster should advise the train dispatcher of the departure time, description and destination of all trains coming onto Panhandle tracks; and that the crews of all "foreign" trains running on Panhandle tracks should be supplied with Panhandle timetables and rules and instructed as to the condition of tracks, signals, switches, signals and so on.[376]

The jury could obviously have been proud of having done a very good and thorough job.

Mühlheim am Rhein (later Köln-Mühlheim) station, Germany was the scene of a major disaster on Wednesday 30th March 1910. Special train No. 40, a military leave train from Dortmund to Metz, carrying soldiers of the XV1 Army Corps, was stopped at signal A H because of another train still in the station. Meanwhile following it was Lloyd Express 174 from Hamburg to Genoa, a luxury service of first-class only coaches, which consisted behind the engine and tender of two sleeping cars, a dining car, a brake van (?) and a luggage van. The troop train consisted in contrast mostly of old Prussian third-class non-corridor carriages.

The troop train had stopped at signal A H at 1.48 a.m. and at 1.53 the signal cleared to admit the train into the station, but there was some delay in restarting the train on a gradient. Meanwhile signal A v and its vorsignal (associated distant signal) were off for the Lloyd train, but this only gave the express permission to proceed as far as signal F v, which with its vorsignal was at danger to cover the troop train. The express also ran past the stop signal at stellwerk (signal box) Mnt which the signalman could not have cleared without having cancelled the troop train to stellwerk Mbt, and because the signal was in any case locked by a track circuit occupied by the troop train. The express then ploughed into the troop train, the seventh carriage of which was thrown onto the opposite track while the third from the rear completely telescoped the one in front of it. Nineteen of the soldiers were killed outright, 39 seriously and 17 slightly injured. Three of those seriously injured died in hospital. Two conductors on the troop train and a passenger and a waiter on the Lloyds express were slightly hurt. There had been only 15 passengers on the express, which resumed its journey at 5 a.m.[377]

Questions regarding the disaster were raised in parliament and there were calls for the introduction of some form of automatic train control.[378] It seems quite likely that the driver of the express had dozed off.

At 4.20 p.m. on Saturday 18th June 1910 omnibus train No. 467 of the Chemins de fer de l'État left Paris Invalides station for Dreux with Conductor Bardet and Driver Sèrons. It called at Champ de Mars, Versailles-Chantiers and Saint-Cyr and at 5 p.m. arrived at Villepreux-les-Clayes, France. Here the driver discovered something wrong with his engine, the exact nature of which varied according to different sources, but at any rate it

---

[376] *Logansport Daily Reporter*, Logansport, Indiana, United States, 26th October 1909
[377] *The Standard*, London, England, 31st March 1910
[378] *Eisenbahnunfälle bei der Deutschen Bahn* by Erich Preuß pgs. 70-3 (in German)

must have been a serious problem. After spending three-quarters of an hour attempting to fix the problem and having made several unsuccessful attempts to restart his train the driver eventually gave up and it was decided to telegraph Versailles-Marelots depot for a replacement locomotive. Pending the arrival of the replacement engine the railwaymen tried to remove the carriages of the local train from the main line; the train was divided, and an attempt was made to draw forward the front carriages to put them into a siding, but the engine only managed to move them a short distance. By that time most of the passengers had alighted onto the station platform but some of the women passengers preferred to remain in the rear four carriages of the train out of the summer heat. The station signals were kept at danger and Stationmaster Cozic had also sent a man back with a red flag.

Meanwhile an express for Granville, train No. 577, had left Paris Invalides at 5.14 p.m. It was formed of 13 carriages, the leading one a restaurant car, drawn by a practically new engine, No. 2794, which had been delivered to the Chemin de fer de l'Ouest just before its takeover by the state company. In charge of the engine was Driver Henri Leduc, with Fireman Lecordier. The first stop of the express was at Versailles-Chantiers at 5.45, where Driver Leduc was instructed to make an extra stop at Nonant-le-Pin to discharge passengers. Two minutes were lost here because of the large amount of luggage to be dealt with, and two more minutes due to adverse signals at Saint-Cyr. Driver Leduc was now anxious to run hard to regain the lost time, but then the injector on the driver's side refused to work and there was an escape of steam into the cab. Driver Leduc made several attempts to get the stubborn injector to work and it was while he was thus preoccupied, and his fireman was busy breaking up lumps of coal that the Villepreux avancé (distant) signal was passed. A warning bell sounded in the cab but Driver Leduc, aware that the only train in front was the omnibus working which should by now be at Dreux, ignored it. A kilometre later, running downhill, his train was running at 102 kilometres an hour (exceeding the maximum permissible speed of 80 kilometres an hour) when rounding a curve he simultaneously saw the Villepreux entrée (home) signal at danger and a man waving a red flag. He then made an emergency brake application, but his train was still travelling at an estimated 70 kilometres an hour when it ran into the rear of the local train.[379]

The passengers standing on the platform at Villepreux thought at first that the approaching express was the relief engine coming to take their train forward but wondered why it was coming in so fast. Then the collision occurred. The express engine overturned to the left, its tender ended up lying next to the door of a crossing keeper's cottage, the brake van at the end of the local train was reduced to matchwood, the restaurant car of the express became 'so much wood for kindling', and the remainder of the express was piled up above the crushed remains of the local train. At least two of the passengers on the platform were reportedly knocked down and killed by fragments of flying wood and metal. Very shortly after the accident burning coals from the firebox of the express engine set fire to the broken woodwork of the carriages while a damaged petrol tank (according to a newspaper account, but a gas cylinder seems more likely) belonging to the restaurant car fuelled a second blaze.[380]

Driver Leduc reportedly fled the scene after the collision, 'and was making for the

[379] Wikipedia – Accident ferroviaire de Villepreux-les-Clayes
[380] *The Standard*, London, England, 20th June 1910

woods, with half a dozen peasants at his heels; and M. Cozic, the stationmaster, after an alternation of objurgations and appeals for mercy, had locked himself in his office, the terrible scene having momentarily deprived him of his reason. The driver of the train whose engine had broken down also became insane, and it was as much as three young farm lads could do to bind him and prevent him from committing suicide.' Fortunately M. Lépinay, stationmaster at Colleville, chanced to be at Villepreux and was able to take over his incapacitated colleague and organize the rescue operation.

Local farm people joined uninjured passengers from the express in the rescue work, while others hastened to Villepreux village, two miles away, to raise the alarm and organise the village fire brigade. Efforts to put out the fire soon proved futile, however, as there was no water supply at the station, and shortly the leader of the rescuers, the young Abbé Bignon, who had been a passenger on the express en route to his Normandy parish, turned his efforts to attempts to remove the dead and injured from the wreckage before the advancing flames reached them. Having thus directed his fellow rescuers he then turned to give absolution to the dying and comfort to the injured. Eventually the blaze became so fierce and intense that efforts to reach the dead and injured pinned beneath one carriage had to be abandoned and it proved impossible to search other carriages.

Only at sundown, with the arrival of hundred military engineers from Versailles and a detachment of cadets from a military school at St Cyr was it possible to make more effort to fight the fire, but it was not until dawn that it was finally extinguished. Before then doctors and nurses had arrived from Paris on an ambulance train, and from Versailles and elsewhere by motor car. The doctors tended the injured through the night in the cramped station buildings, lit by motor car headlights. Most of the injured were eventually taken to Paris though some went to Versailles and three dangerously ill patients were taken to a nursing home near the scene. By the morning one line had been cleared but the track was damaged, and passengers had to change trains and walk 400 metres between trains.[381] Single-line working was opened up at 4.30 p.m. on the day after the accident and both lines brought back into use at 1.20 a.m. the following morning.

The number of victims who were killed and injured in the accident remained uncertain, especially as some of the bodies of those killed were totally consumed by the fire, but at the trial of Driver Leduc in January 1911 figures of 23 dead and 80 injured were given. All the dead but one were presumably French; the exception was an English governess who had a situation in the family of Count Lambert at Néauphle-le-Châtéau. She was known to have boarded the 5.14 train at Paris Invalides and was never seen again.[382]

Driver Leduc managed somehow to get to his home at Argentan where he was arrested the next day. He was said to be a physical wreck and admitted that he was primarily responsible for the accident but said that he had been busy working on the injector, having had no news of any hold-up ahead and that the line was normally clear for many miles. A newspaper report commented that the local train had first broken down ten minutes before the express left Paris, but its crew were not warned there, nor at the stop at Versailles, nor even close to the accident at St Cyr. This is true, of course, but then that is the function of the fixed signals and does not really excuse the driver. The same

[381] Ibid.

[382] *The Times*, London, England, 22nd June 1910

report noted several of what were doubtless many tragedies, for example, 'M. Leriche, a civil servant, who was travelling by the slow train, has lost his wife, his mother, and one of his daughters. When the train stopped, he got out with his second daughter. The express dashed up, the slow train was telescoped, and M. Leriche saw no more of the rest of his family. He is unable to recognise any of them from the remains now lying in the *chapelle ardente* at Villepreux.' [383]

In view of the fact that the accident had been characterised by problems with both locomotives M. Millerand, the Minister of Public Works, was questioned as to the condition of the engines and rolling stock acquired by the État from the Ouest company. He admitted that the material and workshops of the company had been in very poor condition but said that the government was doing its best to rectify matters.[384]

Driver Leduc was tried for manslaughter at the Versailles Criminal Court in January and early February 1911. The court found him wholly responsible for the accident, refusing to consider any defects in the equipment as a mitigating factor, but in view of his good record sentenced him only to two years suspended imprisonment and a fine of 500 francs.

The accident could have been worse still, for the express from Paris was timed to pass its corresponding working from Granville at Villepreux, but because it was slightly late the opposing train had already gone by and a potential double collision was averted.[385]

That common factor in railway collisions, fog, played a part in the next accident, on Monday 18th July 1910. The scene was Richmond, Victoria, Australia, in the suburbs of Melbourne. The accident occurred during the morning rush hour. The 8.20 a.m. passenger train from Brighton was standing in the platform at Richmond when it was run into by the 8.37 a.m. passenger train from Elsternwick, both trains heading for Melbourne Flinders Street station. The fog was so thick that passengers at the rear of the Elsternwick train sat talking for 10 minutes after the impact, not realising there had been a collision. The Brighton train was formed from the 7.41 a.m. down train from Melbourne, which was itself delayed by the fog, and hence the up train did not leave Brighton until 8.25 instead of 8.20 and Elsternwick at 8.36 instead of 8.32, immediately in front of the 8.37 departure from there. Timed to arrive at Melbourne at about 9 a.m., both trains were crowded.

Guard Davitt of the Brighton train was standing on the Richmond platform next to his van and passengers were hurrying up the stairs to catch the train when the Elsternwick train came around the corner and the collision occurred with a deafening roar. The guards van of the Brighton train was demolished by the engine of the Elsternwick train, the carriage ahead of the van was little damaged but two compartments of the succeeding first-class carriage were demolished, while an old Hobson's Bay Railway carriage in the centre of the second train was 'splintered into matchwood' by the stronger, more modern carriages on either side of it.

Almost at once after the accident the fog began to clear, and the scale of the

---

383 *The Standard*, London, England, 21st June 1910

384 *The Times*, London, England, 21st June 1910

385 Wikipedia – Accident ferroviaire de Villepreux-les-Clayes

destruction was revealed. There were soon plenty of people at hand as people in the streets of Richmond had heard the noise of the collision and rushed to the scene, while passengers on an arriving Caulfield train were also witness to the destruction. For a moment the eye-witnesses were too horror stricken to do anything, but then railwaymen and labourers began to hack and cut at the coaches with axes and saws to release the injured. Doctors had been telephoned for from all directions and 'Nurses in their neat uniforms appeared to bandage wounds and broken limbs, the people from neighbouring shops sent along piles of blankets, numbers of men ran to and fro with bottles of brandy for the revival of the white-faced men and women saved from the ruins, and some ransacked bags belonging to passengers in search of suitable material for dressing. The onlookers clapped their hands and cheered as someone not as seriously injured as the rest of the victims was lifted out into the sunlight, which was growing brighter and brighter, bringing out in lurid details the ghastly spectacle of crushed and bleeding bodies. The bodies of two dead men lay on the platform-one with the head crushed into the shoulders. On a seat lay a young man swathed in bandages. On another form close by was a collection of books - books which were being read by their owners when the crash came. A spot of blood blotted an open page. A girl's slight parasol lay disconsolately against a man's Gladstone bag. On the permanent way lay a little heap of bread and jam sandwiches broken loose from a piece of brown paper. A sparrow fluttered down towards them, but, frightened by the bustle and the shrieks, flew back to its post of observation on the station roof.

'A constable on duty, with dozens of comrades, paced up and down the platform, eagerly listening for the names of dead and injured. His daughter was on the train. She was not at her work in the city, and she had not gone home. What had become of her? The father asked everyone around him, but they could only shake their heads. He went home to find her safe.

'A young man ran down the platform breathless, scanning all the faces and looking in vain. His brother was not there, his sister was not there. With his face working with emotion, he asked if anyone knew the names of those that had been killed. Bur no one knew just then. "That is the carriage my young brother would sure to be in," and he pointed to the splintered smoker, "What name is it?" inquired a kindly doctor, in his shirt-sleeves. The brother faltered out the name. "I never heard that name," said the doctor, "You must hope for the best." Yet as the doctor spoke the body of the young man's sister lay at the Morgue.' [386]

Nine people were killed outright in the accident, while by 22nd July the number injured had climbed to 407.[387]

The inquest on Sydney Dauncey Wellington, one of the passengers killed in the collision, into the accident opened on Wednesday 20th with an inspection of the wrecked trains, Richmond station and signal box by the jury. With various adjournments it ran until Friday 26th August, when the jury passed a verdict that, 'We say that on July 18, at the Richmond railway station, Sydney Dauncey Wellington died from injuries caused by a collision of the 8.20 a.m. up Brighton and the 8.37 a.m. up Elsternwick trains. We also find that the accident was caused by these two trains being allowed on the section

386 *The Argus*, Melbourne, Victoria, Australia, 19th July 1910
387 Ibid, 22nd July 1910

between South Yarra and Richmond at the same time during a heavy fog, but no evidence has been tendered to show that this was due to any act of criminal negligence.'

Run more or less concurrently with the inquest was an investigation held by a Parliamentary Board of Inquiry, which opened on 28th July. On the opening day of the inquiry it was noted that the Brighton train had only been stopped at the Richmond platform for a short time when the collision occurred, that it had taken a considerable time to travel between South Yarra and Richmond, and that both the Brighton and Elsternwick trains had been in the South Yarra – Richmond block section at the same time. The South Yarra and Richmond signal boxes were 1,751 yards apart and it was 1,828 yards from the South Yarra up home signal to the Richmond home signal. It was indisputable that there was a thick fog at the time of the accident. There were three sets of line at Richmond – to Brighton, Caulfield and Camberwell and two sets at South Yarra – to Brighton and Caulfield. Richmond box had two signalmen and a train recorder, South Yarra one signalman and a train recorder. Fogsignalmen and groundmen were available at both stations, the fogsignalmen were placed at the distant signals to repeat the indication of the signal to the drivers, placing detonators on the line when the distant was at danger. The groundmen were employed at the signal box during fogs to inform the signalman that trains had passed complete. The line was worked on the absolute block system, with distant, home and starting signals, also at South Yarra an advanced starter which admitted trains to the South Yarra – Richmond block section and when cleared permitted drivers to proceed as far as the Richmond home signal. According to the regulations, trains were not to be let down to the advanced starting signal to await "Line clear" from Richmond during foggy weather.[388]

On the second day of the inquiry evidence was given by Richard Yolland Smith, block recorder at South Yarra. He said that it had been a little foggy at 7 a.m. but had become thick by 9 a.m. when he could not see the station. He had heard of the accident at 9.15 via the switchboard at South Yarra and had told Signalman M'Donald, who had replied, "Well, it can't be any fault of ours." The entries in his book were then gone through in detail and various mistakes in it were pointed out to him, such as Richmond box clearing out the Brighton train at 9.3 a.m., at which time it was still standing at the South Yarra advanced starter. He alleged that he had given Richmond the "Train waiting signal" (3-2-1) at 9.0½ and that Richmond had then cleared back the Brighton train at 9.03. At 9.10 M'Donald had given Richmond the 3-2-1 for another Elsternwick train when Richmond had rung them and said, "Don't send any more trains up, we have two Brightons here." At the same time as the Brighton train was leaving South Yarra an up Oakleigh (Caulfield) train was departing on the other line but he denied any confusion between the two block bells and insisted that Signalman M'Donald had not made any mistake.

Frederick Hollands, block recorder at Richmond, then testified. He said there were two signalmen at Richmond, Signalman Bull operated the Brighton line and half of the Oakleigh line, Signalman Mullins the other half of the Oakleigh line and the Camberwell line. It had been foggy, and he could not see the station platforms at 9 a.m. His book showed the Brighton train arriving at 9.05½. There was no entry regarding the Elsternwick train. At 9.02 Signalman Bull asked him to ring South Yarra to ask them what they had sent the 3-2-1 signal for? He had done so but could not remember any answer. When they heard the noise of the collision Signalman Bull had said, "I hope it is

---

388 Ibid, 29th July 1910

not serious." [389]

Hollands' examination continued at the next sitting of the inquiry. He denied that the Brighton train had been signalled clear at 9.03 and the Elsternwick train accepted at that time as stated by South Yarra and insisted that Signalman Bull had sent the danger signal to South Yarra at 9.11 (denied by them). After much questioning block recorder Hollands was excused and it was the turn of the South Yarra signalman.

Signalman Alexander M'Donald stated that he had arranged with the Assistant Stationmaster, Mr Taylor, to have the fogsignalmen called out at 7 a.m. They were appointed at all the distant signals and groundmen at the down Oakleigh and Brighton homes but not always at the South Yarra advanced starter. If he could not see the advanced starter his practice was to leave it half a minute or a minute after a train should have passed it before replacing it. He had come on duty at 8 p.m. on the Sunday evening and should have gone off at 7 a.m. but the man due to relieve him was sick with lumbago and he knew he could get no relief at that hour so had agreed so had agreed to stay on until 11 a.m. The Brighton train had arrived at 8.52½, he had sent the 3-2-1 signal to Richmond, the Sandringham train had been cleared back and the Brighton train accepted and had departed for Richmond at 8.53½. He had cleared his signals for the Brighton train after getting it accepted by Richmond and replaced them behind it. He had allowed, he estimated, a minute and three-quarters for the train to pass his advanced starter, which was out of sight due to the fog, before replacing it. The Elsternwick train had been stopped at his home signal before it was admitted to the station. He had sent the "Train waiting" signal 3-2-1 to Richmond at 9.0½ and had received the "Arrival" signal for the Brighton train at 9.03. He admitted that the arrival signal had been received for an Oakleigh line train at 9.03 but not at exactly the same time. He was positive that the Elsternwick train had then been offered and accepted by Richmond. He denied any query from Richmond regarding the 3-2-1.[390] Signalman M'Donald was subject to a long, rather grilling inquiry which went on until the next day but would not admit that he had made a mistake.

Signalman Warburton Bull was next examined. He said that the Brighton train had been about to leave when the collision occurred. He had received the 3-2-1 signal from South Yarra at 9.02 and considered that it applied to the Brighton train. He had sent the danger signal, 6 beats, at 9.12 or 9.13 (the book stated 9.11). According to him outside of the Morgue on 20th July Signalman M'Donald had apologised with tears in his eyes for saying that he had had "line clear" from him (Bull) and the following Sunday M'Donald had said that he would cut off his right hand sooner than drag anyone else into his trouble. Signalman Bull persisted in stating that he had believed that the 3-2-1 signal sent at 9.02 applied to the 8.20 Brighton train. He was asked about Porter Power being sent to him at 8.30 by the Richmond stationmaster, Mr Oliphant, to ask if he had all his fogsignalmen. He said he did not remember this and that there were fogsignalmen, as far as was aware, at all his signals except his up Brighton home.

Stationmaster Henry Oliphant was next examined, then Block Recorder Yolland was recalled. He denied receiving any enquiry from Richmond about the 3-2-1 signal.

Clerk Thomas Lavez stated that he was employed checking train register books. He

---

[389] Ibid, 30th July 1910
[390] Ibid, 2nd August 1910

had found errors and discrepancies in the times booked for trains in the Richmond book but no alterations or erasures.[391]

The next day (5th August) Signalman John Zebedee Mullins gave evidence. He remembered his fellow signalman asking why South Yarra had sent him the 3-2-1 at 9.02 when the Brighton train accepted at 8.53 had not yet arrived. He could see that the signals were lowered for the Brighton train and that the instruments were locked because that train had not passed over the treadle. He was sure that Bull had mentioned 8.53 when he had asked the recorder to ring South Yarra. Signalman Mullins said that it had then occurred to him that the train had gone on to Jolimont, but Bull had forgotten to clear it back to South Yarra and he (Mullins) had contacted Jolimont but they said that the last Brighton train had reached them at 8.57. He could not explain entries made by Recorder Hollands for a non-existent Camberwell train. He had known recorders to make incorrect entries and to omit entries, leading to potentially dangerous situations. It would be courting disaster, he said, for a signalman to give line clear assuming that a train had gone through without his being sure that it had. He was prepared to swear that Signalman Bull did not clear back the Brighton train and accept the Elsternwick one at 9.03. He remembered Porter Power asking about the fogsignalman and thought that they were all in their places.[392]

Assistant Stationmaster Alexander Taylor gave evidence on Monday 8th August after Block Recorder Yollands had been briefly recalled again and admitted making incorrect entries in the train register regarding an imaginary Camberwell train. The assistant stationmaster thought that the driver of the Brighton train stopped at the South Yarra advanced starter should have whistled to alert the signalman of his presence and stated that there had been no alternative to permitting Signalman M'Donald to work on until 11 a.m. (when he would have been on duty 15 hours) as it was impossible to obtain another signalman before that time.[393] He stated that the Brighton train had only been stopped at South Yarra for a minute or so. Stationmaster Oliphant was then recalled and questioned as to why there had been no fogsignalman at the Richmond up Brighton home signal. He said that Porter Power had been told by the Richmond signalmen that all the fogsignalmen were at their posts.[394]

Signalman William White, who was relieved by Signalmen Bull and Mullins at 7 a.m., told the inquiry on Tuesday 9th August that he had sent a message to the stationmaster that fogsignalmen were required at 6 a.m. At 7.25 he had gone out to act as fogsignalman himself at the down Camberwell home (after 11 hours working the box). He had told Signalmen Mullins and Bull, who were busy at the time, that there were fogsignalmen at the distants but not the homes, though they clearly misunderstood him. He was followed by Porter William Power, who testified that following the stationmaster's orders he had called up all the fogsignalmen, going to all their houses, then arranged their meals, taken fog signals to the men on the Jolimont side, taken out more meals, then at the stationmaster's request he had gone to the box to check that all the fogsignalmen were at their posts, and Signalman Bull had told him that they were. He had then gone to the restaurant for more meals and when the collision happened had not

391 Ibid, 3rd August 1910
392 Ibid, 6th August 1910
393 Ibid, 8th August 1910
394 Ibid, 9th August 1910

yet had time to visit the up signals. He had been delayed with the meals because the woman in the restaurant had insisted on payment before delivery.

Driver Peter Allen of the 8.20 a.m. train said that he had left Brighton Beach at 8.25 with Fireman Gillman. It was clear there; they first ran into the fog near Windsor. They exploded two detonators at the South Yarra up distant and two more at the home, which was at danger. He had whistled, it was lowered, and he went forward into the station. They stopped there a little longer than usual and it was so foggy that he could not see his guard until the latter came along the platform to flag him off. He then set off, past the starter, which was off, and stopped at the advanced starter, which was on. He gave a fairly long whistle at the advanced starter. It took him nearly three minutes to get from the station platform to the advanced starter and he waited there about four minutes before it was cleared. Asked why the guard or fireman had not gone back to the signal box in accordance with Regulation 75 (equivalent to the British Rule 55) to remind the signalman of their presence he gave the usual excuse that he had been expecting the signal to be lowered at any moment. An up Oakleigh train had passed when his train had been standing about a minute. When the advanced starting signal had cleared, he had set off, crossing the bridge over the Yarra river at four or five miles an hour, then accelerating to about 12 miles an hour. He saw the fogsignalman at the Richmond up distant holding out a green flag and knew it was off and he had glimpsed the home signal off from a short distance away. He had stopped at the platform of Richmond station and had been there about half a minute when he felt his train being run into from behind and it was driven forward.

Fireman Peter Gillman confirmed his driver's evidence and could add little to it.[395]

Guard Michael Davitt of the Brighton train gave evidence before the inquiry the following day. He said his train was stopped at South Yarra about a minute. He could not see the starter through the fog, but the assistant stationmaster gave him the word to go on and he walked down the platform to show his driver a green flag. Three minutes after setting off the train had stopped again, at 8.57½ by his watch. He had then got down and lit his tail lamp and then the side lamps. They expected delays on foggy days and Regulation 75 had not entered his head. It was the densest fog he had ever experienced and difficult for him to look out for signals anyway because there were 12 or 13 passengers in his van. Asked then of his recollections of the actual collision Guard Devitt was overcome by his memory of it and it was some time before he could continue, when he related that he had spoken to the driver of the Elsternwick train and said, "This is a sad affair, isn't it?"

"It is," the driver had replied; he had not said much more.

The guard was pressed about carrying passengers in his van and especially about Regulation 75. He said that in 11 years as a guard he had never known a fireman go back when a train was stopped at the advanced starter. He himself had never gone back in such circumstances. He would have gone back if he considered that his train had been stopped an excessive amount of time. It had not occurred to him that his train should not have been held at the advanced starting signal in foggy weather.

When Guard Davitt's questioning was finally concluded it was the turn of Driver John Melville of the Elsternwick train. The latter said that his train had not been checked

---

[395] Ibid, 10th August 1910

coming into South Yarra. He had stopped in the station about two minutes. The starter had been at danger but was cleared before his guard gave his train the signal to depart. He saw the advanced starting signal off from a short distance. He gave his engine a little more steam and they were going at 12 or 14 miles an hour as they crossed the Yarra River bridge. Three or four hundred yards further on he exploded a detonator and saw a fogsignalman who appeared to be waving a green flag at what he presumed was the Richmond distant signal. He had reduced speed from 15 to 9 miles an hour. The next thing that he saw was the guard's van of the Brighton train about an engine's length in front, when he had made a full application of the Westinghouse brake. He and his fireman had been looking out intently for the Richmond home signal but did not see it. Seeing a green flag waved rather than held steadily and exploding only one detonator instead of two he assumed it was in order to proceed at 15 miles an hour. He expected a detonator at the home signal had it been at danger.

Fireman James Burke of the Elsternwick train testified that at the Richmond distant he had remarked to his driver, "There's only one," (detonator) and Driver Melville had replied, "It's all right; I have got a green flag." He had then been looking out for the home signal but could not see it. The next thing that happened was his driver exclaiming, "Good heavens! (or very likely something stronger) There's a train in front of us." [396]

Fireman Burke gave a little more evidence on 11th August when he estimated the speed of their train just before the collision as between six and eight miles an hour.

Guard William Hargraves of the Elsternwick train stated that he had only been a guard for three weeks, previous to which he was a porter. He was criticised for only having looked out once or twice for signals during the journey and leaving signal observation to the driver.

Thomas Ryan, driver of the 8.30 a.m. up Oakleigh train, said that the Elsternwick train had started from South Yarra just in front of him. He did not see it again. One detonator was exploded at the Richmond distant and he was stopped at the Richmond Oakleigh line home about a minute. He could not see the position of the Brighton line home. In 20 years as a driver he had never sent his fireman back in accordance with Regulation 75. Driver Albert Crocker, acting as Driver Ryan's fireman, corroborated the latter's evidence.

Driver George Johnson of the 8.25 up Caulfield train gave evidence as to the thickness of the fog and stated that in 28 years as driver he had never had occasion to observe Regulation 75.

Porter Edward Emmett stated that he had been on duty since 10 p.m. Sunday as night watchman and had been sent to the Brighton line up distant as fogsignalman by the stationmaster at 6 a.m. He had signed on at the signal box and proceeded to his post with 24 detonators, flags and lamps. The Brighton train passed him, and he had laid one detonator on the rail when another train appeared out of the fog as he was about to put down a second. He imagined for a moment that it must be an up Oakleigh train on the other line, then saw that it was on his line and had to jump out of its way. The train was only 10 or 15 feet away. While he was holding the detonator in his right hand, he had a red flag unfurled in his left. How exactly he had been holding it he could not say but he did not wave it. He did not see the engine driver. He tried to call out to the guard, but his

[396] Ibid, 11th August 1910

voice would not carry. The second train was going rather faster than the first; they were about half a minute apart. Immediately after the passing of the second train he had put down two detonators and then saw that the signal was at danger.

Fogsignalman Wright at the Oakfield line up distant had called out to him, "They're very close together. I am afraid they're be an accident."

He had replied, "I think there will be." He heard of the collision a quarter hour later.[397]

The inquiry continued the next day with Fogsignalman Peter Wright stating that he had had red and green flags in his hands, but he did not see how Driver Melville could have mistaken his flag for a green one on the Brighton line. The inquiry also heard evidence that day as to the rescue work and the telescoping of some of the carriages.[398]

On Monday 15th August Superintendent Edward Blazer gave evidence as to the signalling. Train recorder boys as young as 14 were, he said, used in Great Britain. He said that in that country there were between 900 and 1,000 miles equipped with lock and block signalling in Great Britain, in Australia there were 17 sections equipped with it in Victoria and three sections in New South Wales. In the United States a form of lock and block supplemented by track circuiting was in use on 485 miles out of a total of 275,000. The large expenditure involved had prevented other countries from adopting the system. As regards the Richmond accident Signalman M'Donald should not have accepted the Elsternwick train until he was sure that the Brighton train had proceeded 440 yards beyond his home signal, if necessary, by awaiting it being cleared back by Richmond. The superintendent was then examined on a number of other matters and stated that the wholesale breach of Regulation 75 would be looked into.[399]

Continuing his evidence the next day, Superintendent Blazer commented that he did not believe that either Signalman M'Donald or Signalman Bull had committed perjury but that each was convinced that the other had made the mistake. As for the conflicting evidence between Driver Melville and Fogsignalman Emmett as to whether the latter's flag had been red or green, he could not explain it.

John Mars, assistant superintendent for New South Wales lines, stated that there they had Preece's, Tyer's one and three-wire and lock and block instruments. He considered the Tyer's three-position instruments in use in New South Wales to be superior to the Winter's two-position instruments in use in Victoria. Three-position instruments were used on lines with 10 or 12 trains following one another at three minute intervals. Very rarely had there been two trains in the section together in New South Wales on lines using three-position instruments and he did not know of any accidents. He thought that a repeater lower down the arm of the Richmond up Brighton line home would be an advantage and agreed with Superintendent Blazer that if the signalman could not see the position of his advanced starter, he should not replace it until the train had been cleared back by the box in advance. He did not think that signalmen should be kept on duty as long as Signalman M'Donald had been but noted that he had had little to do during the night.

Signal Engineer Charles Pilkington of the South Australian Railways stated that in his

---

[397] Ibid, 12th August 1910
[398] Ibid, 13th August 1910
[399] Ibid. 16th August 1910

state they had found the Winter's block satisfactory but he thought that those in the South Yarra and Richmond boxes put too great a tax on the signalman's memory (presumably because of the intensity of the train service and the number of lines) and that a three-position instrument would be better.[400]

More witnesses were called the following day, including Block and Signal Inspector Hugh Lynch, who testified that he had visited South Yarra box at 9.58 a.m. on the 18th and had said to him that it was a bad business, to which M'Donald had replied, "I have not let any train up without line clear," and, "We have had nothing to do with it here." The signalman had appeared cool and un-ruffed.[401]

The inquiry continued with questions regarding the safety of the rolling stock, the positioning of the signals and the lock and block system.[402] Block and Signal Inspector Alexander Mathieson stated, somewhat surprisingly that he preferred the Winter's two position blocks to Tyer's three-position instruments.

After several more days of evidence the long-winded inquiry finally concluded on Wednesday 24th August and its results were published on 5th September. In its findings the inquiry found that 'the accident was caused by the succession of contraventions named hereunder of the Railway regulations by a number of employees of the department during an extraordinarily dense fog, and, therefore, at a time when they should execute extreme vigilance and caution.'

The contraventions were: -

(a) The premature replacement by Signalman M'Donald of his up advanced starting signal before he could be sure that the Brighton train had passed it instead of, if need be, waiting for it to be cleared back by the Richmond signalman. Regulation 64(b) regarding the replacement of an advance signal might be made more explicit regarding the signalman's actions in foggy weather.

(b) The contravention by Signalman M'Donald of Regulation 4 in accepting the Elsternwick train from Prahran before he knew the Brighton train had advanced 440 yards beyond his home signal.

(c) The failure of the crew of the Brighton train to make the South Yarra signalman aware that they were stopped at his advanced starter in accordance with Regulation 75(a). As, however, the regulation was uniformly disregarded they could not be blamed.

(d) The infringement of the block regulations by either Signalman M'Donald or Signalman Bull as a result of which the Elsternwick train was admitted to the South Yarra-Richmond section while it was still occupied by the Brighton train, together with the failure of Signalman Bull to come to a clear understanding with Signalman M'Donald as to the meaning of the 3-2-1 bell signal sent at about 9 a.m.

(e) The failure of Driver Melville to exercise extreme caution after exploding a detonator at the Richmond up distant signal so that he overran the up home signal at danger. He had some excuse inasmuch as he might have expected to explode detonators at the home signal, but he had not exercised the vigilance required under the difficult and abnormal conditions.

---

[400] Ibid, 17th August 1910
[401] Ibid, 18th August 1910
[402] Ibid, 19th August 1910

Stationmaster Oliphant was found not to be to blame for the misunderstanding between himself and Signalman Bull which had resulted in the lack of a fogsignalman at the Richmond up Brighton line home signal.

The inquiry noted that 'The contraventions and misunderstandings above set out afford ample proof of the truth of the well-known maxim of railway management that it takes two men to make an accident, for, even had one of the men concerned in the calamitous occurrence of 18 July fully complied with the railway regulations no accident would have occurred.' [403]

On 9th November Driver Melville was charged with 'having wilfully omitted and neglected to observe certain regulations of the Railways department.' At the conclusion of the Crown case the jurors consulted for a few moments and then their foreman announced that they did not wish to hear the defence case; they found Driver Melville 'Not Guilty'. [404]

Durand, Michigan, United States, the location of a major disaster in August 1903 (see above) was close to the location of another bad accident, again a rear collision, seven years later on Wednesday 24th August 1910. Grand Trunk Western Railroad train No. 4, an eastbound Chicago – Montreal express, left Durand at 10.01 p.m. but came to a stand after about three miles at Duffield due to an engine problem. Eastbound train No. 14 left Durand at 10.35 p.m. According to its fireman, George Wilson, it was well under way and travelling at about 45 miles an hour when they ran into the rear of the first train, the only warning of which had been the explosion of one torpedo at the last moment. The rear Pullman car (*Nebraska*) of the first train was split in two by the collision and its passengers thrown in every direction. The wreckage caught fire immediately after the collision and burned fiercely for hours. There was considerable confusion the following day as to the number of people killed. By then eight bodies had been recovered but rescuers who had been early on the scene were positive that 19 had been killed. Those who lost their lives included Engineer George Mitchell, who was under his locomotive making repairs at the time of the collision and died on the way to hospital. Other railroad men who were badly, possibly fatally hurt were Engineer Charles Spencer of train No. 4 and Brakeman George Nelson of train No. 14.[405] [406]

After the accident a relief train with doctors, nurses and hospital supplies was sent from Battle Creek and wreckers from both Battle Creek (the same train?) and Durand. Farmers from the surrounding countryside whose attention had been attracted by the flames lighting the sky also lent assistance.

Responsibility for the collision was not immediately clear, with the crew of train No. 14 stating that a flagman had been sent back half a mile as soon as the train had stopped.[407] An investigation by Grand Trunk officials, however, concluded that Engineer Spencer, of train No. 4, was at fault, that train No. 14 had been properly flagged, that the

403 Ibid, 6th September 1910

404 Ibid, 10th November 1910

405 *Bridgeport Evening Farmer*, Bridgeport, Connecticut, United States, 25th August 1910

406 *New Albany Public Press*, New Albany, Indiana, United States, 30th August 1910

407 *Ogden Evening Standard*, Ogden, Utah, United States, 25th August 1910

engineer's mental condition had been at a low ebb due to domestic troubles and that his failure to stop after being properly flagged had been the result of "moral idiocy" as they expressed it.[408]

The inquest on the victims of the accident, the number of whom was eventually given as 10, concluded, in contrast to the findings of the railroad's own investigation, that the person principally to blame was Rear Brakeman George Graham, who had been grossly negligent in not going back a safe distance to lay torpedoes on the rails and use other means in his power to stop train No. 4. Engineer Spencer they censured 'for not watching the track more closely'. [409] [410]

The first of two bad rear collisions in England in December 1910 happened on the Monday 5th of that month in the northern suburbs of London, at Willesden, on the London and North Western Railway. The 8.27 a.m. up local passenger train from Watford to London Euston was standing at No. 4 platform, on the up slow line, when it was struck, at 8.55 a.m., by the 8.30 a.m. local passenger train from Watford to London Broad Street. The two rear carriages of the stationary train were wrecked and telescoped, and five others damaged; five passengers were fatally hurt and 235 complained of injury.

The entrance to platform 4 was controlled by Willesden No. 5 box, which was normally manned by one signalman and a booking boy, but on the day of the accident a second signalman and a second booking boy were in the box training. It was a mistake by Signalman George Blundell, the regular man, which led to the accident. The 8.27 was a heavy train and it lost four minutes on the journey from Watford to Willesden, perhaps partly because its engine was running bunker first and steaming badly. It arrived at Willesden at 8.51 instead of 8.47 and was then further delayed because of a special examination of season tickets. It had been accepted by Willesden No. 2 box at 8.43 by Relief Signalman Harry Smith, who was learning the box, at 8.43 and the starting signal at the end of the platform had been cleared for it.

The 8.30 train had left Watford on time and arrived at Willesden on time at 8.55. Willesden No. 7 box had offered it to No. 5 box at 8.51 and it had been accepted as the 8.27 train had passed the rear end of the platform, 340 yards in advance of the up slow line home signal, which in accordance with local instructions was the clearing point (the point beyond the home signal which a train had to pass before a following train could be accepted up to that signal – usually 440 yards). It could not be offered forward to No. 2 box because the 8.27 train was still in the section and the up slow distant and home signals should have been kept at danger, but as the 8.30 train was approaching the home signal at about 25 miles an hour it was lowered by Signalman Blundell. Driver George Hill had no warning of an obstruction as he passed No. 5 box and because of an overbridge and a curve he only saw the train in front at the last moment, when he jammed on the brakes.

The Board of Trade inquiry into the accident was conducted by Lt-Col H. A. Yorke, who noted that the accident was clearly due to a mistake made by Signalman Blundell, but that the latter's explanation of how it had occurred did not entirely agree with other

---

[408] *Manitoba Free Press*, Winnipeg, Manitoba, Canada, 2nd September 1910
[409] *The Evening Democrat*, Fort Madison, Iowa, United States, 12th September 1910
[410] *Pulaski County Democrat*, Waynesville, Missouri, United States 10th September 1910

evidence. Blundell stated that after the 8.30 train had been accepted, he had been preparing to bring an empty carriage train out of No. 2 siding to the up fast platform to form the 8.58 departure to London Euston. To do so he needed to pull levers 16 and 19, operating the points, and then 53, working the associated disc, and must by mistake have pulled lever No. 55, the up slow home. Signalman Blundell said that he had then asked Signalman Smith, "Is it all right for the up train?" or words to that effect.

The latter had replied, "Yes."

Signalman William Selfe in Willesden No. 4 box on the other hand stated that the shunting disc for the empty carriage train had been cleared immediately, while when Stationmaster C. L. Sharratt had come into the box Signalman Blundell had remarked, "We have made a mistake between us," implying that rather than pulling the wrong lever he had misinterpreted Signalman Smith's reply to his question as to whether all was right for the up train as meaning that the 8.27 train had gone. The inspecting officer thought it possible that Signalman Blundell, anxious not to delay the 8.30 train and being unaware of the season ticket inspection being conducted had assumed that the 8.27 train must have gone, although he had not heard the "Train out of Section" signal for it.[411]

Lt-Col Yorke noted that as Signalman Blundell had in any case assumed full responsibility for the accident the exact nature of his mistake was perhaps not relevant. According to the signalman's own admission he could not be absolved from great carelessness, which had resulted in the deaths of five persons and injuries to many others. He was an experienced man who had been in the box 10 years. He had either pulled the wrong lever, showing that he was paying scant attention to his duties, or he had violated the block regulations.

It was often urged, the inspecting officer continued, that having two men in the box would increase safety, but this accident did not confirm that, and it was quite possible that the division of responsibility between the two men had something to do with the accident. Nor had excessive hours anything to do with the accident as the two signalmen had only been on duty about three hours after adequate rest. The pier of an overbridge interrupted the signalman's view of the platform and removal of the pier or a re-siting of the signal box would improve matters, but a better solution would be the installation of a track circuit, locking the home signal when it was occupied by a train standing at the platform. A problem in England was the use of passenger rolling stock having Mansell wooden wheels, but this could be overcome by bonding the tyre to the hub with a copper strip. The London and North Western Company were already doing this but all rolling stock had not yet been altered. Meanwhile an electric lock had been fitted to the up slow line home signal such that it could only be cleared when the signalman in No. 2 box had given "Line Clear" on the block instrument to his colleague in No. 5.

Booking Boy Sidney Shepherd had recorded the 8.30 train as having been accepted by No. 2 box at 8.53 whereas in fact it had never been offered. He had only been training four days and his mistake must be put down to inexperience. When asked about his mistake he had naively replied that every column had to be filled in, implying that he had not been properly instructed in his job. The value of train register books depended entirely upon their accuracy and if they could not be relied upon, they were worse than useless and could in fact be a source of danger by misleading a signalman (though

---

[411] Board of Trade report

Signalman Blundell did not suggest that he had consulted the book before pulling off for the 8.30 train). If train registers could not be relied upon, they were better dispensed with. Booking boys also often became signalmen and for them to start with the impression that the books merely needed to be filled in haphazardly was the worst possible education for them. Lt-Col Yorke said that he had long been of the opinion that in busy boxes where the signalman as too busy to do the booking some sort of automatic recording device working with the block instruments could be used to record every block signal given and received on tape. (This idea was never adopted in Great Britain, train register books and booking boys remaining in use at least until the 1970s, when the elimination of the busiest boxes in modernisation schemes meant the end of booking boys, though not train register books, in manual boxes). Meanwhile the train register books might be simplified by reducing the number of columns so that the signalmen themselves could fill them in, eliminating the need for booking boys.

Turning to other matters the inspecting officer remarked that the fatal results of the accident were largely due to telescoping, a feature which appeared to be on the increase, probably due to the adoption of longer coaches with wooden bodies resting upon steel underframes. In the United States 'anti-telescoping' and 'anti-climbing' devices had been introduced to combat the problem, and in England the Metropolitan District Railway had attached a steel buffer block to the front of its cars. In the United States the carriages had a short rigid centre coupling and buffing arrangement without side buffers but in Great Britain the long flexible coupling and side buffers in general use increased the risk of telescoping. It might be well for the railway companies to consider modifications to coach design to reduce the risk of telescoping.[412]

At the inquest on the five passengers who died Signalman Blundell admitted his responsibility for it and the jury returned a verdict of 'Accidental Death' adding that it had been due to the signalman inadvertently pulling the wrong lever. They asked whether, in view of the bridge obscuring the signalman's view of the platform, it might not be possible to devise some means whereby the signalman could see whether the lines were clear.[413]

The safety record of the Midland Railway of England was relatively good, but it did suffer two major disasters, both rear collisions, in less than three years on its Settle-Carlisle main line in 1910 and 1913, the accidents occurring within a few miles of one another. The first accident was an indirect result of the 'small engine' policy of the company, which meant that any train other than the lightest had to be assisted over the steeply graded line crossing the Pennines from either Leeds or Hellifield if it was northbound or from Carlisle if it was southbound. The assisting engines were detached at the summit of the line at Ais Gill and then returned whence they had come. There were a signal box and lay by sidings at Ais Gill, but the nearest turntable was at Hawes Junction, three miles on the Hellifield side of Ais Gill. All assisting engines in both directions therefore had to run from Ais Gill over the up line to Hawes Junction for them to be turned; a necessary operation to avoid many miles of tender running back to Hellifield, Leeds or Carlisle.

---

412 Ibid.

413 The Times, London, England, 6th, 7th, 8th, 10th, 12th, 13th, 15th, 19th and 20th December 1910

On duty in Hawes Junction box during the early hours of Christmas Eve, Saturday 24th December 1910, the day of the accident, was Signalman Alfred Sutton, who had worked as signalman at Hawes Junction for about 20 years. With the extra Christmas traffic he had a busy turn. He described what had happened from about 4 a.m. at the Board of Trade inquiry into the disaster. At 4.06 a.m. two light engines, Nos. 448 and 548, had arrived from Ais Gill and he had put them onto the Hawes branch platform line on the up side. A down express ran through at 4.14, he crossed the engines over to the down side where the turntable was, they were turned and then moved back to the branch platform line on the up side until there was an opportunity to send them back to Carlisle. He could not send them at once as he had been advised by Hellifield of a down fitted goods train. Four more engines – Nos. 247, 249, 313 and 314 – arrived from Ais Gill at 4.36. The down fitted goods passed at 4.41, the four engines were crossed to the turntable and three more engines, Nos. 42, 312 and 317, which he had accepted from Ais Gill under the 'Warning Arrangement', arrived at 4.47. He then accepted an up through goods train from Ais Gill under the 'Warning Arrangement', crossed the three last arrived engines to the turntable and the through goods passed at 4.59. Ais Gill then offered him a Class 'A' goods, which he had refused, thinking the two Leeds engines on the turntable, Nos. 247 and 249, were ready to start. Finding that they were not he had accepted the Class 'A' goods under the 'Warning Arrangement'. A special down express passenger train passed at 5.20 and he then crossed engines 448 and 548 from the branch platform to the down main. Engines 247, 249 and 317 were crossed from the turntable road to the branch platform line, the Class 'A' goods was allowed to proceed towards Dent, and he was offered, but refused, another Class 'A' goods by Ais Gill. Finding that there was no time to get rid of the two Leeds engines before this goods he then accepted it from Ais Gill, and it passed him at 5.43. Ais Gill had cleared back the down special express at 5.26 and he had meant to ask for the two engines (448 and 548) for Carlisle at that moment but preoccupied with other work he forgot to do so. He had been very busy at that moment as one of the Leeds enginemen had come to the box asking about relief and he had been on the telephone to Hellifield about their engines. At 5.39 the midnight down express from London St Pancras was offered to him and he accepted it, offered it to Ais Gill and cleared his down main signals, having forgotten all about engines 448 and 548. Dent sent the express on (gave the 'Train Entering Section' signal) at 5.44 and it passed him at 5.47. He had looked towards Ais Gill before lowering his signals but there had been a heavy gale with driving mist and he could see nothing of the two light engines, which he thought must have moved further forward than was usual for engines waiting there to do. If the engines had whistled, he might not have heard it with the wind. Nor had the enginemen carried out the provisions of Rule 55 and come to the box to remind him that they were there. The similarity to the Richmond collision in Australia a few months earlier (q.v.) is obvious. He then made a couple of other engine movements, when Driver George Tempest from one of the engines came into his box and asked about the Carlisle engines. He looked in his train register book and saw that they were not recorded. He then rang Signalman Benjamin Bellas at Ais Gill and asked what time he had offered the engines. "You haven't given me any, they haven't come," was the reply.

He then knew that he had forgotten the engines and that an accident had probably occurred, but he rang Ais Gill once or twice more to ask if the engines had arrived. "No," was the reply.

With that his relief, Signalman John Simpson, arrived. He looked out of the north window, towards Ais Gill, and saw that there was a reflection of light in the sky. Turning to Signalman Simpson, he said, "Go to Bunce {Stationmaster William Bunce} and tell him that I am afraid that I have wrecked the Scotch express." [414]

Drivers Edwin Scott and George Bath had been in charge of engines 448 and 548, with their firemen George Worley and James Carruthers. When their engines had been crossed over to the down main line, they knew that they were only waiting for the previous express to clear Ais Gill before they should get the right of way. They did not whistle and neither fireman was sent back to the box in accordance with Rule 55. They estimated that they had only been stopped at the signal for six or seven minutes, but it could be fairly assumed, Major J. W. Pringle, the Board of Trade inspecting officer, stated, that they had actually been detained 20 minutes. When the starting signal did finally clear, as they thought for them but really for the midnight express, they had whistled and set off and reached a leisurely speed of 25 miles an hour. Driver Bath stated that they had passed through the short Moorcock Tunnel and were a quarter mile further on, near the footbridge known as Grisedale Crossing, when looking back he saw the headlights of the express in the tunnel. He had only had time to open his regulator, wave his lamp to Driver Scott and tell Fireman Carruthers to open the whistle when the express caught them up. He was briefly knocked unconscious in the collision but coming to walked 1½ miles to Ais Gill box to report the accident, then back to the wreck.

Driver Richard Oldcorn and his fireman (name apparently unrecorded) were in charge of the pilot engine of the express with Driver Henry Wadeson and Fireman William McVittie on the train engine. Driver Oldcorn said that they had run through Hawes Junction under clear signals at about 60 miles an hour, being about 16 minutes late due to signal checks south of Skipton. The speed of the train had increased to about 65 miles an hour as he emerged from Moorcock Tunnel and saw a red light (the tail lamp of the rear engine) immediately in front. There had only been time to apply the continuous brake before the collision occurred.

The two front coaches of the express were telescoped by the violent collision and completely wrecked. Fire broke out immediately in the wreckage and eventually spread to the whole of the train except the rear two brake vans. Twelve passengers died in the accident and nine were injured. The four drivers and four firemen were also injured.[415]

An especially distressing feature of the disaster was that several of the passengers were burned to death, trapped in the wreckage with rescuers unable to save them, as was noted in the London *Times*, which told of the arrival of a relief train in Glasgow. 'The relief train which was sent north with about 20 passengers from Hawes Junction arrived in St Enoch's Station, Glasgow, at 2 o'clock on Saturday afternoon. They were met on the platform by Mr C. E. Cockburn, general manager of the Glasgow and South Western Railway, Mr R. Gilmour, stationmaster, and a number of friends. All the passengers were suffering from the effects of their exposure to wind and rain on the bleak moorland, and several of them had bruised limbs and bleeding faces. But most distressing of all was the appearance of Mr and Mrs James Gray, who had to endure the awful agony of seeing their infant daughter burned to death while they were helpless to save it. The couple were travelling from Eastleigh, Hants, to spend Christmastide with friends in Partick. Mr

[414] Board of Trade report
[415] Ibid.

Gray's face was bruised, his right arm was in bandages, and his legs seemed to have been severely crushed. He was carried to an invalid's chair on the platform. Mrs Gray walked behind. Her face was swollen, her right eye was cut and bleeding, and she looked the picture of despair. Mr Gray, in sobbing accents. Explained to a friend that the child had been burned to death before his eyes. The train, he said, crashed into the engines and split them like matchwood. "Oh, it was horrible! Don't speak about it," he cried.' [416]

Having detailed the circumstances of the accident, Major Pringle considered some points which had been mentioned in newspapers and letters. The primary responsibility for safe working, he wrote in his report, was placed upon the signalman. In accordance with Rule 59 the latter must, by observation, assure himself that there was no obstruction upon the line within his immediate control before lowering his signals to permit a train to approach. Some railway companies provided devices to remind the signalman that the line was blocked, such as lever clips (collars). The Midland Company did not use them on the grounds that they were likely to make the signalman less likely to practise careful observation. Where the signalman's view was restricted train-waiting indicators operated by contact bars or treadles, or track circuits, were provided. But these precautions were not taken by all railway companies, nor were they required by the Board of Trade, at simple block posts where the signalman's view was unobstructed. On the other hand to guard against failures by signalmen to observe obstructions and when the weather made such observation difficult Rule 55 had been introduced. The meaning of the rule was plain enough but it was not always enforced so as to ensure safety. After a train had been detained long enough for a preceding train to have cleared if the train was still detained then someone should go to the signal box to remind the signalman that the train was there. To say that the signal was momentarily expected to fall, and that delay would result from going back was no good excuse. If one of the firemen had been sent back after the engines had been waiting five minutes, he would have reached the box in another three, eight minutes before the express had passed Dent, and no collision would have resulted.

The inquest jury had been unhappy that Hawes Junction North and South boxes had been replaced by one box six months earlier but there was nothing unusual in the replacement of two boxes by one, which improvements in signalling and the working of points permitted to be done with security. The starting signals were not exceptionally far from the box and the work in the box was not too much for one man, Sutton himself had said in his evidence that he had found no more difficulty in working one box than formerly, in fact it was easier in several ways.

The inspecting officer then went on to consider the different railwaymen and their responsibility. Signalman Sutton, he said, had admitted that he had forgotten the two engines and lowered his signals for the express, probably because he had been preoccupied with trying to get the two engines away to Leeds. The tail lamp of engine 548 must have been 270 yards from the box. Sutton should have seen it. Again, Sutton could have noted in his train register book that he had put the engines on the down line. The signalmen (Sutton and Simpson) said that they would insert a poker or a piece of telegraph wire in the latch of a lever to remind them not to pull it, but they were not used for light engines as they were so frequently dealt with. Also, there was no reason why the engines could not have been kept in the branch platform until they were despatched or close to the box on the down main. Even though the number of light engines to be dealt

[416] *The Times*, London, England, 26th December 1910

with added to the work the box was not excessively busy, Sutton having dealt with 58 trains during his ten-hour turn.

With regard to the enginemen, like Signalman Sutton they had given their evidence in a very straightforward manner. Scott and Bath had failed to carry out Rule 55 and excuses that it was not usually carried out there, that they were every moment expecting the signal to come off and that they did not seem to have been waiting that long could not be accepted. They were guilty of a serious dereliction of duty.

Major Pringle did not think it would be reasonable to have expected Driver Bath to realise any quicker that the express was behind them or Driver Oldcorn to have seen the tail lamp of the engines any sooner.

It seemed likely from the evidence of railwaymen, passengers and Arthur Ashton, a shepherd who lived near the scene of the collision, that five minutes or more must have elapsed before the passengers in the express realised what had happened and that the train was on fire. It was clear that the fire had originated from escaping gas from cylinders beneath the first two coaches. Sparks caused by the grinding together of the steel underframes or the passage of wheels over broken rails and ballast could have ignited the gas. The wind caused the fire to spread to the rest of the train and as its vehicles were mostly derailed only the rear two vehicles could be drawn clear and saved.

At about 5.55 a.m. a light engine with Driver John Judd and Fireman Charles Jenkins arrived from Ais Gill on the up line. They were able to release a couple of passengers and then used the water from the tender of their engine to try and fight the fire until the supply was exhausted.

Shepherd Ashton and Extra Labourer Robert Field, who lodged with Ashton, heard the noise of the collision from their house and hurried to the scene. Field had the keys to a permanent way hut from which they obtained some tools and along with uninjured passengers tried to free the trapped passengers. Some of the passengers had complained that with more axes they could have rescued more of those trapped but Major Pringle doubted it unless the fire could have been kept under control.

In conclusion the inspecting officer placed the responsibility for the accident on Signalman Sutton, for failing to assure himself of the position of the Carlisle engines and observe that the line was clear. He found Driver Bath equally responsible for failing to carry out Rule 55, and Driver Scott responsible to a lesser degree. He did not attach any blame to Drivers Oldcorn or Walsden.

Major Pringle recommended that in view of the number of light engines dealt with the up and down lines should be track circuited in the rear of the advanced starting signals; Rule 55 would then not need to be carried out there. Secondly the signalman's hours at Hawes Junction should be reduced to eight and thirdly the attention of enginemen generally should be drawn to the importance of carrying out Rule 55.

The inspecting officer then considered matters relating to the aftermath of the accident. It could not be disputed, he stated, that gas had been the cause of the fire. The substitution of electricity for gas would not entirely eliminate the risk as fires could originate from gas used in the cooking ranges of restaurant cars, as had been the case at Grantham (q.v.) and possibly Cudworth (q.v.). Gas, he said, had been used as an illuminant since 1868 and there had been no previous cases of lives being lost in a fire caused by it. (This might have been true at this date as far as Great Britain was concerned but the immunity would in any case not last much longer). Gas, Major Pringle continued, with the latest improvements in for example incandescent mantles, had many advantages over electricity; it required no expert staff, was nearly as easily controlled, was more economical and less liable to failure.

To improve the safety of gas cylinders several suggestions had been made; the use of valves, the strengthening of the ends of cylinders, a decrease in their diameter, positioning them above the roof rather than under the framing and the use of some incombustible material such as uralite (a mineral, originally described in the Ural Mountains) around them.

The inspecting officer was not in favour of all-steel carriages, which he thought might cause more injury to passengers and make the use of rescue tools impossible.

Having apparently spoken very much in favour of gas lighting Major Pringle then conceded that electricity was safer and more desirable and should be adopted wherever possible! But he considered that gas would be mainly used for a long time to come.

Regarding the telescoping of the carriages the inspecting officer noted that this had been a feature of a number of recent accidents. He suggested that attention should be drawn to:

(a) The need for the use of rolling stock of uniform weight, for example not marshalling light four or six-wheeled vehicles between modern heavy bogie vehicles.

(b) The design of buffers could perhaps be altered to prevent telescoping.

(c) The deepening of the headstocks by the addition of a steel frame below and possibly above.

The addition of a small number of tools on long distance services was desirable.

It was doubtful the provision of fire extinguishers would have made any difference to tackling a fire caused by gas under pressure.

The external doors of the corridor coaches were adequate in number but a central external door in the sleeping cars was desirable. Windows should not be so small as to prevent egress and barred windows were undesirable.

Railway servants should receive training not only in carrying out protection and obtaining assistance but in looking out for fires and dealing with them.

The general public were mostly unaware that many railwaymen were trained in ambulance work and first aid and that many railway centres contained ambulance stores.[417]

The jury at the inquest into the 11 identified victims of the disaster returned a verdict of 'Accidental Death' with a rider drawing attention to the violation of Rule 55 and recommending that the regulations should be made more effective, and light engines should start from the signal box.[418]

---

417 Board of Trade report

418 *The Times*, London, England, 27th, 28th, 29th, 30th and 31st December 1910 and 13th January 1911

Eastbound Northern Pacific Railroad passenger train No. 42 was stopped on the main line at Cheney, Washington, United States, early on the morning of Wednesday 4th January 1911, when it was run into from behind by train No. 2, the *North Coast Limited*, from Seattle, Washington, to Chicago, Illinois. There was ample time for the first train to have gone into the siding but for an initially unexplained reason it was stopped for five minutes on the main track.

The location was preceded by a mile of straight track and when the engineer of train No. 2 saw the obstruction he made an emergency application of the air brakes, but the speed of the train was estimated to have still been as high as 35 to 40 miles an hour at the moment of the collision. The locomotive of the colliding train ploughed through the standing train for three car lengths before it could be pulled up. Five people were killed, and a large number injured. The death toll would have been higher had it not been for the fact that there were only six passengers in the smashed Pullman cars.[419]

The state railroad commission investigated the accident and found that the crew of train No. 42 had not properly protected their train, the engineer of train No. 2 had not had it properly under control and the Northern Pacific rules regarding trains entering Cheney yard were misleading. In summary, the commission stated that, 'Gross negligence on the part of Conductor J. B. Le Rue and Rear Brakeman F. C. Titman of Northern Pacific train No. 42, was the direct cause of the wreck Wednesday morning at Cheney, five people being killed, and a number injured when train No. 2, the North Coast Limited, collided with the rear of train No. 42.' [420]

On Friday 13th January 1911 New York Central and Hudson River Railroad westbound train No. 49, the *Boston and Buffalo Special*, was standing in Batavia station, New York, United States, when it was run into by train No. 23, the *Western Express*, also westbound. The rear Pullman car of the first train was telescoped and the engine continued into the day coach beyond. Four passengers were removed from the wreckage, so mangled and scalded as to be only identifiable with difficulty. They were identified as Cassius C. Perrin, assistant general manager of a detective agency; Fred J. Speaker, of the Dodge Publishing Co, New York; A. R. Brown, of 809, Tallman Street, Syracuse, representative of the George C. Batchelor Co, New York; and probably Leon Doncourt, representative of the Cary Brothers, Philadelphia (identification not positive). A fifth passenger, Robert McFarland, of 2,989 East Grand Boulevard, Detroit, Michigan, died in hospital and Conductor F. Van Valkenburg of train 49 and a passenger, H. J. Cade, of 6,211, Lexington Avenue, Chicago, were probably fatally hurt.[421]

Weather conditions at the time of the accident were bad, with a heavy fog and drizzling rain which at times became heavier and turned into ice the moment it touched anything. Train No. 49 had arrived at Syracuse first and thus preceded train No. 23 onwards towards Buffalo. Engineer Lydell apparently admitted passing a caution signal 4,000 feet from Batavia station, and also a stop signal, only applying the brakes when he saw the tail lights of the first train through the fog. He also reversed his engine and shouted a warning to his fireman, then jumped.

[419] *New York Times*, New York, New York, United States, 4th January 1911
[420] *The Brainerd Daily Despatch*, Brainerd, Minnesota, United States, 7th January 1911
[421] *New York Times*, New York, New York, United States, 14th January 1911

'Patrolman Donald Elliott, on duty at the station, was an eye-witness to the accident. He stood beside the ill-fated train and saw the Western Express bear down upon it. It was impossible to give any warning, and the officer got out of the way of the wreckage just in time to save his own life. "It was not yet daylight," the officer said in describing the scene. After the crash there was a red glow from the firebox of the overturned engine, and in the pile of wreckage, half as high as a house, on top of which the wrecked locomotive steamed, little fires started here and there. Passengers in the Pullman who were able were endeavouring to crawl out of the wreckage. There were cries and groans from those who were pinned down. One man succeeded in getting his head and shoulders to the side of the car and drew himself as far as he was able on top of the wrecked engine. He cried frantically to the railroad men scurrying about with lanterns and to the people who came running to the station from all directions: "My God, help get me out of here. I'm caught and I'm burning to death." Several railroad men went quickly to his rescue, clambering on top of the hot locomotive with axes. Portions of the car were chopped away, and the man, who lost consciousness, was hurried to the hospital.

'Among the uninjured passengers in the Pullman cars forward in the Boston special were Dr Elsie Seeley Pratt of 1.434, Glenarm Street, Denver, Col, who was bound for Cleveland, and Miss Winters, of Yonkers, a trained nurse, who was on her way to Tacoma, Wash. Learning that men and women had been killed and injured, the physician and nurse dressed hurriedly and went back to the wrecked coach. They were of great help in giving first aid to the injured as they were taken from the demolished car. In bandaging wounds the two women tore some of their clothes to shreds.

'Presence of mind undoubtedly saved the life of James O. Moore, of Buffalo, former United States District Attorney. As it is, he is in the hospital with his back injured, a possible fracture of a rib, and his entire body bruised. His physician, Dr Joseph Burke, of Buffalo, who came to Batavia to attend him, says Mr Moore told him jets of live steam escaping from the boiler shot through a portion of the wreckage where he lay. He pulled towards him all the bedding he could reach and wrapped it around his head and chest, thus protecting the vital portions of his body. He was lifted from the wrecked car.' [422]

Superintendent Everett, of the Western division of the company, made a statement regarding the accident. 'From the preliminary investigation it is shown that J.B. Lydell, of Buffalo, engineer of train 49, is responsible for the accident in that he disregarded the signals. Lydell has been in the employ of the company for 40 years, and has been an engineman since 1888, with a perfect record.' [423]

H. H. Cade died in hospital on 16th January making the sixth death in the collision.[424]

The Interstate Commerce Commission pointed out that the Bavaria accident and a similar one on the Pennsylvania Railroad near Nevada, Ohio, again demonstrated the need for a device to stop trains automatically when signals were overrun, something upon which little progress had been made.[425]

---

[422] Ibid.

[423] *The Brunswick News*, Brunswick, Georgia, United States, 14th January 1911

[424] *The Fairfield Daily Journal*, Fairfield, Iowa, United States, 16th January 1911

[425] *Cheyenne State Leader*, Cheyenne, Wyoming, United States, 24th January 1911

The Taff Vale Railway of South Wales suffered the third and last of the major disasters to occur on its system on Monday 23rd January 1911. Like the Willesden accident the month before it was a rear collision resulting from a signalman's error. In this case the mistake was made by one of two signalmen and it was not entirely clear who was at fault.

The accident occurred at Coke Ovens, near Pontypridd, and involved a coal train and a passenger train, the latter running into the back of the former. The signal boxes concerned were Gyfeillon Lower and Rhondda Cutting Junction, situated 986 yards apart. There were three down lines between them: - No. 1, used by passenger and mineral trains and Nos. 2 and 3 relief roads, used only by mineral trains. All three down lines converged into one approaching Rhondda Cutting box. The lines were signalled with Tyer's two position block instruments.

Signalman Albert Hutchings told the Board of Trade inquiry into the accident that he had come on duty at 9 a.m., relieving Relief Signalman Phillip Dipper, though the latter remained in the box as assistant. According to him he accepted a through mineral train from Gyfeillon Upper box at 9.23 and had it accepted by Rhondda Cutting at the same time. He sent the "Train Entering Section" signal for it at 9.31 and received the "Train out of Section" from Rhondda Cutting at 9.36. A coal train from the sidings at his box for the Barry Railway had been accepted by Rhondda Cutting at 9.29 on No. 3 road, sent on at the same time and cleared back at 9.39. The next train had been a passenger train from Treherbert which Rhondda Cutting accepted on No. 1 road at 9.43 and Gyfeillon Upper sent on at 9.47. It was passing and he was just about to send it on to Rhondda Cutting when the latter signalman gave him the 3-3 "Blocking Back" signal.

Signalman Dipper said that he had been getting his breakfast at the time. He remembered, he said, Hutchings asking for the coal train but not his sending it on. He also remembered Rhondda Cutting accepting the passenger train. As he failed to remember the blocking-back signal and thought that Hutchings had sent on the passenger train, which Hutchings himself denied having done, the Board of Trade inspecting officer, Lt-Col E. Druitt, considered that he clearly had not been taking much notice of what was going on and that his evidence should be treated with caution. Dipper said that as the passenger train was going by Hutchings had shouted, "He has blocked the road," and had thrown back his signals, but too late to be of any use, while he (Dipper) had gone to wave a red flag but was also too late. Hutchings had then looked at his train register and exclaimed, "Good God, there is a mineral on No. 1". Either before or after going to the book Hutchings had gone to the telephone and he had then collapsed in a faint and asked Dipper to take over, which the latter did.

Also in the box from about 9.31 had been Groundman W. H. Merchant, who had been checking the traffic in the sidings. He could give no information as to the signalling but remembered Hutchings saying, "Good God, what is down No. 2?" and dropping his pen and collapsing.

Signalman Sidney Quick at Rhondda Cutting Junction said that the first train he knew about at the time of the accident was the Barry mineral train on No. 3 line which (allowing for a three-minute difference in the box clocks) he had accepted and taken on at 9.29. He was positive that he had never accepted a mineral train on No. 1 road and there were no entries in his train register regarding it. He had accepted the passenger train at 9.43. Then at 9.47 Fireman George Lewis of the coal train had come into his box and said, "We are right away to Roath line, standing on No. 1." Quick immediately went to the block shelf and belled the blocking back signal, 3-3, to Gyfeillon Lower.

District Mineral Train Inspector H. Tuck had come into Rhondda Cutting box at about 9.30 or 9.35. He said that he recalled Fireman Lewis coming into the box and moaning to Signalman Quick, "We are on No.1, you let the Barry train out in front of us." And, "We are right away on No. 1 to Roath."

To which the signalman had replied, "I have no record of you at all," but the inspecting officer remarked that Tuck did not seem to have a very good recollection of what had happened.

Fireman Lewis stated that when their train had come to a stand at the Rhondda Cutting outer home signal he had commenced to fill the lubricators. This had taken him five or six minutes, then another four or five minutes had been consumed in walking to the box, as he had had to wait for an up mineral train before crossing the line. In the box he had told Signalman Quick that they were right away on the Roath line. The signalman had not taken it in for a moment, and then had asked him which road he had said he was on and he had repeated they were for Roath from line 1. Quick then told him, "I did not know you were there," and had rushed to the telephone shouting, "Albert! Albert!"

Inspector Tuck had remarked, "You had better get the road for this man as quickly as you can, or you will stop the passenger train."

Driver A. Davison of the mineral train said that he had whistled as he approached and stopped at the Rhonda Cutting outer home, bringing his train to a stand at 9.34.

Driver A. Sellars of the passenger train said that the Gyfeillon Upper and Lower signals had all been clear, the Rhondda Cutting distant had been at danger and he had applied the vacuum brake, preparing to stop at the home signal, when he had seen the brake van of the mineral train about 20 yards away. He had seen it earlier but with the curve and so many lines had assumed it was not on his track. He had braked hard but with greasy rails his train's speed was little reduced, and his train was travelling at 27 or 30 miles an hour when it ran into the van.

Fireman John Jones corroborated his driver's testimony.

In the collision the brake van of the mineral train was demolished, and two waggons were wrecked, and others derailed. More significantly in the passenger train the leading brake van was telescoped into a third-class carriage and another brake van telescoped a second third-class carriage. Eleven passengers were killed and four passengers seriously and 17 slightly injured. Goods Guard George Gould of the mineral train was seriously injured and later died.[426] The passengers who lost their lives included three members of the South Wales Miners' Federation, a Nonconformist minister, a nine-year-old boy and a ten-year-old girl.

A newspaper reporter wrote, 'One of the most interesting of the many accounts of the accident which have been given to-day by passengers on the wrecked train is that of Mr William Jenkins, a miners' agent, of Cymmer. He explains that was usually his practice to leave his carriage at Porth and to join the Rhondda Fach miners' agents in the front part of the train. He did not do so to-day, as he had a friend with him. When they reached

---

[426] Board of Trade report

the Coke Ovens, they felt the brakes being suddenly applied, and the train instantly came to a standstill. Both he and his friend were hurled to the side of the compartment, but they had no difficulty in getting out. He was in the fifth coach, and when he got out, he saw that the rear part of the fourth coach had mounted the front part of that in which he had been. The fourth coach was smashed to ribbons. He saw in front a long train of 30 or 40 waggons loaded with coal. His first thoughts were for his colleagues, and he was horrified to find three of them dead. He thought the fourth coach had suffered most and described the rescue work as amazingly prompt. Mr William Phillips, registrar, of Porth, was sitting, with two other passengers in one of the telescoped coaches. Both were killed, but Mr Phillips was only slightly injured. A great deal of difficulty was experienced in reaching those who were imprisoned in the wrecked coaches. Some of the carriage doors are said to have been locked, and many had to be hammered in before the unfortunate people within could be got out. The guard of the passenger train had his van telescoped and succeeded in climbing out by a hole in the roof.'[427]

Lt-Col Druitt stated in his report that it was hard to understand how the mistake had been made by one or other of the signalmen as there had been little traffic at the time. A possible way in which the mistake had been made, he said, though he did not say it was the correct one, was that Hutchings, after having offered and had accepted the mineral train, had omitted to send the "Train Entering Section" signal. Quick, for his part, had failed to book the mineral train in his train register.

Hutchings had then offered the Barry train on road 3 and when Quick had cleared this train back at 9.36 Hutchings had thought that Quick was clearing back the mineral train on road No. 1. With Quick having failed to take on the mineral train the indicators of the two-position Tyer's instruments would have shown "Train Arrived", their normal position, and there was then nothing to prevent Hutchings offering and Quick accepting the passenger train into the occupied section. Quick's surprise at Fireman Lewis telling him that the mineral train was on road 1 appeared to be genuine whereas if he had accepted the mineral train and then forgotten it he would have probably made a very different remark (such as, "Shit! I forgot all about the mineral!"). Quick should, the inspecting officer thought, have seen the mineral train on road No. 1 when he pulled off for the Barry train, even though he was not expecting it to be there. His explanation for giving the "Blocking Back" signal, 3-3, rather than the "Obstruction Danger" signal, 6 (that it was the first one that he thought of and the one which would be more familiar to Hutchings), was not entirely convincing, but Hutchings had understood in any case that the line was blocked.

Although he said that he did not wish to draw any conclusions from it, it was clear that Lt-Col Druitt considered that the fact that Signalman Hutchings had completely lost his head when he received the blocking back signal also pointed to it having been him who had mainly caused the collision. The inspecting officer also disputed Hutching's assertion that he had not had time to throw back his signals in the face of the passenger train. He noted that it was to be regretted that Hutchings had failed to seize the opportunity given to him.

Turning to the effects of the collision the inspecting officer noted that it had been aggravated by the passenger train having been pulled up very suddenly when it ran into

---

[427] *The Times*, London, England, 24th January 1911

the mass of coal in the waggons of the mineral train, and by the telescoping of the second and fifth coaches of the passenger train by the rear ends comparatively light brake vans in front of them jumping up so that their steel underframes cut through the following coaches. The carriages had been lit by oil gas but there had been no leakage of gas and there was no fire.

The two position block instruments in use gave less information to a signalman than the three position instruments in more general use and should a signalman fail to give the "Train Entering Section" signal and then wished to offer a second train there was nothing in the position of the instrument to remind him that he had already offered one, or to remind the man in advance other than the entry in his train register. Three-position block instruments should be substituted for two-position ones in all busy boxes and where there were several running lines. (Two-position instruments remained in use in some quieter boxes until at least the 1970s.)

Traffic Inspector Tuck had come to the box to question Signalman Quick about some delay to Great Western empties the week before. The inspecting officer did not think that inspectors should come to the box to examine train registers and interrogate the signalmen while they were on duty. As regards Marchant he was a groundman whose duties were to check the numbers of waggons leaving the yard, assist mineral guards and convey instructions between signalmen and guards. It was more convenient to give him a desk in the box than in the traffic office in the yard.

Driver Davison excused his not sending his fireman to the box immediately by saying that it was customary for the fireman to fill the lubricators, which could only be done while the train was stationary, and that he was not expecting to get away until after the Barry train. Lt-Col Druitt did not think these reasons very convincing, and that the fireman had either taken longer than he said in filling the lubricators or had dawdled making his way to the box. It was however a dangerous place to carry out the rule and in such locations the installation of a track circuit would be useful, both to obviate the need for the fireman to go to the box and as an aid to the signalman. But the rule was an important one, and should be strictly carried out, with the difficulties at specific locations reported.

Finally the inspecting officer pointed out that if Guard Gould had kept a look out he could have seen the signals cleared for the passenger train five minutes before the collision, in time to go back and lay detonators on the line if not to reach Gyfeillon Lower box [428] (and would have saved his own life).

At the inquest into the victims of the accident the coroner's jury, unable to decide which of the two signalmen had been at fault, returned an open verdict, which censuring Driver Davison for not sending his fireman to the box sooner.[429]

A feature of American railroads in the years after the First World War was the running of so-called silk trains. The high price of silk during this period justified the running of special trains carrying raw silk skeins from West Coast ports through St Paul and other

---

428 Board of Trade report

429 *The Standard*, London, England, 1st February 1911

mid-West centres to distribution points in New York, New Jersey and Philadelphia.[430] One of these trains was involved in an accident on Monday 18th December 1911. The first section of train No. 18 of the Chicago Milwaukee and St Paul Railroad, known as the *Columbian,* running from Tacoma, Washington, to Chicago, Illinois, was run into by the second section, a silk train, at Odessa station, Minnesota, United States.[431] The first section of the train consisted behind the locomotive of a mail and baggage car, two coaches, a tourist car, a dining car and the Pullman sleeping car *Beverley.* It was in charge of Conductor Gillick with Engineer Eddington.

Odessa station was situated on a single-track main line, worked by a manual block system, supervised by a dispatcher at Aberdeen, North Dakota, 114 miles away. It was formerly open day and night but a couple of months earlier the arrangement was altered to save money and at it was now closed between 1.30 a.m. and 8 a.m. each night. During this time the block section extended from Junction Switch to Appleton. When the Odessa office was open the block signal was normally kept at danger but when it was closed the signal was left clear. However, when train first No. 18 arrived at Odessa at 3.52 a.m. the signal stood at danger. After the train had stopped, Conductor Gillick instructed Head Brakeman Miles to go to the house of the operator and rouse him, and Flagman Law to go back and set the switch at the beginning of the passing track for the other line and then go back further to flag second No. 18. Operator Adams duly appeared (doubtless not best pleased at having been woken up in the early hours) tried the block signal several times and left it in the clear position. He then asked Train Dispatcher Noonan if the latter wanted to alter the meet of first No. 18 from Appleton to Correll (the next station to Odessa) obviously because first No. 18 was now delayed. The dispatcher asked if first No. 18 was protected by flag and having been assured that it was arranged with the Appleton operator to alter the crossing of first No. 18 and No. 17 to Correll. The dispatcher then told Adams that second No. 18 had left Junction Switch.

Operator Montel told the Interstate Commerce Commission (ICC) investigators that before second No. 18 arrived at Junction Switch at 4.8 a.m. he had asked Operator Kaiser at Appleton if first No. 18 had reached there and he claimed that Kaiser had told him that it was coming close (Operator Kaiser strongly denied this). Operator Kaiser said that the Junction Switch operator had twice asked him if first No. 18 was coming and both times he had replied, 'Not coming.'

Dispatcher Noonan said that when Operator Montel had first asked him if he could clear second No. 18 he had replied, 'No.' Contacted by Montel again five minutes later about releasing second No. 18 at 4.14 he had contacted Appleton and asked if first No. 18 was coming and, like Operator Montel, he said that Appleton had told him, 'Coming right close.' Second No. 18 had a meet with No. 17 at Correll and he was sure that the Junction Switch-Correll section must be clear so he had told Montel, 'OK, clear,' (The regulations permitted the dispatcher at his discretion to issue a permissive card to a train permitting it to enter an occupied block section before the previous train had cleared).

Soon after Dispatcher Noonan had released section No. 18 from Junction Switch Operator Adams advised him that in fact first No. 18 was still at Odessa, having stopped there because the block signal had accidentally gone back to danger (as had happened

[430] *Silk Trains on the Great Northern Railway* by Gordon L. Iseminger
[431] *Cook County Herald*, Arlington Heights, Illinois, United States, 22nd December 1911

before). Dispatch Noonan then asked Adams if there was a flagman out to protect first No. 18 and Adams confirmed that there was. He then altered the crossing of first No. 18 with No. 17 from Appleton to Correll, then got on with other work.

Unfortunately, Flagman Law failed to carry out his conductor's orders. He did not reset the switch at the beginning of the loop to divert second No. 18 to the other track, nor did he go back very far to flag it. Instead he hung around about 500 feet from the rear of his train. He defended his actions at the ICC inquiry by saying that he depended on the block to protect against following trains, and that with the straight track he could see a coming train for five miles. When Operator Adams cleared the block signal he thought that his train was ready to proceed and returned to its rear, confident that the second section of No. 18 was still at Junction Switch. It was only when Head Brakeman Miles had shouted that there was a train coming that he knew otherwise. He had then run back but could not get as far as the switch to divert second No. 18 into the loop before the train reached it.

Operator Montel agreed that after speaking to Dispatcher Noonan a second time he had given Conductor White of second No. 18 a clearance card, indicating that the block was clear, without a permissive card, though he knew first No. 18 had not cleared the block. Asked why, he said that it had been done before. Conductor White likewise admitted having left Junction Switch with a clearance card but without a permissive card, Montel having told him that the Appleton operator had said that first No. 18 was 'coming close' to there.

Engineer Sorby of second No. 18 said that the Odessa block signal had been set at clear when he first saw it. Whistling for a road crossing, he had then given a long whistle for the station and soon afterwards saw the outside marker of first No. 18 and applied the air brake. A moment later he saw both markers and made an emergency brake application. Only then did he see Flagman Law a few car lengths behind the stopped train. Engineer Sorby said that if Flagman Law had gone back as far as a cattle-guard about 100 feet west of the house track switch he could have been seen in time to prevent the collision. As it was, the engine of second No. 18 penetrated several feet into the *Beverley* sleeping car while the forward end of the car was telescoped into the dining car. In the collision eight passengers, the conductor of the *Beverley* car and a cook in the dining car were killed and 13 passengers and five railwaymen were injured. The collision would doubtless have been more serious had not Engineer Eddington, who was out on the platform, shouted to his fireman to start the train. The latter managed to move it forward about a car length before the collision.

Not surprisingly, perhaps, the ICC laid the principal blame for the accident on Flagman Law for having failed to go back and protect his train as instructed by his conductor. Others, though, were also at fault. Dispatcher Noonan should not have directed Operator Montel to clear second No. 18 when he knew first No. 18 was still in the block. Operator Montel at Junction Switch should not have allowed second No. 18 to enter the section with a clearance card, knowing that the block was occupied, without a permissive card and the same criticism applied to Conductor White. Finally, Operator Adams at Odessa should not have left his block signal at clear while taking orders from the dispatcher and before first No. 18 was ready to start.

Noting that this was the second accident on the Chicago Milwaukee and St Paul within a few weeks the ICC remarked that steps should be taken to ensure that company employees had a working knowledge of the rules and that they obeyed them. The long (20.6 mile) section between Junction Switch and Appleton provided, the ICC considered, a great temptation to employees to take chances to expedite the running of trains.

The fact that the casualties had occurred in steel cars worried the ICC, given that the use of such cars was often recommended as a means of preventing loss of life, but it was concluded that no design of car could have successfully withstood the shock of such a collision.[432]

A collision at Paris Austerlitz station on Friday 5th January 1912, which injured about 40 passengers, was followed by a worse accident the very next day, Saturday 6th January 1912, on the Chemins de fer de L'Est, only a few miles out of the French capital. A local train, 535 bis, left Paris Est station at 7.25 p.m. It was booked to run non-stop as far as Bondy and then call at all stations to Gargan. The line between Bondy and Gargan was still single-track, a second set of rails having been laid the previous year but not yet brought into use. Having travelled as far as Pont-des-Coquetiers, beyond Bondy, train 535 bis thus had to wait there for the 7.36 train from Gargan to clear the single line, and this evening the latter train was somewhat delayed. Behind train 535 bis was train 535, which left Paris Est at 7.30, called at Est-Ceinture, ran through Pantin, called at Noisy-le-Sec and Bondy and was due at Pont-des-Coquetiers at 7.59.

Train 535 bis was crowded with workmen going home to their families who according to a newspaper report were laughing and singing. It should have been fully protected by a stop signal as well as by its red tail lamps. Driver Louis Gerschel of train 535 apparently saw the tail lamps of the train in front but thought that they were those of a train stabled on the other line and did not at once attempt to pull up. By the time he had realised his mistake it was too late, though, as the newspaper reporter put it, the low speed of his train prevented a bloodbath. The effects of the collision were serious enough though. The rear guard's van and following third-class carriage of the front train were completely wrecked and the colliding train did not escape either, for although its leading guard's van escaped with little damage the carriage behind was driven into it. Eight passengers were killed outright, two fatally injured and almost 20 less badly hurt. Driver Gerschel, his fireman, Jean Petitpas, and the guard in the leading van all escaped unhurt. Driver Gerschel was arrested after the accident.[433] [434] [435]

On Thursday 4th July 1912 Delaware Lackawanna and Western Railroad westbound passenger train No. 9 from Hoboken, New Jersey, to Buffalo, New York, consisted of locomotives Nos. 1,052 and 973, a buffet car, seven Pullman sleeping cars and two coaches, all of wooden construction except the second car from the rear. It was in charge of Conductor Staples with Engineers Still and Schick. The train had left Elmira, New York, at 4.37 a.m. and approached the freight train at East Corning, New York, United States, at about 5 a.m. There it was stopped by the flagman of westbound freight train extra No. 393, which had stopped because the draft sills of a car had broken. The pilot

---

[432] Interstate Commerce Commission report

[433] *Le Petit Parisien*, Paris, France, 7th January 1912

[434] *The Times*, London, England, 8th January 1912

[435] *Le Figaro*, Paris, France, 13th January 1912

engine of train No. 9 was then detached and pushed the rear end of extra freight No. 393 into a siding. This had been done and the engine was returning to its train when the rear end of the train was run into by train No. 11.

Train No. 11 was also running from Hoboken to Buffalo. It consisted of locomotive No. 1026, seven express cars and a combination mail and express car, in charge of Conductor Sullivan and Engineer William Schroeder. It had left Elmira at 5 a.m., run past block signals at 'Caution' and 'Danger' and the flagman of train No. 9 and collided with the rear of train No. 9 250 feet west of automatic signal No. 2771.

The rear coach of train No. 9 was completely destroyed. Both ends of the next steel coach were crushed and that coach telescoped the wooden sleeping car *Esthonia*, third from the end, for two thirds of its length. The rest of train No. 9 was little damaged. Thirty-nine passengers were killed, and 86 passengers and two railroad employees were injured, many seriously.[436]

Miss Marie Falco, a passenger on train No. 9 who escaped injury, related her story to a reporter. "I had been awake for some time before the crash. When it came our car was crumpled up into a mass of wreckage before anyone realized what had happened. I imagine many were killed while asleep. After the shock the car continued shivering from end to end. The sides and roof, which were ripped wide open when we struck, fell inwardly. I occupied a lower berth, which was underneath my brother's. When I gained possession of myself, I saw that his berth had fallen partly from the wall. He was beneath it, gripped tightly in the wreck of the interior of the car. I could see that the weight of the berth and part of the wall of the car were resting on him and gradually settling. I could see that, if he were not dead already, the life was being gradually crushed out of him. He was unconscious, and the blood was flowing from a wound on his head. I could not reach him. I was cut off from him by the berth. The interior side of it had fallen, but its hinges held it firmly to the wall. It sheltered me from the wreckage and undoubtedly saved my life. My knee was pinned under one edge of the berth, but I managed to get it loose and I turned so I could look out of the window.

"Our car was one that remained standing. Out of the window I saw two men and shouted to them. One of them came to the window, The window was shattered and full of jagged edges of glass, but I was in too much of a hurry to notice it. I got through all right and reached the ground by stepping on the man's shoulder and balancing myself with his hand. By this time many others had gotten out and were running around excitedly. Women who had reached safety were screaming, and they were answered by the cries inside the wreckage for almost the whole length of the train. But I had no time to think of that, I was so intent on getting to my brother. The passage into our Pullman had somehow suffered little damage, and I intended to go through it and crawl under the debris till I reached my brother.

"As I started into the train some men came out carrying a dead man on a mattress. I thought it was my brother. I tried to control myself and to give directions to the men who were carrying him away. When I said he was my brother one of the men who bore the mattress said, 'No, Madam, this is not your brother. This is my chum.' Then I looked at the body and saw the mistake I had made. I ran back to the train. This time I saw men who really were carrying out my brother, but though he was bleeding and seemed to be

[436] Interstate Commerce Commission report

badly injured, I knew he was not dead. I spent most of the time with him after that. I had so much to be excited about, that my recollection of the things I saw is very vague, and I am not now able to describe them." [437]

Engineer Hemphill of train extra 393 told the Interstate Commerce Commission (ICC) inquiry into the accident of his train parting due to the breaking of draft sills on one of the cars as he was shunting out of the way of train No. 9.

Conductor Halloran of train extra 393 stated that it had been so foggy At East Corning that he could only see a few car lengths. He had instructed Flagman O'Connor to protect the rear of his train.

Flagman O'Connor said that it had been foggy when he had dropped off train extra No. 393. He had put two torpedoes on the rail and flagged train No. 9, which had pulled up.

Engineer Still on the pilot engine of train No. 9 stated that it had been very foggy, and he could only see signals at a distance of a car length. Signal No. 2773 had been at caution, he had run over the detonators laid by Flagman O'Connor and stopped at signal No. 2781. He had immediately whistled out the flagman. Then he had cut off his engine to push train No. 393 into the siding.

Engineer Schick on the train engine agreed that the weather had been very foggy but said that he had been able to distinguish all of the signals.

Conductor Staples of train No. 9 spoke of seeing Flagman Lane go back. It had been foggy, but he thought he could see a distance of 20 or 30 car lengths.

Flagman Edward Lane of train No. 9 testified that as soon as the train had stopped he had gone back with a red flag and one green fusee to a point midway between signals Nos. 2781 and 2773. It had been a little foggy, but he had been able to see 30 or 40 car lengths. After 10 or 12 minutes he had heard train No. 11 coming and had lit the green fusee and placed it on a tie on the engineer's side of the line. He had not used any torpedoes as required during foggy or stormy weather. Train No. 11 had approached him travelling at 60 or 65 miles an hour and he had flagged it with his red flag until forced to step out of its path. There had been no response to his flag signals and the engine was still working steam. He could see the engineer, who appeared to be looking over the boiler towards the left side of the engine rather than out of the window.

Conductor Sullivan of train No. 11 said that its air brakes had been tested at Elmira where they had been working normally. It had been foggy. There had been no application of the brakes before the collision and the first that he knew of it was feeling the shock of it.

Flagman Sweet of train No. 11 confirmed his conductor's evidence. He said that he had gone back as soon as his train had stopped, had seen the green fusee put down by Flagman Lane still burning and had met Flagman Lane midway between signals Nos. 2781 and 2773. Continuing back to signal No. 2773 he had found it in the "Stop" position.

Engineer Schroeder stated that he had arrived late at the roundhouse that morning and so had had to take his locomotive out without oiling or inspecting it. There had been no

[437] *New York Times*, New York, New York, United States, 5th July 1912

fog at Elmira, but it had become very foggy near East Corning, so that signals could only be seen at a car's length. He had been running at 60 miles an hour. Soon after leaving Elmira there had been a problem with one of the injectors, which had been feeding too fast. He had tried to get it to work properly without success and had then found that the other injector could not be made to work properly either and had had to go out onto the running board of his engine to try and adjust it. While he was still out on the running board his fireman had spoken to him about the water getting low. He admitted that while he had been working on the injectors and then talking to his fireman he had not been continuously on the lookout for signals and that he had seen neither signal No. 2773 nor the flagman's fusee. He had seen signal No. 2781 at red about 150 feet away and beyond it the rear of train No. 9, when he had made an emergency brake application and thought that he had closed the throttle. At the coroner's inquest it had been suggested that he had been under the influence of alcohol the night before but this he denied. He conceded that he had had two drinks of gin between 10 and 11 p.m. and had visited a number of other saloons but said that he had had no more to drink in any of them. He had slept at home between 2 and 6 p.m. on the afternoon of the 3rd, had gone out at 9 p.m. and returned home and gone to bed at 12.10 a.m. He had answered the call boy at 3 a.m. but had not got up and it had been necessary to send the call boy again at 4.10 a.m. He had then got up and arrived at the roundhouse late at 4.48 a.m.

Fireman Mac Huntley of train No. 11 testified that he had gone out onto the running board and spoken to his engineer about the level of water in the boiler sometime after leaving Big Flats, a point midway between Elmira and East Corning. He thought that this must have been when they were nearing signal No. 2773 as he had seen neither the signal, the fusee, nor the flagman. He had still been on the running board when he had seen signal No. 2781 and the train ahead.

The ICC report went on to explain that engine No. 1026 was of the Wootten type, with separate engineer's and fireman's cabs, linked by a running board. The water glass and injectors were in the engineer's cab with only gauge glasses in the fireman's cab so that if the fireman wished to see the water glass or if the injectors were working properly he had to make his way to the engineer's cab.

The ICC concluded that the accident had been caused by the failure of Engineer Schroeder to observe the automatic block signal indications and the indications of Flagman Lane. It pointed out that a greater degree of protection would have been afforded had the signals been properly overlapped so that an obstruction was protected by a 'Caution' and two 'Stop' signals. Such overlapping would reduce the capacity of the line by increasing the headways between trains, but it was used by a number of railroad companies and was recognised as a much safer method of train operation. Beyond this there was a need for an automatic train stopping system, such as would automatically stop a train should the engineer fail to obey a stop signal due to his being asleep, dead or physically incapacitated, or distracted by other duties, or if the signal was obscured by fog, snow or smoke. The commission had made the same call for automatic signalling in reports on the accidents at Bridgeport, Connecticut; Hoosac Tunnel, Massachusetts and Fort Wayne, Indiana. In the absence of such measures trains should be prohibited from running at high speed in foggy or stormy weather.

A contributory cause of the accident was the failure of Flagman Lane to use torpedoes. The commission noted that Lackawanna's rule 99 left it to the judgement of the flagman how far back he should go back and whether he should use torpedoes; these being required only in foggy or stormy weather. A further complication was that the rules prohibited the use of torpedoes in clear weather (doubtless to avoid annoying lineside residents with the noise, like the regulations prohibiting excessive whistling approaching level crossings). In this case Flagman Lane's judgement had not been at fault inasmuch as he had gone back far enough but he had not considered it foggy enough to use torpedoes. Definite distances for the flagman to go back should be laid down and the use of torpedoes made mandatory under all circumstances. In addition Rule 583, which required flagmen to have not less than four torpedoes attached to their flagstaffs, had been habitually disregarded by employees on the line for many years.

Train No. 9 normally ran with a steel car at its rear but because of the extra 4th of July traffic four sleeping cars and a coach had been added to its composition, with wooden car No. 86, built in 1904, at its rear. It was in this car that most of the fatalities had occurred. In order to provide for the safety of the travelling public steel cars should be substituted for wooden ones in high-speed trains at the earliest practicable date.[438]

The jury at the inquest on the victims of the disaster found both Engineer Schroeder and Flagman Lane guilty of negligence and warrants were issued for their arrest on charges of second degree manslaughter. The jury recommended the employment of two engineers on 'Mother Hubbard' (Wootten) type locomotives, the use of torpedoes at all times and the abolition of wooden coaches.[439]

In late September the Grand Jury indicted Engineer Schroeder but not Flagman Lane.[440] The engineer eventually stood trial in June 1913 and was acquitted, a verdict which did not please the New York District Attorney, whose assistant, James Sebring, wrote to a newspaper that the acquittal had been a failure of justice, and that the jury's verdict had been influenced by sympathy for the engineer, whose wife had just died, his good character and long service, the dismissal of the first indictment against him and the long delay before his trial, but most importantly, by 'the tremendous and overpowering' influence of his union, the Brotherhood of Locomotive Engineers and Firemen.[441]

On the morning of Sunday 14th July 1912 there was a rear collision at Western Springs, Illinois, United States, on the main line of the Chicago Burlington and Quincy Railroad between Chicago and Aurora. At this point the main line had three tracks, one for trains in each direction and the third for use by trains in both directions. The line was worked by a manual block system, with telephones. Lower quadrant stop signals, showing *danger* and *clear* indications, were in use, without preceding distant signals. The block stations directly involved in the accident, from east to west, were Congress Park, Western Springs and West Hinsdale.

The sequence of events leading up to the accident began at 6.10 a.m. An eastbound freight train, No. 74, was at Congress Park and had work to do which would involve fouling the southernmost, eastbound track. The Congress Park operator therefore telephoned Mrs Julia Wilcox, the Western Springs operator, to ask her to let him know

---

438 Interstate Commerce Commission report

439 *The Syracuse Herald*, Syracuse, New York, United States, 24th July 1912

440 *New York Times*, New York, New York, United States, 28th September 1912

441 Ibid, 28th June 1913

when eastbound train No. 4 was passing West Hinsdale. Mrs Wilcock understood the message, however, as to mean that she should stop train No. 4, and kept her stop signal on. Approaching at high speed, and with a dense fog prevailing, the engineer of train No. 4 was unable to stop at the Western Springs signal and overran it by 250 or 300 feet. The flagman of train No. 4 came back, Mrs Wilcox was informed that the line was clear at Congress Park, the flagman rejoined his train and it proceeded. Before train No. 4 had cleared Congress Park train No. 2 arrived at Western Springs and likewise overran her signal, this time by about 850 feet. Train No. 2, the *Overland Express* from Denver to Chicago, consisted of engine No. 2816, two baggage cars, two coaches, a chair car, two tourist sleepers and three standard sleepers, in charge of Conductor Frank Hughes and Engineer Eno. Engineer Eno said that the fog had been so dense that he had not been able to see the signal until he was right on top of it.[442]

What exactly happened next is not entirely clear, and depended on whose account was accepted, but it would seem that Flagman John Woodworth of train No. 2 started back to protect his train. Mrs Wilcox then received clearance from Congress Park for train No. 4, received acceptance for train No. 2 and cleared her signal for it. According to her she told the flagman that it was now clear for his train to proceed and he went back towards his train, then came back a second time and told her that there was something wrong with it and it could not proceed. According to a newspaper report one of the rear Pullman cars of train No. 2 had developed a 'hot-box'. [443] Flagman Woodworth stated that he had only come back once and had already known that there was a problem which would delay his train when Mrs Wilcox told him that it could proceed. At any rate, Mrs Wilcox, under the impression that train No. 4 would shortly be underway, then cleared it back to West Hinsdale. In doing so she was strictly not breaking any rules, as the regulations permitted operators to report a block clear once a train had passed 150 feet beyond their signals. Given that another fast train, No. 4, the *Fast Mail*, from U.P. transfer, Omaha, to Chicago, was close at hand and the previous two trains had overrun her signal in fog this was, at any rate, very unwise.

Train No. 8 consisted of engine No. 2521 and seven mail and four baggage cars, all wooden but with three of the mail cars and one baggage car having steel underframes, in charge of Conductor Johnson with Engineer George Brownson. It approached Western Springs travelling at 50 or 60 miles an hour, ran past the block signal at danger, past Flagman Woodworth, past Brakeman Wicke of train No. 2, without any apparent reduction of speed and ploughed through the rear Pullman sleeping car, *Hamlet*, and partially demolished the next car, *Fernwood.* All of those who lost their lives other than Engineer Brownson were in the *Hamlet* car. Brakeman Wicke stated that the brakes had not been applied nor had steam been shut off when the train passed him but Fireman H. Crain said that Engineer Brownson had applied the brakes in emergency and reversed the engine when the engine ran over the first torpedo.[444]

Apart from those killed about 17 people were injured, including Conductor Hughes and Fireman Crain. A newspaper reporter wrote, 'The story of one real heroine comes from the wreck. She was Mrs E. G. Pohlman, bride of eight months of a Santa Rose travelling salesman. When rescuers reached her, she insisted that rescuers care for her

[442] Interstate Commerce Commission report
[443] *The Daily Commonwealth*, Fond du Lac, Wisconsin, United States, 15th July 1912.
[444] Interstate Commerce Commission report

husband first. He was seriously but not fatally hurt. When rescuers reached her, they found her with both limbs and her spine broken. She died on her way to the hospital.' [445]

The Interstate Commerce Commission (ICC) found that Flagman Woodworth was principally to blame for the accident by failing to go back far enough to protect his train although he was well aware that train No. 8 was likely to be following close behind and there was a dense fog. He had also erred in putting down two torpedoes – a caution signal – instead of one – a danger signal. The rules of the company, and many other railroads, were also not sufficiently definite in that they merely stated that the flagman must go back a sufficient distance without stating what that distance was.

Contributory to the accident was the failure of Engineer Brownson to approach Western Springs with his train under control so as to be able to stop at the block signal if he found it at danger.

The signals in use were another factor in the collision, the ICC contended. Stop signals without corresponding distant or caution signals were inadequate to control the movement of trains running at high speed, especially during foggy or stormy weather and the overrunning of signals was not uncommon. In the present instance three successive trains had run past the same signal at Western Springs. Distant signals were now apparently being provided on this division, meanwhile trains should not be permitted to run at high speed during foggy or stormy weather.

Rule 319 of the Burlington company, permitting an operator to clear the block once a train had passed 150 feet beyond the stop signal, was liable to create dangerous situations. In the present case, although Mrs Wilcox had exercised poor judgement, she had violated no rule.

Finally, although passenger trains were limited to 60 miles an hour over the division of the railroad where the accident occurred there was no limit to the speed of mail trains, which frequently ran at speeds of 75 or 80 miles an hour, and which had to run at high speed to maintain their schedules.[446]

On Wednesday 30th October 1912 five persons reportedly died and 40 were injured in a rear collision at Jannowitzbrücke station, Germany, on the Berlin Stadtbahn involving a train from Fredrichshafen and a Berliner Ringbahn train. The accident was blamed on 'faulty signalling'. [447] [448]

An accident near Montz, Louisiana, United States, on Monday 11th November 1912 had several all too familiar features, for example it involved an excursion train and the initial accident was followed by a devastating fire. The train involved was a return working from New Orleans, Louisiana to Woodville, Mississippi, over the Yazoo and Mississippi Valley Railroad. It consisted of engines Nos. 2103 (pilot) and 85 (train), a baggage car and nine wooden coaches and was in charge of Conductor Stinson with Engineers Montgomery and Wright. Also accompanying the train to ensure its safe

---

[445] *The Daily Commonwealth*, Fond du Lac, Wisconsin, United States, 15th July 1912.

[446] Interstate Commerce Commission report

[447] *The Daily Times*, New Philadelphia, Ohio, United States, 31st October 1912

[448] *The Sterling Daily Standard*, Sterling, Illinois, United States, 31st October 1912

movement was Assistant Train Master McBurney.

The train left New Orleans at 11 p.m. on Sunday 10th November (the Interstate Commerce Commission report states that the train left New Orleans on 11th November and that the accident occurred early on 12th November but as it was already reported in several newspapers of 11th November this is clearly not correct).[449] [450] Kenner Junction, 11 miles from New Orleans, was reached at 11.28, and here Conductor Stinson was given orders with a schedule for his train. The excursion left Kenner Junction at 11.38, 13 minutes late on its schedule. The train then travelled another 14½ miles and was within three quarters of a mile of Montz station when it came to a halt, an eccentric blade on engine No. 85 having become disconnected from its link. The train came to a stand at 12.05 a.m. and was still stationary half an hour later when it was run into behind by train No. 58, a regular manifest freight (a manifest freight was one carrying different types of cars with different loads for multiple destinations).

Freight train No. 58 was in charge of Conductor Rodney with Engineer Dreher and Fireman Taylor and was only a short one, consisting of engine No. 404, eight loaded cars, two empty cars and a caboose. In the collision the freight engine penetrated halfway through the rear coach of the excursion. The following two coaches were not badly damaged but the two coaches ahead of them, fifth and sixth from the engines, were telescoped for half their length. Fire broke out after the collision, caused by hot coals from the firebox of engine No. 404, and the three rear coaches of the excursion were entirely consumed, while the two telescoped coaches were partially burnt. The five front coaches of the excursion were undamaged and none of the cars of the freight train were affected. Fifteen passengers died and no less than 249 passengers and three railwaymen were injured.

Engineer Montgomery of the pilot engine of the excursion told the Interstate Commerce Commission (ICC) investigators that upon receiving a signal from Engineer Wright of the train engine he had stopped the train and had immediately whistled out the flagman. Conductor Stinson and Assistant Train Master McBurney both said that they had seen the flagman going back and had then themselves gone forward to find the cause of the stoppage.

Flagman Cunningham had indeed gone back straight away, taking with him a red and white lantern and three torpedoes, but no fusees, as he said the weather as he thought at the time was clear and he did not think he would need them. He said he went back 1,000 feet from the rear of his train and there affixed one torpedo to the rail. From that point he could, most of the time as there was an intermittent fog, see both the rear markers of his train and the switch lights at Sellars station, three-quarters of a mile from where he stood. He saw the headlight of train No. 58 as it was passing Sellars, at which point he took up the torpedo he had put down and started towards the freight train, intending to replace it further back. He signalled to the freight train engineer but got no response until the engine was almost upon him and he had no time to replace the torpedo.

Engineer Dreher, in contrast to Flagman Cunningham's statement that there had only been a slight patchy fog, stated that the night had been very foggy. He had not seen the flagman until he was close to both him and the rear of the excursion train. He had been

[449] *The Daily Herald*, Biloxi, Mississippi, United States, 11th November 1912
[450] *The Daily Ardmoreite*, Ardmore, Oklahoma, United States, 11th November 1912

travelling at 25 miles an hour and had immediately shut off steam, opened the sanders and applied the emergency brakes. His statement that he had braked was supported by the fireman, conductor and Flagman Forbes of his train.[451]

Engine No. 85, the failure of which was the indirect cause of the accident, was a freight locomotive which had been used on work trains until it worked the excursion alone to New Orleans on Saturday 9th November. That trip had not gone well; the train had had to stop at Ethel for five minutes because the right go-ahead eccentric had overheated, then again for 35 minutes five miles further on at Slaughter because the eccentric was slipping. On arrival at New Orleans Engineer Montgomery had reported multiple faults on the engine which needed attention, to which an engine inspector had added more maintenance work to be done. The repairs were duly carried out, but not apparently very efficiently, for it was the loss of the loose clevis pin which was one of the faults noted by the engine inspector that caused the failure at Montz. Engines Nos. 85 and 2013 worked the return excursion from New Orleans; engine No. 85 was to be detached at Baton Rouge for the sugar cane service with engine No. 2013 taking the excursion on to Woodville.

The ICC investigation found that the direct cause of the accident was the gross violation of the rules and absolute neglect of his duty by Flagman Cunningham. According to Rule 99 the flagman should place one torpedo on the rails a quarter mile in the rear of his train and two torpedoes half a mile in the rear. During foggy or stormy weather, he must go back a further distance and under no circumstances allow a train to pass him without exploding a torpedo. Should a train approach him before he had gone back the required distance he must, at night or during foggy or stormy weather, continue towards the train displaying a red fusee. By his own admission Flagman Cunningham had not gone back far enough, had removed the torpedo from the rail and had taken no fusees with him. The flagman was 19 years old, though the date of birth entered on his flagman application form wrongly implied that he was 31 years old.

The failure of Flagman Cunningham to do his job properly was by no means the only thing that the ICC found fault with. They also found that:

(a) Rule 100 stated that when the flagman went back to protect his train the conductor must take the place of the flagman at the rear of the train, while Rule 817 stated that the conductor must not let other duties interfere with the protection of his train and must ensure that the flagman was carrying out his duty properly. Conductor Stinson had disregarded both these rules. Had he been at the rear of the train he could have seen that Flagman Cunningham had not gone back far enough.

(b) Assistant Train Master McBurney was specifically placed on the excursion train to ensure its safe movement, yet both he and the conductor were unaware that freight train No. 58 was following them as they had neglected to check the train register at Kenner Junction. It was incumbent on the assistant train master to see that the flagman was properly performing his duties but not having gone to the rear of the train he was unaware that the flagman had not gone far enough back, or that he had no fusees with him.

(c) The severity of the collision showed that Engineer Dreher had been travelling considerably faster than the 25 miles an hour that he stated, while the time which had

[451] Interstate Commerce Commission report

elapsed between that when his train left Kenner Junction and that when his train collided with the excursion gave an average speed of 34.8 miles an hour. Given the foggy conditions and the fact that he knew that the excursion was in front of him he should have especially careful to observe the rule limiting the speed of his train to 25 miles an hour. He had been disciplined in 1906 and 1908 for being responsible for collisions.

(d) An average of 20 and sometimes as many as 35 trains were run daily over the 111 miles between Kenner Junction and Wilson. There were 15 intermittent telegraph offices, but of these only Gramerey and Baton Rouge, 30 and 78 miles respectively from Kenner Junction, were regularly open at night. It was therefore impossible to abide by the time-card rule obliging trains running in the same direction to keep 10 minutes apart, except in the vicinity of telegraph offices. The automatic block system had been installed where the traffic was most dense in the New Orleans area, but the company argued that they could not afford to install it everywhere. There was, however, the ICC maintained, no reason why a few more night offices could not be kept open and a manual block system introduced.[452]

Three weeks after the Montz accident there was another bad rear collision on the Pennsylvania Railroad near Dresden, Ohio, United States, again precipitated by the first train stopping out of course because of a problem with its locomotive. On Tuesday 3rd December 1912 westbound passenger train No. 125 consisted of an engine and only two coaches, a combination mail, baggage and express car and an ordinary coach, with Conductor Sapp and Engineer Zeisloft. It left Trinway, Ohio, at 5.40 p.m., 22 minutes late, and Dresden at 5.49 p.m., 27 minutes late. After covering another 3½ miles it then came to a dead stand at 5.55 p.m. due to the breaking of the pipe supplying the air whistle from the main air drum, the consequent loss of air causing the brakes to apply.

Behind train No. 125 was passenger train No. 43, another short one, consisting of an engine, a combination mail, baggage and express car and two coaches, in charge of Conductor Evans and Engineer James Bryant. This train left Trinway at 5.47, five minutes late, and Dresden at 5.54, eight minutes late and only five minutes behind train No. 125. As soon as train No. 125 stopped Flagman Koon set off to protect it, equipped with a lantern and torpedoes but without fusees. The latter were kept in the train box in the smoking compartment of the coach rather than at the rear and knowing that train No. 43 was likely to be close behind he did not stop to get them. Partly walking and partly running he managed to get back probably 1,000 or 1,200 feet from the rear of his train before train No. 43 appeared. Because he was on the outside of a curve the engineer only saw him at the last moment. The fireman could have seen him earlier had he been looking out, but he was busy firing. Engineer Bryant applied the brakes when he finally saw the flagman and made an emergency brake application and reversed his engine when he saw the markers of the train in front.

The unavoidable collision which followed telescoped the coach of train No. 125 for two-thirds of its length. Nine passengers and two railroad men were killed and three passengers, three railroad employees and a mail clerk were injured.[453] Apart from Engineer Bryant all those who died were in the smashed coach, escaping steam from the

[452] Ibid.

[453] Interstate Commerce Commission report

colliding engine contributing to the deaths and injuries.

The other railroad employee killed was Brakeman L. H. Blaney. Seven of those who died were from Zanesville, Ohio. Another was Max L. Harris, a bond salesman for Tillitson & Wolcott Co, of Cleveland, Ohio. 'Just before he died Max Harris, of Lodi, a travelling salesman, asked if he had any chance of life and when told there was no hope, told his attendants to notify his sweetheart, Julia Rodman, of Paola, Kan., to whom he was engaged.' [454]

The Interstate Commerce Commission (ICC) investigation found that the direct cause of the accident had been the inability of Flagman Koon to get back far enough to protect his train and his failure to use every available means to do so because he had had no time to equip himself with fusees. Conductor Sapp was blamed for failing to ensure that his flagman had all the means of protection available for emergency use in accordance with Rule 703.

Engineer Bryant, whose train had been travelling at an estimated 55 miles an hour before the collision, was criticised by the ICC for failing to obey the time-card rule limiting the speed of trains to 40 miles an hour over this section. It appeared, however, that this rule was not generally observed.

The air pipe which broke had been reported as needing repair the day before, but when the necessary material for making the repair was found to be unavailable the engine was put back into service in its defective condition, something for which the investigators considered there could be no reasonable excuse.

Pennsylvania Railroad Rule 91 dictated that on lines, such as that on which the accident had occurred, which were not worked on the block system trains must be kept five minutes apart except when closing up at stations. The ICC noted that this interval could only be maintained at open stations, and if, as in the case of the accident, the open stations were far apart, a fast train could easily overtake a slow one. The rule was entirely inadequate to ensure proper protection and the time interval should be increased.[455] To a British observer this would seem to be the crux of the matter, the time interval system having been long discarded in the United Kingdom by 1912.

A major disaster occurred at Borgaon station, near Nagpur, India, on the Great Indian Peninsula Railway, on Wednesday 30th April 1913. On a dark and misty morning, a goods train was stopped by signals outside the station when it was run into by a third-class passenger train running at 35 miles an hour. Three third-class carriages of the passenger train were completely wrecked and three others badly damaged. Thirty-three passengers were killed and 20 seriously injured. The engine of the colliding train was 'twisted into a shapeless mass of metal' but its driver somehow escaped.

Commenting on the fact that only 'natives' had lost their lives a newspaper reporter explained, 'That there was little loss of European life in the tragedy recorded in the cable is due to the fact that white people in India never travel in third-class trains, which are reserved wholly for the use of the natives. The presence of a native in a first-class

---

[454] *The Newark Advocate*, Newark, Ohio, United States, 4th December 1912

[455] Interstate Commerce Commission report

carriage is often resented, even if he happens to be a person of some note.' [456] In fact the English interlocking inspector of the railway and his wife were travelling on the train in the coach next to the brake van; they escaped injury but two of their servants were among the fatalities.

The passenger train was en route from Nagpur to Bhusawal.[457] The accident was blamed on 'the defective working of the block instruments which would, if in order, have prevented the admission of the second train on the section until the goods train had reached Borgaon.' [458]

An accident at Stamford, Connecticut, United States on Thursday 12th June 1913 was in terms of casualties considerably less serious than, say, that which had happened the previous November near Montz, Louisiana, but probably partly because it occurred in New England rather than the Deep South it received much more attention in the press. Additionally, and perhaps even more important, was the fact that it had occurred on the line of the New York New Haven and Hartford Railroad, a company which had suffered a series of bad accidents in the last year or so. Involved in the accident were the first and section sections of train No. 53, the *Boston Express*. Train first No. 53, from Springfield, Massachusetts to New York, New York, consisted of a smoking car, three day coaches and the Pullman parlor car *Skylark*, with Conductor Vincent and Engineer Worthington.

In contrast to the American rear collisions at Montz and Dresden described a little earlier there was no failure of equipment involved in the Stamford accident; train first No. 53 had stopped just east of Stamford station so that its steam locomotive could be replaced by an electric one for the run from Stamford to New York. Also, in contrast to the previous two accidents, the line was worked by what was known as a controlled manual block system, and first No. 53 was protected by semaphore home and distant signals.

Second No. 53, from Boston to New York, consisted of an engine, a postal car, a baggage car, the Pullman cars *Centredale, Pristina* and *Foxboro*, two Boston and Albany Railroad coaches and a New York New Haven and Hartford Railroad coach, with Conductor Dunn, Engineer C. J. Dougherty and Fireman Smith.

At the time of the collision train first No. 53 was stopped 171 feet in advance of its protecting home signal, which in turn was 1,800 feet from the corresponding distant signal. The weather at the time of the accident was clear. Train second No. 53 approached the distant signal at probably over 60 miles an hour, having covered the 4.5 miles from Darien to Stamford, according to the train sheets, in four minutes, giving therefore an average speed of 67.5 miles an hour between those points. It failed to pull up in time and was still travelling at an estimated 15 or 20 miles an hour when it ran into the rear of the first section, telescoping the wooden Pullman car *Skylark* almost the whole length of the engine. Six passengers died and another 20 and two mail clerks were injured.[459]

The New York Times carried a very full account of the accident, which included the

---

[456] *The Sun*, Sydney, New South Wales, Australia, 2nd May 1913

[457] *The Times*, London, England, 1st May 1913

[458] Railways in India – Administration Report for the year 1913-14 Volume 1

[459] Interstate Commerce Commission report

story of a passenger whose wife was one of those killed. 'Of the carnage in the *Skylark* as the big locomotive ate its way through it Mr Kelley of the Canadian Pacific gained an impression typical of those who were in the wooden car. He occupied seat No. 18 in the *Skylark*, while his wife, Mrs Ada Kelley, who was killed, occupied No. 19. "I did not feel any lurch forward," said Mr Kelley, "it was so soft that when I saw what had happened the crash of the *Titanic* into the iceberg came to my mind in a flash. I was sitting facing my wife, who was across the car from me. We had been watching a girl in a blue dress toward the front of the car and I was joking with my wife about a boy who obviously was flirting with the girl. I felt something. It seemed as if someone had struck the back of my chair. I bent forward a little from the shock, and then the back of the chair struck me a harder blow. I looked for my wife. I could not see her. I looked right where she should have been. I could not make her out and I tried to reach for her. Some object took form before me. It looked like a crumpled heap of something human. I picked it up, but it was nothing – some torn wreckage of some kind.

"'It was just kindling wood all around. A hand came through the splinters of wood. It grasped mine and a big, gruff voice said I had to come out. I said to leave me alone, that my wife was in there with me and I must find her. But the big voice grew more commanding, and I felt the pressure of the hand on my wrist. I struggled, but I was dragged out. I saw that the locomotive had scooped our car right up on the boiler. There was a ladder running up to a window. I struggled to reach the ladder and begged them to let me go up it to get my wife. But they held me fast. I saw the end of the Pullman was open – the end away from the engine. I tried to make to that place. They said they would attend to my wife all right and I must go to a hospital. When I was struggling with them I saw the lower part of my wife's body hanging from a window. Two men who held me walked me to the window, but just before we reached it, my wife's body fell to the ground.

"'They picked me up and walked with me towards a little signal tower of some kind. I held her hand all the way. Her pulse still was beating when they laid her down. A doctor from the Stamford hospital came and said she was dead. The locomotive had simply jammed her against the ceiling of the *Skylark* and held her there until she was crushed to death.

"'I am a railroad man with a long railroad experience. It is a mystery to me how these New England railroads can continue to operate in defiance of legislation and of all sane rules of railroading."'[460]

Engineer Docherty testified at the Interstate Commerce Commission (ICC) investigation that the brakes of his engine had not been working very well on the day of the accident and not stopping the train as quickly as they should have done. Approaching Stamford, he had shut off steam before reaching the distant signal. Seeing the distant signal at caution he had made a 15-pound reduction in train line pressure. The brakes had not seemed to act properly, and he had made a further reduction of 10 pounds, then when he was 300 feet from the rear of the first section an emergency brake application. He had also attempted to reverse his engine, but the reversing lever was stiff, and he could only move it slightly.

Speaking of his experience, Engineer Docherty said that he had worked almost 10 years as a fireman and about a year as an engineman, but up until two days before the

---

[460] *The New York Times*, New York, New York, 13th June 1913

accident only on switch engines or slow freights. On his first day working fast trains he had overshot the station at Bridgeport and had reported that the engine brakes were 'no good'. The locomotive he had driven on his first day and on the day of the accident was of a type he had never driven before. He had received no instruction on working the engine other than off-hand comments regarding other engineers having problems with the brakes and water glasses. Although the change from steam to electric engines had been made east of Stamford on previous trips he had made he was not aware that it was customary for engines to be changed there.

The statements of Fireman Smith and Conductor Dunn did not add much to the events immediately preceding the collision.

The ICC report on this accident is long and there is much evidence regarding the efficiency of the brakes, different brake tests carried out and so on. The conclusion was that the direct cause of the accident was the failure of Engineer Docherty to apply his air brakes in time to stop at the home signal, as a result of an error of judgement due to his inexperience. The responsibility for placing an inexperienced and uninstructed engineer in charge of a fast train rested with the higher officials of the company. Contributory causes were the fact that the protecting distant signal was not placed far enough out and that the brakes of the train were not as efficient as they needed to be.

The report went on to criticise the New Haven Company for their slow progress in replacing wooden with steel rolling stock, steel cars being both better to withstand collisions and avoiding the fire risk associated with wooden vehicles. A detailed criticism followed of the fact that Engineer Docherty was allowed to take out a fast train without being supervised by a travelling engineer.

The report ended with the following conclusions.

(1) The engineman of second No. 53 did not have the special experience and instruction required for the operation of such a train.

(2) The New York New Haven & Hartford Railroad had no reliable method for determining the capacity of enginemen before placing them in charge of high-speed passenger trains.

(3) There was negligence on the part of the corporation in putting in charge of such a train as second No. 53 an engineman inexperienced and uninstructed for this class of work.

(4) This railroad has no proper system of checking work reports so as to determine whether or not defects on locomotives are actually repaired. This lack of supervision is dangerous, and it ought not to be permitted to continue.

(5) The distant signal at Stamford is too near the home signal and should be at least 2,500 feet therefrom.

(6) Modern steel equipment for high-speed passenger trains should be installed at the earliest possible time, as recommended in previous reports of this Commission, and legislation fixing such a time should be enacted without delay.

(7) Establishment of safer and more efficient operation of this railroad is immediately necessary if congressional legislation extending the scope of governmental

regulation of railroads is not to be called for and justified in the interest of public safety.[461]

If the methods of working some of the American main line railroads at the beginning of the twentieth century were primitive by British standards, then those of working some of the interurban electric railways might be termed prehistoric, as demonstrated by an accident at Vineyard, near Los Angeles, California, United States on the Venice Short Line of the Pacific Electric Railway, on Sunday 13th July 1913. This double-track line was operated by time-card without either train orders or signals, trains being despatched by starters, as they were called, from the Hill Street station, Los Angeles, and Venice. The times of extra trains were entered on the time-card, but no record was kept of their times of passing stations. The spacing of trains was solely by vision. With such an antediluvian system it seems almost inevitable that a disaster would happen sooner or later and in mid-July 1913 one did.

On the day of the accident train extra No. 532 left Venice at 8.55 p.m. and train extra No. 874 left North Beach, Santa Monica, at 8.53 p.m. and supposedly passed Venice at about 9.05 but in accordance with the normal method of operation its passing time was not recorded. Meanwhile at 9 p.m. an overhead trolley wire broke at Vineyard, a junction 5½ miles out of Los Angeles. Only about 50 feet of wire was affected but it took some time for the broken wire to be secured and tied up and a succession of trains came to a halt behind the broken section. Once the wire had been secured trains were able to coast past the broken section, provided that they started far enough back to gain sufficient momentum to climb the 1.7 percent (about 1 in 59) gradient at that point.

Extra No. 532 reached Vineyard at 9.15 p.m. and was duly flagged by the flagman of the preceding train. Stopping to pick him up it then proceeded to a point about 15 feet from the rear of the preceding train. At 9.20 it was run into by extra No. 874.

Motorman Clark of extra No. 532 told the Interstate Commerce Commission investigators that he had whistled the flagman out as soon as his train came to a stand, had recalled him when the train ahead began to move forward, then whistled him out a second time when the preceding train stopped again after going only a short distance.

Conductor Van Dorn, in charge of the first car of extra No. 532, agreed that the motorman had whistled out the flagman immediately the train had stopped and said that he saw the flagman run back and his signalling to extra No. 874 when he was 600 or 700 feet back. Conductor Hart, in charge of the second car of extra No. 532, confirmed that the flagman had gone back at once with red and white lanterns as far as the beginning of a curve 180 feet in the rear of his train and had not started back when the motorman whistled him in.

Conductor Emil Bartholomai, in charge of the rear car of extra No. 532, was also acting as its flagman. He said that he had been signalled out at once when the train stopped at Vineyard. He had seen extra No. 874 approaching when he had got back about 500 feet when he saw extra No. 874 approaching and ignoring the recall whistle had continued to walk and run towards it, signalling with his lantern, but the motorman of extra No. 874 did not respond until the train was close to him when he was just beyond

[461] Interstate Commerce Commission report

the slow board for the curve. He had torpedoes and fusees with him but did not use either as he wanted to get as far back as possible. He thought extra No. 874 was travelling at 35 miles an hour.

Motorman Joseph Forster of extra No. 874 said that he saw the flagman near the slow board after he had shut off current and whistled for the junction. Seeing the brakeman, he applied the brakes and thought his speed at that point was 30 miles an hour. When he realised that extra No. 532 was in front he had made an emergency brake application. At the moment of collision his train was running at 10 or 15 miles an hour. The brakes of his train had been in good condition and working properly.

Conductor Shafter, in charge of the head car of extra No. 874, said that his train had made normal stops between Venice and Vineyard. He did not think that the motorman was running unusually fast. He put the speed of the train at about 30 miles an hour approaching the slow board and said that he had noticed no reduction in speed before the collision.

Conductor Sexton, in the second car, said that he had had passengers for Vineyard and had signalled the motorman to stop there. His signal to the motorman was not acknowledged but speed was reduced.

One of four passengers riding in the vestibule of extra No. 874 related that power had already been shut off when he saw the flagman running towards the train and the motorman then applied the brakes. He put the flagman's position as about halfway between the rear of the train in front and the slow board and estimated the speed of extra No. 874 at that moment at 20 miles an hour.[462]

The trains were 'crowded with happy passengers returning from the beach' and the results of the collision were disastrous, 'the impact of the collision caused the rear car of the standing train and the forward two cars of the moving train to telescope-that is, to ride on top of one another. It was under these telescoped cars that most of the victims were trapped. Before all of the dead and injured could be taken out it was necessary for the wrecking derrick to be brought into action to raise the coaches and swing them to one side. A wait of more than an hour was necessary before the derrick could be gotten into position where it could grapple with the wrecked cars and move them without endangering the lives of many of those pinioned beneath but still breathing. Within an hour the first trainload of the injured, made up of mostly of persons who were taken out of the wreck unconscious, was on its way to the city. These injured were taken to the emergency hospital in the Pacific Electric building at Sixth and Main streets, and from there distributed among the various hospitals of the city.

'G. C. Norman and wife, who was a bride of a few weeks, both terribly injured, occupied adjoining rooms at the receiving hospital. When surgeons examined the bride, they said her chances of recovery were but one in a thousand, and that was through an operation. The bride begged to say farewell to her husband. Both were tenderly carried to the operating room, where the surgeons and nurses turned tearfully away as the young couple said farewell. Norman was then carried away. The wife was operated on but failed to rally and died soon after. Norman asked to be carried once more to his wife's side, and

[462] Ibid.

after viewing the body was returned to his room. Norman will probably recover.' [463]

In all 14 passengers died in the accident and some 200 were injured.

The Interstate Commerce Commission (ICC) noted that the speed of trains approaching junctions on the Pacific Electric line was limited to 10 miles an hour, while a bulletin order directed that trains must be operated under complete control around sharp curves. Regarding flagging the rules directed that the flagman must go back 500 feet and put down two torpedoes 60 feet apart. Should a train appear before the flagman had gone back that far he must put down one torpedo and if conditions required display a fusee.

After carrying out braking tests the ICC concluded that if Motorman Forster had obeyed the speed limit the accident would not have occurred despite the short flagging.

The investigation determined that Conductor Bartholomai had entered the service of the Pacific Electric Railway only 24 days before the accident, having had no previous railroad experience, nor had he been examined. Motormen Clark and Forster had several years' experience but most of the other employees on the two trains had only a few months.

From Ivy Junction, just over 3½ miles west of Vineyard, trains to and from Venice were joined by trains to and from Redondo Beach, and between 9 a.m. and 9.20 p.m. on the day of the accident 123 trains had passed over the eastbound track from Ivy Junction and Vineyard, an average of one train every six minutes over 12 hours.

In conclusion the ICC declared that the accident was due to the failure of Flagman (Conductor) Bartholomai of extra No. 532 to properly protect his train, the failure of Motorman Forster to properly control his train and the failure of the company to provide an adequate method of operation to ensure the safety of their trains.

To prevent the recurrence of such accidents, trains should not be run with crews composed entirely of inexperienced men and an inexperienced man should not be appointed to the responsible position of rear conductor and flagman. The rules restricting the speeds of trains at dangerous points should be strictly adhered to and employees should be impressed with the absolute necessity of obeying rules and orders relating to the safe operation of trains.

Even, however, if accidents resulting from recklessness or disregard of rules could be entirely eliminated frequent errors of judgement could be expected to occur under the existing method of operation, the ICC concluded, and the density of traffic indicated that an automatic block system was urgently required, especially in view of the inexperience of many of the company employees. Such a system was already installed and successfully operated on a number of other electric railroads around the United States, the commission noted.[464]

Two days after the accident a coroner's jury placed the blame for the accident squarely on the Pacific Electric Company for operating trains too closely together, the cars being overcrowded and the employees being unacquainted with the traffic rules.[465]

Two major disasters, both rear collisions, occurred on 2nd September 1913, one in

---

[463] *Santa Ana Register*, Santa Ana, California, United States, 14th July 1913

[464] Interstate Commerce Commission report

[465] *Manitoba Free Press*, Winnipeg, Manitoba, Canada, 16th July 1913

England and the other in the United States. We will consider the one in England first. It happened on the Settle and Carlisle main line of the Midland Railway, near Ais Gill, only a couple of miles away from the Christmas Eve, 1910, disaster. This time two up trains were involved, and it was a driver rather than a signalman who was responsible for the accident.

On the morning of the accident an express from Glasgow, with a portion from Stranraer Harbour, to London St Pancras left Carlisle at 1.38 a.m., three minutes late. It was followed at 1.54 a.m. by a second express from Edinburgh to London St Pancras, this train also having portions from Aberdeen and Inverness and leaving five minutes behind time. The expresses were thus 16 minutes apart at that point.

The Glasgow express consisted of 4-4-0 engine No. 993 and tender, a Midland and Glasgow and South Western Joint (M & GSW Jt) brake composite, two Midland sleeping cars, an M & GSW Jt third-class coach, an M & GSW Jt brake van, an M & GSW Jt brake composite, a Midland third-class coach, a Midland sleeping car, an M & GSW Jt third-class coach and an M & GSW Jt brake van. It was in charge of Guard Oliver Whitley, riding in the rear brake van, with Assistant Guard David Donnelley, who was in the fifth vehicle, Driver William Nicholson and Fireman James Metcalf.

The Edinburgh express was drawn by engine No. 446, another 4-4-0, and tender and the following vehicles, all Midland and North British Joint stock, a third-class brake composite, a sleeping car composite, a brake composite, a sleeping car, a brake composite and a composite coach, in charge of Guard William Walker with Driver Samuel Caudle and Fireman George Follows.

From Carlisle the line climbed gently for 33 miles to a point near Ormside, but from Ormside it climbed for 15¼ miles at an average of 1 in 125 to the summit at Ais Gill box, where there were lay by sidings but no station. Kirkby Stephen, the last station before Ais Gill, was almost 7 miles short of the summit, almost exactly midway between the two places was a block post at Mallerstang.

The Glasgow express passed Ormside box at 2.20 a.m. and the Edinburgh train at 2.39, so that it was now 19 minutes behind the first train, having taken three minutes longer on the easier stretch from Carlisle. Beyond Ormside the crew of the Glasgow express had difficulty maintaining steam pressure and the speed of the train began to fall, despite Driver Nicholson assisting with the firing. Kirkby Stephen was passed at 2.37 and Mallerstang box not until 2.46, the train having averaged only 20 miles an hour over the three miles from Kirkby Stephen. The Edinburgh express passed Kirkby Stephen at 2.51, thus now 14 minutes behind the Glasgow train, and reached Mallerstang box at 2.57.

On duty in Mallerstang box was Signalman George Sutherland, who had worked for 24 years in this lonely moorland cabin with only the occasional sheep for company. He told the subsequent Board of Trade inquiry into the accident that expresses usually took six minutes to run between his box and Ais Gill. The Glasgow express having passed him at 2.46 he therefore expected Ais Gill to clear it out about 2.52. At about 2.55 or 2.56, this not having happened and with the Edinburgh express close at hand, he rang his mate, Thomas Clemmet, at Ais Gill, to enquire where it was, but Signalman Clemmet could only say that he had not yet seen it. The Mallerstang signals had all been put back to danger after the first train. While Signalman Sutherland was on the phone he saw the Edinburgh express coming up to his home signal. Its engine did not appear to be steaming and taking it for granted that it was slowing down, having missed the distant signal, he had lowered his home signal to let it up to the starting signal to await "Line clear". He had then taken his eyes off the train briefly and when he had looked again he could see that its locomotive was in fact steaming hard. He had then thrown the home signal to danger, though possibly not before the engine had passed it, and grabbed his red handlamp, but by then the engine was past his box and the guard was not looking out either. He had not been sure if the train was going to stop at his starter but when it did not he had sent the "Train Running on Right Line" signal, 4-5-5, to Ais Gill, which had been immediately acknowledged, and had then spoken to Signalman Clemmet on the phone. Signalman Clemmet had said that there was still no sign of the first train and Signalman Sutherland said that he had replied that there was nothing more that he could do to stop the second train. At about 3.03 or 3.04, Signalman Sutherland continued, he had heard engines whistling and then a rumbling noise which he had presumed was the sound of the collision. He had then rang Signalman Davis at Kirkby Stephen, whom he had already informed of the Edinburgh express running through his signals, and had asked him to call out Thomas Moss, the Kirkby Stephen stationmaster, which had been done at once.[466]

Driver Nicholson of the Glasgow train said that they had lost time all the way up the steep gradient because of poor steaming, due he thought to small coal and a heavy train rather than any fault of his fireman. The coal came from the Blackett Colliery, South Tyne, and the Naworth Coal Company, Cumberland. It was of reasonable quality, but too small. He had had the regulator wide open, but the train nevertheless came to a stand just north of the Ais Gill distant signal. He would have stopped at Mallerstang box had he not thought that he could make it to Ais Gill. After they had come to a stand, at 2.55 by his watch (Guard Whitely said 2.57 which the Board of Trade inspector thought was more accurate) Guard Donnelley had come up and asked how long they were likely to be stopped. "A few minutes," he had said. At this moment Fireman Metcalf had been about to go forward to Ais Gill box to carry out Rule 55 but he (Driver Nicholson) then glimpsed the light of the following train and told his fireman to try and run back and stop it. He had then sounded his whistle repeatedly and tried to get his train moving, but without success. He had had no difficulty in seeing the Mallerstang signals.

Fireman Metcalf agreed that the problem with the engine fire had been caused by the small size of the coal. Speed had been lost immediately after passing Mallerstang. There he had been sure they could make it to the summit, but the steam pressure quickly fell from 120 to between 80 and 90 lbs and they came to a stand. He had been on the point of proceeding to Ais Gill box when his driver pointed out the light of the second train, when he had seized a red lamp and set off to try and stop it but had not got as far as the rear of his own train when the collision occurred. After the accident he had run straight back to his engine and then on to Ais Gill box to inform the signalman, reaching there probably a few minutes before 3.15 (Signalman Clemmet said he had arrived at 3.10).

Driver Caudle said that the tender of his engine had had 30 cwts of Blackett coal and 20 cwts of Naworth coal. Both classes were very small. Up until a month previously a certain amount of Yorkshire coal had been mixed in, which had made it easier to keep a better fire. He had complained about the quality of the coal, but only verbally. When he saw the coal, he thought that if his engine, which was a good steamer, could manage with that quality of coal, it could manage with anything.

[466] Board of Trade report

Describing the journey Driver Caudle said that things had gone well as far as Appleby, and even beyond there at Crosby Garrett, but then the steam pressure had begun to drop and there had been difficulty in keeping enough water in the boiler. He had seen all the Kirkby Stephen signals at clear. Two miles further on, approaching Birkett Tunnel, he had left the engine footplate to oil the engine, going around the front of the smoke-box before returning to the footplate on the opposite side. It was a practice, Driver Caudle said, to go around the engine once between Carlisle and Leeds to oil the auxiliary boxes. He had been expecting to get back to the footplate before reaching the Mallerstang distant signal, but the wind had been stronger than he had thought. As they were leaving Birkett Tunnel he had glanced at the Mallerstang up distant signal and thought it was off. He usually checked by looking a second time but getting back to the footplate he was anxious to check the water level and so failed to look at the signal again. His fireman was struggling trying to get the left-hand injector to work properly and had not been able to get the right-hand one to work at all, so he (Driver Caudle) had busied himself with that. By the time that he had got the injector to work they had run past the Mallerstang home and starting signals without either he or his fireman having seen them. According to the rules having failed to see the signals they should have stopped. They kept going and were losing speed somewhat with the gradient and his engine's poor steaming. He had told his fireman to put a little more coal on, and was watching the latter doing so, when suddenly the fireman sounded the whistle and shouted, "Look out, Sam, there's a red light in front of us!" There was just time for him to close the regulator, reverse the engine and apply the brakes before the impact.

Fireman Follows said that he had been busy working on the injectors after Kirkby Stephen and then when Driver Caudle had returned to the footplate after going around the engine he had done some firing. He did not see the Mallerstang distant himself but did not tell his driver that he had not, it was not the practice to do so unless asked. Busy firing, he had likewise not seen the Mallerstang home and starting signals and he pointed out that even had he looked the glare from the fire would have made it difficult for him to see them. When he had ceased firing approaching Ais Gill he had looked up and saw two red lights, which he thought at first were the Ais Gill distant and home signals, so he had opened the whistle to alert the signalman. Then he had seen another red light close to the ground, like that of someone stepping down from a carriage with a handlamp and had shouted to his driver. They had by then been only about 50 yards from the rear of the Glasgow train.

The third red light seen by Fireman Follows would have been that of either Guard Donnelley or Fireman Metcalf, as they were on Follows' side of the train. Guard Whitley said that he had got down from the train once it had stopped, had gathered from Guard Donnelley that they should not be stopped long and then, about three minutes after they had stopped, he had heard the sound of another train coming. He had immediately begun to run back, waving a red lamp and whistling. He had detonators but did not stop to put one down. He estimated the Edinburgh train was travelling at 40 miles an hour. After the collision he had gone back to the rear of the Edinburgh train, told Guard Walker what had happened, and Walker went back to provide protection for both trains.

In the accident the engine of the colliding train crashed through the brake van at the rear of the Glasgow train and buried itself midway along the third-class carriage beyond. The Edinburgh express did not escape either as the roof of the wrecked van of the first train cut through three passenger compartments of its leading composite carriage. Fire broke out immediately or very shortly after the collision and the rear three vehicles of the Glasgow train were consumed. Fourteen passengers were killed outright, and their bodies were wholly or largely consumed by the fire. Two more passengers died of their injuries and 38 were more or less seriously hurt or suffered shock or from exposure. All of those who lost their lives had been passengers in the Glasgow train.[467]

The accounts of one or two of the surviving passengers featured in the newspapers the following day. 'Mr Ivan Thomas, of Cardiff, who worked manfully with the rescuers, says that he saw at least three people under the burning coaches whom they failed to get out. They had to leave one poor fellow several times in order to stagger up the bank away from the smoke and flames and get fresh air. They finally succeeded in getting him out, badly injured, but still alive. Mr Thomas, who was travelling in the same compartment with Miss Nan Clark, of Upper Norwood, fortunately saw the express approaching in time to jump out with his fellow passenger. They scrambled up the bank of the cutting, which is about 10ft high. For a moment or two, they expected the train to rock over on to them. It was swaying from side to side, and finally gave one great shudder and came to rest. Then the two, joined by others, set to work to help their less fortunate fellow passengers. Some of the windows in the wrecked coaches were actually not broken. Nobody knew how to break them until Mr Thomas thought of his golf clubs and started rescue work with them.

'Miss Clark makes it clear that the fire originated with the engine. The line was littered with burning red-hot coal, and the flames were right under the engine. The first man rescued was a man in mechanic's overalls, either the driver or the fireman of the second train. He was frightfully injured. I must give one fine incident in Miss Clark's own words: - *The next they got out was a little boy of 14 years, but I want you to say he was one of the bravest boys imaginable. He was very much injured and had an arm or his collar-bone broken, but he just said, "Oh, I'm all right. Leave me alone and look after the others." He was splendid. He couldn't stir a limb, and yet he said, "I can stand all right just now. Help the others."*

'Miss Clark was not only brave, but she showed a woman's resource. Before any rescue had been made she collected as many towels as she could from the train to relieve the sufferings of the injured and spread all the cushions and pillows she could find by the side of the line. One of those she had helped to rescue came to her later, grasped her hand, and said, "You are a plucky little girl. I only wish every man had been as brave as you were." Many of the passengers were asleep in the sleeping cars at the time of the collision. One of these, Mr H. Smallwood, says that he owes his life to Captain Hill, who smashed the woodwork and got him out of a carriage which was afterwards burnt to ashes. Captain Hill would not give the name of his regiment. All he would say was by way of compliment to a brother officer in the 57th Sikhs, who rescued three passengers who would otherwise certainly have been burned to death. Finally, I must tell you of a clergyman who knelt and prayed for one poor man who could not be released.' [468]

Signalman Clemmet at Ais Gill testified that he had seen a red glow in the sky just after 3 a.m. and at 3.10 Fireman Follows had arrived at his box, as he remembered out of breath, and had told him of the collision. He had already phoned Hawes Junction box and

---

[467] Ibid.

[468] *The Times*, London, England, 3rd September 1913

asked Signalman Simpson to alert the stationmaster, William Bunce. A down goods train standing at the Ais Gill home signal was shunted into the siding so as to clear the road for the stationmaster.

Stationmaster Bunce related that Signalman Simpson had woken him at 3.05, he had quickly dressed and reached the box at 3.10, where Signalman Simpson told him that there had been a collision. The 9.30 p.m. down express from London St Pancras was then nearing Hawes Junction. Having himself spoken on the phone to Ais Gill and confirmed what had happened Stationmaster Bunce instructed Signalman John Simpson to ring Carlisle for the breakdown vans, then proceeded to Ais Gill on the detached pilot engine of the 9.30 p.m. express, wondering, perhaps, if this was going to be as serious as the Hawes Junction disaster and thinking, *Here we go again.* The pilot engine left Hawes Junction at 3.26 and Ais Gill was reached at 3.35. There, presumably not having thought to do so before, he asked Signalman Clemmet to ring Hawes Junction and call out the platelayers, and he was at the accident scene at 3.40. He then sent the engine back to Hawes Junction, telling its driver to go down the branch to Hawes to get medical assistance. The forward part of the Glasgow express was despatched to Hawes Junction at 4.20 a.m. while from Ais Gill box he had spoken to the Kirkby Stephen stationmaster and arranged for an engine to come from there to draw the rear part of the Edinburgh train back.

Meanwhile Stationmaster Moss at Kirkby Stephen, roused by his signalman, had tried to telephone Ais Gill and Mallerstang but the lines were not working properly so he called out the platelayers and the first one that appeared he despatched to rouse local doctors Gibson and Highet. A Carlisle to Manchester train then arrived and as there was a doctor on the train he decided to proceed on the express engine with this doctor rather than waiting for the local ones. Telephone communication having been restored he also spoke to Stationmaster Bunce and arranged to draw the rear part of the Edinburgh train back to Kirkby Stephen. He left with the doctor on the engine at 4.20.

Major J. W. Pringle of the Board of Trade reported at length on the disaster. Having detailed the circumstances of the accident, as outlined above, he next considered three secondary circumstances: (a) the non-provision of a pilot engine for the Glasgow train (b) the quality of the coal and (c) the practice of oiling engines in motion.

(a) At 243 tons the first express was 13 tons heavier than the specified load for the engine puling it. Driver Nicholson had thereby asked for a pilot but was told none was available. General Superintendent Cecil Paget had explained that trains were permitted to run with loads in excess of that laid down without an assisting engine but in such cases their drivers would not be penalised if they failed to keep to the scheduled times because of the heavy load.

(b) The Blackett coal had a high calorific value but formed clinkers too easily. The Naworth coal was equal to the best Welsh coal and did not form clinkers but tended to be small. Following complaints about small coal the colliery had tightened up on its screening and matters had improved but the situation had clearly deteriorated again. But the small coal, and the problems with the injectors of the engine of the second train, should not themselves have been a source of danger.

(c) Recent improvements meant that it was really no longer necessary for drivers to apply lubrication during a long journey, but old habits died hard. The practice carried risks to the drivers themselves of falling off or coming into contact with lineside obstructions as well as the risk of their failing to observe signals.

Turning to other matters the inspecting officer considered Signalman Sutherland had been incautious in lowering his home signal when he had, and his action had been inconsistent with the intention of Rule 40(a). A recent collision on the Great Western at Yeovil Pen Mill showed the danger of a signalman lowering his home signal before he was sure that the train was stopping. As, however, Driver Caudle stated that he did not see the home and starting signals at all the incident had no bearing on the present collision. (It seems that Signalman Sutherland must have felt, wrongly, that he was partially responsible for the accident. Had he merely denied lowering the home signal at all no one would have been any wiser).

Rule 217(a) decreed that should a train be stopped in a block section because of engine failure or some other cause the rear guard must immediately go back to protect his train. It was unfortunate that Driver Nicholson had not been more accurate when asked by Front Guard Donnelley how long they were likely to be stopped. According to Nicholson he had said to Donnelley, "Only a few minutes."

The guard remembered the driver saying, "Go back to your van."

Donnelley had then shouted to Rear Guard Whitley that they would be, "Only a minute or so," giving the latter reason to think that there was insufficient reason to go back. Estimates as to the interval of time between the stoppage of the Glasgow train and the collision varied but the most reliable estimate was seven minutes. It was difficult or impossible for guards to always literally abide by the rule but in the present instance both guards must have realised from experience that the stoppage would last some time, and both must therefore share some responsibility for failing to protect their train sooner.

Rule 6 stated that, 'The safety of the public must under all circumstances be the chief care of the servants of the Company.' The safe working of a railway depended entirely on drivers obeying signals and every driver would admit that if he did not see a signal it was his duty to act as if it had been at danger and to proceed, if at all, with caution. Caudle stated that emerging from Birkett Tunnel he had "formed the impression" that the Mallerstang distant signal had been clear. He had not, as was his usual custom, looked at the distant signal again and then had been so intent upon working on the injectors that he had likewise failed to see the Mallerstang home and starting signals, not had he asked his fireman to look for them. Having finished working on the injectors he recognised that he had run past the Mallerstang signals without seeing them. This was the critical point. He never expected to find a train stopped in the section so close to the summit and instead of proceeding with great caution he had taken a chance that the line was clear. Nor had he been keeping a look out, as the night was clear and the tail lights of the Glasgow train were visible for 200 yards from the fireman's side. The difficulties which Driver Caudle had experienced with low water in the boiler and problems with the injectors notwithstanding there could be no excuse for his carelessness in missing the Mallerstang signals and then failing to reduce his speed and keeping a keen lookout and the main responsibility for the accident must rest with him.[469]

The responsibility of Fireman Follows was in any case secondary to that of his driver and given his difficulty with the small coal, his unfamiliarity with the engine and the fact

[469] Board of Trade report

that the glare of the fire had impaired his vision his responsibility was small.

As for the railwaymen in charge of the first train, it did not seem as if they had had more than two minutes' warning of the approach of the second train and it did not seem as if they could have done any more than they had. Suggestions had been made that Guard Whitley should have attempted to evacuate the passengers from the rear of the Glasgow train but with the very limited time, the darkness and the absence of a platform this would probably just have resulted in more casualties.

The inspecting officer then considered recommendations for the safer working of railways. As regards drivers overrunning signals various suggestions had been made. The idea of having a third man in the cab had been considered in other reports and rejected for various reasons. Automatic train control, with cab signalling, had been adopted by the Great Western Railway at a number of locations, but needed further experimentation and tests before it could be recommended as a panacea for all difficulties. Another method in its infancy was wireless control of trains. On the London Underground tube railways automatic train stops were in use but the idea had not yet been tried on steam railways. The railway companies should unite more fully in experimenting with it. Meanwhile detonator placers should be installed in all signal boxes, at least on main line and express routes. For the protection of trains stopped in sections various ideas had also been suggested. Major Pringle suggested that at night flares should be adopted, as used on American railways and had been recommended by other inspecting officers in previous reports. (This suggestion seems to have fallen on deaf ears – as far as the author is aware flares have never been employed on British railways). The question of a daytime equivalent to flares was more difficult.

The next part of the report dealt with the mitigation of the results of accidents. No real telescoping had occurred in the accident because the two rear carriages of the first train had had wooden underframes. Instead the engine had smashed their headstocks and split open their frames. Coaches with steel underframes were more liable to telescope but this could be prevented with shock-absorbing buffers, steel castings and double cases. Older wooden coaches should be withdrawn as quickly as possible and replaced with modern stock. Other improvements to be considered were a modification of carriage doors to prevent their jamming, that bars to windows if fitted should be removable and that the fire-proofing of rolling stock should be looked into.

The trains appeared to have been supplied with an ample number of tools and items of equipment. All the fire extincteurs (extinguishers) were apparently used and were effective in subduing the fire at first though probably not used to the best advantage. Once the fire had progressed the extinguishers were ineffectual. Little use was made of the fire buckets and although the train had actually come to a halt over a stream (Far Cote Gill) no one realised this in the darkness. More safety lamps on the train would have helped provide much needed lighting. Short hooked ladders and small screw jacks would have been useful.

Witnesses were divided as to the origin of the fire. The railwaymen, except Driver Caudle, and a couple of the passengers were of the opinion that it originated from live ash thrown among the baggage and wreckage; the remainder of the passengers attributed it to the ignition of escaping gas. Accounts agreed that there were two separate outbreaks of fire, the first smaller one was extinguished, the second more serious outbreak came some minutes later. The inspecting officer concluded that the initial outbreak was due to gas and was put out but a second outbreak caused by baggage and luggage piled upon the engine had smouldered out of sight of witnesses and gaining in intensity until it reached the roof of a carriage resting on the engine chimney when there was a sudden burst of violent flame.

Since 1898 there had been 11 accidents in Great Britain on steam railways in which the wreckage had caught fire. Of these in one the origin of the fire was doubtful, in two it arose solely from the engine, in two both gas and electricity were involved and the remaining six were due solely to gas. It was thus clear that gas was less desirable than electricity as an illuminant, a point addressed to the railway companies by the Board of Trade in February 1911. By January 1913, following the Ditton Junction accident, the railway companies were making some change of attitude 'in the desired direction', but progress was slow. Fires due to gas would feature in several further accidents in Great Britain before gas lighting was finally eliminated, the last at Charfield, also on an ex-Midland Railway line, in October 1928.[470] (The last major accident anywhere in the world involving a gas-fed fire was probably that at Dugald, Canada, as late as 1947.)

Because two of the passengers had died later three separate inquests were held on the deaths of the victims. At the inquest on 14 of the victims the jury had to retire four times, having repeatedly come out with a verdict which the coroner told them he must interpret as one of manslaughter against Driver Caudle and Fireman Follows, and the jury saying that they did not want their verdict so interpreted. They finally agreed on a verdict that the deaths had been caused, "accidentally and by misadventure". At the second inquest, at Leeds, on the death of stockbroker George Lindsay, the jury returned a verdict that the accident had resulted from the failure of Driver Caudle and his fireman to observe the danger signals, but they did not consider that under the circumstances it amounted to culpable negligence. The final inquest was held at Carlisle on the death of Sir Arthur Percy Douglas. This time the jury were less forgiving and took only half an hour to return a verdict of manslaughter against Driver Caudle.[471]

Having thus had to go through the ordeal of three separate inquests Driver Caudle stood trial at Carlisle in late October and he was found guilty of manslaughter with a strong recommendation to mercy. Mr Justice Avory sentenced him to two months' imprisonment. As might be expected the verdict and sentence provoked a strong reaction from the National Union of Railwaymen and there were calls for the Home Secretary to release Caudle immediately. He was in fact granted a Royal Pardon and released on Saturday 1st November, 10 days after his conviction.[472]

The second disaster to occur on 2nd September 1913 happened near North Haven, Connecticut, United States, on the New York New Haven and Hartford Railroad. The second portion of the southbound *Bar Harbor Express*, train second No. 91 was run into by the first portion of the southbound *White Mountain Express*, train first No. 95. The results of the collision were disastrous. The engine of first No. 95 cut through the rear two wooden Pullman cars of the first train and lifted the third car, occupied by boys on

---

470 Ibid.

471 *The Times*, London, England, 4th, 5th, 6th, 8th, 10th, 15th, 16th, 17th, 18th, 19th, 20th, 23rd and 24th September 1913

472 Ibid, 9th, 10th, 21st, 22nd, 27th and 30th October and 1st and 3rd November 1913

their way to a summer camp at Monmouth, Maine, up into the air, when it then fell onto its side. The passengers in the rear two Pullmans, alive or dead, were thrown to the sides of the track together with the wreckage of the cars. Early reports spoke of 22 dead and 50 injured; the final figures were 21 passengers dead and 33 passengers, and two railroad employees injured.

The *New York Times* covered the accident very fully. In one column it wrote of the experiences of the summer camp boys. 'Harry Rich Mooney, who lives in the Elks Club in this city, was in charge of fifty-eight boys who were returning on the second section of the Bar Harbor Express from Camp Cobbassee, in Maine. Mr Mooney and most of his charges were in the *Chisholm*, the car third from the rear. A few, however, were in the car back of the *Chisholm*, and two of these, William Altschuler and Albert Green were killed. "The most heroic thing that I know connected with the wreck," said Mr Mooney, "was the work of Walter Schuster, a sixteen-year-old boy of Mount Vernon, in saving two of his companions. The three boys were on the platform in the rear of the third car in the wait before the crash came. They were looking back through the mist to see if they could find out what was delaying the train, when a brakeman who was standing on the ground near them shouted, 'Jump, or you'll be killed.' Schuster acted instantly, shoving off both his companions, Joe Hennessey and E. Bauman. Then he jumped himself. All three boys were rolling down the embankment when the oncoming train struck. Another instant on the platform would have been sure death to them.' [473]

Train second No. 91 consisted of a locomotive followed by a Maine Central Railroad baggage car and 10 wooden Pullman sleeping cars; *Salvador, San Ardo, Messenia, Omena, Eleaya, Volta, Tournay, Chisholm, Kasota and Chancellor.* It was in charge of Conductor Adams with Engineer Wands and Fireman Barnfather. Behind it was train first No. 95 in charge of Conductor Fowler and Engineer Augustus Miller and Fireman Elbert Robertson.

The line through North Haven was double track and worked by train order and timetable rights. It was equipped with automatic disc signals known from their appearance as banjo signals, dating from about 1888. The signals were spaced a mile apart. Some of them were provided with associated distant signals and at other locations overlap track circuits were provided so that the second signal behind a train did not go to clear until a train had cleared a track circuit extending some distance in advance of the first signal behind the train. In the vicinity of the accident, however, there were neither distant signals nor overlap track circuits. If a stop signal remained at danger a train was allowed to pass it and proceed as far as the line was clear. Notwithstanding the automatic signalling General Rule 99 required the flagman of any train stopped at a signal to go back a sufficient distance to protect it.

Engineer Wands of train second No. 91 told the Interstate Commerce Commission inquiry into the collision that after leaving Springfield, Massachusetts, he had had a clear run as far as an automatic signal at Wilson, 23 miles to the south. Because of a fog he could only see the signal when he was 200 feet from it and overshot it despite applying the brakes in emergency. It took him five minutes to recharge the air brakes and he then restarted his train, without recalling the flagman. His train passed through Hartford and continued as far as signal 23, which again was at danger and which he again was unable

---

[473] *New York Times*, New York, New York, United States, 3rd September 1913

to see through the fog in time to stop at it despite an emergency brake application. Again, it took three or four minutes to recharge the train line and he then restarted but was stopped by the conductor as the flagman had not rejoined the train. He had whistled to recall the flagman and two minutes later the collision occurred. Engineer Wands testified that in foggy weather it was impossible to avoid running past automatic signals and he had never been disciplined for it. To avoid doing so an engineer would need to run at only 10 or 15 miles an hour in fog.

Fireman Barnfather corroborated his engineer's account in all details, including the fact that trains frequently ran past signals when it was foggy.

Conductor Adams of train second No. 91 stated that it had left Springfield at 5.13 a.m. At Wilson the flagman had gone back and stopped first No. 95 which had pulled up behind them. At signal 23 he saw someone alight when his train stopped whom he had assumed must be the flagman going back, though he could not be sure because of the fog. When the train restarted he stopped it again to recall the flagman. The recall whistle had been closely followed by the sound of exploding torpedoes and then the collision. He had witnessed Flagman Murray 'short flagging' before and had spoken to him about it and mentioned it to the crew dispatcher. Trains frequently ran by stop signals in fog.

Flagman Charles Murray testified that he had been with the railroad a year. He had never taken the flagman's examination nor was he qualified as a flagman. He had gone back a short way at Wilson but denied having flagged first No. 95 as stated by the conductor. At signal 23 he had been on the platform of the rear car and had started walking back when it stopped, carrying two lanterns, torpedoes and a fusee. He kept walking until he heard the recall whistle, when he had placed two torpedoes on the rail. As he was about to light a fusee he heard a train approaching and started back, swinging his red lantern and trying to remove the cap of the fusee with his teeth, but had then had to jump aside to avoid being run down. He did not remember a conductor ever criticising the way he was carrying out his flagging, he had not been given any instructions, and he did not have a watch, though qualified flagmen were meant to have one. Other men acting as flagmen on passenger trains were also unqualified, he added.

Engineer Miller of first No. 95 testified that he had left Springfield at 5.33. It had been very foggy. He had not been flagged near signal 83 as stated by Conductor Adams and did not see second No. 91 before the collision. Signal 25, just over a mile north of signal 23, had been green, indicating that the line was clear as far as signal 23. He had made a slight brake application before looking out for signal 23. When he saw the flagman, he made a 10lb brake reduction, quickly followed by an emergency application when he saw both signal 23 at danger and the rear of the train ahead. The signal had only been visible for 150 or 200 feet, the flagman was 100 feet from the rear of the train and 50 feet north of the signal and the torpedoes were 50 feet further north again. Engineer Miller said that he could only recall running past a danger signal once before, nine years ago. He attributed his running past signal 23 to the fog, the lack of a distant signal and of landmarks in the vicinity which would have enabled him to locate the signal. Because he had been covering some of the runs of a colleague who was off sick as well as his own and had to deadhead 40 miles to and from him home between shifts he had had little sleep in the days preceding the accident, but he testified that he had not felt sleepy even so.

Fireman Robertson agreed that train first 95 had not been flagged at all at Wilson and that signal 23 was the first red that they had encountered. After the accident he had gone back and found the torpedoes 60 or 70 feet north of signal 23. He too stated that cases of trains running past signals were numerous.

Conductor Fowler said that the brakes on the baggage car had been cut out (inoperative) when he took over the train at Springfield. Passenger trains occasionally ran with the brakes cut out on a vehicle, though not commonly. The brake application, the explosion of the torpedoes and the collision had all happened very close together in time.

Train Despatcher Kelly testified that trains first No. 93, second No. 93, first No. 91, second No. 91 and first No. 95 had passed Wallingford, Connecticut, a few miles north of the accident scene, at 6.19, 6.21, 6.35, 6.43 and 6.51, together with a Hartford – New Haven local train at 6.32. All the trains had been delayed by the fog. Train first No. 95 had been running rather faster than any of the other trains. The collision itself occurred at 6.55. Finding that there were five trains between Wallingford and Air Line Junction already he told the Wallingford operator to hold first No. 95 there but Air Line Junction had then reported that first No. 93 was passing there so he had cancelled the order to hold first No. 95. There were seven automatic signals between the two points, and he wanted to keep the traffic moving.

Signal Engineer Morrison stated that there were about 154 signals between Springfield, Massachusetts and New Haven, Connecticut, of which 25 had distant signals. Following two accidents earlier in 1913 the disc signals were being replaced by semaphore signals with distant indications. Over 1,000 schemes for automatic signalling had been submitted, two of which had been selected for experimental tests.

The Interstate Commerce Commission (ICC) concluded that the direct causes of the accident were the failure of Flagman Murray to properly flag his train, the failure of Engineer Miller to drive at a speed such that he could stop short of signal 23 and the failure of Conductor Adams to ensure that his flagman was doing his job properly. A contributory factor was the failure of Engineer Wands to stop his train before passing signal 23.

Had Flagman Murray at once run back and lit a fusee the accident would probably not have happened. Instead he had negligently loitered with his train for six or seven minutes and at no time did he go back more than 200 or 300 yards. He had limited experience but was familiar with his duties.

Engineers Miller and Wands should have run their trains at greatly reduced speed in view of the fog to enable them to stop short of signals at danger. Despite the shortage of staff there was no need for Engineer Miller to have done a double duty for a week and his faculties must have been impaired by fatigue.

Conductor Adams was a man of long experience and his first duty when the train stopped was to ensure that it was protected, especially in view of his lack of confidence in the flagman, but he had flagrantly neglected this duty.

While the accident was thus directly due to the failure of several employees to do their duty the signalling system was entirely inadequate for a line with such a large amount of traffic, while there was a deplorable lack of supervision of their employees by higher officials.

The ICC was, predictably, less than happy with what seemed to be an endless series of bad accidents on the New York New Haven and Hartford Railroad and blamed the management of the company. Its report went on, 'The recurrence of disastrous accidents on this railroad and the large numbers of deaths and injuries resulting therefrom makes advisable a less narrow treatment of the whole subject than that followed in mere inquests or in the ordinary official investigation of an accident… In reports by this commission of the investigation of previous accidents on this railroad defects of management, neglect of obvious precautions for safety, ineffective rules and these not systematically enforced, high-speed trains closely following one another without adequate safeguards, dereliction of duty on the part of certain employees, as well as a general weakness of discipline, have been denounced and condemned. It is astounding that this state of affairs is allowed to continue to exist.'

Engineer Miller and Flagman Murray were arrested after the accident on a charge of being criminally responsible for it but were freed a day or so later. The coroner's inquest was held in private by Coroner Mix and the findings of it appear to have been overshadowed by the very speedy publication of the ICC report, less than a month after the accident and apparently while the inquest was still in progress.

The Great War, later to become known as the First World War, having broken out at the end of July Germany invaded Belgium and declared war on France at the beginning of August 1914. German troops advanced rapidly across Belgium and part of northern France until their progress was halted in September at the First Battle of the Marne. Meanwhile there was a large movement of civilian refugees fleeing the fighting, heading for either Brittany or across the Channel to England. Although the front had stabilised by the end of September the movement of refugees seems to have still been in progress in mid-October, for on probably 18th October 1914 a train carrying 2,000 Belgian refugees was involved in a major disaster between Marquise and Wacquinghen, France, on the Chemins de fer du Nord. With the war dominating the news the accident was presumably not given much attention by the press, or possibly details of it were censored at home, though it is mentioned very briefly in a handful of American and Canadian newspapers. The refugee train was involved in a collision, most likely a rear collision, with either a supply train or a troop train according to different sources. Some reports stated that over 50 died, many of whom were never identified, and more than 200 injured were taken to hospitals back in Calais.[474] Another, possibly more reliable report, spoke of 31 dead and 81 injured, gave the destination of the train as Boulogne rather than Rouen and added that many of the refugees had come from Antwerp. The collision happened at 8 p.m.[475] It was blamed on a wrong side signal failure, a broken wire causing a signal to wrongly show 'clear'.[476]

A disaster at Miechów, Russian Poland, on Tuesday 3rd November 1914 involved trains of the Austrian army, which was withdrawing in front of the advancing Russians. To stem the advance of the Russian forces a tunnel nine kilometres east of Miechów on the Kielce – Kraków line was blown up at 8 p.m. that day as soon as the last Austrian

[474] Webpage – Médecins de la Grande Guerre – Comments les habitants de Calais aidèrent les réfugiés et blesses belges en Octobre 1914
[475] La Métropole (Being temporarily published in London, England), 22nd October 1914
[476] *Tyrone Daily Herald*, Tyrone, Pennsylvania, United States, 22nd October 1914

train had passed through it, while in preparation for the retreat the telephone and telegraph apparatus at Miechów station had already been removed. Between Miechów station and the tunnel at 5.30 p.m. had stood, in order from the station, trains of the Eisenbahn-Betriebs-Kompanie No. 26, Reserve-Eisenbahn-Bau-Kompanie No. 1, Reserve-Eisenbahn-Bau-Kompanie No. 16 and Festungs-Eisenbahn-Bau-Kompanie No. 4. Two hours later Captain Ermlich ordered Driver Schönfelder of Festungs-Eisenbahn-Bau-Kompanie No. 4 train to proceed. With all communications out of use trains could only be driven on sight. Unfortunately Driver Schönfelder appears to have assumed that the three trains in front must have long left Miechów and did not exercise sufficient caution. Rounding a curve with visibility limited to 500 metres he suddenly found Reserve-Eisenbahn-Bau-Kompanie No. 16 train in front of him and could not avoid running into the back of it. His engine was running tender first and the tender climbed up onto the rear waggon of the train in front, tender and waggon were derailed and forced into one another. The waggon was loaded with a motor car, supplies of petrol and 4,500 kg of explosives.

Driver Schönfelder and his fireman were killed in the collision. Fire broke out in the wreckage and injured soldiers cried out for help. As their comrades came to their assistance the explosives in the rear waggon detonated. The official report on the accident stated that there were 60 dead and 52 seriously injured; other sources state that over 100 died.[477]

On 20th November 1914 an up passenger train was brought to a stand at Sambhu station, India, on the North Western State Railway for its driver to obtain a permit to proceed to the next station. The permit should have been in a pouch picked up mechanically by the engine but was missing. There was some delay due to the assistant stationmaster misdating a second permit and the train was set back to the centre of the station. In this position it was run into by a following mail train. Three vehicles were wrecked, eight passengers killed and 25 injured.[478]

At 4.50 a.m. on Saturday 5th December 1914 'accelerated' (an all-stations passenger train which was slightly faster than the omnibus trains) train No. 1809 from Rome Termini to Naples was being held by signals Riardo-Pietramelara station, Italy. The problem was congestion at the next station at Teano and the need to give priority to *direttisimo* (high speed) train No. 102 bound for Rome, freight train No. 6581 in front of train No. 1809 and a train of empty coaching stock which was to form train No. 1782, the 4.46 a.m. from Vairano to Isernia and Sulmona, which was thus already late in starting. After 25 minutes had elapsed train No. 1809 was still waiting at Riardo when a decision was made to advance optional (pick-up?) freight train No. 7179 to Riardo and put it onto a side track there.

Freight train No. 7179 from Rome to Caserta consisted of two engines and 30 refrigerated waggons with hand brakes. It was duly despatched to Riardo but on reaching there it failed to come stop despite its drivers whistling for brakes and collided at considerable speed with the rear of train No. 1809, still stood at the signal. The passenger

477 *Eisenbahn-Unfälle in Europa* by Erich Preuß pgs. 65-6 (in German)
478 Administration Report on the Railways in India for the year 1914-15

train consisted behind its locomotive of a sleeping car, five first and second-class carriages, four third-class carriages and a cattle waggon loaded with mules, the latter being the last vehicle. The engine of the freight train demolished the cattle waggon, killing some of the mules and injuring a number of the soldiers accompanying them. Those mules that were not killed ran about wildly, kicking their minders, who were forced to shoot the beasts. The third-class carriage next to the cattle waggon was smashed and the next two carriages were badly damaged. Some of the passengers in these carriages were trapped and injured, a few fatally so. These were mostly soldiers and carabineri (police). The remainder of the train was undamaged and after the injured passengers had been transferred to the forward coaches it left for Naples three hours after the collision. It arrived there at 9.30 a.m. and the injured passengers were taken to various hospitals.

Passengers killed in the collision were Ghiselli Emilio, from Cuneo, a soldier in the 32nd infantry, who had been accompanying the mules, an assistant electrician, a sailor in the Regia Marina (Italian navy) and two others who could not immediately be identified. Antonio Munlam, the Riardo signalman, also lost his life. Of the 50 people injured there were 10 seriously hurt, two of whom shortly died of their injuries. Most of those injured were soldiers and sailors, a few were railwaymen and their families going to Naples.

The initial railway departmental inquiry into the accident had theorised that Signalman Munlam had caused the accident by failing to reset the road to divert the freight train into the side track, but in answer to a question in Parliament on 7th December Augusto Ciufelli, the Minister of Public Works, stated that the home signal had been at danger but the train had run past it for a distance of 700 metres despite its driver calling upon six brakemen to put down brakes, and had then run into the passenger train. The decision to forward the freight train to Riardo when the line ahead was blocked with traffic and the passenger train was still waiting there, was also questioned.[479]

A return excursion train on the Baltimore and Ohio Southwestern Railroad featured in a rear collision at Orient, Ohio, United States on Thursday 12th August 1915. The train had been organised by the Mount Sterling Knights of Pythias and was carrying 300 men women and children home from their annual picnic at Cedar Point, which doubtless included a visit to the amusement park on the shore of Lake Erie.[480]

Excursion train extra No. 1467 consisted of engine No. 1467, a combination car and five coaches, all six vehicles being of wooden construction with Conductor Schefler and Engineer Glancy. It left Sandusky, Ohio, at 7.40 p.m. on Wednesday 11th for Bloomingburg, Ohio. Considerable delay to the train was caused when its engine sustained a broken trailer truck, and at Newark it had to be exchanged for engine No. 1474. Scheduled to leave Columbus, Ohio, at 10.40 p.m. on the 11th the excursion did not leave there, after further delay, until 1.28 a.m. on 12th, having received order No. 49, giving its times to Bloomingburg, order No. 3, instructing it to run 2 hours 30 minutes late on the times given in order No. 49, and orders Nos. 51, 54 and 57, instructing it to run slowly at different points due to the danger of track washouts following heavy rain. It passed Gn tower, 1.4 miles from Columbus, then was flagged by the brakeman of freight

[479] Wikipedia – Incidente ferroviario di Riardo
[480] *Xenia Daily Gazette*, Xenia, Ohio, United States, 12th August 1915

train No. 199 at Broad Street siding, a mile further on, from where Engineer Glancy telephoned Gn tower and was given further instructions about track conditions. Freight train No. 199 was in the siding at Broad Street and the excursion then proceeded.

Freight train No. 199 from Newark, Ohio, to Midland City, Ohio, was in charge of Conductor Hagan with Engineer Stephen and Fireman Davidson and had arrived at Columbus yard at 9 p.m. It received the three slow orders at Gn tower, ran to Broad Street siding and then its engine returned to Gn tower for order No. 3, which was a message that the excursion train was now extra 1474 instead of 1467 because of the change of engines, and a clearance card instructing it to follow 10 minutes behind the excursion. The engine of the freight train then returned to Broad Street siding where the excursion train passed it as described above. Leaving Broad Street 10 minutes after the excursion, the freight train consisted from that siding of Engine No. 2669, 20 loaded and 14 empty cars.

At a point about 1¼ miles east of Grove City the freight was stopped by the flagman of the excursion, and it caught the excursion up again a mile further on. The excursion was then lost sight of until the freight train ran into the rear of it about 6 miles further on, 500 feet east of Orient station. Although the freight train was only travelling at an estimated 5 or 10 miles an hour the rear coach of the excursion was demolished and about 10 feet of the following car was crushed. The third car from the end of the excursion train escaped with a broken corner post and slight damage was caused to the front of the colliding engine. Six passengers were killed outright in the accident and 32 were injured. [481]

Those who lost their lives at the scene of the collision were Logan Holler, a Pickaway County farmer; Margaret Sollars, aged 18; Thornie Neff, aged 25; Thad Mitchell, aged 14; Carl Reichelsderfer, aged 16, and Fred Gerhardt; all from the Mount Sterling area.[482] Following the accident Orient, described as a small village of a few hundred people, was 'practically turned into a field hospital', with practically every resident rushing to the scene to offer assistance. The private hospital of Dr J. A. Knight was filled to capacity and some of the injured were taken to private homes, while a relief train with nurses and physicians was sent from Columbus.[483] One of the injured, Frank Miller, a Mount Sterling merchant, died in hospital on 17th August, making a seventh fatality.[484]

Engineer Glancy of the excursion told the Interstate Commerce Commission (ICC) inquiry into the accident that he had stopped his engine at the water tank at Orient but on account of the height of the tank he had had to move his engine back a foot or so before he could take water. He had done so and was just about to whistle out the flagman when someone shouted for him to go ahead. His engine had just begun to move when the collision occurred. Between Grove City and Orient he had run at about 15 miles an hour.

Conductor Schefler testified that he did not consider that freight train No. 199 had waited the full 10 minutes before following his train, but Flagman Fuller had afterwards told him that he (the flagman) had a full understanding with the engineman of the freight

[481] Interstate Commerce Commission report.

[482] *Washington Court House Daily Herald*, Washington Court House, Ohio, United States, 12th August 1915

[483] *Xenia Daily Gazette*, Xenia, Ohio, United States, 12th August 1915

[484] *Washington Court House Daily Herald*, Washington Court House, Ohio, United States, 17th August 1915

train as to the work the excursion train had to do. When the excursion stopped at Orient the flagman had been somewhere in the train, but he had come out of the rear after about 15 seconds and started back the train. Shortly afterwards, the conductor continued, he had heard the roar of the freight train approaching and had shouted to the flagman to hurry. When the headlight of the locomotive of the freight train appeared, he thought it was going to stop; when it became apparent that it was not he had stepped onto the rear platform of the car and pulled the air whistle for his driver to go on. He thought that Flagman Fuller had got back 400 feet. Given that the freight train had identical slow orders to his train and its engineer had also been spoken to by his flagman, he had not thought it necessary to use fusees as per Rule 99.

Flagman Fuller said that he had told the engineer of the freight train that the excursion train would be stopping at all stations and also be taking water at Orient. When his train stopped at Orient he had stopped briefly to help several passengers disembark before collecting his flagging equipment and starting back. Hearing train No. 199 approaching he had begun to run. Three or four telegraph poles from the rear of his train he had seen the headlight of the freight a quarter mile away and had waved his red lantern but received no acknowledgement. Like his conductor he said that having spoken to the engineman of the freight train he had not thought he needed to drop a lighted fusee when they stopped at Orient. He considered that his first duty as a flagman was to attend to the passengers, then do the flagging.

Engineman Stephens of train No. 199 agreed that the flagman had told him that the excursion would call at all stations. The excursion had departed from Grove City at 2.50 a.m., his train had not left for another 12 minutes. Approaching Orient, he had made a 20 pound brake application and reduced his speed to about 12 miles an hour before releasing the brakes. Then, rounding a corner the back markers of extra No. 1474 had come into view about six car lengths away. He had made an emergency brake application and applied sand, reversing his engine when about a car length away. He insisted that he had had his train under control, but that heavy rain had interfered with his vision and washed the sand from the rails, plus he could have pulled up had the excursion been standing at the station rather than the water tank. The oil headlight of his engine was ineffective in lighting up the track for any great distance. His air brakes had been tested before leaving Broad Street and were in good working order.

Fireman Davidson of the freight train more or less agreed with his engineer's statement and could not add very much. The fireman and Brakeman Rodgers of the freight thought that the flagman had only been a car length or so in the rear of the excursion train.

Superintendent Brooks of the Baltimore and Ohio Southwestern stated that trains on the Midland subdivision were operated under a permissive manual block system. Under this system no train was permitted to follow a passenger train from an open telegraph office until the passenger train had cleared the block section, except that a train starting from a non-telegraph office, such as Broad Street, could follow a passenger train after an interval of 10 minutes. He admitted that this block system did not provide adequate protection.

The ICC report concluded that the accident had been caused by the freight train following the passenger train through the block section without its being run 'under control' so that it could be stopped within the engineer's range of vision, as per General

Rule No. 108. Engineer Stephens was therefore required to run his freight train through a 20-mile block section at a speed such that he could stop within his range of vision, with a feeble oil headlight, on a dark, stormy night with wet slippery rails. He was aware, of course of the weather and the poor condition of the track, and had already twice overtaken the excursion train, indicating that the freight train was running faster than the passenger train. Under such circumstances he should have used extreme caution, but he had not, and was therefore principally responsible for the collision.

Contributing to the collision was the failure of Conductor Schefler and Flagman Fuller to properly protect their train by using fusees thrown down when their train was running slowly and after they became aware of the approach of the freight train in accordance with Rules Nos 99 (a) and 99 (c) and of the conductor to ensure the flagman was carrying out his duties properly in accordance with General Rule No. 803.

The commission noted that while the direct cause of the collision was the failure of the employees involved it was the system in operation which rendered the accident possible. The 70-mile stretch of line on which the accident occurred was divided at night into five block sections, three 10 miles in length, one 15 miles and the other 21 miles long. The exception to Rule 108 requiring a positive block (that is what in Great Britain would be termed an absolute block) to be maintained after passenger trains at all times was supposedly provided to meet the case where a freight train had entered a section in front of a passenger train and had had to be shunted at an intermediate siding where there were no means of communication. Such a situation had not existed on the night of the accident, but the superintendent stated that such an interpretation was permissible, even though he admitted that in such a case the block system provided no additional protection. The night offices at Derby and Madison Mills had been done away with in April 1914 and those at Grove City and Sabina in October 1914, in the interests of economy. If such economies were considered expedient where the safety factor was reduced, as here, by the running of night passenger trains either the absolute block system should have been maintained over the longer distances or the closed night offices temporarily opened.

Finally, the ICC remarked that the rear coach of the excursion train, in which all the deaths had occurred, had been an all wooden one with open platforms dating from 1883. The superiority of cars of all-steel construction had been repeatedly demonstrated and had such cars been used the number of fatalities would doubtless have been less.

On Thursday 7th October 1915 the supervising signalman at Bischdorf (now Satopy Samulewo), in East Prussia, Germany, an intermediate station on the Allenstein-Insterburg line, instructed his assistant to admit fast goods train No. 6212. At the same time, he received a report from Bergenthal that Berlin-Eydtkuhnen express D 55 was passing there. Presumably forgetting at this point that he had not telegraphed back the arrival of the goods train to the next block station at Schönenberg he then telephoned Schönenberg and advised the signalman there that the line was clear for the express. The signals were duly cleared at Schönenberg for train D 55 and it passed there at speed and collided with the goods train as the latter was entering Bischdorf. Fifteen people died and 55 were injured. [485] [486]

---

[485] *Eisenbahn-Katastrophen in Deutschland* by Hans Joachim Ritzau pg. 134 (in German)

Early on the morning of Tuesday 28th December 1915 a Rimini-Bologna passenger train, carrying soldiers returning to the front after Christmas leave, sailors travelling from the ports to spend the New Year with their families and civilians travelling on business, collided with the rear of a stationary goods train near San Lazzaro di Savena station, Italy, killing about 20 passengers and injuring over 70, some of them very seriously. There was a dense fog in the area at the time.

Goods train 6050 from Faenza to Bologna was running an hour and a half late. It reached San Lazzaro a few minutes after midnight and warned by the explosion of three detonators its driver brought it to a stand at a stop signal. After waiting some time without the signal having been cleared Driver Fanelli advanced his train to signal box No. 7 to inquire as to the reason for the delay. Driver Fanelli then heard the whistle from the engine of *direttisimo* train 54 coming up behind him from the direction of Mirandola. He immediately opened his regulator in an attempt to avoid the collision but in vain.

Train No. 54 had slowed up a little at Mirandola but its driver finding the signal there clear had then continued towards San Lazzaro. Passing signal box No. 7 Driver Armando Sinigagliesi saw a red light a short distance ahead. He closed the regulator and applied the Westinghouse brake but was unable to avoid running into the rear of the freight train.

The engine and tender of the colliding train were derailed and ended up buried in the earth and ballast. Behind them a waggon loaded with postal packages, a mail van, a luggage van and two third-class carriages were piled up in a heap of wood, metal, glass and the bodies of passengers mixed up with their possessions. The following second-class carriage and another third-class carriage were derailed and badly damaged, and the last four waggons of the goods train had been crushed and thrown to the side of the tracks.

The rescue work went on all night and through to the following afternoon, made more difficult by the darkness and the continued fog. Ambulances removed the injured as they were released, and the bodies of the dead were provisionally laid out in a long line alongside of the track. Some of those killed were so badly mutilated as to be unrecognisable.

The conclusion of the preliminary inquiry into the accident saw the arrest of San Lazzaro stationmaster Mario Conti, Guard Attilio Malaguti and Rear Brakeman Ferdinando Pizzi of the freight train and Pietro Caprara, the signalman in No. 7 box. At their trial at Bologna in mid-May 1916 Stationmaster Conti was found guilty, with extenuating circumstances, and sentenced to 3 years 2 months and 26 days imprisonment and a heavy fine, in addition to which he was ordered to pay damages and expenses. Guard Malagutti, Rear Brakeman Pizzi and Signalman Caprara were found not to have been criminally liable and were acquitted.

The sentence on the stationmaster seems rather harsh while the acquittal of Signalman Caprara, if, as seems likely, the collision resulted from a signalling error, seems strange. The latter probably thought himself very lucky. [487]

---

[486] Webpage – Zugunglück Deacademic (in German)
[487] Webpage – 1915 Disastro ferroviario a S. Lazzaro by Pier Luigi Perazzini (in Italian)

CHAPTER 5

SIDELONG, HEAD-ON, DOUBLE AND FOULING COLLISIONS

Trains can come into collision with one another in a variety of ways. Rear collisions, that is ones involving one train running into the back of another, trains meeting one another head-on on a single track and trains colliding on rail/rail level crossings are detailed in other chapters. This chapter will deal with other types of collision such as those involving trains running into one another at junctions, meeting head-on on multiple tracks and colliding with vehicles left in sidings but which ended up foul of running lines.

On Wednesday 18th September 1901 the Cape express of the New York New Haven and Hartford Railroad which had left Boston South station at 1.08 p.m. was passing through the station at Avon, Massachusetts, United States, at about 1.37 when it was run into by a freight train which had just been made up in the station yard. The engine of the freight train struck the express and threw part of it off the line.[488] The first stop of the express was down the line at Brockton, where it was due at 1.43, and it was running through Avon at 40 miles an hour.

Local physicians were quickly on the spot after the accident and were later joined by doctors from Brockton, Randolph and Stoughton. By 3.30 all the injured had been taken to Brockton hospital and the blocked tracks had been cleared by 4.15.

The freight train was in charge of Conductor Brown, with Engineer William Sheldon and Fireman Arthur Cooper, and had arrived at Avon a short time earlier from Braintree. Some of the 33 cars of the train were to be left at Avon, where in any case it had to shunt out of the way of the express. The shunting operations had been partly completed before the arrival of the express but as the freight train engine moved forward to make another shunt move it was diverted back onto the southbound main line via a set of points which had been left open. The express consisted only of a combination baggage and mail car, a smoker and a day coach in charge of Conductor A. W. Cushman with Engineer Lewis

---

[488] *Boston Daily Globe*, Boston, Massachusetts, United States, 18th September 1901

Tirrell and Fireman Jesse Lincoln. The freight engine struck the first car a few feet from its rear end, scraping it slightly, struck the smoker, in which there were 15 or 20 male passengers, hard enough to tear out part of the side of it, and the day coach, which contained both men and women passengers, was struck harder still. The second and third cars were thrown over on their sides and into a ditch on the right-hand side of the track.

A newspaper reporter described what happened after the accident. 'Station agent Edgar L. Crocker, happened to be down near the crossing, about halfway between the station and the wreck at the time. He rushed toward the scene. About the same moment, Harry C. Lake, the young telegraph operator at the station, saw the cars strike the freight engine and topple over. Without waiting to see them land in the ditch he rushed from the platform where he was standing into the office and tried to communicate with the train dispatcher's office in Boston. The circuit happened to be open, however, and while he was still trying to get connected agent Crocker rushed in and ordered him to get up to Avon Center with all speed and notify the local physicians. Young Lake mounted his bicycle and started on the uphill journey of a mile through the rain and mud. Such a mad race against time was never before performed by the wheelman in Avon and the young man was pretty well tuckered out when he reached the office of Dr R. A. Elliot and found he was in the barber shop getting his hair cut. Without an instant's delay Dr Elliot started for the scene, and with him went Dr E. P. Linfield, who had been notified by telephone.'

Six passengers were killed outright in the accident and about 14 seriously and 28 slightly injured. Conductor Cushman and Brakeman Joseph Pickford of the express were also hurt. Four of the passengers who lost their lives were quickly identified as Everett Joyce, of Joyce and Fletcher, Brockton shoe manufacturers; Gustavus Hay Jr, a Boston lawyer; William Coughlin, a prominent real estate operator, and Miss Winfred Tomkin, senior at the Howard seminary, Bridgewater. Two women, one aged about 60 and the other younger, remained unidentified.[489]

An inquest on the accident was held on Monday 30th September and an inquiry by the Massachusetts railway commissioners was held the next day. The key witness at the commissioner's inquiry, who did not give evidence at the inquest, supposedly because he was not given notice of it in time, was Brakeman Warren Murphy, who admitted that he had forgotten that the switch was still open so as to let the freight train back out onto the main line.[490]

The verdict of the inquest was announced a week later, on 8th October, when Judge Marden announced that he found both Engineer Sheldon and Brakeman Murphy guilty of causing the collision through negligence.[491]

On the morning of Saturday 21st December 1901 a train from Milan collided with a second train at Melegnano, Italy. Six people were killed and 15 seriously injured. Both trains were completely wrecked.[492]

---

489 Ibid, 19th September 1901
490 Ibid, 30th September and 1st October 1901
491 Ibid, 8th October 1901
492 *Evening News and Evening Mail*, London, England, 21st December 1901

An accident at Bloemfontein, Orange River Colony on the evening of Sunday 24th August 1902 involved a train of refugees bound for Johannesburg. By this date the (Second) Boer War had ended so the refugees were very likely Boer families who were being repatriated from British concentration camps (this was the phrase used at the time, though it did not then have the connotation that it acquired after the Nazi atrocities of the Second World War). A shunting engine collided with the carriages at the front of the train. Several carriages were 'smashed to pieces', several women and children were killed outright, and others were injured and had to be removed to hospital.[493]

A westbound Big Four (Cleveland Cincinnati Chicago and St Louis Railroad) passenger train, No. 25, which left Cleveland, Ohio, at 9.30 p.m. on Monday 23rd February 1903 collided with an eastbound freight train, No. 74, a mile east of Berea, Ohio, United States, on a double-track section of line. The passenger train was in charge of Conductor Michael White with Engineer William Fish and Fireman E. Glesman and was carrying many school supervisors and teachers on their way to Cincinnati to attend a meeting.[494] In charge of the freight train was Conductor Burns with Engineer John Ferguson and Fireman Havre. Both engineers and firemen survived the accident, though those on the passenger train had no time to jump and both engines were demolished. Not so fortunate were four mail clerks riding on the passenger train and two tramps, concealed in a stock car on the freight train, all of whom were killed. The track was piled high with the wreckage of the trains, which caught fire after the collision and burned so fiercely that the lineside telegraph poles were consumed, cutting off communication. Express messenger Joe Coulter was trapped, pinned between boxes and express packages, and would have burnt to death had not passengers thrown snow onto the wreckage to check the fire until he could be freed.

The mail clerks killed were named as G. W. Smiley, L. W. Walworth and W. H. Blackwell, all of Cleveland; and F. H. Kreissman of Cincinnati, Ohio. The names of the tramps were not known. A relief train with doctors was sent from Cleveland just before midnight.[495]

Engineer Ferguson of the freight train made a statement to a newspaper reporter after the accident. 'We had been in the siding to let train No. 16 pass and I saw the light of the dwarf signal change from red to green and I pulled out, for that gave me orders to go on to Lindale. Had the red light not have been changed I would have been on the siding yet, for no man will willingly jeopardize lives and property by running contrary to plain orders and signals. We had scarcely reached the eastbound light when I saw a glaring light in front of us. I called to Fireman Havre, 'For God's sake that headlight is on our track! Unload (jump) for your life. Meantime I had put on the brakes and brought the engine to a standstill. I jumped as far away from the engine as I could and plunged into a deep bank of snow on the side of the embankment. Engineer Fish of No. 25 (the passenger) had his orders to go to Lindale on the eastbound track. I do not know why such orders are given but they have been given many times lately. Perhaps this accident will stop such business. As for my headlight not being lit as the passenger crew say, it

493 *The Times*, London, England, 26th August 1902

494 *The Daily Huronite*, Huron, South Dakota, United States, 25th February 1903

495 *The Newark Advocate*, Newark, Ohio, United States, 24th February 1903

was burning clear and strong. Besides that my train was an "extra" and had two large white blizzard lights on the front of the engine. There were really three headlights on our pilot, so I don't see how the engineer on the passenger train could have made such a mistake.' [496]

At the inquest on the victims of the accident on Thursday 26th February the evidence offered was conflicting, but Deputy Coroner Burke decided that the weight of it was in favour of the crew of the freight train, three of whom testified that the signal had been cleared for them to proceed, and he placed the responsibility for the accident upon Berea Operator L.J. Peonteck, who denied having cleared the signal.[497]

Early on the morning of Monday 20th April 1903 a long double-headed westbound freight train broke down near Red House, New York, United States, on the Erie Railroad. The freight train was due to meet the eastbound train No. 4, the *Chicago Flyer*, at Red House, and the leading engine was detached and went forward to the siding with a flagman to warn the express. The eastbound train consisted of locomotive No. 545, six, a combination baggage and smoking car, two day coaches, three two sleeper cars and two private cars, and was in charge of Conductor George Thomson with Engineer Samuel Cook and Fireman Fred Bell. One private car was occupied by W. J. Murphy, General Manager of the Queen and Crescent Railroad, his wife and two company officials, and the other private car by J. L. Fraser, General Superintendent of the Toledo St Louis and Western Railroad.

Repairs having been made, the freight train was restarted and had just reached the siding when the passenger express appeared. The second engine and two or three cars had actually entered the siding when the passenger train crashed into the side of it. The force of the collision telescoped the leading vehicles of the passenger train and fire, fed by gas, broke out quickly and consumed the front five cars of the passenger train and several freight cars, though the two rear cars were detached and drawn free. Seven passengers were killed, along with Brakeman Robert Hotchkiss of the freight train, who was buried under the passenger engine. The bodies of the passengers, three men, three women and a child, were so badly burnt as to be unrecognisable.

After the accident eastbound passenger train No. 2 was backed seven miles down the line from Salamanca and subsequently went forward to New York with 30 of the uninjured passengers.[498] The work of clearing the line took all day and meanwhile trains were diverted via Dayton on the Salamanca – Dunkirk branch, involving a reversal at Dayton.

Red House siding was about a mile and a half long and there was a tower at its west end, manned by 17-year-old operator Lawrence Vale. It was believed that Operator Vale, seeing the headlight of the light engine, had wrongly assumed that the freight train had entered the siding. Acting on this assumption he had shown a white all-clear signal to the engineer of the passenger train, and the latter, concentrating on the signal from the operator, had failed to see that of the flagman vainly trying to warn him.

---

[496] *Defiance Crescent-News*, Defiance, Ohio, United States, 25th February 1903
[497] Ibid, 28th February 1903
[498] *Allegany County Reporter*, Wellsville, New York, United States, 21st April 1903

A surviving passenger, Harold Pulsifer of New York, gave a reporter a detailed account of his experiences. "I was in the forward sleeping car, and when I went to bed was the only passenger in that car. During the night, and unbeknown to me, a family party – an old woman, her daughter and granddaughter – got aboard, and later a man. I was awakened with a feeling as if my neck had been driven into my body, and then I could feel the car bumping over the ties. Almost immediately I smelled smoke. I knocked out the ventilator and opened the window. The car was tilted at an angle of thirty degrees, and I was on the upper side. As I opened the window flames burst from beneath it.

"Thinking I was the only person in the car, I flung out my belongings and was about to climb out after them when I heard a child's voice calling, 'Come and help us get out.' I got out in the aisle, where the smoke was now thick, and tried to get at my imprisoned companions. That side of the car was crushed against the engine, the upper berth was jammed down upon the occupants, and I could do nothing to help them. By this time the fire had gained such headway that my own life was in peril and I crawled out of the car through a window into a little space between my car and the wreck of the freight train. The engine was piled upon a flat car and beneath the flat car was the mangled body of the brakeman, R. S. Hotchkiss.

"While I stood between the two trains the engineer of the passenger train came running up, crying that there were women in the car burning to death. Together we tried with irons from the wrecked freight cars to break the windows of the car, but the flames were too much for us. The other man in my car had escaped, as had also the negro porter, Cook. Both were only slightly injured. The two women and the child were burned to death. Both the fireman and the engineer were thrown from their engine, neither knew how, and the fireman, though his right eye was closed, his face fearfully gashed, and his right leg badly torn, was the foremost among the rescuers. It was he who rushed down the line and opened the vestibules of the coaches. He finally fainted, after all was over, as he was carrying a woman's valise to the Reed {sic} House. As he was carried off on a stretcher he asked for a smoke and a last sight of his engine. He was the hero of the day."[499]

Reports of missing persons who had supposedly been passengers on the train led to a statement on 27th April that 10 or possibly 12 passengers had died in the accident, their bodies consumed by the fierce fire.[500]

Fireman W. S. Porter of the freight train stated a few days after the accident that this was the second wreck in which he had narrowly escaped death and he was giving up working on the railroad for a safer occupation.[501]

As No. 13 up mixed goods train was entering the loop line Sukheke station, India, on the North Western State Railway, on 13th June 1903, it was run into by No. 110 down goods train, whose driver had overrun signals. Eleven passengers were killed and seven injured while there was 'considerable damage to rolling stock.[502]

[499] *The New York Times*, New York, New York, United States, 21st April 1903
[500] *The Daily Palladium*, Oswego, New York, 27th April 1903
[501] *Allegany County Reporter*, Wellsville, New York, United States, 1st May 1903
[502] Administration Report on the Railways in India for the year 1903

There was a major disaster at Rockfish, Virginia, United States, on the Southern Railway, at 3.12 p.m. on the afternoon of Tuesday 7th July 1903. Passenger train No. 35 from Washington D.C. to Atlanta, Georgia, with cars from Boston and New York, was in charge of Conductor Mays with Engineer Charles Davis and had about 120 travellers. It consisted of a locomotive, an express car, a baggage car, two day coaches and two Pullman cars. The passenger train collided head-on with local freight train No. 68, on its return trip from Lynchburg to Charlottesville,[503] which was standing in a siding, into which the passenger train was diverted because the switch had been left open. Severe telescoping occurred between the baggage car and the following day coach and by the following day it was known that there were 22 dead and 14 injured. Those who died comprised Engineer Davis, Fireman Charles Gray of the freight train, Front Brakeman Thomas Shepherd of the freight train, Porter James Arnold, Dining Car Waiter Charles Laitch and 17 passengers. One of the passengers who lost their lives was another engineer, James McCormick, who was dead-heading to Monroe, the others were a mixture of coloured passengers and Austrian and Italian immigrants, the immigrants being 'bound for points as distant as California'. About 10 young immigrant children survived the accident but being unable to speak English they were unable to give any information as to their dead parents. They were being cared for by Charlottesville people, presumably as a temporary measure before they were consigned to an orphanage. Engineer E. C. Hale of the freight train was among the injured.

One of the passengers on train No. 35, and the only physician on hand, was Dr W. A. Lambeth, professor of hygiene at the University of Virginia, who gave his story to a reporter. "I was probably the first man out of the train. I found one engineer dead and the other painfully hurt. The engines were piled up together. The baggage coach had crashed through the passenger coach next to it, carrying off its top, and between these coaches, the one on top of the other, I could see that many persons had been crushed to death. Arms and feet stuck out at different points and in some cases crushed heads were in view. We got axes, cut through the panelling of the coach, and for an hour and a half worked like beavers to remove the dead and injured. In that time we took out 21 dead, not to speak of those that were injured. There was no panic among the ladies. On the contrary, they were extremely helpful, washing the blood from the faces of the wounded. It was a horrible sight and the work of relief so exhausting that when it was well over I keeled over in a somewhat collapsed condition and in turn was the object of the kind attention of one of the lady rescuers."

A relief train was sent from Charlottesville and returned there at 8 p.m. with some of the dead and almost all of the injured, 13 of whom were taken to the University of Virginia hospital.

The accident resulted from a simple mistake by Conductor W. B. Brubeck of the freight train, who had misread his orders as reading that the passenger train was an hour and 20 minutes late instead of just 20 minutes late. He was said to be half insane from mental anguish over the disaster. Engineer Hale was, of course, equally at fault as he had stuffed the orders into his pocket without actually bothering to read them, taking the conductor's word for what they said.[504]

---

[503] *Weekly Defiance Express*, Defiance, Ohio, United States, 16th July 1903
[504] *The Sun*, Baltimore, Maryland, United States, 8th, 9th, 10th and 11th July 1903

Buffer stop collisions, which are usually the result of driver error, have occurred from time to time at many terminal stations but are rarely that serious. An accident on Monday 27th July 1903 at St Enoch station, Scotland, the Glasgow terminus of the Glasgow and South Western Railway, was an unfortunate exception, and was also by an easy margin the worst accident to happen at any time on one of that company's lines.

The train involved in the accident was the 6.50 a.m. special passenger train from Ardrossan. It was the close of the Glasgow fair holidays and the train was carrying holidaymakers returning from Douglas, Isle of Man, on the steamer *Tynwald.* The train consisted of engine No. 221, three third-class carriages, a bogie composite, a first-class carriage, a third-class carriage, a first-class carriage, two third-class carriages, two first-class carriages, a third-class carriage and a brake van, and was in charge of Guard Hugh McCallum, with Driver Henry Northcote and Fireman Hamilton Kerr. Glasgow St Enoch station had been recently rebuilt and enlarged, so that at the time of the accident it had 12 platforms. The six new platforms were shorter than the old ones, and that into which the Ardrossan train was routed was, at 617 feet, the shortest platform of all but one. The signals controlling the immediate entrance to the platforms were placed on a gantry and each signal post carried a sign showing the number of the platform to which it applied. There was a speed restriction applying to all trains entering the station and the weekly circular detailing the running of the special train included a warning from the Superintendent of the Line that drivers must adhere to the speed restriction of 10 miles an hour between Clyde Junction signal box, 230 yards from the end of Platform 8, and the buffer stops, while Rule No. 5 (f) required drivers to enter terminal stations at such a speed as to be able to stop their trains short of the buffer stops using only their hand brakes.

The special train from Ardrossan failed to pull up before it reached the buffers, and there was a violent collision in which the rear ends of the carriages near the front of the train were thrown upwards, with the result that the underframe of the second coach got below that of the first coach, and the forward momentum of the train then resulted in the complete telescoping of the second carriage, in which many passengers were killed or severely injured. In all 13 passengers were killed outright, three fatally injured, 17 seriously enough hurt to need to be taken to hospital and 47 less badly hurt.[505]

'Samuel Knox, one of the injured, said that the *Tynwald* left the Isle of Man shortly after midnight with a large number of men, women and children returning from their holidays. A smart passage was made, and Ardrossan was reached at ten minutes to seven. The train ran at very high speed, and the carriages swayed much when St Enoch Station was being entered. He remarked to his wife that the train was certainly going too fast and had no sooner spoken than he found himself hurled into the next carriage. As further indicating the violence of the crash, he said that the dust was blinding. "One end of the second carriage was sticking high up, and I saw three men held fast to the seat by the broken woodwork of the compartment. They were unable to move, and, before they were extricated, it was necessary to saw through the woodwork. Mr Knox, with his son, was at Ibrox Park when the terracing collapsed. On that occasion he fell a distance of 75 ft and lay for three months in the infirmary.' (25 had died and over 500 were injured in the

[505] Board of Trade report

Ibrox stadium disaster on 25th April 1902.)

'Michael McGheen, who was in the last compartment of the second carriage, stated that he was sitting in a corner seat, when he was jerked violently through the open door onto the platform. He went to the assistance of the injured and helped to take six from the wrecked carriages. One woman, whom he was holding, died in his arms. The porters procured water and brandy, and this was poured down the throats of those who were pinned in the wreckage and could not be extricated at once. McGheen, although bespattered with blood, escaped with only slight injuries.

'A railway guard stated last night that many of the passengers in the ill-fated train took their seats at Ardrossan in a train which was to follow the express with met with disaster. At the last moment, however, two score or more of them changed trains, crossing the platform and entering the two first carriages of the express. But for this circumstance the telescoped carriages would have run practically empty, and the death list might have been nil.'[506]

Driver Northcote told the Board of Trade inquiry that his train was fitted with automatic vacuum brakes which had been tested before leaving Ardrossan and had been working properly. His train had left Ardrossan nine minutes late. There was only one booked stop before Glasgow, at Paisley, which was made normally. The signals between Paisley and St Enoch had all been off. He had shut off steam at Gorbals Junction and applied the brakes shortly afterwards. At Gorbals Junction his train had been running at 15 or 16 miles an hour. At Clyde Junction had had released the brakes and was going at 10 miles an hour there. Realising that his train was being directed into Platform 8 he had applied the brakes again. He had told the fireman to look out for any obstacles, meaning carriages, at the platform and when the latter said it was clear he had released the brakes again almost completely. His fireman had then shouted, "Stop Harry, there's the buffers." He had then applied the brakes with full force, reversed his engine and applied steam, while his fireman had put down sand. He also alleged that the vacuum brake on the engine had failed as the train reached the platform, with both needles showing zero. He had only been in St Enoch four times since the rebuild and never before into platform 8, which he had thought was 100 yards longer. He knew that Rule No. 5 required him to enter a terminal platform at such a speed that he could stop his train using the handbrake alone and was sure that he could have done so had the platform been as long as he thought it was.

Fireman Kerr said that he had only been into St Enoch station three times and was not aware that platform 8 was a short one. Until he saw the buffer stops close ahead as he was looking out for obstructions. He had then shouted to his driver, applied the hand brake hard and put down sand. He had thought just before the collision that the train would strike the buffers, but not that it would be a severe collision.

Guard McCallum said that he usually worked goods trains in the winter and passenger trains in the summer. His train had been rather light, with only 400 passengers and room for 600; most of the passengers had been seated towards the front. Everything had gone normally until they arrived at St Enoch, with normal stops at Paisley and Eglington Street. The train had been running at no more than 15 miles an hour between Eglington Street and Clyde Junction and somewhat more than 10 miles an hour at Clyde Junction,

---

[506] *The Standard*, London, England, 28th July 1903

where the vacuum brake had been applied and then released. His train was travelling at 8 miles an hour when they reached the platforms, which he thought was rather fast. He had applied his hand brake when they were going over the Clyde Bridge and it had been hard on as they were entering the station. He had not applied the vacuum brake because he thought that his driver had full control of the train.

Assistant Stationmaster John Douglas at St Enoch said that he had seen the Ardrossan train coming in and thinking that it was running too fast had signalled to its driver for him to pull up, but he did not think that the driver, who in any case was doing all he could to stop, had seen him. After the accident he had at once sent for doctors and telephoned for the ambulance vans, to the Central Police Station and to the locomotive sheds for the breakdown crew.

Station Inspector John Mathie was in St Enoch box arranging for the return of empty trains when the excursion train arrived. Seeing that it was coming in unusually fast he had remarked to the two signalmen that its driver did not seem to realise which dock he was running into. Both signalmen had come to the window and one of them had remarked, "I'm afraid he will strike the buffers." He estimated that the train had still be running at 20 miles an hour as it passed the station box.

Lt-Col H. A. Yorke, the Board of Trade inspecting officer, concluded that the disaster must be attributed to want of care in coming into the station too fast and inattention in not realising that he was going into a short platform on the part of Driver Northcote. Six witnesses had testified that he had entered the station at too high a speed to be able to pull up using the hand brake alone but even then he could have pulled up safely had he applied the continuous brakes earlier. His statement that he had been unaware that he was entering a short platform was doubtless true but hardly an excuse. He could have asked for a pilotman and even had he been directed into a long platform there might have been vehicles standing at it which would have given him no more distance to stop than a short platform. Tests carried out after the collision showed no evidence to corroborate the driver's suggestion that the vacuum brake on his engine had failed at the last moment, and in any case had he been braking using the hand brake alone this would have been immaterial. Driver Northcote, who was a strictly sober man with 18 years' railway service, must therefore be held responsible to the accident.

Suggestions had been made, Lt-Col Yorke added, that to reduce the consequences of such accidents one or two empty vans should be added to the front of all trains, but this would hardly be practicable and would result in the need for longer platforms and much wasteful running of empty vehicles, especially if as, to be logical, an empty vehicle or two was to be added to the rear of trains to guard against the possibility of rear collisions.

The inspecting officer considered that the rule requiring drivers to enter terminal stations 'at such a speed' as to enable them to pull up using only hand brakes should be amended to include an actual speed, say six miles an hour.

In conclusion Lt-Col Yorke called for the adoption of quick-acting brakes of the kind which had been developed by both the Vacuum Brake and Westinghouse Brake Companies, but which were not yet in use in the United Kingdom.[507]

---

[507] Board of Trade report

A special train carrying 954 passengers, including several hundred students of Purdue University, met with disaster on the outskirts of Indianapolis, Indiana. United States, on Saturday 31st October 1903. On that day Purdue University was playing Indiana University at Indianapolis in the Indiana State American Football Championship and the Cleveland Cincinnati Chicago and St Louis (Big Four) Railroad put on two special trains to carry the fans to the game, hauled by engines Nos. 188 and 330. It was extra No. 330 which was in the accident.

The accident had its origin in confusing rules and orders and a bad habit that train crews had fallen into of disregarding the rule regarding cautious running in the vicinity of Indianapolis. The operation of special or extra trains was covered by rules which stated that they had no rights and must keep out of the way of regular trains, running on orders issued by the dispatcher or superintendent. As regards the approach to Indianapolis, Rule No. 23 stated that, 'Trains not scheduled, when permitted to run between North Indianapolis and shops, must keep under control, expecting to find the track occupied by yard engines.'

Although according to the rule book specials trains had no rights Trainmaster P. J. Maloney issued an order stating that engines 188 and 330 would run as two extra trains from Lafayette to Indianapolis with rights over all trains. Train extra 330 was to leave Lafayette at 8 a.m., arrive at North Indianapolis at 9.29 a.m., North Street at 9.35 a.m. and arrive at Indianapolis at 9.40 a.m. The order was presumably intended to supersede the rule which stated that extra trains had no rights. The rule as to extra trains running through the yard limits at Indianapolis under control had also been overridden by the normal procedure of the Indianapolis yardmaster notifying switch crews of the movements of extra trains so they could keep out of the way of them. Aware of this, the crews of extras had got into the habit of running through the yards at a speed considerably in excess of that which was compatible with having their trains under control. The engineers were also under considerable pressure to keep time, as should their trains arrive late they were expected to go personally to the superintendent's office to explain the delay.

When extra 330 left North Indianapolis that morning it was nevertheless 23 minutes late, so it is probably not surprising that Engineer W. H. Shumaker 'stretched the rules a little' when he continued towards Indianapolis.[508] Tragedy now followed because the yardmaster had failed to advise switch crews of the extra because he had not himself been advised of it by the dispatcher. As a result, at a gravel pit switch near 18th Street a switch engine pulling a cut of cars collided with the front coaches of the extra. The first coach was crushed to splinters, the second telescoped and thrown down a 15-foot-high embankment and the third coach was overturned across the track. The coal cars were thrown through the first two coaches, which were filled with Purdue University players and substitutes, nine of whom were killed. In all 16 passengers died.

'The scenes following the rescue of the dead and injured were distressing. Strong men wept and cried aloud when they looked upon the faces of their dead classmates or at the sufferings of their classmates and brothers. Here and there someone would stop and lift from the face of one of the dead a handkerchief or piece of garment which had been

---

[508] *A History of Railroad Accidents, Safety Precautions and Operating Practices* by Robert B. Shaw, pgs. 295-6

placed over the features by some friend. Many of the girls rolled up their sleeves and knelt at the sides of the stricken men and bathed their wounds until the services of a surgeon could be secured. The women on the train did everything possible that their limited facilities would permit. Nearly all of them assisted in the work and their soiled dresses when the work was over was mute evidence of their willingness.

'All the young men killed were in the first coach, which was reduced to kindling wood. The coach held about 70 persons. Those who retained consciousness when laid on the grass urged and in some cases demanded that their friends hasten to the telegraph office and inform their parents that they had escaped with a few bruises. Several of those suffering fellows piteously pleaded that their names not be published because of the anguish it would cause their mothers or fathers. There were genuine heroes among the wounded boys. Several of them requested that the physicians first attend to the others and then return to them.

'Down town, three miles away, three hundred cheering students had arrived on a special train from Bloomington, Ind., with the sturdy boys of the Indiana university eleven eager for the contest on the grid-iron. They poured out of the train with band playing and colors flying, to be met with the news of the tragedy to their rivals from Purdue. Instantly bands ceased, colors were lowered and hidden beneath coats. Tears followed laughter and college yells faded into mournful expressions of sincere and heartfelt sympathy. The incoming Bloomington crowd broke into sad and silent groups that huddled at street corners, visiting the morgues and hospitals, searching for relatives and friends, or boarded trains of street cars that began carrying thousands to the scene of the wreck. The game has been indefinitely postponed. The injured at the hospitals are being well cared for. Surgeons have already amputated several limbs and are still hard at work over the most dangerously hurt.' [509]

At the close of the inquest on the victims of the disaster Coroner Tutewiler placed the responsibility for it upon Chief Dispatcher B. C. Byers of Kankakee, Indiana, who had failed to advice Indianapolis General Yardmaster John Q. Hicks of the running of the extra. Had Hicks received notice of the special he would, according to custom, have advised the operator at North Street, the switch tenders at Market Street and Senate Avenue and the assistant yardmaster at Shelby Street. Operator I. H. Bishop, of Shelby Street, Indianapolis, who had received notice of the extra the previous day but was not expected to post it, was not found to be at fault. Also specifically exonerated from responsibility by the coroner were Engineer Shumaker and Conductor F. M. Johnson of the extra, on the grounds that although their train had exceeded the speed limit through the yards they were warranted in doing so by custom and by orders inconsistent with the rule. This was in direct contrast to the findings of the railroad company, who had found the engineer and conductor responsible for the accident.[510]

What was described as the worst accident in the history of the Wabash Railroad occurred on the evening of Sunday 3rd July 1904 at Litchfield, Illinois, United States. At 5.36 p.m. passenger train No. 11 was wrecked a mile north of Litchfield depot.

Train No. 11 was in charge of Conductor James Crawshaw with Engineer James

[509] *The Fort Wayne Weekly Sentinel*, Fort Wayne, Indiana, United States, 4th November 1903
[510] *The Indianapolis News*, Indianapolis, Indiana, United States, 14th November 1903

Sanford and Fireman Sam Smith and consisted of a locomotive, a combination baggage and smoking car, two day coaches, a Cincinnati Hamilton and Dayton Railroad coach, a parlor car of the same company, a Wabash parlor car, an observation café car and the private car *Texas*. The accident was caused by it being diverted through an open switch onto the house track whilst running at a speed of at least 60 miles an hour and the results were disastrous. Standing on the house track was a string of freight cars and the express collided with these. Its leading coaches were shattered and almost immediately the wreckage caught fire. The Litchfield fire brigade were on the scene within 10 minutes, but the fire was fanned by gale force winds and they were unable to prevent it from consuming the whole of the train with the exception of the two rear coaches, which were uncoupled and drawn clear.

All the doctors in Litchfield attended the accident together with others from nearby Staunton, Mt Olive and Taylorville. Every bed and cot in Litchfield hospital was occupied by injured from the accident, some lying on makeshift beds on the floor while less seriously hurt cases were sent to a hotel or to their homes. A reporter wrote, 'A score of physicians labored among the injured and nine sisters were never so busy in their lives. The air was freighted with the odour of chloroform and the groans of the injured could be heard on every side. The smell of warm blood and the sight of raw and gaping wounds and bloody bandages was trying to unaccustomed nerves.' Despite the best efforts of the doctors five of the injured had died before midnight.[511]

A guard was placed on the switch after the accident while Coroner Gray opened an investigation.

The mechanism of the accident was unclear. It was thought highly unlikely that Engineer Sanford and Fireman Smith would have failed to notice the red target at the switch had it been showing which suggested the possibility of sabotage, but this too was thought to be unlikely. The third section of train No. 19 had passed through the switch safely only a few minutes earlier. It was believed, at least immediately after the accident, that the cause must have been a defective switch, the rod connecting the points having broken with the passing of the previous train.

Wabash Superintendent Cotter stated, however, that he believed that someone had tampered with the switch, while Wabash President J. Ramsey Jr remarked, "Someone turned that switch at Litchfield. Whether it was one who had a spite against the railroad or one who was morally crazy I don't know. Certainly it seems that no human spite against the railroad could be so bitter as to be willing to sacrifice the lives that were lost or to cause the suffering that has resulted." Others pointed out that the accident had happened in broad daylight at a spot where people were passing frequently, making an act of sabotage unlikely.[512]

By 8th July the official number killed in the accident was 21, with three others 'missing', their bodies possibly totally cremated. About 30 other persons were injured.[513]

By the close of the inquest the number of dead was put at 25, including four 'missing'. The jury found a verdict that the deaths of Isaac Mills and an unknown person had been caused by the train running into an open or misplaced switch. 'We further find from the

[511] *The Daily Review*, Decatur, Illinois, United States, 4th July 1904
[512] Ibid, 5th July 1904
[513] Ibid, 8th July 1904

evidence that the switch was disconnected, and that the keeper on the switch bridle, which held the connecting rod on the bridle, was broken off. We further find from the evidence that Margaret Steiner, Benjamin E. C Connell, Fannie Connell and Mary Irene Connell were killed and burned in said wreck, and that their bodies were never recovered.'

The verdict did not, of course, clarify, whether or not the switch had been deliberately damaged.

A return excursion train on the Chicago and Eastern Illinois Railroad, carrying largely children who were returning from a Sunday School picnic organised by the Doremus Congregational Church, was returning from Momence, Illinois, when it collided with a backing freight train near Glenwood, Illinois, United States, at 6 p.m. on the evening of Wednesday 13th July 1904. The passenger train was running at 30 miles an hour when it collided with the freight train which was coming towards it. The leading coach and the baggage car of the passenger train were telescoped in the collision and the other coaches were smashed and splintered. The lights of the train also went out in the collision and even the uninjured passengers were thrown temporarily into a panic, especially as they feared that fire would break out. As soon as the men on the train recovered themselves they set about the rescue work, which was made more difficult by the fact the bodies of the dead and injured were pinioned beneath the wrecked cars. Men ran up and down the track in both directions, to Glenwood and Chicago Heights, and in a short time trains carrying physicians came from both directions, first one from Chicago Heights with six physicians and then one from Glenwood with more physicians and a number of nurses.

Bonfires were lit to assist the work of rescue in the darkness and just before 10 the last of the injured had been taken from the wreckage. Half an hour later the train started for Chicago, stopping at Sixty-first Street, from where the injured were hurried to Englewood hospital and the bodies of the dead were taken to various undertakers.

Describing the rescue, a reporter stated, 'Physicians hurriedly dressed wounds and administered stimulants. Field surgery was practiced. Women and children were drafted into the work of relief. Car doors were torn off and used as stretchers and everything possible that could be done was accomplished. Out along the wire fences were laid 24 heads, and somewhere under the rough covering were the arms, legs and trunks. The sight was sickening. It seemed a grateful fact that families had not been separated. The lack of water caused much delay and a dozen mangled little ones breathed their last under the scrub oaks on the seat cushions. Fires were built to heat water. The farmers brought water in milk cans. The field hospital was operating among the piles of dead until eight o'clock, when the last bodies which could be found were taken out and placed on a special train for Chicago.' It will be noted that this report implies that the relief train to Chicago left somewhat earlier than stated in the previous account.

The children had been participating in the annual picnic of the Doremus chapel on Thirty-first and Butler Streets, which always took place at Island Park, Momence. The excursion was returning to Chicago as the second section of a passenger train and was carrying 750 passengers, mainly women and children from poor families, attracted by the cheap fares. Preceding the excursion had been a freight train which broke in two between Chicago Heights and Glenwood. Engineer Frank Hoxie had not realised that he had lost part of his train when he arrived at Glenwood and was in the process of crossing over onto the southbound track there when the excursion train ran into it on that track, to which it had been crossed over at Chicago Heights.

Eighteen passengers died in the accident, most of them small children, together with some women. Eighty-seven passengers were injured.[514]

Divisional Superintendent J. E. Russ investigated the accident and finding Engineer Hoxie to blame ordered his arrest a day or so later pending the coroner's inquest.[515] Another report stated that although Conductor Frank Cooper and Brakeman Charles Wright of the coal train had reported to the offices of the railroad company the engineer had disappeared.[516]

The inquest was duly held on 26th July, without Engineer Hoxie who remained a fugitive. Conductor Cooper explained, weeping several times as he testified, how the accident had happened. He said that when the train had become divided at Chicago Heights he had continued with the front of it, riding on the 15th car. He had not seen the red train order board at Glenwood and had ordered the engineer to proceed to Thornton. At Thornton he had given the engineer three distinct signals to cross to the other track and go into the siding. Then the Thornton operator had come up and asked if they had received any orders at Glenwood and was told they had not. The operator told him that he had heard the dispatcher tell the Glenwood operator to hold the coal train there. As he had been on the ground talking to the operator he had seen his train backing up and had said to the operator, "What is that fool engineer trying to do with the train?" The conductor said that he had then gone with the Thornton operator to the office and told him to tell the Glenwood operator to send Hoxie back to Thornton, but the dispatcher had intervened and said to leave the head end of the freight train at Glenwood.

He (the conductor) had then said to the operator, "They will strike, tell them that."

The operator worked the telegraph key and shortly afterwards looked up and exclaimed, "They have struck."

Earlier Brakeman Wright had admitted that had he remained at the rear of his train he might have seen the excursion train coming and flagged it, but he said that he had left to look at the orders on the bulletin board, which he had no right to do. He had only been a brakeman six months.

After five hours' deliberation the coroner's jury found five employees guilty of criminal carelessness, while criticising the company for overworking its employees and using dilapidated passenger coaches. The four employees taken to jail to await the pleasure of the grand jury were the conductor, the brakeman, Dispatcher Frederick Whiteman, who had ordered the excursion train to use the southbound track between Chicago Heights and Thornton and Frank Costello, the fireman of the coal train.[517] [518]

At the end of September, the grand jury exonerated Dispatcher Whiteman and

---

514 *The Daily Review*, Decatur, Illinois, United States, 14th July 1904
515 *The Goodland Herald*, Goodland, Indiana, United States, 23rd July 1904
516 *The Wahpeton Times*, Wahpeton, North Dakota, United States, 22nd July 1904
517 *The Moline Daily Despatch*, Moline, Illinois, United States, 27th July 1904
518 *Jefferson County Republican*, Fairfield, Iowa, United States, 29th July 1904

Fireman Costello, the other three being charged with manslaughter.[519] Engineer Hoxie appears to have remained at large and nothing further to have come of the matter.

Train No. 1, the westbound Canadian Pacific Railway *Transcontinental Express*, was run in two sections on Thursday 1st September 1904, the first section consisting of the baggage, express, mail and second-class cars and the second section of the tourist, dining and Pullman cars and two private cars carrying Lord Minto, the Governor General of Canada, and his party. Lord Minto was going on a farewell tour of the country before raking up a new post as Viceroy and Governor General of India. The second section was in charge of Conductor Battle, with Engineer Wilson and Fireman Delbridge. As the express made its way westwards coming in the opposite direction was special through freight train No. 118, which had an order to meet the express at Sintaluta, North-West Territories (it became part of Saskatchewan in 1905). There was already a westbound freight train in the crossing track and there were two cars in the other siding, so the freight train had to run forward past the siding on the single main track and back into it. In accordance with the usual procedure the front end brakesman went forward to flag the passenger train should it appear before the freight train was in clear and once the freight was in advance of the siding connection the rear brakesman set the road for the siding and then returned to the caboose. The train having backed in clear of the main line the engineer whistled to recall the front brakesman while the conductor had gone into the station buildings. No one thought to reset the switch.

There was apparently no target light at the switch, so the express engineer had no warning until his train was suddenly diverted into the siding, when he applied the emergency brakes just before the large express and freight locomotives collided head-on. Driven forward by the weight of the following Pullman cars the first tourist car of the express was forced against the tender of the locomotive and its leading end was telescoped. The second tourist car was also somewhat damaged, but all the deaths and injuries occurred in the first car. The collision occurred at 9.10 p.m.

Killed in the accident were Mrs J. C. Dowsett, of Edmonton; Mrs Warren, of Sand Point, Ontario; Miss Agnes Shirley, of Ottawa and two unidentified women supposed to be sisters named Hewitt, dressmakers en route from Toronto to Calgary. Mrs Dowsett was going to join her husband in Edmonton and had her three children with her; they escaped unhurt. The injured passengers were Mrs Back and Miss Golden or Goldman, en route from New York, United States, to Manila, the Philippines; Miss Murray, of the China Inland mission, bound for Shanghai, China and Ira Ross, of Blenheim, Ontario, going to Victoria, British Columbia. Porter L. P. Hart was also badly hurt, while Engineer Wilson of the express, who was hurled through the cab window, suffered two broken ribs.

Fortunately, there were soon three physicians on the scene to treat the injured, Dr Bouju of Sintaluta being joined by Drs Kemp and Harvey, who were brought on a light engine from Indian Head. Relief and wrecking trains were also sent for, information of the accident reaching Winnipeg at 1 a.m. Initially it was reported that there were a large number injured, it being difficult to assess casualties in the darkness. The vice regal party in their cars at the rear of the express escaped injury and continued their journey at 3 a.m.

[519] *The Attica Daily Ledger*, Attica, Indiana, United States, 29th September 1904

As was common at this time the inquest into the victims of the accident was held already the afternoon after it had happened. It found Conductor J. A. Irwin and Engineer Thomas Johnston of the freight train to have been responsible for the collision and they were arrested on a charge of criminal negligence amounting to manslaughter and taken to Regina to stand trial.[520]

The prosecution of the conductor was eventually dropped. Engineer Johnston went on to stand trial at Regina in late February 1905 and it took the jury an hour to find him 'Not Guilty'. [521]

On 4th September 1904 No. 35 up mail and No. 64 down goods trains collided at Machhpara station, India, on the Eastern Bengal State Railway. Five passengers died, six were injured and there was considerably damage to engines and rolling stock. The accident was blamed on the non-observance of shunting rules by the assistant stationmaster and the driver of the mail train running past signals.[522]

Having already been put off of a freight train three times a group of tramps re-boarded it and five of them lost their lives in an accident on the Wabash Railroad at Sadorus, Illinois, United States, on Monday night, 24th October 1904. The freight train on which they were riding became divided and the two portions then came together again, whereupon the train was derailed. One of the tramps who died was identified as J. Francisco, a cigarmaker of Bloomington, Illinois. Another tramp, Martin Tracy, of Boston, also a cigarmaker, hung suspended for four hours in the wreckage close to the bodies of his dead companions. He reportedly tried to commit suicide by strangling himself but was found and released by trainmen.[523] [524]

The unintended diversion of a train into a siding caused a major disaster in England in 1905, this time as the result of a mistake by a signalman. The location was Hall Road, on the Lancashire and Yorkshire Railway Liverpool – Southport branch, a busy commuter line which had already been electrified at this early date.

Many trains on the branch from Liverpool terminated and started from Hall Road and the station was provided with a siding on the down side and a middle siding situated immediately north of its up and down platforms where trains could be stabled on arrival from Liverpool pending their return there. Specifically, on weekdays there was a 10-minute interval service of trains between Liverpool and Hall Road, plus a 20-minute interval service between Liverpool and Southport. The two sidings were reached by facing points from the down main line. Mechanical detectors fitted to the points made it impossible for the down main home signal to be cleared unless the points were set for that line, an electric lock on the signal lever meant that it could not be pulled unless the switch blocks of the facing points were in the exact adjustment and mechanical locking

[520] *Manitoba Free Press*, Winnipeg, Manitoba, Canada, 3rd September 1904
[521] *The Stonewall Argus*, Stonewall, Manitoba, Canada, 22nd February 1905
[522] Administration Report on the Railways in India for the year 1904
[523] *The Goodland Herald*, Goodland, Indiana, United States, 29th October 1904
[524] *The Iowa Recorder*, Greene, Iowa, United States, 2nd November 1904

between the levers meant that the signal lever for the down main home could not be pulled if either of the points levers were over, nor could the distant signal be cleared unless all the main line stop signals were cleared. Full signalling protection was therefore provided, in contrast to the situation at Sintaluta, but human error caused an accident all the same.

On duty in Hall Road signal box on the evening of Thursday 27th July 1905 was Signalman William Boote. He gave evidence at the subsequent Board of Trade inquiry into the accident, where he testified that he had been offered the 6.20 p.m. passenger train from Liverpool at 6.36, it had been given "On line" at the same time and had reached him at 6.39. This was one of the short workings, and after its passengers had alighted it was put into the middle siding at 6.40. He was then offered the 6.30 p.m. express from Liverpool and reversed the facing points into the middle siding and went to clear his signals for the express, but he found that he was unable to pull the lever working the inner home signal. He had then thrown back to danger the outer home signal, reversed the lock bars of the siding points and another set of facing points and set the road for the siding again. He had then pulled the lock bars again and cleared his outer home and starting signals but had found that he still could not clear the inner home. This signal was of course now mechanically locked because he still had the points set for the middle siding. Clearly flustered at this moment, afraid that he would delay the express unnecessarily and thinking that the problem was a recurrence of a fault with the electric locking, he had then seized a green flag and displayed it to the motorman of the express, which was just slackening speed at the outer home.

Being an electric train the express had been able to accelerate sharply, much more quickly than a steam train, and by the time that its motorman had realised that he was being diverted into the middle siding it was going fast enough for there to be a violent collision with the stock of the 6.20 train which was standing there. The leading car of the express and the first car of the empty stock were telescoped into one another and two bodies of the empty train were driven beneath the second one. Twenty passengers in the leading car of the express were killed outright and two seriously and 45 slightly injured. One of the seriously injured passengers later died. The motorman of the express was also seriously injured.[525]

A passenger related his experiences. 'The train was nearing Hall-road when the passengers felt two successive jolts. He believed the train was travelling from 38 to 40 miles an hour. The whole front of the third-class compartment, in which he thinks there were about 17 or 18 passengers, was telescoped. The bottom of the carriage fell out and the whole of the passengers were pinned beneath it. The train came to a sudden stop and he was the first to get out. The first thing he saw was a man in blue overalls hanging from the front carriage. The motor had blazed up and the man's clothing became ignited. This fire, with the assistance of another man, was put out. They tried to gain admission to the compartment by the carriage door, but this was found to be impossible. They had to tear the boards from the sides of the carriage, and immediately he found five bodies grouped together. In his opinion the whole of the passengers were lifeless. He saw 13 dead bodies. He believed the deaths were due to the concussion, and impact with the debris. The bodies were absolutely wrapped up in the framework at the bottom of the carriage. In the case of a young man they got the head and body clear. Everybody able to do so rendered

[525] Board of Trade report

the utmost assistance in rescue work.'[526]

It was over a fortnight before Motorman Willian Rimmer was sufficiently recovered from his injuries to be interviewed. He said that his train had been checked at the Seaforth and Waterloo distants and had then had a good run to Crosby. The Hall Road outer distant had been against him but the inner distant had been cleared as he had approached it. What he called the outer home had been kept at danger, but the signalman had exhibited a green flag from his box. Not being satisfied, he had applied the brakes and popped his whistle, but the signalman had shouted from the window, "Right away driver," whereupon he had applied the current. The platform starter or inner home had been off when he first saw it but had been thrown back when his train was close to it. He then saw that the points were set for the middle siding and had shut off the current and applied the brakes. He had never been signalled past a signal at danger with a green flag before as a driver, though he had had his driver do so when he was a fireman.

Guard Sam Aldred of the express said that he had seen the signals between Waterloo and Crosby and had then been walking through the train counting the passengers when he had been suddenly thrown to the floor. He saw both the Hall Road distants at danger and through the train window a green flag being waved from the signal box. He took the green flag waved horizontally from side to side by the signalman as a right-away signal. He had not seen such a signal before. He did not notice his train accelerating from the fifth and rear carriage of the train, it just seemed as if his motorman had released the brakes.

Recalled, Signalman Boote explained that he had feared that he would be fined if he had delayed the express without good reason. He had never been fined previously, though he had heard of other signalmen who had been. He denied that the inner home had been off as stated by the motorman. No troubles had been affecting him that day and he had had no reason for making his mistake other than being afraid of causing delay. The work of the box was very heavy at times, if straightforward, and, along with his mate, Signalman Smith, he had applied for the assistance of a booking boy.

The Hall Road stationmaster, George Hartnup, told the inquiry that the first he had known of the accident was hearing a crash as he came out of the office. He had seen the train in the middle siding and a great deal of arcing going on, which had subsided as he ran down to the siding, but the woodwork of the carriages had been smouldering. Worried about the danger of fire he had run back the station to ask the power house to switch off the current. He had then telephoned the Crosby Fire Brigade, advised Mr Shaw at Liverpool, who said that he would contact Bootle Hospital, and contacted the locomotive department at Sandhills for the breakdown train. Returning to the scene of the accident he had found the guard playing on the fire and uninjured passengers attempting to remove a body. Motorman Owen Humphries had pointed out that it would be better if they had an engine to pull one coach from the top of the other, so he had gone back to the station again to get the locomotive department to send out an engine at once. At about 7 he had gone to the signal box and had asked the signalman what he had been doing and the signalman had said, if he remembered correctly, that he had forgotten to reverse. He had not asked the signalman any more questions. Thinking that it would be best not to leave the signalman alone and unable to find anyone else he had asked a motorman to

[526] *The Times*, London, England, 28th July 1905

stay with the signalman in the box. He had not given the signalman any general instructions regarding flagging trains past signals but on a couple of occasions he had flagged them past himself after having carefully checked the lie of the points.

Lt-Col E. Druitt, the Board of Trade inspecting officer, pointed out that the causes of the collision were clearly shown by the evidence of Signalman Boote and Motorman Rimmer. Their statements as to the positions of the signals were contradictory, but that of the signalman was clearly correct and the motorman's recollections had been affected by his severe head injury.

Despite Guard Aldred's statement that his train had not accelerated after his motorman had responded to the signalman's green flag Lt-Col Druitt believed that he had and had been travelling at 40 to 50 miles an hour when he saw the road set for the middle siding. It was impossible for the inner distant to have been clear as stated by the motorman, but denied by the guard, and the inner home signal had certainly been at danger thanks to the mechanical and electric locking. The only explanation of the motorman's statement was that his memory had been affected by his injury. It also seemed doubtful if the motorman had shut off the current before the accident as he said, or that the signalman had shouted, "Right away," which Boote denied.

Signalman Boote said that he was well aware of Rule 73, which stated that should a signal become defective a competent person should be placed at it with hand signals, to work under the instruction of the signalman, and when there were facing points it must be checked that they were safe and secured for the line on which the train was to run. Boote had inadvertently left the points set for the siding and then in order to prevent delay had disregarded the rules. His statement that the box was busy the inspecting officer did not accept and in any case that had nothing to do with the collision.

Motorman Rimmer had likewise disregarded the rules and had driven recklessly.

The only lesson to be learnt from the accident was the very old one of the necessity of always complying exactly with all the rules and regulations and that punctuality and fast running must always give way to safety.

The high number of deaths on an electric railway had, the inspecting officer stated, attracted much attention, and he concluded his report by referring to several points raised in the press and elsewhere. These were as follows:

(a) The supposed danger of facing points. With modern safeguards to ensure that a signal could not be lowered unless the points were correctly set there should be no danger. Facing points were a great advantage in permitting a slow train to get out of the way of a faster one without the time-consuming operation of bringing the whole train past the connection, stopping and then setting back. At Hall Road they were the quickest and simplest way of getting turnback trains off of the down road and therefore the safest.

(b) Fears that open corridor trains such as that used on the Liverpool express were more liable to telescope than the older type of coach with separate compartments were unfounded. At Hall Road the collision had been a very violent one between the express running at high speed and the empty train weighing 170 tons standing with its brakes on, so that some telescoping was inevitable regardless of the type of vehicle. Regarding the old suggestion that all trains should be run with an empty carriage or van next to the locomotive this overlooked the fact that rear collisions were as common as head-on collisions, requiring an empty vehicle at each end of the train, entirely preventing the use of single-coach steam railcars such as several companies were now using, and greatly adding to the running costs. Much greater security could be obtained by a strict observance of the rules.

(c) As regards the increased danger of fires from the running of electric trains there was no fire in the carriages at Hall Road, though there was a slight charring of the woodwork and some smoke, nor had any of the passengers suffered an electric shock. The current had been cut off four minutes after the accident when the stationmaster telephoned the sub-stations, something which he should have done when he first became aware of the accident. Iron bars were carried on the trains and supplied to the stationmasters to set up a short circuit and blow out the circuit breakers at the sub-stations in an emergency. It might be noted that there was a serious collision the previous winter involving a steam train where several coaches were destroyed by fire (at Cudworth in January).

(d) Though there was only one motorman on an electric train as opposed to a driver and fireman the guards were instructed how to stop a train in an emergency should the driver (motorman) be suddenly incapacitated.[527]

On 10th August the inquest on the accident victims was concluded, the jury returning a verdict that, 'the accident was caused by a mistake by signalman Boote and driver Rimmer, which was censurable, but not criminal. They recommended that the facing points at Hall-road and everywhere possible should be abolished. They also strongly suggested that, especially on express trains, the guards should be conversant with the working of the motor in case of accident, and they thought that all signalmen and drivers should be examined periodically as to their knowledge of the rules appertaining to their work.' [528]

A third collision due to a train being accidentally diverted onto the wrong line occurred on Monday 25th September 1905 at Paoli, Pennsylvania, United States, on the Pennsylvania Railroad. Paoli lies only a short distance from the city of Philadelphia, was the terminal station suburban trains and was provided with a large yard on the north side of the line for the storage of rolling stock and engines between trips. One of the trains involved in the accident on 25th September was a local train, No. 140, which had just left the yard, the other was the eastbound St Louis – New York express. The latter train, one of the fastest on the system, usually ran through Paoli on track 1 at 2.50. On the day of the accident it was five minutes late, and at 2.52 the local train was taken from the yards across to track 1, the signals being placed against the express, which was to be re-routed to track 2 after the local had crossed from the yard. Perhaps because the local train was longer than usual its engineer pulled up prematurely, so that the rear truck of the special car was still standing on the end of the crossover switch, thus preventing the crossover being reset for the main line. The express received a green caution signal at the first Paoli tower, where it was crossed over to track 2. The signal at the second tower was red, but the express engineer kept going, despite the signal and the blowing of a whistle by Operator J. Bryan. The express was diverted through the crossover and into the back of

[527] Board of Trade report

[528] *The Times*, London, England, 11th August 1905

the local train.

All those killed and some of those injured in the collision were riding in the private car of Pennsylvania Railroad General Manager Atterbury which had been attached to the rear of the local train. Mr Atterbury was on vacation in Maine until 1st October, and in the meantime the car had been overhauled and then sent out in the morning from the West Philadelphia yards on a test run. It had been attached to the afternoon local train to take it back to Philadelphia and thence to Maine. In the collision the engine ploughed 10 feet through the special car, which contained about a dozen men who had been overhauling it. The special car was in turn forced about halfway through combination passenger and baggage car ahead of it, from which Baggagemaster Ross jumped just in time to escape. Five of those in the special car were killed or fatally injured and all the rest hurt. The half-dozen ordinary passengers in the front of the local train escaped injury, as did the driver and fireman of the express, but about half a dozen express passengers were hurt by being thrown against tables or the sides of cars. After the collision Engineer Ulysses Bromall of the express reportedly placed his head in his hands and cried, “My god, why wasn’t I killed?”

Those who lost their lives in the special car were George Pennypacker, foreman car inspector at West Philadelphia, George. Walton, steam heat inspector at Altoona; Frank Brastow, Philadelphia general agent of the Safety Car Heating and Lighting Company and brother-in-law of Mr Atterbury; Charles Danbauer, of Philadelphia, described simply as a railroad employee in the West Philadelphia yards; Richard Garland, general foreman of the West Philadelphia railroad shops, who succumbed to his injuries in hospital the same evening, and coloured cook Norman Redmond, who died the next morning.

Hundreds of railroad men from the adjoining yards were quickly on the scene after the accident. Most of the injured were loaded into the undamaged forward cars of the local train, which was hurried into Philadelphia as a special, stopping at Bryn Mawr station, half a dozen of the injured being taken thence to Bryn Mawr Hospital.[529]

A travelling salesman named S. G. Weber witnessed the accident. “I was standing at the station when I noticed the express coming up the track. Simultaneously, it seemed, with the sounding of the whistle from the signal tower came the crash of the oncoming train. The crash was deafening, and I do not know whether the passengers realized the danger in time to scream there was such a din. I rushed across the tracks, and as I did so those of the passengers who were not hurt crowded down from the local train.” [530]

The inquest on four of the victims of the accident was held the next day. Operator Harry Stevens of Paoli station tower testified that he had told the man at the levers (not named but presumably Operator Bryan) to cross the local train onto track 1, which there should have been plenty of time to do. However, the rear car of the local did not go six feet beyond the switch and was resting directly on the guard rail so that the crossover could not be reversed.

Engineer Bromall stated that just before Paoli station he had slowed his train down using the air brakes and crossed safely from track 1 to track 2. There had been a freight train in the station, the steam of which had obscured the next signal. He was just passing it when his fireman had shouted, “You are passing a red.” He had applied the brakes in

[529] *The Washington Post*, Washington, D.C., United States, 26th September 1905
[530] *The Philadelphia Inquirer*, Philadelphia, Pennsylvania, 26th September 1905

emergency and got its speed down to about eight miles an hour, but with a train of six heavy Pullmans it was impossible to stop. Had the local not been standing with its brakes on there would have been no wreck.

Fireman Frank McCormick said that he had been looking out for signals approaching Paoli. His side of the train had been clear, but the engineer's side had been clouded with steam. He had shouted when he saw the signal at danger. He was positive that the train was going no faster than eight miles an hour and had not jumped as he had not expected there to be a collision.

Several other witnesses declared their surprise at their severity of the collision.

After several hours' deliberation the jury found Engineer Bromall and Fireman McCormick responsible for the collision for running past a red signal and the crew of train 140 responsible for not taking proper measures to ascertain that their train was standing on the switch.[531] A second inquest on the deaths of Messrs Garland and Redmond returned a similar verdict.[532] No prosecutions for manslaughter seem to have followed the verdicts.

At the end of 1905 the 1905 Russian Revolution was in full swing and the railways were severely affected. It was stated that 'A condition of anarchy prevails on the Siberian railway line. All the stations have been pillaged, and the military trains follow one another indiscriminately, in charge of unqualified engine drivers. Thirty soldiers were killed, and many injured through a collision between two military trains at Lebedin.'[533] This accident occurred on Sunday 31st December 1905. *The Times* in London stated that the victims in the Lebedin accident were recruits and railway delegates and mentioned a second collision in Russia, also involving two military trains, on about the same date, between Znamenska and Trepovka stations. Twenty soldiers were reportedly killed, and several others were injured in this accident.[534]

On Sunday morning, 17th June 1906, a party of 'pleasure seekers' wished to get from their homes in Patty, Tennessee, United States, to a nearby town, and probably on the suggestion of one of them, J. H. Brinkley, who was employed by the Louisville and Nashville Railroad, they decided to borrow a hand car and use that to take them there. The hand car, carrying men, women and children, was on its way and crossing a bridge over the Hiwassee River when it as in collision with engine No. 558, in charge of Conductor T. R. Kelley with Engineer C. Simpson, coming in the opposite direction. The light engine met the hand car on a curve and was upon it before any of its riders could jump. The engine was pulled up as quickly as possible but four of those riding on the car were killed outright and three others seriously injured, two of whom died later. J. H. Brinkley was one of those who lost their lives.[535]

---

[531] Ibid, 27th September 1905
[532] Ibid, 5th October 1905
[533] *The Warwick Examiner and Times*, Warwick, Queensland, Australia, 6th January 1906
[534] *The Times*, London, England, 2nd January 1906
[535] *The Times Dispatch*, Richmond, Virginia, United States, 18th June 1906

A troop train collided with a goods train near Marianovka, Russia, 'on the Siberian Railway' on about Sunday 7th October 1906. Seven soldiers were killed and 29 injured.[536]

A serious collision occurred at Epernon station, France, on the Chemins de fer de l'Ouest, in the early evening of Sunday 14th October 1906. Omnibus train No. 510 which had left Rennes at 7.40 a.m. was due to leave Chartres at 3.50 p.m. and to arrive at Paris Montparnasse at 5.50. It left Chartres 22 minutes late at 4.12 and had been shunted into a siding at Epernon out of the way express train No. 514 from Brest, due to pass Chartres at 4.13, stopping between there and Paris only at Versailles.

The latter train duly passed but Driver Reverdy noted that it appeared to be running that day in two sections as there was a green marker on the rear of the express indicating that another train was following. He asked the stationmaster if he must wait for that train as well. No, the stationmaster replied, the omnibus train could now proceed, and Driver Reverdy noticed that the yellow carré signal had been cleared for him. He opened his regulator and was just drawing his train out onto the main line when it was struck violently by a light engine, No. 980, coming from Le Mans with Driver Luc Louvet and Fireman Le Breton. The engine was travelling at 80 kilometres an hour and crushed five carriages of the omnibus train. The leading mail vans and the rear vehicles of the omnibus train escaped, two first-class, two third-class and one second-class carriage in the middle of the train taking the impact and being crushed. Driver Louvet was only slightly hurt and his fireman escaped altogether.[537] The collision occurred at about 5.15 p.m.

News of the accident quickly spread through the town of Epernon and a crowd soon assembled at the station and began the rescue work, assisting a crew of firemen under a Lieutenant Goussard. Nine passengers were killed outright in the collision and about 32 others were injured, three of whom subsequently died. The train carrying some of the injured arrived in Montparnasse at 10.10 p.m., others having been left at Essarts-le-Roi, Versailles and elsewhere.[538] Those who lost their lives in the accident included Dr Charles Floquet, Chief Medical Officer for the Palace of Justice; Mme. Henriette Deschamps; 15-year-old Henry Deschamps; M. Gaston Mongeot, owner of a women's clothing store; M. Georges Laurent, a commercial traveller; Mme. Jeanne Moreau, aged 15, and M. Alfred Brisard, a painter.

On being informed of the accident M. Barthou, the Minister of Public Works, went to the scene accompanied by the Director of the Ouest Company, M. de Larminat, and an investigation into the cause was opened at once by the judicial magistrate. There was a conflict of evidence between Driver Louvet, of Vaugirard depot, in charge of the light engine, and the Epernon signalman, the driver claiming that the signal had been off for him. Driver Louvet was said to be a very good driver with a previously clean record but the inquiry nevertheless found him responsible for the accident and he was charged with manslaughter through negligence.[539]

Both Driver Louvet and his fireman were found guilty at their trial in March 1907 and

---

536 *The Times*, London, England, 9th October 1906

537 *Le Petit Journal*, Paris, France, 16th October 1906

538 Ibid, 15th October 1906

539 *Le Petit Parisien*, Paris, France, 16th October 1906

the driver was sentenced to a year's imprisonment. His fireman was clearly thought to be much less responsible as he received only 50 francs suspended fine.[540] The Ouest Company were found to be civilly liable. In November the Court of Appeal reduced Driver Louvet's sentence to one of six months' imprisonment.[541]

Westbound Minneapolis St Paul and Sault Ste Marie (Soo) Railroad passenger train No. 106 collided with a switch engine a mile west of the passenger station at Enderlin, North Dakota, United States, at 2.10 a.m. on Sunday morning, 23rd December 1906. Enderlin was a division point and the switch engine was engaged in making up a freight train to be sent east. The string of cars was too long for the sidetrack and the engine, which was at the west end of them, had come out onto the main line during the shunting operations. Prior to the fouling of the main line by the engine a flagman had been sent out who testified that he had gone back half a mile, put down two torpedoes and waited there. When the passenger train had appeared, he had waved his lantern at it but there had been no reaction to either his hand signals or the torpedoes, which he presumed had exploded. The night had been very foggy.

The passenger train had already whistled for Enderlin station, where it was booked to stop, but was still travelling at considerable speed when it struck the engine. Nine passengers were killed outright in the accident and about 35 seriously injured. Engineer Frank Barnes and Brakeman Harry Dizard of the passenger train were also hurt, while the deaths of two of the injured passengers after they had been taken to Minneapolis brought the number of fatalities up to 11. All the deaths and most of the injuries were to men riding in the smoking car, which was completely telescoped by the baggage car. Only two of the occupants of the smoker escaped unhurt. Many of its passengers had got to their feet to alight at Enderlin upon hearing the engine whistle for the station. A few of the passengers in the day coach were hurt but no one in the sleeping cars, which were not derailed. Most of the passengers on the train had been on their way to visit relatives or friends for Christmas.[542]

With Enderlin station close at hand help was soon on hand after the accident and a special train was sent from Valley City with physicians and nurses. Overturned stoves set fire to the wreckage so that there was a race to extricate the dead and injured before the advancing flames, but with great effort they were all freed. It was not thought worthwhile to try and fight the fire after that and the cars were left to burn themselves out. The little hospital at Enderlin was soon overwhelmed with casualties and the Hotel Hilton and other establishments were converted into emergency hospitals.

The train concerned in the accident was an accommodation working from Moose Jaw, Saskatchewan, Canada, to St Paul, Minnesota, due there at 11.45 p.m. but almost three hours late. It was reportedly running fast in an effort to make up a little of the lost time.[543] Earlier in the same journey the train had collided with a hand car near Hecker, North Dakota, killing one man and seriously injuring the other.[544]

540 *L'Écho Nogentais*, Nogent-sur-Seine, 17th March 1907
541 *L'Écho Nogentais*, Nogent-sur-Seine, 24th November 1907
542 *The Fargo Forum and Daily Republican*, Fargo, North Dakota, 24th December 1906
543 *Logansport Daily Reporter*, Logansport, Indiana, United States, 24th December 1906
544 *Wahpeton Times*, Wahpeton, North Dakota, 3rd January 1907.

Following the accident warrants were issued for the arrest of Engineer John Moore, Yard Foreman Charles Acker and Flagman J. E. Walsh pending a hearing which was expected to result in their being charged with second degree manslaughter for fouling the main line in the path of the passenger train and/or not taking sufficient steps to warn its engineer of the obstruction.[545]

It was Engineer Moore who first stood trial at Lisbon, North Dakota, at the end of February 1907, when it emerged that he had had orders to go out a mile and a half with the flagman but had in fact only gone half that distance. Only a white flag (according to the newspaper account but given that it was 2 a.m. clearly a lantern) was shown and, despite the initial testimony of the flagman that he had put down two torpedoes it was stated that there had been no torpedoes on the switch engine, nor presumably any fusees.[546] Nevertheless, after about three hours' deliberation the jury found him "Not guilty". Upon this becoming known the charge against Flagman Walsh, who had been still awaiting trial, was immediately dropped and he was discharged. It must have already been decided not to proceed against Yard Foreman Acker.[547]

On about Sunday 17th March 1907 a passenger train and a goods train collided at Turnshiche station, Manchuria. Seventeen people died and 35 were seriously injured.[548]

On the evening of Saturday 24th August 1907 express train No. 624 which had left Bordeaux at 10.15 p.m. for Paris via the Chemin de fer de Paris à Orléans came into violent collision with locomotive No. 1002 which was performing shunting operations about a kilometre from the passenger station at Courtras, France. The express consisted of a locomotive and 13 vehicles; eight vans, a mail van, destined for Paris Austerlitz, and two first-class, one second-class and one third-class carriages Both engines were severely damaged, all the carriages were more or less broken up and 12 goods waggons were smashed. Eight passengers and Driver Pihet and Fireman Gatinel of engine No. 1002 were killed outright and about 30 passengers were injured. Driver Limouzin of the express and his fireman saw the danger, jumped in time and escaped. The collision occurred at 11.05 p.m.

Train No. 624 from Bordeaux Bastide ran only between 8th July and 15th October and was a relief train to the regular express, train No. 24, which left Bordeaux Saint Jean at 10.30 p.m. Train No. 624 provided a connection with the Midi line and was specially intended for passengers going through to Paris, where it was due to arrive at 7.27 a.m. It was timed to run at an average speed of 70 kilometres an hour between Bordeaux and Angoulême.

On hearing the noise of the crash railwaymen from the station and workers from the adjoining Calvet oil mill hurried to the scene and began the work of trying to extricate the approximately 150 passengers from the wreckage with the aid of lanterns. The rescue work continued all night, and for some reason it was not until 6 a.m. that Dr Bouchard

545 *The Fargo Forum and Daily Republican*, Fargo, North Dakota, 28th December 1906
546 Ibid, 20th February 1907
547 Ibid, 22nd February 1907
548 *The Times*, London, England, 20th March 1907

arrived from Libourne and began treating the casualties, who had been lying on cushions taken from the carriages since midnight. The injured were then loaded onto a special train which had been sent from Bordeaux. They arrived back in Bordeaux at 8.30 a.m. At an unstated hour express train No. 31 from Paris to Bordeaux was stopped at Coutras because of the accident. By chance Dr Henri de Rothschild was on this train, travelling with two French and one Spanish doctor to Tangier, and he (and presumably the other doctors) on hearing of the accident hurried half-dressed to the scene to assist in treating the injured.

Single-line working was established at 6 a.m. and both tracks were back in use at 10 a.m. M. Barthou, the Minister of Public Works, visited the accident site the next day.

The accident was attributed in a French newspaper report to a mistake by a signalman, without further explanation.[549] A report in *The Times* newspaper in London from a correspondent at the scene spoke of the faulty operation of a set of hand points which had been recently installed prior to the construction of a marshalling yard at Coutras. The same newspaper mentioned that there had only been two English passengers on the train, Messrs W.A. Bradley and Egbert James of Wellingbourough, who were returning from a three-week holiday in the Pyrenees.[550]

A mistake by an operator was also blamed for a head-on collision on the Baltimore and Ohio Railroad at Bellaire, Ohio, United States, at 2.30 p.m. on Saturday 28th September 1907. A westbound freight train had arrived at Bellaire and was slowly moving along the siding. Eastbound Chicago and Wheeling express passenger train No. 14 was also due and Operator William Morgan in the Bellaire tower had directed Operator Buckhannon in Schick tower, a mile further west, to put the passenger train onto the main track there. Operator Buckhannon misunderstood what he was to do, the switch was not thrown and the passenger train continued along the siding to collide head-on with the freight train. The express was running three hours late. The trains met on a sharp curve which prevented either engineer from seeing the other train much before they met. The two locomotives were badly damaged and the smoker was completely telescoped by the baggage car, several of its occupants being killed and the remainder seriously injured. The passengers riding in the day coach and the two Pullman cars were thrown from their seats but not badly hurt. The passenger train engineer, H. A. Liscomb, was trapped beneath his engine. In an attempt to release him surgeons amputated one of his legs, an ordeal which he reportedly bore bravely despite the pain not only of the amputation but of scalding steam passing across his face. His ordeal was in vain as he died from his injuries. The freight train engineer, I. N. Galbreith, was likewise scalded to death and Conductor E. E. Motz of the freight train was the third railroad employee to die in the accident. Five passengers were killed outright and about 15 were more or less injured. All but one of the 75 members of the "The Spring Chicken" Comic Opera Company escaped injury because their private car was the last one in the train. The company was due to perform at Wheeling in the afternoon and evening; instead in the evening they held a thanksgiving service to mark their escape.[551]

---

549 *Le Petit Journal*, Paris, France, 26th August 1907

550 *The Times*, London, England, 26th, 27th and 28th August 1907

551 *The Sun*, Baltimore, Maryland, United States, 29th September 1907

After the accident a newspaper reported that, 'Many of the women on the train tore up their linen to provide bandages and families living in the neighbourhood threw open their houses which were converted into emergency hospitals. Several of the injured were carried down the bank to a houseboat, where two of them died.

'An affecting scene took place when an uninjured passenger asked if any in the houseboat were Catholics, and receiving an affirmative reply, presented a crucifix and prayed with them. One man murmured that he was not a Catholic, but would like to hold the crucifix. His wish was granted and five minutes later he expired.

'A score of physicians from Bellaire and Wheeling were early on the scene and the work of taking the injured from the wreck and caring for them was prosecuted vigorously. Some of the men in the wreck of the smoker were pinned beneath the timbers and it was found difficult to extricate them.'

The passengers who had lost their lives at the time of the initial report on the accident were Michael Heinze and William Shaw of Wheeling, West Virginia; Carl Bereron, of Milwaukee, Wisconsin; F. I. Rose, of Cleveland, Ohio, and newsboy Harry Seitz, of Massillon, Ohio. In addition it was feared that passengers Harry English, of Connersville, Ohio; Curtis Lafferty, of Cambridge, Ohio and John Hawk of Pittsburg, Ohio were fatally injured.[552]

The express had left Chicago, Illinois, at 9.30 p.m. the night before, and was divided at Chicago Junction, part of it going to Pittsburg via Akton and part to Wheeling via Newark, Ohio.

The seriousness of an accident near Budapest, Hungary, on Thursday 10th October 1907, varied considerably between different reports. The reports agreed that an express train from Semlin collided with a goods train and that a fire broke following the accident. Beyond that they differed. A Reuter's report published in the London *Times* stated that only three railwaymen had lost their lives, one of them the driver of the goods train, and that four passengers had been seriously and 15 slightly injured. This report added that a large number of passenger carriages had been smashed and that the fire involved the setting fire of a bridge by sparks from the locomotive.[553]

Contrasting with this report was one from Vienna which stated that in fact 10 persons had died and 30 had been injured, many of them seriously. This account stated that the collision had occurred on a high embankment and that both trains had been thrown into a deep ditch. Fire, it was stated, had broken out in the passenger carriages and some of their occupants had roasted to death in front of their fellow travellers who had been fortunate enough to escape.[554]

It is, of course, impossible to say which account was the more accurate, though if, as the Reuters report states, a large number of passenger coaches had been smashed to pieces, it seems surprising that all the passengers in them should have escaped with their lives.

---

552 *The Titusville Morning Herald*, Titusville, Pennsylvania, United States, 30th September 1907
553 *The Times*, London, England, 11th October 1907
554 *Manitoba Free Press*, Winnipeg, Wisconsin, Canada, 11th October 1907

Eighteen people died and 20 were injured in a morning collision between a passenger train and a goods train near Rabshtyn station, Russian Poland, on the Vistula Railway. The accident happened on Tuesday 22nd October 1907.

On Thursday 24th October 1907 No. 9 up passenger train collided with No. 154 down goods train at Kot Lakhpat station, India, on the North Western State Railway. Eleven passengers lost their lives and 25 passengers and four railwaymen were more or less injured, while considerable damage was done to rolling stock. The accident was attributed to the driver of the goods train having 'started his train with a wrong line clear and against signals'. [555]

Twenty-eight people died and 15 were seriously injured in a collision on the Ferrocarril Central Mexicano at Gargantua Siding, Mexico, 8 kilometres west of Maltrata, at 2.5 p.m. on Friday 24th April 1908. The trains were the second section of a heavily loaded passenger train carrying Roman Catholic pilgrims who were returning to Mexico City from the shrine at Guadalupe, having started out from the capital early that morning, and an extra freight train. Both engines, one passenger car and two freight cars were totally destroyed. Escaping steam from the wrecked locomotives added to the horror of the scene and probably was responsible for many of the casualties. The freight train had been backing into the siding at the time of the accident. Medical aid was sent after the accident from Orizaba and the injured were taken there.[556]

The accident was blamed on the failure of the crew of the freight train to provide flag protection for it while it was being put out of the way.[557]

By coincidence the next accident in this chapter also involved a train carrying Roman Catholic pilgrims, although in this case it was not a special train and they only made up some of the passengers. The accident, at Contich, Belgium, was at the time it happened the most serious ever to have taken place in that country. On Thursday 21st May 1908 a passenger train left Antwerp for Turnhout, taking the Brussels main line at Contich. Here it had to wait for a connecting train from Brussels before proceeding to Turnhout and was stopped at the beginning of the branch. The passenger train was filled with pilgrims from Notre Dame du Sacré Cœur in Brussels, destined for the shrine of Notre Dame de Montaigu, near Lierre, who had just joined the train, as well as soldiers from Antwerp and peasants. It consisted of five coaches, with three relatively modern vehicles at the front and two old third-class carriages at its rear, packed with working-class passengers.

Behind the Turnhout train a train for Brussels had left Antwerp at 8.49 a.m. and was due at Contich at 9.01. It arrived at Contich on time and about 120 metres from the station was suddenly diverted onto the Turnhout branch and into the rear of the branch train. The two flimsy third-class carriages at the end of the front train were completely telescoped and the following second-class carriage was overturned. The driver and fireman of the branch train reportedly saw what was going to happen at the last moment

[555] Administration Report on the Railways in India for the Calendar Year 1907

[556] *The Daily Ardmoreite*, Ardmore, Oklahoma, United States, 26th April 1908.

[557] *The Franklin Reporter*, Franklin Grove, Illinois, United States, 30th April 1908

and ran towards the oncoming train waving red flags. The driver of the Brussels train did his best to pull up, then jumped from his engine and left the scene, though he could have done no more to prevent an accident which was not his fault. It was initially stated that 38 passengers had been killed in the collision and 68 injured, the number of dead later rose to 40 and of injured to 132. Most of those who lost their lives had been in the two third-class carriages, a few had been in the adjoining second-class coach.

For a few moments the station staff were too shocked by what had happened to do anything but they then began the rescue work in earnest. A message was sent to a nearby barracks and a squad of soldiers arrived at the station and began the work of releasing the victims, carrying the injured to the station buildings, where they were treated by doctors from Contich and Antwerp. Many of the bodies of those killed, which included several children, were badly disfigured and some had been decapitated. Others in the rear carriages escaped death, if not injury, when the carriage floor collapsed and they fell to the track with it. Contich station was transformed into a temporary hospital and later ambulance waggons and litters, with a staff of nurses and attendants, arrived and the injured were transferred to Antwerp Hospital, though some went to Brussels and others to a hospital at the barracks. The occupants of the colliding train suffered little or no injury, one passenger merely wondered why his train had stopped rather suddenly.

A passenger who had come from Brussels described his experiences. 'We left Brussels by the 8.2 train, which gives a connection at Contich for passengers from Brussels for Lierre and Turnhout. The train was very full, the third-class carriages being packed. The porters were just starting to close the doors prior to our leaving Contich when there was a fearful shock. Passengers were thrown through the open doors onto the platform, and many, terror-stricken, rushed off into the fields. I, fortunately, was in one of the forward carriages. Rushing to the back of the train, a terrible scene met my eyes, while shrieks of pain filled the air. In one compartment, lying amid several companions terribly injured was a man whose chest had been torn completely open, the heart and lungs being exposed. Others had had their heads crushed. In fact, the whole spectacle was one of such horror that I had to turn away, and had not recovered from its effects yet.'

Initially it was theorised was due to a fault with the points, but it soon emerged that the interlocking in the signal box was being renewed, a procedure which involved removing the levers from the frame in the box, one by one, the jobs being expected to take about three hours. At the time of the accident the lever for the signal protecting the points had been removed and a man was stationed outside of the box who caught hold of and pulled the signal wire to clear the signal when requested to do so by the signalman. The Contich stationmaster said that he had known nothing of this arrangement. The signalman without thinking instructed his helper to clear the signal for the Brussels train on the assumption that the facing points were correct, forgetting that they were still open for the branch as he had not reset them for the main line, as he could have easily seen, had he looked, by the position of the point and facing point lock levers in his box.[558] The signalman fled the station after the accident, with much more reason of course than the engine driver. He soon turned up at his home and it was said that he was confined to bed and was expected to die of a weak heart.[559] [560]

558 *Railway Gazette 1908* 8 pg. 545 and 9 pg. 361
559 *The Times*, London, England, 22nd, 23rd and 25th May 1908

India suffered a disaster on Friday 26th June 1908, on the line of the Bombay Baroda and Central India Railway. An express train, carrying mainly 'natives', ran into a goods train. Fire broke out in the wreckage from burning coal or wood from the fireboxes of one or both engines and four carriages, four vans and four goods waggons were consumed. The all too frequent horror of passengers trapped in the wreckage being burnt to death before they could be released was repeated.[561] Initial reports spoke of 'many dead', then of 15 dead and 270 injured and the final official report put the number of fatalities at 26. According to these reports the trains should have crossed at Baroda, but the goods train was late. The passenger train nevertheless pulled out of the station and rounding a curve encountered the goods train travelling at speed.

A Bombay reporter added that, "The road officials will be called before a government investigating board, and it is likely that a new method of running trains will be enforced as a result. The government has long been planning regulations forbidding the overcrowding of cars, and the wreck will make such action certain."[562]

As noted above newspaper reports of the disaster made it appear that it was a single line collision, but this was apparently not the case, for the official report states that down express train No. 5 collided with down goods train No. 47. It gives the location as Itola station, and states that the collision was due to the driver of No. 5 train having run past signals and the disregard of rules by the station staff. (Baroda and Itola are places very close to one another). Twenty-four of the dead were passengers and two railway servants, while according to this report 34 passengers and five railway servants were injured.[563]

The elevated section of the Berlin U-Bahn suffered a major disaster on Saturday 26th September 1908. At this date the first part of the U-Bahn, the Stammstrecke, consisted of branches from Zoologischer Garten and Warschauer Brücke to Potsdamer Platz. There was also a loop line allowing direct running between the two branches and forming a triangular junction, the Gleisdreieck. Despite the name of the railway the tracks at this point were elevated above the streets rather than being carried underground. The passenger trains involved in the collision were one from Zoologischer Garten to Warschauer Brücke which was run into by one from Potsdamer Platz to Warschauer Brücke. The three junctions were fully signalled from a central signal box located within the triangle.

At about 1.45 p.m. Motorman Karl Schreiber was approaching the junction of the Potsdamer Platz and Zoologischer Garten lines with a train from Potsdamer Platz. He glimpsed a siding signal at clear and somehow took it as applying to his train, missing his own signal, which was at danger. A passenger sat behind Motorman Schreiber described what happened. 'When we drove to the Gleisdreieck I saw the train approaching from the west (from Bülowstraβe) and assumed that this train would pull up as usual, so that we had a clear route through the junction. Then, to my horror, I saw that the other train was heading for the track that we were on.'

---

560 *The Standard*, London, England, 22nd and 23rd May 1908

561 *The Argus*, Melbourne, Victoria, 29th June 1908

562 *Edwardsville Intelligencer*, Edwardsville, Illinois, United States, 27th June 1908

563 Administration Report on the Railways in India for the year 1908

The signalman could also see what was going to happen. He threw his signal against the train from Zoologischer Garten but it was too late, the train was past it. A passenger on this train related, 'As we were sat with our backs to the other line we could not see that another train was coming into us. Then, suddenly, the passengers sat opposite to us jumped from their seats with cries of horror and fear. At that moment our car received a terrible blow. The other train had run into our car. Because it was struck from the side, our train was lifted from the track and car No. 3 broke through the railings bordering the viaduct and fell ten metres to the ground. The car also landed on its roof so that the passenger compartment was crushed by the 17 ton undercarriage.' [564]

The second car of the train might well have followed the leading one over the edge of the viaduct had not the couplings between the two broken. Instead it ended with half of its body suspended over the viaduct. The third and final car remained on the line and its occupants escaped with cuts and bruises. The colliding train suffered little damage, although its second car was derailed.

Almost all of the occupants of the car which fell from the viaduct were either killed or injured. The appearance of some of the bodies suggested that a proportion of those killed had been electrocuted. Initially it was stated that there were 21 dead and 18 seriously injured, with other less badly hurt victims having left the scene. On Monday the number of fatalities was stated to have been only 16 but it seems that two of the injured subsequently died bringing the final total to 18. Most of those killed and injured were men and girls returning to their homes at the end of their day's work, having finished early because it was a Saturday.

After the accident the police and fire brigade were, as might be expected given that it had happened in the middle of the city, quickly on the scene. Once the dead and injured had been removed work began to stabilise the second car which was threatening to follow the first into the courtyard below. Traffic was suspended until the evening, when single-line working was brought into use over the undamaged Zoologischer Garten line. Meanwhile the Prussian Minister of Railways had ordered an immediate investigation into the collision. The provisional inquiry determined that Motorman Schreiber was at fault, and that he had somehow overrun not just one but two signals. He was arrested. The signalling itself was also checked and found to be working properly. Kaiser Wilhelm was at his shooting box at Rominten, where he was informed of the disaster. He expressed his profound grief at what had happened and desired to be kept informed of the progress of the injured.[565]

Motorman Schreiber was subsequently tried and found guilty of manslaughter and sentenced to a year and nine month's imprisonment. When the Great War broke out in 1914 he was called up and was one of those who did not return.

On 1st October the railway was extended from Potsdamer Platz to the Spittelnarkt in the centre of the city and it was said that public confidence in the working of the railway, which had naturally been shaken by the disaster, was returning.[566]

In 1913 the layout at Gleisdreieck was altered, with the two lines now crossing one another on different levels with an interchange station allowing passengers to transfer

---

[564] Webpage – Tagesspeidel: Tod am Gleisdreick (in German)

[565] *The Times*, London, England, 28th and 29th September 1908

[566] Ibid, 2nd October 1908

between them.[567]

The Caribbean island of Haiti was the somewhat unlikely location of a serious accident on Sunday 4th July 1909 in which the then president of the country, General Antoine Simon, had a narrow escape. The evening train in which he was riding, probably on the line of the Chemins de fer de la Plaine du Cul-de-Sac, was in collision with a freight train near Beudet. Several cars were wrecked and 10 soldiers were killed and others injured. The president himself was unhurt and having organised the clearing away of the wreckage and the care of the injured he returned to the capital, Port-au-Prince, on horseback.[568]

A handful of American newspapers contained details of a report from Rome of an accident at Rezzato, Italy on Saturday 2nd October 1909. It was stated that 20 persons had died and many others had been injured when a Milan express had collided with a train waiting in the station.[569] These reports appear, however, to have been greatly exaggerated in the way that such reports often were (and are). A presumably more reliable report from a Swiss newspaper stated that a passenger train from Verona had collided with a goods train in the station. One waggon was smashed and some passengers were slightly injured.[570]

On the morning of Christmas Day, Saturday 25th December 1909, an accident occurred at Ubersko station, near Chotzen, Bohemia. Owing, it was said, to a mistake in signalling, an express train bound for Vienna collided with a goods train. Three passenger carriages were reportedly smashed to pieces and both locomotives, six passenger carriages and two guards' vans were badly damaged. Eleven persons died and 20 were injured.[571]

Around 1910 during the summer months the Chemins de fer de l'État in France ran a special excursion train known as a G train from Bordeaux to the seaside resort of Royan, the train starting at 8 a.m. and taking the main line to Paris as far as Pons, from where it took a single-line branch to Saujon and thence the branch to Royan. Sunday 14th August 1910 was the eve of Assumption Day, a public holiday in France, and with the two-day break the train was exceptionally well patronised. It was made up to the maximum load of 24 vehicles (19 third-class and three second-class carriages plus two guards' vans) and was carrying well over a thousand passengers, crammed 15 or 20 to a compartment or standing in the corridors. To provide more capacity benches were added to the two vans to accommodate passengers but even then when the train set off, 20 minutes late, over a hundred disappointed passengers had to be left behind due to lack of space.

At Saujon station freight train No. 1512 from Royan to Pons had arrived from Royan

[567] Webpage – Tagesspeidel: Tod am Gleisdreick (in German)
[568] *The Lincoln Daily Star*, Lincoln, Nebraska, United States, 5th July 1909
[569] *New Castle News*, New Castle, Pennsylvania, United States, 2nd October 1909
[570] *Journal de Genève*, Switzerland, 3rd October 1909
[571] *The Times*, London, England, 27th December 1909

at 10.35. It had to shunt the goods yard there and also to wait for the late-running special from Bordeaux and to be overtaken by train No. 98, a Royan – Paris express, before continuing its journey towards Pons.

The Bordeaux excursion train approached Saujon, made the regulation stop at the advance signal at La Traverserie halt and then came around the curve into the station. Driver Dieuaide found the entrance disc showing clear for track 1, but then noticed that engine No. 1512 of the freight train, at the head of a rake of waggons, was obstructing the line. He made an emergency brake application and reversed his engine but struck the engine of the freight train, which overturned onto the right-hand side of the line. The engine, leading van and first passenger carriage of the passenger train were derailed and the second and third carriages were pushed up against them. The next three carriages were also telescoped to some degree, but not seriously damaged, while the remainder of the train, although derailed, escaped.

M. Cros, the Saujon stationmaster, fortunately had the presence of mind to get the home signal from the Royan branch put back to danger, thus preventing a possible second collision, as the Royan – Paris express was close at hand. The initial panic having passed, the rescue work was begun by station staff and uninjured passengers, assisted by local inhabitants who had been alerted by the noise of the collision, which witnesses described as sounding like cannon. Medical assistance was provided by doctors from a health centre in the town, joined later by others from the surrounding area and from Royan, from where a relief train was despatched. The three leading carriages of the train had been partly occupied by 70 passengers from Barsac, members of a friendly association of former pupils of the state school on their annual holiday. The leading and rear compartments of these three coaches more or less escaped damage but the remainder formed a tangled pile of wreckage, and 33 people in them were killed outright. Their bodies, 12 of which were unrecognisable, were first taken to the station waiting room and then to the goods shed, which had been transformed into a chapel ardente. Also killed was Fireman Papillon of the freight train, crushed beneath the overturned tender. Driver Dieuaide of the passenger train and Fireman Ploquain jumped and escaped with bruises. After treatment on the spot the injured were taken to local hospitals and some were later conveyed by train to Saintes and Bordeaux. With the death of several of those badly hurt the final toll was 38 dead, with a further 103 injured.

Although there was a lack of heavy equipment to clear the wreckage, most notably of a breakdown crane, workmen managed to get the line open for traffic again in 48 hours, clearing away the wreckage of the carriages and waggons and righting the overturned locomotives with the aid of jacks. After the accident there were the usual visits to the scene by politicians and expressions of sympathy from them and assorted foreign royalty, including Kaiser Wilhelm II of Germany and King Victor Emmanuel III of Italy.

Simultaneous inquiries into the accident were begun by a judge from the Saintes Criminal Court and by the Minister of Public Works. The freight train had to perform various shunting operations at Saujon, attaching and detaching traffic. These had been completed but before the engine as recoupled to its train its driver, Fernand Bouissou, had briefly left his engine in charge of his fireman and gone to the pump for a drink of water. Foreman Shunter Drouineau admitted permitting the driver to leave his engine but denied responsibility for the accident. There was briefly speculation that the deceased fireman had mistakenly taken the train foul of the main track but the position his body was found in made it unlikely that he had been at the engine control at the moment of the collision. Driver Bouissou did not deny leaving his engine but had doubtless assumed that it would be protected by the signalling, as the signals should not have cleared until the signalman had been advised by the foreman shunter that the main line was not obstructed. The station signalling was controlled from a cabin at the far end of the station, in charge of Signalman Firmin Desaffit. He was judged to be directly to blame for the accident, as having seen the rear of the freight train in the siding near his box he had wrongly assumed that the whole of it as clear of the main line, and to avoid further delay to the excursion train he had taken off the signal at the Pons end of the station without waiting to be advised by Foreman Shunter Drouineau. The train should have only been travelling at 15 kilometres an hour as it neared the station but according to the Flaman speed recorder it had approached at a speed 50 kilometres an hour, Driver Dieuaide doubtless trying to regain some of the lost time, though he managed to reduce the speed to 15 kilometres an hour at the moment of impact.[572] [573]

The circumstances of the accident were discussed in Parliament in February 1911, when criticisms were made of the composition of the excursion train, made up because of the pressure to obtain sufficient rolling stock to cope with the heavy traffic of a heterogenous jumble of vehicles, including light carriages of 8 to 10 tons which had been crushed by heavier carriage of 18 or 20 tons. Another criticism raised in the House was the failure of the state railway to install a more modern signalling system at Saujon, with the interlocking of points and signals, and the position of the signal cabin, from which the east end of the goods yard was out of sight around a curve. It was finally pointed out that had the freight train been equipped with automatic couplers, as called for by the House a year earlier, the shunting operations would have been expedited.[574]

The examining judge chose to ignore the wider issues which had led to the disaster and instead to place the responsibility for it entirely on Signalman Desaffit. He stood trial in April 1911 charged with causing death and injury through clumsiness, recklessness and non-compliance with the rules and regulations. In his defence it was argued that the station staff had been overworked and conditions on the day of the accident had been difficult. The prosecution called for a prison sentence but announced that it would accept a suspended one. In the end he was found guilty and sentenced to a year's suspended imprisonment and a 200-franc fine, while the State was found to be civilly liable for the accident.[575] [576] [577]

Following so soon after the disaster at Villepreux-les-Clayes (qv) which had also occurred on the State system, the accident gave new ammunition to those who opposed the public ownership of the railway system, which had first come about in France in 1878 following the bankruptcy of some private railway companies, and there was a considerable debate about the issue in the Press.[578]

The accident also caused a reaction from Catholic clergy who had been opposed to the

[572] *Le Matin*, Paris, France, 15th and 16t August 1910
[573] *Le Petit Parisien*, Paris, France, 15th, 16th and 17th August 1910
[574] *Journal Officel de la République française*, Débats parliamentaires, 2nd February 1911
[575] *Le Figaro*, Paris, France, 6th April 1911
[576] *L'Humanité,* Paris, France, 12th April 1911
[577] *La Presse*, Paris, France, 13th April 1911
[578] Wikipedia – Accident ferroviaire de Saujon

Separation Law of 1905 which had finally resulted in a secularisation of the state. For example, Archbishop Andrieu of Bordeaux penned a letter in which he railed against the perilous path of a world seeking the mirage of earthly pleasures.[579]

Seven passengers died and 40 were injured in a head-on collision in Rottenmann station, Austria, on the state railway system, on Tuesday 20th September 1910. The driver of a Rome – Vienna express overran signals and his train collided head-on with the corresponding Vienna – Rome express. It was presumed that he had misread the signals. He perished in the collision.[580]

The French state railway, the Chemins de fer de l'État, suffered yet another disaster at Courville, France, on Tuesday 14th February 1911. At 6.07 p.m. on the evening of that day freight train 6330 had arrived at the station from Le Mans, three hours and fifteen minutes late. It was immediately put onto track ter, the only siding on the up side, so as to leave the main up, track 2, free for the following freight train, a so-called *détail* freight train, presumably a parcels train, which had to unload and load at the station platform. A third train was following the two freight trains, Guingamp – Paris omnibus train No. 516. Stationmaster Raymond had to free track 2 for the omnibus train and with the ter siding occupied by the late running Le Mans freight his only option was to shunt train 2338 into siding 1 ter on the down side, which meant taking the train across the down main line track 1,300 metres from the station. Stationmaster Raymond had to deal with the tickets for the omnibus train passengers and the job of shunting the freight train was thus delegated to Head Porter Launay.

It would appear that there was no signal box at Courville, for before commencing to shunt the freight train Head Porter Launay telephoned a signal box 3.7 kilometres on the Paris side of the station to ascertain the whereabouts of *rapide* train No. 513 from Paris Montparnasse to Brest and Nantes. Told that it was not yet about he began the shunt, only to be shortly advised that it was now to hand. He at once operated a lever placing the *disque d'entrée* signal a kilometre from the station at danger and in addition sent Shunter Chailloux up the line with a red lamp to stop the *rapide*.

Despite these measures *rapide* No. 513 failed to stop and its Pacific locomotive collided with the middle of the shunting train, demolishing 10 vehicles of train 2338. The engine overturned to the right, ploughed along the ground for 100 metres and came to a stand 200 metres from the station. Behind the engine lay the tender and leading van. Next came a third-class carriage, a postal composite carriage and a dining car and the first two of these three vehicles were telescoped by the heavier dining car. The sudden stop caused by the automatic application of the breaks caused a coupling to break behind the dining car and the remaining vehicles of the train came to a stand little damaged, though derailed.

Train No. 2338 having been crossed onto the down line prior running into the siding, omnibus train No. 516 had run into the station and was on its way again towards Chartres when the collision occurred. Driver Georges Béquille saw what was happening in front of

---

[579] *L'Univers*, 19th August 1911

[580] *The Times*, London, England, 21st September 1910

him and did his best to pull up but his engine ran into the pile of wreckage and overturned, adding to the chaos of the scene.

The force of the collision had ruptured gas cylinders beneath the wrecked leading carriages of the *rapide* train, and fire broke out at once. The staff and all 19 passengers who had been in the dining car were able to make their escape and uninjured passengers from the *rapide* and omnibus trains, together with station staff, tried desperately to rescue the passengers who had been in the leading two carriages, but only a few could be released before the intensity of the fire drove the rescuers back. About the same number of passengers were injured and were taken to Courville hospital.[581]

A good many of the dead were members of a wedding party who had been returning from Paris to Sable. One of the only members of the party who escaped said, "We were 11, the bride and bridegroom., the best man, and other guests. We were singing merrily after a delightful day when a terrible crash occurred. I was thrown into the corridor and without knowing what I was doing sprang through the open door and fled into the night. All around were screams and groans and jets of flame. Presently I came back to the scene. Of the tour party of 11 I have already counted five dead. One of them is the young bride, whose body was burned almost beyond recognition. As to her husband, he has one leg torn off, and has been taken to hospital, where he keeps saying that he wishes to die, and calling for her who was his wife for one day." [582]

A relief train was sent to the accident scene from Chartres with doctors and a company of the 102nd Infantry, Courville firemen fought the fire and more soldiers from the 13th Cuirassier Regiment arrived by road, so that eventually there were 200 rescuers. The Orléans company lent a large breakdown crane to raise the two engines and M. Louis Puech, the Minister of Public Works, arrived on the scene on another special train. The line was reopened, with single-line working, at 7.30 a.m. the next morning.

Driver Bourcier of the *rapide*, aged 46 with 17 years' service, was arrested and charged with manslaughter, but released on bail. In his defence he pleaded that smoke blowing down from the chimney of his engine had prevented him from seeing either the *disque d'entrée* signal or the hand signal of the shunter. Crossing from the left-hand to the right-hand side of his cab he had seen a white light which he had taken to be the *disque d'entrée* signal, though in fact it was the headlight of the engine of the omnibus train. After a trip on a locomotive the new Minister of Public Works, Charles Dumont, considered this explanation plausible.

Also arrested were Stationmaster Raymond and Head Porter Launay, for fouling the running line within five minutes of the expected arrival time of the *rapide* train (6.12 p.m.) in defiance of the rules. According to the regulations this could only be done under special authorisation, or in an cases of absolute emergency. They pleaded that the *rapide* had not been announced when they had first contacted the signal box, and that they had expected that, as on previous days, it would be running late because of three separate permanent way restrictions in operation between Paris and Chartres. In addition they argued that the need to shunt train No. 2338 out of the way of the omnibus train constituted an absolute emergency, which seems a very doubtful assertion, especially as it risked delaying a *rapide* train to avoid delaying an omnibus working.

---

581 Wikipedia – Accident de ferroviaire de Courville (in French)
582 *The Times*, London, England, 16th February 1911

At the subsequent trial of the three the prosecutor treated the stationmaster and the head porter leniently, while 'throwing the book' at Driver Bourcier, who apparently had a record of previous driving transgressions. The court found all three guilty. They held that the two station staff had not had any special authorisation for shunting train 2338 when the *rapide* was due within five minutes and an urge to avoid delay to the omnibus train did not constitute an emergency. Both were sentenced to six months imprisonment and fined 100 francs. Driver Bourcier was sentenced to a year in prison and fined 500 francs for neglect of duty overrunning signals. The prison sentences were all suspended ones. On appeal the sentence on the driver was reduced to two months, but it was no longer a suspended sentence. He was then briefly imprisoned before being released following the intervention of his union.

Predictably yet another accident on the state railway system led to further discussion in parliament and much criticism of the whole idea of nationalisation by sections of the press.

Measures to prevent a repetition of the accident focussed on two areas; the fouling of running lines by shunting operations when through trains were due and the observation of signals by drivers. With regard to the first M. Dumont sent out a circular to all companies on 15th March 1911 asking them to amend the regulation to read that trains should not be shunted within the five-minute deadline unless there was imminent danger to them remaining in the station. In the absence of a means of preventing smoke from locomotives obscuring signals measures were to be made to make signals more visible by increasing their heights and changing the shapes and colours of their indications, as well as supplementing the danger indication with the automatic application to the rails of detonators. Pending these alterations staff giving hand signals with flags or lamps were also to supplement them with detonators.[583]

Many persons were killed and injured in a collision near Constantinople (now Istanbul), Turkey, on Saturday 7th October 1911. Since 29th September Turkey had been at war with Italy and the Turkish authorities made the dubious claim that the accident had been caused by Italians who had formerly been employed by the railway but had been discharged upon the outbreak of the war and had removed a rail in revenge.[584] Quite how removing a rail could cause a collision, as opposed to a derailment, was not explained.

An accident at Malmslätt in Sweden on Sunday 16th June 1912 had its origin in a misunderstanding between different railwaymen. On that day it was initially reported that international train No. 2 from Berlin, Germany, via the port of Trelleborg was delayed due to delay to the ferry with which it connected and would not leave there until 11.16 p.m. instead of 9.04 p.m., and that international train No. 12 from Copenhagen via Malmö would therefore precede it instead of following it. Train No. 12 normally passed through Malmslätt at 4.32. a.m. The ferry arrived a little earlier than expected and train No. 2 was less delayed than expected so at 5 a.m. a further telegram was sent arranging for train No. 2 to cross mixed train No. 751 at Bankeberg. Due to a misunderstanding Porter Carlsson, who was in charge of the signal box at Malmslätt, was informed that

[583] Wikipedia – Accident de ferroviaire de Courville (in French)
[584] *The Syracuse Herald*, Syracuse, New York, United States, 7th October 1911

train No. 751 would cross train No. 12 at Bankeberg rather than at Malmslätt and he according set the road for train No. 751 to depart, setting the points at the Bankeberg entrance to the station right for the departing train. Train No. 12 then suddenly appeared, was diverted due to the wrongly set points onto the track on which train No. 751 stood, and the two collided head -on in the station. Several carriages were severely damaged in the collision, including the leading luggage van in the express; the following through carriage from Copenhagen and third carriage from Malmö were telescoped into one another. The roofs of these two carriages had to be cut off during the rescue operation to reach the trapped passengers. Fire also broke out in the wreckage, though the blaze was fortunately quickly extinguished. The collision occurred at 5.12 a.m., at a time therefore when many of the passengers were asleep.

After the accident a relief train was despatched from nearby Linköping with doctors, nurses and equipment and by 7.30 a.m. the passengers were brought there. The track was then cleared so that train No. 2, with through carriages from Berlin, could pass through at already at 11 a.m. and continue its journey to Stockholm.

Driver O. Bonavier of the Malmö express was somewhat hurt but was able to give an account of the accident to a newspaper reporter from the *Ostergotland Dagblad*. He said that his train had been travelling at its maximum permitted speed of 50 kilometres an hour. The semaphore signal at Malmslätt had been clear and he had only been 40 or 50 yards from train No. 751 when he had seen that the points were lying wrong. He had shouted to Fireman Olsson, applied the brakes, and the engines had met. After the accident Driver Bonavier was said to have been distraught and to have asked repeatedly if it had been his fault.

At the investigation into the accident Conductor Sahlin of train No. 751 stated that he had asked at Malmslätt where train No. 12 was and had been told that it would meet his train there as normal. There were two clerks on duty at the station at the time of the collision named Eriksson, the one who had spoken to the conductor, and Kjellsson. According to the evidence Eriksson received a message by telegraph from Bankeberg that train No. 12 was leaving there and told Kjellson. The latter had then left the office and Eriksson had got on with other work. Asked as to the position of the points when he had accepted train No. 12 from Bankeberg Erikson could not answer. It would seem that earlier when Kjellson entered the office Erikson had told him that train No. 2 was late and that they would probably receive orders regarding it but that it was Kjellsson had not understood that train No. 12 was proceeding to Malmslätt, so that when Porter Carlsson had asked him, "Has it left Bankeberg?" meaning train No. 12, Kjellsson had replied, "No, this train (No. 751) must cross it at Bankeberg." The signalman had then reset the points and shortly afterwards the collision had occurred. After the accident Signalman Carlsson said that he had gone to Kjellsson and asked him what was that order he had given? Kjellsson had been lost for words and had not replied but had gone to the office, shortly returning and saying that it was train No. 2 which the mixed train was to have been crossed at Bankeberg.[585] [586]

In all 22 people died and 12 were injured in the accident; naturally enough most of the victims were Swedes but there were also Danes, Germans and two Englishmen; a Mr

[585] *Dalpien*, Falun, Sweden, 18th June 1912

[586] *Kalmar*, Kalmar, Sweden, 19th and 22nd June 1912

Clibborn, who lost his life, and the Rev Osborne, who was injured. Another person who lost her life was Greta von Philp, daughter of August Strindberg, a famous Swedish writer and playwright. Her husband Henry survived.[587]

Reports of a collision near Lauromuller, Brazil, on Wednesday 31st July 1912 were perhaps more than usually conflicting. According to one report the accident occurred in the late evening, news of it only reaching Rio de Janeiro early the next day when a special train carrying doctors, nurses and a hospital field outfit was rushed to the scene. This report, which seems as reliable as any, stated that two suburban passenger trains filled with passengers on the Central Brazilian Railway (Estrada de Ferro Central do Brasil) had collided head-on on a curve due to a mistake in signals. 'The heavier of the two trains ploughed halfway through the other and almost immediately the wreckage caught fire. Many of the injured will die and railroad officials here {Rio de Janeiro} express the fear that it will prove the worst railroad disaster that Brazil has ever known.' At the time of the report it was said that at least 100 persons were killed and twice that number injured.[588]

The somewhat complex sequence of events that led to a collision at Irvington siding, on the outskirts of Indianapolis, Indiana, United States, in the early hours of Wednesday 13th November 1912 began with the departure of eastbound Cincinnati Hamilton and Dayton Railroad freight train No. 95 from State Street yard, Indianapolis, at 1 a.m. The train consisted of an engine, 26 loaded and four empty cars and a caboose and was in charge of Conductor Hines with Engineer Yorke. Just before reaching Irvington the train stalled on the 1% (1 in 100) gradient. It was then divided, and the engine went forward with 14 cars and placed them in the siding at Irvington. Once this had been done Head Brakeman Carl Gross cut off the engine, which proceeded out onto the main track at the east end of the siding. The head brakeman left the engine at this point to go forward to flag westbound passenger train No. 36, while the engine went back to collect the remainder of its train. This portion was left at the west end of the siding, the engine returned via the main track to the east end of the siding, set back onto its train and the two portions were re-joined. The intention had then been to wait at Irvington for passenger train No. 36, but they then received an order giving them until 2.50 a.m. to reach Julietta siding, six miles further east along the single track, and they set off. They had not left Irvington, though, before Engineer Yorke told Conductor Hines, who was riding on the engine, that he thought that the train was too heavy for them to reach Julietta in the time allowed in the order. The conductor then directed him to back into the siding once more and returned to his caboose. As they were setting back Head Brakeman Gross asked what they were doing, and Engineer Yorke explained that they did not have time to reach Julietta and invited the head brakeman to ride back into the siding on the engine. Gross said that he must stay out and protect against train No. 36.

The train was clear back in the siding at 2.38 a.m. Engineer Yorke then directed his fireman to cover the engine headlight and whistled Head Brakeman Gross in. The engineer said that he had then set off towards the east switch to check that it was closed

587 Wikipedia – Järnvägsolyckan I Malmslätt (in Swedish)
588 *The Daily Reflector*, Jeffersonville, Indiana, United States, 3rd August 1912

but about 150 feet from his engine he had met the head brakeman coming in so had asked him if the east switch was closed. "Yes, it's closed all right," Gross had replied and the two had returned to the engine. On the way they had met Rear Brakeman Cox, sent up by Conductor Hines to check that the east switch was right., and Gross had likewise reassured Cox that it was, whereupon the rear brakeman had returned to the rear of the train, while his conductor had gone to the telegraph office for orders.

Passenger train No. 36 from Cincinnati, Ohio, to Chicago, Illinois, consisted of a locomotive, a mail car, a combination baggage and smoking car, a day coach and two Pullman sleeping cars, all of wooden construction. It was in charge of Conductor Earl Wiggins and Engineer William Sharkey and was 52 minutes late when it passed Reedville, the last open telegraph station, 13 miles east of Irvington, and it reached Irvington at 3.17 a.m. Despite Head Brakeman Gross's assurances the east switch had not been closed train No. 36, travelling at 45 miles an hour, was diverted into the siding where it collided head-on with the engine of train No. 95. Both engines were badly damaged, and four freight cars were destroyed, the mail and combination cars of the passenger train were damaged and worse, the combination car telescoped the following day coach, in which most of the casualties were found. Conductor Wiggins, Engineer Sharkey, Fireman C. Burg and 11 passengers on train No. 36. Along with Brakeman Bert White of train No. 95 died, while six passengers and five employees were injured.[589] Fire broke out in the wreckage but was put out before much damage was done.[590]

The accident was the subject of an Interstate Commerce Commission (ICC) report, but their investigators do not themselves seem to have questioned any witnesses, referring instead to the investigations carried out by the railroad company and the Indiana Railroad Commission. Brakeman Gross told the company inquiry that when he was returning after being whistled in he had seen a light near the switch and had assumed that someone was closing it. He had not looked closely at the switch stand as he passed it and pleaded that the red and white lantern that he had been carrying did not throw a very good light. He admitted telling Engineer Yorke and Rear Brakeman Cox that the switch was closed because he thought it was. Rear Brakeman Cox told the same inquiry that he had asked Gross if he, Gross, had closed the switch and Gross had replied that he had.

Engineer Yorke told the Indiana Railroad Commission inquiry that train No. 36 had been approaching at speed, its engine using steam and fire flying from the stack.

It also transpired that the lamp at the switch stand had been out. It was a long-burning signal lamp, meant to stay alight for eight days, had been attended to the previous Saturday and had been burning at 5.10 p.m. the previous evening, but between then and the accident three train crews had reported that it was out. There had been oil in the reservoir after the accident and it seemed likely that a strong wind had blown the lamp out.

The ICC found Engineer Yorke and Head Brakeman Gross equally responsible for the accident. The engineer that part of Cincinnati Hamilton and Dayton rule No. 104A which read, 'When a train backs in on a siding, to be met or passed by another train, the engineman, when his train is into clear, must see that the switch is properly set for the main track.' Head Brakeman Gross had failed to ensure that the switch was closed and

[589] Interstate Commerce Commission report

[590] *The Monmouth Daily Atlas*, Monmouth, Illinois, United States, 14th November 1912

then erroneously stated that it was.

The commission considered that Engineer Yorke had also exercised poor judgement in whistling the flagman in before the switch had been closed and that he had had ample time to close it himself. He also should not have permitted his engine headlight to be covered before the switch was closed. Furthermore the entire crew of the freight train had displayed a lack of alertness in not relighting the switch lamp which all of them must have seen was not burning.

Rule No. 27 of the company stated that, 'A signal imperfectly displayed or the absence of a signal at a place where a signal is usually shown, must be regarded as a stop signal and the fact reported to the superintendent.' A footnote to the rule added that switch signals came within the definition of 'fixed signals'. The superintendent of the division stated however that switch lights were not regarded as coming within the rule inasmuch as train crews were not expected to stop at ones which were unlit, but only to report them. Engineer Sharkey of train No. 36 had therefore been following the customary practice in not stopping. The investigation had disclosed that switch lamps were frequently out, showing that they were either inadequate or poorly maintained.

Records showed that trains were frequently required to double into Irvington, including on 16 days out of the 60 preceding the accident. Most of the trains were not fully loaded and the problem appeared to be the use of poor coal (cf. the rear collision at Ais Gill, 1913).

The accident had occurred within the city limits of Indianapolis, where there nominally a 30 miles an hour speed limit, but the investigation showed that this was frequently exceeded.

The engine of the passenger train had been equipped with an electric headlight, but this had been removed for repair two months previously and not replaced. Had the engine had an electric headlight the engineer might have seen the wrongly placed switch in time to avoid the collision.

The commission concluded that, 'The operating conditions disclosed by this investigation should be materially improved in an effort to avoid a recurrence of such accidents.' It noted that measures should be taken to keep switch lamps burning at night, employees should be required to obey the rules regarding stopping at switch lamps which were out and to relight any which they found extinguished. The commission was also critical of the fact that trainmen of the freight train were all newly promoted or inexperienced and of the frequent need for trains to be doubled to Irvington. Finally it pointed out that had automatic block signals been provided the open switch would have been detected and the accident averted.

On the evening of Sunday 15th December 1912 as-required freight train No. 913 (number also stated as 9113) to Catania was stopped in the third track at Guardia Mangano station, Italy (on the island of Sicily) awaiting the passing of through passenger No. 116 from Syracuse to Rome (via Messina and the ferry to Villa San Giovanni). Train No. 116 was double-headed. It was due at 17:53 but was considerably delayed and did not arrive until 18:40. When it did the driver of its pilot engine, Giovanni Raspaolo, found that the points at the Acireale end of the station had been left in the wrong position, despite the disc signal at the entrance to the station being clear. The passenger train was travelling at a considerable speed and a violent head-on collision with the freight train was unavoidable. The freight engine and the pilot engine of the passenger train were almost destroyed, the leading baggage van, a third-class carriage and a first- and second-class composite carriage were badly damaged and the restaurant car was telescoped by the following Pullman car and broke in two. The waggons at the head of the freight train doubtless also suffered in the collision. Thirteen passengers lost their lives in the accident; most of them had been in the restaurant car at the time. Driver Raspaolo and his fireman, Salvatore Durante, were also killed and over 30 people were seriously injured, including the crews of the freight locomotive and the train engine of the passenger train and the staff of the restaurant car.

Rescuers, ambulances, doctors and police were soon on the scene after the accident and relief trains were sent to the scene. The injured were taken to hospital in Acireale, and some to the Garibaldi Hospital in Catania. The next morning 160 soldiers arrived to assist with clearing the wreckage.[591]

Fireman Durante was killed almost instantly but Driver Raspaolo, who was trapped in the wreckage of his engine, wedged between the reversing wheel and the tender, lingered for about six hours in extreme agony, begging his rescuers to shoot him as they worked feverishly to get him out. Between his cries he asked to be remembered to his parents, his wife and his five children. Finally he was freed, but by then he was lifeless. Driver Raspaolo and his firemen were mourned as keen trade union members, especially Driver Raspaolo, who had been reduced to working shunting engines, with consequent loss of pay, for taking part in a strike back in 1905. Their funeral in Messina on 18th December was attended by over 5,000 mourners.

As after the Sujon disaster in France questions were asked in parliament following the accident and it was declared that there were various factors behind the accident beyond the simple mistake of the signalman who had left the points wrongly set. It was mentioned that the local press had repeatedly pointed out the need for the line to be doubled, given that it was saturated with traffic, with the addition of 24 extra pairs of trains, putting severe pressure on the railwaymen, while the extra traffic was blamed by those on the left on Italy's military adventures in Tripolitania, which were supported by the banks, the industrialists and the Vatican. The poor lighting at night of the disc signals, which used rapeseed oil and which after being lit for a few hours gave a very dim light, also came in for criticism, with signals only being visible from a few metres away and yet drivers being heavily disciplined should they pass such a poorly lit signal at danger.[592]

A livestock and goods train (No. 41 down) which was due to leave Toowoomba at 8.50 p.m. was derailed a short distance down the line at Murphy's Creek, Queensland, Australia, at about 11 p.m. on Wednesday 29th January 1913. Seventeen waggons, including five cattle trucks, left the line and six cattle were killed, while others escaped from the wrecked trucks and wandered about the tracks. The single line was badly blocked. To maintain some sort of train service until the wreckage could be cleared away passenger trains were worked up to the point of obstruction from either side.[593] The

[591] Wikipedia – Incidente ferroviario di Guardia Mangano. (in Italian)

[592] *Cub Rail Altervista*, 14th May 2014

[593] *The Telegraph*, Brisbane, Queensland, Australia, 30th January 1913

derailment was according to Driver Joseph Doyle of the goods train due to excessive speed, a result in turn of the Westinghouse brake pump failing to work.

So far, therefore, a relatively minor accident, and hardly one worthy of inclusion in a book on world railway accidents. The initial derailment was, however, not the end of the matter, for a considerably more serious collision was to follow. A special train was empty carriages was sent down from Toowoomba on Thursday morning, 30th January 1913, to pick up the passengers who had left Brisbane on the 6.30 a.m. and 7.50 a.m. trains, which were stopped on the Murphy's Creek side of the obstruction. When the empty train reached the site of the derailment it failed to pull up in time and collided with an engine and breakdown van, throwing the engine of the breakdown train onto the derailed waggons and the gang of men who were working there clearing the line. As a result Lengthsmen Frederick Freise and James Little and Temporary Labourers Patrick McCarthy and Charles McGregor were killed outright, Temporary Labourer Walter Claydon and Ganger Alexander Matthews were fatally injured and four other men were less badly hurt.[594]

The accident was the subject of a prolonged magisterial inquiry. On its opening day Driver George Thompson told how he had worked the train of empty stock from Toowoomba, consisting of an engine, eight coaches, two waggons and a second engine at its rear. Guard William Hall and Pilotman James Rafter had also ridden on the engine. He saw no danger signals and exploded no detonators before he saw the engine of the breakdown train an engine length away. Guard Hall had then cried out, "Steady up, you're on her!" and his engine had struck the breakdown van as he was braking and putting down sand. Had there been detonators 400 or 500 yards back he could have pulled up but there had been no protection of any sort.[595]

Further into the inquiry the crews of various trains gave evidence that the protection provided at the accident site had been patchy, some trains being warned and others not. Towards the end of the inquiry General Traffic Manager Arthur Lloyd said that he had acted as guard on an engine going to the accident site and that the protection for that engine had been inadequate, but admitted he had not said anything to Acting Traffic Manager Michael Carmody about the lack of protection or checked that permanent protection was being provided, although he had later suspended Guard John Scanlon, guard of the first breakdown train, for failing to provide protection. Three other guards had also been suspended, including Guard John French of train No. 41 down, for failing to protect his train by permitting himself to be called away by a superior officer for another task.

The last day of the inquiry Acting Traffic Manager Carmody gave evidence and admitted that he had taken Guard Scanlon away from his job of protecting his train to do another job, on the assumption that he (Carmody) would be able to get to Murphy's Creek, obtain food for the men and be back in time to take over the protection duties himself before another train appeared. Some of his further replies to questions seem contradictory, such as stating that he was acting traffic manager but did not do traffic manager's duties, and that permanent protection of the obstruction was the business of the engineers but that did not relieve the traffic department of the responsibility for

---

594 Ibid, 31st January 1913

595 Ibid, 25th February 1913

providing protection.[596]

The final conclusions of the inquiry do not seem to have been made public but on 27th March the Railway Commissioner announced that General Traffic Manager Lloyd and Acting Traffic Manager Carmody had been suspended for an unspecified period. The four guards had presumably been previously reinstated.[597]

The Attorney General announced on 9th May that no legal action was contemplated, so Messrs Carmody and Lloyd escaped facing manslaughter charges.[598]

The Second Balkan War had just begun with an attack by Bulgaria on Serbia and Greece when two trains, one of them carrying Bulgarian troops, met in a collision near Jenikeul, Bulgaria, between Drama and Janthi, on about Thursday 19th June 1913. Five officers of the reserve forces and 50 soldiers were killed and a number of others were injured.[599]

A number of tramps were killed in a collision on the Chicago Milwaukee and St Paul Railroad between Delmar Junction and Riggs, Iowa, United States, at 10 a.m. on Sunday 22nd June 1913. A work train was at Riggs, east of Delmar Junction and a flagman had been sent back to the junction to protect it. It was supposed that he had cautioned the crew of freight train first No. 70, in charge of Conductor LaRug with Engineer Stickney but that Engineer Stickney either misunderstood the location of the work train or that the work train had moved closer to the junction. The work train had its engine at the rear. Its crew could not see the freight train coming but heard it and tried to get out of the way by pushing their train forward, but the freight train was coming fast downhill and a collision could not be avoided. Eight or 10 cars behind the engine of the freight train were wrecked, the next 10 or 15 remained on the line and 10 or 12 cars behind them were wrecked. It was in these last cars that the tramps who lost their lives were riding. According to one newspaper report three tramps were killed outright and a fourth fatally injured; other reports spoke of seven or eight dead. Neither the engineer and conductor of the freight train nor Conductor Mellish and Engineer Leguire of the work train was hurt, and other than the tramps only one person, a Japanese labourer, was hurt. When it became obvious that a collision was about to happen the foreman of the work train had shouted to his men to jump. All had done so except the injured man,[600] possibly he did not understand what the foreman was shouting if his English was not very good.

The collision occurred near a bridge and a car of butter was shot off an embankment and grain cars scattered all about. The wrecker was sent from Savanna and much time was spent in clearing the wreckage as every car had to be picked up carefully lest their be men beneath them. As most of the wrecked cars were situated in a cutting they could also not just be swung to the side of the line. A special train was also sent from Marion to pick up the dead and injured but for some reason returned without them. The line was cleared

596 Ibid, 26th, 27th, 28th February, and 1st, 3rd, 4th, 5th 6th, 7th, 11th, 12th, 13th and 14th March 1913
597 Ibid, 27th March 1913
598 Ibid, 9th May 1913
599 *Journal de Geneve*, Geneva, Switzerland, 20th June 1913
600 *The Quincy Journal*, Quincy, Illinois, United States, 23rd June 1913

by the evening.[601]

A number of senior officers in the Serbian army were killed or injured in a collision between two trains near Usküb (now Skopje), Serbia, on a branch of the Chemins de fer Orientaux on Monday 8th September 1913. Eight persons were reportedly killed outright and 30 mortally injured. Part of the Serbian general staff had been riding on one of the trains.[602]

Fourteen passengers died and 28 were injured in a collision between an express train from Kiev to St Petersburg and a stalled locomotive near Dvinsk station, Russia, on Monday 6th October 1913. The victims of the accident included Alexander J. Quchkoff, the former president of the duma, the Lower House of the Russian legislative assembly.[603]

During a thick fog a passenger train collided with and empty train on a siding at a station in Warsaw, Prussian Poland, on Wednesday 29th October 1913. Seven people were killed and 37 seriously injured.[604]

Two express trains reportedly collided near São Paulo, Brazil, on or about Monday 3rd November 1913, killing 50 people and injuring many others.[605] The accident happened on the line of the Companhia Mogiana de Estradas de Ferro. No more information on this accident has come to light.

France suffered another major disaster on Tuesday 4th November 1913, this time a sidelong collision not too far out of Paris at Melun, on the Chemins de fer de Paris â Lyon et â la Méditeranée (P.L.M.). Melun was the junction of two routes from Villeneuve-Saint-Georges, the original route via Brunoy and the newer route via Corbeil-Essones. At Melun the old main line continued to Montereau via Fontainebleu, while the newer line ran to Montereau via Hêricy, while a double line connection allowed trains to and from Paris via the old line to cross to the new one towards Hêricy. The collision occurred at 9.20 p.m. and involved train No. 2, an express from Marseilles to Paris and train No. 11, a mail train travelling in the opposite direction.

Train No. 2 had let Marseilles at 9 a.m. and behind the Pacific class engine consisted of two baggage vans, a mail van from Nice, seven first- and second-class carriages and a van. It carried over 150 passengers, including English and Dutch travellers who had disembarked at Marseilles from the Dutch steamship *Willis*, operating the Malle des Indes (Indian mail) route, plus passengers who had boarded at intermediate stops,

---

601 *The Cedar Rapids Republican*, Cedar Rapids, Iowa, United States, 24th June 1913
602 *Logansport Pharos-Reporter*, Logansport, Indiana, United States, 9th September 1913
603 *The Washington Post*, Washington, D.C., United States, 7th October 1913
604 *Journal de Geneve*, Geneva, Switzerland, 30th October 1913
605 *The Age*, Melbourne, Victoria, Australia, 5th November 1913

including about 80 who had joined the train at Tarascon from the Midi line.[606] The train had been in charge of Head Conductor Charles Vernet with Driver Jules Dumaine and Fireman Louis Nicolle.

Train No. 11 was exclusively a mail train, and consisted of an engine, a van, two carriages for the storage of mail, seven sorting carriages and another van. The sorting carriages were designated according the area which their mail was destined for: *"Lyon II"*, *"Marseille"*, *"Méditeranée"*, *"Lyon rapide"*, *"Paris-Besançon"*, *"Paris-Mont Cenis"* and *"Paris-Pontarlier"*. Mail was also sorted for Switzerland, Italy, Greece, the Far East and North Africa. About 68 postal workers were on the train. This train had left Paris at 8.40 p.m.

Train No. 11 was due at Melun at 9.14 p.m., arriving from Paris via the old main line, but from there it was booked to cross onto the new line via Hêricy, which involved crossing the line on which train No. 2 was booked to arrive at Melun via the Fontainebleu line at 9.21 p.m. On this day the postal train (No. 11) was running seven minute late, and thus approached Melun at exactly the same time as the express. The Melun signalman decided to give the postal train precedence through the junction and the signals for the Marseilles express were therefore necessarily at danger. Despite this the driver of the express took no action until his train was within 150 metres of the crossing, when he finally initiated an emergency brake application, and his train was still travelling at 80 kilometres an hour when it rammed the mail train, smashing into four vans behind the leading one. The express engine overturned onto its left side in the collision and buried itself in the earth and its tender formed an obstruction to the following vehicles. The leading baggage van and mail van climbed up onto the tender before falling over onto their sides and the following second baggage van, second-class and composite carriages were telescoped against it and crushed by the vehicles behind. Gas reservoirs beneath the carriages and postal vans exploded as a result of the collision, setting off a fire which quickly spread through the wreckage.[607] [608]

The accident happened close to the station and the town, so that fireman, together with soldiers from the 13th regiment of dragoons and the 31st infantry regiment stationed at Melun. The rescuers managed at the last moment to remove a waggon containing 1,400 square metres of compressed gas from the scene, thus averting a potential explosion, but they were forced to retreat from the burning inferno of the wreckage without being able to save victims trapped in them, who were burnt to death. In a tragedy reminiscent of that at Guardia Mangano the previous December one passenger, the wife of a Captain Amie, was pinned beneath the locomotive of the Marseilles express. Breakdown cranes sent from Paris were unable to shift the engine and like Driver Raspaolo the year before she lingered in agony, imploring rescuers to end her misery, until finally doctors were able to diminish her suffering with an injection of morphine. She kept asking for her husband, whose body had been one of the first to be found, and towards 5 a.m., abandoning all hope of rescue, gave farewell messages for her relatives and passed out the jewellery she had been wearing to be handed to her mother. She died, still trapped, 8½ hours after the accident. The blaze was eventually brought under control and extinguished, when the remains of passengers and their valuables were collected and the operation to clear the

---

[606] *La Croix*, Paris, France, 6th November 1913
[607] *Le Petit Journal*, Paris, France, 5th November 1913
[608] *Le Figaro*, Paris, France, 5th and 6th November 1913

line was begun. Traffic could be resumed after 22 hours.[609] [610]

Three months after the disaster the Ministry of Public Works put the final death toll at 41, with 57 injured. The dead included Professor Mathieu Jaboulay of Lyons University, a prominent surgeon. However, the *Times* in London was able to state that 'it can be confidently stated that no English people have been killed. The only foreigners among the dead and injured are a Dutchman and a German.' Many of those killed were, in fact, postal workers from train No. 11, 15 of whom lost their lives. The loss to the postal service in mail was also heavy, that of the departments of Ain, Côte D'Or, Jura, Doubs, Savoie, Haute-Savoie, Drôme, Vercluse, Gard and Hêrault being lost in the fire, along with international mail for Switzerland (including the diplomatic bag destined for the French embassy in Berne), Italy, Greece and Egypt.

Condolences for the accident were sent from numerous national and international sources and the Chamber of Deputies unanimously agreed to grant of 12,000 francs to the families of the postal workers killed and injured and of 14,000 francs to cover the cost of the funeral of the 15 who lost their lives.

The signalling at Melun was naturally checked immediately after the accident and found to be working normally. The express would first have encountered a distant signal, placed 1,150 metres from the junction and displaying a single red light; 180 metres further on was a repeater signal, displaying two green lights, and warning that the junction lay 970 yards ahead. Finally, 150 metres from the junction were the two red lights of the carrê, which was an absolute stop signal. The driver of the express had not begun to brake until he reached the latter. The decision of the signalman to give priority to the mail train might be debated from an operating point of view but from a safety point of view could not be found fault with, and the responsibility for the accident lay wholly with those operating train No. 2.

According to the rules and regulations and their interpretation by the courts not only the driver of the express but its conductor were responsible for monitoring the signals. Thus Conductor Vernet, riding in the leading brake van, should have operated the brake valve when he found that his driver had failed to react to the signals. Conductor Daudans, riding in the rear van, was not required to look out for signals, but stated that he had applied the brake when he saw the red lights of the carrê. Driver Dumaine admitted that he had not seen any signals before the carrê, when he had applied the brakes in emergency. He attributed his failure to the fog and to steam beating down from the front of his boiler.[611] [612] Fireman Nicolle merely confirmed his driver's statement. Driver Dumaine was initially arrested, then released, then re-arrested along with his conductor.

At the trial of the driver and conductor, which began on 18th March 1914, the question of the visibility of the Melun signals was relegated to a secondary issue when evidence emerged that although Driver Dumaine had previously had an exemplary record, during the past year it had deteriorated and he had overrun signals on three occasions. Witnesses stated that the day before the disaster he had failed to stop at Joigny while in charge of a freight train and had had to set back, and that he had persuaded fellow railwaymen to

[609] *The Times*, London, England, 6th November 1913
[610] *Le Petit Journal*, Paris, France, 6th November 1913
[611] *Le Figaro*, Paris, France, 7th November 1913
[612] *The Times*, London, England, 7th November 1913

keep quiet about the incident. One wonders if Driver Dumaine was suffering from a medical condition which was causing him to suffer mental absences, such as transient global amnesia, or if he had suffered a head injury. On 27th March the court at Melun sentences Driver Dumaine and Conductor Vernet to four and one month's imprisonment respectively. At the same time the court considered the case of one of the widows of the deceased postal workers, the only one who had not agreed to the compensation offered by the railway company. She was awarded 4,000 francs plus an annuity of 1,400 francs a year.[613] Driver Dumaine alone appealed against his conviction, with the support of his union, the Fédération des cheminots, and in May 1915 the Court of Appeal in Paris reduced his sentence to a suspended one.[614]

The disaster led to fresh calls from the railway and postal unions for improvements in railway safety.

Immediately after Driver Dumaine had been gaoled his union, the *Fédération générale de Groupements de Méchaniciens et Chauffeurs des* had demanded his release and the case had been debated at length in parliament. The new Minister of Public Works, M. Fernand David, maintained that the signalling at Melun was adequate for safety but this was hotly disputed by Socialist M.P. Jean Colly and the *Fédération générale de Groupements de Méchaniciens et Chauffeurs des* underlined its previous vain repeated efforts to have the signals repositioned further from the junction, given their poor visibility in fog and increasing train speeds. Effectively admitting this was a problem, M. David persuaded the railway company to move the advance signal to a point 2,420 metres from the junction after the accident.

The accident at Melun, following those at Villepreux-les-Clayes, Saujon and Courville, emphasised the general problem of relying upon train drivers observing signals at increased train speeds. Proposals in parliament for the provision of a third man in the engine cab, charged solely with observing signals, came to nothing (and would certainly have been resisted as worse than useless by Board of Trade inspectors in Great Britain and their counterparts in the United States). The introduction of automatic train control devices, warning the driver that he was passing a signal at danger and still in the experimental stage, was given a boost. At the end of November 1913 the Minister of Public Works issued a circular giving railway companies two months to issue draft proposals for the installation of the "Crocodile" system, which caused a horn to sound in the cab when a train passed a danger signal, on fast and express trains.

Another issue was the continued use by the P.L.M. company of oil gas for lighting, the abandonment of which had already been called for in numerous ministerial circulars following previous accidents. The railway companies pleaded that the switch to electric lighting was expensive and could not be undertaken at once. In response, the previous Minister of Public Works, M. Thierry, had issued a circular warning railway companies that no new rolling stock proposal which included gas lighting would be approved. His successor, M. David, thought it worthwhile to reissue the circular in January 1914.

The layout of tracks at Melun was altered after the disaster; an alteration to the layout had already been proposed back in 1906. This reduced the risk of a repeat of the accident,

613 *L'Aurore*, Paris, France, 19th, 20th and 25th March 1914
614 *L'Humanité*, Paris, France, 16th March 1915

but did not eliminate it (as shown by a further disaster in October 1991).[615]

At 19:45 on the evening of Saturday 29th November 1913 the 18:00 express from Rome to Naples was diverted onto the wrong line at Ceccano, Italy, due to a mistake by a signalman, and collided head-on with a standing goods train. The express had fortunately reduced speed due to adverse signals at Frosinore but the leading third-class carriage of the express was badly damaged and five passengers in it were killed and another 17 seriously injured. In contrast none of the occupants of the following dining car, which was more solidly constructed, was even badly hurt. The engine crew jumped and escaped but the signalman was crushed by the wreckage. The regular signalman had reportedly delegated his duty to another man, with insufficient explanation of what to do.[616 617]

Very brief reports appeared in several American newspapers of a collision between a passenger train and a freight train near Costesti, Romania, south-west of Bucharest, late on Friday 5th December 1913. One hundred people were reportedly killed or injured.[618]

Thirty men were reportedly killed and 85 seriously injured near Darno, Austria, on Friday 15th December 1913 when an express train overran a signal and collided with a work train carrying 800 men.[619]

A German troop train was involved in a serious accident at Woippy, near Metz, Germany, on Friday evening, 2nd January 1914. This was part of Alsace-Lorraine, previously part of France but which since the end of the Franco-Prussian war had been incorporated into the German Empire. Elberfeld – Trier – Metz train No. 44 was carrying soldiers returning to their infantry regiments in Alsace-Lorraine from their homes in Rhenish Westphalia. It was to have been shunted at Woippy station out of the way of a following express but owing in different accounts either to carelessness on the part of its driver or to a wrongly set pair of points it was diverted into a dead-end road and collided violently with a set of buffers. The engine was lifted into the air by the force of the collision, demolished a corrugated iron shed standing behind the buffer stops and then buried itself in the ground. The leading vehicle of the train was completely demolished and the rear luggage van crashed into the third-class carriage in front of it. Seven soldiers died and three were seriously injured, two being described as in a very critical condition.[620 621]

What at the time was the worst accident ever to happen in the state of New South Wales, Australia, occurred at Exeter at 11.50 p.m. on Friday 13th March 1914. Exeter was

---

[615] Wikipedia – Catastrophe ferroviaire de Melun du 4 Novembre 1913 (in French)
[616] *The Times*, London, England, 1st November 1913
[617] *Eisenbahnunfälle Band I* by Ludwig Stockert pg. 12 (in German)
[618] *Washington Daily Herald*, Washington Court House, Ohio, United States, 6th December 1913
[619] *The Lima Daily News*, Lima, Ohio, United States, 17th December 1913
[620] *The Times*, London, England, 5th January 1914
[621] *Salzburger Wacht*, Salzburg, Austria, 3rd January 1914

a crossing place and there was a long refuge siding at the east (Sydney) end of the station. Train No. 5, the Temora Mail, had left Sydney at 8.10 p.m. and was due to arrive at Exeter at 11.40 p.m. From the opposite direction train No. 12, an up fast goods train, had arrived at Exeter. It usually crossed the mail train at Moss Vale, 6 miles further on, but because it was late there was insufficient time for it to reach Moss Vale without seriously delaying the down mail. It was also too long to fit into the crossing loop, so there was no option but to take it forward out onto the single line, under the protection of the down home signal, and back it into the refuge siding. It was while it was performing this operation that it was run into head-on by the mail train, which had overrun the hone signal.

The collision between the trains was a violent one. Both train crews escaped with bruises and the passengers at the rear of the mail train were not too badly hurt but two carriages at the front of the mail train were severely telescoped.

Rescue operations were begun after the accident by railwaymen on the scene and uninjured passengers, who were soon joined by local residents who had been awakened by the sound of the crash, and by navvies from a nearby camp. Breakdown trains were sent from both Sydney and Goulburn at 1 a.m. Saturday morning, followed by another train from Sydney at 2 a.m. carrying Superintendent of Lines Hodgson, Assistant Superintendent of Lines Mars, Chief Mechanical Engineer Lucy and Dr Taylor. The rescue work went on all night in mist and rain, with light provided by lineside fires. The last victim was removed at 7 a.m. and the line was reopened to traffic at 7.30 a.m. The dead and injured were taken first to the station buildings and thence to Bowral.[622]

Thirteen passengers and Mail Guard Alfred Bray died in the accident, though only five were killed outright. Three died before they could be released, one as he was on the point of being released, two on the station platform and the other three in a sleeping carriage converted into a makeshift hospital. Others though had lucky escapes. 'Extraordinary escapes from sudden death characterised the tragedy. One man and his wife escaped with a slight shock, while Mr and Mrs Minnis on the opposite seat were killed. Another passenger was sitting still and two dead bodies were thrust on top of him. Mr Peter Cassanarti only arrived in Australia last week from Europe, where he spent 14 months fighting with the Greek army against the Turks. The passengers on either side of him were killed. The driver and fireman of the Temora mail were particularly lucky to escape. Despite the force of the impact and the damage to the engine they only suffered slightly from the quantity of the coal which was heaped on their legs. The driver and fireman of the goods train also escaped. Several children in the fatal carriage were thrown into corners and on the racks and were also lifted out little the worse for their startling experience. Just as the concussion of the collision threw the passengers from their seats in the rear portion of the train one man sat up, rubbed his bruised head, and exclaimed, "I knew something would happen. Serves me right for travelling on the unlucky Friday, the thirteenth."' [623]

The fate of one family was especially sad. Mr John Heaver of Parramatta was taking his wife's body to Cootamundra for internment. Accompanying him were his daughters Alice, Lilian and Viola, his son Arthur, his daughter-in-law and her sister, Mrs Pitts, who

---

[622] *The Sydney Morning Herald*, Sydney, New South Wales, Australia, 16th March 1914
[623] Ibid.

alone of the family was riding in the sleeping car. Mr John Heaver, Alice Heaver and Mrs Arthur Heaver were all killed and Arthur Heaver, Lilian and Viola were all injured. 'With their father, sister and sister-in-law killed outright, and their brother pinned in and injured among the wreckage, the two young ladies who escaped were in a pitable state of distress, accentuated by the shock of the collision itself. Being in the solidly constructed sleeper, Mrs Pitts suffered no ill-effects from the bump, and slipping on a kimona over her night attire immediately ran to the smashed carriages, calling out to her relatives. The carriage in which the latter were was crumpled up, and one of the ladies who escaped was thrown into the hat rack.

'Miss Viola Heaver was the first to recover her self-possession after the shock and grief attendant upon the disaster and its fatal results. She pleaded with Mrs Pitts to go back to the sleeper and get dressed lest she should catch cold in the rain, and then did what she could to help the others. Mrs Pitts also stayed on the spot as long as she could be of service.

'Mr Arthur Heaver bore himself bravely, "Look after the girls first," was his request, when the rescuers began to extricate him.' [624]

Almost a week after the accident Night Officer Alexander Callan told the inquest that he had been on duty at Exeter on the night of the accident. The goods train had arrived at about 11.48 and the mail train was due at 11.44. He had commenced to shunt the goods train into the siding when the mail train arrived and the collision occurred. Just about the time of the collision a fog had come up which had obscured the backlight of the down distant signal. Questioned as to whether it would not have been wise to call out the fogsignalman he stated that he had not thought it was necessary, as the fog had not been that dense and in his experience fogs often came and went again in a few minutes. He said that after the accident he had met the driver of the mail train, who had looked in a very bad state, excited and upset. He had said to the driver, "You passed the signal at danger. How did you manage it?"

The driver had replied, "I don't know. It is something terrible; something awful."

Later he had spoken to the fireman, but the latter had said that he had not seen the signals because they were on the driver's side. He said that he was aware that the driver of the mail had alleged that the distant signal had been green but that the driver had said nothing to that effect to him at the time.

The Exeter stationmaster, Alfred Pavey, next gave evidence. He said that the noise of the crash had awoken him. He saw the guard of the goods train who told him that the mail had run into them. Night Officer Callan had told him that the mail had overrun the signals. He had not liked to ask who was to blame at the time. The locking would prevent the distant signal being off with the home signal at danger. It had been foggy but not bad enough, he considered, to require the services of the fogsignalman.

Driver George Barlow, in charge of engine No. 738 of the goods train, testified on the other hand that it had been very foggy that night and he had only been able to see the up distant signal an engine length away. He had seen that the down home signal was at danger. The loop at Exeter was always too short to accommodate his train and should be extended.

[624] Ibid.

Fireman Raymond O'Neill said that at the previous station at Bundadoon the fog had been too thick to enable them to see the guard signalling from his van on the engine. The fog might have been slightly less thick at Exeter, but not much.

Guard Walter Punley said that his train had been 61 minutes late at Bundadoon. Because of the fog there the night officer had had to walk half the length of the train to give the signal that they might depart. He considered that the fog had been much the same at Exeter.

Driver William Morgan said that he had been in charge of the engine of No. 11 mail train, which was stopped at Moss Vale because of the accident. He had reached Exeter at about 1.35 a.m. and had spoken to the driver of the Tamora Mail. "This is a bad accident, Peter. How do you account for it?"

The driver had replied, "I don't know. The distant signal was showing a bright green light when I passed."

He (Driver Morgan) had repeated this statement to Night Officer Callan, who had replied, "That could not be. It was never pulled by me." [625]

The next day (Friday 20th) Guard W. J. Denneby gave evidence at the inquest. He testified that he had been the guard of the Temora Mail that night. He knew the position of the Exeter down distant and had been looking out for it but owing to fog, smoke and steam had failed to see it. His train was then travelling at 20 miles an hour and had gradually slowed down due to the driver applying the brakes so that he did not think that it was travelling at more than six or seven miles an hour at the moment of the collision. He did not see the home signal either, though he was looking out for it.

The inquest was concluded on Tuesday 24th March. Neither Driver Peter Irwin nor Fireman Harold Terrett was permitted by their solicitor to give evidence. The inquest jury found that the disaster had been caused by Driver Irwin overrunning the Exeter down home signal, adding a rider that the crossing loops should be extended to accommodate any train, or alternatively that refuge sidings should be provided at each end of Exeter station, also that more precautions should be taken to ensure the safety of the travelling public in foggy weather. The coroner duly committed Driver Irwin to stand trial for manslaughter. [626]

On the first day of Driver Irwin's trial Night Officer Callan, the Exeter stationmaster, the driver and fireman of the goods train and others gave evidence similar to that tendered at the inquest. Night Officer Callan admitted that he had been 'removed' (temporarily suspended perhaps) as night officer at Springwood 15 months earlier for falling asleep on duty and delaying a mail train. [627]

At the resumed trial the following day Driver Irwin himself gave evidence. He said that at Werai (the station on the Sydney side of Exeter) he had run into very thick fog and the fog had become denser as they neared Exeter. From landmarks he knew where the Exeter down distant should be, but did not see it until he was within 65 yards of it, when he saw it showing a full green light. When he saw the distant off he had taken it that the remaining signals were off and that the section was clear but on account of the fog he had

[625] Ibid, 20th March 1914
[626] Ibid, 25th March 1914
[627] Ibid, 28th April 1914

not taken it for granted that the home signal would be green. He followed this oxymoronic statement by saying that he had been travelling at about 20 miles an hour then but had shut off steam and applied the brakes approaching the home signal. Before seeing the home signal he had seen two white lights through the fog and had realised that there was danger. He had then seen the white post of the home signal and its red light and had applied the brakes in emergency as his fireman had shouted, "Whoa, mate." He had also tried to reverse his engine but the wheels had skidded on the greasy rails; then he knew it was too late and the collision occurred.

Guard Denneby and Fireman Terrett were then questioned, the fireman corroborating the evidence of his driver and agreeing that the Exeter down distant signal had been showing a green light.

Driver Irwin was found "Not Guilty". [628]

A point which does not seem to have been raised at all during the inquest or trial was the fact that Night Officer Callan had accepted the mail train from Werai with the goods train being shunted only a few yards in advance of the protecting home signal. Clearly the signalling regulations in New South Wales were not as strict as in the United Kingdom, where this would not have been permitted even in clear weather unless the "Warning Arrangement" was authorised, in which case the mail train would have had to have been cautioned at Werai that the section was clear but the station was blocked.

On the evening of Monday 20th July 1914 two trains came into sidelong collision at Empalot, France, a few kilometres outside of Toulouse on the Chemins de fer du Midi. Empalot was the junction for a branch to Auch and train No. 456 from that place and bound for Toulouse was three-quarters of the way through the junction when it was struck by the engine of train No. 510 from Bayonne. The engine of the Bayonne train was thrown onto its side, the last two carriages of the Auch train were telescoped and the others partially forced into one another.

Police lieutenant Guillame Dane, who was a passenger on the Bayonne train, described the accident. 'It was 8 o'clock. We were going fast. Suddenly there was a terrible noise. We were thrown one on top of the other as our train suddenly stopped. Some of us got down onto the track. Cries and appeals for help were coming from the Auch train. Distraught men and women were running here and there. There were soldiers travelling on the Bayonne train. I called them and they rendered invaluable aid in the rescue work.'

Six passengers died in the accident. They were: Mme Mousseron, aged 75, of Brax-les-Guévin; Mme Ferrèrre, 56, mail attendant; Saliquet, re-engaged soldier of the 14th line, identified by General Poline; Jean Abeille, 36, employee at the Midi station; his wife, Mme Abeille, aged 32, and their daughter Juliette, 3½ years old. Fifty or more other passengers were injured.

An investigation into the accident was immediately opened by the public prosecutor. Driver Jean Bart, aged 47, of the Auch train, stated that approaching the junction he had read the signals as clear for him. It was only when his train was a few metres from the semaphore immediately adjoining the junction that he had realised that it was at danger.

[628] Ibid, 29th April 1914

He had then jammed on the brakes, reversed his engine and applied steam, but it was then too late.

Signalman Dupaul at the junction stressed that he had maintained the disc signals from the Auch branch at danger so as to give the express from Bayonne precedence through the junction.[629]

The newspaper report on the accident noted that an identical, though much less serious accident, had occurred at the same junction five or six years previously.

The Minister of Public Works, M. René Renoult, ordered the Midi company to make its own full investigation and advise him of the results.

France entered the First World War on 3rd August 1914. Less than three weeks later there was a major catastrophe on the Chemins de fer de L'Est, on 23rd August 1914 at the junction at Saint Laurent, near Epinal. Due to wartime censorship the accident was not mentioned in the press at the time, but it is recorded in the archives of the Société Nationale des Chemins de fer français. An ammunition train and a troop train were in collision, leaving 79 dead and many injured. [630]

An accident occurred at Schönhauser Damm, Germany, on Sunday 22nd November 1914. A wrongly positioned set of facing points diverted passenger train Zug D 6 into the rear of goods train No. 5130. Five persons lost their lives, two were seriously and 11 slightly injured.[631] The goods train had been routed onto track 3 at Schönhauser Damm and the signalman replaced the signals and attempted to reset the points for track 2 for the passenger train. Meanwhile train D 6 was stopped at Blockstelle 29. Although the Schönhauser Damm signalman had replaced the stop and distant signals at danger the signals failed to properly return to danger with the result that he was unable to reset the points. The passenger train, having been stopped at Blockstelle 29, set off again towards Schönhauser Damm and, given that it had had to restart from a dead stand, should not have been travelling at more than about 50 or 55 kilometres an hour when it reached Schönhauser Damm. Driver Perl of train D 6 stated, however, that he only saw the Einfahrsignal (home signal) at danger when it was about 20 metres away. He had then applied the brakes in emergency but had been unable to avoid colliding with the goods train with sufficient force to telescope three waggons. It seemed possible that in fact he had mistaken the Ausfahrsignal (starting signal), which had already been cleared for track 2, for the home signal.[632]

While the First World War raged in Europe in Central America Mexico was embroiled in a civil war of its own, during the course of which there were a number of major railway disasters, many of them the result of sabotage. On Friday 27th November 1914, however, lives were lost not in a derailment due to sabotage but in a collision, presumably accidental. A train carrying General Pancho Villa, on his way to Mexico

---

[629] *Le Matin*, Paris, France, 22nd July 1914

[630] Wikipedia – Liste des accidents ferroviaires en France au XX siècle

[631] *Eisenbahnunfälle Band I* by Ludwig Stockert pg. 31. (in German)

[632] *Eisenbahn-Unfälle in Europa* by Erich Preuß (in German) pgs. 80-2

City, collided with a train carrying General Chao 'above Tula'. According to some reports 32 soldiers and camp followers were killed and 40 injured, but the general himself escaped (other reports state that there were only 25 killed). It was clearly his lucky day, for a mile beyond the scene of the accident his train ran over six mines connected to a battery. The mines failed to explode and were discovered the following morning. General Chao was slightly hurt.[633 634 635]

The French press contained reports at this time of several train accidents in parts of Belgium and France occupied by the Germans, information received second-hand from neutral countries such as the Netherlands and Switzerland. Thus it was stated in early December 1914 that two German trains had collided near Liége-Guillemins station, Belgium, presumably on about Saturday 28th November 1914. A large number of soldiers were killed and injured, including six generals and a number of officers, plus there was much material damage.[636 637]

A French colonial newspaper reported on 23rd January 1915 a 'terrible collision' which had allegedly occurred at Kalitz on 27th December 1914 involving a packed German troop train and an equally well-filled hospital train. Over 20 coaches were said to have been wrecked, 400 people killed and 500 injured. A number of railwaymen had been arrested following the accident, accused of treason.[638] Another report from Varsovic via Petrograd dated 23rd December probably relating to the same accident, stated that two Austrian trains were involved and there were 1,000 dead and injured. The location was here given as Kalish. If this report relates, as seems likely, to the same accident it clearly must have occurred before 27th December, perhaps on about Sunday 20th.[639] The accident probably occurred in Russian Poland but it has not been possible to identify the correct location.

The year 1915 was only a few hours old when England suffered a serious railway accident. It occurred on the Great Eastern Railway, at Ilford, at about 8.40 a.m. on Friday 1st January. The 8.20 a.m. up suburban passenger train from Gidea Park to London Liverpool Street, consisting of an engine and no less than 18 carriages, was crossing from the up local line to the up through line, when it was struck by the 7.06 a.m. express from Clacton to London Liverpool Street, consisting of an engine and tender and nine carriages. The front of the Gidea Park train had already reached the up through line and the engine of the Clacton train struck the seventh and eight carriages, the first of which was very severely damaged, while the eighth was practically destroyed. The next four carriages were badly damaged, as were the engine and two leading vehicles of the Clacton train. Ten passengers, mostly from the Gidea Park train, were killed outright, and

633 *Moberly Daily Mirror*, Moberly, Missouri, United States, 2nd December 1914
634 *The New York Times*, New York, New York, United States, 29th November 1914
635 *The Galveston Daily News*, Galveston, Texas, United States, 29 November 1914
636 *Courrier de la Rochelle*, Rochelle, France, 3rd December 1914
637 *Gaulois*, Paris, France, 3rd December 1914
638 *La Dépêche Malgache*, Antananarivo, Madagascar, 23rd January 1915
639 *L'Homme Enchaîné*, Paris, France, 24th December 1914

over 500 reported being injured or suffering from shock. The driver and fireman of the Clacton train were scalded and otherwise hurt.[640]

A newspaper reporter interviewed a local resident, Mr George Mills, of the White Horse, Ilford, who was on the spot soon after the accident, 'The intense cold added to the plight of the sufferers. It was the first thing they complained of when we regained consciousness. We took the cushions from the first-class carriages and with some blankets rigged up a sort of shelter to protect them until they could be taken to hospital. It was terribly difficult to extricate the injured, but there was no lack of medical help when we once got them out. Half a dozen doctors were there in a quarter of an hour and worked unsparingly. Bandages had been procured at once, and officials of the Great Eastern Railway were at once on the spot with all sorts of survival appliances.'

The same newspaper's report of the accident continued. 'The station was immediately closed to the public. Outside the gates there had gathered a little group of people waiting for news of friends and relatives. The names of the injured and of the hospitals to which they had been taken were posted in the windows of shops nearby without unnecessary delay. A touching incident was told by a gentleman who had been travelling in the wrecked train. He said:- My father was in the first-class carriage which was shattered in the collision. I was in the second-class carriage immediately behind it. Our compartment was crowded; there were five people standing. The first warning we had of anything amiss was a dull grinding sound, followed by two or three crashes. We were all pitched on top of one another. One man was cut by the splintered glass of the window, but the rest of us had escaped with a shaking. I got out and, of course, my first thought was to look for my father, but I could not find him anywhere and feared he might be still under the debris. I continued to search, but at last I was compelled to continue my journey to the City. There, to my great joy, I found that a telephone message had been received that my father, slightly injured and shaken, had been taken to the emergency hospital at Ilford.

'Dr Fraser, of Belgrave-road, Ilford, whose house was close to the railway line, was in bed when the accident took place. Hearing the sound of a collision, he feared that something was amiss on the line and hastily dressed in order to go and offer his services. As he left home he was met by a messenger who had been sent to summon him. With the other doctors he was soon engaged in binding up wounds and otherwise attending to the injured. It is unpleasant to record that while he was engaged in this act of mercy his bag containing surgical instruments and a pocket case was stolen.

'A kindly thought prompted a number of the residents near the railway line to bring out hot coffee for the victims of the accident.' [641]

Four running lines ran through Ilford station and there were double junctions between the through and local lines at the west (London) end of the station. Two signal boxes, Ilford East and West, controlled the station area. The next box towards London was Aldersbrook, and towards Ipswich it was Ilford Carriage Sidings. The lines were signalled by Sykes Lock and Block.

The Board of Trade inquiry into the accident took the evidence of a number of witnesses, beginning with that of Signalman Alfred Nicholls of Ilford East box, who stated that the Gidea Park train had been offered to him by Ilford Carriage Sidings at

[640] Board of Trade report

[641] *The Times*, London. England, 2nd January 1915

about 8.34 a.m. He had accepted it at that time and offered it to the West box, who had accepted it at 8.40. He had then lowered his up local home signal and the train had passed him, running at about 10 miles an hour. The Clacton train was offered to him and accepted by the Carriage Sidings signalman at about 8.37 a.m. He had offered it forward to the West box but it was not accepted, so he had not lowered any of his through line signals. The Carriage Sidings signalman sent the Clacton train on immediately he had accepted it. Watching the Clacton train approaching he saw that it was not stopping at his home signal. He had picked up a red flag, opened the box door and gone out onto the landing, from where he had waved the red flag and shouted as hard as he could to the driver as the train passed him. He was sure that the detonators which were placed automatically on the line when the home signal was at danger had exploded.

Relief Signalman Henry Nicholls was on duty at Ilford West box. He said that he had not accepted the Gidea Park train from the East box until he had got his booking boy to find out from Chadwell Heath how the Clacton train was running. Finding that it was running four minutes late he decided to give the Gidea Park train precedence, had reset the road from the up local to the up through, accepted it from the East box and cleared his signals for it. The Clacton train had been offered to him at 8.39 when he had cleared out a Southend train when instead of accepting it he had sent the "Blocking Back" signal. All his signals for the up through line were at danger. As he was engaged supervising some shunting operations in the yard he suddenly saw the Clacton train dash past his box. He had had no emergency running away signal (as the East box signalman had been out on the landing trying to stop the Clacton train and the boxes were only 250 yards apart he doubtless had no time to send one). He saw the collision occur.

Driver John Buttress testified that he had been the driver of the Gidea Park train. He testified that he had been severely checked at the Ilford East signals but not quite stopped. The signals at the West box had been lowered for him to cross from the up local to the up main and he had passed that box at a speed he estimated as 15 miles an hour. The first he knew of the collision was his engine being forced forward and his being knocked down on the footplate. Getting up he saw that another train had run into his. It had been a very clear morning and he had had no difficulty in seeing the signals. He had frequently driven the Gidea Park train and they were often crossed over in front of the Clacton one.

Acting Fireman E. M. Hicks of the Gidea Park train corroborated his driver's evidence. He said that when the collision occurred his mate had asked, "What's up?"

He had looked over the side and replied, "The Clacton is into us."

Second Guard John Mason said that he had been acting as head guard of the Gidea Park train. He saw the signals lowered for crossing from the local to the through line at the West box. He had known nothing of the impending collision until he was knocked down in his van.

Riding in the third-class brake composite near the rear of the Gidea Park train was Second Guard Charles Turner, who thought that his train had passed the West box travelling at 10 or 15 miles an hour. He said that the first he had known of the impending collision was hearing the whistle of the Clacton train, when he had got to his feet and exclaimed to the other men riding in the van, "What is that express whistle?" Immediately afterwards he had seen the engine of the Clacton train passing on his left and had flown to the Westinghouse brake a moment before the collision. His van had not left the line and he had only felt a slight vibration.

Head Guard Henry Cook of the Clacton express testified that he had been riding in the third-class brake composite sixth from the engine. His train had stopped at Thorpe-le-Soken, Colchester, Chelmsford and Shenfield. It had left Shenfield two minutes late and Chadwell Heath three minutes late. Smoke and steam from his own train had prevented his seeing the signals at Ilford. He had felt the continuous brake being applied near Ilford East box. He had been about to look out of the window to try and see the signals when the collision happened.

Fireman George Adams of the Clacton express said that he had been firing and had not seen the Ilford East distant signal. He saw the home signal at Ilford East box at danger but did not hear any detonators explode. He had not said anything to his driver because the latter was already applying the brakes. He saw the stop signal at the west end of the station at danger too but said nothing to his driver about that one either. They had been travelling at, he thought, 50 miles an hour past the first danger signal and were still going at 30 or 40 miles an hour past the second. Halfway along the station platform he saw the Gidea Park train crossing over and knew that there must be a collision. He did not see the actual collision as his driver had told him to stoop down, which he had done. That was the only thing that his driver had said. He did not think his driver had reversed the engine.

Driver Frederick Bloomfield of the Clacton express said that his brakes had been in quite good order. He had sighted the Ilford East distant signal and had taken it as off. It had not been a doubtful signal. It had been nice and clear. Steam beating down had made it difficult to see the Ilford East home signal but close to it he had seen that it was at danger. He had then shut off steam and applied the brakes fully. He heard no detonators go off. Halfway through the station he had seen the Gidea Park train crossing over and that the Ilford West home signal was also at danger. He had whistled going through the station and knew when he saw the Gidea Park train that there would be an accident. He had not reversed his engine. When he first saw the Ilford East home signal at danger he had not grasped its meaning.

Various other witnesses gave evidence at the inquiry regarding the events leading up to the collision and whether or not the detonators at the Ilford East home signal had exploded.

In his conclusion to the inquiry the Board of Trade inspecting officer, Lt-Col P. G. von Donop, summarised the evidence of the various witnesses and then pointed out that there could be no doubt as to the position of the up through line signals at the two Ilford boxes. With the road set for the Gidea Park train to cross over both the interlocking and the Sykes Lock and Block would have prevented the lowering of the up through line signals. The only person who alleged that any of these signals had been clear was Driver Bloomfield, who maintained that the Ilford East up through distant had been off. It was clear that he was mistaken, and that the collision was primarily due to this mistake. The evidence as to where steam was turned off and the brakes applied was contradictory, with witnesses such as Driver Bloomfield, Fireman Adams, Signalman Alfred Nicholls, Guard Cook, Conductor Challow and Chef Smith in the restaurant car of the express and a passenger, Mr Carne, with nine years' railway experience on the Great Western, all stating varying locations. From the balance of the evidence steam was turned off and the brakes applied 300 yards short of the point of collision. The speed of the train must have been at least 20 miles an hour at the moment of impact.

The driver, fireman and guard of the express all denied hearing any detonators explode at the Ilford East home signal but exploded detonators were found in the machine by a platelayer shortly after the accident which could not have been detonated by any other train. Furthermore explosions had been heard at the time that the Clacton train was passing by a foreman porter, an off-duty signalman in bed at home and two lineside residents. If Driver Bloomfield's statement that he saw the home signal at danger be accepted, whether or not the detonators went off was immaterial. It was, however, strange that Driver Bloomfield had not immediately reacted upon seeing the home signal at danger. If in fact it had been the report of the detonators that had first alerted the driver to the fact that the signal was at danger that would explain the delay in his stopping his train. At any rate the accident must be entirely attributed to Driver Bloomfield's lack of care.

The Great Eastern Railway Company had, the inspecting officer stated, provided a large number of emergency detonator placers at home signals, but it was at the distant signals that the warning was especially needed, and he urged the Company to consider trialling such devices, as several other companies had done.[642]

The inquest on the deaths of nine of the victims of the disaster returned a verdict that the accident had been 'due to an error of judgement on the part of the Clacton express driver in misreading one signal for another, near the scene of the accident, thus working on a false presumption, he being ignorant of any error at the time.' Driver Bloomfield thus escaped a possible manslaughter charge. The jury added a rider that 'In our opinion all guard's vans which have back-view look-outs should be at the rear of all trains. We agree that the guard of the local Gidea Park train should be commended for so promptly applying his brakes, and, in our opinion, saving the lives of hundreds of other passenger.'[643]

The jury at the inquest on the tenth victim presumably returned a similar verdict.

A French newspaper reproduced on 6th January 1915 a report from the Amsterdam *Algemeen Handelsblad* newspaper regarding a collision on the line between Antwerp and Esschen in Belgium. The accident happened on Sunday 3rd January 1915 when a passenger train from Antwerp came into a sidelong collision with a German military train 'at the junction of the various lines leading to the forts' – presumably on the outskirts of Antwerp. The engine, baggage car and one coach of the passenger train were completely destroyed and other carriages derailed. Eight passengers were killed and 25 others more or less seriously injured. At least some of the passengers were still unidentified on Tuesday.[644]

A paragraph in a French newspaper of 15th January 1915 stated that information has been received from Basle in neutral Switzerland of a sidelong collision involving a German military train running without any lights near Batilly station in Lorraine,

---

642 Board of Trade report

643 *The Times*, London. England, 25th January 1915

644 *L'homme enchaîné*, Paris, France, 6th January 1915

Germany, presumably on about Tuesday 5th January 1915. An official note added that 19 soldiers and two railwaymen had been killed.[645]

Two military trains collided at Konitz, Germany, on the Dirschau – Schneidemühl line on Wednesday 3rd February 1915. Six died and 20 were injured.[646]

An engine ran into a passenger train at Dirschau, Germany, on Thursday 23rd March 1915. Six died and 20 were injured.[647] (Given that these last two accidents apparently occurred not far from one another and resulted in exactly the same number of deaths and injuries there is a possibility that the accounts relate to a single event.)

The pressures on the railways caused by the First World War led to a series of accidents in most of the countries involved. In Great Britain the year 1915 was the worst in railway history in terms of passenger fatalities in train accidents. The first of these occurred as we have seen on the first day of the year and on 22nd May in Scotland there was an overwhelming catastrophe which resulted in more than twice as many deaths as any other accident in Great Britain before or since up to the present time (February 2020).

The location of the accident was Quintinshill signal box, just over 10 miles north of Carlisle on the Caledonian Railway main line. There was no station here but up and down loops were provided, together with a mains crossover. The signal boxes on either side were Gretna Junction and Kirkpatrick. On the morning of the accident the 11.45 p.m. and 12 midnight sleeping car expresses from London Euston were running late. In accordance with normal practice, when the late running of the 12 midnight express meant that it could not leave Carlisle by 6.20 a.m. the 6.10 a.m. local train was allowed to precede it, as otherwise the local train would also have to wait for the 6.30 a.m. Glasgow and South Western express from Carlisle to Glasgow and the 6.35 a.m. North British express from Carlisle to Edinburgh, much delaying the local train and causing disruption down the line. On such days the local working travelled as far as it could without further delaying the express following it and was then shunted out of the way of the express at a convenient location.

On Saturday 22nd May 1915 the down local train preceded the express as far as Quintinshill box. Here the usual and easiest procedure would have simply to run it into the down loop, but this was already occupied by the 4.50 a.m. down goods train from Carlisle, which had arrived at 6.14, and the only option was to cross it over onto the up main line, which was done. Once it had been crossed over a third train, an up empty coal working, was admitted to the up loop. The first of the delayed down expresses (due to leave Carlisle at 5.50 a.m.) was then accepted from Gretna Junction and passed Quintinshill safely at 6.38. Meanwhile in the signal box the early turn signalman, James Tinsley, who had ridden down from Gretna on the engine of the local train, was just taking over from his colleague, Signalman George Meakin. One or other of them cleared back the empty coal train to Kirkpatrick, accepted an up troop train, although the up main

[645] *La Charente*, Périgeux, France, 15th January 1915
[646] *Eisenbahn-Katastrophen in Deutschland* by Hans Joachim Ritzau, pg. 132 (in German)
[647] Ibid.

line was blocked by the shunted local train, and cleared the up line signals. The second down express (due off Carlisle at 6.5 a.m.) was also approaching. The troop train collided head-on with the engine of the local train and before it could be stopped the down express collided with the wreckage. Fire, fed by gas from damaged cylinders, broke out afterwards and eventually consumed all or part of not just the local train, the troop train and the express but also the two goods trains. In total 223 passengers (214 troops and nine civilians) and three railwaymen were killed and 242 passengers (191 troops and 51 civilians) and four railwaymen were injured. The bodies of many of those who died were entirely consumed so that they were 'missing' and the exact number of troops killed was uncertain as the battalion roll had been lost in the accident.[648] The troop train was one of two carrying officers and men of the 7th Royal Scots from Larbert to Liverpool, en route not to the Western Front but to Gallipoli to fight the Turks.

Although the accident involved a troop train in wartime news of it was not censored and was duly reported in the press, though with all the war news perhaps in less detail than would have been the case in peacetime. Thus, 'Splintered wood caught fire from the engines, and fanned by a slight breeze the flames quickly swept along the carriages. Before rescue work had been carried far the troop train was blazing from end to end, and soldiers, helpless through injury or pinned down in the shattered carriages, were burnt in a furnace from which there was no release. Where there was a life to be saved the rescuers, many of them injured, worked splendidly and bravely to free men from the fire, but into the hearty of the flames no one could go. Fire engines arrived from Carlisle and jets of water were poured into the blazing pile. The flames, where their hold was not too intense, were quelled, but round a spot where the three engines were locked together water could not stay the fire.

'Medical assistance was secured by telephone calls sent from Gretna village, and doctors and nurses travelled as quickly as they could to the scene. There was work for every one of them. The casualties among the passengers in the express and the local train happily are not believed to be numerous, but hardly one man in 10 of those crowded in the troop train escaped death or injury.

'First-aid to the sufferers had to be given in a long green field on the east side of the line, and soon this field, enclosed by hawthorn hedges, became a temporary mortuary as well. The injured were laid on mattresses, the dead – and release from the fiery trap came too late for many pulled clear of the debris - were placed side by side on the grass, and reverently covered with white sheets. Later the bodies were removed to a farm at the end of the field or to the tiny village hall at Gretna. Before sundown over 100 corpses or little bundles of charred bones and flesh, which were all that remained of Scottish soldiers who had hoped to fight in France, had been carried to their resting places. Every hour during the afternoon and night a motor-car took down two or three more of the pitiful collections of burned human fragments.

'Witnesses of the morning scenes tell stories of horror. Men, they say, were crying aloud to be rescued. Some of the shouts came from imprisoned victims who could not be seen through the smoke and flames. When the passengers had scrambled out of the express train, or the few carriages of the troop train which remained intact, and had turned themselves into rescuers, tried to answer these calls for help, the heat of the fire

[648] Board of Trade report

drove them back. The cries from this centre, where the wreckage of the three trains was joined, soon ceased. Where the workers could make their efforts more successfully the tragedy was almost equally great. In two cases the legs of men were amputated to save them from the fire. Many soldiers were taken out of the wreckage only to die in the field or on the way to Carlisle.

'Among the first people to get to the scene of the accident were Mr and Mrs Dunbar, who act as caretakers of the marrying blacksmith's shop at Gretna Green. Mrs Dunbar told me that the crash of the collision was so tremendous that she exclaimed at once, "The Germans have come." She and her husband went outside, and when they realized that something had happened on the railway they hurried over the fields.

'"It was a terrible sight," Mrs Dunbar said. "There were three trains all piled up. A fire had begun, and the poor things in the carriages were calling out. There seemed hardly anybody to help at first, but some of the men who got out of the trains uninjured fetched some bars and began to force open the doors. Then there were no doctors, and I came back to the village to see what could be done. The young lady at the post office telephoned to a lot of doctors in Carlisle, and then I got some food and drink ready. We haven't anything left in the house tonight."

'Mrs Dunbar described how the bodies had been arriving all the day at the little hall in the village. Talking outside the door of a Border building which romance has made famous, it seemed impossible to associate the charnel-house a score of yards away with the name of Gretna. "It was only yesterday we had a wedding here," Mrs Dunbar told me.

'Mr Dunbar all through the morning helped with the work of rescue. He said that numbers of the soldiers had broken limbs and thighs, and were quite unable to get out of the carriages without help. The accident would have been terrible without the fire, but the fire brought with it such scenes as a man would remember with a shudder to the end of his days.' [649]

Signalman George Meakin was the first witness to give evidence at the subsequent Board of Trade inquiry. He testified that for the last two or three years he and his mate had been changing shifts when he was on nights at 6.30 a.m., writing down the times of bell signals received after 6 a.m. on a scrap of paper to be copied into the train register book by his mate when he reached the box, so that it appeared that they had changed over at the proper time. His mate, Tinsley, had arrived on the local train at 6.30 a.m. and had reached the box a couple of minutes later. Shortly afterwards the fireman of the local train had arrived at the box to sign the train register book in accordance with Rule 55. Signalman Meakin said that he had told his mate that the up empty coal train had been put into the loop because Kingmoor yard at Carlisle could not take it, and that a special troop train was behind it and had been telephoned from Lockerbie as passing there at 6.32. His mate had then taken over and he had sat down in the corner of the box to read a newspaper which had come down on the local train. The paper was addressed to Tinsley and was thrown off the local train on days when it did not stop. He had sat in the box for 20 minutes reading the paper and perhaps discussing the news with others in the box; he had not taken any notice of what his mate was doing. The brakesman of the down goods train had come into the box for 10 minutes after it had arrived, later the brakesman of the empty coal train came to the box and was still there at the time of the collision. Having

[649] *The Times*, London, England, 24th May 1915

finished looking at the paper, Signalman Meakin said he was just leaving the box to go home when the troop train arrived and collided with the standing local train. He immediately turned to Tinsley and shouted, “Where’s the 6.5?” Tinsley had rushed to the levers but with that the second collision occurred. He (Meakin) had run down the steps and seeing that a serious accident had happened he had returned to the box and wired to District Superintendent Blackstock, Inspector McAlpine and the Gretna Junction and Kirkpatrick stationmasters for assistance.

Signalman Meakin said that the last bell signals he had given were clearing out the local train to Gretna and accepting the 5.50 a.m. express. He had not cleared out the empty waggon train to Kirkpatrick. He had not completed the operation of restoring the facing points leading into the up loop or clearing out to Kirkpatrick and hence had not sent the “Blocking Back” signal to Kirkpatrick or used a lever collar.

Signalman James Tinsley gave evidence next. He stated that although he had travelled down from Gretna on the local train all recollection of it had then passed out of his mind. The first bell signals he recalled giving were in connection with the 5.50 a.m. down express passing the box. He admitted accepting the troop train from Kirkpatrick at 6.42 and taking it on (receiving the ‘Train entering Section’ signal) from Kirkpatrick at 6.46. He had then obtained the road for it from Gretna Junction and cleared all the up signals, entirely forgetting that the local train was standing on the up main line outside of the box.

According to Signalman Tinsley his mate was still sat in the box when the first collision occurred and said to him, “Whatever have you done, Jimmy?”

To which he had replied, “Good heavens! What can be wrong? The frame’s all right and the signals are all right.”

Meakin had exclaimed, “You’ve got the Parly standing there!”

Gretna Junction had sent on the 6.5 a.m. express at 6.48 and it was approaching the box at that moment. He had at once sent the “Obstruction Danger” signal to Kirkpatrick and Gretna Junction and the down express had run into the wreckage. While he was copying out the train times there were at one time both his mate and the two brakesmen in the box, but they had not interfered with what he was doing. The fireman had signed the train register book and left the box again. He was sure that he had not cleared out the empty waggon train and his mate must have done so. He had not looked out of the box window when he accepted the troop train. The fireman of the local train had followed him into the box but had not reminded him where it was standing (which since he had just ridden down on it was hardly surprising). He had not thought to block back behind the empty waggon train.

Brakesman Charles Leggat said that he had been the guard of the troop train. He had received a blow on the head in the collision so could not remember much of what happened afterwards, but he recalled seeing a tongue of flame two yards high which he thought was burning gas, and remarking to a soldier, “Look at that now, and there’s no water.”

Driver David Wallace of the local train said that his fireman had gone to the box as soon as they had crossed over to the up line. After about five minutes he had returned and said that they would be kept there until the 6.05 a.m. express had passed. The fireman had got out his piece-box, then had suddenly noticed that the up signals were off and had drawn his attention to the fact. With that, they had seen the troop train approaching and only had time to jump clear before the collision. Following the initial collision soldiers were getting out of the troop train and running across the down line so he had shouted to try and warn them of their danger. Then the express ran into the tender of the troop train engine, which along with some of its telescoped carriages was foul of the down line. Shortly after the second collision fire broke out in the wreckage. He had discharged four or five fire extinguishers at the blaze, but without, he thought, much effect. Some persons, probably soldiers, had got a hand-cart and force pump and played water from a ditch on the flames until the fire brigade arrived at about 11 a.m. An explosion or two, apparently from gas tanks, had also occurred. All the tools and crowbars had been used with great effect by a group of naval men and soldiers. He thought that a number of soldiers had been run down by the down express.

Fireman George Hutchinson of the local train stated that he had gone to the box almost at once, and had been closely followed by the brakesman of the Welsh coal empties, so that at that point apart from himself there were two signalmen and two brakesmen in the box. He had signed the train register book, Signalman Tinsley had told him that his train would be held for the 6.05 express, and he had returned to his engine after about four minutes. Back on the engine he had noticed that the signals were off, and had warned his driver, but then the troop train had been close upon them and there had been no time to do anything other than jump for their lives. Fireman Hutchinson admitted that he had not carried out Rule 55 in its entirety by checking that there was a lever collar on the up home signal.

Guard Douglas Graham of the local train spoke of being thrown across his van when the troop train had run into his train. Stunned for a moment, he had then thought of the 6.05 a.m. express and had run up the line, whistling and holding up his hands. He had had no time to grab a red flag or detonators. The brakes of the express had been applied. He had then gone back to Gretna to tell the signalman and stationmaster there, followed by the guard of the express going back to protect his train.

Driver Andrew Johnstone of the train engine of the 6.05 express said that the Quintinshill distant signal had been clear for him but he had then seen the guard holding up his arms, had shut off steam and braked but there had been no time to reverse his engine. He had not seen the Quintinshill down home signal as his fireman had shouted that there was something wrong and he had been applying the brakes. He described the fire and explosions which had happened after the accident.

Fireman John Graham, Driver Johnstone's fireman, and Fireman David Todhunter on the pilot engine gave similar evidence. Driver John Cowper on the pilot engine was badly injured. Various other witnesses gave evidence, including Brakesman Thomas Ingram, of the down goods train, and Brakesman William Young, of the Welsh coal empties, both of whom had gone to the box but had not taken much notice of what was going on, nor had either of them signed the train register book in accordance with Rule 55.

Signalman Robert Kirkpatrick at Gretna G&SW Junction said that Quintinshill had telephoned him at about 6.26 and told him that the "Boy would get a ride today," and he had indicated to Signalman Tinsley, who was coming down the road, that he would be able to get a lift to the box.

Stationmaster Alexander Thorburn at Gretna said that Platelayer Richard Broach had called him out at 7 a.m. and had advised him that there had been a bad smash at

Quintinshill. He had told Signalman Kirkpatrick to phone Border Union box and get them to phone Longdown for doctors and assistance. He had then hurried to the scene. He found that three trains had come into collision and fire had broken out. Going to the signal box he had found that all the local stationmasters had been advised and asked to send assistance. There were many rescuers at work so he had gone to Kirkpatrick on the engine of the 4.50 a.m. down goods train to open up single-line working from that end, then returned to Gretna, where District Superintendent Blackstock had arrived and after that he had been busy with single line working, removing undamaged portions of trains and so on until relieved.

Stationmaster Thorburn denied knowing anything of the Quintinshill signalmen changing over at unauthorised times.

Lt-Col E. Druitt had the task of conducting the Board of Trade investigation. Having described the accident in his report and given in full the evidence of the various witnesses he went on to outline how it had happened. It was clear, he said, that either Meakin or Tinsley must have cleared the up Welsh coal empties back to Kirkpatrick, though each man had denied doing so; and that one or other of them must have released the indicator lock of the (Tyer's two position) block instrument, allowing the troop train to be accepted from Kirkpatrick. Neither signalman had given the "Blocking Back" signal to Kirkpatrick in accordance with Block Telegraph Regulation No. 13 (a) and (c), nor had a lever collar been applied to the lever of the up home signal.

The inspecting officer considered that the responsibility for the disaster lay entirely with the two signalman. Having crossed the local train over Signalman Meakin should unquestionably have given the "Blocking Back" signal for it and used a lever collar. He should also have finished dealing with the coal empties by reversing the points and clearing it out to Kirkpatrick. Also had the indicator lock of the instrument to Kirkpatrick not been released the troop train could not have been accepted. Preoccupation with copying out the times into the train register book, and dealing with the two down expresses, were probably the factors which led to Tinsley automatically accepting the troop train when it was offered by Kirkpatrick, forgetting that the local train was on the line, only 62 yards from the box.

Fireman Hutchinson had gone to the box in accordance with Rule 55 and signed the train register book but had not obtained an assurance from Signalman Tinsley as to the use of a lever collar. Likewise Brakesmen Ingram and Young, although they might have reasons to go there, such as carrying out Rule 55, asking for relief or enquiring how they were to be kept there, should not have stayed so long. Nor should Signalman Meakin have remained in the box reading the newspaper once he had handed over to his colleague.

Lt-Col Druitt noted that men engaged in routine work sometimes got into loose ways of working and habitually neglected regulations laid down for theirs or others' protection. There had been other cases in which the use of lever collars had been neglected, and Meakin and Tinsley admitted that they had not thought them necessary for a train crossed over to the other running line, and rarely used them. It was also hard to understand how the Gretna stationmaster could have remained unaware of the unauthorised change over times when Signalman Tinsley lived next door to him.

Collisions with light engines or vehicles standing on the lines had led to the installation of track circuits at a large number of signal boxes, but the provision of track circuits at such a location as Quintinshill would mean that almost every signal box on a main line would require them, something which was not practicable at the present time, and was not likely to be for a great many years, due to the need for equipment and the cost. Given that it was little more than a block post Quintinshill would be one of the last places to be so fitted.

Due to its high speed the damage to the troop train had been excessive and the escape of live coals from its engine due to damage to the firebox and the proximity of wreckage all around had made a fire inevitable, the blaze then being intensified by the escape of gas from damaged cylinders, though a fire would have doubtless occurred even had all the vehicles been electrically lit.

Lt-Col Druitt stated that it seemed likely that in view of the excessive damage to the troop train all the victims of the accident had been either dead or unconscious before the fire reached them. This stands in contrast to newspaper reports of victims being burnt alive and other reports of officers shooting trapped men in 'mercy killings'. It seems possible that this may have been an attempt by the inspecting officer to lessen the horrors of the disaster, though on the other hand newspaper reporters have a tendency to sensationalise events to make a better story.

The circumstances of the accident had, said the inspecting officer, been altogether exceptional, but as to preventing the terrible results of a fire arising in the wreckage after a collision, he need only refer to the recommendations made in previous reports:

(a) Coaches should as far as possible be steel, with shock-absorbing buffers. Any timber used should be non-flammable. Doors should be constructed so as not to jam and any window bars should be easily removable and windows openable to the fullest extent.

(b) All new rolling stock should have electric lighting and gas lighting should be done away as opportunity arose.

(c) All brake vans, sleeping cars and so on should be provided with tools and appliances for fire extinguishing.

Some of the Companies were, Lt-Col Druitt said, using non-flammable wood, most were using electric lighting in new rolling stock and all were providing fire extinguishing apparatus as per his third recommendation.[650]

The surviving uninjured soldiers eventually continued their journey to Liverpool. There they were given a medical examination and one officer and all 58 men were declared to be unfit for foreign service. Colonel Peebles and six officers embarked for the Dardanelles.[651]

An inquest on 27 of the victims who had died at Carlisle was held there and its jury found not only the two signalmen but Fireman Hutchinson guilty of manslaughter, and they were committed for trial. The situation was complicated by the fact that the accident had taken place in Scotland, where inquests are not held, and Signalman Tinsley had been arrested there, charged with culpable homicide in the deaths of Driver Francis Scott and

[650] Board of Trade report

[651] *Gretna: Britain's Worst Railway Disaster* (1915) by John Thomas, pg. 60

Fireman John Hannah of the troop train and three officers.[652]

The trial of the three railwaymen was held in Edinburgh on 15th September. No evidence was offered against Fireman Hutchinson and the jury were directed to find him Not Guilty. It took the jury only five minutes to find the two signalmen guilty, and Meakin was sentenced to 18 months' imprisonment and Tinsley to three years.[653]

Following their release the Caledonian Railway offered to re-employ the two men and Tinsley went back on the railway, taking a job as a lampman at Carlisle. Meakin declined the offer and took a job as a coal merchant.[654]

The Quintinshill disaster was followed on Tuesday 17th August 1915 by an accident due to a signalling error at Dietikon, Switzerland, just outside of Zurich. This time the results were much less serious; six people died and a good number were injured.

Train No. 122 was a Zurich-Bienne-Geneva express which departed from Zurich at 5.55 p.m. At Dietikon local train No. 2837 should have left for Zurich at 6.05 p.m. and having had to make a shunt move was before departure standing on track II, that on which the express as due to pass through without stopping at 6.08 p.m. The departure of the local train was slightly delayed, but it started at 6.07 and was in the process of crossing over to track III, the main line to Zurich, when the express arrived and collided with it on the crossover. The express had been offered to Dietikon a little early and accepted, whereupon Station Assistant Steiner had thoughtlessly cleared the home signal at the Zurich end of the station.

The express engine struck the local train between its engine and leading carriage. The engine of the local train was thrown off the track, its leading carriage demolished and the second badly damaged. The fireman on the engine of the local train, named Brandenberger, had time to jump clear but Driver Jacob Blumer was fatally injured. Four passengers, all on the local train, were killed outright. They were named as Mme Behrli, an office apprentice; M. Henri Ungricht and M. Oefell, house painters and M. Andréas Koller. Six passengers were seriously injured, one of whom, Oscar Lauber, died later.[655] A relief train left Zurich for the accident site at 6.30, carrying doctors and company officials. Some of the injured were taken by motor ambulance to Zurich, others were cared for at Dietikon and the most seriously injured were taken on the relief train to Zurich, arriving there at 8 p.m., and thence to the cantonal hospital.[656]

After the accident trains were diverted over the Oerlikon-Seebach-Wettingen line. Work to clear the line went on all night and normal working was resumed the next morning.[657]

Driver von Felber of the express told the inquiry into the accident that the disc signal at the entrance to Dietikon station had been clear and he had passed it at about 75 kilometres an hour. As his train exited the curve on the approach to the station he had

---

652 *The Times*, London, England, 26th and 31st May and 4th and 25th June 1915

653 Ibid, 16th September 1915

654 *Gretna: Britain's Worst Railway Disaster* (1915) by John Thomas pg. 121

655 *Journal de Genève*, Geneva, Switzerland, 19th August 1915

656 Ibid, 18th August 1915

657 Ibid, 19th August 1915

seen the local train crossing the line in front of him. He had immediately made an emergency brake application and his fireman, named Kamber, had applied the hand brakes as hard as he could. They had not tried to jump and had been buried in coal when the trains collided. According to a newspaper report Fireman Kamber was more badly hurt than his driver but from another report it would seem that both were badly hurt; the passengers on the express all escaped injury.

It was quickly determined that Station Assistant Steiner was alone responsible for the accident. He was arrested and taken to gaol,[658] but was subsequently released under caution.

Station Assistant Steiner came to trial for negligence at the beginning of May 1916. The court took a lenient view of the case, stating that in view of the fact that Steiner had a record as a conscientious employee and to some extent had been the victim of circumstance, they would not sentence him to three years' imprisonment but let him off with a 50-franc fine plus costs.[659] It seems surprising that there was no interlocking to prevent the signal being cleared for the express with the route set for the local train to cross its path.

On Monday 20th September 1915 a military train and a goods train collided at Plüschow, Germany, between Lübeck and Bad Kleinen. Five died and 22 were injured.[660]

The final British railway disaster of 1915 took place on Friday 17th December at St Bedes Junction, England, on the North Eastern Railway, almost opposite, a newspaper reporter noted, the ancient church of the Venerable Bede. The accident occurred early on a foggy morning. The 7.05 a.m. passenger train from South Shields to Newcastle ran into a light engine standing on the line near St Bedes Junction box at 7.20 a.m. Both engines were derailed foul of the opposite running line and were almost immediately run into by the 6.58 a.m. down empty coaching stock train from Hebburn to South Shields. The light engine and the engine of the up passenger train were thrown down the embankment and the two leading carriages of the passenger train were telescoped and caught fire. Eighteen passengers died and their bodies were subsequently consumed by fire; 76 passengers and six railwaymen were injured. One of the injured railwaymen later died as well.[661]

The accounts of two passengers who survived the accident may be noted. 'One of the survivors, William Watson, a Sunderland carpenter, employed on munition work at Palmer's, Jarrow, has given the following story of the accident:- With other workmen he joined the 7.5 at Tyne Dock. The compartments were fairly well filled, and when the train was well on its journey from that station there was a crash. The doors jammed and the passengers could not get out. The first carriage quickly caught fire, and there were screams and groans from the imprisoned passengers. Those who got free from the other compartments ran to the first carriage to render assistance, and they found one man at the door of a blazing compartment. Every effort was made to rescue him, but so fierce was

---

[658] *Gazette de Lausanne*, Lausanne, Switzerland, 20th August 1915

[659] Ibid, 3rd May 1916

[660] *Eisenbahn-Katastrophen in Deutschland* by Hans Joachim Ritzau, pg. 91 (in German)

[661] Board of Trade report

the fire that he had to be left to his fate. Several men were released from the first carriage, but it was impossible to get at others, and before long the second carriage had caught fire and was burning furiously.

'John George Eales, a munition worker living at Sunderland, who travelled in the fourth carriage, has given the following account of his escape:- I had a marvellous escape after the collision. I found myself unable to move. It was pitch dark. I could feel broken wreckage all round me and the men from the other parts of the compartment were groaning fearfully. I struggled in vain to get free from the seat which pinned me. Then the compartment filled with gas and one end burst into flame. It was the end opposite to where I was pinned and I gave up hope of ever getting out alive. The flames attacked the passengers, who shrieked fearfully in their agony. I do not think any of them can have escaped. I was almost overcome with the gas and the heat, and as the fire burst and came closer to me I do not know precisely what happened, but the seats lost their grip upon me and I forced myself free. I climbed out of the roof of the carriage and fell down the embankment, where shortly afterwards I was picked up. Only two other men I know of made their escape from that compartment.' [662]

St Bedes Junction was situated between Jarrow and Tyne Dock stations on the South Shields branch and was the junction for a double-track branch which descended to Tyne Dock Bottom. The signal box also worked a trailing crossover between the main lines. It was customary for goods trains to be banked up the incline from Tyne Dock Bottom to St Bedes Junction, from where the engine returned whence it had come via the down branch line. It was while the engine was waiting at the junction to return down the branch that it was run into.

Acting Fireman Robert Jewett told the Board of Trade inquiry that he had been acting as a fireman for seven months. His engine had joined the 6.50 a.m. goods train from Tyne Dock Bottom in the New Yard without being coupled up to it and had pushed it up the bank. At St Bedes the goods train had just kept going and their engine had come to a stand. Going up the bank there had been a green light on the bunker of the engine which he had then changed for a red one. A down goods train passed, going down the branch, after which his driver had whistled. The signals were then cleared for a train on the down main, and his driver had explained Rule 55 to him, and said that if the signalman did not let them away after that train had passed he had better go to the signal box. A down passenger train passed, the signals were not cleared for them and his driver gave him a hand lamp and told him to proceed to the box. Near the box he heard the up passenger train approaching. Presumably he had shouted to the signalman for he said that the latter had come to the window and asked what was the matter. When he had explained that his engine was waiting to go back to Tyne Dock Bottom the signalman had told him to show a red light to the passenger train and he had done so.

Acting Fireman Jewett did not think that more than five minutes had elapsed before he had set off to the box. He said he had never carried out Rule 55 before and had not been entirely sure what to do. After the accident the signalman had run down the steps and told him that he had not known that there had been an engine at the rear of the goods train.

Signalman William Hodgson said that he had been on the railway 42 years and a signalman 40 years. He testified that he remembered the 6.50 a.m. goods train coming up

[662] *The Times*, London, England, 18th December 1915

the bank and passing his box at 6.55. He saw a red tail lamp on the brake van at the end of the train but nothing of a banking engine. He was sure that it was a red light that he had seen. He knew nothing of any engine coming out onto the main line and did not hear any whistle. A down goods train had gone down the branch at 7.02 and a down main passenger train passed at 7.10. Harton box had offered the up passenger train at 7.08 when he had accepted it. Jarrow box had cleared out the Tyne Dock goods train at 7.11, and accepted the passenger train, when he had cleared his signals for it. At the moment that the fireman had appeared and told him of the engine the up passenger train had been passing his home signal and there was no time to do anything. He had accepted the empty stock from Jarrow at 7.12 and it had been sent on at the same time. He had not lowered his signals for it as the line to Harton was blocked. The second collision came a minute after the first. It had been dark and foggy at the time. Even after the fireman had told him that the engine was on the line he could not see it when he looked from the box. He was sure that he had seen a tail lamp and two side lights at the rear of the goods train. The only way that he knew that there was an assisting engine was by seeing it. The 'engine assisting in rear of train' 2-2 bell code was not used on the branch.

Fireman Frank McArdle stated that he had been on the engine of the 7.05 a.m. passenger train. He had not seen the signals at St Bedes as he had been breaking up coal and his driver had shouted no warning. He did not remember the engine leaving the rails but suddenly found that they were at the foot of the embankment. The collision had taken him completely by surprise. After the accident a soldier had taken him to a house for treatment; it was not until after he had left the house again that he had realised that a down train had also been involved.

Guard Alexander Mitchell of the passenger train said that he had not seen the St Bedes signals because of the fog. His train had been travelling at 25 or 30 miles an hour and its speed had not been checked before he was suddenly thrown from his seat. After the accident he had gone forward to Jarrow to protect his train from that direction; when he returned its front carriages had been on fire.

Guard Robert Taylor of the empty stock train said, like Guard Mitchell, that he had been unable to pick out the St Bedes signals because of the fog. His train had been travelling at 12 or 15 miles an hour when it stopped very suddenly and he was thrown to the floor. After the accident he had met Guard Mitchell who had said that as he, Mitchell, was the younger man and could move more quickly he would protect towards Jarrow. Instead he went down the branch towards Tyne Dock till he met a platelayer who told him that the line was protected from that direction. By the time that he had returned to the junction the fire was burning furiously.

Pilot Guard Michael Vernon said that he had been in charge of the Tyne Dock goods train. He said that arriving at St Bedes his van had had two red side lights but no tail lamp while the pilot engine (banker) had a green light at its rear. The pilot engine had not left the train until they had passed the crossover. Sometimes the pilot engines carried red lights and sometimes green. He had put a red light on his van after they had left the pilot engine behind. He had worked the branch for two years and the 6.50 a.m. train sometimes went up the bank without an assisting engine, but not often.

Driver Samson Tolliday said that he had been riding as a passenger on the 7.05 a.m. train. The train's speed was not checked before there was a very severe jerk. He had been riding in the second compartment of the second carriage. After a second jerk the coach seemed to go to pieces. The doors would not open so he got out through a window and promptly fell down the embankment. Regaining the line he had gone towards the signal box to check that the lines were protected and met a guard (Mitchell) who assured him that the lines were protected and asked if he was an ambulance man. "Expert," he had replied. The guard asked him to attend to the passengers. Driver Tolliday said that first he went to the box to send a message to the Locomotive Superintendent at Tyne Dock to send all the ambulances and doctors possible. Then he set to work with other volunteer passengers to free those trapped in the wreckage. They had released three and were working on a fourth when fire broke out. He obtained some extincteurs and attempted to fight the fire but without success. They went back to trying to free the trapped passengers but the fire made it impossible.

Driver William Hunter stated that he had been driving the banking engine. His engine had not been coupled up. It had had a green light at its rear which he was aware was not correct but had been the usual practice. His engine had been close to the rear of the goods train when they passed St Bedes box. He had stopped the engine at the customary spot. When he had not immediately been signalled back towards Tyne Dock Bottom he had told his fireman to change the green tail lamp to a red one, which was done. A goods train had gone down the branch, and he had then whistled. When the down signals came off again for the main line he had told his fireman that he would have to go to the box to carry out Rule 55. His fireman had asked what that was and he had had to explain it. The fog had suddenly come down thickly and he had thought it best to wait until the down passenger train had passed before sending his fireman to the box. When he had realised that the up passenger train was approaching he had tried to get his engine moving and had just done so when it was struck, throwing him down the embankment and rendering him unconscious. He was aware that the fireman should have immediately gone to the box.

Driver William Smith of the passenger train stated that the St Bedes up distant and home signals had been off for him. His first warning had been seeing what appeared to be the tender of a locomotive about an engine's length away. He had shut off steam and gone to apply the brakes but was not sure they had applied at all before the collision. His engine began to oscillate after the collision and as it finally was coming to rest there was another shock. He was dazed by the accident.

Driver George Taylor said that he had been working the engine of the down empty stock. The St Bedes down distant signal had been at danger and he had been bringing his train to a stand at the home signal when he saw a sudden flash of fire and steam and remembered nothing more. He had come to lying on the embankment.

Fireman Thomas Winter of the stock train was fatally scalded.

Lt-Col P.G. von Donop investigated the accident for the Board of Trade. He concluded that three causes had led to the accident. The first, for which the signalman could not be blamed, was that the 'engine assisting in rear of train' bell signal was not used on the Tyne Dock branch. This was apparently because although assisting engines had been employed for some time on the branch their use had never been officially authorised by the Company. Clearly this bell code needed to be authorised and used.

The second cause was want of care on the part of Signalman Hodgson. He was aware that the engine assisting bell code was not used and that the train was more often than not assisted up the bank. He should therefore have been keeping a careful lookout. The green light on the engine should also have alerted him.

The third cause was the delay in carrying out Rule 55. The engine had been standing for 17 minutes, from 6.55 until 7.12, before the fireman had set off for the box. For this Driver Hunter must be mainly responsible, and he could not accept the driver's excuse that he had been worried about the fireman reaching the box safely, as the fireman could have reached the box without going near any line. Some degree of responsibility also attached to Fireman Jewett as after 3½ years' service and seven months as a fireman he should have been acquainted with Rule 55.

A track circuit would have prevented the accident and although the view from the signal box was good the fact that light engines were frequently left standing ahead of the mains crossover made the provision of one desirable.

Lt-Col von Donop then went on to consider the fire. All the witnesses agreed, he stated, that the fire had not broken out until some minutes after the accident. Statements as to where it had originated differed, but the most reliable evidence was that of Driver Tolliday, who, the inspecting officer added, had shown great resource and energy in attempting to rescue trapped passengers and deserved much credit for the way in which he had risen to the occasion.

It seemed certain that the fire had originated with the ignition of a jet of gas escaping from one of the cylinders of one of the leading vehicles, the gas having then set the woodwork alight. Hot cinders or ashes from the engines had probably set the gas alight. Fires accompanying recent accidents at Witham, Grantham, Hawes Junction and Ditton Junction had been found to have originated from escaping gas, while in other accidents such as Ais Gill and Quintinshill escaping gas had intensified the fire. In the St Bedes Junction accident the role of escaping gas was more definite than in any previous case, and furnished therefore a very conclusive objective lesson of the additional danger in an accident of a train having gas lighting.

The reports on the Ditton Junction, Hawes Junction and Ais Gill accidents in particular had recommended that railway companies substitute electric for gas lighting as quickly as possible. The North Eastern Company had made some progress in the matter but had not, like several other companies, undertaken to fit electric lighting to all new rolling stock, nor had they taken any steps to convert existing gas-lit stock. The company said that it was considering matters such as the provision of safety valves, strengthening of cylinders and using non-flammable materials for coaches, but such expedients could only be regarded as temporary ones, and not as permanent substitutes for conversion to electricity. In accordance with Board of Trade recommendations the North Eastern had also supplied 770 out of a total of 836 brake vans with rescue and fire appliances. In the St Bedes accident the tools had proved very useful although the extincteurs had been ineffective against a well-established gas fire. The company had also made a start in modifying their coaches to prevent telescoping.[663]

The inquest on the victims of the disaster concluded on 25th January 1916, when the jury returned a verdict that there had been neglect by certain officials, but that the negligence was not culpable.[664]

---

[663] Board of Trade report

[664] *The Times*, London, England, 26th January 1916

CHAPTER 6

SINGLE LINE COLLISIONS

The operation of trains in opposite directions over a single line of rails carries the obvious danger of head-on collisions between them. By 1901 there were many hundreds of miles of single track railways throughout the world, especially in places where the density of traffic did not justify the provision of a second line. Most of them were worked with the aid of the electric telegraph, with decisions as to the meetings between trains being either made locally or, as in United States, by a central dispatcher. In Great Britain some lines were worked telegraph supplemented by a staff and paper tickets, while an improved method of single-line control had been introduced 1880 after the invention by Edward Tyer of the electric train tablet system but this and similar methods were not used extensively outside of the British Empire. The invention of a failsafe track circuit in 1872 by the American William Robinson[665] led to the introduction in the United States of automatic block signals, by which train movements on a single line could be safeguarded by signals warning of the approach of an opposing train, but the cost of installing such signals was high, and as late as 1906 the Interstate Commerce Commission was stating that out of 194,727 miles of railroad there were still only 41,916 miles of track protected by them.[666] With the protection of trains running over a single line left solely to the exchange of telegraph messages, rather than by the more secure electric train tablet or its later equivalents, and not supplemented by some form of automatic signalling, collisions due to human error in the transmission or interpretation of telegraph messages became all too common.

Perhaps appropriately, since the country was to have a good many more in the years to come, the first serious single-line collision in the twentieth century happened in the United States, on the Pennsylvania Railroad, at a location variously referred to as Ruslin's Siding, Bordentown and Hamilton, New Jersey. One of the trains involved was the fastest train on the Pennsylvania, known as the *Nellie Bly* after a woman journalist who famously made a round the world trip in 72 days by steamship and railway in 1889.

---

[665] Wikipedia – Track Circuit
[666] Wikipedia – Automatic Block Signalling

The other train in the accident was accommodation (local) train No. 330.

The *Nellie Bly* express from Jersey City to Atlantic City, otherwise train No. 495, usually ran in two sections, but on Thursday 21st February 1901, for the first time, it was run in three sections to cope with heavy Lent holiday traffic and a large contingent of Italian immigrant labourers on their way to work on the construction of a reservoir at Atlantic City. It had left Jersey City at 3.14 p.m. with Engineer Walter Earle and Fireman Michael McGowan and consisted behind the locomotive of a coach, a baggage car, a combination smoker and baggage car, and three coaches. Meanwhile the accommodation train, bound from Camden to Trenton, had left the former place at 3.20 p.m. This train was in charge of Conductor Edward Sapp, with Engineer Benjamin Thompson and Fireman Edward Garwood and consisted of an engine, a combination smoker and baggage car and two coaches.

The accommodation train crossed the first section of the express at Bordentown and here Conductor Sapp was handed a train order reading: 'Trains No.330 and second (2nd) and third (3rd) No. 495 meet at Ruslin's Siding. Train No. 330 will take siding.' Accustomed to only meeting two sections of the express and having, as he had afterwards alleged, covered the part of the order relating to the third section with his thumb (a common excuse used when train orders were misread) Conductor Sapp signalled his train to proceed once they had crossed the second section at Ruslin's.

The trains met on a curve three miles south of Trenton. The express, filled with a mixture of excursionists and labourers, was travelling at high speed when the passengers of both trains heard the grinding of the air brakes, and the crash followed. The cars piled up in all directions and two went over an embankment and into the bed of a canal, which had been drained of water. As soon as everything had stopped those passengers who had not been killed or injured and were not trapped in the wreckage clambered out and many set to work to rescue those trapped and hurt.

A newspaper reporter wrote, 'The most terrifying feature of the wreck was the fire that broke out immediately after the shock of the trains. The stoves were overturned and while the passengers in the rear coaches looked on helpless and miles from aid of any kind, the wretches pinned down beneath the burning cars lay and roasted to death. Others who did not meet death in this manner were crushed outright between the colliding cars, or died on the way to hospitals in Trenton, Camden or Atlantic City.'

An early report on the accident spoke of 11 dead, 5 missing and 20 injured. Two railwaymen were among those killed; Engineer Earle of the express, who was decapitated, and Baggagemaster James Birmingham of the accommodation train. However, 'It was on the 'Nellie Bly' express, in the forward coach between the engine and the baggage car, that the greatest number of fatalities occurred. There, crowded on the seats, were about thirty Italians on their way from New York to begin construction work in Atlantic City. Few of these came out alive, and of the number that escaped death outright, many were so severely injured that they are not expected to live. In this car the scene was sickening. The frail coach was crushed like an eggshell when the engine and baggage car closed in on it, and a moment later it blazed up like tinders when the stoves were overturned. Behind the baggage coach, which climbed the car where the Italians lay pinioned, was a day coach crowded with excursionists on their way to spend Washington's birthday in Atlantic City. This car was turned as if it were a feather and thrown into the canal which runs on both sides of the high embankment. It was crowded with passengers from New York and points on the New York Central, the Belvidere division of the Pennsylvania Railroad and as far west as Michigan and Illinois. Half the passengers were women. The panic here was terrifying, half the women having to be dug out of the windows by members of the train crew, armed with hatchets. There were several casualties in this car, not a few being crushed beneath the car as it fell down the embankment.'

After the accident there were naturally calls for physicians but the rescue effort seems in this respect to have been somewhat disorganised until one of the conductors reportedly walked (more likely, one would have thought, ran) the mile and a half to Bordentown, and for over two hours there was only a lone medical student, who must have been a passenger on one of the trains, to tend the injured.[667]

Brakeman Frank Bolden was one of several of the injured who subsequently died.[668]

The inquest on those killed in the accident was opened a few days later but then postponed until mid-March. At the resumed inquest Dr Shipps of Bordentown testified that he had been advised of the wreck by a man named Fitzpatrick at Bordentown station, and had arrived at the wreck with other Bordentown doctors at 6.15. The chemical engine (fire engine) had come with them but was supposedly not used because railroad men feared that directing a chemical stream upon the wreck might injure some of the trapped victims.[669] However, the next day the inquest heard from P. Brakeley, the foreman of the chemical engine company, who stated that they had set out at 7.30 with the engine loaded on a flat car but had met a train returning from the wreck and had to return to Bordentown to let it pass. That done, they did finally reach the scene, but with only 200 feet of hose could not get close enough to be any use.[670]

Engineer Thompson died of erysipelas on 27th March at the St Francis hospital in Trenton, the 17th fatality and the fourth railroad man to die.

Brakeman Theodore Wright told the inquest on 29th March that Conductor Sapp had remarked to him at Bordentown that they were to wait for two sections of the 'Bly' at the shipyard siding (Ruslin's Siding) and that the conductor had not pulled the bell cord at the siding for the train to proceed but that Engineer Thompson had pulled out onto the main track on his own responsibility.[671]

Despite his brakeman's testimony Conductor Sapp admitted on the final day of the inquest on Thursday 4th April that he had forgotten about the third section of the express, although he had read the order twice. He was not the only one at fault, though, for Fireman Garwood stated that Engineer Thompson had read the order as soon as they were given to him, but he too had failed to appreciate there was a third section. The fireman said that Conductor Sapp had mentioned to him, the fireman, that there was a third section and that he had mentioned it to his engineer but then forgotten about it until he had seen the express coming around the corner. "Here comes the third *Nellie Bly*!" he had exclaimed to Thompson and jumped. Thompson had stuck to his post.

Fireman McGowan of the *Nellie Bly* testified that he had said to Engineer Earle, "My

---

667 *Trenton Times*, Trenton, New Jersey, 22nd February 1901
668 Ibid, 26th February 1901
669 Ibid, 18th March 1901
670 Ibid, 19th March 1901
671 Ibid, 29th March 1901

God, here comes a train!"

Engineer Earle had answered, "Jump and save yourself! Hurry, hurry, hurry!" He had got off as quickly as he could, seeing his engineer applying the brakes and reversing the engine.[672]

The jury predictably found Conductor Sapp and Engineer Thompson responsible for the collision and the conductor was arrested on a charge of involuntary manslaughter. The jury also censured the Pennsylvania company 'for not maintaining a double track with the block system or automatic signals.'[673]

In the end Conductor Sapp did not have to stand trial as in mid-June it was reported that the Grand Jury had refused to indict him.[674]

Shortly after 7.30 a.m. on Wednesday 10th July 1901 the Chicago and Alton Railroad's westbound limited passenger train No. 7 from Chicago to Kansas City, in charge of Conductor Robert Dunsmore, collided head-on with a fast livestock train (the second section of freight train No. 88)[675] on the single track between Marshall and Norton, Missouri, United States. The two engines were completely wrecked and pushed to either side of the track. The baggage and chair cars of the passenger train and the freight cars of the livestock train were piled up on the engines. The dining car of the passenger train overturned and one end of it was telescoped. Steam and scalding water from the boilers of the locomotives poured on the passengers trapped in the wreckage and spilt coals from the engines quickly set the wooden coaches alight. The forward Pullman car and the tourist sleeper in front of it were burned.

The injured passengers were removed from the wreckage as soon as possible but not before many of them had suffered terrible burns and scalds. Few of them suffered broken limbs, most of the injuries being from scalding and the inhalation of steam. Several of those injured were in a more serious condition than an initial examination had indicated so that three died soon after admission to the University hospital and a fourth soon after admission to St Joseph hospital, both in Kansas City. A relief train had left Kansas City at noon and arrived back there with the injured and the bodies of those who had died on the journey at 6 p.m., when the injured were divided between the two hospitals. Six people who had been killed outright were left at Slater, Missouri. By the end of the day there were known to be 16 dead, including Engineer Frank Briggs of the passenger train and Conductor Daniel Mc'Anna, Engineer P. J. Anderson and Fireman Daniel Donnelly of the livestock train.

The passenger train was running in three sections because of many passengers wishing to go to San Francisco for an Epworth League (a Methodist Church youth organisation) convention. The train involved in the collision was the first section, which was carrying mostly ordinary passengers and very few leaguers.

The collision apparently had its origin in an error made by Conductor Mc'Anna of the livestock train. He had received an order to meet the second section of the passenger train

[672] Ibid, 4th April 1901
[673] Ibid, 6th April 1901
[674] Ibid, 11th June 1901
[675] *Fort Wayne Weekly Sentinel*, Fort Wayne, Indiana, United States, 17th July 1901

at Slater, the next station east of Norton, and had overlooked the fact that his train had not yet met the first section of the train, which was running 55 minutes late. According to the front brakeman of the livestock train, who was riding on a car about four back from the engine, it was the conductor himself who had been in charge of the engine between leaving Marshall and the point of the collision, which had happened on a curve and high embankment two miles west of Norton.[676]

By 13th July the number of dead was 21 and when Miss Ora Tallman, of Valparaiso, Indiana, an Epworth Leaguer, died at the university hospital on 21st July the death toll had risen to 23; 19 passengers and four railroad men.[677]

Following the disaster Superintendent Barrett of the Chicago and Alton issued a notice that thereafter every train order must be read by every member of the train crew.[678]

Initial, probably exaggerated, reports spoke of 100 to 150 people having been 'sent into eternity' as a result of a collision on 27th November 1901 on the Wabash Railroad near Seneca, Michigan, United States. The trains involved were No. 4, the eastbound *Continental Limited*, in charge of Conductor George Martin with Engineer Aaron Strong and Fireman Bastien and consisting behind the engine of a baggage car, a combination car and a sleeper; and train No. 13, a double-headed westbound service carrying emigrants and in charge of Conductor Charles Troll with Engineers Samuel Work and Robert Parks. The trains met on a straight stretch of track a mile east of Seneca. Five cars of the emigrant train were demolished, and the combination car of the express was telescoped. The boiler of one of the locomotives of the emigrant train exploded and the engine of the limited overturned into a ditch. As in the Chicago and Alton collision in July the wreckage then caught fire.

A newspaper correspondent wrote, 'It is not known how many people there were on the emigrant train, but the number of dead, injured and burned will be anywhere from sixty to one hundred and fifty. The people that were on the train were caught like rats in a trap and crushed. Then the wreck caught fire and those that were not instantly killed were slowly roasted to death and none of the few spectators who hastily gathered from the farmhouses nearby were able to afford aid. The whole emigrant train was soon consumed by flames and every person on that train it is reported now, was killed. Farmers residing along the track rushed in on the blazing mass to rescue those whom they thought might be alive. The bodies hauled out of the wreck were taken to nearby farmhouses which were filled with dead, and a large number of injured were taken to a hospital at Peru, Ind.

'Along the track, long lines of burned bodies lie covered with blankets, presenting a gruesome sight. It may be possible that the exact number of killed or who they are will never be known, At present it is impossible to get anything resembling a list of injured or dead from Seneca.'

When news of the accident reached Division Superintendent George Burns at Detroit, he ordered the dispatch of special trains to the scene from Adrian, Michigan; Montpelier, Ohio; and Peru, Indiana, while a fourth special train, carrying 32 physicians and

[676] *The Herald Democrat*, Leadville, Colorado, United States, 11th July 1901
[677] *Brooklyn Daily Eagle*, Brooklyn, New York, United States, 22nd July 1901
[678] *The Weekly Pantagraph*, Bloomington, Illinois, United States, 26th July 1901

surgeons, was sent from Detroit. The first relief train to leave the scene with injured was that from Adrian, which set off back to there at 10.45 p.m. Earlier the train from Montpelier, in the opposite direction, had arrived at 9 p.m. but could not get close because the heavy vestibule cars of the limited stood in its path.

The two trains were timetabled to cross at Montpelier but the emigrant train, which was timetabled to leave Chicago at 2.30 p.m., did not get away until 4.20., and an order was issued for the to meet at Seneca, where the limited was due at 6.43. It was stated soon after the accident that Engineer Strong of the limited had misread the order, thinking that it read Sand Creek, a station about five miles further on. Conductor Martin of the special was said to have read the order correctly and thought that his engineer was stopping and by the time that he had realised that he was not, and had applied the air brake, it was too late.[679]

Describing the scene, the morning after the disaster another reporter wrote, 'With the wounded having been taken to the nearest point where adequate medical attendance could be procured, the train crew went on with the work of clearing the track and taking out the bodies. In the fitful glare of torches, coupled with the glow from the red-hot engines and tenders full of coal, still dimly burning, the task was one of extreme difficulty. In the darkness little could be done, and all hope for those left in the wreck being set aside, the crews cleared off the splintered beams of what had once been passenger coaches from the outer edges of the massed confusion.

'Dawn broke, and brightened into a clear, cold day, and as daylight showed up the wreck many features mercifully hidden by the darkness were brought to view. The clearing away of the wreckage, and the taking out of the bodies proceeded apace and by noon the greater portion of the track was cleared, and the men were at work repairing the track for traffic. But upon the banks of the railroad, little heaps covered with blankets, some the size of a man, some merely a small heap, told eloquently of the tragedy that had happened the night before and of the sleeping passengers destined not to see the sunrise.

'Farmers for miles around drove to the collision yesterday morning, in a majority of cases bringing their wives and families with them. After viewing the debris scattered along the track they would walk through the fields to the house where Giovanni Folonaro, one of the emigrants who escaped from the burning cars, was tossing on a bed and muttering in his native tongue. When these spectators arrived at the scene of the holocaust, they saw a bleak and desolate sight. Strewn along both sides of the track for 500 feet were car wheels, brake beams and such other parts of the wrecked coaches as would not burn. Engines 88 and 609 were lying in the ditch on the right-hand side of the road, veritable scrap heaps. Broken cars, many of them daubed with the blood of their last occupants, were scattered about and being used as cushions by those huddled around the open fires about the scene.'

Meanwhile controversy reigned as to the number of victims. The day after the accident the Wabash Company was listing only 10 dead and 48 injured, but, 'The Detroit newspaper men who were at the scene of the wreck and talked with the survivors of both trains say that the official list does not bear out the statements of the passengers nor the evidence of the horrible loss of life at the spot where they witnessed the accident which had occurred.

[679] *Fort Wayne Weekly Sentinel*, Fort Wayne, Indiana, United States, 28th November 1901

'The Free Press today will say that the loss of life was in round numbers 100 and that the statements made by the Italian immigrants in train No. 13 bear out this claim. These immigrants say that there were about 125 of their nationality on the train and that only about 25 of them escaped. While there are only ten of them known to be dead, it is asserted that the other ninety were completely incinerated and that with the removal of the wreckage the bodies crumbled to dust, which was scattered by the wind which swept across the country. If the names of those dead are ever known it will be when friends make inquiry for them, and in many instances, this seems highly improbable.'

Superintendent Burns, however, disputed this and said that he did not think that the total number of dead would exceed 20.

Another newspaper report spoke of the sad scenes at Detroit, 'At the Union station in this city during the night were many sad scenes. Relatives and friends of passengers who were known to have been on the wrecked train besieged the railroad officials with entreaties and inquiries. One little girl waited until 2.30 a.m. for her sister, Vida Degas, of Tupperville, Ont., and was then informed that the sister was among the dead. "Dead! Dead! Dead!" she shrieked, fastening her arms around the operator who had announced the names. The grief of the child exhausted her so that she had to be carried away.

'Job Witchell, of the firm of Witchell, Sons & Co of Detroit, was expected home on No. 4. His son was at the depot to meet him. At midnight a despatch came numbering Job Witchell among the dead. The son was in the waiting room when the news was broken. The terrible reality was too much, and he swooned, falling against a radiator. He was assisted to his seat and remained in that attitude, stunned in the reality. His father had been absent in Dallas, Tex., for some time and had just telegraphed his intention to return on the ill-fated train.'

Engineer Parks on the rear engine of train No. 13 (the emigrant train) stated, 'We had orders to meet train No. 4 at Seneca and were running about thirty miles an hour. The track is straight for several miles and I could see the headlight of No. 4 coming when we were about three miles west of Sand Creek. I put on the air brakes and the shock came before I had time to jump. The three engines all piled up together. I was thrown through the window or roof of the cab. I don't know which. The engine of train No. 4 fell to the south and our two piled up together. I picked myself up on the bank and then another man crawled up to me. It was Engineer Work. Both our firemen must have been killed.'

Engineer Strong, of train No. 4, the limited, stated, 'I read my orders that I was to stop at Sand Creek, and evidently the other crew read theirs to stop at Seneca. I know that there were four of us that read my orders the same-the fireman, the conductor, the brakeman and myself. I lost my orders in the confusion of the wreck, but the records will show what they were and what the other crew's orders were. When I first saw the headlight of the other train, it was far enough away, I presume {that} I could have stopped, but you know how it is with these electric headlights-you cannot tell anything about how far away they are. It was on a straightaway track and I thought they were waiting for me at Sand Creek. They doubtless thought we were stopped at Seneca, and so both of us came on at full speed. We must have been running at fifty miles an hour. When we saw that they were on top of us I saw I must jump. There was a bridge right ahead and I did not want to fall in that, so I waited until we were past it and fell. There was only a slight embankment and the crash came right after I let go. When I came to there was my engine on one side of me and the trucks of the second car, which was telescoped, on the other.'

Superintendent Burns confirmed that it was Engineer Strong who had been at fault. The relevant order, which was delivered to train No. 4 at Montpelier, Ohio, and to train No. 13 at Holloway, ten miles east of Sand Creek, read:

"Order No. 82, trains Nos. 13 and 3 – No. 4, engine 609, will meet No. 13 at Seneca, and No. 3, engine 623, at Sand Creek. Engines Nos. 88 and 151 will double-head No. 13." [680]

Conductor Martin made a statement a couple of days after the accident. 'My orders, which I took to Engineer Strong and read over to him, and which agreed with his orders, read, 'No. 4, engine 609, will meet No. 13, engines 88 and 151, and No. 3, engine 623, at Sand Creek. After receiving these orders, I remember asking the agent at Montpelier whether the new siding at Seneca would contain my train. I was toward the rear of the chair car, fixing up a passenger's ticket, when I saw the station lights of Seneca flash by and immediately felt the air-brake applied. I felt no alarm, supposing the engineer was slowing down for the east switch, in order to see the numbers of 13's engines, which I supposed had cleared. Just then the brakeman, who had been on the lookout, came hurrying into the chair car and pulled the cord twice. "What is it Tony?" I called. "We're going through Seneca," he answered, and an instant later the crash came.'

Confronted with the conductor's statement Engineer Strong refused to admit that he alone was to blame. He stated, 'I have not denied at any time that we received such an order; I only say that if we did get such an order I misread and misunderstood it and I should have been set right by the conductor. My fireman will tell you that he, too, understood that we were to meet both trains at Sand Creek instead of at Seneca. If I made an error I am not alone to blame. The other three members of the crew must have understood as I did or else, they would have corrected me.' [681]

At the resumed inquest on the victims of the accident on 3rd December Engineer Strong came up with the excuse that as he had been reading the train order the wind had curled up the edge of the sheet of paper and obscured the word Seneca – a variation on the more usual excuse of the thumb of the reader obscuring the name.

At the conclusion of the inquest on 4th December the jury found Engineer Strong negligent for failing to understand and obey the train order; Conductor Martin negligent for failing to apply the air brake; Fireman Bastien negligent for failing to alert his engineer when the latter did not give the proper signal and slow up at Seneca; Brakeman A. W. Dittman negligent for failing to signal the engineer to stop and applying the air brake when the train passed Seneca; and the Wabash Company negligent for failing to provide a head brakeman and proper signals at Seneca.[682]

At about 11.28 p.m. on the night of Saturday 14th December 1901 eastbound Illinois Central Railroad passenger train No. 4 collided head-on with a westbound freight train first No. 51 between Perryville and Irene, Illinois, United States. The passenger train should have been in Chicago at 9.40 p.m. but was running about three hours late. The freight train was due in Rockford, Illinois, just west of the scene of the accident, at 12.36

---

680 *The Evening Gazette*, Burlington, Iowa, United States, 29th November 1901

681 *Logansport Reporter*, Logansport, Indiana, United States, 3rd December 1901

682 *The Brookfield Gazette*, Brookfield, Missouri, United States, 7th December 1901

a.m. and hence was running early. Both trains were running fast when they met.

Passenger train No. 4 was in charge of Conductor J. H. Quinlan with Engineer Richard Ormsby and Fireman James Reardon and consisted behind the locomotive of a combination baggage and express car, a smoker, day coach, parlor car and diner. The freight train was in charge of Conductor William Osten with Engineer David Behan and Fireman Edward Casey.

The collision of the two trains was very violent and was reportedly immediately followed by the explosion of one of the engine boilers and a minute later by the explosion of an oil car which had formed part of the freight train. As a result, either of the initial impact or of one of the explosions a casting was thrown 100 feet while a roof of one of the passenger cars was hurled into a field some distance from the track.

To make matters worse, 'Fire followed immediately, and aided by the ignited oil furrowed its way through the wreckage, reducing it to ashes in less than an hour. By heroic efforts the way car alone was pushed back from the short string of freight cars left standing on the rails and was saved from the fire. The injured were hatless and coatless. The wounded, with broken bones, bleeding from deep cuts and gashes, with limbs dismembered or partially so were stretched out on the frozen ground. The fortunate ones divested themselves of their open clothing in order to wrap the wounded. The smoking, express and baggage cars were piled up on top of the locomotives, penning in the occupants of the smoker. Of the half-dozen passengers in the smoker only three escaped. The three victims, if not immediately killed, were roasted to death, and their bodies, with those of the engine crews, were entirely consumed.'

The freezing conditions proved an ordeal for the survivors. 'The temperature stood at 20 degrees below zero. The bitter cold wind swept over the prairie surrounding the wreck. No protection was afforded the injured or those who escaped. Only the caboose was saved from the fire. The wounded were placed in this and two hours later, when a switch engine arrived, they were hurried three miles east to Irene, where first aid was afforded them. A wrecking train left Rockford at 1.40 a.m. yesterday morning carrying a number of physicians and a railroad wrecking crew. This relief train was met by the engine and caboose. The other survivors were taken into Rockford a few hours later. There attaches of the hospitals and physicians who were not available when the relief train departed were in readiness to aid the injured.'

Some passengers were very lucky. 'Rev P.E. Henson, pastor of the First Baptist church, had a narrow escape from death and was subjected to the most thrilling experience of his life. After being reported to his wife as dead, Dr Henson reached the city {Chicago} just in time to reach his pulpit and preach at the evening service. "I was in the parlor car," said Dr Henson. "The crash came without warning. In fact, it seemed to be a succession of shocks, three in all. The cars surged like a boat on water during a storm. Every chair was torn from the floor and the pieces flew everywhere. I was thrown over several chairs, but fortunately sustained only a few bruises and a severe shaking up. I assisted two women from the car and we quickly saw it was useless to attempt any further work at rescue. The explosion of the oil car threw great sprays of flame over the entire mass of debris and the flames fought back the fortunate ones as they endeavoured to lend assistance to the maimed and injured who were crying for assistance. We returned to Rockford, where the injured were well cared for. It was a terrible experience and a night of horror and suffering for the uninjured as well as for the less fortunate.

"'I'm alive," gasped Bernhard R. Ahrendt, one of the Chicago survivors of the wreck. As he staggered into his mother's store at 669 West North avenue yesterday noon and fainted on the threshold from the fatigue and nervous strain he had been compelled to endure. He has no remembrance, he says, how he got out of the hospital at Rockford, boarded the train, and on his arrival at Rockford found his way out to the northwest side of the city. When he had been revived in the afternoon, he told his family this story of his experience:

"'I left Waterloo, Iowa, yesterday on No. 4 which was four hours late. I could not get a berth in the sleeper, so I rode in the chair car. I had asked the porter for some pillows to make myself comfortable for the rest of the night and was just placing them in position when the crash came. I was knocked senseless for a moment and the first I knew the steam from the burst pipes was enveloping me. I broke the window with my fist and climbed out, cutting and bruising myself badly.

"'A fellow passenger who sat in the chair next to me jumped out after me and landed on his head. The wreck caught fire almost immediately, and in ten minutes was reduced almost to ashes. The oil from the tank car and the high wind that prevailed helped the blaze. There were a great many miraculous escapes. A young woman with her baby jumped out of one of the cars and escaped without a scratch. We were all obliged to stand out in the bitter cold for two hours waiting for the relief train. I was taken to the Rockford hospital and my injuries were dressed and then I got out and came home, but I don't know how I got here. It all seems like a terrible dream.'" [683]

The final number of dead was probably 13. Five of them were railroad men; Engineers Ormsby and Behan, their Firemen Reardon and Carey and Brakeman J. W. Funk of the passenger train. Of the others seven passengers were confirmed dead and a Frank Rideout of Paxton, who had boarded the train at Freeport, was 'missing'.

On 17th December Illinois Central Division Superintendent Horace Baker issued a statement in which he gave his opinion as to how the accident had occurred. He stated:

"On Saturday night, December 14, passenger train No. 4 received at Freeport order No. 71, reading: 'No. 4, engine 227, will run three hours, fifty minutes late, Freeport to Chicago.' This order was also delivered to the first section of freight train No. 51, engine No. 29, at Hawthorne.

"By the time card train No. 4 was due at Rockford at 7.03 p.m., at East Rockford at 7.08 p.m., at Perryville at 7.17 p.m., at Irene at 7.23 p.m., at Colvin Park at 7. 33 p.m., and at Hart at 7.40 p.m. On order No. 71, which was the only order held by train No. 4, that train could not pass Rockford before 10.53, East Rockford 10.58, Perryville 11.07, Irene 11.13, Colvin Park 11.23 and Hart 11.30.

"As shown by the registers and time sheet, train No. 4 left Freeport at 10.27 p.m., and Rockford at 11.10 p.m., seventeen minutes later than the time at which it could have left under the order No. 71

"The first section of train No. 51, engine 29, left Colvin Park at 11.04 p.m. Colvin Park is as far as the first section of train No. 51 should have gone under the order. The accident occurred according to the best information we can get, between 11.25 and 11.30 p.m. The accident was due to the first section of train No. 51 having used more time than

---

[683] *Waterloo Daily Courier*, Waterloo, Iowa, United States, 16th December 1901

it received. The first information we received of the accident came at 12.04 a.m. from Irene."

Also, on 17th December Conductor Osten, suffering badly from frostbite, gave evidence at the inquest on the victims of the accident. He stated at he had said to his engineer that they might be able to reach Rockford. At Irene he had looked at his watch. It had been 10.57 and he had figured that No. 4 would reach Perryville at 11.11. His train had been going at 50 miles an hour and should have covered the five miles to Perryville in five or six minutes, giving them fully five minutes to make the siding there. The head brakeman had been back in the caboose eating his lunch and he had been in the cupola when the flash of No. 4's headlight had appeared in the frost covered windows only a second or so before the crash.

Testimony was also given by Operator Herman Reimer, of Rockford, who produced his train register with the signature of Conductor Quinlan of train No. 4 timed at 11.10. Operator Reimer stated that that time was correct both by the depot clock and by his own timepiece.

Conductor Quinlan was badly injured as his testimony had to be taken in hospital. He simply produced his orders from Freeport, showing that he had left there four hours late, and testified that he had registered in Rockford at 11.05.

The newspaper reported that Conductor Osten was evidently in error in stating that his train should have been able to reach the siding at Perryville by 11.02 as they had in fact not been reported out of Colvin Park until 11.04, and it was 12 miles from there to Perryville. It seemed likely that he and Engineer Behan had somehow miscalculated and imagined that they could reach Rockford before train No. 4 left there. The inquest jury agreed, and their verdict found that the responsibility for the accident lay with the crew of the freight train.[684]

In one of several bad railway accidents which happened during the Second Boer War but which were not the result of Boer sabotage, on 12th April 1902 troops of the 8th New Zealand Mounted Infantry were involved in a single-line collision near Machavie, in the Transvaal. On the day of the accident an empty eastbound goods train from Klerksdorp to Johannesburg was due to cross two westbound troop trains at Machavie. The Machavie acting stationmaster was said to have been young and inexperienced. One goods train arrived, and there was also a ration train standing in the station yard which the acting stationmaster mistook for the second troop train. He duly despatched the empty goods train towards Potchefstroom. Shortly afterwards he realised his error and telephoned to the next blockhouse to try and stop the second troop train, but it was too late.

The troop train was descending an incline and the empty train was rounding a curve when they met. The drivers saw one another, braked and then with their firemen jumped off. The trains reportedly dashed into one another at a speed of 12 to 15 miles an hour. In the collision the first truck of the troop train was smashed to pieces and the second truck completely overturned. After the accident a Corporal Hunter, of Invercargill, New Zealand, who had been riding in the second truck and escaped injury, mounted a horse and rode to Potchefstroom for assistance. On the way he passed more than one

[684] *Dubuque Telegraph Herald*, Dubuque, Iowa, United States, 18th December 1901

blockhouse at which he was mistaken for a Boer and fired upon, but he reached Potchefstroom unscathed and gave the alarm, whereby an ambulance train as despatched to the scene.

Nine soldiers were killed outright, four died shortly afterwards and 14 were injured, three of whom died from their injuries later. Driver Robert Conchie and Fireman Robert Margison were also hurt (which engine they were on was not stated). The injured soldiers were taken to Potchefstroom Hospital while the bodies of the dead were wrapped in blankets and conveyed to Klerksdorp. It was 8 a.m. when the accident happened, and it took until 5 p.m. to clear the line. Learning the consequences of his mistake the Machavie acting stationmaster attempted to shoot himself but was prevented and placed under arrest.[685] The dead soldiers were buried at Klerksdorp in an impressive funeral, with a procession led by a firing party, pipers of the Seaforth Highlanders playing *The Land o' the Leal* and a band playing the *Dead March*.[686]

Boxing Day, Friday 26th December 1902, saw a major disaster in Canada, on the Sarnia branch of the Grand Trunk Railway from Toronto and Hamilton to Sarnia, Windsor and the United States. Train No. 5, the westbound Chicago express collided head-on with an eastbound freight train at Wanstead, Ontario. Blame for the accident rested between a despatcher and an operator.

On the evening of 26th December Despatcher James Kerr in London, Ontario, had to arrange a routine meet between the express and the freight train. The initial plan was for the two trains to pass one another at Wyoming but the express was behind time, running an hour late because of the heavy Christmas traffic and a blizzard affecting the area through which it was running. Despatcher Kerr therefore decided to arrange for the trains to cross at Wanstead, one station to the east of Wyoming, and at 9.48 p.m. he sent order No. 93 to the station agents at Wyoming and Watford, Watford being the second station east of Wanstead. This order read: "Order No. 93 No. 5 Engine 980, and extra east Engine 773 will meet at Wanstead." The order was duly repeated back to Kerr by the station agents at Watford and Wyoming and acknowledged by him. Up to this point the statements of the three men involved, and more particularly of Despatcher Kerr and Operator Andrew Carson at Watford as to what transpired agreed. It would seem that having sent this order Despatcher Kerr had misgivings as to the wisdom of having the already late Chicago express wait for the freight at Wanstead. Quite apart from the fact that it was already well behind time it had District Supervisor W. E. Costello as a passenger and were it to suffer further unnecessary delay the despatcher would be likely to be asked to explain his actions. Unfortunately, Despatcher Kerr did not just let these thoughts run through his head but communicated them via the telegraph to Station Agent Carson. What exactly was telegraphed was afterwards a matter of dispute between the two men. The despatcher maintained that he had simply mentioned to Carson that he *might* annul the order that he had transmitted and let the express proceed to Wyoming. According to Carson on the other hand Kerr had telegraphed, "Bust the order."

At about this time the express reached Watford and Conductor John McAuliffe went into the office to see what orders there were, as the train order board had been against his

[685] *Otago Witness*, Otago, New Zealand, 28th May 1902
[686] *The Times*, London, England 15th and 18th April 1902

train. Operator Carson told the conductor that there had been an order, but that it had been “busted”. He then gave the conductor clearance to proceed and the express set off at 10 p.m., expecting to cross the freight at Wyoming. Meanwhile the despatcher had been on the telegraph to Wyoming asking if the freight was about ready to go.

“Yes, it’s all ready,” Operator J. A. Lamonte had replied. “Shall I stop it?”

Deciding that there would be time for the freight to reach Wanstead after all and not risk delaying the express Kerr told Lamonte, “No, let her go.” The despatcher then realised that he had still not been advised by Watford that the crew of the express had received order No. 93. He contacted Watford to be told that the express had just left and that order 93 had not been delivered because the despatcher had said that it was “busted”.

Precious seconds were wasted by the despatcher and the operator engaging in a futile argument before Despatcher Kerr thought to telegraph Wyoming to stop the freight. The operator there said that the freight was just pulling out. He ran out onto the platform and waved a red lantern at it. A rear brakeman spotted the signal through the falling snow and signalled for the train to stop, thinking that perhaps the conductor had been left behind. Conductor J. A. Graham indicated to the brakeman that he was safely aboard. The train went on.

There was still another chance to save the day for between Watford and Wanstead and two miles short of the latter was the small station of Kingscourt Junction. The despatcher desperately telegraphed there but for seven or eight minutes he could get no reply. On duty at Kingscourt was 17-year-old Operator James Troyer, doing only his third day in the service and his first night duty. He was conscientiously studying the timetable and with little more than a rudimentary knowledge of the telegraph. The machine had been busy for several minutes as Kerr contacted Watford and Wyoming and it was some time before Troyer realised that it was now his station, code “KC”, which was being called. He finally answered it at 10.08, just as the express was passing through at 50 miles an hour, too late for him to put his semaphore signal to danger.

Even then the accident might have been avoided for the freight train reached Wanstead just before the express. As it neared the end of the siding Brakeman James Smith jumped off and ran forward to operate the points to let the train into the loop. The deep snow slowed him up, however, and when he reached the switch, he found that it was frozen. Anticipating that the brakeman would have had time to operate the points Engineer James Gillies of the freight train failed to pull up before he reached them and overshooting the switch by half a car’s length had to back up again. By now the brakeman had got the points to work but just as the freight engine was moving forward again the Chicago express was upon them. Brakeman Smith vainly tried to flag it down. It was 10.10.

Engineer MacKenzie on the express glimpsed the headlight of the freight engine through the snow but assumed that it was on the siding, only discovering too late that it was on the main track. Conductor Graham and the fireman of the freight train managed to jump clear in time, but Engineer Gillies of the freight and Fireman Arthur Ricketts of the express were amongst those who lost their lives. The leading baggage car of the express and the following day coach, packed with passengers, were telescoped into one another, the floor of the baggage car slicing through the day coach at an angle so that all the fatalities were on one side of the coach. Baggagemaster Alexander Caulfield, sat in the baggage car reading a newspaper, survived with painful but not serious injuries. Twenty-six passengers were less lucky and were killed.

Despatcher Kerr, waiting and hoping against hope that the freight train had been able to get into the siding at Wanstead in time, instead received a telegraph message from there from District Supervisor Costello telling him to send auxiliary trains out. Trains were sent from London and Sarnia.

Costello had reached the Wanstead office at the same moment as the off-duty Wanstead operator, Wesley Young, who had witnessed the collision from his house across the road from the station.[687]

Fire broke out in the wreckage but was extinguished before it could properly take hold by Conductor McAuliffe, using fire extinguishers from the two Pullman sleeping cars, assisted by a passenger, A. C. Clarke of Sarnia. Other passengers threw snow onto the flames. The Pullman sleepers were then turned into first-aid stations and two doctors and three passengers riding on the train did what they could to treat the injured. As in the Perryville accident the survivors suffered much from the cold.

The most serious damage done to the freight train was the destruction of two cars of chilled beef, which was heaped into a useless mass, though a number of other freight cars were wrecked.

A passenger described the accident. 'We were running about forty miles an hour, when without the slightest warning the two trains met with terrific force. On examination it was found that the two engines were in the ditch. The baggage car was thrown on top of the first-class coach, instantly killing a great many and pinning about fifty other passengers in the debris. The screams, moans and prayers of the injured were heart-rending. One poor woman begged that her child be saved as she was dying. The little one was carefully taken from the wreck and will probably recover. The mother was afterward released, but only to die in a few minutes.' [688]

The inquest on the victims of the accident dealt specifically with the death of Fireman Ricketts, he being, of course, taken as representative of all the victims. At 11.40 p.m. on Tuesday 30th December, after hours of deliberation, the jury returned a verdict that: 'We find that Arthur W. Ricketts was killed in the collision at Wanstead on the evening of December 26, 1902; that said collision was caused by wrong orders being given No. 5 at Watford, and the responsibility for this issuance of wrong orders we are not agreed upon as between Operator Carson and Despatcher Kerr.

'That after No. 5 had left Watford by the issuance of wrong orders, we consider that the accident could have been averted by the operator at Wyoming or Kingscourt Junction, had the railway company had more experienced operators at these points, one being but a boy of sixteen, at each of these places the despatcher having ample time to do it, endeavoured to get the opposing trains stopped.'

In the face of the conflicting testimonies the inability of the jury to decide who was to blame is probably not surprising, though really both operator and despatcher were at fault. Operator Carson should not have taken such a vital step as annulling an order merely on the basis of an informal use of the words, "Bust it," if indeed such a telegram

---

687 *Marquette Reporter*, Rapid City, Manitoba, Canada, 1st January 1903

688 *Manitoba Free Press*, Winnipeg, Manitoba, Canada, 29th December 1902

was sent, which the despatcher strongly denied. Carson claimed that it was not the first time that an order had been cancelled so informally, something which the despatcher also denied. On the other hand, there was no justification for Despatcher Kerr informing the operator of what he was merely considering doing and thus confusing him, and of course he was Carson's superior, and possibly intimidated him.[689]

At 2.53 a.m. on Wednesday 28th January 1903 two express trains collided near Esmond, Arizona, United States, on the Southern Pacific Railroad. The trains involved were No. 9, the westbound *Sunset Limited*, and the eastbound *Crescent City Express*. The latter was normally designated as train No. 8 but due to a wreck in Texas it was running no less than 12 hours late. The schedule of train No. 8 had therefore been annulled and instead it was running as the first section of train No. 7. At Vails an order should have been delivered to train No. 9 to meet train first No. 7 at Esmond, three miles further west of Vails. Conductor G. W. Parker of train No. 9 went into the office at Vails but failed to pick up the order. Shortly after the express had left Night Operator George Clough saw what had happened and rushed to the door but the train was already some distance away. Reportedly a little later he could see the light of the fire which had broken out in the wreckage and wired the dispatcher's office in Tucson that there had been a wreck. Another source stated that the first news which Tucson had of the wreck was the arrival of the rear Pullman sleeper of train first No. 7, which the shock of the collision had detached, and which had rolled back down the bank for 15 miles to the yard at Tucson, where it collided with a switch engine, severely damaging both.[690]

On news of the accident being received in Tucson a hospital and wrecking train was quickly made up and hurried to the scene, carrying four physicians and undertakers (!) 'When the relief train reached the wreck the sight which met their gaze was horrifying. The entire wreck was ablaze, and the cries of the unfortunate victims pinned in the burning cars could still be heard. The injured were scattered along the side of the wreck and the passengers who escaped from the wreck in their night-robes were assisting the less fortunate to places where they could be safe from the fire. Day dawned shortly after the relief train reached the scene of the wreck. The work of rescue began in earnest. At daybreak a large gang was put to work building a shoofly around the burning wreck and all trains were held at Tucson and Benson, the relief train making frequent trips back and forth to Tucson with the dead and injured.' [691]

John Harmons, of Caribou, Maine, a passenger on the Sunset Limited, was interviewed later in a Bakersfield, California, hotel, and described the accident. 'I was in a tourist sleeper, the fourth car from the engine, and when the collision occurred and fortunately for myself and perhaps for others I kept my head. Everything was in confusion in an instant, women screaming and rushing in terror. I mounted on the seat and warned everybody to keep cool. The car was thrown off the track and caught fire a few minutes after the collision. Some of those who were nearest the doors succeeded in getting out there. I waited too long, looking for my dressing case, and when I started for the door the car was already ablaze in parts and so filled and choked with smoke that I

---

[689] Ibid, 30th and 31st December 1902

[690] *The Weekly Gazette*, Colorado Springs, Colorado, United States, 29th January 1903

[691] *Sedalia Weekly Sentinel*, Sedalia, Missouri, United States, 30th January 1903

could not find my way. I was completely lost. After groping around in the dark a minute or two I managed to find a window and, kicking out the glass, crawled out to safety. I was fortunate in not receiving any injuries beyond cuts and bruises, which were painful enough but not serious. There was a relief train from Tucson early on the scene and another passenger train from the east was only an hour behind us, so we received attention as early as could be expected under the circumstances, as the scene of the accident was on the desert far from any town.

'The fire was caused by the explosion of the gas generators and started scarcely two minutes after the collision. When we left the scene of the wreck coming to Los Angeles it was still burning and so hot as to make it impossible to do much towards clearing it away. I understand that there had been twenty-four taken out dead then and there were many more in the ruins. Quite a number were hurt were hurt and killed in the car where I was. Some were caught in the wreck and burned to death.' [692]

Immediately after the accident Operator Clough blamed Conductor Parker. Clough then disappeared and could not be found. It soon emerged that in fact the blame lay not with the conductor but with Operator Clough himself. He had been sent two orders to deliver to train No. 9, the first to meet train first No. 7 at Esmond and a second order to meet an eastbound freight train at Wilmot, the station beyond Esmond. These were both "31" orders, which had to be signed by the conductor or another member of the train crew in the operator's office after having been read back to him and the signing of the order then reported to the dispatcher. The order regarding the Esmond meet was received first and Operator Clough succumbed to the dangerous temptation of telegraphing it back to the dispatcher as having been signed by the conductor before the train had even reached Vails. Then, when he received the second order relating to the Wilmot meet, he quite forgot about the first one. Conductor Parker signed for the second order, knowing nothing of the first, and the trap was set.

By the evening of the day of the accident 20 bodies had been recovered, including those of Engineer Jack Bruce and Fireman Joe McGrath of train No. 8 and Engineer R. Wilkie of train first No. 7. One of the passengers was known to be capitalist J. M. Hilton of Cambridge, Massachusetts; 16 others were unidentified.[693] Two days later the number of known dead had risen to 38, including four coloured dining car porters and three trespassers, while there were 45 injured. Of those killed only six had been identified, so charred were the bodies.[694]

On 31st January it was reported that Tucson Divisional Superintendent Stoufe had completely exonerated Conductors Parker and Schriven and Engineers Bruce and Wilkie from responsibility for the collision and placed the entire blame for it upon Operator Clough.[695]

In mid-February the inquest jury at Tucson, Arizona, found Operator Clough solely responsible for the wreck. His whereabouts remained a mystery. The jury added that more of the victims might have been identified had souvenir fiends not looted the

---

692 *The Daily Californian*, Bakersfield, California, United States, 30th January 1903
693 *Daily Times*, Dubuque, Iowa, United States, 29th January 1903
694 *Defiance Crescent-News*, Defiance, Ohio, United States, 4th February 1903
695 *The Brookfield Gazette*, Brookfield, Missouri, United States, 31st January 1903

wreck.[696]

Flood damage, which had completely blocked the main line of the Atchison Topeka and Santa Fe Railroad, was the indirect cause of a collision on the Missouri Pacific Railroad near Stilwell, Kansas, United States, on Thursday 4th June 1903. Santa Fe trains were being temporarily diverted over the Missouri Pacific tracks and the trains involved in the collision both belonged to former company.

According to W. J. Black, the Santa Fe passenger agent, the northbound train concerned in the accident, designated as the first section of train No. 2, had been made up at Ottawa, Kansas, from trains from the west and south. It was running through to Chicago and was carrying 180 passengers.

The southbound train, special No. 301, in charge of Conductor A. J. Rader with Engineer H. A. Pierson, had left Kansas City at 11 a.m. The trains were due to cross at Stilwell but Conductor Patton, in charge of the northbound train, allegedly misread his orders and the train failed to stop there. A mile north of the station it met the southbound one. Nine persons were killed outright and 28 more were injured, 10 it was feared fatally. All of those killed had been on the northbound train.[697] Those who lost their lives included Pullman car conductor Frank Johnson and two tramps riding on the bumpers of a baggage car.

Newspaper reports on the accident were even more confused and conflicting than was usually the case, for example another report identified the northbound train as the third section of train No. 4 and the southbound as train No. 1, also stating that all the fatalities except those of the tramps had occurred to passengers in the chair car of the southbound train.[698]

The mistake which led to a collision at Raymond, Iowa, United States on the Illinois Central Railroad at 2.40 a.m. on Friday 19th June 1903 was quickly identified. The trains involved were the first section of eastbound passenger train No. 2, in charge of Conductor Thomas Quinlan with Engineer John Griffen and Fireman Daniel Bantz; and westbound freight train No. 87, in charge of Conductor Charles Judd, with Engineer Frederick Stoneman and Fireman Irwin Mills. Passenger train No. 2 normally originated as two trains starting from Sioux City. Iowa and Omaha, Nebraska, which were combined into one train at Fort Dodge, Iowa for the run to Waterloo, Iowa, where they were combined with a train from Minneapolis, Minnesota, from where they continued eastwards as one section, arriving in Dubuque, Iowa, at 5 a.m. On Thursday evening, however, the Fort Dodge train arrived in Waterloo on time, but the Minneapolis train was about an hour and three-quarters late. The Fort Dodge train was therefore allowed to leave Waterloo on time and without orders relating to it, while the Chief Dispatcher at Dubuque sent an order that the second section was running an hour and 30 minutes late. Herein lay the cause of the accident. The order was delivered to the freight train at Independence, Iowa. Both Conductor Judd and Engineer Stoneman somehow misread the order as stating that train

[696] *Brainerd Daily Despatch*, Brainerd, Minnesota, United States, 13th February 1903

[697] *The Iola Daily Register*, Iola, Kansas, United States, 5th June 1903

[698] *The Eau Claire Leader*, EAU Claire, Wisconsin, United States, 6th June 1903

No. 2 (which did not normally run in two sections) was an hour and a half late, thus giving them ample time to reach Waterloo in front of it.

When the trains collided at Raymond both were going at high speed, as the place was located in a dip approached by falling gradients from both sides. The trains met with a terrific crash and both engine No. 1008 of the passenger train and No. 44 of the freight train were damaged beyond repair, as were the baggage and mail cars of the passenger train and seven cars of the freight train. The remaining vehicles of the passenger train, two day coaches and two sleepers, were only slightly damaged and only three passengers were injured. Killed in the accident were both engineers, both firemen, Brakeman Erwin Stickney of the freight train, Mail Clerk F. L. Groom, emigrants W. H. Ricketts and Frank Kline of Epworth,[699] and three men who had been stealing a ride. Two of the latter gave their names as John O'Neil, of Syracuse, New York and Earl Book, of New York City. They were trapped in the wreckage and smothered before they could be released. Apart from the passengers two mail clerks and two tramps, Charles Moran and John Hurt, were injured, the latter very seriously.[700]

The trains met 30 car lengths west of the depot at Raymond. The latter was a day station only but the operator, who lived close to the station, was naturally awakened by the noise of the collision and hurried to his office, from where he notified the divisional headquarters at Dubuque and also Waterloo. Wreck trains were rushed from both places with physicians on board. That from Waterloo was the first to arrive and the injured were taken on it back to a hospital at Waterloo. The work of clearing the line then began and was reported to be a stupendous task.

Conductor Judd afterwards made a statement fully admitting his fatal mistake. "I didn't notice the fact that the train was running in two sections and neither did Engineer Stoneman. We both received copies of the same order at Independence and had No. 2 been running in one section as we supposed we would have been in Waterloo long before it left there."

Conductor Quinlan of the passenger train also made a statement. His train finally reached Dubuque at 11.30 a.m. with a new baggage car and engine crew, having reversed to Waterloo and thence travelled over the Chicago Great Western to Dyersville before regaining Illinois Central tracks there. Conductor Quinlan stated, "We were just out of Waterloo and were running along smoothly when the crash came. I was in one of the coaches getting the passengers ready to alight at Independence, and surprising as it may seem the concussion was not as great as one would expect after seeing the wreckage. The passengers did not become greatly alarmed and there was little or no confusion, many not knowing what had happened and others merely thinking that we had made a sudden stop. I immediately went through the train to see whether or not any of the passengers had been hurt, for as soon as the crash came, I knew there had been a collision. Some of the passengers in the sleepers were nor awakened by the impact, which surprised me very much. I did not know what the cause of the wreck at this time was, for I was certain that we were running alright, and I knew just where we were, as the engineer had whistled for Raymond." The conductor went on at length to describe the rescue operation.[701]

---

[699] *Waterloo Daily Courier*. Waterloo, Iowa, United States, 19th June 1903

[700] *Cedar Rapids Republican*, Cedar Rapids, Iowa, United States, 20th June 1903

[701] *Dubuque Telegraph-Herald*, Dubuque, Iowa, United States, 19th June 1903.

The inquest on the victims of the accident was concluded the day after the accident, the jury merely finding that the collision had occurred 'because the conductor, engineer, brakeman and fireman on said freight train No. 87 mistakenly misread the order they had received for the moving and handling of said train.'

A Waterloo newspaper reported that 'Thousands from this city and neighboring towns viewed the scene of the Illinois Central wreck near Raymond yesterday morning. During the day the liverymen and hackmen of the city were taxed to the utmost of their establishments to supply those who wished to visit the wreck. Every rig was engaged for another trip before it had returned from the scene.' Speaking of itself, the newspaper added, 'The *Daily Courier* owes its thanks to Frank Skinner, the newspaper artist who is at present visiting in the city. He was roused out of bed in the early hours of the morning, taken to the scene of the catastrophe, and a number of good sketches were hastily made by him which were later converted into chalk plates for readers of the *Courier*. Besides the drawings made a number of Kodak views were taken of the wreck. In addition to the regular edition of the *Courier* yesterday, a 1,000 extra copies were printed, all of which were disposed of in a short time. Out of town news dealers placed many orders for from 25 to 100 copies, all of which were supplied. Several hundred were sold at the scene of the wreck last evening.'

Despite the magnitude of the clearance operation the line was back in use by 6 p.m. on the day of the accident.[702]

In late August 1903 the city of Udine, Italy, was reportedly celebrating a rare visit from King Victor Emmanuel III and Queen Elena but the celebrations were cut short by news of a serious accident on the main line to Treviso and Venice. At 22:30 on Thursday 27th August 1903 a troop train coming from Udine collided head-on with a freight train between Pasian Schiavonesco (Basiliano from 1923) and Codroipo stations. The two engines met with an immense noise and several carriages were piled up upon one another and derailed.

After the accident railwaymen from the nearest station and uninjured soldiers began the rescue work and a special train was sent from Udine with doctors and medical equipment, arriving at the scene of the accident about midnight, and a second relief train was despatched at 02:00. Fifteen soldiers and a captain of the 14th Regiment were killed outright and by the following morning 34 soldiers and nine officers, including a Colonel Bona,[703] had been admitted to hospital at Udine. The driver and fireman of the troop train also lost their lives, but the engine crew of the freight train seem to have escaped, or at least were not listed among the casualties.

The King and Queen, who by then were at Treviso, went to Udine on the Friday afternoon to visit the injured in hospital, the king shaking hands with all the injured soldiers, as did General Ottolenghi, the Minister of War, and Archbishop Zamburlini. Meanwhile Stationmaster Léonard Valenti at Pasian and Conductor Bisolfi of the troop train were arrested. The newspaper report commented, however, that the accident was probably due rather to the extra work caused by the concentration of troops than by

[702] *Waterloo Daily Courier*, Waterloo, Iowa, United States, 20th June 1903

[703] *The Times*, London, England, 29th August 1903

neglect.[704] Italy was not involved in any wars in 1903, other than a naval blockade of Venezuela, so the reason for the concentration of troops is unclear, probably they were just routine military exercises.

An investigation into the accident found that it was the fault of the driver of the troop train, who had been completely intoxicated, and had driven through Pasian station, where he was supposed to stop, at high speed. As he was killed he had paid for his mistake with his life.[705]

A wreck near East Paris, Michigan, United States, on Boxing Day, Saturday 26th December 1903, was, according to General Superintendent Smith of the Pere Marquette Railroad Company, the first occasion in which a passenger, as opposed to an employee or a trespasser, had been killed in an accident to one of the company's trains. Severe weather conditions, with a blizzard and gale-force winds, were apparently the main cause of the accident.

The trains involved were westbound passenger train No. 5, in charge of Conductor Neill, with Engineer Frank Waterman and Fireman Moon and eastbound passenger train No. 6, in charge of Conductor Edward Weigel with Engineer Charles Stoddard and Fireman Burns. The westbound train was reportedly running down grade at 60 miles an hour, while the eastbound was labouring up the hill at 40 miles an hour, with a raging blizzard greatly restricting the visibility of the two engine crews.

Describing the collision, a newspaper reporter wrote, 'The westbound engine drove through engine No. 6 like a wedge, separating the gear from the boiler and standing the latter on end like a broken, twisted shell. The running gear was crumpled up like so much twine. Great cast steel rods were bent into many shapes and even the steel tires of the big driving wheels were split and sprung from the wheels, landing in some cases ren feet from the engine. The boiler of engine No. 297 turned over several times, and the very force of the twisted impact seems to have been the means of saving the life of Engineer Waterman. He was flung forty feet over the fence at the edge of the right of way. Moon, his fireman, was sitting on the opposite side of the cab, and he too, was thrown clear of the pile of wreckage.

'No. 6, eastbound, was composed of a mail and baggage car, combination, smoker, buffet parlor car, day coach and diner. The baggage car caught between the engine and heavy train, crumpled up like pasteboard, only the roof remaining partially unbroken. The big blind mail coach, heavily bulkheaded, took the brunt of the blow, but withstood the crash and communicated it to the sleeper next in the rear. The smoker was swept clean, as the mail car telescoped its entire length and turned over at the north side of the track. The smoker in turn crashed several feet into the parlor car, which was at the end of the smoker, and forward end of the cafe car, where the passengers were killed and injured. From a space six feet square the bodies of two women, a boy and a man were taken out.

'The westbound flyer, with its heavy bulkheaded cars, stayed on the rails, back of the baggage and mail cars, and the passengers in this train escaped serious injury. As one man they rushed out of the train to assist the less fortunate ones on the train from this city

---

704 *Journal de Genève*, Geneva, Switzerland, 29th and 30th August 1903
705 *Gazette de Lausanne*, Lausanne, Switzerland, 31st August 1903

{Grand Rapids, Michigan}. Many stripped warm overcoats from their backs and covered the suffering injured as they were borne to the less shattered cars of No. 5 and into the remains of the parlor car of No. 6. Nearby farmers, hearing the crash, turned out with blankets and did valiant service.'

The day after the accident it was stated that 22 people had died and another 38 were injured, several probably fatally. Those killed included Engineer Stoddard, Fireman Burns and Baggagemen Austin Wager and William Helmrich of train No. 5.

Three relief trains were sent to the scene of the disaster. The first returned with the injured and the bodies of five of the dead which could be extricated without wrecking equipment. The second included the tool car and steam crane, and within 20 minutes its wrecking crew had dug out eight bodies, each frozen stiff in the bitter cold. The dead were taken to the morgue at Grand Rapids, where just before noon a third relief train came in with the bodies of Engineer Stoddard and Fireman Burns, who had been found beneath the wreckage of their locomotive.

The accident came about because train No. 5 failed to stop at McCords, where Operator Booth stated emphatically that as soon as he had received orders to hold No. 5 he had placed the train order board to danger, showing a red light, and that the lamp had been burning brightly only two minutes before the train arrived. He stated that he had heard the rumble of the approaching train and when he realised that its brakes were not being applied, he had rushed to the door of his office to flag it just as its rear coach was going past. Engineer Waterman on the other hand, supported to a degree by his fireman and the conductor, maintained that the signal lamp was burning but showing a white light, and that had no light been displayed he would have stopped. There were three white switch lights at McCords, but these were three car lengths from the order light and placed considerably lower down. Engineer Waterman did not think that he could have mistaken one for the other. Operator Booth had been with the Pere Marquette for 35 years and had never had an accident before.

General Superintendent Smith outlined what he thought had gone wrong. "As stated last night, from the information I had then, it appeared the collision resulted from the fact that the high wind had blown out the order board light and before the arrival at McCords of train No. 5 and too late for the operator on duty at McCords to flag No. 5. This morning I talked with the engineer, fireman and conductor of train No. 5. They are positive that the order lamp did not show red and the engineer, on whose side the order board is, was positive that it did show white, for the reason that he would have stopped it there had been no signal light burning.

"The only explanation I can see is that the engineer on account of the high wind and the snow storm mistook the light on the switch target for what he supposed to be the white light of the order board and that the order lamp was actually not burning.

"A meet order was given to the train to meet at Oakdale Park and one of the trains was delayed more than expected and the meet order was changed to read Fox, which is between Oakdale Park and McCords. The orders were issued in the regular form provided for by the rules an in accordance with the rules promulgated by the American Railway Association.

"As soon as the operator at McCords found that No. 5 had passed he called on the dispatcher at Plymouth who in turn called up the operator at Grand Rapids yard and the yard master at Grand Rapids, F. Neubecker, who happened to be in the office, realized the difficulty and called up the postmaster at East Paris by 'phone to try to flag No. 6, but all too late." [706]

Two points come to mind in relation to this accident. One is the continued use of white as an all clear signal as late as 1903. Clearly there were not yet that many other external light sources which might be mistaken for a clear signal. The other is the sole reliance on the train order signal, even when visibility was much reduced. It seems surprising that the regulations did not require the operator to put down torpedoes in such conditions to provide an additional layer of protection.

On 27th December 1903 No. 15 down passenger and No. 80 up goods train met on the single line at mile 533-9 between Sini and Gomharria, India, on the Bengal – Nagpur Railway. Both engines and five carriages were smashed, 12 passengers and three railway servants were killed, and nine passengers and six railwaymen were injured. The goods train driver had left Gomharria on an 'incomplete conditional line clear ticket.[707]

Less than a fortnight after the East Paris disaster there was a collision at Willard, Kansas, United States on the Chicago Rock Island and Pacific Railroad. The trains involved in this collision, just after 1 a.m. on Wednesday 6th January 1904, were the California and Mexico express which had left Chicago at 11.30 p.m. on Monday and Kansas City at 10 p.m. on Tuesday, and a special stock train. The express consisted behind its engine of a combination baggage and mail car, a baggage car, smoker, tourist sleeper and standard sleeper, and was in charge of Conductor J. C. Nagle with Engineer Benjamin.[708] The trains were due to meet one another at Willard. When the passenger train reached there, Engineer Benjamin saw a train standing in the siding. It was theorised that assuming that this was the stock train referred to in his train order, the engineer had opened his throttle and pressed on. Had he taken care to check the number of the freight engine with that given in his train order for the stock train he would doubtless have realised his mistake. His train was declared to have accelerated to fully 65 miles an hour when the two trains met, though another report stated that it had been travelling at 40 miles an hour, which seems a more likely speed. That a wreck was likely to occur was known by the Topeka dispatcher eight minutes before it happened when the passenger train was reported to have passed Willard and the stock train to have gone through the next station at Maple Hill.

With the wreck occurring in the darkness in an isolated location there was considerable delay in caring for the injured and getting out the bodies of the dead, but a newspaper reporter recorded that there had by chance been a doctor on the express. 'The hero of the wreck was a Dr Bell, of New York city, a young physician, himself crippled, and walking on crutches. Dr Bell, although slightly hurt in the collision, was the first man to leave the Pullman sleeper, which was not damaged. Struggling heroically forward on his crutches, he immediately assumed charge of the rescue work. First, he ordered the chair car and sleeper cleared. Then, after directing the removal of the injured to these

[706] *The Herald Democrat*, Leadville, Colorado, United States, 28th December 1903

[707] Administration Report on the Railways in India for the year 1903

[708] *The Morning Herald*, Lexington, Kentucky, United States, 7th January 1904

cars, he allayed their pain as much as possible without medicine or instruments, bandaged broken legs and arms with strips torn, at his direction, by others, from sheets and pillow slips, administered what whiskey there was at hand to deaden the pain until the relief train should arrive, and saved the life of the fireman by tying an artery with the aid of a penknife and a piece of string.

'After working till nearly exhausted Dr Bell only gave way when the physicians arrived from Topeka with medicine and instruments. Then he steadfastly refused to disclose his name and it was only partially learned from his fellow passengers. Dr Bell was on his way to El Paso, Texas, for his health. Soon after returning to Topeka from the scene of the wreck he boarded another train for the south, refusing either to reveal his full name and address, or to submit to an interview.

'The relief train hurrying back to Topeka with the injured, left the dead in the wreck, to be cared for later.' [709]

Seventeen persons were initially reportedly to have died in the disaster with 37 others injured, all but one of whom were expected to recover.[710]

A day or so after the accident Conductor Nagle made a deposition from his hospital bed. "We had orders to wait for a special stock train at Valencia until 12.58 and to pass another stock train at Willard. We were late and passed through Valencia at 1 a.m. and no stock train was in sight. At Willard we saw a stock train on the siding and made no stop. I suppose the engineer thought the two stock trains had been combined. I could not see the number of the engine on the sidetrack on account of the steam and because we were running so fast. We were going 50 miles an hour. Engineer Benjamin and I had duplicate orders."

The doctors would not at this time permit Engineer Benjamin to be interviewed in view of his condition.[711] A few days later, though, he did give an interview, in which he said, "I was to wait until 12.55 a.m. at Valencia for the extra east (the stock train with which we collided) but we were late, and the train not being there we proceeded west. At Willard we passed 489., as we were ordered to do, and ran on until we met the extra stock train in the collision. I would not undertake to say where the blame for this wreck lies. It is not for me to say. My belief was that I was doing everything in the line of duty." [712]

It seems possible that Engineer Benjamin was under the impression that there was only one stock train to be crossed instead of two.

Simple human forgetfulness was the cause of a collision near Sand Point, Ontario, Canada, at 4.53 a.m. on Tuesday 9th February 1904. Two Canadian Pacific Railway expresses were the protagonists; train No. 7 westbound and train No. 8 eastbound. Train No. 7 was in charge of Conductor James Nidd with Engineer John Dudley and Fireman Ernest Dubois and had passed through Ottawa over an hour late. It had been issued with clear orders that it was to meet train No. 8 at Sand Point, a siding 60 miles beyond Ottawa and between five and six miles west of Arnprior. When train No. 7 reached

[709] *The Advertiser-Courier*, Hermann, Missouri, United States, 13th January 1904
[710] *Austin Daily Herald*, Austin, Minnesota, United States, 7th January 1904
[711] *The Pulaski County Democrat*, Waynesville, Missouri, United States, 15th January 1904
[712] *The Chillicothe Constitution*, Chillicothe, Missouri, United States, 14th January 1904

Arnprior Engineer Dudley was having difficulty maintaining steam pressure and distracted by this problem he clean forgot about the meet with train No. 8 at Sand Point. When he failed to stop there Conductor Nidd or Brakeman Lamour should have signalled for him to stop but they apparently forgot about the meet as well. Fireman Dubois perhaps had more excuse in view of the poor running of the engine.

Train No. 8 was in charge of Conductor Forester with Engineer Joseph Jackson and Fireman Pike and was also running late. The trains met on a curve. Train No. 7 was travelling at 30 miles an hour, train No. 8 was probably going faster. Despite the terrific impact neither of the trains was derailed, but the baggage cars of each were telescoped into the tenders of their engines and a second-class coach of train No. 7 telescoped into the baggage car. This coach was filled with shantymen[713] (lumbermen), seven of whom lost their lives. Engineer Dudley probably jumped but his fireman was fatally injured and died on the train taking him to Ottawa. Engineer Jackson stuck to his post, applying the brakes, reversing his engine and whistling. He was killed outright, while his fireman was badly hurt. Also killed were Baggageman John O'Toole, Express Messengers Nelson Robertson and Robert Thompson and News Agent W. McMullen. Engineer Dudley, Fireman Pike, Mail Clerks Beach and Black and about 20 passengers were injured. By chance a Catholic bishop was travelling on one train and a Catholic priest on the other; the two went among the casualties administering the last rites to the dying.

After the accident the operator at Sand Point was roused from his sleep and telegraphed a message to the divisional superintendent at Ottawa, Mr H. B. Spencer.[714] A special was quickly made up and sent off and wrecking trains were sent from Carleton and elsewhere. Doctors in Arnprior and Renfrew were summoned and later an ambulance train was despatched. The rescue work was hampered by the terrible cold and a lack of tools. The injured were taken by train to Arnprior.[715] The inquest on the victims of the disaster finally concluded on 19th March. The jury returned a verdict that, "From the evidence submitted we find that the said collision was caused by the temporary loss of memory of John Dudley, driver of the engine of No. 7 train, in running past Sand Point station, at which station said driver had orders to cross No. 8 train, and that in the opinion of the jury Conductor Nidd, Brakeman Lamour and Fireman Dubois were in a measure to blame in not protecting said train, No. 7, on the morning of the collision. It is the opinion of the jury that the Canadian Pacific Railway company should not have issued crossing orders at such a distance as they did in this case and that a collision might have been averted had a red Nunn signal been displayed at Sand Point." [716]

The Nunn signal mentioned by the jury was simply a form of train order signal set by the operator.[717] As there was no night operator at Sand Point anyway it is hard to see how this would have made any difference. On the other hand, the rider about train orders not being delivered so far away from where they applied seems eminently sensible, unless there were no night stations open closer to Sand Point.

---

[713] *Manitoba Free Press*, Winnipeg, Manitoba, Canada, 10th February 1904
[714] *The Citizen*, Ottawa, Canada, 10th February 1904
[715] *Wreck!* by Hugh H. Halliday, pgs. 109-112.
[716] *Manitoba Free Press*, Winnipeg, Manitoba, Canada, 21st March 1904
[717] Webpage – Railroadiana: Adams and Westlake catalog 1907

Canada suffered another bad single-line collision on Wednesday 31st August 1904, this time on the Grand Trunk Railway near Richmond, Quebec. The trains involved were a special excursion running from Montreal to Sherbrooke Fair, and passenger train No. 5, a regular working from Island Point, Vermont, United States, to Montreal. The excursion train was running as the first section of the Portland express, which normally crossed the Island Point train at Richmond, but Conductor Norman Atkinson failed to hold the excursion train at Richmond until train No. 5 had arrived. He reportedly disappeared after the accident.

The excursion train consisted behind its engine of a baggage car and 10 coaches, into which were crammed about 1,000 excursionists, while train No. 5 consisted only of an engine, a baggage car and five coaches and was only carrying a few passengers. The excursion train had only just cleared Richmond yard and was rounding a curve when the Island Point train was met, running at high speed. Both engineers reversed their locomotives and then along with their firemen jumped, all four escaping with slight injuries. A mile away in Richmond the sound of the collision was plainly audible. The two engines were locked together as a result of the collision with the baggage car of the excursion train reared up and telescoped the following smoker. It was in these two the latter that most of the casualties occurred. Nine people were killed outright and 23 hurt, including Engineer Paquette of the Island Point train. [718] One of those who lost their lives was J. E. Blanchet, Liberal M.P. for St Hyacinthe.[719] The number of passengers who died later rose to 10.

Another report theorised that Conductor Atkinson had become confused between the Island Point train and another train and believed that train No. 5 had arrived. Two trains bearing the same number apparently arrived at the station on the same track at about the same time. As the excursion train began to pull out Operator Charles Withrow shouted to the conductor that he must wait for train No. 5. The conductor nodded as if he understood but the train kept going. Station officials and passengers stood on the platform, anticipating disaster, followed the train and found that the inevitable had happened only a mile down the line.

A reporter wrote, 'The two engines were a mass of scrap. The engine of the excursion train lay half through a barbed wire fence while the engine of the other train lay at the same angle but was far more seriously damaged than the other.

'News of the accident soon spread, and the townspeople and uninjured passengers hurried to the assistance of those pinned beneath the wreckage. The worst sight was the wreck of the excursion train. The baggage car of this train had telescoped the first passenger coach. The baggage car was lifted clear of its trucks and jammed completely inside the passenger coach to within three feet of its rear. In this coach most of the dead and injured were found, and it is a wonder that a single one escaped alive. It required more than half an hour for the wrecking apparatus to render possible the removal of the dead and injured. To do so the side of the car had to be cut open. Then the horrible sight was laid bare, and many of the spectators were badly affected.

'While the gang of men and passengers were clearing away the debris of the special train, assistance was also rendered to the local train. The smoking car of the local train

718 *The Times Dispatch*, Richmond, Virginia, United States, 1st September 1904
719 *Durham Review*, Durham, Ontario, Canada, 8th September 1904

was telescoped by the mail coach, and the three who were killed in this train were riding in the former car. Of the passengers in the other coaches of this train very few were injured, the full weight of the impact having fallen on the excursion train.' [720]

On 9th September a coroner's inquest at Richmond found Conductor Atkinson and Engineer Frank Schwager of the excursion train responsible for the accident but not criminally negligent. The jury also recommended that the Grand Trunk company be more careful in the handling of train orders. [721]

It was the Southern Railway of the United States which featured in an accident on Saturday 24th September 1904, near Hodges, Tennessee. Westbound passenger train No. 15 from Bristol to Knoxville, due in Knoxville at 11 a.m., had received orders at Morristown to meet train No. 12 from Chattanooga to Salisbury, North Carolina at New Market, 17 miles west of Morristown. No other trains were mentioned in the order, the crew of train No. 15 were experienced, and it was midmorning on a clear day. Nevertheless train No. 15 failed to stop at New Market and proceeding towards Knoxville was a mile and a half beyond Hodges when it collided with train No. 12.

Train No. 15 was a local one, with only three cars, and carrying 140 passengers mostly bound for Knoxville for shopping or to attend a fair there. Train No. 12, known as the *Carolina Special,* was a heavier one, consisting two mail cars, three wooden day coaches and four steel Pullman cars. It was carrying 210 passengers, many bound for the World's Fair at St Louis, Missouri.

When it was realised that train No. 15 had left New Market prematurely telegraph messages were sent to both Hodges, the point where the two trains normally met, and to the next station at Strawberry Fields. For some reason there was no one at Hodges to answer the message, and it went unanswered. Train No. 12 was just leaving Strawberry Fields when the message was received there, and efforts made to stop it were unsuccessful. According to one account train No. 15 was running at 70 miles an hour down grade while train No. 12 was travelling at 60 miles an hour. At any rate the collision was an extremely violent one, the boilers of both engines reportedly exploding and fire breaking out in the wreckage.[722] Even so, the only fatalities on the local train were Engineer William Kane and his fireman. On train No. 12 Engineer Richard Parrott and his fireman were also killed, together with at least 62 passengers, the high number of deaths being due to the heavy Pullman steel cars crushing the lighter day coaches in front of them. One account of the accident puts the number of dead as high as 113 but this, like the alleged speed of the trains, seems well too high.[723] About 160 others were injured. One of those to escape with relatively slight injuries from train No. 12 was Congressman Henry R. Gibson, of Knoxville.

A wrecking and relief train left Knoxville as soon as possible after news had been received there of the disaster. Meanwhile countless local people had gone to the scene upon hearing the sound of the crash and did what they could to assist. Jessica Whittaker, the sister of Joseph Whittaker, on whose farm the accident had happened, tore up all the

---

720 *Manitoba Free Press*, Winnipeg, Manitoba, Canada, 1st September 1904

721 *The Daily Colonist*, Victoria, British Columbia, Canada, 10th September 1904

722 *The Evening News*, Cumberland, Maryland, United States, 26th September 1904

723 Webpage: History of the New Market Train Wreck

fine fabric from her hope chest for use as bandages. The dead and injured were taken by train to Knoxville, the injured being taken to the hospital there or to their homes or those of friends. A local undertaker received an order to have 150 cots ready.[724] All day Sunday 150 men toiled to clear the line.[725]

Why train No. 15 failed to wait at New Market, or even at the usual crossing place for the two trains at Hodges, has never been explained. One theory was that Engineer Kane of the local train had fallen asleep, though as the collision occurred at 10.08 a.m. this hardly seems likely. The conductor of the train, whose name does not seem to have been recorded in any newspaper report, was of course equally at fault. At least one of the firemen. James Mills was coloured. It is not clear which train he was firing on; if it was train No. 15 he might well have been reluctant to query the actions of his white engineer.

The St Louis World's Fair, which featured in the Hodges accident, was also a factor in a collision near Warrensburg, Missouri, United States, on the Missouri Pacific Railroad, on Monday 10th October 1904. The accident occurred at about 4.10 a.m. between Warrensburg and Montserrat and involved an eastbound passenger train carrying passengers bound from Kansas to the World's Fair and a westbound freight train. The passenger train was very crowded, with passengers standing in the aisles and crowded into the vestibules. Most significantly, as far as the number of casualties was concerned, there was no baggage car behind the locomotive, instead the leading vehicle was a well-filled passenger car. This was because the heavy train had had to be divided at Pleasant Hill because of its weight and the baggage car had gone forward with the first section, while passenger train No. 50 was interspersed between the two sections, though the first section was carrying the proper markers to indicate that another section was following.[726] The second section was a short train, consisting only of two coaches, a Pullman car and a freight caboose, which because of the lack of space was also filled the passengers. Reportedly the passenger train was travelling at 60 miles an hour when the trains met. The engine of the freight train demolished the passenger engine in the collision and went on to plough three quarters of the way through the leading coach. The only slightly fortunate feature of the accident was that the impact burst the water tank of the passenger engine and the escaping water put out the fire of the locomotive, thus preventing the wreckage catching fire, though passengers were scalded by escaping steam.

Immediately after the accident trainmen ran to Warrensburg to raise the alarm but it was almost two hours before any relief reached the scene. Meanwhile trainmen and uninjured passengers did what they could to tend the injured. 'It seemed almost impossible for them to them to know where to begin first, as everywhere were cries for assistance and heart-rending appeals for aid. Most of the dead and injured were in the first coach, next to the engine, and they were hemmed in by the wreckage and seats until it seemed almost impossible to extricate them, but with undaunted courage and tremendous energy the work was pursued and with axes ways were hewed for the escape of the wounded and the bringing out of the dead. Many were pinned beneath the debris and it required herculean efforts to rescue them.'

---

[724] *The Boston Sunday Globe*, Boston, Massachusetts, United States, 25th September 1904
[725] *The Evening Herald*, Montpelier, Indiana, United States, 27th September 1904
[726] *Moberly Daily Monitor*, Moberly, Missouri, United States, 11th October 1904

By daybreak large numbers of local residents were at the wreck, having been notified in some cases by telephone calls. At 7 a.m. almost all the dead and injured had been removed from the wreckage and laid out along the right of way, the bodies of the dead being covered with whatever was at hand and the injured being administered to. Many of the victims were children. Relief trains arrived from Sedalia in the east and from the west at about the same time. The bodies of the dead were placed on the train for Warrensburg and the injuries on the train for Sedalia, though more than one of the injured died before the Sedalia train left the scene and the body was transferred to the Warrensburg train. By the evening of the day of the accident 28 passengers and one railroad employee were known to have lost their lives.

'Percy N. Allen, of Coffeyville, who was among the injured, having his left leg very badly mashed, was in the first car, where nearly all those who were killed and injured were. When talking to a Sentinel reporter he said that at the time that the two trains came he was asleep in the first car and was awakened by the fearful crash. The dead were piled all around him and over him and his legs were pinioned by the wreckage and it was about thirty minutes before he could be extricated and then only by having a hole cut in the side of the car and being pulled out of his dangerous position. He said that the thirty minutes that passed while confined in the car were the most terrible that he had ever endured and were most terrible to contemplate, he being covered by the dead and dying, many of them dying within the time before his rescue was accomplished. Mr Allen, although badly injured, was perhaps the happiest man on the train coming into Sedalia and expressed himself over and over again in thankfulness for his miraculous escape from instant death.

'Little Ruth Stewart, aged 10 years, a pretty little girl, who was accompanying her uncle and aunt to see the World's fair, was among the very bravest of those who were injured. Although this little girl was confined in the wreck for over two hours after it had occurred and was only rescued by tremendous endeavour, she kept up her courage through it all and was an example for those even older and of more mature years. Her fortitude and sweetness under the distressing circumstances won the admiration and love of everybody on board the relief train coming into Sedalia. There were many on the train who would have been only too glad to have taken the little girl home with them. Her aunt was badly injured and died a few moments after being taken from the wreck, and her uncle is one of the worst injured now in the hospital. The little girl will probably be sent to her home in Independence, Kas.'

In charge of the passenger train was Conductor Charles Barnes, with Engineer Ed Rosson, Fireman W. W. Young and Brakemen P. G. Gray, C. H. Bright and C. J. McCabe. The freight train was in charge of Conductor J. W. Preston with Engineer John Horton, Fireman E. G. Dawson and Brakemen W. Seidle, Eli Zeigler and L. L. Haynes. Brakeman Seidle was killed and Engineer Horton and Fireman Young were injured.

Engineer Horton told a reporter, "Well, I suppose it was our mistake. We knew we had orders to meet the second thirty at Knob Noster, and I would swear even now that she went through, and when I thought she had gone I whistled for off breaks and left the station. Then the collision came. It looks to me like as if we made the mistake."

Engineer Rosson also made a brief statement. "I had no time to jump. I simply applied the brakes and the crash came. The force of the impact threw me off the engine and along the side of the track." [727]

The inquest on those who had died in the accident was concluded already the day after it happened. Evidence was offered that Conductor Preston and the two brakemen of the freight train had fallen asleep while their train was in the siding at Knob Noster. The conductor stated that when Engineer Horton had indicated that it was in order to proceed, he had asked the engineer twice if all the opposing trains had passed and had been assured that they had. When they reached Montserrat, the conductor said that he had checked for danger signals from the rear caboose but seeing none had allowed his train to proceed. His train had left Jefferson City that morning and Sedalia at 11.50 p.m. with orders to run extra to Pleasant Hill, taking the siding at Montserrat for trains Nos. 9, 19, 30 and 50, but his train had developed a 'hot-box', hence the need to stop over at Knob Noster. Conductor Preston had been warned that evidence he gave might be used against him but stated that he wanted to tell the whole story. Night Operator A. T. Fisher at Warrensburg stated that he had had no orders for the passenger train, but it had stopped there for a minute to drop off three passengers. A minute after he had notified the dispatcher of its departure the latter had asked him to go out onto the platform and see if he could still see the lights of the train.[728] The Montserrat night operator, W. W. McVey, stated that the dispatcher at Sedalia had ordered out the wrecking crews, section men and doctors before the collision had taken place. This was doubtless because McVey had reported the passing of the freight train and Warrensburg the passing of the passenger train.[729] The jury found the crew of the freight train responsible for the collision but did not recommend their prosecution. County Prosecuting Attorney Morrow stated nevertheless that he would begin such a prosecution. In addition, Brakemen Haynes and Zeigler were charged with stealing from the body of their fellow brakeman and Mrs Julia Wood, of Oxford, Kansas, died of her injuries, making a thirtieth fatality.[730]

Prosecuting Attorney Morrow subsequently filed an affidavit charging Conductor Preston and Engineer Horton only with fourth-degree manslaughter.[731] It is not clear if the case ever came to trial.

A simple mistake by an operator led to a collision near Azusa, Wyoming, United States, on the Oregon Short Line Railroad, shortly before midnight on Friday 11th November 1904. The train order involved was delivered to a freight train by the Operator J. E. Miller at Granger and should have given the eastbound freight train 30 minutes to reach Azusa and there cross westbound passenger train No. 3. Unfortunately, the night operator delivered what was referred to as a "bulled" order; the order as delivered reading 'fifty' instead of 'thirty' so that the crew of the freight train were under the impression that they had 20 minutes more than was actually the case to reach Azusa and was hurrying there, rather than waiting at Granger for the passenger. Even so the freight train almost made it, but the trains collided at speed a mile and a half short of Azusa. It was supposed that it must have been somewhat foggy as the trains met in the middle of a straight stretch of line several miles in length and the passenger engine at any rate was

727 *Sedalia Evening Sentinel*, Sedalia, Missouri, United States, 10th October 1904
728 Ibid, 11th October 1904
729 *The Mitchell Capital*, Mitchell, South Dakota, United States, 14th October 1904
730 *The Herald Democrat*, Leadville, Colorado, United States, 12th October 1904
731 *Cynthiana Argus*, Cynthiana, Indiana, United States, 19th November 1904

equipped with an electric headlight. Both engines were demolished, while in the passenger train the mail and baggage cars were telescoped and the day coach badly damaged, though the Pullman cars did not leave the rails. Six refrigerator cars from California, loaded with fruit, raisins and nuts were also piled up in a heap of wreckage.[732] Fire reportedly broke out in the wreckage.[733] Initial reports on the accident spoke variously of nine people killed but the final number of dead was probably 12. Those killed were Engineers Ben Eccles, Conductor Joseph Lowham and Brakeman John Redmond of the freight train; Engineer William Murray and Fireman William Combstock of the passenger train, Express Messenger J. J. Winslow, Mail Clerk H.M. Sherman, Car Repairer S. S. Epperson and passengers Roy Chamberlain, James Frew, Carl McNeil (a private in Company K, U.S. Cavalry) and B. Tucker. A passenger named Martin Samuelson spoke of seeing a woman and a child killed a few feet from where he sat but their bodies had not been recovered. Several others who were injured were taken to hospital at Rock Springs, among them Fireman Fred Wilson of the freight train, whom it was feared was fatally hurt.[734]

There was apparently almost a double collision as passenger train No. 3 was being closely followed by train No. 5, whose engineer was only just able to pull up when he saw the wreckage blocking the line in front of him, finally bringing his engine to a stop when its cowcatcher was within 20 feet of the obstruction.[735]

Probably because it happened in one of the sparsely populated Mountain states the accident did not receive much coverage in the national newspapers, but a Salt Lake City, Utah, reporter managed to locate a couple of passengers to interview. 'Arthur Howell and his sister, Mary C. Howell, of Washington Court House, O, now at the Wilson hotel, were in the Azusa railroad wreck, and barely escaped with their lives. Miss Howell was found this morning at the hotel and stated that she and her brother were on their way to the Pacific coast, and at the hour of the accident were occupying seats at the rear of the chair car. Like a bolt of lightning, at one minute before midnight, there was a crash, a roar, and a rush, as if of escaping steam, the front end of the car seemed to rise in the air and swerve to one side, and then all was darkness and a horrible confusion. Mr Howell's first thought was to save his sister, and at the instant of the shock he threw himself in front of her to save her from threatening injury. In doing this he was caught in the dislocated chairs and his jaw was knocked out of place, besides suffering bruises and several cuts, but fortunately, none of these were serious and he is able to get around. Miss Howell's feet were at the time caught under the chair in front of her, but her brother's prompt action enabled her to escape being pinned down in the wreckage of the chairs, with the probable loss, as she believes, of one of her legs. The chair car was not badly broken, though the fixtures inside were thrown all over the cars, and passengers and light luggage and chair fixtures were all in confusion. The car in front was smashed to pieces, and its occupants killed or injured as already detailed in the press accounts.

'Miss Howell said as the two trains came together, the engines rose right up in the air, and then fell over to one side, and four cars on the freight, and three cars on the passenger train, were crushed like eggshells. Immediately after the collision, as the passengers and

[732] *The Morning Echo*, Bakersfield, California, United States, 13th November 1904
[733] *The Cheyenne Daily Leader*, Cheyenne, Wyoming, United States, 13th November 1904
[734] Ibid, 15th November 1904
[735] Ibid, 16th November 1904

uninjured trainmen recovered their senses, big bonfires were lighted and active efforts made to succour the wounded and rescue those in danger of further injury. Miss Howell looked after her brother's injuries and then took an active hand in doing everything a woman could do on such an occasion and made herself generally useful. She says the dead were dreadfully mangled and crushed, so that it was sickening to look at them. The special with the surgeons was on the spot as soon as it was possible to get them there, but it was 17 hours before traffic could be resumed. She does not believe there were any persons left under the wreckage, as has been reported, and to her best knowledge and belief the responsibility of the dreadful accident lay with the crew of the passenger train rather than the freight, as the passenger was I hour and 50 minutes late, and its crew misread the order relative to meeting the eastbound extra at the siding.

'Mr and Miss Howell were brought on to this city yesterday afternoon, and they will remain at the Wilson for a week, as they are much pleased with the town, and wish to see the city and its surroundings, before continuing on to Los Angeles, their destination.'

According to a passenger on a following train the load of the freight train was scattered all over the track, and Italian labourers who were busy clearing up the mess quickly gathered up the nuts and loading their pockets with them 'made quite a meal'. [736]

The subsequent inquest placed most of the blame on Operator Miller, who fled after the accident, allegedly boarding a westbound freight train, though it had been rumoured that he had shot himself dead upon learning of it. He apparently initially took the order correctly from the dispatcher but made a transcription error in tracing the original copy, substituting a '5' for a '3'. A Cheyenne newspaper reporter commented, 'Miller was not present to testify in his own behalf before the coroner's jury, having fled shortly after the terrible wreck. Little is known of him, as he was a new man at Granger. He was a fair operator, but the pride which tempted him to re-write the train order that he might put out a prettier copy, cost the lives of 12 persons, and caused one of the most frightful railroad accidents that ever happened on the Harriman system.' [737]

Operator Miller was, however, not the only person to blame, the jury found. Their verdict read in part, "We further find that the cause of death was due to injuries received in a head-on collision between Union Pacific passenger train No. 3, westbound and Union Pacific freight train extra No. 1661, eastbound, at about 11.59 p.m., November 11, at a point four miles east of Granger, Wyo.

"We further find that the said collision was caused by the careless and gross negligence of J. E. Miller, the night operator at Granger, in furnishing a wrong order to the conductor and engineer of said freight train and to carelessness of the said conductor and engineer in leaving Granger, Wyo., under such short order."

The freight train were found to be partly to blame because they had already received a correct copy of the same order at Altmont. The order read, "No. 59 to C.S.E. extra east 1661. No. 3 (three) will run 1:30 late Green River to Evanston."

Operator Miller had, it emerged, drawn his time (resigned) from the railroad three days before the accident and was awaiting transportation elsewhere, but had agreed to

---

[736] *Deseret Evening News*, Salt Lake City, Utah, United States, 14th November 1904

[737] *The Cheyenne Daily Leader*, Cheyenne, Wyoming, United States, 16th November 1904

take the place of the regular operator who was ill.[738]

By April 1905 the Russo-Japanese war had been in progress for over a year. The Trans-Siberian Railway, the construction of which had begun in 1891, had been a major factor in precipitating the conflict, because it made possible Russian economic and military expansion into eastern Asia, and into areas which the Japanese had eyes on themselves. When war came, however, the task of transporting large numbers of Russian troops and great quantities of supplies over the many miles of single track placed a great strain on the railway, as was demonstrated by an accident on about 6th April 1905. A report transmitted via St Petersburg stated that a troop train coming from Kharbin (now Harbin) had neglected to go into the loop at No. 9 Siding. The explosion of detonators and shouts from the lineside had failed to alert the engine crew and the train continued along the main line until it collided head-on with another troop train coming in the opposite direction. Eight soldiers died in the accident and 26 others were injured. The engine driver at fault was severely injured. The driver of the colliding train, his assistant {?} and the fireman had all been sound asleep due to exhaustion.[739]

On Wednesday 14th June 1905 a train wreck occurred at Mount Hope, just outside of Baltimore, Maryland, United States, on the Western Maryland Railroad. The accident was not a serious one in terms of casualties, but there was a considerable amount of wreckage to clear up afterwards. The railroad company had a dedicated team of workers to deal with such incidents, who went from place to place as their job demanded and hence were known as “floaters”. By late afternoon on Saturday 17th June their work was almost finished, and they boarded a train from Baltimore Hillen station to Hagerstown to return to their homes in western Maryland, most of them living near Thurmont. The passenger train, which had left Baltimore at 5 p.m., consisted of only a baggage car, a mail car and three coaches and was already crowded. It was in charge of Conductor George Buckingham, with Engineer George Covell and Fireman John St Ledger. The 35 workmen crammed themselves into the baggage car of the train with some of them also riding on the buffers between the baggage car and the tender of the engine and others on the buffers between the baggage car and the mail car. A few minutes later these men would be the principal casualties in a major disaster.

Coming in the opposite direction along the single track to Hagerstown was a heavily laden extra freight train No. 43 consisting of 17 cars, drawn by two engines, Nos. 41 and 43, and bound for Port Covington. This train was in charge of Conductor Vernon Derr, with Engineer John Crouse and Fireman William McNamee on engine No. 41, Engineer Lewis Rice and Fireman J. Knipple on engine No. 43, Flagman George Lynch, Brakeman O. R. Nyman and Trainee Brakemen E. R. Stott and H. W. Berger.

The trains collided at a location known as Ransom, a mile or so on the Baltimore (east) side of Patapsco station, Maryland. They met on a curve and just west of a bridge over the Patapsco River. The three engines piled up on one another and the baggage car of the passenger train was crushed between the engine tender and the mail car, and demolished. The workmen who had been riding on the buffers were killed outright and

---

738 Ibid, 17th November 1904

739 *The Times*, London, England, 10th April 1905

those within the baggage car were killed or injured and trapped in the wreckage. Fortunately, there was no fire and no leakage of steam from any of the locomotives. The passenger coaches suffered only broken windows and doors and their occupants nothing worse than bruises and cuts from glass fragments. The number of dead was initially reported as 29 or 30; it included the three engineers, the three firemen, Conductor Derr and Trainee Brakeman Stott.[740]

The noise of the crash and the cries of the wounded brought the inhabitants of nearby farmhouses to the scene and farmers' wives made bandages of their clothing and household linen and 'worked heroically among the blood and grime'. One or two physicians who had been riding on the train directed the rescue efforts. Meanwhile, a messenger hurried to the nearest telegraph station to summon help from the nearby town of Westminster. All the physicians in the town assembled at the station and by chance the fast mail train for Baltimore shortly arrived. The train passengers were requested to disembark, the physicians climbed aboard, and the train set off at record speed for the accident scene. It reached Patapsco at about the same time as another relief train from Baltimore, carrying more physicians and supplies. The less seriously injured victims were taken to Westminster and Union Bridge, the more seriously injured to Baltimore. One of the victims had died before the train arrived at Baltimore Union station.[741]

How the accident had happened was explained to some degree by the testimony of Flagman Lynch, which was supported by the other two surviving trainmen from the freight train, Brakeman Nyman and Trainee Brakeman Berger. Lynch stated: "We got into the siding at Gorsuch about 4.25 p.m. and waited there considerably more than an hour. While we were waiting, we all got down from the train and sat on a pile of ties near the track. The two engineers and the conductors had their time cards and schedules and we talked for a while about the time we were making, how long we had to wait for No. 5 and where we would run to after she passed. No. 17, the Union Bridge Accommodation, came past on time, and then No. 11, the Blue Mountain Express, which was making her first trip of the season, followed. We talked about what a fine train she was, and, so far as I know, there wasn't any misunderstanding about No. 5. We had seen the other two trains pass and were waiting for the regular.

"I took a bucket and went to a spring nearby to get some water. When I got back to the track the engineer was in his cab and the train was just pulling out. They called to me to jump aboard if I was going, and I did. I looked at my watch and saw that No. 5 was not due for a few minutes. I yelled out to the fireman ahead: 'Where are you going to pass No. 5?' He yelled back, as near as I could catch, 'At Lawnsdale.' Again, I looked at my watch, saw the time, and shouted, 'For God's sake look at your watch.' He waved his hand at me, and I could only think that my watch much be wrong. There were the two engineers, the two conductors and the fireman, five men, all of whom had the time and knew the schedule as well as I did.

"For a minute or so I thought of putting down the airbrakes, but I thought again I must be wrong, and these five men could not be mistaken. If I had put down the brakes it would have almost torn the train to pieces. We were going down grade, and I was back at the rear of the train when we struck. There was a jar and then a succession of bumps, but

740 *The Morning Herald*, Hagerstown, Maryland, United States, 19th June 1905
741 *The Philadelphia Inquirer*, Philadelphia, Pennsylvania, United States, 19th June 1905

I was not thrown down. Why we left the siding when we did, I cannot tell." [742]

A partial explanation for the accident might be that the freight train was only scheduled to pass train No. 11, the Blue Mountain Express, at Gorsuch, and trains Nos. 17 and 5 at Lawnsdale, six miles further east. However, after train No. 11 had passed a loose bolt was found on the forward engine of the freight, which took time to fix, and so the freight waited at Gorsuch for train No. 17, and then overlooked train No. 5, or perhaps believed that they had time to reach Lawnsdale before it left there. When Flagman Lynch protested, they apparently joked that his watch was wrong. They were only about a mile from Lawnsdale when they met train No. 5.[743]

The final number of dead in the accident was fixed at 26, eight of them members of the train crews.

Manager Clark and Superintendent Ennes of the Western Maryland Railroad conducted an exhaustive inquiry, without coming to any new conclusions. At the conclusion of the inquiry Mr Ennes stated, "There was nothing new developed as to the cause of the disaster. Although there was a thorough and searching examination made, nothing what is already known to the public was brought to light." [744]

No inquest was held in Carroll county, where the accident occurred. State Attorney Guy Steele stated that it was evident that the accident was attributable to the crew of the freight train. The responsible persons and witnesses were dead. If an inquest jury found the freight crew guilty of criminal neglect of duty no criminal action could be brought against them.[745]

In what has been described as a classic single-line collision passenger express train No. 112 from Görlitz collided head-on with express passenger train No. 113 from Berlin at Schrankenbude (crossing hut) 7 near Spremberg, Germany, on Monday 7th August 1905. The two engines were destroyed and the two leading coaches of the Berlin express telescoped one another, all their occupants being killed or injured. Nineteen people died and 40 were injured.

Station Assistant Gustav Stullgys began his duty at Spremberg at 4.30 p.m. in an intoxicated condition. There then took place the series of telegraph messages which led to the collision. The stations involved, in sequence, were Bagenz, Spremberg, Schleife and Weißwasser.

First Spremberg asked Schleife, "Where is train 112?"

Schleife replied, "Train has not been offered yet."

Then Bagenz asked Spremberg, "Where is train 112?"

Spremberg replied, "Train is not yet at Weißwasser."

Bagenz suggested to Spremberg that the crossing of trains 112 and 113 be moved to Spremberg.

742 *The Sun*, Baltimore, Maryland, United States, 19th June 1905

743 *The Morning Herald*, Hagerstown, Maryland, United States, 20th June 1905

744 Ibid, 21st June 1905

745 Ibid, 22nd June 1905

Spremberg agreed.

Bagenz approved the change of crossing place from Bagenz to Spremberg and offered Spremberg train 113.

Schleife then offered Spremberg Hauptzug 112, that is the first part of train 112, and Spremberg accepted it.

Bagenz offered Spremberg train 113 again, Spremberg accepted it and Bagenz reported its departure.

Schleife telegraphed Spremberg, “Train 112 passed at 17.35.”

Spremberg acknowledged the message and offered train 113. The offer was repeated. Schleife answered.

Schleife offered Spremberg Nachzug 112, the second part of train 112. “Will you accept Nachzug 112 when Hauptzug 112 arrives?” No answer.

Schleife asked again, “Will you accept train 112?” And then a third time, “Will you accept train 112?” He did not specify ‘Nachzug’ as the ‘Hauptzug’ had already been accepted and was on its way to Spremberg.

Spremberg answered, “No, why?”

Schleife offered Nachzug 112 a fourth time. “Take train 112 on.”

Spremberg neither accepted nor refused train 112 but offered train 113.

Schleife replied, “We have the train here already; Nachzug 112 has to come to you, I can’t let the work train go in front of the express. There was also a work train at Schleife, so Nachzug 112 and train 113 could not cross one another there.”

Spremberg, “Yes, but.”

Schleife, “Please, is train 112 there?”

Spremberg, “No, not yet.”

Schleife, “Please report back as soon as you can.”

Stullgys at Spremberg now intended to get in touch with Bagenz, but in error telegraphed Pointsman Paul Wiedermann at Schleife instead, so that Schleife received the message, “Will you accept train 112, when train 113 arrives here?” This of course was a message not meant for Schleife.

Schleife replied, “Is train 112 there?”

Spremberg replied, “Yes, it is here.”

With this, Schleife let Nachzug 112 proceed towards Spremberg. Meanwhile, Stullgys repeatedly tried to call Bagenz, but he was still connected to Schleife, so Bagenz did not answer. He sent, “Train 113 here. Will you accept train 112?”

He repeated the message.

Schleife replied, “Schleife here, train 112 through 17:43.” Spremberg acknowledged the message.

Station-assistant Terpo was now at the telegraph apparatus at Spremberg, and Bagenz answered. Spremberg reported the arrival of train 113 and offered Hauptzug 112, which

Bagenz accepted.

Schleife again informed Spremberg that Nachzug 112 had passed, but was interrupted by Spremberg, who believed that Schleife was referring to Hauptzug 112. Stullgys telegraphed to Schleife, “Yes, train 112 is here, will you accept train 113?”

Schleife answered, “Send train 113 on, Nachzug through 17:43.”

Stullgys still did not understand that Nachzug 112 was still approaching from Schleife, despite Wiedemann referring to Nachzug 112, and Schleife having accepted train 113 on the assumption that Nachzug 112 was at Spremberg, he let train 113 depart towards Schleife and telegraphed its departure. “Yes, train 112 here, train 113 left at 17:47”

Schleife called Spremberg again, and when Stullgys finally answered, asked, “Please, has train 113 left?”

At this point Terpo took over the Morse key and asked Schleife, “Where is Nachzug 112?”

Terpo knew that train 113 had been reported out at 17:47 and asked Schleife, “What time did you report Nachzug 112?”

Schleife answered, “17:43.”

Spremberg asked, “What time did Hauptzug 112 leave?”

Schleife replied, “Please, don’t you know what trains you have on the line?”

Spremberg then telegraphed to Schleife that train 113 should come back (!) At that moment the collision occurred.

Hauptzug 112 had had a white disc at its rear (Signal 17b) indicating that it was running in two parts but even without this signal the running of the train in two sections was a regular occurrence. It would seem however that Stullgys, in his intoxicated state, overlooked this and was thinking only in terms of a single train.

Stullgys, Wiedemann and Schmidt, the Spremberg Weichensteller (pointsman), all subsequently stood trial at Cottbus for manslaughter. Stullgys was sentenced to sixteen months’ imprisonment and Schmidt to one month; Wiedemann was acquitted.[746]

The Nickel Plate (New York Chicago and St Louis Railroad) had an accident early on Sunday 13th August 1905 at Kishman, Ohio, United States. A fast eastbound passenger train collided head-on with a westbound freight train. First reports stated that 12 people had died and 25 were injured, of whom it was feared eight would die. One of those killed was Charles Poole, the engineer of the passenger train, the other 11 were mostly foreign labourers who had been riding in the smoking car at the head of the passenger train. They were employed by the Standard Oil Company and were on their way from Fort Seneca, Ohio, to Brookfield, Ohio. Engineer Poole had died as he reached for the air brake upon seeing the headlight of the freight train; his fireman, named Haefner, had jumped and survived. The engineer and fireman of the freight train also jumped and escaped.

The high speed of the passenger train threw its engine and first three vehicles over the engine of the freight train and the smoker and following car were telescoped, while the

[746] *Eisenbahn-Unfälle in Europa* by Erich Preuß, pgs. 116-8

forward cars of the freight train was described as having been 'smashed to fragments'. None of the occupants of the smoker escaped injury and many of them were pinned down by heavy timbers and unable to get out without assistance from rescuers. Fortunately, there was no fire. The injured were hurried to a hospital in Lorain and the bodies of the dead taken to a morgue there.

Almost all the passengers in the smoker had been dozing when the crash came and the force of the impact hurled them out of their seats into the aisle or against the roof and sides of the car, 'which seemed all at once to crunch in like a paper bag, pinning down each passenger in the position in which he had been caught.' Most of the passengers in the following car were also hurt to a lesser degree, suffering broken bones, sprains or bruises. The two rear sleepers escaped serious damage and none of their passengers was hurt.

The uninjured trainmen did what they could to help those injured, but lacked tools, and were assisted by the few local residents in the sparsely populated area. Some of the injured were placed in the undamaged sleepers and others accommodated in nearby houses.

Nickel Plate President Caniff stated that from information he had received from officials the crew of the freight train had had orders to go into the dead-end siding at Kishman for the passenger train. Why, if they had not been able get clear of the main line a flagman had not been sent out to warn the engineer of the passenger train was not clear. It was rumoured that the watch Engineer Kirby of the freight train had been slow, and that he had believed he had eight minutes more than he actually had to get in clear. The freight train had stopped and was about to set back into the siding when the passenger train arrived, travelling at 45 miles an hour.

Passenger Roy Winch was badly hurt but was the first person to reach the wrecked smoker. He was very critical of the behaviour of the surviving passengers in the smoker. 'Terrorized by the death of their companions, the hiss of escaping steam and the fear of death in the darkness, the Italians lost control of their scattered nerves, and stamping over the living and dead fought their way out of the car. Several of them were almost as badly injured by fighting each other as by the wreck and to what extent their pounding and walking on the wounded is responsible for many of the injuries cannot be known. Whether it was that the people were dazed or that it was a case of brutal selfishness, I do not know, but I never so people so unwilling to help. They would come up and take a look at the car, shudder and walk away.' [747]

The mind of Engineer Kirby was said to have been deranged by the accident.[748]

On 6th September Coroner French of Lorain placed the blame for the accident both on the crew of the freight train and on the dead Engineer Poole, blaming the latter on the grounds that the accident had occurred on a piece of straight track and the headlight of the freight train had been burning and in plain view. The coroner submitted his findings to Prosecutor Stroup, but the latter declared that under Ohio law the conviction of a person responsible for a death by negligence alone was practically out of the question.[749]

---

747 *The Evening Independent*, Massillon, Ohio, United States, 14th August 1905

748 *The Waterloo Times-Tribune*, Waterloo, Iowa, United States, 15th August 1905

749 *The Daily Reflector*, Norwalk, Oho, United States, 6th September 1905

The York and Dallastown Electric Railway had the only serious accident in its history on Saturday 9th September 1905. A passenger car was approaching Stabley's Switch, Pennsylvania, United States, when Conductor Snyder and Motorman Meisenheider saw a freight car approaching at a rapid rate and not more than 100 feet away. The motorman applied the brakes and jumped for his life, followed by the conductor; both escaped serious injury. The passengers inside the car were unable to escape; they became panic stricken and made desperate efforts to jump off but not one was able to do so. The car had been crowded, the register showing 102 fares, and two passengers were killed outright, three were so badly injured that they died later in hospital and another 75 were less seriously hurt.

After the accident 20 of the most seriously injured were placed on the freight car and taken to York, six miles away. The noise of the accident attracted all the local farmers and they rendered all the aid possible, while some of the less seriously hurt were cared for in nearby farmhouses.

The accident was 'supposed to have been due to a mistake in signals'. [750]

Another single-line collision occurred in the American state of Wyoming shortly before midnight on Wednesday 6th December 1905. A report on the accident from Rock Springs spoke of 16 dead. The accident involved the eastbound Union Pacific Railroad *Overland Limited* flyer (passenger train No. 2) to Chicago and a westbound freight train. The two collided in a cutting between Ah Say and Wilkins. The passenger train was according to one report travelling at 70 miles an hour and the freight train at 40 miles an hour, attempting to reach the next siding to clear the flyer. The passenger train was running considerably behind time. The freight train included several cars loaded with steel rails which were forced through and over the engine and then ploughed their way through the leading vehicles of the passenger train. Fire broke out in the wreckage soon after the collision. Nine dead bodies had been removed from the wreckage before the first assistance arrived.

By noon on 7th December the identified dead were Engineer A. E. Brink and Brakeman Earl Smith of the freight train, Electrician Warren Stigers and his assistant Student Electrician Frank McKenna, two mail clerks – names unknown, Chief Cook James Busby, Second Cook Ed Rosenbaum and Fourth Cook John Lawless. Engineer Anthony Jordan of the passenger train was said to be dying and Fireman Oscar Peterson of that train was also thought to be fatally hurt.[751] Fireman J. H. Alfers of the freight train suffered a broken leg.

Another report stated that the baggage dynamo car and dining car of train No. 2 had been demolished, and named three dead mail clerks: Jacob Newson, J. F. Phillipar and Frank Peterson.[752]

One report stated that the collision had been due to a mistake made by the operator at

[750] *The Vienna News*, Vienna, Georgia, United States, 12th September 1905
[751] *The Durango Wage Earner*, Durango, Colorado, United States, 14th December 1905
[752] *The Daily Journal*, Telluride, Colorado, United States, 7th December 1905

Wilkins, who had subsequently gone crazy.[753] Another, probably more reliable report, stated that there had been an error on the part of the conductor of the freight train, who had overlooked a train, having been given orders against four passenger trains, three from Ogden and off the Granger branch of the Oregon Short Line.[754] A Cheyenne newspaper explained this more fully, stating that the freight train had had an order to wait at Ah Say for the first and second sections of mail train No. 10, passenger train No. 6 and train No. 2, the Overland Limited. When the first three trains had passed the conductor gave the signal for his train to proceed. It covered a mile before the collision. The four passenger trains had been held up by the derailment of eight cars in Echo canyon, east of Ogden Utah, and when this had cleared had followed one another within a few minutes, thus leading to the conductor's confusion.

The Cheyenne paper stated that the baggage dynamo, mail, dining and one Pullman car of the Overland Express and had piled up in a heap, except the Pullman car, which had rolled down an embankment, injuring its passengers. The explosion of a gas tank beneath the dining car had caused the fire, which had consumed all the mail and baggage carried on the passenger train. This was rather ironic given that the Union Pacific had just gone over to using baggage dynamo cars to generate electricity for lighting to do away with the danger of gas explosions in the baggage cars.[755] Gas was presumably needed in the dining car for cooking purposes, electric cookers not being practical at this date.

Automatic block signals, which would probably have prevented the accident, were meant to have been installed on 1st December, but the job had been put back.[756]

What was described as 'one of the most awful wrecks in the history of western railroading' occurred near Abode, Colorado, United States, on the Denver and Rio Grande Railroad, at 2.20 a.m. on Friday 16th March 1906. Two passenger trains were involved, train No. 3, the westbound Utah and California express, and train No. 16, a Colorado-New Mexico express. The engineer of the westbound train made every effort to stop his train, 'and he went to his death with his hand on the throttle, faithful to his charge. This much is vouched for by his fireman, who, seeing the uselessness of remaining in the face of sure death, jumped and saved himself. Of the conduct of the entire crew of the eastbound train it can only be said that they died at their post, for no one lives to tell the story of their heroism.' According to Fireman J. H. Smith the trains were running at the usual speed despite the fact that they were running through a blinding snowstorm. He said that when Engineer Walter Cosslett of the westbound train had seen the headlight of the eastbound one, he had applied the emergency brake and its speed had been checked momentarily, but slippery rails and the momentum of the train had carried it forward.

Surviving trainmen stated that the impact of the collision had been scarcely noticeable, and it was thought that the helper engine at the front of train No. 16 had cushioned the impact, minimising the force and weight of the heavy mountain locomotives hauling the trains. The helper engine was badly crushed, and the two main

753 *The Durango Wage Earner*, Durango, Colorado, United States, 14th December 1905

754 *The Ogden Standard*, Ogden City, Utah, United States, 7th December 1905

755 Webpage: Utah rails.net – Early Head End Power

756 *The Cheyenne Daily Leader*, Cheyenne, Wisconsin, United States, 8th December 1905

engines locked themselves together. The baggage car of train No. 3 broke in two and the leading coaches were squeezed together. The baggage and mail car and coach of train No. 16 buckled but there was no telescoping. As in so many wrecks fire broke out almost immediately. 'In the forward coach of the westbound train every seat was occupied by the passengers, most of whom were homeseekers on their way to the northwest. A number of foreigners were among them, and in their terror, they gave up life without making any attempt to reach safety outside of the burning car. Paralyzed with fear, and with prayers upon their lips, they sank to the floor of the car and were roasted alive. The cooler ones in the car, seeing their danger, rushed for the windows and doors, and with the aid of the passengers in the rear of the train and those members of the train crew who were unhurt, managed to reach the open air. Many of them were injured more or less seriously by the rough handling they received or from flying glass and timbers. Although many were willing to undertake the risk, efforts to rescue those who remained in the burning cars would have been suicidal, as the heat was unbearable. When the occupants of the two standard and two tourist sleepers of the westbound train saw that nothing could be done to check the flames, they aided the trainmen in pushing back the sleepers and those cars were not damaged in the least. The sleepers on the eastbound train were also pushed back, and soon they were placed out of the reach of the leaping flames. The wrecked cars were soon reduced to a mass of smouldering ruins.'

An early report noted 19 dead in the accident, including Engineer Cosslett, his counterpart Engineer William Hollis and Fireman H. D. Sudduth of train No. 16 and Express Messenger Enos McFarland. Among the passengers killed were Deputy Sheriff of Denver and his prisoner, Archibald Whitney, who the sheriff was taking to the penitentiary at Canyon City. The sheriff was reportedly killed outright in the accident and his prisoner was unhurt but handcuffed to the law official was unable to escape the fire.[757] Twenty-two passengers were listed as injured.[758]

The cause of the collision was quickly established as having been an error on the part of Operator Frank Lively at Swallows. The two trains normally crossed one another at Canyon City, further to the west, but train No. 3 was running late, and the dispatcher wished to alter the meet of the two trains to Abode, 12 miles east of Canyon City. Having checked with Lively that train No. 3 had not yet passed him, the dispatcher issued an order to Lively at Swallows and to the operator at Canyon City altering the meet to Abode. Operator Lively, it transpired, had agreed to work the beginning of the night shift after having finished his regular day shift, to accommodate the night operator, Van Deusen, who had gone to Pueblo, but Van Deusen did not return when expected, and by 2 a.m. Lively had been on duty for 19½ hours, since 6.30 a.m. He had thus dozed off, perhaps only for a minute or so, but during that time train No. 3 had passed unnoticed. Unaware of this he had accepted the order to deliver to it.[759]

Operator Lively reportedly said after the accident, "I was dead for rest. I went to sleep and did not give No. 16 {sic, but this must be No. 3} its orders. I am responsible for the wreck. I will not run away. God knows I wish I knew how I happened to go to sleep." [760]

---

[757] *A History of Railroad Accidents, Safety Precautions and Operating Practices* by Robert B. Shaw, pg. 147
[758] *The Advertiser-Courier*, Hermann, Missouri, United States, 21st March 1906
[759] *The King City Chronicle*, King City, Missouri, United States, 23rd March 1906
[760] *The Advertiser-Courier*, Hermann, Missouri, United States, 21st March 1906

Operator Lively must have had second thoughts about not running away for he left Swallows two days after the accident, joined his brother, who was also an operator, at Portland, and both later departed on an eastbound train for their home in Munfordville, Kentucky. A subpoena was issued for his appearance at the inquest[761] but although he was subsequently located and agreed to return to Colorado[762] the inquest was eventually held without him.

The inquest on the accident concluded on Monday 25th March, it having been postponed in the hope that Lively and Van Deusen would attend it. It was stated at the inquest that Van Deusen had gone on the train to Pueblo to cash both operators' pay cheques, so they could pay their bills, there being no bank closer. According to the arrangement made between the two operators Van Deusen would return on the train arriving at Swallows at 6.45, so Lively would only have had to work an extra quarter hour. For reasons which were not explained, although Van Deusen was at the postponed inquest, he was not on the 6.45 train, nor any of the next three, so 19-year-old Lively was forced to stay on duty.

The jury returned a verdict 'that said collision was the result of negligence on the part of one S. F. Lively, but that said negligence was not wilful, malicious or felonious, and is not of such a nature as to render the said Lively or the said railroad company liable to criminal prosecution under any statute in this state.' The jury condemned Operator Van Deusen for deserting his post[763] and made the sensible recommendation that in future train order signals should be maintained at danger until whistled off by the engineers of approaching trains.[764] The jury also recommended the installation of automatic block signals, or the doubling of the line.[765]

As a number of bodies were totally consumed in the fire the final number of dead in the disaster could not be stated but was said to have been about 40.[766]

Eastbound Chicago mail train No. 18 and train first No. 21 a westbound Chicago and St Louis express, both running at full speed, met in a head-on collision near Clover Creek Junction, Pennsylvania, United States, on the Petersburg cut-off, 18 miles from Altoona, on the Pennsylvania Railroad, at about 10.45 p.m. on Friday 4th May 1906.

A few hours earlier 27 freight cars had been wrecked in a head-on collision at Union Furnace, between Petersburg and Tyrone. Before this became known the Chicago mail had left Altoona for Tyrone, but when it reached Elizabeth Furnace it was found that the line was completely blocked at Union Furnace it had to be taken back to Altoona to take the alternative route via the Petersburg cut-off. To save time the end was cut off the front of the train and attached to its rear for the journey back to Altoona and onwards. In the opposite direction the Chicago and St Louis express was also diverted via the cut-off.

Train No. 18 consisted of an engine, a coach, four horse cars and six mail and express cars and was in charge of Conductor J. H. Collins with Engineer J. L. Fickes, Fireman A.

---

761 *Converse County Herald*, Lusk, Wyoming, United States, 22nd March 1906
762 *The Shelby Sun*, Shelby, Nebraska, United States, 29th March 1906
763 *The Anaconda Standard*, Anaconda, Montana, United States, 27th March 1906
764 *The Salida Mail*, Salida, Colorado, United States, 30th March 1906
765 *Centralia Daily Sentinel*, Centralia, Illinois, United States, 28th March 1906
766 *Fairplay Flume*, Fairplay, Colorado, United States, 30th March 1906

T. Cook and Brakemen H. Hambright and H. T. Harder. Train No. 21 consisted of an engine, a baggage car, an express car and five coaches and sleeper cars, and was in charge of Conductor C. B. Snyder with Engineers John Lehr and James Dougherty, Fireman J. C. Baumgardner, Baggagemaster J. W. Herr and Brakemen R. W. Shatto and C. T. Fisher.

Engineer Dougherty, of train first No. 21, had orders to wait at the tower at the end of the double track for train No. 18, but misunderstood his orders and proceeded, meeting the mail train near Clover Creek. Both trains were running at their usual speed, believing that they had a clear road. The express car of train No. 18 telescoped the flimsy coach ahead of it, crushing it like an egg shell. Of the 22 passengers in the coach four were killed outright and all of the remainder injured. A postal car of train No. 18 was also badly damaged and four mail clerks in it were killed instantly, though three others were only slightly injured. Max Carlowe, in charge of the horses, was also killed when the end of the horse car he was riding in was crushed and three horses were injured. The heavy vehicles of train No. 21 withstood the collision better than those in train No. 18, but its baggage car was badly damaged and Baggagemaster Herr was fatally injured.

After the accident the part of train first No. 21 which was not derailed was dragged back and took the bodies of the dead and some of the injured to Altoona via Petersburg. On the opposite side train No. 10 had followed train No. 18 and the remainder of the injured were taken to Altoona on this train. Previous to this the engine of train No. 10 had been detached and run down the Clover Creek branch to Williamsburg, where the four physicians of the town had been awakened from their beds and taken back to the wreck within 40 minutes. In addition to the 10 dead 37 people were injured.

'Mr D. P. Schaeffer of 1304 Seventh-avenue {Altoona} who was pilot on No. 18, speaks of the wreck as follows: "We were running at about fifteen or twenty miles an hour when suddenly the headlight of the other train burst upon us. I hollowed jump, which the fireman and myself did, one on each on each side of the engine. The engineer for some reason did not. We were rounding a curve when I jumped, and I rolled down a steep embankment and fell into a hole which had been dug for a culvert. After taking in the situation I crawled back up the embankment and helped to extricate the engineer, who was pinned into the corner of his cab, unconscious. The coach next to the engine, where most were killed, was shattered to splinters and the cries of the injured were awful. Several other cars were badly damaged, but none upset, and fortunately fire did not break out.

""At this juncture I thought of the engine and saw that water was kept up to prevent any further trouble. On the other train the baggage car ran partly under the mail car. Mail Clerk Wagner was caught in here and killed. Everybody who was able set immediately to the work of rescue and everything possible was done until the arrival of the wreck train an hour later."

'In spite of the fact that Mr Schaeffer was injured in the hand and knee he nervily refused to give up but assisted wherever possible and retained self-possession enough to remember that the water supply in the engine must be kept up. It was a case of the kind of nerve needed in an accident of this kind.' [767]

The circumstances of the accident were gone into very thoroughly at the inquest, with

---

[767] *Altoona Mirror*, Altoona, Pennsylvania, United States, 5th May 1906

evidence from all the witnesses involved.

Dispatcher C. G. Troxell testified that because of the wreck at Union Furnace he had issued orders for train No. 18 to proceed east over the Petersburg branch to Springfield Junction, thence to obtain orders to continue to Petersburg. He added that trains were often run with coaches next to their locomotives.

Engineer Shaeffer stated that he had piloted train No. 18 over the branch and had received written orders at Springfield Junction to proceed to Petersburg.

Conductor Collins of train No. 18 corroborated Shaffer's testimony and said that the order had been a telephone order.

Dispatcher J.M. Speddy testified that he had issued an order for train first No. 21 to run over the Petersburg branch.

Engineer Lehr stated that he had been in charge of the engine of first No. 21. He had never driven over the branch so at Petersburg Engineer James Dougherty had taken over to pilot the train to Altoona. The operator at Petersburg tower had given them an order to run as an extra west to ST tower. After leaving Petersburg he (Lehr) had no control whatever of the train. He added that he had asked for Dougherty as pilot because he known him for 33 years as a careful man. Recalled, Lehr said that as far as he was aware no orders were received after leaving Petersburg except a slow order from ST tower. Having passed ST tower Dougherty had sent his fireman back for orders, then gone back himself. Someone at the tower gave the signal to go ahead and Dougherty had returned and started the train. He (Lehr) had seen no further orders but assumed all was all right.

Fireman Baumgardner of train first No. 21 testified to their getting orders to proceed to ST tower and receiving a slow order from the operator at ST but no running orders. They had stopped just beyond the tower and his engineer had sent him back for running orders. He had done so and coming back had told his engineer there were none for them. Dougherty had then gone back himself and returning had told him that their orders were to run to SJ tower, the other end of the single-line section. They had then set off, the man in the office operating the switch for them to pass onto the single track.

Engineer Dougherty read a prepared statement. He began by protesting that the dispatcher, knowing that his and other trains were running on the branch in both directions, should have taken the precaution of positively blocking his train at ST tower. At the latter place he had been given a clear block and the switches had been set for his train. He had been given two orders there, but they were not the ones he required. He sent the fireman back to the conductor, thinking that the latter might have the right orders, but the fireman came back and said they were for the track. Engineer Dougherty said that he himself had then gone back to the conductor, who had taken out his schedule book, looked at his orders, and decided that they gave them the right to run to SJ tower. Going back to his engine he had concluded that he had erred in thinking that the order only took him to ST tower, and had said this to his fireman and Engineer Lehr, to which the fireman had remarked that it was a poorly written "J" on the order. They had then set off at low speed. When he had seen the headlights of the other train, 10 lengths away, he had almost been able to stop his train before the collision. He admitted that he had misunderstood his orders but repeated that if the dispatcher had blocked that part of the road the accident would not have happened. After the accident the conductor had pointed out to him that ST tower did not appear on his schedule. Questioned by jurors, he could give no reason for taking his train onto the single track without having a positive order to do so.

Conductor Snyder of train No. 21 stated that he was not familiar with the road. He could not see ST tower on his schedule and thought that it must be SJ tower and had told Engineer Dougherty so.[768]

Several further witnesses were examined on the final day of the inquest. Operator W. H. Withers at ST tower told of receiving an order by telephone for train No. 21, but it was only an order to go slow west of the tower. He admitted opening the switch and allowing the train to proceed. Asked about this, he stated that that night he had received orders that three trains would run over the Petersburg branch, all to run as extras. Train No. 29 on the main line had had through orders on the branch and not having received any other orders he assumed that train No. 21 had the same orders. He had received no orders regarding any eastbound trains. As soon as train No. 21 had passed he had notified Altoona by telephone and had then been asked by the dispatcher if trains Nos. 18 and 21 had passed. He had been at ST tower 17 months.

Operator Warren Gosnell at SJ tower told of receiving an order by telephone that: "No. 18 will run from SJ to GI," GI tower being near Petersburg. Operator Gosnell said that he had only been employed at the tower a week before the accident. He wrote down all his telephone orders upon receiving them and repeated them back to the person sending them. He had known nothing of any westbound trains.

Passenger Trainmaster W. Brooke Moore testified that learning of the Union Furnace wreck he had arranged for trains to be diverted over the Petersburg branch, each to be provided with a pilot. Train No. 29 had through orders to run over the branch to Hollidaysburg and thence over the New Portage road. Train No. 21 was to run to Hollidaysburg and thence to Altoona. In the opposite direction trains Nos. 18, 14 and 6 were to run eastwards over the branch to Huntingdon. Train No. 21 had received no orders to proceed beyond ST tower. The signals at ST and SJ towers had nothing to do with it. There was nothing to prevent trains being marshalled as train No. 18 was, with a coach leading.

Conductor Snyder recalled stated that he had taken the order to read ST when he had first seen it but later thought that it might be SJ as ST was not on his schedule.

Engineer Dougherty recalled stated that he had first taken the order to read ST and it was not until he had spoken to the conductor that he had had his doubts. He repeated that had the SJ operator had an order to stop his train the collision would not have happened and went on to relate previous occasions when he had prevented accidents through a presentiment that something was wrong.

Trainmaster J. R. Fisher testified that the Petersburg branch was run on block signal only as far as Hollidaysburg, after that train orders were used. On the night in question the line was being worked in accordance with the rules and train No. 21 had no right to run past ST tower without further orders.

Superintendent C. A. Preston, asked if he wished to make a statement, remarked that the actions of Engineer Dougherty and Conductor Snyder had been a flagrant violation of the rules. In running beyond ST tower, the engineer had done exactly what he should not have done. He understood the orders and knew better. As for the conductor, if he had not

---

768 Ibid, 11th May 1906

understood the order he should have gone to the operator and made himself positive as to their meaning.

The jury returned a verdict that, "Death was caused by a head-on collision between trains known as No. 21 west on the main line of the Pennsylvania railroad and running as extra No. 459, and train No. 18 east on the main line, running as extra No. 276, about 11 p.m., on May 4, 1906, at a point 150 yards east of Clover Creek Junction, on single track of the Petersburg branch of the Pennsylvania railroad, between eastern tower ST and western tower SJ. We find that the cause of the accident was due to the pilot in charge of extra train No. 459 west, in passing tower ST, which was in violation of his running orders and the rules of the Pennsylvania railroad company in governing the movement of trains." [769]

Train No. 44, the eastbound regular mail train and a westbound extra freight train, running as the first section of train No. 39, a passenger train, collided about two miles west of Hamlet, South Carolina, United States, on the Seaboard Air Line Railroad, at 8.40 p.m. on Sunday 22nd July 1906.

The passenger train was in charge of Conductor John Bowen with Engineer Frank Lewis, Fireman Thomas Hill, Baggagemaster H. S. Byrd and Porter John. Bogan and was running from Rutherford, near Charlotte, South Carolina, to Wilmington, North Carolina. The freight train was running from Raleigh, North Carolina to Monroe Junction (Hamlet) and was in charge of Conductor Hunter with Engineer Owen Bundy. The baggage car and 'Jim Crow' coach of the passenger train were wrecked, and eight cars of the freight train were derailed and some crushed to pieces.

Early reports regarding the accident stated that 20 persons had been killed and 24 injured, the dead including the entire crew of the passenger train. The dead were reportedly eight whites and twelve coloureds and the injured five whites and 19 coloureds. The names of the dead and injured whites only were listed in some newspapers, though others named all the casualties who had been identified.[770]

Blame for the accident was immediately placed by the railroad company on the operator at Rockingham for failing to hold train No. 44 for the freight extra.[771] However, on 26th July a local newspaper stated that the Rockingham operator claimed to be entirely innocent in the matter and that the blame lay at Hamlet.

An eye-witness described the accident. 'Five minutes after the train pulled out of Rockingham the engine of No. 44 darted into a deep curve and swiftly passed up to what is known as one of the sharpest curves on the Seaboard's system. Before the brave engineer of No. 44 knew what was ahead of him, the headlight of the extra freight darted from around the opposite end of the curve. There was a terrible crash – a fearful hissing of steam and the two ponderous steeds reared straight up by the force of the collision.

'During the few minutes that followed in which the escaping steam and the roar caused by the collision there could be heard the wails of the dying and those more fortunate.

---

769 Ibid, 17th May 1906

770 *The Galveston Daily News*, Galveston, Texas, United States, 24th July 1906

771 *Fort Wayne Weekly Sentinel.*, Fort Wayne, Indiana, United States, 25th July 1906

'As quickly as possible, Conductor Bowen and those who had escaped from the terrible collision, went forward to where the engine was standing in an upright position and called for Engineer Lewis. There was no answer to his call for the hand that guided the fast flying passenger engine, while still holding to the throttle of his engine, was stilled by death. Through the mist of escaping steam and the pile of iron and steel, could be seen a portion of the brave engineer's body as it was being roasted alive by the boiling flood of water that was pouring from the engine boiler.

'Wedged in between the wreckage of the baggage car and that of the Jim Crow car could be seen the mangled remains of Expressman Bird. He must have met an instant death as ponderous portions of iron and wood held his body in deep embrace.

'The scene where the Jim Crow car and the baggage and express car were telescoped, was one of intense horror and suffering. There, in the mangled mass of flesh and blood, were 19 torn and dismembered bodies of negroes who were ushered into death without a moment's warning. Their bodies were piled, one upon another, in a frightful mass and now and then one could hear the dying wails of a lost soul.

'As the accident took place at nightfall and as there were few if any lights available, the scene was one of weirdness as well as horror. The ones aboard the train who escaped injury did all that was possible to relieve the dying and wounded, but their efforts were curtailed by the many disadvantages incident in the wreck.

'By 9 o'clock a wrecking crew from Hamlet reached the scene and as fast as possible the injured were taken from their perilous position. All through the weary hours of the night men and even women worked to relieve the suffering and care for the dead. The bodies of the dead were taken out one by one and the lifeless forms were laid beside the track until preparations could be made to move them to Rockingham.'

Another paragraph in the newspaper explained the escape of the engine crew of the freight train. 'The fireman of the freight train says that a few minutes before the collision he saw the reflection of the electric headlight of No. 44. He went over to the engineer's side and told him to look. The engineer looked and the next moment he reversed his engine and with a signal to the fireman to jump he reversed his engine and leaped into the darkness. This explains why both the engineer and the fireman of the freight train saved their lives.' [772]

No report of an inquest or inquiry into the accident has come to light.

Monday 12th November 1906 saw another disastrous single-line collision in the United States with all too familiar features; it involved a train carrying immigrants and would have been much less serious had a fire not broken out in the wreckage. A newspaper reporter described the tragedy of the immigrants who had come across the Atlantic in the hope of a new life and instead had found death in a train crash. 'In the gray hot ashes that night were the withered hopes of almost half a hundred souls. In America, they saw the light. In America, too, they awakened in the fire of destruction. Escape was impossible. They had been trapped and they died in the flames because of the neglect of a train crew. All the way from the East the immigrants had watched the shining lights in the homes of the Americans. Cheerfully they wrapped their thin blankets around them,

[772] *The Hickory Democrat*, Hickory, North Carolina, United States, 26th July 1906

thinking that when they awoke, they would be at the entrance to new homes – their homes. Dreaming peacefully of the morrow and its joys they were rudely aroused by an awful crash, like the roar of a cannon, and the cars plunged down an embankment into a heap.'

The accident occurred near Woodville, Indiana, on the main line of the Baltimore and Ohio Railroad, about 46 miles from Chicago. The immigrant train was from Locust Point, Baltimore, Maryland. It was running as the second section of passenger train No. 47 and had 167 persons on board, mostly Russians, Serbians and Poles.[773] It was in charge of Conductor W. H. Brooks with Engineer Rensman and Fireman A. Culler. The first section of train No. 47 had passed westbound freight train first No. 98, in charge of Conductor Samuel Moste, at a blind siding at Babcock, five miles on the Chicago side of Woodville. Engineer Anthony Burke was unaware that the passenger train was running in two sections. According to the regulations when a train was running in more than one section all, but the last section had to display green flags or lights and when passing an inferior train must sound one long and two short whistle blasts. The first section was carrying green lights but, in the darkness and blizzard conditions with gale-force winds and driving snow, the green lights, encrusted with snow and ice, if displayed, were not seen and the whistle, if sounded, went unheard. Thinking all was clear Engineer Burke pulled out onto the main line and set off eastwards with his 30 cars. Meanwhile the immigrant train, travelling 15 minutes behind the first section, had received clear signals at Suman, five miles on the other side of Woodville.

Fireman Culler was the only one of the four engine crew to lose his life in the collision but both engineers and Conductor Brooks were badly hurt. An unidentified man, perhaps a passenger, ran to some farmhouses at Woodville siding and aroused the occupants. Telephone messages were sent to Chesterton, McPool, Porter, Suman and Valparaiso and telegrams for wrecking trains and physicians were sent to Chicago and Garrett. Riders were sent from the farmhouses to other farms in the vicinity and within half an hour 50 men were attempting to save the lives of the passengers trapped in the wreckage of the immigrant train. In the collision the leading two coaches of this train had rolled down a 20-foot embankment on the north side of the line and instantly caught fire. The blaze spread very quickly and despite the best efforts of the rescuers only about a dozen of those who had not been thrown clear could be saved. The fire had reportedly originated from the contents of the engine fireboxes.

The collision had occurred at about 3.20 a.m. At 5 a.m. a train was made up at Chicago carrying a score of physicians while nurses were picked up at Englewood and South Chicago. Meanwhile two hundred men from Chesterton, three miles away from Woodville, had been awakened by fire whistles and bells and came to join in the rescue work. The injured were taken to the only nearby farmhouses, those of Daniel Cole and James Dorgan, and 10 of them to the general store of B. J. Lindaman at Woodville siding.

Engineer Burke, related his story, between spells of unconsciousness, to Baltimore and Ohio Superintendent Bachelder at Cole's farmhouse. "They had no signals out. I watched for them and there wasn't a light on the tender. I always listen for a whistle and there wasn't a sound. I was sure it was the second division and pulled onto the track. I don't remember anything more until after the smash up. It was so dark I don't think either

773 *Weekly Breeze*, Monroeville, Indiana, United States, 15th November 1906

engineer saw anything until they came together."[774]

Conflicting evidence was given at the inquest at Valparaiso. Conductor Moste stated that the first section of train No. 47 had not whistled, nor had it shown the correct lights to indicate that another section was following. In this he was contradicted by Head Brakeman Daniel Woodward, who stated that the first section had whistled but he had not understood what the whistles meant. According to him he had discussed the whistling with Engineer Burke. Fireman Herbert Schaffer of the first section of train No. 47 said that he had been told by his engineer to check that their lights were in proper shape, but he could not say if the correct signals had been displayed. His engineer had whistled at Babcock, but the signal had not been acknowledged.[775]

President Smietkna of the Polish Alliance questioned Conductor Brooks as to how many children there had been on the immigrant train. The conductor, it would seem, was reluctant to answer, but Brakeman J. H. Clemens said that as far as he could remember there had been between 25 and 40 children under five years of age. He also admitted that the cars had been old and rickety.[776]

The most dramatic evidence given at the inquest came from Engineer Frank Galnauer, in charge of the engine of the first section of train No. 47, who testified that he had given the proper whistle at Babcock that another section was following but the engineer of the freight train had not acknowledged it and he had repeated it, but had still had no response. He admitted that when his whistle had not been answered he should have stopped and at this point he broke down and wept, sobbing like a child. He added, however, that the conductor of the freight train should have showed him a hand signal. Down the line a short way at McCool he had stopped, he did not know why, and found that his lights were out.

Conductor Porter of the first section of train No. 47 said that the lights had been burning properly just before the accident.

Flagman Niles Baum of the freight train testified that the first section had given no signals as it passed.

Superintendent Bachelder told the inquest that 13 persons reported to have died had been found to be still alive and that the total number of victims who could be checked up on was 62, though there might be some more infants.

At the close of the inquest Coroner Carson ordered the arrest of Engineer Galnauer, Conductor Moste and Head Brakeman Woodward on a charge of manslaughter, to await the decision of the Grand Jury on 12th December. President Smietkna of the Polish Alliance protested that higher officials of the railroad company should also have been charged.

Despite Superintendent Bachelder putting the number of dead at 62 the same newspaper report stated that a memorial service had been held at St Patrick's Roman Catholic church, Chesterton, at which a requiem mass had been sung for 75 victims.[777]

Eventually it was only Engineer Galnauer who stood trial and on 13th April 1907 it

---

[774] *The Reflector*, Jeffersonville, Indiana, United States, 16th November 1906
[775] *The Lake County Times*, Hammond, Indiana, United States, 21st November 1906
[776] Ibid.
[777] *The Fort Wayne Sentinel*, Fort Wayne, Indiana, United States, 22nd November 1906

took a jury at Valparaiso only 30 minutes to find him "Not Guilty". [778]

The year 1907 had barely begun when there was a disaster on the Chicago Rock Island and Pacific Railroad near Volland, Kansas, United States. As was often the case, most of the victims were what might be termed second-class citizens, in this case not blacks or poor immigrants but Mexican labourers. This was doubtless because such passengers were expected to travel in the most inferior rolling stock, the coaches that were the least likely to be able to withstand the rigours of a collision and, because of their location, those most likely to catch fire.

The trains involved in the accident early on Wednesday 2nd January 1907 were the westbound train No. 29, the southbound *California Fast Mail*, and its northbound counterpart, train No. 30, the *Chicago Fast Mail*. Train No. 30 should have been held at Volland for a meet with train No. 29, but it was not and the two met west of Volland and between there an Alta Vista. The southbound train, No. 29, consisted of 10 cars, having been strengthened for the heavy homeseekers excursion traffic, and it was also carrying a party of Mexican labourers who were going to El Paso to a job on the Rock Island. It had left Chicago, Illinois, at 8.30 a.m. the day before and was in charge of Conductor William Gane with Engineer Slater and Fireman Sweeney. The northbound train, No. 30, also had 10 cars and had left El Paso, Texas, on Monday night. Its engine was manned by Engineer McMahon and Fireman Brown.

The accident occurred on what was known as Alta Vista Hill and train No. 30 came down the hill at high speed when near the bottom Engineer McMahon saw the other train. He barely had time to brake before the two trains collided, but the two engineers and two firemen were able to jump and save themselves. It was train No. 29 which suffered most in the terrific collision. Its smoking car was literally split in two and the leading chair car was telescoped for half its length. Fire broke out in the wreckage and the two baggage cars, mail car, smoker, chair cars and tourist car were all subsequently consumed by the fire, which was fed by gas from broken cylinders. The fire also involved the foremost tourist car of train No. 30, but not until its occupants had left it. All the deaths and serious injuries were in the smoker and chair car, the passengers in the other cars escaping with sprains and bruises, caused by their being thrown forward by the sudden violent braking of the train. In the smoker the Mexicans, most of whom had been curled up asleep, were pinned under their seats, while the doors of the car were jammed. Many of the passengers in the chair car of train No. 29 were likewise pinned under their seats. The passengers from the rear cars of the train escaped in their nightclothes and dressed in the mud beside the track, then alerted by the cries of the trapped passengers hurried forward to try and rescue them. Every man and woman was said to have participated in the fight to save lives, but the heat near the smoker soon made it impossible to approach it, though the passengers in the chair car were probably all got out alive.

Train No. 30 had been plodding uphill at 10 miles an hour. It was little damaged and only a few of its passengers were hurt and none killed, though some were badly burnt in the attempt to rescue those trapped in the coaches on fire.

Conductor Gane, who suffered a broken collarbone and bruises, stated, "The train was just pulling clear of the Volland yards and was travelling about thirty miles an hour. I

[778] *The Plymouth Tribune*, Plymouth, Indiana, United States, 18th April 1907

dropped into the front seat and dozed. There were 32 Mexicans, sitting two in a seat, back of me, Al Link, a negro porter; an interpreter, a foreman, and five paid fares.

"I was awakened by the setting of the brakes. I jumped head first out of the window and struck on my shoulder. When I turned and looked at the train, I saw flames leaping fifty feet high from the smoker."

The first relief train arrived at the scene an hour and a half after the accident from McFarland, Kansas, bringing surgeons and helpers. Prior to the arrival of the relief train assistance was given by a professional nurse, a Miss Sadler, from McIntyre, Iowa, who chanced to be a passenger on train No. 29 and escaped with slight bruises.[779] A second relief train was sent from Topeka. Two of the injured Mexicans died on the way to Topeka and Porter Link died in the hospital there shortly after his arrival.[780]

Later the day of the disaster Superintendent Rourke of the Rock Island company stated that there were 35 dead, 10 of whose bodies had been recovered. Apart from the Mexican labourers the dead were said to have included their interpreter, Clarence Dauchy, Porter Al Link and William Thomas Miller, an Iowa soldier on furlough and a tramp riding on the baggage car of train No. 30. Some passengers also spoke of a woman and child trapped and killed in the chair car of train No. 29.[781]

Responsibility for the accident rested largely on Operator John Lynes at Volland. Initially train No. 29 had been meeting northbound trains Nos. 14 and 30 at Alta Vista, the next station, but it was running somewhat late, so Dispatcher J. H. Shoemaker at Topeka decided to change the meets with the two northbound trains to Volland. Operator Lynes was arrested after the accident but before being taken to gaol made a statement. "I had been awake all night and was sober. About 4 or 5 o'clock this morning the dispatcher gave me four orders. One was for the southbound train to meet two trains at Volland instead of at Alta Vista as previously arranged. The southbound train headed into a switch and let one train pass, backed out of the switch, and headed down the main line without waiting for the other train. I thought it was going to stop to take water, as trains have been doing, but instead it went by at about ten miles an hour. The southbound board was at 'danger' according to the lever in the office, but the train did not stop, and I ran out with my lantern, but with two swings across the track it went out. I then ran to the pumphouse, grabbing the pumper's lantern, waved it a few times and it also went out. I also called the pumper, telling him what had happened. I went back and told the dispatcher that the southbound train was by, and we waited to see if it was going to come back or hit the northbound train. I told the dispatcher that I was coming to Topeka and was trying to get there when arrested." [782]

Rock Island General Superintendent and Division Superintendent Rourke held a conference on 5th January as a result of which they placed the whole blame on Operator Lynes and completely exonerated Dispatcher Shoemaker.[783] The inquest jury, however, a few days later, blamed both the operator and the dispatcher. It was not thought likely that

---

[779] *The Helena Independent*, Helena, Montana, United States, 8th January 1907
[780] *The Weekly Pantagraph*, Bloomington, Illinois, United States, 4th January 1907
[781] *The Lincoln Daily Star*, Lincoln, Nebraska, United States, 2nd January 1907
[782] *The Washington Post*, Washington, D.C., United States, 3rd January 1907
[783] *Logansport Reporter*, Logansport, Indiana, United States, 8th January 1907

any prosecutions would follow.[784] The jury's verdict seems reasonable inasmuch it was generally held to be unwise to issue a hold order to a train at the point at which it applied, and if done special precautions should be observed. As far as the operator was concerned, he should clearly have called Conductor Gane into his office and ensured that he understood that his train was be held there before accepting the order from the dispatcher. Operator Lynes had also apparently claimed to be several years older than he actually was and to have had six years' service on other railroads when in truth he had none.[785] The final number of dead was stated to have been 32, with about 55 injured.

Fire was again a feature of a collision at Fowler, Indiana, United States, on the Cleveland Cincinnati Chicago and St Louis Railroad (the 'Big Four') on 19th January 1907. Eastbound train No. 38, known as the *Queen City Special*, had left Chicago at 11.30 p.m. the night before. It consisted of a locomotive, a combination baggage and day coach, two Pullman sleeping cars for Cincinnati, a Pullman sleeping car for Indianapolis and the private car of Vice President Charles Schaff, occupied by his wife, Mrs Schaff. It was in charge of Conductor Peter Hiddinger, Engineer Edward Tripp and Fireman Henry Olcott.

At about 2.17 a.m. on Saturday 19th January train No. 38 was approaching Fowler, where it had to cross train No. 95, a westbound fast freight. The line was worked under the telegraph block system. The block signal at Fowler was displaying a bright red light against the passenger train but there was a thick fog. The operator at Fowler, fearful that the engineer of train No. 38 might miss the signal, reportedly waved his lantern frantically at the train and fired off half a dozen shots of his revolver but the train kept going, supposedly at 50 miles an hour, and collided head-on with the freight train 500 feet east of the switch. The force of the collision was sufficient to cause the tender of the passenger train engine to telescope the combination car for almost its entire length and of its occupants, of whom it was officially stated there were 13, only two escaped death, and they were seriously injured. The combination car caught fire after the collision and the dead and injured who could be removed were placed in the first Cincinnati sleeper, moved when that caught fire to the Indianapolis sleeper and finally to Vice President Schaff's private car, in which they were taken to Kankakee. 'Arising from her comfortable berth, she {Mrs Schaff) surrendered her private car to the injured and with her own hands helped to dress wounds and to care for the injured.' Only a few people in the sleeper cars were slightly hurt and no one in the private car was injured. The combination car and Cincinnati sleepers were entirely consumed by fire, but the Indianapolis sleeper must have been uncoupled before the fire reached it for it escaped undamaged.

The noise of the collision awakened all the residents of Fowler and along with the glare of the burning wreckage brought them to the scene. At first, they formed a bucket brigade, dipping into water lying in pools and ditches alongside the railway. Their efforts had little effect and were abandoned when the fire spread to the sleeping cars.

Not all the dead or injured passengers could be extricated from the combination car

---

784 *Moline Daily Despatch*, Moline, Illinois, United States, 15th January 1907

785 *A History of Railroad Accidents, Safety Precautions and Operating Practices* by Robert B. Shaw, pg. 148

before the intensity of the fire drove the rescuers back and yet again some of the trapped victims probably burnt to death. At daybreak it was reported that the smouldering skeletons of some of the victims could be seen in the wreckage.

The number of dead was officially stated soon after the accident to have been 13, but other reports placed it higher. Those killed included Conductor Hiddinger, Fireman Olcott and Baggagemaster Magee of the passenger train. Engineer Tripp of the passenger train and Fireman Minor Griffin of the freight train were seriously injured. Engineer John Koble of the freight train was not hurt. A draft for $200,000 in the name of Henry A. Price, of Long Beach, California, was found among the wreckage. He was known to have been on the train and it was believed that his body was one of those cremated.[786]

Commenting on the disaster, E. G. Hall, the local prosecuting attorney, stated, 'It is evident that the railroad men and officials blundered at the coat of lives of eleven persons, who were crushed and burned to death in the wreck of the Queen City Flyer. The Queen City disaster will be probed to the bottom and every effort exerted to bring those who are culpable to justice. The rigid examination already begun will be pushed vigorously. So far as I can see now, at least three persons blundered. The charge, in case of conviction, would be involuntary manslaughter.'[787]

At the inquest into the accident the Kankakee assistant dispatcher, George Sears, stated that at 2.01 a.m. he had issued order No. 38 to Fowler for the passenger train to be held there until 2.17 a.m. and to Swannington for the freight train to proceed to Fowler. The freight train had left Swannington at 2.05 and would take eight minutes to reach Fowler and four minutes to clear the siding, making the full time to 2.17. He said that no order had been issued to the passenger train before it reached Fowler because of a company rule that passenger trains should be advanced as far as possible and not given orders until the last possible point.[788] It was, he said, the freight train crew's own affair whether or no they attempted to reach Fowler in the time allowed and that "manifest" or fast freights such as No. 38 were considered to be as important as first-class passenger trains and must be advanced as rapidly as possible.

Chief Dispatcher C.A. Borchers expressed the opinion that Dispatcher Sears and the crew of the freight train were slightly at fault in their anxiety to advance the freight train on such a foggy night.[789]

Coroner Charles Comley returned his verdict at the conclusion of the inquest into the victims of the disaster on 26th February. He found Assistant Dispatcher Sears and Conductor Charles Mills and Engineer John Koble of the freight train, also the head of officials and management of the Big Four, guilty of gross negligence. Engineer Tripp did not escape criticism either, the coroner finding that he had been running his train at excessive speed. The Fowler operator, apparently coincidentally named Ross Fowler, was completely exonerated. Despite the earlier statement of the prosecuting attorney the coroner did not find any evidence of criminal negligence and no arrests were ordered. Attorney Hall conceded that he did not think there was sufficient evidence to warrant a charge of criminal negligence. The coroner had called over 30 witnesses and his report

---

[786] *The Ogden Standard*, Ogden City, Utah, United States, 19th January 1907
[787] *The Logansport Times*, Logansport, Indiana, United States, 25th January 1907
[788] *The Muscatine Journal*, Muscatine, Iowa, United States, 24th January 1907
[789] *The Cannelton Enquirer*, Cannelton, Indiana, United States, 16th February 1907

covered over three hundred typewritten pages. It was thought that the inquest would cost over $300.[790]

The coroner stated that the number of fatalities in the accident, including one unidentified person and two passengers who died later, was nine. Earlier reports had stated that about 20 people were injured.[791]

Separate from the inquest the state railroad commission conducted their own investigation, which concluded that the crew of the freight train had violated the rules in not clearing a superior class of train by five minutes.[792]

In contrast to the Fowler wreck the next accident, in Canada, on the Canadian Pacific Railway, happened in broad daylight. At 5.10 p.m. on Friday 28th June 1907 westbound train No. 97, the *Pacific Express*, came into head-on collision with eastbound train No. 95 which was carrying 232 Chinese emigrants in bond to Montreal. (As an aside, it is not quite clear what the term 'emigrants on bond' used in the newspaper report means, but the author believes that these were emigrants whose right to stay in Canada was under review. The immigration of Asian emigrants was being restricted at this date and since 1900 each person had been required to pay a $100 head tax – the only ethnic group so penalised.)[793]

The trains met 'four pole lengths' (22 yards) east of Butler, Manitoba. It is not entirely clear from the various newspaper reports if the collision actually occurred on the single line or at the crossing loop, but it has been assumed to be a single-line collision. The two locomotives were badly damaged, but the engineers and firemen all jumped and escaped injury. The heavier passenger cars of both trains were little damaged, and none was derailed; some of the passengers in them received a shaking up but none was injured. Much less fortunate were the Chinese in the two tourists cars in train No. 95. These lighter vehicles were broken up, crushed by the engine and heavier cars in front of and behind them, and eight Chinese were killed, together with a guard, M. E. Connor, of St John, New Brunswick, and one Japanese man. Guard Connor was likely not a railway guard (who in any case were called conductors in Canada) but one of several security officers in charge of the emigrants. The Japanese man was stated to have been a passenger on train No. 97 but was he another emigrant? Nine other Chinese were badly hurt and arrived in Winnipeg aboard a special car attached to train No. 1, the *Imperial Limited*, at 11.40 a.m. the next day, being taken to Winnipeg hospital. The bodies of the 10 dead were left at Kenora.

A Winnipeg newspaper reporter wrote that, 'The sympathy expressed for the Chinese by the officials of the Canadian Pacific was very deep. The fact that almost all the dead were from the far east did not affect the spirit in which the catastrophe was viewed and everything possible was at once done to relieve the sufferings of the wounded.

'A number of local Chinese residents were at the depot at the time of the arrival of the train from the east and eagerly sought for information in reference to the dead and the

790 *Goshen Weekly News-Times*, Goshen, Indiana, United States, 1st March 1907
791 *The Kokomo Daily Tribune*, Kokomo, Indiana, United States, 27th February 1907
792 *The Plymouth Tribune*, Plymouth, Indiana, United States, 19th February 1907
793 Webpage – The Canadian Encyclopaedia: Chinese Canadians

wounded. None of the Chinese on the train had relatives in the city so far as could be learned.' [794]

It was reported that train No. 97 had acquired a bad reputation and was being referred to by railwaymen as the "Hoodoo". Just over a week earlier, on Thursday 20th June, it had been in another accident 40 miles west of Chapleau, Ontario. A forest fire had set fire to a wooden trestle over a creek in the pulp forest and the engine of the first section of train No. 97 had gone through the damaged trestle into the creek. After that passengers, mail and baggage had to be taken across the stream on a corduroy bridge to an eastbound train stopped on the opposite side, an operation which at night took 12 hours, with the sky lit up by the blazing forest. With the apparent jinx attached to train No. 97, Winnipeg passengers were using other services instead wherever possible.[795]

The inquest on the victims of the accident, who by then numbered 12 with the death of two more Chinese, was held at Kenora, Ontario, on Saturday 30th. Its jury returned a verdict that "death occurred at Butler through the failure of train No. 95 to carry out the orders and running past the meeting place." Crown Attorney Mackenzie immediately ordered the arrest of Engineer George Wrighton and Fireman Charles Green, who were subsequently released on bail of $2,000 each.[796]

It was stated on 2nd July that the eight Chinese in Winnipeg general hospital were progressing satisfactorily and that they should be released in 10 days or a fortnight. However, another victim, Ching Sing, died in Winnipeg hospital on Saturday 20th July, the 13th fatality in the disaster.[797]

Engineer Wrighton alone stood trial and on 10th October was found guilty and sentenced to three months' imprisonment. A petition was immediately drawn up requesting that he be granted parole on the grounds that he was a cautious, sober and industrious engineer, one of the best on the division; that he had not been alone in misunderstanding the orders and that he had a wife and child depending on him. [798]

Train No. 155, a special eastbound excursion on the Pere Marquette Railroad, collided with westbound freight train No. 71 near Salem, Michigan, United States, on Saturday 20th July 1907. There were 800 passengers on the excursion train, many women and children, crowded into 11 old wooden coaches and bound for Detroit. The passengers were all Pere Marquette employees from the company shops at Ionia, together with their families, going on their annual trip. The freight train was a local one, consisting of only an engine and seven cars. The three leading cars of the excursion train were telescoped in the collision and the fourth, fifth and sixth cars smashed. The accident happened at about 9.10 a.m. and by noon 22 bodies had been recovered. Fire broke out in the wreckage but does not to have played a significant part in the accident. Although there were estimated to have been 150 children on the train few of them were killed or hurt, most of those killed being men.

After the accident messages were rushed to Salem and neighbouring villages and all

[794] *Manitoba Free Press*, Winnipeg, Manitoba, Canada, 1st July 1907
[795] *The Winnipeg Tribune*, Winnipeg, Manitoba, Canada, 29th June 1907
[796] Ibid, 2nd July 1907
[797] Ibid, 20th July 1907
[798] *Manitoba Free Press*, Winnipeg, Manitoba, Canada, 14th October 1907

the doctors available went to the scene. A relief train was sent from Detroit, rather more than 30 miles away, as quickly as possible with doctors, nurses, surgical equipment and cots. By the time that it reached the scene several score of injured were waiting to receive treatment and the bodies of the dead had been laid out on the grass beside the track.[799]

By Monday 22nd the number of dead was said to have risen to 33. The fatalities included Conductor E. E. Bixley and Fireman Knowles of the excursion.[800] As the excursionists had been railway folk travelling on free passes rather than fare-paying ordinary passengers the Pere Marquette company could have avoided paying any damages to the relatives of the victims but speaking on behalf of the company General Counsel Stevens stated that this immunity would be waived, and the excursionists treated as if they had been ordinary passengers.

Responsibility for the accident had already become apparent, as it had become clear that the freight train had been trespassing on the time of the passenger train, and was not only on the main line at a time when the passenger train was due but actually going towards it.[801] The freight train had left Plymouth, 6½ miles from Salem, at 8.49 a.m. and would take 21 minutes to cover that distance, arriving at Salem at 9.10 a.m., which was exactly the time that the excursion train was due to leave there. Conductor Hamilton of the freight train had in fact admitted his responsibility for the collision to the general superintendent and the company placed the blame upon him and freight train Engineer W. I. Rogers. It transpired that there had been a simple mistake in the reading of a train order. The order in question (No. 3) had been sent to the crew of the freight time advising them of the running of the excursion train and detailing its scheduled times at different stations, including Salem and Plymouth, thus:

Salem	9.10 a.m.
Plymouth	9.25 a.m.

As copied by the operator the lines ran uphill to the right, so that the time 9.25 stood almost opposite the word Salem, although connected by a dotted line to the correct word Plymouth. Although the operator said that Conductor Hamilton had read it back to him correctly afterwards the conductor, the engineer, head brakeman and flagman all managed to misread it, while it was apparently not shown to the fireman at all, as it should have been according to the rules. The freight train thus proceeded from Plymouth thinking that it had 15 more minutes to reach Salem safely than was really the case.[802]

During the inquest the many duties of operators at small stations were revealed. It was questioned whether, after the freight train had been reported as leaving Plymouth, the dispatcher could not have got the Salem operator to hold the excursion train, but Operator Dean Perkins at Salem testified that as well as attending to the telegraph he had to sell tickets, attend to the switch lamps, deal with the freight and baggage, handle the mail and deliver it to the post office. By chance he had been at the post office when the excursion train had passed.[803]

---

[799] *Dubuque Times-Journal*, Dubuque, Iowa, United States, 20th July 1907

[800] *The Weekly Sentinel*, Fort Wayne, Indiana, United States, 24th July 1907

[801] *The Marshall Daily Chronicle*, Marshall, Michigan, United States, 23rd July 1907

[802] *A History of Railroad Accidents, Safety Precautions and Operating Practices* by Robert B. Shaw, pg. 124

[803] *The Evening Review*, East Liverpool, Ohio, United States, 1st August 1907

Dispatcher Marcus Bonsell stated at the inquest that he had had such absolute faith in the crew of the freight train reading and obeying their orders that the idea of holding the excursion train at either Salem or South Lyon (the station before Salem) had never entered his mind, while the double precaution of notifying the excursion train of the presence of the freight train was practically unheard of.

Ii was agreed that the order involved had been illegible but stressed that a careful trainman would not have accepted such an important "31" order without assuring himself of its full meaning.

Other evidence was that the average length of railroad employees was less than two years and that three of the excursion train cars were 35 years old and three more 24 years old.

Trainmaster Everett Kane stated that under similar circumstances the operating department would have acted just the same. Asked sarcastically if the company had learnt nothing from the wreck he replied, "Not a thing." An honest answer, perhaps, but not one likely to appeal to the court.[804]

S. Laverne Laughlin, brakeman on the wrecked excursion, testified that it was his first run, while Rear Brakeman Sweeney said that he had been in the railroad business about four weeks.

Flagman Frank Becker of the freight train testified that when Conductor Hamilton had handed him the order, he had got tangled in reading it, but declared that properly handled the order would have been a safe one.

General Superintendent W. D. Trump was clearly not impressed with the proceedings, and stated from the witness box, 'Our system of dispatching trains is the work of the best railroading minds of the country and I don't have to come before a coroner's jury to be instructed on the question of conducting a railroad.'

On the final day of the hearing of evidence Conductor Hamilton unhesitatingly admitted to having misread the order. Engineer Rogers on the other hand testified that he did not think that one man in a hundred would not have been confused by the order. Pressed by the prosecutor he still insisted that he still preferred the double order system used by the Pere Marquette to the single order system which the prosecutor tried to make him concede was safer. (The single order system was the older one. In it the order was first given to the superior train, for example train No. 5 would be given an order, 'Meet train No. 2 at A, train No. 6 at B and train No. 7 at C.' Train No. 2 would get an order, 'Meet train No. 5 at A,' train No. 6 would get an order reading 'Meet train 5 at B' and train No. 7 an order reading 'Meet train No. 5 at C.' With the introduction of the Standard Code in 1887 the double order system was made standard. In this system trains Nos. 2, 5, 6 and 7 would all get an identical order reading, 'Train No. 5 meet train No. 2 at A, train No. 5 at B and train No. 7 at C.' [805] The double order system was generally held to be superior, whatever the prosecutor might think.)

At 10.30 p.m. on 2nd August at Northville, Michigan, the inquest jury, after an absence of six hours, returned a sweeping verdict on the death of Mrs Abraham Eddy and 30 other persons in the collision. The verdict read in part '...that they do further say that the said

[804] *Pinckley Dispatch*, Pinckley, Michigan, United States, 1st August 1907
[805] *Brotherhood of Locomotive Firemen and Engineman's Magazine* Vol 57 (1914) pgs. 305-6

head-on collision was the result of a misreading of order No. 3 on the part of Conductor Hamilton, Engineer Rogers, Head Brakeman Briggs and Flagman Becker, of the crew of train No. 71, and that the said misreading of the said order No. 3 was due to the imperfect and improper manner in which the order was prepared by Operator Sayre and delivered by Operator Cassaday, of the Plymouth station; that they do further say upon their oath that they find the operating system of the said Pere Marquette Railroad and the rules and regulations governing the same are defective in that they do not provide that all trains which are running under a special schedule not in the regular time card of the said railroad should be absolutely obliged to stop for orders at all stations which have been indicated by orders for other trains as a meeting point, that they do further say upon their oaths that the absence of such a rule and regulation from those governing the operation of trains on the said Pere Marquette Railroad was as equally a direct and proximate cause of and equally responsible for the said collision and the failure of the crew of the said train No. 71 to properly read order No. 3 or the imperfect and improper manner in which the same was prepared and delivered.'

Assistant Prosecutor Robison had already announced that no prosecutions would follow the case as no evidence to justify any had emerged. The newspaper reporter saw the verdict as a victory for him even so, as although the jury had ignored his concerns about the double train order system and ruled or unruled train order blanks, they had agreed with his main contention that the Pere Marquette system of train operation was defective.[806]

The last week of August 1907 was the week of the Coles county fair at Charleston, Illinois, United States and 30th August was "Big Friday" when the largest crows attended. The fair officials had prepared a special program and expected banner crowds. The weather was good and all morning the cars of the Central Illinois Traction company had been crowded with merrymakers. To cope with the traffic a trailer was attached to the heaviest electric car on the line. The interurban car and trailer left Mattoon for Charleston at 11.15 a.m. From Urban Park, midway between the two places, to near Charleston the line followed the Ambraw River with sharp curves and heavy grades. Halfway down one of these grades and just beyond one of the sharpest curves the eastbound passenger train collided with a westbound express car loaded with freight.

Motormen Charles Botts, on the passenger car, and Ben McClara, on the freight car, saw one another at the last moment, applied the brakes and jumped. The speed of the cars was only slightly reduced, and the frail upper portions of the cars were cut from the bogies 'like so much cheese'. The bodies of the passengers were thrown high into the air among splintered woodwork and broken glass. There had been 99 passengers on the eastbound train and the majority of them were killed or injured. A reporter wrote, 'The crash of the collision was followed by screams of the dying, which attracted the attention of farmers working in fields nearby, and they came to the rescue. Superhuman strength seemed to be there as they lifted the broken timbers from the injured.'

Some of those injured, mad with pain and fright, reportedly fled into the fields while others gave temporary aid to those more seriously hurt. The unnamed nine-year-old daughter of Farmer John Linden ran over a mile down the track to the fairgrounds to

---

806 *The Marshall Daily Chronicle*, Marshall, Michigan, United States, 3rd August 1907

summon help and thousands from the fairgrounds walked to the scene to lend assistance. It was almost an hour before the first relief car reached the scene with physicians and nurses. Telephone messages to Charleston and Mattoon also brought ambulances from both places and the casualties were taken to the Methodist Memorial hospital and the Montgomery Sanitarium in Charleston. The hospital facilities proved inadequate though to deal with the large number of casualties and farmers wives transformed their homes into emergency hospitals. An early report on the accident spoke of 16 passengers being killed outright, four dying shortly after removal to hospital and two score more injured. Conductor Gucker escaped with slight bruises.

The same newspaper account added that, 'This is the third serious collision that has occurred on the Mattoon-Charleston interurban line within the last three years. In the first 20 persons were injured in a collision near the State Odd Fellow's Old Folks' home in this city {Mattoon}. The second took place Sept 4 last year, at the same place, and one was killed and 18 injured. Both were head-on collisions. As a result of the last collision, John A. Backus, who was in charge of the dispatching system, committed suicide from excessive brooding. He was a protégé of Judge Peter A. Grosscup of Chicago, who is the heaviest stockholder in the road.' [807]

At the inquest on the victims of the accident the number killed had been revised down to 14. The verdict of the jury was that both Motorman Botts of car No. 14 and Motorman McClara were to blame, McClara for having started from Charleston without knowing where car No. 14 was and Botts for disobeying orders in not having his train under control when rounding the curve where the accident occurred. Both were arrested. The jury also found that the Central Illinois Traction company employed incompetent men, that it had no person employed in its principal office from whom employees could get positive instructions as to the movement of cars and that the instructions to employees on the days when the regular schedule was not in force were misleading.

Following the accident, the railway company was placed in the hands of the receiver in anticipation of the large number of claims for damages which were likely to be received. [808]

Following the findings of the Grand Jury at the end of October President Edgar Potter, Superintendent Fred Moore, Directors Francis Peabody, Marshall Sampsell, Arthur Underwood and Judge Peter Grosscup were all arrested on a charge on involuntary manslaughter in connection with the accident.[809]

At the end of February 1908 the indictments against all the accused were quashed. State Attorney McNutt announced in April that he intended to re-indict them.[810] However, Judge Thompson, in instructing the grand jury, ignored the case, and it was believed that the matter would be dropped.

According to one source this was the end of the affair.[811]

Another source states that the matter did come to trial, but this appears to be a reference to the inquest. Evidence was given at this proceeding that the superintendent of

---

[807] *The Daily Free Press*, Carbondale, Illinois, United States, 31st August 1907
[808] *Cook County Herald*, Arlington Heights, Illinois, United States, 13th September 1907
[809] *Logansport Daily Reporter*, Logansport, Indiana, United States, 31st October 1907
[810] *The Des Moines Capital*, Des Moines, Iowa, United States, 17th April 1908
[811] *The Daily Review*, Decatur, Illinois, United States, 21st April 1908

operations, who was also a circuit clerk of Coles County, had only been doing the job a short time, and that although he was regarded as an experienced railwayman he had in fact only worked for a short time as a brakeman on the Clover Leaf (Toledo St Louis and Western Railroad) many years earlier. Due to the heavy traffic the usual hourly schedule had been altered to a half-hourly schedule. The superintendent testified that he had told Motorman McClara and his assistant of the change in schedule, they testified on oath that he had not, but had said that the trains would run the same as yesterday. The freight car had then departed westwards into collision.[812] Train orders were usually transmitted by telephone, with phone booths at sidings along the 12-mile line. It would seem that neither car involved in the accident had been issued with any orders regarding the other.

The final number of fatalities in the collision was probably 18, all of them passengers.

Visitors returning from the fair at Sherbrooke, Quebec, Canada, were the main victims of a collision north of Canaan, New Hampshire, United States, on the Boston and Maine Railroad, early on the morning of 15th September 1907. The trains involved were train No. 30, a Quebec to Boston express, and train No. 267, a northbound freight. The express was running down the grade towards Canaan station while the freight was trundling uphill towards the siding at West Canaan, four miles on the White River Junction side of Canaan, which its engineer expected to be able to reach in time. As the express rounded a curve the engineers saw one another's headlights emerge from the fog, jammed on their brakes and jumped. In the collision the engines were said to have locked themselves into an embrace and they then rolled over into a ditch. The leading baggage car of the passenger train was forced back through the following day coach, tearing it asunder from end to end. The rear of the car climbed up on the top of the smoker, which only suffered a few broken windows, while the Pullman cars behind the smoker were undamaged. There were more than 50 passengers riding in the day coach. Just before the accident a few of the men had gone back into the smoker, leaving the women to try and get a little sleep in the straight seats. A surviving passenger stated that as the train rounded the curve someone at the front began to 'sing' (cry out?) so that almost all the passengers were awake at the moment of the collision. The uninjured passengers from the smoker and sleepers jumped down onto the track after the accident and hurried forward to the demolished coach. The fact that the engines had gone off to the side of the line probably prevented the wreckage from catching fire.

Trainmen and passengers began the rescue work. The bodies of the dead were laid alongside the track and the injured were dragged out and placed on the backs of broken seats or on blankets taken from the sleeping cars. Wounds were bound and cuts staunched using strips of bedding from the sleepers. 'The little band worked diligently in the dawning light before the doctors came.' There were only a few farms in the vicinity of the accident site, but the farmers were aroused and came and lent their help. Meanwhile calls for assistance had been sent to White River Junction, Concord and Hanover and a large number of physicians hurried to the scene. Passengers from the following Montreal express also assisted.

One of the first West Canaan farmers to arrive at the scene was a man named Benjamin Briggs. Two hours after he had arrived, he retrieved a body from the wreckage

---

[812] Webpage – Central Illinois Traction: Hick's Car Works

and looking at its face found that it was that of his own mother, who had been visiting friends in the north and was returning to Boston. He almost collapsed with grief.

Brakeman Frank Ryan of the express was badly injured, with supposedly a severed artery, and lost consciousness for a quarter hour. When he revived his first thought was for the safety of the following Montreal to Boston express. He made someone aware of the danger and the fact that probably no one had gone back to protect the line. Someone did so and the express was stopped a quarter mile short of the obstruction.

Initial reports spoke of 24 dead and dying passengers and 27 others hurt, most of them seriously. The number of fatalities later rose to 26.

A statement from the Boston and Maine company outlined the circumstances of the accident. Train No. 34, the Montreal to Boston express, normally preceded train No. 30, the Quebec to Boston express. On the day of the accident both trains were running late, and in reverse order. Train No. 30, running 40 minutes late, left White River Junction at 3.40 a.m., with train No. 34 20 minutes behind. J. A. Crowley, the night dispatcher at Concord, New Hampshire, had to arrange altered meets between the two delayed expresses and opposing inferior trains and sent an order to White River Junction and Canaan advising northbound trains that train No. 34 was 1 hour and 10 minutes late. According to the dispatcher's records this was the order which he sent to Canaan and which was repeated correctly by John Greeley, the Canaan operator. The order as delivered to freight train No. 267, however had '30' on it instead of '34', making the conductor and engineer of that train believe that they had 30 minutes more than was really the case to reach West Canaan siding.[813] Conductor Lawrence of the freight train was able to produce his order after the accident proving that it read 30 rather than 34.

Dispatcher Crowley told the state railroad commission inquiry into the accident that the first he knew of anything being wrong was when he had contacted Canaan to ask them to notify the freight train crew that the train order board at Lebanon (near White River Junction) was faulty. Operator Greeley had then replied that the freight train had already gone. The dispatcher said he had told Greeley that the freight train would not have enough time to get to West Canaan, to which the Canaan operator had answered that the Quebec train was an hour and 10 minutes late, so there was plenty of time. Dispatcher Crowley said that he had told Greeley that it was the following Montreal train, which was that late and that, knowing that a collision was inevitable, he had wasted no time in informing the superintendent and calling out the wrecking train. He was positive that he had made no mistake in transmitting the order to Canaan or in hearing it verified.

Operator Greeley identified the incorrect order and declared that that was how he had received it and repeated it back to the dispatcher. He said that he was on duty from 7 p.m. to 7 a.m. daily and that in addition to working the telegraph he dealt with the baggage, mail and freight for three trains. He had been in his office and near the telegraph key all night before the wreck and had not been asleep. He was sure that the records showed the message as he had received and repeated it.

Operator Maher at White River Junction testified that the order as he had received it corresponded with that sent by the dispatcher. It also came out that the orders had not been sent to Canaan and White River Junction simultaneously, as they should have been.

---

[813] *The Evening Record*, Greenville, Pennsylvania, United States, 16th September 1907

Superintendent Ray of the Concord division stated that the investigation had narrowed the blame to either the dispatcher or the operator and it was clear that neither train crew was in any way at fault. Both dispatcher and operator were careful, prudent and responsible men.[814]

Regardless of who was responsible the disaster showed yet again the vulnerabilities of the train order system.

The next bad single-line collision happened in Mexico, near Encarnación de Diaz, early on Thursday 19th September 1907. The express train from El Paso, Texas, United States to Mexico City via the Ferrocarril Central Mexicano (Mexican Central Railway) was rounding a curve in the foothills near the town when it collided head-on with a freight train. A report from Mexico City stated that every car of both trains had been wrecked and that two coaches of the express had plunged down a steep gully, where many bodies still lay at the time of the report. The report complained that many of the injured were practically uncared for, Encarnación de Diaz being only a small place with two physicians and no hospital. The railway authorities admitted that there had been a serious accident and a relief train was on its way to the scene from Mexico City. The initial report spoke of 30 dead and twice as many injured.[815] By the next day a further report from Mexico City stated that the number of dead was 68, with 42 injured, many of whom were expected to die.[816] Those who were killed and injured were all Mexicans except Engineer McFarland of the passenger train, who was probably fatally hurt.[817]

A later and presumably more accurate report stated that the leading four coaches of the express had been telescoped and demolished but that the Pullman cars at the rear of the train were undamaged and none of the Pullman passengers was hurt. The express was two hours late and trying to make up time.[818] It was crowded with people returning from the national fiesta at Aguascalientes.[819]

The accident reportedly resulted from the gross negligence of the American crew of the freight train, Engineer Brock and Conductor Randleman. The collision was due to the freight train running past the siding where it should been shunted out of the way of the passenger train. At the moment of the collision only the Mexican fireman and the coal pusher were on the engine, the engineer having abandoned the locomotive to join the conductor in the caboose, where they occupied themselves drinking together. The fireman was killed outright in the collision, being found on the right-hand side of the cab with his head crushed, and the coal pusher died shortly afterwards from his injuries.

Back in the caboose the conductor and engineer escaped death or serious injury in the accident and disappeared after the wreck. Mounted soldiers and locals scoured the countryside for them but the two evaded their searchers and made their way to the city of Aguascalientes. This did them no good, however, for they were arrested there while trying to hide in the American quarter and it was believed that the chances of their

---

814 *Concord Enterprise*, Concord, Massachusetts, United States, 25th September 1907
815 *The Ogden Standard*, Ogden City, Utah, United States, 20th September 1907
816 *Anaconda Standard*, Anaconda, Montana, United States, 22nd September 1907
817 *The Sun*, Baltimore, Maryland, United States, 21st September 1907
818 *The Daily Huronite*, Huron, South Dakota, United States, 23rd September 1907
819 *Santa Fe New Mexican*, Santa Fe, New Mexico, United States, 20th September 1907

avoiding long prison sentences for their negligence were exceedingly slim.[820]

Only five days after a serious collision at Kot Lakhpat station the North Western State Railway of India had another bad accident, on Tuesday 29th October 1907, this time at milepost 1042-18 between Jagadhari and Kalanour stations, when No. 27 up passenger train collided with No. 3 Rajpura down special goods. Nine passengers and no less than 13 railwaymen were killed, and six passengers and the same number of railwaymen were injured. Like the earlier accident this one was due to driver error, the driver of the goods train having run through the signals at Jagadhari station. 'The rolling stock was considerably damaged.' [821]

In what was presumably a single-line collision No. 18 down passenger train and No. 61 up goods train collided between a temporary block hut at mile 230-6¼ and Dadpur block hut on the East Indian Railway on Tuesday 17th December 1907. Eight third-class ('native') passengers were killed and 18 passengers were hurt. The guard of the goods train was slightly hurt and there was considerable damage to rolling stock. The accident was blamed on the failure of the staff at Jhajha to give a caution order to the driver of the up goods train. [822]

In the third collision in India in quick succession, No. 5 up and No. 12 down passenger trains collided at mile 935-2, between Ludhiana and Lahowal on the North Western State Railway on Christmas Day, Wednesday 25th December 1907. This time 15 passengers and six railway servants were killed, and three passengers and four railway servants were injured. Those killed included four Europeans. The trains were reportedly travelling at 30 miles an hour when they met. A stationmaster was stated to have been arrested, bail being refused. This was probably the assistant stationmaster at Ladhowal, whose disregard of the rules was blamed for the accident.[823] [824]

An American interurban line was in trouble again on Tuesday 28th April 1908 when the Detroit United Railway had a collision at Denton, Michigan, United States, four miles east of Ypsilanti, on what was known as the Ann Arbor branch. Nine people died when a limited eastbound car from Ypsilanti collided head-on with a regular westbound car from Detroit. Hardly any of the 50 passengers who were riding in one or other of the cars escaped some sort of injury, those who were not badly hurt suffering bruises and cuts from broken glass. The cars were reportedly smashed to pieces. The dead included Motorman Isa Fay of Jackson. Following the collision telephone calls were made to Ann Arbor and Detroit and a special car was rushed to the scene and took the casualties to Ypsilanti; the seriously injured passengers being then all taken on to the university

[820] *Abilene Daily Reporter*, Abilene, Texas, United States, 24th September 1907
[821] Administration Report on the Railways in India for the Calendar Year 1907
[822] Ibid.
[823] Ibid.
[824] *The Times*, London, England, 28th December 1907

hospital in Ann Arbor. The accident occurred at about 2.25 p.m.[825]

A more detailed report on the accident stated that the limited car, driven by Motorman Fay, had left Ann Arbor at 1.59 p.m. and Ypsilanti at 2.17 p.m. for Detroit, comfortably filled with about 40 passengers. The running times of the car had been both changed and reduced that day, so that the car had started 10 minutes earlier than before, a factor which might have led to the mistake which caused the accident. The limited and regular cars should have passed one another at Harris switch, two miles short of the scene of the collision, but Motorman Fay continued past there at high speed. Rounding a curve two miles further on the local car came into sight, also travelling at considerable speed, and a collision was unavoidable.

Motorman Wingrave of the local car jumped at the last moment but Motorman Fay, perhaps realising that he was at fault and trying to do his best to minimise the effects of the collision, did not, and was crushed to death in the vestibule of his car. The regular, local car was considerably higher than the limited car so that in the collision the heavy body of the local rode up over the flooring and ironwork of the limited car, which was telescoped for 30 feet, half its length, though it was not derailed. [826]

'There were screams of fright from the passengers on the limited as they saw the local train loom before them and then they were silenced in the crash. Survivors say that there was a moment of deathlike stillness following the collision and then the uninjured and slightly injured passengers crawled from the wreck and after a few dazed moments began the work of rescuing those who were pinned down under the local car.' [827]

The day after the accident no further deaths had occurred, but one passenger was said to be in a critical condition and another passenger and Conductor George Meade of the limited car were in a serious state.[828]

Early in May 1908 the London *Times* carried a very brief report on an accident in India in which it was stated that ten 'natives' had been killed.[829] The next day it stated that 15 had died [830] and on 1st June it was stated that the official death toll had risen to 120.[831] This official figure was later reduced to 79, though an almost contemporary author claimed a death toll of 400.[832]

The accident occurred on the Oudh and Rohilkhand Railway, on Wednesday 6th May 1908, on the single line between Ghaziabad and Dasna stations. Two passenger trains, each reportedly travelling at 40 miles an hour, met on a curve where forward visibility was further restricted by a clump of trees. The leading six vehicles of each train were completely wrecked and piling up upon one another then caught fire. The blaze, fed by gas from damaged reservoirs, spread very rapidly due to a high wind and was further fed by the coal from both engines. The carriages were full, and the fire spread so quickly that

[825] *The Weekly Pantagraph*, Bloomington, Illinois, United States, 1st May 1908
[826] *The Evening Independent*, Massillon, Ohio, United States, 29th April 1908
[827] *The Warren Mail*, Warren, Pennsylvania, United States, 30th April 1908
[828] The Weekly Sentinel, Fort Wayne, Indiana, United States, 6th May 1908
[829] *The Times*, London, England, 8th May 1908
[830] Ibid, 9th May 1908
[831] Ibid, 1st June 1908
[832] *Engines of Change: The Railroads that made India* by Ian J. Kerr, pg. 100

anyone unable to move was burnt to death. The bodies of 13 passengers were recovered but a very large number, perhaps 100, were consumed. Fifty other passengers were known to have been injured but a large number of others were removed by their friends. Both drivers and both firemen were killed.

What made this accident exceptional, apart from its high death toll, was that fact that since 1902 the section involved had been worked by the Tyer's electric train tablet system, in place of the written "line clear" system generally in use in India. Neither of the train crews was at fault, two tablets having been taken out at the same time for the Ghaziabad – Dasna section, something which should have been impossible unless one of the instruments had been interfered with. An expert committee was set up at Delhi to investigate how the accident could have happened and how to prevent a repetition.

The accident subsequently featured in civil court case in which Mrs E. C. Blanchett, the widow of one of the dead drivers, who had been left with six children to bring up alone, made a claim for damages against the Oudh and Rohilkhand Railway company. From this case it emerged that the investigation into the disaster had been unable to determine where the blame for the disaster lay. It was stated that the tablet instruments could be tampered with and the names of three Oudh and Rohilkhand railwaymen, Mr Collett White, Jawahir Lal and Sander Lal came under the spotlight. It could not be determined which train had started first and which, therefore, could have legitimately been given a tablet. The tablet instruments had been thoroughly tested after the accident and there was no defect in them. Dasna was a simple crossing place while Ghaziabad was a more important station, also used by the East Indian Railway. It was stated that there was nothing to connect Mr Collett White with the accident. Sander Lal at Dasna had previously been suspended following allegations that he had not learnt by heart the bell codes used in connection with the tablet instrument, but this allegation was later shown to be false and he proved himself perfectly capable of working the instrument. The company subsequently discontinued the use of the tablet system, presumably going back to the old written "line clear" working instead. [833]

The Washington Baltimore and Annapolis Railway, another interurban line, was the location of the next accident on Friday 5th June 1908. Two electric cars collided just at about 7.39 p.m. near Camp Parole, Maryland, United States. The collision involved a regular car from Annapolis, driven by Motorman James O'Neal, who was one of those killed, and a special car driven by Motorman George Wadsworth. One car was going at perhaps 60 miles an hour and the other about 35 miles an hour when the motormen caught sight of one another's cars and realised that a collision must happen.

'A second later,' a Baltimore newspaper reported, 'the heavy green cars, with their passengers, were twisted like shots from a catapult into each other and were knocked and broken to splinters. An awful scene ensued. Several men were killed by the side of their wives. Among them were Mr J. W. McDaniel, of Baltimore, and Mr A. H. S. Schultz, both of whose wives were injured.

'As soon as anything could be done, the work of rescue began, and the Emergency Hospital at Annapolis was told to prepare for the injured. As rapidly as they could be, they were taken there, and Drs S. S. Hepburn, George Wells, B. B. Brown, Walron

---

[833] Webpage – Mrs E. C. Blanchett vs Secretary of State for India

Hopkins, W. S. Welch and J. J. Murphy worked like Trojans to save lives. Mr McDaniel died at the hospital after reaching it. All the doctors had to operate and work at the greatest speed.'

Extra services were being run during the week of the accident to handle the large crowds in connection with the week's festivities at the United States Naval Academy at Annapolis, and especially on the day of the collision a farewell ball being given to the graduates. The regular car driven by Motorman O'Neil was largely filled with passengers going to attend the ball. Many of the passengers on the car saw the special from Annapolis coming towards them and retreated to the back of the car, thus saving themselves. Both cars were described as having been 'smashed almost to smithereens' in the collision but that of the local service seems to have suffered the worse, the car of the special having ploughed almost entirely through it before it rolled off into a bank, though its rear bogie remained on the rails.

'The scenes following the collision were terrifying. Passengers were hurled hither and thither and several were terribly injured by being jammed between the seats and by flying bits of timber and glass. The screams of the women and groans of the injured could be heard at a good distance, and residents of the neighbouring country within a radius of a mile hurried to the scene and gave assistance to the injured by providing pillows and other comforts. Motorman O'Neil's body was not badly mangled, but it was hurled a distance of 50 feet, and it was some time before it was found in a clump of bushes at the base of an embankment.

'A relief train with several physicians was sent from Annapolis, and eight of the injured, with the bodies of seven who were killed outright, were brought to Annapolis. The injured were hurried to the Emergency Hospital in ambulances. In the meantime, several of the injured, including General Traffic Manager Slaughter, Judge James R. Brashears and others were brought to Annapolis in carriages. Judge Brashears suffered considerably from shock. He was taken to the Maryland Hotel, where he was given treatment, but soon recovered himself, and returned to his home, near Eastport.

'The bodies of the killed were viewed by Police Justice John N Davis, acting coroner, and, upon their arrival at Annapolis, were turned over to local undertakers to be prepared for burial.'

Upon becoming aware of the accident General Manager and Vice President Shanahan set off for the scene but had to wait for an hour at Naval Academy Junction (near Odenton) as the electric current had been cut off. Arrangements were then made to run a car to the scene and passengers who had been waiting at the junction for about three hours were also were also taken aboard. At the accident site they walked around the wreckage and on the far side another train took them to Annapolis, while Mr Shanahan remained at the scene.

The death of 19-year-old Harry Jacobs in the Emergency Hospital on Saturday night brought the final number of dead to nine.

By Sunday 7th blame for the accident was being placed on Motorman Wadsworth and his conductor, Edward Bernasco. It was stated by Dispatcher B. C. Scott at Naval Academy Junction that Wadsworth had been given an order at 7.30 p.m. to work an extra car from Annapolis to the Naval Academy Junction without passengers and to sidetrack at Camp Parole for the regular car to Annapolis.[834] The next day it was said that Motorman Wadsworth and his conductor had been instructed at Annapolis to proceed to Naval Academy Junction, "calling in" at Best Gate, the station on the Naval Academy Junction side of Camp Parole, and to watch out for regular car 133 – the one with which they collided – which was due at Annapolis in 11 minutes. One of the regulations stated that extra trains must be in clear of regular trains five minutes before the departure time of the latter at every siding. Motorman Wadsworth told a newspaper reporter that he and Conductor Bernasco had checked their watches and the time table when they arrived at Camp Parole and they had concluded that they had enough time to press on to the next siding at Best Gate. In his own words Motorman Wadsworth said, "My only salvation rests on whether Bernasco, my conductor, preserved the orders we received last night. My own orders were lost, and I am still too confused as to what happened to recall a word of them. When a man has gone through what I have, when he has looked almost certain death in the face, details don't count for much.

"One thing I do feel sure of, though; that was that I had no orders regarding Camp Parole. I believe I should have stopped at Best's Gate, but I had no time to get there. The first warning I had after rounding the curve after Parole was the headlight of the approaching special. {This reference to a special appears to be an error either of the motorman or of the reporter who wrote down his words. It was not a special but the regular train.}

"One of the first things a motorman learns is to switch off his currents suddenly, and this I did from force of habit and then jumped off into the night. I am not at all clear, but it seems to me that the crew of the special, judging from what my orders were, could have carried out their instructions.

"No, I don't think I'm seriously injured, but the doctor says I won't be able to put my foot to the ground for at least two weeks and maybe for a month.

"I'm not trying to withhold anything. If I could just remember clearly, I would gladly tell anybody all I know of the circumstances of the wreck, but it was all over in a minute, and I can't tell just why the collision occurred, unless, as I said, the other crew disregarded orders. When I threw on the brakes and heard them catch that was all I could do, and I couldn't have made things any better by not jumping." [835]

Motorman Wadsworth had only been employed on the line for about a month but had previously been a motorman on the Ohio Traction Company for about seven years. Conductor Bernasco had considerably more experience, including 18 years on the United Railways of Baltimore.

The inquest on Mrs Mary Green, representing the nine people who lost their lives in the accident, concluded on 22nd June, without having heard the evidence of Motorman Wadsworth, who had disappeared after the collision, it was supposed to his home in Ohio, and could not be found. The basic finding of the inquest was that the extra car had left Annapolis at 7.30 with sufficient time to cover the three miles to Camp Parole and be in clear there five minutes before the arrival of the regular car, as per the regulations. It was because they had pressed on to Best Gate that there had been an accident.

---

[834] *The Sun*, Baltimore, Maryland, United States, 6th June 1908
[835] Ibid, 7th June 1908

Conductor Bernasco testified that they had left Annapolis at 7.30 but said that he had no recollection of the time that they had passed Camp Parole. It appeared that he had misconstrued the despatchers order at Annapolis, which were to call at Best Gate for further instructions, as giving the extra right to the track as far as Best Gate, instead of looking out for the regular car at some point short of there. This was not, however, the whole story. It developed that the order as delivered to the crew of the extra was void and should not have been accepted. The original order to extra No. 21, issued by Chief Dispatcher Bernice Scott at Naval Academy Junction, read, "Leave Annapolis for the Junction, and meet the regular and the work engine at Best Gate." Later Dispatcher Carson Schumacher at Annapolis, on instruction from Scott, had altered the order by crossing out the words "and meet the regular and the work engine at Best Gate", and had substituted, "report at Best Gate". Instead he should had annulled the order and issued an entirely new one.

Dispatch Schumacher, giving evidence, emphatically denied having given any verbal order to Conductor Bernasco. The conductor stated that not only on this occasion but on several previous occasions he had taken a car from one siding to another on the basis of such verbal orders. Scott and Schumacher reiterated their denials that such verbal orders had ever been given.

Conductor Isaac Walraven of the regular train admitted that his car had been late leaving Naval Academy Junction. It had started from there at 7.28, passed Waterbury station, five miles on, at 7.33, and the collision had occurred at 7.38 or 7.40, at which time he estimated his car had been moving at 60 miles an hour. The car was running at high speed because it was 12 or 18 minutes behind schedule. He had received no orders and knew nothing of the extra being in the block.

After discussing the matter for over three hours the inquest jury found that the causes of the collision were:

(a) By Motorman Wadsworth and Conductor Bernasco receiving a defaced or scratched order.

(b) By the inexperience or ignorance of Dispatch Schumacher in issuing a defaced and scratched order in direct violation of rule 334, Book of Rules, governing the movement of trains and the duties of the employees.

(c) That the conductor and motorman of car 21 should have taken precautions to clear train 133 five minutes, according to rule 320.

(d) That there was negligence in the movement of trains on June 6th, 1908.

The jury stopped short of finding any employee guilty of criminal negligence but Conductor Bernasco had already been arrested and it was expected that his case would be considered by the Grand Jury in October.[836]

No record has been found of any further developments in the matter.

Two Missouri Pacific Railroad passenger trains, the California special from St Louis and the St Louis special from Kansas City, collided two miles east of Knob Noster, Missouri, United States, at 5.20 a.m. on Wednesday 1st July 1908. Both trains were

[836] Ibid, 23rd June 1908

travelling at full speed. The California limited consisted of two mail cars, a baggage car, two chair cars, two sleepers and an observation dining car. The Kansas City train had eight cars and was behind time, having been held two hours for the addition of two cars of soldiers discharged from Fort Leavenworth.

The two engines were smashed in the collision and thrown from the track but did not overturn. Four of the cars from St Louis and three of those from Kansas City left the rails and piled up. The two mail cars on the Kansas City train were of the new steel type; both were derailed and badly damaged. Part of the wreckage caught fire. A thick fog prevented the engineers having much warning of the impending accident, but they had time to jump. A special train was run to the scene from Sedalia after the collision, carrying all the available physicians from that town and another wrecking train was sent from Kansas City. Until the line could be cleared trains were diverted over the Lexington branch. The unhurt passengers on the two trains involved were transferred to trains on the opposite sides of the obstruction. The injured and the bodies of the dead were taken to Sedalia.

Nine persons were killed and 36 injured. Those who lost their lives were Baggageman Campbell and two assistants, three Western Union linemen named Fred Storey, John Hood and Michael Burke, coloured mail clerk W. H. Harding, W. J. Frisbie, salesman for Roberts Rand Shoe Company and S. R. Inglish, a lumberman.[837]

The trains had been originally arranged to cross one another at Knob Noster but F. W. Strange, the Sedalia dispatcher had subsequently issued new orders changing the meeting point to Lamonte, seven miles east of Knob Noster.[838] The accident was, it soon became clear, caused either by the failure of the dispatcher to issue the new order to an operator or the failure of an operator to deliver the said order to the Kansas City train, which thus instead of waiting at Lamonte continued towards the original meeting point at Knob Noster.[839]

At the inquest on the victims of the accident Dispatcher Strange stated that he had issued an order to Lamonte and Knob Noster to hold train No. 3, the California special, at Lamonte, for train No. 12, the late-running St Louis special, and that Operator Charles Taylor at Lamonte had replied ‘O.K.’ but the train had already left the station. In this he was corroborated by his written record and by Conductor Williams of train No. 12 who had read the same message off the wire while his train was waiting at Knob Noster. The jury found both Strange and Taylor responsible and Taylor was subsequently arrested on a charge of manslaughter. Strange was not but he was discharged by the railroad company, perhaps for issuing a holding order at the point at which it applied.[840] In a somewhat surprising development at the preliminary hearing of Taylor’s case at Warrensburg on 17th July he was exonerated from all blame, despite the verdict at the inquest.[841] [842]

The narrow gauge Flensburg – Rundhof line suffered a bad accident on Sunday 9th

837 *Checotah Enquirer*, Checotah, Oklahoma, United States, 10th June 1908

838 *The Cheyenne Daily Leader*, Cheyenne, Wyoming, United States, 3rd July 1908

839 *The Lincoln Daily Star*, Lincoln, Nebraska, United States, 2nd July 1908

840 *Sedalia Democrat-Sentinel*, Sedalia, Missouri, United States, 10th July 1908

841 *The Laclede Blade*, Laclede, Missouri, United States, 25th July 1908

842 *Sedalia Democrat-Sentinel*, Sedalia, Missouri, United States, 23rd July 1908

August 1908 at Groß Tarup, Germany, when passenger train No. 32 collided with a special train. Two coaches of the train coming from Sörup were telescoped into one another, nine people lost their lives and eight were seriously injured. The driver of the passenger train had a written order to cross the special train at Groß Tarup but the latter train was running ten minutes late and so had not arrived there. Not seeing the special train the driver of the passenger forgot about it and proceeded. As it was a lightly used minor railway there was neither a signalman nor fixed signals to prevent such a mistake.[843]

Young's Point, Montana, United States, was the scene of a disaster at 8.02 a.m. on Friday 25th September 1908 on the Northern Pacific Railroad, though it involved a Chicago Burlington and Quincy Railroad train running over foreign metals, which had left Helena, Montana, at midnight. Eastbound express train No. 16, was known as the *Burlington Flyer* and consisted of an engine, baggage car, smoker, day coach and two Pullman cars. It was running down grade at a reported 50 miles an hour when it crashed head-on into a heavily loaded westbound freight train. Both engines were wrecked, and the baggage car and smoker of the flyer were completely telescoped, the smoker ending up next to the tender of the engine and being 'reduced to a heap of kindling wood'. None of the other vehicles was derailed.

The freight train, which was described as being exceptionally heavy, had left Park City, four miles east of Young's Point, with only six minutes in hand to get to Young's Point. It had got partly into the siding there when its engineer heard the passenger train coming. Realising that there was no time to get in clear he had, instead, backed his train out onto the main line again and begun to reverse towards Park City, while sending a flagman forward to stop the passenger train. The engineer had got his train back two passenger train lengths from the siding when the collision occurred. The flagman hurried towards the oncoming express, but it was foggy, and snow was falling heavily so that the red flag that he was frantically waving was not seen in time by Engineer Joe Bessinger of the passenger train. As a last resort the flagman threw a fusee through the window of the engine cab but even this was not noticed. Engineer Bessinger did finally see the freight train in time to throw on the brakes in emergency and leap for his life. He escaped with a few bruises. Less fortunate was his fireman, Ora Babcock, who jumped through the cab window and struck his head on a switch rail. His skull was fractured, and he died in a short time. Head Brakeman Milo Halloway of the passenger train was also killed but Express Messenger R.W. Ledue got away with a few bad bruises and Conductor T. M. Griffin was unhurt. The crew of the freight train jumped as well but according to one report Fireman Phillips was killed.[844] None of the 35 occupants of the smoker were said to have avoided being killed or injured. Eighteen of those in the smoker were killed outright and three more died later. The day coach passengers only sustained cuts and bruises and the Pullman passengers were unhurt. The heavy, wet snow did at least prevent the wreckage from catching fire. Among the passengers on the train was the Spokane delegation to the National Irrigation Congress at Albuquerque, New Mexico.[845]

---

[843] *Eisenbahn-Unfälle in Europa* by Erich Preuß, pg. 12 (in German)
[844] *Sanders County Signal*, Plains, Montana, United States, 1st October 1908
[845] *Morning Oregonian*, Portland, Oregon, United States, 26th September 1908

Railroad men and volunteers from Livingstone and Billings helped treat the injured and clear the wreckage and a relief train was sent from Billings, 30 miles to the east, with Dr Watkins and his assistants and several nurses, more physicians being picked up at Laurel and Park City, while farmers living nearby supplied clothing and blankets and received many of the injured into their homes. The injured were brought around the wreck and taken on to Billings on the relief train.[846]

The day after the accident an inquest on those who lost their lives had already been held and completed, the jury placing the blame for the accident on the crew of the freight train for 'running on short time' or 'stealing time' as it was termed.[847] Prosecution of the crew of the freight train was to follow.

On 3rd December 1908 up mail train No. 3 and down mail train No. 4 collided on the single line at mile 1029 between Mustafabad and Barara, India, on the North Western State Railway. Fifteen passengers and seven railwaymen lost their lives and 20 passengers and four railwaymen were injured, with considerable damage to rolling stock. Line clear was given for the down mail after line clear had already been given for the up train.[848]

Westbound Denver and Rio Grande Railroad passenger train No. 5 left Denver on the morning of Friday 15th January 1909 and was due at Glenwood Springs at 10.30 p.m. It never reached its destination for at 9.35 p.m. it was in a head-on collision with an eastbound double-headed freight train near Dotsero, Colorado, United States, while running through the darkness at 45 miles an hour. The passenger train was expecting to meet the freight train five miles further on, while the freight train was heading for the blind siding at Dotsero. As a result of the collision the engines were smashed, the combination baggage and express, smoker and coach were telescoped and crushed against the engine by the three heavy Pullman cars and dining car.

The uninjured passengers from the rear of the train did what they could to help their less fortunate fellow travellers and were credited with saving many lives. Fire broke out in the wreckage but was quickly extinguished by some of the passengers using snow lying alongside the line. News of the accident was sent to Glenwood Springs and a relief train was sent to the scene with physicians and nurses.

'The most harrowing conditions of the Dotsero wreck were the lack of light for the rescuers, there being but two trainmen's lanterns available; the wreckage of a long freight obstructing the arrival of rescuers, doctors and nurses; the narrowness of the embankment besides the wreckage, where a mishap meant probable death in the river. The sufferings of the injured were prolonged, and this, with exposure to cold, probably means death for some who otherwise might have survived. Even after the living victims were brought here there were not enough surgeons to attend them and it was more than twelve hours before it was possible to give them more than the usual first aid.'

First reports were that 21 persons had been killed and another 30 injured. Among the

846 *Heppner Gazette*, Heppner, Oregon, United States, 1st October 1908
847 *The Dillon Tribune*, Dillon, Montana, United States, 2nd October 1908
848 Administration Report on the Railways in India for the year 1908

injured were Engineers Gus Olson, Sig Olson and W. H. Jeffery; Fireman Harry Cardinal and Express Messenger Phil Peters Jr.[849]

Engineer Gus Olson had apparently checked his watch as he approached Dotsero and read the time as 9.45 when it was really already 9.50. Thinking that he had time enough to make it to the next siding he opened the throttle and continued past Dotsero. Realising that his train was not pulling up at Dotsero, Conductor Al McCurdy jumped to the bell cord and signalled to his engineer to stop, but by then it was too late.[850]

The inquest on the victims who had lost their lives in the collision, the number of which had by then risen to 26, was held at Red Cliff, Colorado, on Saturday 23rd, and duly found Engineer Gus Olson responsible. He was unable to attend the inquiry, having suffered a nervous breakdown as a result of the accident.

It developed at the hearing that the passenger train had been given an order at Wolcott to wait at Dotsero until 9.55 p.m. for the third section of train No. 66, the one with which they had collided. That they had not done so was indicated by the fact that the trains had met 1,414 feet west of the siding at 9.47 p.m.

Assistant Superintendent Hugh Wilson testified as to the point at which the collision had occurred and stated that the gradient there was 26 feet to the mile (about 1 in 200). The passenger train had had eight coaches and the freight train, which had had two engines, 27 cars. Gypsum, the next station east of Dotsero, was 6.8 miles therefrom and the ordinary running time between the two places was 13 minutes at an average speed of just over 30 miles an hour. The train had in fact been going at 40 miles an hour, which was permissible as it was running late. All the cars of the passenger train had been fitted with air brakes, enabling the conductor or any member of the train crew to stop it at any time.

Dispatcher C. A. Whiteman said that train No. 5 had had orders to wait at Dotsero until 9.55 for third train No. 66. He had also ordered it to meet second No. 66 at Gypsum. It had been a "31" order which must be receipted by the conductor and read by the conductor to the engineer. Conductor McCurdy's signature had been repeated back to him by the operator as per the regulations.

Operator Roy Crouch at Shoshone testified that he had delivered a "19" order to the conductor of third No. 66, instructing it to take the siding at Dotsero for train No. 5. A "19" order was one issued to an inferior class of train.

Operator Clark Argabright at Wolcott stated that he had delivered the order to Conductor McCurdy to meet third No. 66 at Dotsero. The conductor replied that they were already 23 minutes late and a wait at Dotsero until 9.55 would mean further delay.

Conductor McCurdy took the stand, noticeably pale and looking weary and careworn. He testified to repeating the orders to the operator and said that the orders were the wrong way round, with that regarding Dotsero before that regarding Gypsum, and he had so remarked to Engineer Olson, but he understood that they were to meet third No. 66 at Dotsero. Brakemen F. J. Roberts and Orville Wilson had been shown the orders and said that they understood them. He said that Dotsero switch was on a curve so a train at the west end could not be seen at the east end. He had looked out, thinking that they were at the east end. Then Brakeman Roberts had opened the door and shouted, "Al, they are not

849 *The Express Messenger*, Salida, Colorado, United States, 22nd January 1909
850 *The Bulletin*, Linneus, Missouri, United States, 21st January 1909

there." He had immediately signalled to Engineer Olson to stop and felt the air brakes come on. The speed had dropped to 20 miles an hour, then the crash had come. The conductor corrected Mr Wilson's statement. The train had had nine vehicles – a baggage car, smoker, chair car, two tourist sleepers, a diner, two standard Pullmans and an observation car.

Brakeman Roberts corroborated his conductor's testimony. Asked to explain Engineer Olson's mistake he suggested that he had misread his watch by 10 minutes, misread the order or mistaken one station for another.

Conductor McCardy said that his and the engineer's watches had synchronised at Minturn, about 15 miles east of Wolcott.

Assistant Superintendent Wilson related the names of the crews of trains Nos. 5 and third No. 66. No. 5 – Conductor Al McCurdy, Engineer Gus Olson, Fireman D. N. Wheeler, Front Brakeman F. J. Roberts, Rear Brakeman Orville Wilson. Train third No. 66 – Conductor J. N. Cope, Engineers Sig Olson (brother of Gus?) and W. H. Jeffries, Firemen John Anderson and Harry Cardinal, Brakemen Harry Flynn and Carl Walker.[851]

No legal action appears to have been taken against Engineer Olson, presumably because of his mental breakdown. The only further news of him was that he had left Grand Junction on 8th February with his wife to spend a few months in Los Angeles to recuperate.[852]

Ten people died and 28 were badly injured late on the night of Saturday 19th June 1909 when two interurban cars collided head-on 'at a terrific speed' near Dune Park, Indiana, United States on the Chicago Lake Shore and South Bend Railway. About halfway between Gary and Michigan City was a sharp curve at the top of a slight summit in the line. The trains were both running at speed in order to easily surmount the incline. The westbound train was near the summit of the bank and its motorman caught sight of the eastbound train and managed to reduce speed a little, but the eastbound train came around the curve at full speed. In the collision the westbound car was lifted into the air, saving the lives of everyone in that train. The eastbound car ended up beneath it and the passengers in the forward smoking compartment were crushed and buried in the mass of debris. It was three hours before the last body could be released.

The collision occurred in an unsettled area and it was an hour before the first assistance arrived. At first the uninjured passengers spent some time trying to help their screaming and hysterical fellow travellers before someone thought to telephone for help, when men set off in both directions in search of a phone. The nearest one was found in the farmhouse of Fred Carrigan, three quarters of a mile away. The first news of the accident was received in Gary at about 10.30 p.m. and other calls were made to Michigan City and to Chesterton, to the south of the line. Attempts to run a relief train from Michigan City were frustrated when it was found that the electric current had been cut and the Lake Shore and Michigan Southern Railroad offered to run a train to their Chesterton station and have them carried across country to the South Shore line to the north. Meanwhile two electricians had gone to the site on a small gasoline-powered

851 *The Herald Democrat*, Leadville, Colorado, United States, 24th January 1909
852 *The Salida Record*, Salida, Colorado, United States, 12th February 1909

railcar and they were able to restore the current, which had cut because the leading car of the westbound train had fouled the overhead lines when it was pushed upwards in the collision. After this had been done relief trains could be sent to the scene. Meanwhile physicians had reached the scene in automobiles from Gary and Chesterton and when the relief trains reached the scene the injured were laid out in rows respectively alongside the track ready to be loaded onto them.[853]

'Parts of the wrecked cars hung over the brow of a hill {which} ascended sharply from the other side of the track so that even those who were uninjured found it difficult to move about, and a matter of great risk to attempt to remove the injured beneath the splintered trains. The accident occurred in Porter county, and Deputy Coroner Willfong, who was one of the first outsiders to arrive at the wreck, promises a strict investigation. The motorman of the east bound train is among the dead, but the conductor will be able to testify.' The injured were taken to the Mercy hospital in Gary.'

David Crawford, a South Bend chauffeur, was a passenger on the eastbound train. He related, "There were about 50 of us on our car. Most of us got on at Hammond, having been at the auto races at South Bend. Tired out by the long day of excitement, many of the passengers were asleep. About 9 o'clock, when we were running between 50 and 60 miles an hour, there was a terrific crash, smashing of timbers and creaking of steel. Then suddenly everything was dark and deathlike stillness followed. But it was for a second only. Immediately afterward the stillness was broken by the cries of the injured and dying. Shrieks and groans horrified the passengers who had escaped. The cars telescoped each other."

In a statement the day after the accident General Manager H. V. Wallace stated that the crews of east and westbound trains had been issued with orders at Gary and Michigan City respectively by Dispatcher Lambke to cross one another at Wilson Siding. The records indicated that these orders had been delivered and the fault lay with Conductor Delmar Kenney and Motorman George Reed of the eastbound train in not waiting at that point.[854]

Motorman Shimmel, of the westbound train, stated, "It was not my fault. I had orders to make the Wilson siding and was nearly there when I saw the reflection of a headlight against the trees at the Shady Side curve. I shut off power and jammed on the brake. My car had practically come to a standstill when the other struck us. My car was lifted straight up in the air. Everyone was so dazed that it was hard doing anything for the men under the car that was beneath mine. Many of the injured were so held down so we had to await for the arrival of the axes and crowbars before we could get them out. Some of the men went to look for a telephone and they were thoughtful enough to bring lamps from all the houses they visited.

"There was one woman. A Mrs Stutman, I think she said he name is. She was walking around carrying a baby. Her face was cut by glass, but all she could think of was to say to everyone she met, 'I saved baby all right.' When we finally pulled out Kerrigan, we had to lift out four bodies that were resting upon his. All these were dead, but he, the last man, was unhurt."' [855]

---

[853] *The Weekly Sentinel*. Fort Wayne, Indiana, United States, 23rd June 1909
[854] *The Lake County Times*, Hammond, Indiana, United States, 21st June 1909
[855] *The Weekly Sentinel*, Fort Wayne, Indiana, United States, 23rd June 1909

The verdict at the inquest on the victims of the accident at Valparaiso, delivered by Coroner Carson, was that "The cause of said collision was a disobedience of a train order by train No. 59 eastbound which read, 'Meet train No. 58 at Wilson.' I would, therefore, place the entire responsibility of said wreck upon the mismanagement of train No. 59 for disobeying said order in not taking the siding at Wilson, as the order directed."

No action was recommended by the coroner.[856]

Train No. 5, an extra, left Coeur d'Alene on the Spokane and Inland Empire Railroad interurban line at about 4.35 p.m. on Saturday 31st July 1909. It had only travelled two miles when it collided head-on with regular passenger train No. 20 at La Crosse, Idaho, United States. The two leading cars of each train were telescoped, the light smoker of the extra was practically demolished and 12 passengers riding on the platform of it were killed outright while Motorman Edgar Campbell was very seriously injured. Motorman James Delaney of train No. 20, had time to almost bring his car to a stand and to leap to safety, but Motorman Campbell had no time to get clear and the brakes of his overloaded train failed to work, so that it was still travelling at 20 miles an hour when the cars met.

Help came quickly after the accident with doctors coming from Coeur d'Alene, assistance from the workmen at the local Stack-Gibbs lumber mill and a special train with doctors, nurses and medical supplies sent from Spokane. The more badly hurt were taken to hospitals in Coeur d'Alene and Spokane and the less seriously injured cared for in local homes. There had been a total of about 600 passengers on the two trains.

By Monday the number of dead had risen to 13 and at least 100 people were said to be injured. It had also become apparent by then that a mistake by Motorman Campbell had led to the disaster, as after receiving his orders he had started from Coeur d'Alene with the extra train on the time of the regular train. The dispatcher there had assumed that he was just taking his to the end of the station yard, but he had kept going. Motorman Campbell, who at the time had believed that he was speaking with his dying breaths, stated that he had thought that he had the right of way.

A newspaper placed part of the blame for the accident on an unprecedented amount of passenger traffic in the days preceding the accident putting a heavy demand on staff who had to run up to 240 miles a day shuttling between Spokane and Coeur d'Alene, though this was denied by the company. The extra traffic load was the indirect result of a decision by the United States Congress in February to make available to the public 668,000 acres of land on three Indian reservations. Applicants for parcels of land had to register at Spokane, Coeur d'Alene or Missoula between 15th July and 5th August, giving rise to a so-called land rush.

In a few days the inquest on those who had lost their lives in the collision, the number of which finally reached 15, had been held, but without hearing the testimony of Motorman Campbell, still lying weakened and wasted from his many injuries in a Coeur d'Alene hospital.

The inquest found that Conductor Whittlesey had violated the company rules in several respects. To begin with he had registered out of Coeur d'Alene station both for himself and for the motorman. He admitted that this was something that he had done

856 *The Monon News*, Monon, Indiana, United States, 2nd July 1909

frequently before, and Superintendent R. C. Bowditch declared from the witness stand that had he been aware of this the conductor would have been suspended for violating the rules. In a further violation of the rules the conductor failed to show his orders to Brakeman Trudell, although the latter had asked to see them, but instead said to Motorman Campbell, “All right, are you ready?” According to the newspaper report the orders made no mention of train No. 20 but merely read “Meet extra No. 5 at Allen.” This clearly cannot be correct as Conductor Whittlesey’s train *was* extra No. 5 but at any rate the train order clearly gave train No. 5 no rights against the incoming regular train which, as the newspaper correctly pointed out, the extra had to wait for, in theory up to 12 hours if need be. The regular train was late, doubtless due to the pressure of traffic; it should have arrived at Coeur d’Alene at 4.12 p.m. The newspaper stated it was Conductor Campbell’s first trip working an extra and charitably suggested that for that reason he had forgotten the provisions in the rule book regarding extras giving way to regular trains. It seems more likely that it just did not occur to either him or Motorman Campbell that the late-running regular train had not yet come in. The train having started, the conductor busied himself with collecting the tickets from the numerous passengers while his train passed the Y outside the station and entered the single line. Brakeman Trudell at the back of the train could have stopped it but not having seen the orders was unaware that anything was wrong.

In a final twist of fate when Motorman Campbell saw the regular train coming towards him, he had immediately reversed his train and made an emergency brake application. The air brakes applied but then the air pressure was lost, and the brakes released and all he could do was reverse his train. The cause of the apparent brake failure could not be determined later because of the damage to them in the accident. It was also alleged that during the period of exceptional traffic caused by the land rush it had become the standard practice for the extra to run from Coeur d’Alene as far as La Crosse siding to meet the local train but that on this occasion Motorman Campbell had overrun the siding as well.

The jury found Conductor Whittlesey of train No. 5 and Motorman Campbell mainly to blame for the accident, with presumably some blame attached to Brakeman Trudell for not insisting that he be shown the train orders. Attorney Will Graves, counsel for the railroad company, stated that they would hold a further investigation once Motorman Campbell had recovered sufficiently to testify.[857] No further report of such an investigation has, however, been found.

Motorman Campbell eventually recovered to a degree but had to have a leg amputated. He subsequently sued the Inland Revenue Company for damages for his injuries, claiming that a brake failure had partially caused the accident. He was awarded $7,500.

None of the victims of the accident was selected in the lottery held for the homesteads. The Inland Revenue Company made much money from the from the extra land rush traffic, only to have to pay out most of it in damages to accident victims.[858]

---

[857] *The Spokane Press*, Spokane, Washington, United States, 7th August 1909

[858] Webpage – historylink.org: Two Spokane and Inland Empire trains collide northwest of Coeur d’Alene, Idaho with disastrous results on July 31, 1909

Saturday 14th August 1909 saw another head-on collision on the Denver and Rio Grande Railroad, this time near Husted, Colorado, United States, 12 miles north of Colorado Springs. Train No. 8 northbound had orders to cross southbound second train No. 1 southbound at Husted. As train No. 8 reached Husted its crew saw a light engine in the siding there and their view obscured by undergrowth, they mistook it for train No. 1 and continued northwards. Half a mile further on train No. 1 came around a curve and its conductor saw train No. 8 approaching only 300 yards away and applied the air brake. Train No. 1 was brought to a standstill but train No. 8, which was double-headed, kept going, its brakes being applied only at the very last moment.

'The impact was terrific. Three giant locomotives came together, built a pyramid, and then toppled over. The tender of No. 8's second engine forced its way into the combination baggage car and smoker of No. 1, putting it on end and injuring only two persons. The combination baggage and smoking car of No. 8 was split in twain, its human cargo was hurled through windows, pinioned under seats, shot through the roof. Death and destruction were everywhere. In that car the carnage was great. There were two coaches, an observation car, a chair car and two sleepers behind the combination smoker and baggage car on train No. 8. They carried upwards of 278 passengers. Behind the first coach on train No. 1 were two observation cars, the train carrying in all 137 passengers.

'Nearly three score were injured. As many had miraculous escapes and the brown and red of the hillsides ran rivulets of blood. Bodies were hurled through closed windows, across the railroad right-of-way and into enclosed pastures; others fell down the embankment and the groans of the dying and the shrieks of the injured rang in the ears of the more fortunate. It all happened in the twinkling of an eye. Passengers saw their seat companions carried into eternity as a candle was snuffed out, while they themselves were left blinking and only half conscious of what had taken place.

'It was a wreck without reason. One little girl, name unknown, was thrown as from a catapult through an open window up in the air ten week or more and alighted at the bottom of the embankment, none the worse for her experience. A man, D. W. Johnson of Denver, engineer of No. 1, was pulled from under the body of J. A. Gossage, fireman of the other train, with only his left hip and back injured. Gossage was killed outright.

'There are forty-four men and eighteen women in the list of injured. Not one woman was fatally hurt. Some little children escaped without a scratch. One boy, half grown, received some bruises on his leg. His father, who sat beside him, had his leg broken.

'There were other strange features and many coincidences. Fireman Gossage met death almost in his front yard. His wife and four children, living on a ranch at Husted, were the first to arrive on the scene. Mrs Gossage swooned in the arms of sympathetic passengers. The little ones clapped their hands and gleefully shouted, "Papa!" They did not know their father was beyond the range of tiny voices! The scenes were heartrending.

'With 400 men, women and children on the escape list there was a state of panic aboard the relief train. Dr Henry W. Hoagland and Dr C. F. Stough were among the first physicians to reach the scene. Miss Margaret Kredel, a trained nurse, happened to get aboard at Palmer Lake, and she was kept busy during the short ride to the Springs. The dead and dying had the right of way of Colorado Springs streets all afternoon and night. Every dead waggon was pressed into service, and carriages and automobiles were used for ambulances.'

Fireman Gossage and eight passengers were killed outright, and 62 passengers were reported injured.[859] Another passenger, C. M. McCreary, of Dallas, Texas, died of his injuries in hospital the next day.

On 17th August the inquest jury visited the scene of the accident and it was stated that there were only one of two more witnesses to be examined. Engineer O. H. Lessing of the leading engine of train No. 8 was said to be in such a physical condition as a result of the wreck that he could not discuss the accident coherently. Conductor Dalton of train No. 8 stated that 3,000 feet short of the accident he had realised that a terrible mistake had been made and becoming panicky had signalled on the bell cord to Engineer Lessing rather than immediately applying the emergency brake. The engineer had not responded.[860]

Emil Kaupe, of Chicago, Illinois, died of his injuries on 18th August. The same day, following the jury's verdict that the accident had resulted from the criminal negligence of the crew of train No. 8, a deputy sheriff left Denver to arrest Conductor Dalton, Engineers Lessing and Hollingsworth, Wright, the surviving fireman, and Brakeman Melbourne.[861] Engineer Lessing stood trial for manslaughter first and was duly acquitted, whereupon the charges against the other four railroad men were dropped.

The report of the state railroad commission, published on 10th November, came to the same conclusion as the inquest jury, that the blame lay with the crew of train No. 8. By then a 12th person had died of his or her injuries.[862]

An accident near Topeka, Kansas, United States on Friday 8th October 1909 killed at least 17 people but apart from Brakemen Ira Clemons and E. L. Bolte they were mostly labourers, plus 10 or 11 of them were Mexicans, so it did not receive a great deal of press coverage. It occurred on the Atchison Topeka and Santa Fe Railroad between Kilmer and North Topeka, and although perhaps not strictly a single-line collision in the usual sense its inclusion in this chapter seems most appropriate. The victims were riding on the four flat waggons of a construction train which was backing into Topeka. Due, reportedly, to the crew of the work train disregarding orders the construction train was in collision with a northbound freight train. Fifty Mexicans were riding on the flat waggons and 10 of those who were not killed outright were seriously injured. The trains met on a curve. Engineer Edward Ash and Fireman Joseph Corey of the freight engine applied the brakes and then jumped off.

Some of the victims were pinned in the wreckage for several hours before they could be released.[863] [864]

A newspaper editorial wrote sympathetically of the Mexican labourers whose names remained unknown, '"Ten unknown Mexicans dead." This was the line completing the list of fatalities in the Topeka wreck. Just ten little workers wiped out of existence and no

[859] *Elbert County Banner*, Elizabeth, Colorado, United States, 20th August 1909

[860] *The Bayfield Blade*, Bayfield, Colorado, United States, 19th August 1909

[861] *The Durango Democrat*, Durango, Colorado, United States, 19th August 1909

[862] *Colorado Springs Gazette*, Colorado Springs, Colorado, United States, 11th November 1909

[863] *The Burlington Hawk-Eye*, Burlington, Iowa, United States, 9th October 1909.

[864] *Logansport Semi-Weekly Reporter*, Logansport, Indiana, United States, 12th October 1909

one knows who they are. Foreigners in speech and manners they did not get close enough to their American bosses to hardly be known by name. Now they are dead, and the state will bury them and soon the matter will be forgotten. But somehow, we cannot help thinking of some homes in Sunny Mexico where parents, sweethearts or wives wait for word from loved ones who are silent. There will be many days of anxious worry and then a settled sadness for those who are gone. The little Mexicans will labor all day building and repairing railroads, are not a lazy, worthless lot of men but are among the best of the laborers in Mexico. Not content with the low wages of their own country they venture forth, work hard, make what they can, spend little, and eventually return to their native land with their savings. But ten of them will not go back. Instead they will occupy nameless graves because someone disobeyed orders.' [865] Some sources stated that the freight train was an oil train and that the accident occurred at a location known as Three Bridges.[866]

The *New York-Cincinnati Flyer* of the Cleveland Cincinnati Chicago and St Louis Railroad (the Big Four), diverted onto the tracks of the Cincinnati Hamilton and Dayton Railroad because of a wreck on its own tracks at Gano, Ohio, just before noon, was wrecked in a head-on collision with a freight train 300 yards north of West Middletown station, Ohio, United States, on Monday 4th July 1910. The Big Four train was one of the fastest trains between New York and Cincinnati, and was involved in the accident after having been diverted onto Cincinnati Hamilton and Dayton tracks at Dayton and was running as train second No. 1. In the opposite direction northbound freight train No. 90 had orders to take the siding at Poasttown to cross the Flyer. According to an early report on the accident the engineer, when he received instructions to run to Poasttown, had only seven minutes to get there. He set off but then realising that there was not enough time, he had reversed his train with the intention of setting back into the siding at West Middletown. He failed in this as well, and the flyer came around a curve and running at 50 miles an hour crashed head-on into his engine.

'The impact was terrific. Not only did one engine become imbedded in the other to the depth of two feet, but the combination baggage car and smoker were knocked into kindling wood, thrown across the track and down an embankment. The dining car behind the smoker practically was demolished and the woman's passenger coach was torn about, and the front part smashed to shreds. Even the first of the three heavy Pullman sleepers, "*Duvas*" was torn asunder. The forward vestibule was a mass of broken timbers and splintered woodwork throughout.

'In the smoking car were five passengers at the moment of the collision. Every one was killed. In the diner were 15 passengers, not one of whom escaped injury. Nearly all were killed. Three women met death in the women's coach, as did some of the male occupants. Although given a fearful jolting none of those in the sleepers suffered more than bruises or broken limbs.'

Another account of the results of the collision read, 'The third coach jammed into the locomotive and was reduced to splinters. It was here that the accident claimed most of the victims. No one would have been able to tell what the wreckage had ever formed a

865 *Lawrence Daily World*, Lawrence, Kansas, United States, 11th October 1909
866 *Laredo Weekly Times*, Laredo, Texas, United States, 10th October 1909

minute after the crash. The roof was torn off and hurled many feet up the track and one side was completely shattered. The interior of the car was a terrible mass of wreckage and it was under this that the victims were hurled in a mass. The impact of the locomotive caused the steam box on the passenger engine to explode, and this threw a deluge of scalding steam and water upon the imprisoned victims. It was a scene which beggars description wen the rescuers clambered upon the wreck and attempted to give aid. Most of the victims had been pinned beneath the debris and were dead when carried from the mass of shattered wood and iron. Body after body was taken from this coach, some moaning in the last agony of death.

'The Pullman parlor car apparently checked the momentum of the passenger train, for it remained on the track, and though badly shattered on one side no one was even injured. Charles Hunt, the colored porter, assisted the four women and three men passengers to clamber to safety through a window. In the diner and sleepers no one suffered injury.

'The first car of the freight train was loaded with coal, several tons of it. If workmen with a derrick had lifted the car and its contents from its truck and placed it on top of the next car a cleaner job could not have been done. Little other damage was done to the freight, however, aside from the smashing of a box car when a car load of lumber crashed through the end. Both engines were so badly wrecked that they will be sent to the scrap heap.'

Of the railroad employees Brakeman Frank Golden of the passenger train was killed while Conductor Tom Maloney had a lucky escape. On the passenger engine, No. 6418 Pilot Conductor William Wall was unhurt, but Engineer Ed Weisner was badly hurt and had to have a leg amputated, while Fireman W. P. Lamb was also hurt. Engineer Harry Hamel and Fireman Vernon Wallace on the freight train jumped just in time and escaped unhurt.

Following the collision, the people of West Middleton threw open their homes for the care of the injured and all the local ambulances were pressed into service. Other victims were taken to physician's offices and to police headquarters. Every physician in Middletown attended the wreck and the Dayton Company sent a special train with doctors and nurses from Dayton. The Big Four had made up another train within an hour in Cincinnati but were told by the Dayton Company that it would not be needed.

Several passengers standing on the platform of West Middletown station witnessed the accident, one woman exclaiming, "Oh, the passenger train will run into the freight!" She turned away, not wishing to see what happened.[867]

A former West Middletown railroader, J. M. Foley, saw the trains when they were 200 feet apart and heard the impact. The daughter of one of his neighbours, Thomas Hines, witnessed the accident and the two men hurried to the scene and helped to carry the dead and dying from the wreckage.

One of the passengers, from Dayton, said that he owed his escape to the fact that having gone to the diner and picked up a menu he realised that he had left his hat behind and so had gone back to the coach where he had been sitting to get it. At that moment the collision occurred. He was merely jolted. Another passenger in the chair car was thrown to the roof when the trains collided and at the same instant the roof parted, forming an

[867] *The Hamilton Telegraph*, Hamilton, Ohio, United States, 14th July 1910

immense hole through which he was able to crawl to safety. Celebrating his narrow escape, he later went to the bar of a local hotel where he treated everyone he could see to a drink. He would not give his name.

Fireman Wallace told a reporter, "We had seven minutes to make the siding when we left the Middletown yards, just below the city. That must have given us a full five minute's margin as we passed the station. Yet we had barely cleared the depot with our rear flags when the flyer came around the curve. I yelled to the engineer and jumped, and he did the same. The passenger ran ahead of schedule and that is the way it happened. I did not see my engineer's orders, but from what he said to me shortly after receiving them, evidently from the dispatcher at Dayton, we had a free track after we passed West Middletown. Our train was somewhat late, but we had plenty of time to have made the siding at the scheduled place. Instead of taking a chance he told me he would take a siding in front of the station at West Middletown. We had but four car lengths to pull up to enable the brakeman to throw the switch, when he saw the Limited shoot out from the bend above us. The engineer, Harry Hamel, pulled the whistle and then jumped. He sprained his left ankle in landing. I followed suit, hurting my back."

Pilot Conductor Wall, on the other hand, declared that the flyer was strictly on schedule, and that he believed that the freight train crew must have miscalculated the time they needed to reach the siding.

Dispatcher Albert Smith at Dayton also made a statement. "I am not responsible for the wreck. I told the freight to stay on the siding and I can't understand why it was not there. I don't care to place the blame at this time." The dispatcher had issued a "wait" order to keep the passenger train at Poasttown, three miles north of West Middletown, but had then annulled it and had instructed Operator Crider at West Middletown to keep the freight train there. Conductor John Weaver and Engineer Hamel of the freight stated that they had never received this order at West Middletown.

On the evening of the accident General Manager Van Winkle of the Big Four stated, "The cause of the wreck appears to be problematical at this time, and while I have kept in close touch with the situation, I am not in a position to make an official statement. One train was running as the second section of a regular C.H.& D. train and was about one hour late. It was in charge of C.H.& D trainmen, upon whose report the decision as to the cause will probably be based. I have received reports as to the number killed and injured but have no authoritative or official information beyond this. Our train was being brought into Cincinnati on the C.H.& D. because of an accident to a freight train on the Big Four near Sharon, which interrupted traffic. Until a thorough investigation is made it will be impossible to locate the responsibility."

The initial report spoke of 21 dead and 36 seriously and 30 to 50 slightly injured.

Comprehensive details of the evidence given by the main inquest on the victims of the accident was reproduced in at least one local newspaper. (Two inquests were held, one on the majority of the victims, who died in Butler county, and a second on those who died later at Dayton.)

Pilot Conductor Wall stated that the flyer had left Dayton with orders relating to the sidings at Whitfield and West Middletown being blocked, a slow order and a running order but no wait order applicable to Poasttown. The running orders were for train first No. 1 to run 20 minutes late and train second No. 1 to run one hour late; he knew nothing of any wait orders and had no orders regarding meeting any other train. They had been slowing down to make a booked stop at West Middleton when the accident happened.

Rear Brakeman Walter Kuhlman of train No. 90 testified that they had started from Elmwood Place at 8.46 a.m. that morning, running on time with a heavy train of 35 cars. If they received any orders at Hamilton he did not know of them. They met train No. 620 on the south siding at West Middletown, where they were given an order, and also met train No. 616 there. He did not know what time they set out for Poasttown. He knew that the crash came at 1 p.m. because he looked at his watch then, though he had thought at first that it was merely a rough stop. His engineer was a careful man.

Head Brakeman Percy Bergamyer of freight train No. 90 said that they had received orders at Hamilton, but he did not know what they were. At Busenbark (about four miles south of West Middletown) they had received an order to cross train No. 620 at West Middletown. Train first No. 1 had been running between trains Nos. 620 and 616. At West Middletown the operator had told them that he would give them more time on train second No. 1, and that the passenger train would wait there for them until 1.07 p.m., so that would give them more time to get there and meet the passenger there. They decided, however, that they did not have time and were backing into the siding at West Middletown when train second No. 1 appeared. He estimated that the passenger train was 15 or 16 minutes before its time. The passenger train should have waited at Poasttown until 1.07 and if train No. 90 had not appeared it should have proceeded, keeping a careful lookout. They had not flagged the freight train because they believed that they had plenty time to get in clear. The order they had received at West Middletown had read, "Second No. 1 wait at Poasttown until 1.07." The crews of both trains should have received this order.

Engineer Hamel testified to leaving Glendale at 8.45 and read orders he had received at Glendale, Hamilton, Busenbark, Trenton and West Middletown. At the latter place he had first received an order to meet trains Nos. 616 and 620 and to wait for train second No. 1. Later he received a second order from Operator Leo Crider that second No. 1 would wait at Poasttown until 1.07. He was at West Middletown at 12.45. He had received no annulling order cancelling the second order. He attached no blame to Operator Crider.

Fireman Wallace corroborated his engineer's testimony. He said that he did not know that train second No. 1 had left Poasttown until it was right on top of them.

Operator F. C. Spies at Carlisle (the next station on the Dayton side of West Middletown and three miles therefrom) stated that he had received no order to hold train second No. 1 at Poasttown and when it passed through Carlisle at 12.57 at 50 miles an hour he had given it a signal indicating a clear track.

Operator T. J. O'Neil at Dayton testified to having received no orders to hold the passenger train at Poasttown until 1.07.

Operator Leo Crider at West Middletown testified to receiving an order at 12.58 to hold the freight train there for train second No. 1. The freight train was then passing the station. Twenty-five cars had already gone by, but he just had time to give the conductor the order. The collision occurred a minute and a half later.

Conductor Weaver said that the freight train left Ivorydale at 8.55 a.m. The track had been open to them to Poasttown until 1.07. As his train was pulling out of West

Middletown he saw the operator running along the platform. He jumped off the train and ran back. He did not actually collect the order, but the operator had shouted to him that they wanted his train to wait there. He had shouted, "All right," and signalled for his engineer to stop, sending the head brakeman forward to flag any approaching trains. The train could have got into the siding by 1.04. He did not think the passenger train could have stopped at West Middletown the speed it was travelling.

Conductor Maloney of the passenger train stated that to the best of his knowledge they had received no order to wait at Poasttown. The pilot conductor got the orders, but he saw them. He considered that the train was running at excessive speed. He had told the pilot conductor and engineer to stop at West Middletown because there were passengers to get off there.

Chief Dispatcher Smith pleaded that because of the extraordinary amount of business on 4th July his mind had probably been in an unduly excited condition, and he had thought that he could get the freight to Poasttown by 1.07 and clear of the passenger train. While the West Middletown operator was repeating this order, the thought had flashed through his mind that there not enough time and he had sent a further order annulling the first. He had not sent the order to the passenger train to wait at Poasttown until 1.07 (and so at that point he had effectively issued a "lap order", giving both freight and passenger trains right to the track between West Middletown and Poasttown at the same time). By way of excuse he said that he had thought that the freight train was in the siding at West Middletown and not on the main line, also that he considered that the passenger train had been travelling at an excessive speed, faster than any other Big Four train that day, thought he admitted that as there were no speed limits the engineer could run as fast as he thought fit.[868]

J. J. White of Dayton died on 12th July, the 22nd fatality, and a 23rd person apparently died later.

Delayed awaiting the testimony of a couple of seriously injured witnesses the principal inquest was not concluded until 10th November, when Coroner John Burnett announced a verdict that, "I do find from the testimony given that Train Dispatcher Smith did err in giving order to an inferior train and furthermore if he had intended issuing order No. 60, second No. 1 should have received it at Carlisle.

"From the orders submitted to me, I do find that Smith did give second No. 1 a clear track from Dayton to Cincinnati. Furthermore, from the testimony submitted I do find that second No. 90 should not have been out on the main track for the rules of the railroad company specify that an inferior train should be in the clear five minutes before a superior train is due. The orders issued show that second No. 1 (the Big Four passenger) was running an hour late on first No. 1 which was due at Middletown at 12.01. This would put second No. 1 at Middletown at 1.01 p.m.; therefore, it would have been an utter impossibility for the crew of second No. 90 to have reached Poasttown and been in the clear five minutes, even if second No. 1 had received the order first.

"The testimony also shows that the crew of second No. 90 also erred in not placing a flagman to protect them while on the main track. The evidence of Clarence South, rear brakeman of No. 616, shows that he not only flagged second No. 1, but placed torpedoes on the track for second No. 1. This should have warned William Wiesner, who was in

[868] Ibid.

charge of the Big For train, to run at moderate speed and keep a look out for a train ahead. The testimony of Smith as to Wiesner answering his flag is substantiated by Donald Schaeffer and by the members of the crew of second No. 90. The testimony of various witnesses clearly establishes the fact that second No. 1 was running at an excessively high rate of speed at the time that the wreck occurred.

"Therefore, I am of the opinion that the wreck on July 4th at Middletown on the C.H. & D. line was due to the carelessness and negligence of Train Dispatcher Albert B. Smith, the train crew of second No. 90 and to Pilot Engineer Wiesner of the Big Four train." [869]

Less than three weeks after the West Middletown disaster, on 23rd July, Cincinnati Hamilton and Dayton Railroad train No. 1 collided with a freight train at Poasttown due to a switch having been left open. This time no one was killed, though several were hurt.[870]

The Northwestern Pacific Railroad suffered a serious collision a mile and a half south of Ignacio, California, United States, on Monday 8th August 1910, when regular passenger train No. 6 collided with a work train, consisting of an engine, two flat cars and a caboose. The trains met on a curve, the passenger train travelling at a reported 40 miles an hour and the work train also moving at some speed.

'Suddenly the passengers were hurled forward by the quick shutting down of the brakes, and at the same time there came a terrific impact and the sound of escaping steam. The two engines plunged into each other. The baggage car and the smoker were crushed together so that two-thirds of the smoker was telescoped. When the dazed passengers made their way out of the three rear coaches, which were unharmed, they found a pathetic scene. It was light enough to see plainly. From the crushed smoking car came shrieks and groans, and the waving of bloodstained arms. It seemed as though the passengers in the smoker – between twenty-five and thirty men – had been caught in a sitting position and those that were not killed were penned in so by the weight on their legs that they were unable to help themselves or their neighbors in misfortune.

'A few men made their way out of the panic-stricken throng to the forward end of the smoker, where the greatest damage had been done. They had no tools, but they tried by unaided strength to pull away the windows and jumbled timbers. In that mass of wood and flesh, however, there was little movement. The work on the rear half of the smoker was easier. Some of the rescuers scattered to nearby houses to telephone for help, while others sought to get the injured out of the wreck. The body of one man was lying at the side of the track with his head gone. Another man was sitting in a window of the smoker, dead, an expression of mingled wonder and pain on his face. The messenger, a young man, was found dead.'

Even though it was only 1910 at least 50 automobiles visited the accident site in the hours after the collision had happened and some of the injured were taken away in them. Relief trains also went to the scene from each side with physicians, General Manager W. A. Palmer and Superintendent J. H. Hunter. The bodies of those killed were taken to San

---

869 *Butler County Democrat*, Hamilton, Ohio, United States, 10th November 1910

870 *Xenia Daily Gazette*, Xenia, Ohio, United States, 23rd July 1910

Rafael and San Rosa and the injured to various places.

Eleven passengers were killed outright. One of them was John Wilkinson, a prison guard at San Quentin, who had only been married a month and was on his way to see his wife. Conductor Alexander Shera and Engineer C. Speaker of the passenger train, Engineer J. D. Lytta and Fireman Edward Reynolds of the freight train and half a dozen passengers were seriously injured, and one of the passengers, W. D. Burton, of Berkeley, died in hospital at San Francisco the next day, making a 12th fatality.

The day after the accident Superintendent Hunter placed the blame for it on Conductor George Flaherty of the work train, who he said had ignored a train order to wait at Ignacio for the passenger train.[871] [872] By then the conductor was stated to be in a state of collapse and unable to make a statement.[873] Later, however, when he had recovered himself somewhat, he said, "It was my mistake. I thought the Somona train which passed me as I stood on the siding at Ignacio was the Santa Rosa train. After it had passed I gave the sign to the engineer and we pulled out on the track. When I saw the coming train as we rounded the curve I knew we had blundered. We were running fast and almost instantly I felt the awful jar as we struck the other train." [874]

On 22nd August a coroner's jury found that there had been gross negligence and placed the blame for the accident equally on Conductor Flaherty and Engineer Lytta of the freight train and on the railroad company.[875]

'ARRESTS MAY BE MADE' read the banner headline of *The Fort Wayne News* of Thursday 22nd September 1910. The whole of the front page of the newspaper was devoted to news and photographs of a disastrous collision the day before, Wednesday 21st September, near Kingsland, Indiana, United States, on the Fort Wayne and Wabash Valley Traction interurban line. An empty southbound car collided head-on with a regular northbound car and both cars were telescoped for about half their length. The initial report listed 40 dead passengers. Conductors Del Wilson and E. T. Spiller of the southbound and northbound cars, Motormen Benjamin Corkwell and Charles Van Dyne of the southbound and northbound cars and five passengers were listed as seriously injured.

Describing the accident, the Fort Wayne newspaper wrote, 'The fair city of Bluffton is today a city of mourning. The very flower of its citizenship is lying bruised and mangled in the city's morgues, and the living mourn with a deep grief, that time alone can efface, whenever the awful events of yesterday's tragedy are recounted. In the full glory of a lovely autumnal day, forty-nine men and women, whose names stand for all that is best in the life of Wells county, boarded a local interurban car on the Fort Wayne and Bluffton line. It was a holiday crowd, bent on enjoying a day's relief at the Fort Wayne fair, from the cares and routine of the daily grind. Today thirty-five are dead, and the remainder are either hovering between life and death in the Fort Wayne hospitals, or are nervous wrecks at their homes.

'With never a thought of impending disaster, Motorman Van Dyne whirled his little

[871] *Daily Capital Journal*, Salem, Oregon, United States, 9th August 1910
[872] *The Salt Lake Tribune*, Salt Lake City, Utah, United States, 9th August 1910
[873] *The Weekly Sentinel*, Fort Wayne, Indiana, United States, 10th August 1910
[874] *Dispatch Democrat*, Ukiah City, California, United States, 12th August 1910
[875] Ibid, 26th August 1910

car through the little town of Kingsland, and skirting a lovely little wood half a mile north of that town, drove the coach head-on into a special car of the Wabash Valley company's largest type, before he could realise the awful doom that was rushing down the track to meet him and his human freight.' [876]

The northbound car had left Bluffton at 11.58 a.m. Only one of the passengers on it escaped unhurt. John R. Boyd, of Marion, had just boarded the car at Kingsland and was still standing on the step when he saw the southbound car 200 feet away and he was able get off it 40 feet before the point where the two cars met.[877] Conductor Spiller of the northbound car was stunned and injured but was able to run back the line towards Kingsland and flag the following Indianapolis Limited car, which approached at speed a short time later, and thus prevent a further disaster.

News of the accident soon reached Fort Wayne, and Superintendent of Transportation Frank Hardy of the Fort Wayne and Wabash Valley Traction immediately procured a relief train and set off for the scene with Master Mechanic Jacques and half a dozen doctors. This relief train left Fort Wayne at 1.50 p.m. and picking up a number of section hands on the way arrived at Kingsland at 2.20. Just after the relief train left Fort Wayne James M. Barrett and Henry C. Paul, the company's attorney and treasurer respectively, left Fort Wayne for the scene in an automobile, arriving there at 2.40. A second relief train, carrying more doctors, was sent from Bluffton and reached Kingsland just in front of the Fort Wayne one. By 2.45 the injured were being taken to Fort Wayne on the relief train from there and at 3.20 the Bluffton relief train departed with the bodies of those killed.

There was a great pressure on local communications after the accident. 'During last evening it was almost impossible to get Bluffton over the telephones. The number of messages sent out from that city to newspapers and to relatives of the dead was simply enormous, and no calls were accepted by the local telephone exchanges for that town. It was only by watching the Bluffton exchange and breaking in during the lull that it was possible to get parties in that city. Owing to the interruption of interurban traffic and the impossibility of getting the city on the telephone wires, the only communication between Fort Wayne and the other city was through the telegraph wires, and the Bluffton operators were so busy that it was very hard to get them to take the messages. During the evening and late afternoon, a tremendous number of automobiles went from this city to Bluffton to view the scene of the accident, and a great deal of communicating between the two cities was done by this means.' [878]

Blame for the accident was quickly placed on Conductor Wilson and Motorman Corkwell of the southbound car. Dispatcher Frank Frymeier at Bluffton had issued an order for the southbound car to take Siding No. 106, a mile north of Kingsland, but it did not wait there. Conductor Wilson, however, claimed that he had had an order to continue to Siding 107 on the far side of Kingsland. He stated after the collision, "I had orders to stop at No. 107 and ran right on through. I stopped at Ossian for a few seconds but picked up no passengers. I don't know where those orders are now, as I put them in the little box provided for them in the front vestibule of the car. When I came to after the collision I

[876] *The Fort Wayne News*, Fort Wayne, Indiana, United States, 22nd September 1910
[877] *Decatur Daily Democrat*, Decatur, Indiana, United States, 22nd September 1910
[878] *The Fort Wayne News*, Fort Wayne, Indiana, United States, 22nd September 1910

was on my hands and knees and Corkwell was lying over me, bleeding like a hog. I was standing in the aisle of the car with one hand in my pocket when the cars met. As I had my back to the front of the car I never saw the local until we struck." Officials who heard Conductor Wilson's story of his having an order giving his extra car rights over the regular train said that they strongly doubted the truth of it.

Multiple investigations were made into the disaster, by the traction company, by two coroners and by the Wells county grand jury. The traction company investigation was concluded first, and a three-page report was submitted to the Indiana railroad commission. It read in part, 'We find that train 56, between Indiana Union car No. 233, was being operated in the regular scheduled local service between Muncie and Fort Wayne, and in charge of Motorman C. A. Van Dine and Motorman E. A. Spiller. Train 56 was on time, and due in siding 106 at 12.16 p.m., and at siding 105 at 12.22 p.m.

'We find that car No. 303 was being operated from Fort Wayne to Bluffton as an extra train, leaving Bluffton at 11.40 a.m., in charge of Motorman Benjamin T. Corkwell and Conductor Del Wilson, who were given an order as per the copy herewith:

Form 186-50 M.

FORT WAYNE AND WABASH VALLEY

TRACTION CO

Train Order No. 30 9-21-1910

To Conductor and Motorman

Train No. Motor Car No. 303 at

Siding No. Wayne

Train No. Extra Car No. 303 and

Train No. Extra Car No. 54 will meet

At Siding No. 101

Train No. Extra Car No. 303 will

report at Siding No. Bluffton Will

run extra to Bluffton.

Motorman-Corkwell, Conductor-

Wilson. Train No. – Extra 303. Com-

plete- F.H. Time- 11.40 a.m.

We find that no other order was given by the dispatcher to the men in charge of 303 after the order shown above.' [879]

The Bluffton inquest ended next, with Coroner Thomas ordering the arrest of Conductor Wilson on a charge of involuntary manslaughter.[880]

The Indiana railroad commission found that Conductor Wilson and Motorman

[879] *The Auburn Courier*, Auburn, Indiana, United States, 29th September 1910
[880] *The Fort Wayne Sentinel*, Fort Wayne, Indiana, United States, 29th September 1910

Corkwell had disregarded the rule of the company which laid down that extra trains should take sidings five minutes ahead of regular trains and had attempted to reach a siding two stations beyond the proper meeting point. Conductor Wilson had claimed that he had had an order to go to the siding which he was attempting to reach, but the commission had no doubt that this claim was false. The crew of the extra train lacked proper training, experience and fitness for the service they were performing. There were no block signals on the line, although the daily movement of trains was about 38 on a single track, and although the double order rule, requiring that both conductors and motormen receive orders, was in force, it had been found that it was frequently violated by crews and dispatchers, perhaps daily. It appeared, indeed, that the orders for the extra train involved in the accident had only been given to the motorman.[881]

The grand jury came to similar conclusions as the first inquest and the railroad commissioners, but Coroner Kesler, whose inquest at Kingsland was the last to be concluded, blamed not only Conductor Wilson and Motorman Corkwell but also Motorman Van Dine of the northbound car, who he said could have stopped or at least reduced speed but instead had applied power and then deserted his post when he saw that a collision was unavoidable. It was not thought, however, that any action be taken against Van Dine, whom all the other investigations had held to be blameless and who was back at work.[882]

At the beginning of November, it was reported that Motorman Corkwell was violently insane and would probably be taken to the state insane hospital in Richmond.[883] No report of Conductor Wilson having stood trial has come to light.

The final death toll was 41.

Only three days after the Kingsland disaster, on Saturday 24th September 1910, there was another single-line collision on an interurban line in Indiana. This time it was the Indiana Union Traction Company. At about 10.40 a.m. on 24th September a northbound passenger car in charge of Conductor J. D. Mains, with Motorman Joseph Baker, was rounding a long curve at high speed a mile and a half north of Tipton, Indiana, United States, when it collided with a southbound freight train. As a result of the collision the passenger car was telescoped one third of its length and five men in the smoker were killed, as was Motorman Baker. The passengers who lost their lives were Louie Broo, of Kokomo, Indiana; Dr William Holthausen and his brother William Holthausen, both of Brooklyn, New York; Verdal Railback, of Hymena, Indiana and B. F. Welsch, of Fort Wayne, Indiana. About 15 passengers and Conductor Mains were injured. Dr Holthausen was to have been married to Miss Nellie Coxon of nearby Kokomo only two days after the accident.

The passenger car had been running about 15 minutes late. Had it been a couple of minutes earlier the two cars would have met on a long straight and the collision probably altogether averted. As it was, the view of the motormen where the trains met on the curve was obstructed by a lineside orchard. Both cars were travelling at a good speed, but the

[881] *Galveston Leader*, Galveston, Indiana, United States, 21st December 1910
[882] *Huntington Weekly News Democrat*, Huntington, Indiana, United States, 21st October 1910
[883] *The Logansport Times*, Logansport, Indiana, United States, 4th November 1910

motorman of the freight car had time to jump and escaped with bruises.[884]

After the accident freight car Motorman Dan Lacey admitted, "I guess we overrun the orders," and Conductor Sebree was unable to explain why the freight car had passed the switch where it should have waited for the passenger car.[885]

At the inquest Train Dispatcher Charles Dragoo stated that the extra freight had been ordered to take the siding north of Sharpsville for the passenger train and its failure to do so had caused the collision. Motorman Lacy was not present at the inquest, having disappeared.[886]

Slightly less than a fortnight after the Kingsland disaster, on Tuesday 4th October, there was another interurban disaster of like magnitude near Staunton, Illinois, United States, and again it was a single-line collision. This time the initial count was 35 dead and 31 injured when two Illinois Traction cars collided on Dickerson curve, two miles north of Staunton.

The cars which collided were No. 14 northbound and second No. 73 southbound. Car No. 14 was in charge of Conductor M. A. Leonard with Motorman John Lierman and car second No. 73 had Conductor J. M. Duncan and Motorman E. J. Young. The collision occurred in a dip so that both cars were moving at high speed and their motormen barely had time to jump for their lives before they met. None of the passengers had chance to follow suit, the crash coming immediately after shouted warnings from the train crews. The southbound car was crowded with merrymakers en route to St Louis to view the Veiled Prophets ball (an annual ball held in St Louis since 1878 and sponsored by local business interests).

'The two cars came together with such awful force that they were not only telescoped, but they were battered out of all semblance to their original shape. They were simply a mass of splintered wood and twisted iron and steel 8 feet high, and in this pile of wreckage the dead and wounded were flung in every conceivable form of mutilation. Some of the bodies of the dead were actually torn apart and streams of blood flowed down the debris in a dozen places. The farmers and citizens of Staunton with desperate haste, however, and in a short time had taken from the wreckage all of the living and some of the dead.

'By means of the rural telephone the news of the accident was spread throughout the country, and when it became known that twenty-eight of the killed had been taken to Carlinville, the long distance to that town was literally swamped with messages from anxious inquiries asking for the names of those killed. To all of these, however, but one answer had been given up to 9.30 o'clock, and this was that no identifications had been made. The special train from Staunton bearing the twenty-eight dead reached Carlinville at 6.30, and volunteers offered their services in carrying the torn and mangled bodies to the undertaking parlors of H. C. Heinz & Bros. Hundreds of people crowded about while the officers fought to keep them back to permit the work of identification to progress undisturbed. But relatives of those missing were irresistible in their efforts to see the

---

884 *The Kokomo Daily Tribune*, Kokomo, Indiana, United States, 24th September 1910
885 *The Journal*, Winchester, Indiana, United States, 5th October 1910
886 *Logansport Daily Reporter*, Logansport, Indiana, United States, 29th September 1910

dead, and most intense excitement reigned.' Another report stated that 27 bodies had been taken to Carlinville and eight to Granite City, and the injured to Granite City.

Train No. 14 had been given orders to pass train No. 73 at Staunton, the order stating explicitly that the southbound train was running in two sections. It waited for the first section, then pulled out in the path of the second section.[887]

Both conductors and both motormen were arrested on the evening of the disaster.

Another newspaper reported that after the collision, 'Old employees of the company refused to go out on learning of the wreck. They complained that new men, employed for the state fair traffic, were disobeying orders and running by switches where they were under orders to stop. It is reported that four full crews mutinied and refused to go out. The system has been taxed to its extreme capacity since last Saturday by the immense crowds hastening to the state fair. Many additional men, mostly motormen and conductors employed in various cities of central Illinois, have been pressed into service on the high speed interurban cars.' [888]

The first witness at the inquest on the victims of the disaster (the number of dead having now risen to 37) was Chief Dispatcher Roy Alexander of Staunton, who explained the importance of the time card, on which were written the regular meeting points of trains. He was followed by Dispatcher H. O. Lamb, of Staunton, who had actually issued the relative order involved, identical copies of which were found in the possession of the two motormen after the wreck and which corresponded with the entry in the dispatcher's book.[889]

Upon Motorman Lierman setting off onto the single track Conductor Leonard, at the back of the unit, had, he said, given one stroke on the bell, indicating "stop". The motorman stated, however, that he had given two strokes, indicating "proceed" and he had kept going. Upon this Conductor Leonard assumed that the motorman had received an amended order (though he had not), but proceeded towards the front of the car to clear the point up. On the way down the aisle he was detained by a passenger with a query of some sort and never did reach the motorman's cab before the collision.[890]

Not surprisingly the coroner's jury found Conductor Leonard and Motorman Lierman responsible for the disaster. Their verdict read, "We, the jury, find that the thirty-seven victims of the interurban wreck which occurred October 4, came to their deaths by a head-on collision on the Illinois Traction System of car No. 238, northbound, in charge of Motorman Lierman and Conductor M. A. Leonard, and car No. 239, southbound, in charge of Motorman E. J. Young and Conductor J. M. Duncan, immediately south of Dickerson crossing, on the afternoon of October 4, 1910, about 3.37 o'clock. We find that said collision was caused by the crew of the northbound car not observing train order No. 47, given to them by Train Dispatcher H. O. Lamb, at Staunton, Ill., to take the siding at Wall."

The jury did not make any recommendations as to charges to be preferred against the

---

[887] *Alton Evening Telegraph*, Alton, Illinois, United States, 5th October 1910
[888] *The Ashton Gazette*, Ashton, Illinois, United States, 13th October 1910
[889] *The Daily Free Press*, Carbondale, Illinois, United States, 14th October 1910
[890] *A History of Railroad Accidents, Safety Precautions and Operating Practices* by Robert B. Shaw, pg. 376

conductor and motorman.[891]

The next major single-line collision occurred near Indianola, Nebraska, United States, on the Chicago Burlington and Quincy Railroad, on 29th May 1911. It was the first accident of its type for which a report was published by the Interstate Commerce Commission. The trains involved in the accident were westbound No. 9 from Denver, in charge of Conductor Willetts, with Engineer William Leahy and Fireman Dixon, and eastbound local No.12, in charge of Conductor F. W. Rank with Engineer John Hyder and Fireman Damron.[892]

Mail Clerk O. H. Anderson, who had been riding on the local train and was injured in the accident, described it to a newspaper reporter. ‘We left McCook almost on time. We stopped for a cream car at Red Willow, where we should have sidetracked for No. 9. Conductor Rank had no orders to that effect, and so we went on. We rounded the curve just this side of Red Willow and then the crash came. There was no warning at all. I don’t know how I got out of the car. The first I know is that I was out by the fence, but I must have climbed out of the car, for I was not thrown out. My back was hurt, but in the hurry and excitement I did not notice it much. I found my partner, R. D. Voorhees, almost unharmed. He had been lying on a table against the side of the car, in the safest position that he could have picked.

“As soon as we saw what had happened, we started to see who we could get out of the wreck. Smith, one of the baggagemen on No. 12, told us that his mate, Frazier, was under the wreckage, but that he didn’t have the nerve to pull him out. We went after him and recovered him. A big salesman’s trunk had fallen on him and knocked a hole in the side of his head. An iron rod had run through his face. He was dead when we found him. After we had got Frazier out we went back to the engines. They were smashed together in a fearful mess. Both cabs were crushed in so you could hardly tell where they were. All of the enginemen were found in the cab space. One of them was mashed so that his body could not be removed entire. All were badly mutilated. Apparently they hadn’t had a chance but had died without moving from their seats.

“We went back to the smoker of No. 9. There had been six passengers there, only one got out. He was thrown through a hole in the roof and was found out by the fence, crazy from the shock. I think he was Robert Anderson, but I am not sure. All the others died. The car was a mass of splinters. I don’t believe there was a single stick of timber over two feet long in the whole pile. But the chair car of No. 9 was the worst. It lay on its side, not badly smashed, but with every window closed and the doors shut. Inside the steam pipes had broken and the steam was filling the interior like a boiler. Apparently the engine had been running on full steam for the pressure must have been way up. The steam just poured from the broken pipes into that closed car and the people inside had no chance to get out. As we came up to the car we could hear them crying inside for help. We broke open a few windows in an endeavour to get them out. Whenever we broke a window the steam came pouring out like out of a boiler, deathly hot. Finally, we got enough windows broken to let a lot of it out and then we began to get them out.

“W.O. Euhler, a traveling man of Oxford, and myself, climbed up on to the side of the

[891] *The Weekly Pantagraph*, Bloomington, Illinois, United States, 28th October 1910
[892] *Lincoln Evening News*, Lincoln, Nebraska, United States, 29th May 1911

car, which lay uppermost, and another man went down inside and handed them out to us. They were all fearfully burned. We would lift a man out, and the flesh would drop off his face and hands as we got him into the air. They were all white, like cooked meat, and all horribly puffed out. The steam had just literally cooked them to death. We got them all out and put them on mattresses on the ground or in the Pullmans. Of all of those in the chair car, I don't believe a single woman pulled through. They had all inhaled steam. Several of those we took out were still conscious. A number of them dictated statements to us to be given to relatives. One man, it was either Hilsabeck or Shepherd of Holdrege, died while he was in the middle of a statement to his wife. But none of us realized so much the horror of it then, for we all had so much to do.

"And right here I want to say it was surprising how sensibly everybody worked, despite the terrible situation. The passengers who were injured helped the others and the Pullman passengers helped us all. One man in a Pullman did a Trojan's work. When I asked him for his name he said, "Oh, that's all right. Forget it." No one was leader, however. The trainmen were looking for their fellow workmen and we all worked by ourselves.

"G. G. Peugh, of Indianola, a painter, was the man who did the most of any of us. He was there just a few minutes after the wreck, and he seemed to be everywhere at once. He knew where to get towels and water and he had some medicine with him. Most of the Indianola doctors were still in bed, but they got down in a hurry and worked without food till afternoon. The first relief train was that from McCook, with a lot of doctors. It got there about an hour after the wreck. The Holdedge train was slow."

The injured were taken to Holdedge, where a new hospital had been recently opened. There was only one nurse there, who reportedly worked solidly for 48 hours after the accident, assisted by volunteers. Others of those injured were taken to hospitals at Holbrook and Indianola.

General Superintendent Allen of the Burlington Company made a statement regarding the accident. "The operator at McCook was primarily responsible. He did not give the crew of No. 12 the order for it to meet No. 9 at Red Willow. The crew of No. 12 was also partly responsible. If the trainmen had checked over their orders they would have seen that the meeting order was missing. I don't think there is anything to the statement of people at the wreck who say that the conductor asked the operator about the meeting order and was told that there was none." [893]

Surprisingly it was decided that no inquest would be held into the accident. There was, however, the Interstate Commerce Commission investigation (ICC). This stated that the disaster had claimed 18 lives, with 32 others injured. Killed in the accident were both engineers, both firemen, express messengers Frazier and Freer and 14 passengers.

According to the ICC report Dispatcher D. N. Cobb at McCook issued train order No. 19 at 5.40 a.m. on the day of the accident. This order was to the effect that westbound train No. 9 was to meet eastbound trains No. 2 at Cambridge and No. 12 at Red Willow. The order was sent to McCook (where the dispatcher's and operator's offices were in the same building but separate) at 5.40, Indianola 5.42, Cambridge 5.45 and Oxford 5.46.

Operator R. D. Lyon at McCook claimed that Dispatcher Cobb only issued three

[893] Ibid, 1st June 1911

copies of order No. 19 to him, two of which he gave to the conductor and engineer of train No. 2, which left McCook at 5.40, and the other of which was retained for the office file. Dispatcher Cobb insisted that he had issued five copies. The ICC investigators pointed out that in any case order No. 19 mentioned both trains Nos. 2 and 12, and that as train No. 12 had started from McCook and Red Willow was the next station along the line it could not have received the order anywhere else. Operator Lyon had an hour between the departures of trains Nos. 2 and 12 to have enquired of the dispatcher if the order was meant for train No. 12 as well.

Conductor Rank of train No. 12 signed for orders 875, 300, 302 and 3 at McCook. When Operator Lyon informed the dispatcher that he had done so the dispatcher sent the operator order No. 31, which was what was known as a 'may go' order, which must be delivered to all passenger trains before they left a terminal. This order read, 'On receipt of orders Nos. 875, 300, 302, 3 and 19 No. 12 may go.' Here, the ICC investigators stated, Conductor Rank had made a blunder. He had not checked the numbers of the orders that he had been given with the 'may go' order, which would have shown him that order No. 19 was missing, but only checked them with his clearance card, on which order No. 19 was also missing. Operator Lyon stated that the dispatcher had not included order No. 19 in the may go order, Dispatcher Cobb insisted that he had. The operators at Indianola, Cambridge and Oxford all stated that they had heard the dispatcher address order No. 19 to the McCook operator, but none of them could recollect hearing the operator repeat the address.

The ICC concluded its report by placing the responsibility for the disaster equally on Operator Lyon for failing to deliver the order to the conductor and on Conductor Rank and the deceased Engineer Hyder for failing to check that they had been given all the orders enumerated in the 'may go' order. It was noted that notwithstanding the mistakes regarding orders the accident would doubtless have been prevented had the line been provided with automatic signals.[894]

Within a mile of the July 1906 disaster covered earlier in this chapter another single-line collision occurred at Hamlet, North Carolina, United States, on the Seaboard Air Line Railroad, on Thursday 27th July 1911. An excursion train carrying 912 negroes on a day's outing from Durham, North Carolina, to Charlotte, South Carolina, collided with a freight train coming into the yards at Hamlet from Wilmington. The passenger train came around a curve at a good speed and struck the freight train. The two engines were telescoped and six of the 11 cars of the excursion reportedly 'crumpled like pasteboard' though most of the casualties were in the third and fourth coaches, which were old and frail. A man sleeping with his head out of a coach window was beheaded.

Local physicians set up a field hospital and the less badly hurt were taken later to Charlotte, those too critically injured to be moved remaining at Hamlet.[895] [896] Ten passengers lost their lives and 56 passengers, and four railroad employees were injured, many seriously.

The cause of the accident was unusual and related to local traffic conditions. The

[894] Interstate Commerce Commission report
[895] *The Joplin Daily Globe*, Joplin, Missouri, United States, 28th July 1911
[896] *Warren Mirror*, Warren, Pennsylvania, United States, 28th July 1911

excursion train, running as train second No. 33, was running southwards to Hamlet over the Raleigh District of the railroad. The regular freight train, No. 17, had arrived from Wilmington over the Wilmington District, but passed on to the Raleigh District at Hamlet to reach the freight yards there. Dispatcher Purvis had charge of the Wilmington District, and Dispatcher Witt was in charge of the Hamlet District, the two working from the same Hamlet office.

The Standard Code, used by the Seaboard Company, required that all trains arriving at registering stations such as Hamlet must check the register to ensure that all opposing trains over the single line had arrived. At Hamlet this posed a problem due to the severe gradient from the Wilmington line into the Hamlet yards, and the need, therefore, to avoid, if possible, having to stop freight trains on the incline to register, because of the difficulty they might have in starting. The practice had therefore arisen of the Wilmington District dispatcher, after having checked with the Raleigh District dispatcher that no further trains were expected, notifying incoming freight trains at Laurel Hill, 10 miles short of Hamlet, that 'All overdue first class trains are past Hamlet.' On the day of the accident Dispatcher Purvis had thus sent this message to freight train No. 17 at Laurel Hill, without, however, having first checked with his colleague, Dispatcher Witt, that this was indeed the case, and when in fact train second No. 33 was approaching. Thus, misinformed the conductor and engineer did not stop to check the register at Hamlet, and proceeded with no knowledge of the excursion, while the crew of the excursion had no knowledge of the freight train.

Because of the gradient the freight train had to be 'doubled' (divided and taken up the bank in two parts) and consisted only of a locomotive and nine cars, travelling at five or six miles an hour. The excursion train was travelling at between 25 and 30 miles an hour.

The Interstate Commerce Commission naturally found Dispatcher Purvis responsible for the accident. It noted that the higher officials of the company were fully aware of the practice of advising the crews of freight trains that they need not check the register and that the practice had been followed for years. The commission stated, however, that 'The practice of issuing message instructions to trains relieving them of duties required under the Standard Rules is a dangerous one and it should not be permitted on any railroad. Such authority should be given only by regular train orders, so that all trains affected shall have notice.'

The commission also noted that the manual block system was in use between Raleigh, N.C. and the Hamlet yard office. Had its use extended to Hamlet passenger station the accident would have been prevented.[897]

No reason was given for Dispatcher Purvis's mistake. Possibly, unaware of the running of the excursion train and knowing that the regular passenger train had arrived, he had simply assumed that there were no first-class trains due without bothering to check.

The day after the Hamlet tragedy a return excursion train from Kidders Point to Caribou featured in a collision on the Bangor and Aroostock Railroad at Grindstone, Maine, United States. The excursion train and a regular passenger train from Van Buren

---

[897] Interstate Commerce Commission report

to Bangor met on another on a dark and stormy night at 9.10 p.m. on Friday 28th July 1911.

Running as the second section of passenger train No. 7 the northbound excursion train consisted of engine No. 55, a combination baggage car and passenger coach and four day coaches and was in charge of Conductor Dibble with Engineer F. W. Garcelon and Fireman Gallagher. The southbound passenger train, No. 8, consisted of engine No. 243, a baggage car, two passenger coaches and two sleepers. It was in charge of Conductor Chester Nutter with Engineer William Orr and Fireman Wentworth. The excursion train was carrying 170 passengers and the regular train about 100, many of whom, in the sleepers, were 'thrown from their berths in scanty attire'.

Describing the accident, a newspaper reporter wrote, 'The excursion train had barely left the curve and a small viaduct over the river when the trains crashed. Engineer Garcelon, an instant before the great locomotives leaped head-on into each another, threw off the lever, applied the brakes, and the next instant was crushed to death as the engines hit, and rearing for a moment high in the air, fell in a twisted heap of steel and iron. The dead passengers were riding in the car immediately back of the excursion train's tender. This coach was almost completely telescoped by the coal car, its sides projecting along the outside of the tender to smoking engines and catching fire.

'Grindstone is a hamlet of only 40 or 50 persons. All had heard the crash and before the fire demon had a chance to add to the horror, they had formed a bucket brigade from the nearby river and extinguished the flames.

'Believing that not a passenger in the coach where all the dead were found had a chance to escape, the attention of the hardy men and women of the little backwoods camp turned immediately to the cars in the rear where the cries of the injured were heard. The hands of the rescuers were rough and unskilled, but they worked regardless of fatigue, some with only their hands to tear windows and boards from the cars in which the injured were imprisoned. Others had run to their homes and camps not far distant for axes, bandages and such medicines that they kept in their homes. But even the best that these sturdy backwoodsmen could do offered little comfort to several whose hurts were so terrible that they may succumb to them.' [898]

Grindstone, another newspaper stated, had just 43 inhabitants. Its only means of communication was by the railroad and there was not a single telephone there. There was not even a logging road out of the place, and of course no physicians or medical facilities. Help was, however, summoned by telephone by the nearby house of the foreman of a logging gang.[899]

A relief train was sent from Millinocket and a second reached the scene from Bangor about midnight. The dead and injured were taken to Millinocket, back to Bangor or to Presque Isle, where most of the excursionists lived. Engineer Garcelon, Firemen Gallagher and Wentworth and five passengers were killed more or less outright in the accident and 17 people were seriously injured, including Engineer Orr and Brakeman George Estabrook of the excursion train, the latter of whom died on Sunday morning. The passengers who died were Dr Hugh Pipes, a Presque Isle dentist; Frank Seeley, son of the cashier of the Presque Isle National bank; Harry Clark and Zerne Harris, both of

898 *Bridgeport Evening Farmer*, Bridgeport, Connecticut, United States, 29th July 1911
899 *Daily Kennebec Journal*, Augusta, Maine, United States, 31st July 1911

Presque Isle, and Claude Loomer of Washburn.[900] Some or all of the passengers who lost their lives were members of the Presque Isle Brass Band.

The cause of the collision was clearly documented in the Interstate Commerce Commission (ICC) report on it. Passenger train No. 8 was superior to train second No. 7 by direction, meaning that train No. 7 must keep clear of No. 8's leaving time at all stations. At Millinocket at 8.55 p.m. Conductor Dibble and Engineer Garcelon of the excursion train had been given an order instructing them to run 3 hours and 45 minutes late, that is 3 hours and 45 minutes late on the booked times of regular train No. 7, which was running that day as first No. 7. The regular train No. 8, was not given and did not require any orders regarding the excursion as it was superior by direction and was running on its timetable rights.

The excursion train, second No. 7, departed from Millinocket at 8.58 p.m., Conductor Dibble having asked his engineer if he could get to Grindstone before train No. 8 was due to leave there at 9.09 p.m. According to the rules the excursion train had to be in clear at Grindstone by 9.09 p.m. or if still on the main line to be properly protected. As already mentioned it was a very inclement night with strong winds and heavy rain, which, as the ICC pointed out, should have discouraged any attempt at high speed, but the conductor and engineer nevertheless decided to attempt to get to Grindstone, despite the fact that to do so successfully would require an average speed of almost 50 miles an hour. The trains met at 9.10 p.m. with train second No. 7 still two fifths of a mile short of the points which would have taken it into the siding. Conductor Dibble told the investigators that he had realised at 9.09 that train No. 8 was leaving Grindstone but had made no attempt to flag it or to stop his own train, because he had hoped that No. 8 would see them. He had stood at the door of the baggage car and when he saw the headlight of No. 8 and it had become apparent that No. 8 was not stopping he had jumped.

The commission concluded that the accident had been due to Conductor Dibble and Engineer Garcelon 'running on short time' and that even so the conductor could have averted it had he stopped his train and properly protected it when it became obvious they could not reach Grindstone in time. Had some form of block signalling been in use the accident would also have been prevented.[901]

An accident similar to that at Hamlet in July, inasmuch as it involved as it involved the checking of train registers, occurred on 15th October 1911 near Fort Crook, Nebraska, United States, on the Missouri Pacific Railroad. Northbound passenger train No. 105, consisting of engine No. 7707, a baggage car, a combination mail and passenger car, a day coach, a chair car, a combination mail and passenger car, a day coach, a chair car and a Pullman sleeper, collided with extra fast freight train No. 407, consisting of engine No. 407, 14 loaded cars, three empty cars and a caboose. It was in charge of Conductor F. R. Travers and Engineer John Scott.

Fort Crook was a few miles south of Omaha. For about five miles south from Omaha Missouri Pacific trains used the double track of the Union Pacific Railroad, then at Gilmore Junction they passed on to their own single-track line. Fort Crook was the next station, 1½ miles south of Gilmore Junction.

---

900 *Boston Sunday Post*, Boston, Massachusetts, United States, 30th July 1911
901 Interstate Commerce Commission report

Passenger train No. 105 had left Kansas City at 12.01 a.m., about an hour late. At Union, 30 miles from South Omaha, its train crew was given train order No. 15, which read:

'No. 105, Engine 7707, will run late as follows, leaves Murray 8.16 a.m., Mynard 8.23 a.m., Plattsmouth 8.30 a.m., LaPlatte 8.40 a.m., Gilmore Jct. 8.50 a.m.'

The train then continued towards Omaha and was an hour and a half late at the time of the accident.

Freight train No. 407 was in charge of Conductor L. C. Gross and Engineer E. C. Crawford. It left the yard at Nicholas Street, Omaha, at 7.30 a.m. after having received order No. 19, which read:

'Engine 407 run extra Omaha to South Omaha with right over No. 155.'

The train arrived at South Omaha at about 8.05 a.m. Four loaded cars were picked up, and its conductor and engineer were given train order No. 28, which read:

'Engine No. 407 will run extra South Omaha to Falls City Yard, will meet extra west 2303 at Union and extra west 7710 via Talmage at Nebraska City Jct.'

The conductor and engineer then went to the yard office to check the train register book that all superior trains had arrived. The freight train left South Omaha at about 8.30 a.m., passed Gilmore Junction and entered the single line. It collided with passenger train No. 105 near Fort Crook station at 8.46 a.m. The trains met on a curve of about three degrees. The crew of the freight train saw the passenger train in time to almost come to a stand, but the engineer of the passenger train did not see the freight train as early from his side of the engine while his fireman was busy firing. The passenger train was probably still travelling at 40 miles an hour when the trains met. Both locomotives were telescoped and derailed in the collision, the day coach was telescoped for two thirds of its length by the combination car, which was practically destroyed, and one end of the baggage car was badly damaged. Brakeman O. W. Keeler of the passenger train and six passengers were killed outright, and 29 people were injured, four of them seriously, and one of whom, Mrs Emma Harvey, later died from her injuries.[902] The passengers who were killed outright were; F. W. Petring, a Nebraska City merchant; Mrs Fred W. Rowtmann and her six-year-old daughter, Marcia, also of Nebraska City; A. W. Sprague, an electrician, of St Joseph, Missouri; Miss Frances Lilian Kanka, of Washington, Kansas and an unidentified coloured woman.[903]

By chance a passenger train was due to leave Omaha for the south when news of the accident was received there. It was converted into a relief train and sent to the scene with a staff of Union Pacific surgeons and Dr Klick, the local Missouri Pacific surgeon. This train returned to Omaha with the uninjured passengers and those slightly injured passengers who were well enough to be sent home. The more serious cases were cared for at the army hospital at Fort Crook rather than their being moved.[904]

The conductors of freight trains were required to register the arrival and departure times of their trains in person in the South Omaha train register book. The conductors of passenger trains were not required to register their trains, this being done by the South

[902] *Ottumwa Daily Review*, Ottumwa, Iowa, United States, 19th October 1911
[903] *Sullivan Sentinel*, Sullivan, Missouri, United States, 20th October 1911
[904] *Centralia Evening Standard*, Centralia, Illinois, United States, 16th October 1911

Omaha yard office operator, who received the information in a roundabout manner. When a Missouri Pacific passenger train passed Gilmore Junction the Union Pacific towerman there telephoned the time to the Union Pacific dispatcher at Omaha. The latter notified the Missouri Pacific dispatcher at Falls City, Nebraska, about 100 miles away, and he in turn issued a train order to the operator at South Omaha, about 93 miles from Falls City, to register the passenger train as having arrived at South Omaha at a certain time. The Interstate Commerce Commission (ICC) noted that this system was likely to lead to errors and should be discontinued and instead conductors should be required to register in person at Gilmore Junction. Conductor Gross had stated in his evidence at the inquiry that he often had to call up the dispatcher to find what trains were in and had been told to go out to the yard and look for them.

Missouri Pacific dispatcher David Lane at Falls City stated that he had sent a message to the South Omaha yard office for Conductor Gross, which read:

'Conductor Gross, 407. No. 105 should pass Gilmore Junction about 8.50a.m. with no signals. W.F.K.'

Both Conductor Gross and the South Omaha operator denied receiving this message. Conductor Gross stated that he had checked the register at South Omaha and thought that train No. 105 had arrived, but that he might have mistaken train No. 155 for it, or else checked the wrong date. Engineer Crawford did not check the register, as required by the rules, but relied on his conductor.

The ICC blamed the accident on the negligence of Conductor Gross and Engineer Crawford in not properly checking the South Omaha register and held that there was no excuse for their failures. The commission recommended the maintenance of an adequate train register book at Gilmore Junction and the installation of a block signalling system.[905]

The little Ligonier Valley Railroad was one of what in the United States were appropriately termed 'short lines'. It had single-track main line of a mere 11 miles in length from Latrobe to Ligonier, and there was also the Mill Creek branch, also single track, on which the accident occurred. For 35 years it had a very good safety record, with neither a single employee nor passenger losing their life in an accident. Its operating methods were primitive, however, with no printed rules, no block system, no train registers and no written train orders. Perhaps, therefore, it was no surprise that on Friday 5th July 1912 it suffered a major disaster.

On most days a coal train left Wilpen, near the end of the Mill Creek branch, in the late morning and arrived at Ligonier before noon but on the day of the accident it was unavoidably delayed and its heavy train, drawn by two engines, was not ready to start until 3 p.m. At that hour its conductor, H. H. Knox, telephoned the office at Ligonier, where the call was answered by car distributor Frank Noel, and the conductor asked permission to bring his train down to Ligonier, 3⅓ miles away, before the afternoon passenger train came up to Wilpen. Noel said that General Manager George Senft and Dispatcher Russell Minnich were also in the office. They agreed to let the freight train come down first. Minnich was instructed to tell Conductor Charles Kuhn of the passenger

[905] Interstate Commerce Commission report

train to wait until the freight train had arrived and stated that he had done so verbally. Noel said that he then had some other work to do a short distance from the station and that while he was so engaged he saw the passenger train pulling out. He assumed that it was just being run onto a siding and made no attempt to stop it.[906]

The passenger train consisted of an engine and a single combination baggage and passenger car. There were ample sidings and a "Y" at Ligonier where the engine could have run round its coach, but this was not done, and the passenger train left Ligonier with its engine pushing the car. Its engine was in charge of Engineer Dunlap, with Fireman Frank Byers. It was crowded with about 50 passengers, including a number of children who had been taken on a day's outing by Miss Matthews, a nurse at the home of General Manager Senft. Other passengers had been visiting Latrobe and nearby towns.

The freight train had left Wilpen at 3.20 p.m.. It consisted of engines Nos. 7 and 14 and 14 cars loaded with coal. It was travelling at an estimated 20 to 25 miles an hour at the moment of the collision, while the passenger train was running at 10 to 15 miles an hour. A newspaper reported described what happened when the trains met. 'The freight engines plowed through the wooden coach, crushing it as it would an egg shell. The coach was ripped to pieces. All the occupants were hurled to the roadbed, some fell in the path of the engines, while others were imbedded partly in the cinder and stone ballasting. The first engine of the double header pulling the freight stopped after it had passed through the passenger coach, turned half way around and fell over on its side.

'Conductor Charles H. Kuhn of the passenger train leaped when he saw the impending crash and received a broken leg. Engineer Dunlap and Fireman Frank Byers of the passenger train stuck to their posts and were uninjured. Dunlap and Byers are the only persons known to be on the wrecked trains who escaped injury.

'Farmers near the scene of the accident and men who were at work at the Denny race track went to the assistance of Engineer Dunlap and Fireman Byers, who were pulling the dying and injured from the wreckage and laying them on the bank. A messenger sped with all haste to Ligonier and notified the residents there of the accident. In a few minutes every resident of the town who was able to walk was on the way to the race track. A frantic call for physicians and aid was sent to Latrobe.

'When notified of the wreck the Pennsylvania Railroad company, of which the Ligonier Valley railroad is a branch, sent a special train from Pittsburgh to Latrobe to bring the injured from the overtaxed Latrobe hospital to Pittsburgh hospitals. The rescue train on the Ligonier valley steamed into the Latrobe station about 7 o'clock last night and was met by almost the entire town. Every conveyance in the town was at hand, and the injured were taken with dispatch to the hospital.

'The wreck train of the Pennsylvania railroad from Pitcairn was ordered sent here. It is thought that there may be more bodies in the wreckage. Coroner McMurray, of Westmoreland county, and several deputies came here and took charge of the dead in the local undertaking establishments. Fourteen dead are in Lowrey's morgue here, while the others are in Slater's and Hartman's morgues at Latrobe.' [907]

'Miss Mathews, the nurse who had charge of the children, proved herself a heroine.

---

[906] *The Philadelphia Inquirer*, Philadelphia, Pennsylvania, United States, 11th July 1912
[907] *New Brunswick Times*, New Brunswick, New Jersey, United States, 6th July 1912

Just before the two trains struck Miss Mathews glanced through the door of the coach which was being pushed backward along the single track. A glance was enough to tell her that a wreck was inevitable. Miss Mathews caught up the nearest two children and threw them into the baggage compartment of the coach. With two others in her arms she ran toward the front of the train. Then the crash came.

'When the rescuers arrived Miss Mathews, with the two children clutched in her arms, was buried under a pile of wreckage near the engine. When the men sought to lift the timbers from her body she called their attention to the children and told them to get the little ones first. When Miss Mathews was finally taken from the wreck it was found that she had received a fracture of the left hip and arm, a severe scalp wound and internal injuries. It was due to Miss Mathews that all of the little ones who had gathered around her seat were not killed instantly.' [908]

The accident cost the lives of Engineers Frank McConnaughy and Smith Beatty, Firemen George Byers and John Ankney and probably 22 passengers, including one who succumbed to her injuries as late as 11th March 1914.

At the inquest upon the persons known at that time to have died in the disaster the jury exonerated Conductor Kuhn from blame and placed the responsibility for the accident on the Ligonier Valley Company. Their verdict was that the company had been negligent "in operating their trains without written orders, without certainty of verbal orders, without signal code, without general rules with respect to operating trains, without track signals and railroad instructions to employees, without proper appliances especially for a signal system and without a general, recognized up-to-date method of operation." [909]

Testifying before the Interstate Commerce Commission investigation, Dispatcher Minnich stated that he had told Conductor Kuhn no less than three times that his train must wait at Ligonier for the arrival of the freight train, once when he was in the combination car and twice while he was on the station platform. This was corroborated as regards the combination car by Freight Clerk Clop and as regards the station platform by Janitor and Baggageman William Frothy. The other members of the passenger train crew said that they had no knowledge of any message from the dispatcher regarding the holding of their train.

Conductor Kuhn was badly hurt in the accident and had to be interviewed in the hospital. He denied having been told by the dispatcher to wait for the freight train and said that the only instruction given to him by the dispatcher related to a freight train which ran between Latrobe and Ligonier and which was in Ligonier yard at the time, the dispatcher telling him that his train could go as soon as the freight train had left the yard.

The conclusion of the ICC was that the accident had resulted from either the failure of the dispatcher to deliver or of the conductor to obey or understand an order that the passenger train wait at Ligonier for the arrival of the freight train. With no written order the matter resolved itself into a question of veracity between the conductor on the one hand and the dispatcher and other employees on the other, but the weight of the evidence led to the conclusion that Conductor Kuhl was responsible.

Turning to other matters, the commission stated that the pushing of the combination

---

[908] *Altoona Mirror*, Altoona, Pennsylvania, United States, 6th July 1912

[909] *Altoona Mirror*, Altoona, Pennsylvania, United States, 7th September 1912

car by the engine was an extremely dangerous method of carrying passengers. There was ample opportunity for turning the train at Ligonier and had this been done there would undoubtedly have been a material reduction in the loss of life. The final sentence of the report stated, 'The method of train operation on this road is extremely faulty, and until adequate rules regarding train operation are adopted and confirmed, accidents of this character are liable to occur.' [910]

Thirty-three people were reportedly killed and 70 injured on Monday 7th April 1913 in what was presumably a single-line collision between an express train and a freight train at Ogulin, Croatia, on the Hungarian State Railway. Both locomotives, an express car and a sleeping car of the express were telescoped in the head-on collision. Those killed included both drivers, both firemen and a guard. The telegram giving notice of the accident stated that there might be more bodies hidden in the wreckage.

Ogulin was described as being 23 miles south west of Karlstadt on the Dohert River.[911]

In what was presumably a single-line collision two locomotives were destroyed in an accident between Poroi and Andjista, Bulgaria, on Friday 23rd May 1913. The stationmaster judged responsible for causing the accident was shot on the order of the Inspector of Bulgarian Railways, clearly a drastic punishment reminiscent of those imposed on railwaymen in Russia in the 1930s, during the Stalin era. No other details of the accident have emerged.[912]

On Thursday 19th June 1913 another collision occurred on an American interurban line. This time the accident happened in the far west, near Vallejo, California, United States, on the San Francisco Napa and Calistoga Railway. Vallejo was the southern terminus of the line. An incoming train in charge of Conductor B. E. Catton and Motorman William Juarez collided head-on with an outbound train in charge of Conductor Horace Richmond with Motorman Eugene Hough. The two-car outgoing train was crowded with passengers from the steamer *Monticello* from San Francisco as well as people from Vallejo. The incoming train from Napa was likewise carrying passengers intending to take the steamer to San Francisco. The trains met near the end of a straight stretch of track at One Mile House but were reportedly within a few hundred feet of one another before the motormen realised that they were on a collision course and applied the brakes. The leading cars of each train were telescoped, and it was in them that most of the deaths and injuries occurred. The other cars were thrown from the track and piled up on one another. The uninjured passengers went to the assistance of those less fortunate than themselves.[913]

A surviving passenger from Oakland, California, Mr C.A. Johnson told a newspaper reporter of his narrow escape, "My escape from that car of death was made through the

---

[910] Interstate Commerce Commission report

[911] *Naugatuck Daily News*, Naugatuck, Connecticut, United States, 7th April 1913

[912] *Railway Gazette* Volume 18 pg. 682

[913] *Public Press*, New Albany, Indiana, United States, 24th June 1913

most awful confusion I have ever witnessed. My escape itself was miraculous. I was about to enter the smoker when somehow or other my desire to smoke left me in an instant and I decided to wait until two or more stations had been passed. I was going to seat myself in a front portion of this car when once again Providence guided me to a rear seat near the door of the train. If I had followed my own inclinations I would certainly have been mangled. There was certainly some good influence beside me that guided my movements.

"When the trains met a crash that was hollow and at the same time seemingly slight resulted, and I thought for an instant that we had hit a waggon or some animal that had wandered on to the track. The first faint crunch was followed by a longer and louder one; timbers began to groan and strain and at last I felt the whole car raise itself on end. I do not know what happened next; I think I was struck a glancing blow by a falling timber, but when I did come to a full realization of what had happened I found myself endeavouring to raise myself from a half-reclining position on the arm of the seat. I was bleeding profusely from a cut on the face and a nasal injury. I had raised myself to an upright position when all of a sudden a great splinter shot past me, parting of it lodging in my hat. Without a doubt I would have been killed outright had the whole splinter hit my derby; you see luck – or should I say Providence was still on my side."

After the accident local farmers rushed to the assistance of the victims and firemen from Vallejo were soon on the scene. The casualties were taken in automobiles to hospital.

Thirteen passengers and Motorman Hough died in the accident and Conductor Richmond, Motorman Juarez and about 25 passengers were injured.

The inquest on the accident was held in Vallejo on 26th June. Dispatcher Edward O'Leary at Napa testified that as well as being responsible for train movements over the whole system he was also charge agent for the company there, had to keep the time of all employees and do miscellaneous clerical duties. He said that the southbound train had called him from Collins (the first station out of Vallejo) whereupon he had told the conductor to wait until he had located the whereabouts of the 'flyer' (the northbound train involved in the collision). He had then called the agent for the Monticello Steamship Company at Vallejo and had been told that it was ready to leave but was loading express matter. He had assumed that the northbound train would not set off until he had been contacted and told the conductor of the southbound one to proceed.

Following the dispatcher into the witness box was Traffic Superintendent K. E. Harrington, whom the newspaper reporter characterised as 'a poor witness'. What evidence he gave was not reported.

Conductor Richmond of the northbound train suffered a fractured skull and the inquest had to go to the hospital for him to be interviewed from his bed. He was said to be still very weak. He said that he had been a railroad man all his life, although this had been only his second trip over this line. His train had started about 17 minutes late. He had contacted the dispatcher and had received clearance to Collins, four miles beyond the place where the collision occurred.

Motorman Hough of the outbound train had made a dying deposition which was read out and which corroborated Conductor Richmond's testimony.

There was therefore a direct contradiction between the statement of the dispatcher, which was also supported, and that of the crew of the northbound train.

The verdict of the jury was that the San Francisco Napa and Calistoga Company was itself responsible for the collision, with Dispatcher O'Leary and Conductor Richmond being but the agents of a very lax system. District Attorney Raines was reported to intend to vigorously prosecute the railroad.[914]

The accident was apparently not investigated by the Interstate Commerce Commission, presumably because it did not occur on a steam railroad, though a few years later they were inquiring into significant interurban accidents as well.

On 13th July 1913 two trains collided head-on near Irún, Spain. The accident occurred in a tunnel and is therefore described in the Tunnel Accidents chapter.

The next significant single-line collision anywhere in the world did not occur until just after a year after that in Spain. When it did it brought to prominence a new hazard of train travel. The accident was a collision between a Missouri and North Arkansas Railroad gasoline (petrol) motor car and a conventional Kansas City Southern Railroad passenger train. It happened less than a mile south of Tipton Ford, Missouri, United States, on Kansas City Southern tracks over which the Missouri and North Arkansas had trackage rights (running powers). The motor car, No. 103, was a single gas-electric car, 70 feet long and weighing 47 tons. Its gasoline tank had a capacity of 150 gallons and at the time of the accident it was carrying about 105 gallons. In charge of it was Conductor S. A. Nichols with Motorman Herbert Ratcliff. It had left Joplin, Missouri, at 5.30 p.m. on Wednesday 5th August 1914, an hour and 15 minutes late and was running as train No. 209.

The other train involved was northbound Kansas City Railroad passenger train No. 2, running as the first section of freight train No. 56. It consisted of locomotive No. 805, a mail and baggage car, two baggage and express cars, an express car, a day coach, a chair car and a Pullman sleeper car, in charge of Conductor J. B. Sisk, with Engineer John Brennan and Fireman Guy Walters. It had left Neosho, Missouri, three hours and 15 minutes late.

The trains met on a curve, each travelling at about 35 miles an hour, at 6 p.m. The heavy passenger train overwhelmed the lighter motor car, which was telescoped 20 feet, pushed back 650 feet and totally destroyed. The engine and leading baggage cars of the passenger train separated from the rest of the train. The impact of the collision burst the pipes and tank of the motor car, the escaping gasoline ignited and the whole of the car was immediately enveloped in flames, making the rescue of its occupants impossible.

The local Joplin Daily Globe newspaper not surprisingly gave very extensive coverage to the accident, including the stories of some of the survivors. 'Mrs Fannie Embrey, a negro, escaped with her son, Ernestine, ten year old, and her daughter, Lois, 7, but her husband, W. H. Embrey, was burned.

'One negro mother's prayer was answered, according to W. M. Drury, of Webb City, who, with his family, escaped.

---

914 *Oakland Tribune*, Oakland, California, United States, 27th June 1913

'"Oh God, save my baby!" the woman cried, Mr Drury said yesterday. Flames were sweeping up around her and her agony was intense. A courageous man was helping me, and we rushed to the car and opened the door.

'"As if in answer to the mother's plea, the baby rolled from the car and at our feet," Mr Drury concluded. Mr Drury doesn't know what became of the mother or the child. Possibly the mother perished. At St John's hospital last night, no information could be obtained concerning either. The youngest negro survivor at the hospital was Lois Embrey.

'Mr Drury, with the aid of a "low, heavy set fellow" who aided him, saved eleven persons, working from one side of the motor car. He enumerated the rescued as follows: His son, Forest; a negro woman; his daughter, Pansy; a strange man; three negroes; a young man; his baby Fred, thirteen months old; his wife and an old man.

'"As many more could have been saved if the survivors had not been dazed," Mr Drury said, "We worked from one side of the car only and more could have been taken out from the other side. We cried to the men survivors to help us, but they remained, seemingly dazed, on the embankment above, while the human cargo burned."

'Mr Drury once gave up his wife as lost but determined to make one more effort to free her from timber which pinned her down. The last time he was successful. She was burned and her face was badly cut.

'The speed of the car and train saved the survivors, Mr Drury believes, from a fiery fate. "The passenger train was so heavy it pushed the motor car back several hundred feet," he said, "but gasoline was spilled along the track, so that coaches of the passenger train stood over it and were scorched. The flames followed the path of the gasoline and it gave rescue workers a few moments to get the victims out. When the fire finally reached the car it was filled with flames, which prevented saving many others."

'Mr Drury said the motor car, he believes, was travelling at a rate of thirty miles an hour.

'Engineman Ratcliffe's body fell half through a window of his cab and hung there. Mr Drury saw the flames eat their way up and consume the body. Ratcliffe, unconscious from the first crash, never moved.

'Mr Drury saw an old man who carried a suitcase and wore a straw hat. Clinging to both, the man was catapulted the length of the car. He escaped, still carrying his hat and baggage. He is believed to be the only person who escaped with property. 'H. E. Littleton, of Harrison, Ark., who was badly cut on the head, was made unconscious and he remembers nothing about the accident.' [915]

Killed in the accident were Conductor Nichols, Motorman Ratcliff, Brakeman Edward Bradley of train No. 209 and, according to the Interstate Commerce Commission report two other railroad employees, together with at least 38 passengers. Four railroad employees and 34 passengers were injured.

Before leaving Joplin Conductor Nichols of the motor car had been given several train orders by Operator William Hadley, one of which was No. 84, which read:

"First No. 56 meet No. 209 at Tipton Ford and wait at Tipton Ford until 5.30 p.m.,

915 *The Joplin Daily Globe*, Joplin, Missouri, United States, 7th August 1911

Saginaw 6 p.m. for extra No. 563 south."

The accident was clearly due to the failure of the crew of the motor car to obey the order and wait at Tipton Ford. The problem was, why?

The Interstate Commerce Commission inquiry heard evidence that Conductor Nichols had failed to register his train out of Joplin, but Operator Hadley was positive that the conductor had received his train orders and they were signed for.

Dispatcher W. R. Seberd at Pittsburg, Kansas, remarked that it was very unusual for a conductor not to register. He explained that the passenger train had been delayed by a wreck in Arkansas and there was no passenger schedule open at that time of the day, hence it had to be given the schedule of a freight train.

Superintendent S. O. Lucas stated that he had not looked into Operator Hadley's past record beyond learning that he had given satisfactory service in his last job with the St Louis San Francisco. It transpired that Hadley had previously been dismissed by the Atchison Topeka and Santa Fe Railroad for failing to deliver an order at Holly, Colorado, but Superintendent Lucas remarked that, "If a man has failed to deliver orders once I believe it makes him a better man than before; it makes him more cautious," and that "Operators all over the United States – I might almost state hundreds of them – have overlooked orders and are still in the employ of the railroads. I do not think either the Kansas City Southern or the Missouri and North Arkansas would employ men under the same circumstances as this, but it is my method."

As regards Conductor Nichols he had never known him failure to register. He did not know if the conductor had been under the influence of liquor; he had never known him to be.

The superintendent admitted that it would have been possible for the operator to have failed to give the conductor his orders and then to have signed for them himself, forging the conductor's signature.

Conductor Sisk testified that he had been in the smoking car when the collision occurred and had not been hurt. His first impression was that they had run into a burning bridge and he had set up a bucket brigade. When the real cause of the fire had become apparent he had realised that it would be useless to try and fight it.

Kansas City Southern Operator Charles Keck at Neosho testified to receiving order No. 84 at 4.53 p.m. and to hearing the Joplin operator repeat it back to the dispatcher. Shortly after 6 p.m. he had heard the dispatcher ask the Joplin operator if he had delivered the order to the motor car conductor. The Joplin operator had replied, "Wait a minute, I'll see," and two minutes later, "I sure delivered it."

There were allegations that the signature of Conductor Nichols on the carbon copy of train order No. 84 retained in the office at Joplin was not his writing but there was no real evidence to this effect.

Conductor H. C. Gibson, of train extra No. 563, Conductor Morris, Conductor Hailey and a Joplin taxi driver named L. S. Wells all testified to having seen Conductor Nichols in the telegraph office, but none could actually swear to having seen him receive and sign for order No. 84.[916]

---

[916] Ibid, 8th, 9th and 10th August 1914

With the entire crew of train No. 209, and in particular Conductor Nichols, unavailable to give their side of the story, the Interstate Commerce Commission (ICC) could do little as regards the cause of the accident than state that it was due to train No. 209 failing to wait at Tipton Ford for train first No. 56. Why it had failed to wait could not be determined.

The commission noted that this was the first accident that they had investigated involving a gasoline motor car and stated that, 'The rapidly increasing use of these motor cars, carrying large quantities of gasoline, introduces such an additional element of danger as to demand extraordinary precautions against the possibility of collisions wherever such cars are used.'

The ICC also remarked that no block system was used on the Kansas City Southern Railroad and even the inherently weak train order system was not used at its highest degree of efficiency. Rule 208 of the company provided for the placing of a meet at the designated meeting point (a so-called 'middle order') where practicable. Tipton Ford was one of two meeting points between Joplin and Neosho where no operator was employed, and where, therefore, a middle order could not be placed.[917]

A single-line collision between Nieborow and Lowitsch, Russian Poland, on the evening of Monday 11th January 1915. Telegraphist Kablitz at Nieborow asked Telegraphist Schmidt of Lowitsch if a train was due. Schmidt answered that, "Yes, there is a train due here soon, but it won't be going any further very quickly because the waggons it needs to collect are spread over several sidings." This enquiry did meet the requirements of a request for clearance for a train, which had a mandatory form of wording. Furthermore the telephone message was not recorded in the train register at either place.

Almost two hours after this telephone conversation, at 20:40, Fahrdienstleiter (Signalman) Boikert at Lowitsch gave the train permission to proceed to Nieborow. He claimed later that Schmidt had told him that Nieborow had accepted the train, an allegation which Schmidt vigorously denied.

This mistake would, however, not have led to an accident had a second mistake not followed. At 20:55 Schmidt accepted an empty military train from Nieborow without consulting Boikert. At 21:10 the munitions train hauled by Prussian engine Munster 1006 collided on a curve and in a fog with the empty train pulled by engine Stetten 4901, the latter running tender first. The collision resulted in a big pile of wreckage and the deaths of 19 people.

Boikert and Schmidt were sent back to Germany to stand trial for endangering the running of the railway and causing disruption to traffic. Their trial was set for August 1916, then for November 1916, but each time was postponed due to lack of witnesses and eventually the case was dropped.[918]

On Monday 22nd November 1915 a Central of Georgia Railway passenger collided

[917] Interstate Commerce Commission report
[918] *Eisenbahn-Unfälle in Europa* by Erich Preuß, pgs. 125-7 (in German)

with a special train of the Con. T. Kennedy Carnival Company near Muscogee Junction, Georgia, United States.

Eastbound passenger train No. 2 consisted of locomotive No. 1610, a combination mail, baggage and express car and two coaches, in charge of Conductor J. W. Reichert and Engineer J. L. Fickling. The special train consisted of locomotive No. 1716, 16 flat cars loaded with waggons and other circus equipment, three stock cars and nine sleeping cars, in charge of Conductor Convey and Engineer W. R. Bittick. The show train was en route from Atlanta, Georgia, to Phoenix, Alabama. The accident occurred a few miles from Columbus, Georgia. East of Columbus for just over two miles the line was double track, but at Muscogee it became single. The trains met one another three and a half miles further east. The two engines were almost destroyed in the collision and the flat cars at the front of the train immediately caught fire. By the time of the first report on the accident, the evening that it occurred, 12 of the flat cars had been destroyed by fire and others were still burning. At that time there were known to have been six fatalities and 12 others were missing. Two of the dead were identified as Mr and Mrs Fred Kempf, of Kansas City. They and their daughter, Hazel, had been riding in a specially constructed 'automobile truck' loaded onto one of the flat cars. The two adults were trapped in their car, unable to escape, and burnt to death, but Mrs Kempf managed to hand her daughter out of a window to rescuers, and the child escaped with cuts and bruises. The other four known initially to have been killed were showmen, one known as 'Whitney' and the others unidentified. Two of the stock cars, containing a lion, two bears, six camels, horses and dogs overturned. All of the animals were, it was believed, killed except some of the horses and dogs. The principal performers, riding in the sleeping cars at the rear of the train and were only bruised.

'From ten to twenty attaches of the carnival company were likewise riding in the ten-car section which was destroyed, sleeping compartments being arranged in the various fronts for this purpose. The people were for the most part asleep and not less than twelve are believed to be piled under the burning debris. Only after a careful inspection will it be determined just how many were killed. {There is} hardly a single person on the show train or that part of it well forward toward the engine who did not receive painful injury.

'An employee on the passenger train ran back the track about a quarter of a mile and gave the first news to Columbus. Immediately a special train was dispatched to the scene, bearing every physician that could be found, and a call was sent for more. Another special train was sent half an hour later with these men.

'Dr W. F. Whitehead, who happened to be hunting near, heard the collision and was the first physician on the scene and he set about at once doing what he could. Other physicians began coming on special trains and in automobiles and the injured were taken to one side and loaded on the baggage car of the passenger train and rushed back to Columbus and taken to a side track near the city hospital, where other physicians were in waiting.

'Con T. Kennedy, owner of the shows, travelling in the rearward coaches, was among the first on the ground to give his attention to assisting his people. He it was, when the show train was fired that gave direction for cutting the cars loose and rolling them back up the track. Kennedy, a strong man, flushed with the pride of success that has been accorded his attractions, is one of the hardest hit by the catastrophe. Not only from a financial reason, but because of the loss of life and suffering of those who have weathered the show season with him. Kennedy stated at 3 o'clock that he had no idea as to the damage that had been done his train. Likewise, he did not know the extent of death and injury that had been visited the ranks of his employees.

'The show people generally knew in a few minutes of the death of Mr and Mrs Kempf and few there were who did not shed tears of real grief at their untimely end. That there were others under the burning section of the train was the stressed implication of all those who were in reach. "I saw poor fellows pinned in their sleeping waggons and they could not get out," was a phrase heard on all sides.' [919]

According to the Interstate Commerce Commission report the final number of casualties was seven circus employees killed and 27 circus employees, three passengers and several railroad employees injured. Engineer Fickling was badly hurt.[920]

The circus train had left Macon at 7.40 a.m. and at Reynolds had been given train order No. 28, which read:

"No. 2, engine 1610, will remain at Muscogee until one thirty-five 1.35 p.m. for No. 37, engine 1759 and extra 1716."

Train No. extra 1716 left Reynolds at 11.20 a.m., passed the last open telegraph office (not named, but probably Upatoi) at 1.07 p.m. and the trains collided at 1.26 p.m.

At Columbus passenger train No. 2 had been given train order No. 28, reading:

"No. 2, engine 1610, will remain at Muscogee until one thirty-five 1.35 p.m. for No. 37 and 1716."

Conductor Reichert was also handed a message reading:

"No. 37 has engine No. 1759."

Baggagemaster Frierson of train No. 2 said that he had read train order No. 28 before they left Columbus. They passed train No. 37 on the double track before Muscogee. Approaching there he had sounded three blasts on the air whistle, which the engineer had acknowledged. The train had stopped, and he had held up two fingers to the engineer and lowered one to indicate that one of the two trains they were to meet had passed. Despite this the engineer began to pull out, saying that he had a message. The baggagemaster said that he had sounded the air whistle again, whereby the engineer had stopped, repeated that he had a message and restarted. The baggagemaster said that he had then gone back to the conductor with the order, told him that only one of the two trains mentioned in it had been met and that the engineer had said something about having a message. The conductor replied that the engineer knew his business. Supposing that he must have been mistaken the baggagemaster had then returned to the baggage car.

Engineer Fickling of train No. 2 testified that he had received train order No. 28 at Columbus, and that as far as he remembered it had read:

"Train No. 2, engine 1610, will wait at Muskogee Junction until 1.35 p.m. for train No. 37, engine 1716."

He had glanced at the order in the telegraph office, then gone to his engine. There the train porter had brought him the order, together with a clearance card and a message. The

[919] *The Atlanta Constitution*, Atlanta, Georgia, United States, 23rd November 1915

[920] *Bluefield Daily Telegraph*, Bluefield, West Virginia, United States, 23rd November 1915

order he had read to his fireman, who had been on the running board wiping the locomotive. He said that he had read the order on the engine the same as he had read it in the office. He had also read the message out, which stated that engine No. 1759 was on No. 37. This train they passed before reaching Muscogee and having misread the order he had been under the impression that this was the only train that they had to meet. He recalled getting three blasts on the whistle but thought it was just to indicate that there were passengers to be discharged, Muscogee being a flag station. After his train had stopped there his fireman told him that the baggageman had an order and told the baggageman that he had a message. He denied seeing the baggageman holding up two fingers, hearing him say anything or stopping twice. It was common practice for dispatchers to correct orders with messages and the rule that orders should be given to the engineers by the conductors was commonly not observed. When the fireman had told him that the baggageman had an order he had not bothered to check his own copy, being confident that he had read it correctly.

Fireman Allison confirmed that he had been working around the engine when the train porter had appeared, and that his engineer had called out to him that they were to wait at Muscogee until 1.35 for train No. 37, which the message said was hauled by engine No. 1759 rather than 1716. Like his engineer he had assumed that when the whistle was sounded at Muscogee it was because there were passengers to be dropped off. He saw the baggageman holding up two fingers and asked his engineer if train No. 37 was running in two sections but his engineer replied it was not. At Muskogee someone also came up to the engine and said something and he heard his engineer reply that they had already passed train No. 37. Approaching the scene of the collision he had been firing when he had seen the engineer suddenly reverse the engine and had jumped, as had the engineer from the other side.

Conductor Reichert testified that he had received train order No. 28 at Columbus and had sent the engineer's copy to him via a porter and given his own copy to the baggagemaster. He had been at the rear of the train and remembered the baggagemaster sounding the air whistle twice at Muscogee, he supposed for train No. 1716, but thought that that train had passed, though he did not go out of the car to see what was happening. A train had passed them on the double track, but he had not positively identified it as such. Leaving Muscogee the baggagemaster had come back and told him that there was another train coming and the engineer did not seem to have understood the train order. He had not, however, signalled the engineer to stop, though he could not explain why not. The engineer having started away, he had been satisfied that train extra No. 1716 must have been met. He admitted that he should have delivered a Form 31 train order to the engineer personally and that had he done so the accident would have been averted. Orders were sometimes given to the engineer by himself and sometimes by a porter.

Train Porter Lockert testified as to giving the train order to the engineer and the latter reading it to his fireman to the effect that they were to wait at Muscogee until 1.35 for train No. 37 and that that train had engine No. 1759 rather than 1716.

Train Porter Lee said that he was aware, having been shown the train order by the conductor, that they were to pass two trains and that when they had started from Muscogee without having done so he had intended to speak to the conductor but had not done so because he saw the baggagemaster doing so.

Dispatcher Mercer stated that when he had first issued train order No. 28 the number of the engine of train No. 37 had been accidentally omitted. The Columbus operator then drew his attention to the omission. The operator at Reynolds said that there was room on the order as written by him to insert the missing number so was directed to do so, while the Columbus operator was given a separate message that train No. 37 had engine No. 1759. Dispatcher Mercer said that it had not occurred to him that thus dealing with the problem was not safe, though he had had no instructions permitting him to violate Rule No. 202, which read:

"Each train order must be given in the same words to all persons or trains addressed."

Operator Wilkinson at Columbus said that when Conductor Reichert had read the order back to him he had asked the conductor if he noticed anything unusual with it. The conductor said he did not. He then gave the conductor the message regarding the missing engine number. The conductor read back the order twice and each time named both train No. 37 and train extra No. 1716.

The Interstate Commerce Collision attributed the collision to the failure of Engineer and Conductor Reichert to obey the provisions of order No. 28, despite the attention of both of them having been drawn to it by Baggagemaster Frierson; the engineer because he had been confident that he had read it correctly and the conductor because he had left the safe movement of the train entirely to his engineer. The collision would also probably have been averted had the dispatcher issued identical orders to both trains.[921]

Twelve people died in a collision on the Nashville Chattanooga and St Louis Railway near Rockledge, Tennessee, United States, on Thursday 23rd December 1915. However, all but two of them were negro labourers, so the accident did not receive a great deal of press attention and none of the victim's names was published.

The section of line on which the accident occurred crossed the Cumberland Mountains west of Chattanooga and had severe gradients and many curves. To expedite traffic working over this difficult stretch of line automatic block signals were provided at an early date over the 9 miles between Cowan and Sherwood, with train movements being controlled by signal indications and train orders dispensed with. Between Cowan and Sherwood, near the summit of the line, a siding was provided at Rockledge, with an associated signal tower. At the south end of the loop siding was a short runaway spur. This end of the siding was out of sight of the tower. The siding to spur and siding to main line points were not operated from the tower, but a dwarf signal controlling the exit from the siding was. Its indications were not repeated in the tower.

On the day of the accident southbound fast freight train No. 55, consisting of engine No. 610, 45 cars and a caboose, left Nashville at 11.10 a.m. in charge of Conductor Wiggs with Engineer Charles Gribble and Fireman John Gilmore. In the words of the newspaper report it was 'carrying Chicago business' between Martin, Tennessee, and Atlanta, Georgia. Two assisting locomotives were attached to its head at Cowan station and at Rockcliffe the tower operator put the train into the siding so that its helper engines could be detached and go forward into the runaway spur.

In the opposite direction regular passenger train No. 2 from Atlanta, Georgia, was running late and an extra train was thus set up starting from Chattanooga and running as

[921] Interstate Commerce Commission report

first No. 2. It consisted of engine No. 419 and four coaches, in charge of Conductor Hancock with Engineer Irby Reed and Fireman W. E. Lehr, and left Chattanooga at 1.35 p.m. It reached Sherwood at 3.36 p.m., crossed train No. 95 and left there at 3.47 p.m. Four miles further north, at a location known as Hominy Hill cut, it collided with train No. 55 at 3.55 p.m.[922] The passenger train suffered most; its leading (Jim Crow?) coach being telescoped by the tender of the engine and ten of its occupants killed and 37 injured. Engineer Reed and Fireman Lehr were also killed and Engineer Gribble, Fireman Gilmore and three other passengers were injured. The negroes were going into Nashville from Chattanooga and intermediate stations and were on their way home for the Christmas holidays. The passenger engine left the rails and a car of mules and a carload of merchandise in the freight train were wrecked.[923]

The helper engines had been detached from the front of train No. 55 and gone forward into the runaway spur. The crew of the freight train then saw that the dwarf signal had been cleared and in accordance with the usual procedure they operated the points and the train set off towards Sherwood. Seeing train No. 55 moving along the siding the Rockledge operator at first assumed that it was just moving forward to clear the main line but when it kept going, despite the fact that he had not cleared the dwarf signal, he realised that it was going back out onto the main line and attempted, without success, to alert its crew. He then telephoned Sherwood, but train No. 2 had already left there, and finally a lineside section foreman's house, but was again too late.

Giving evidence at the Interstate Commerce Commission inquiry Conductor Wiggs testified that on arrival at Rockledge he had started to go to the tower when he saw his train moving forward. Soon after passing the tower he had started over the train to confer with the engineer regarding work to be done at Sherwood and had been 12 or 15 cars from the engine when the trains met.

Rear Brakeman Hamrick said that he had ridden on the top of the train up to the time that the train stopped at the collision point, when he went back to flag without realising that there had been an accident. He had not noticed the indication of the dwarf signal.

Engineer Gribble stated that the dwarf signal had been in the clear position for his train to leave the siding. There had been no indication of anything being wrong. He had no thought of impending danger. He had not seen train No. 2 and the first thing he remembered of the collision was his fireman helping him out of the cab.

Fireman Gilmore said that he had told his engineer the dwarf signal was clear. He saw the engine of train No. 2 but not soon enough to warn his engineer.

Head Brakeman Marshall of train No. 55 stated that he had been turning up retainers on the head car when the train had started and saw the dwarf signal at clear.

Operator Christian at Rockledge and Operator Johns at Sherwood agreed that a clear signal had been given for train No. 2 to proceed from Sherwood to Rockledge at 3.47 p.m. Operator Christian stated that he had been instructed by the dispatcher to hold train No. 55 for train No. 2 so he put it into the siding. It had stopped with its rear still on the main line so when it began to move he thought it was just pulling forward. He had written the time on his block sheet, then it struck him that the train was going too far. The ringing

---

[922] Ibid.

[923] *The Brandon News*, Brandon, Mississippi, United States, 30th December 1915

of a single-stroke bell then confirmed that the head of the train had passed out of the siding. He had not cleared the dwarf signal. He had then attempted to stop first train No. 55 and then train No. 2 but had failed to do so.

Engineer Kimmingham of helper engine No. 451 said that he had seen the dwarf signal at clear and his fireman and the helper had operated the switches for the freight train to leave the siding. The first he had known that anything was wrong was when his fireman had telephoned the tower for instructions and learnt that the operator had not cleared the dwarf signal.

Fireman Miller and Helper Sargent of engine No. 451 confirmed the dwarf signal being clear and operating the switches and stated that after the train had pulled out onto the main line the dwarf signal had not returned to danger as it normally did.

Signal Maintainer Hixon stated that he had been told that the dwarf signal had not returned to danger after the train had left. He had then gone to the signal, opened its case and found that the relay was energised. Learning that the operator had not cleared the dwarf signal he realised that it had given a false clear indication. There was nothing he could do to stop train No. 55 so he set about attempting to discover the cause of the false indication. Western Union linemen were at work at that point transferring wires and poles to a new location further from the line. The insulation of the signal wires had also been damaged by the burning, on some previous occasion, of some old ties beneath them. Also, at two places guy wires were touching the railroad company's common wire, causing a heavy ground. He was of the opinion that the false clear had been caused by a cross between the signal wires and the Western Union telegraph wires at the point where the insulation on the signal wires had been damaged.

The Interstate Commerce Commission concluded that the accident had been caused by a false signal indication (a wrong-side signal failure) due to a foreign current. In tests made after the accident it was possible to cause the dwarf signal to move to clear by crossing its control wire with any of the 12 Western Union telegraph wires. The commission noted that this was the first accident that they had investigated which was positively known to have been caused by a false block signal indication. Accidents due to such causes were exceedingly rare. In view of the location of the automatic signalling installation between Cowan and Sherwood even the best modern signal engineering practice would not call for the use of alternating current or other special precautions against foreign current. More complete protection would however have been provided at Rockledge if the exit switch from the loop had been controlled from the tower.

CHAPTER 7

RUNAWAY TRAINS AND VEHICLES

A variety of situations could lead to either a complete train or part of a train running away out of control. If there were railway staff in attendance the train or engine could only run away out of control if there was a failure of the brakes or if the brakes were incapable of holding it on a steep gradient. In the case of unattended vehicles, the action of gravity alone can be sufficient to precipitate a runaway if the train or vehicles are left standing on an incline. In contrast to the situation on a road a train or vehicles can sometimes run for many miles without mishap, an accident occurring only if the runaway meets with another train, comes to the end of the line or gets to a curve too sharp for the runaway to negotiate at the speed at which it is travelling.

A report from St Petersburg spoke of an accident at Saratoff, Russia, on the River Volga, on Wednesday 17th April 1901. Fourteen tank waggons, which had become detached from the engine pulling them, ran back down an incline and collided with a passenger train, wrecking four carriages and a mail van. Five people were killed, six seriously injured and 15 less badly hurt.[924]

An accident occurred on the morning of Saturday 15th June 1901 in France, involving the derailment of a train of workmen on the narrow gauge line then under construction from Laval to Landivy. A ballast train which was carrying workmen was derailed at 6 a.m. that morning near Chailland, a small town 22 km from Laval. Six of the workmen were killed and seven seriously injured.

According to the report of the bridges and roads engineer, who made an investigation at the scene, the accident was due to insufficient brakes and to the wheels of the train slipping on rails made wet by morning dew. An additional factor was the fact that on the Laval side of Chailland the line first climbed fairly sharply but then dropped steeply. It is down this slope that the train ran away. It was only a short one, consisting of an engine and five ballast waggons, on which were riding 18 workmen.

---

[924] *The Times*, London, England, 18th April 1901

The train became derailed entering Chailland. The locomotive hit a house serving as an office and overturned. The waggons mounted one another and most of the workmen were buried beneath the ballast they had contained.

Six horribly mutilated bodies were found, while one of the injured died in the evening, bringing the total number of dead to seven.

The families of all the victims were left destitute. It was said that two more of the injured would not survive, and the prefect of Mayenne, who immediately went to the scene of the accident, had to wire the government to ask for emergency assistance. The inhabitants of Chailland reportedly showed great compassion for the workmen and made their homes available to the authorities.

The dead were named as Driver Brisset, Conductor (?) Vandemael and Mssrs Dalière, Lebreton (father and son), Terrier and Court. The most seriously injured were Mssrs Radoux, Blottière, Frilouze and Clément.[925]

On the morning of Sunday 8th September 1901 Northern Pacific Railroad mixed train consisting of seven freight cars and two passenger coaches and coming from Oakes, on the James River line of the company, got out of control as it was approaching Jamestown, North Dakota, United States, and collided with an engine standing outside of the station buildings. Five men were killed outright, two fatally injured, and four others were hurt, some of whom it was feared at first might die. The men were all farm labourers having no money and travelling without tickets. There were 17 of them in all, riding on an open flat car marshalled second behind the engine and sandwiched between two heavily laden two box cars. The men were all strangers to one another and were coming from La Moure 'to get work in threshing'. Most of the men were asleep, wrapped in blankets, at the time of the accident and were described as having been 'caught like rats in a trap'. In the collision the flat car broke in two and its centre rose in the air, the victims rolling off it and between it and the box cars. None of the ordinary train passengers was hurt. The names of five of those who lost their lives were John Gilley, H. J. Kilpatrick, R. D. Vickers, William Glenny and Frank Howard.[926]

Engineer Nichols claimed that the air brakes would not work. Both he and Fireman Cleveland jumped, the engineer only at the last moment. Neither of them was hurt. There was no one on the light engine, its engineer and fireman being in the station eating room. The stationary engine was little damaged, the colliding engine much more so.

'The injured were taken to the hospital and every care given them.' [927]

A horrific accident involving an express and a train of petroleum occurred near Palota, Romania, close to the border with Hungary, on 20th (possibly the 21st) September 1901. The express train which had left Bukharest (contemporary spelling, now Bucharest) at 6 p.m. the previous evening and passed through Palota at 2 a.m. Closely following it was a train carrying petroleum, consisting of 16 tank waggons, each containing 10,000 litres of oil. Beyond Palota the line descended steeply, and the express was negotiating the incline

[925] *Le Matin*, Paris, France, 16th June 1901

[926] *Bismark Daily Tribune*, Bismark, North Dakota, United States, 9th September 1901

[927] *Brainerd Daily Dispatch*, Brainerd, Minnesota, United States, 10th and 13th September 1901

at the regulation speed of 35 or 40 kilometres an hour. As the express neared the bottom of the descent its guard, at the rear of the train, heard an alarm whistle from the engine of the heavy petroleum train, the brakes of which had failed on the bank, but he reportedly misunderstood its meaning, for instead of attempting to signalling to his driver for the train to accelerate away out of danger the express was brought to a standstill. A moment later the runaway oil train crashed into it. In a scenario reminiscent of the Welsh Abergele disaster of August 1868, 14 of the 16 waggons of the petroleum train lost their contents and the inflammable liquid poured over the carriages of the passenger train and instantly caught fire.

What a newspaper report described as heartrending scenes followed. By some miracle a sleeping car inspector named Mercovich and his wife managed to escape from the express unharmed but then realised that they had left their daughter behind and had to stand and watch as she was burnt to death before their eyes. M. Dinu, a wealthy man from Crajova, reportedly a millionaire, was trapped with his foot caught between two beams. As the flames advanced he begged the guard of the train to cut his foot off and so free him, promising to reward the guard handsomely, but before the guard had time to do anything the man sank into the flames and was consumed. Also caught between two beams was a retired Austrian artillery officer named Schwarz. A Romanian labourer attempted to rescue him and in his agony the officer dug his nails deeply into the neck of his would-be rescuer, tearing his flesh. The rescue attempt failed and the labourer himself had to be dragged to safety.

The guard of the express, together with its driver and fireman, set off for Turn-Severin for assistance, presumably on the detached express engine. 'When the first rescue party arrived, the scene before them was too terrible for words. Parts of the wreck were still in flames, while here and there amongst the smouldering ashes could be seen the charred remains of the victims.'

Six carriages of the express were completely destroyed, and 32 of the 38 passengers on the express and eight railway servants reportedly lost their lives, while damage was estimated at about 1,000,000 francs.[928] [929] [930]

Seven months after the Romanian disaster one occurred in southern Africa, on Easter Sunday, 30th March 1902, between Barberton and Kaapmuiden, in the Transvaal. On 24th March the 2nd Battalion Hampshire Regiment, quartered at Barberton, was ordered to proceed to Johannesburg.

The first train provided for this move, carrying 102 men of 'E' and the Volunteer companies, reached Johannesburg safely, but the following train, carrying 105 men of 'C', 'E' and the Volunteer companies met with disaster.

Between Barberton and Kaapmuiden the railway fell 200 metres in 7 kilometres, thus with a ruling gradient of 1 in 35, and this stretch also included two sharp curves, the first to the right and the second to the left. Beyond the second curve a straight stretch led to a 40-foot-high bridge over a donga (a normally dry watercourse) known as Hyslop's Creek.

---

928 *The Times*, London, England, 23rd September 1901

929 *The Daily Telegraph*, Napier, New Zealand, 15th November 1901

930 *The Clarence River Advocate*, Maclean, New South Wales, Australia, 6th December 1901

The train left Barberton at about 8 a.m. in charge of Guard Ross, with Driver J. Rankin (or Rankine) and Fireman J. Constance. Behind the locomotive it consisted of seven open trucks, a store truck, a 'boy' truck (meaning of this term unclear), a passenger carriage, two more open trucks and a van. The three leading trucks were empty, the fourth and fifth trucks carried the men of the Volunteer company, the sixth and seventh the men of 'E' company and the two rear trucks the men of 'C' company. Riding in the passenger carriage were three officers and a handful of civilians.

Soon after starting the journey Driver Rankin seems to have lost control of the train and the oscillation caused by the excessive speed warned the passengers that an accident was likely, if not inevitable. 'It was impossible for the men to keep their seats even on the floor of the trucks. Dust and stones flew up in all directions. Rifles, kits, etc were thrown violently from one side of the trucks to the other, and a loud grating noise was heard, probably caused by one of the trucks which had already left the rails.' The speed of the train reportedly reached 80 miles an hour. At the first sharp curve the engine left the rails, hurtled into the air, turned a complete somersault and crashed to the ground, its boiler bursting as it did so. About 60 yards beyond the engine one truck rolled over several times and came to rest on its side, while three others piled up in a heap. The trucks carrying E company got as far as the bridge. One of them fell 40 ft into the donga, the other rolled to the edge of the ravine, throwing its occupants out and then falling on them. The remaining vehicles crossed the bridge, though some of them were derailed. Guard Ross screwed the handbrake down hard and was credited with preventing the passenger carriage from going into the donga, though it was badly damaged. A passenger named William White, thinking to save himself, jumped out of the carriage on the bridge, fell 40 ft into the donga and was killed. Only one other person in the carriage appears to have died, variously reported to have been a Dutch (Boer) boy and a Boer woman from Barberton concentration camp.

After the accident the initial assistance was provided by the troops who had been fortunate enough to escape injury. Colour Sergeant C. D., the senior non-commissioned officer on the train, 'exhibited great presence of mind and coolness, and it was largely due to him that no panic ensued.' There was no spare locomotive at Barberton to haul a relief train from there and horse-drawn military ambulances had to be sent out by road, arriving after about an hour. Every doctor, nurse and medical auxiliary of the Royal Army Medical Corps assisted and by 1 p.m. all the dead and injured had been brought back to Barberton. A relief train was also sent from Kaapmuiden. Driver Rankin managed to crawl out of the wreck and murmur a few words, 'Lay me on the s…' then he died.

Fireman Constance, unconscious and severely injured, died while being taken to Barberton hospital. The dead were buried the next day in Barberton cemetery.

Six of those injured in the accident later died and the final death toll was 49, comprising 42 soldiers, the majority from 'E' company, five civilians and the driver and fireman. Forty-one soldiers and an unknown number of civilians were injured.

Two inquiries were held into the disaster, one by the military and the other by the railways board, but their results are unknown. Possibly the accident was due to a simple error of judgement on the part of Driver Rankin with a heavy train.[931] [932] [933]

---

[931] *The Times*, London, England, 1st, 2nd, 3rd, 4th, 5th and 7th April 1902

The 1902 Independence Day celebrations in the United States, on Friday 4th July 1902, were marred in New York state by a disaster near the end of the day on the Mountain Lake Electric Railroad. The Mountain Lake was a short electric line which climbed from Gloversville, New York, up the slope of the Bleeker Mountain for four miles to a popular resort and hotel at the summit. The line climbed about 1,000 feet in four miles with an average ruling gradient of about 1 in 21. On the day of the accident, Friday 4th July 1902, some 1,700 people had taken the railway up to the summit. A few set off homewards already at 5 p.m. but the majority stayed for a fireworks display which was to conclude the day's celebrations. When this finished at about 10 p.m. there was a predictable rush by pleasure-seekers to get home. Just before 10 p.m. an open car, No. 1, was despatched down the mountain with 75 passengers on board, in charge of Conductor James Cameron with Motorman Arthur Perkins. Five minutes later, No. 5, a heavier closed combination baggage and passenger car which had spent most of the day stabled at the terminal, followed, in charge of Conductor F. Yost and Motorman William Dodge.

On its way down the single track car No. 1 met and crossed car No. 4, on its way up the mountain, but Conductor Cameron neglected to tell the crew of car No. 4 that car No. 5 was following them. The outcome was that cars Nos. 4 and 5 met one another between crossing places. There was no collision, but car No. 4 had to set back to the nearest siding so that car No. 5 could pass it, resulting in a delay. Anxious to regain time Motorman Dodge then began to make the descent at too high a speed and his car was reportedly travelling at 60 miles an hour when disaster struck. Car No. 1 reached a sharp curve and Motorman Perkins, aided by a passenger named J. V. Hine, applied its brakes to reduce speed to negotiate the curve. Motorman Dodge likewise tried to brake his car but maintained afterwards, in an antemortem statement, that had not been able to get its brakes to work. Car No. 5 rammed car No. 1 and both cars then began to hurtle down the hill even faster. In an effort to regain control of his car Motorman Dodge put it into reverse, but this action just blew the circuit breakers, plunging both cars into darkness. As the cars approached a reverse curve a few of the passengers in the open car were able to jump off and escaped with a few scratches.[934] There was then a second collision. 'The big double car crashed into the car ahead. With the terrific force with which it was running it raised the rear end of the first car up into the air and over the right embankment on the side. The collision was too much for the baggage car, and for the road bed. The tracks spread, and the baggage car followed closely in the wake of its predecessor. Both cars crowded to their utmost capacity with passengers, were hurled into the ditch. The passengers were thrown from the cars in every direction, and many were killed instantly, while others died from their wounds immediately afterward. After the cars struck the curve it was but a matter of a very few seconds before the open left the track and pitched forward on its side, hurling some passengers to their death by crushing them under the side of the car, maiming others for life, and causing scenes of agony which were terrible in the extreme.

---

[932] Webpage – The South African Military History Society: Military History Journal, Vol 11. No. 5 The
Anglo-Boer War: A Railway Accident near Barberton by Steve A. Watt

[933] Webpage – The Barberton Rail Crash: The Royal Hampshire Regiment

[934] Webpage – The Mountain Lake Railroad

'In the last gasps of the passengers, and the terrible sufferings to which the injured were subjected, some scenes occurred which show that heroes are found in other places than the battlefield, as men who were pinned beneath the weight of the overturned car lay there for over an hour and never made a murmur or complained of their sufferings.

'The injured were taken in wagons to the Nathan Littauer Hospital in Gloversville. All the doctors in Gloversville were pressed into service and assisted greatly in relieving the wounded.' [935]

The accident happened at about 10.30 p.m. some 2½ miles from Gloversville. A further accident was reportedly averted by a passenger, 17-year-old William Berghoff, running back up the line and stopping another descending car.

It has not been possible to find any account of the inquest held upon the victims of the accident.

The final number of dead was 14, consisting of 12 passengers and two railroad employees, one of them Motorman Dodge, who died two days after the accident. Sixty others were injured. Possible factors in the accident were the inexperience of Motorman Dodge and the fact that the cars lacked a more modern and reliable magnetic braking system.[936]

A dispatch from Breslau brought news of a major disaster on the railway between Sucha and Kalvaria, in Austrian Silesia, on Saturday 1st August 1903. Eleven waggons loaded with stone broke away from a goods train climbing an incline and reportedly ran back 20 miles before they finally collided with a passenger train. Thirty persons were said to have been killed and 52 seriously injured.[937]

The Alamogordo and Sacramento Mountain Railway was basically a logging line, constructed to exploit the timber from fir and spruce forests. Climbing into the mountains the line included viaducts and switchbacks and had very steep gradients, as severe as 6.4% (1 in 16).[938] On the afternoon on Monday 19th October 1903, not long after the line had been opened, a serious accident occurred in the Cox canyon, between Cloudcroft and Russia, New Mexico, United States. The line here ran along the side of the mountain, with a steep wall rising on one side and a precipice falling on the other. The victims of the accident were a number of labourers riding on flat cars. As it became obvious that the train was running away the labourers jumped off towards the wall of the mountain but rebounding from the rock face they fell beneath the wheels of the train cars and were crushed. Six of them were killed outright and 11 more seriously injured, at least two of whom subsequently died. The engineer and fireman chose to stay on the locomotive until it left the track on a curve and were only slightly hurt.

Mystery surrounded the reason for the train running away, but it was theorised that one of the labourers on boarding a flat car had accidentally stepped on an angle cock,

---

[935] *The New York Times,* New York, United States, 6th July 1902

[936] Webpage – Tragedy on the Mountain Lake Electric Railroad July 4, 1902

[937] *Ottumwa Daily Courier*, Ottumwa, Iowa, United States, 3rd August 1903

[938] Wikipedia – Alamogordo and Sacramento Mountain Railway

closing it and rendering the air brakes ineffective.[939]

Lebanon suffered a serious runaway accident on or about Friday 8th April 1904. The line of the Chemins de Fer de Damas-Hamah et Prolongements ran east from the port of Beirut in Lebanon to Damascus in Syria. En route it had to cross the Mount Lebanon mountain range and thus included an incline beyond Aley station with a gradient of 7%, or about 1 in 14, worked on the Abt rack system.[940]

On the day of the accident a passenger train left Beirut at 7 a.m. for Damascus. The journey went without problem as far as Aley. Beyond the station there the train began to ascend the incline. It had only covered a mile when a pipe of the steam brake burst and steam entering the cab threw the driver backwards. Either because of the shock of what had happened or because he was blinded by steam he closed the regulator, and the train, which had been struggling up the incline, first came to a stand and then began to run backwards. As well as the steam brake the train had two hand brakes and two Abt brakes, despite which it ran back at increasing speed, shot through Aley station and became derailed at the Beirut end, its carriages smashing onto some rocks. Two carriages were wrecked, and eight passengers were killed and 21 more or less seriously injured.

The German writer in Lebanon who described the accident commented that as well as the driver, fireman and a supervisor who had been riding on the engine there had been another driver and two conductors in the leading brake van, and yet none of them had thought to apply the brakes. Instead, as soon as the train had begun to run back they had jumped off and followed it back to the station on foot. He added that the accident had caused a sensation in Beirut and the surrounding area and there had been an outcry against the railway company.

The driver of the train was a former naval fireman who had only been employed on the railway for a month and had never previously driven an engine. The railway company did not get better staff, the writer remarked, because they did not pay them enough, while their engines were said to be in a dreadful condition.

The chief engineer sent by the Lebanese government to investigate the accident over the objections of the General Manager, had, the German commentator wrote, already repaired the damage but was unable to express an opinion as to how the initial mishap had occurred.

The relatives of the victims of the accident had already begun an action for damages against the company but the company was subject to Turkish law (Lebanon was of course still a part of the Ottoman empire) and the writer did not think that the claims for damages would be very successful.[941]

While a goods train was performing shunting operations at Ferrara in northern Italy, early on the morning of Tuesday 20th September 1904, its driver left it for a while in charge of his fireman. The latter started the engine, presumably to continue with the

---

939 *Santa Fe New Mexican*, Santa Fe, New Mexico, United States, 21st and 23rd October 1903

940 Wikipedia – Rail Transport in Lebanon

941 *Unfall auf der Hauranbahn in Die Lokomotive Illustrierte Fachzeitung* May 1904 pg. 49 (in German)

shunting operations, but then found himself unable, for some reason, to stop it again and the locomotive ran away in the direction of Poggio Renatico. After covering about three kilometres the runaway engine collided with an oncoming through train from Bologna to Venice. Seven people were killed in the collision and another 16 injured. Those who lost their lives included the fireman of the runaway engine and the guard of the express, while the injured included Dr Otto Ritter, Under-Secretary of the Austrian Ministry of Finance. Three carriages of the express caught fire following the collision and a mail van was completely destroyed and part of the mails was lost. The lines were blocked, and the telegraph interrupted. A relief train with doctors and surgeons was immediately despatched to the scene. The driver of the goods train was subsequently arrested.[942]

As noted earlier in relation to the 1901 derailment at Ingolf, Ontario, in what was a sort of Canadian equivalent of the annual excursions of poor Londoners to enjoy a holiday of sorts picking hops in Kent, from the beginning of the exploitation of the prairies in the 1880s until the mechanisation of agriculture in the 1920s every autumn large numbers of people travelled each year from eastern Canada to labour harvesting wheat. To cater for these migrations the railway provided trains known as harvester specials and it was one of these trains which featured in a disaster at Azilda, Ontario, Canada, on Wednesday 12th September 1906. On this day, to cope with the volume of passenger traffic caused by large numbers of harvest excursionists bound for the prairies Canadian Pacific Railway train No. 1, otherwise known as the *Imperial Limited,* was run in three sections, leaving Toronto at 1.45 a.m., 2.30 a.m. and 4 a.m. It was the third section which was to meet with disaster.

Train third No. 1 consisted, behind its locomotive and tender, of a baggage car, seven lightly constructed colonist cars and a heavy Pullman car at its rear. It was in charge of Conductor Kirby, with Engineer P. Limoge. At Azilda it had to cross an eastbound express, train No. 2, and at 7.30 it was pulled up just short of the east switch of the siding there to await the arrival of train No. 2 before it headed into the siding.

Eastbound train No. 2 had at its head, immediately behind the tender of its locomotive, a refrigerator car loaded with salmon, followed by the usual baggage car and then the passenger coaches. It was in charge of Conductor Thomas Kehoe with Engineer W. Thurlow and Fireman D. Roche. The train was running late, but Engineer Thurlow had managed to make up 11 minutes in the last 28 miles. Approaching Azilda from Cartier, 20 miles back, there was a falling gradient almost all the way, and the morning was wet, with slippery rails. Engineer Thurlow told the inquest on the victims of the accident that he had a regular train order advising him of the meet with train third No. 1 at Azilda. Coming down the bank from Cartier he had applied his air brakes several times in order to steady the train on the descent and they had worked well. About three quarters of a mile from Azilda he had shut off steam and a quarter mile further on he applied the brakes to bring his train to a stand at the siding. His gauge had showed 70 pounds of pressure in the train pipe and 110 pounds in the reservoir, but suddenly the brakes would not work except on the engine and tender, the refrigerator car and the baggage car. The train was still travelling at 40 mile an hour on the wet rails and its speed was dropping very little.

942 *The Times*, London, England, 21st September 1904

Conductor Kehoe was on the alert and realising that his train was not pulling up applied the emergency brake from his van at the rear of the train but still to no effect. When it became apparent that he could not stop his train Engineer Thurlow gave the "back up" signal to the harvester special, then he and his fireman jumped. Engineer Limoge on train third No. 1 heard the "back up" whistle but had no time to do anything as the express came towards him and he and his fireman likewise jumped to safety just before the collision. The refrigerator car marshalled behind the tender of the eastbound express absorbed much of the force of the collision and although the following baggage car was also smashed the coaches behind were little affected and the express passengers escaped. In the harvester special it as a different story, for the leading baggage car telescoped the following colonist car and it was here that all the deaths and injuries occurred.

Some passengers though, even in the ill-fated colonist car, had lucky escapes. James Bartley, of Murray Township, Ontario, told a reporter, "I was three seats from the front of the car, and I saw that ponderous luggage car come right into our car with hardly a moment's notice. I crouched down, but two men, who occupied the double seat with me were killed. I was pinned down for a few minutes, but releasing myself I crawled through a window, wondering at my escape. The baggage car seemed to chew everything to pieces that was in its path. I lost all my baggage but was very thankful to get away so luckily."

A passenger from the rear of the harvesters special also spoke to reporters. 'Sitting at breakfast, James F. Kennedy and his daughter, of Lindsay, who also returned {whence they came, having abandoned their journeys after the accident}, were interrupted by the shock, followed by a shower of broken glass from the chandeliers, while the victuals were dashed to the floor. Mr Kennedy was in the seventh coach of the westbound special. "I went outside to investigate," said Mr Kennedy, "but for fully five minutes we could not distinguish anything because of escaping steam. Subsequent sights accompanied by groans of the injured were simply horrible. Many of the bodies were fearfully mangled, some of them being crushed flat." [943]

Notice of the accident was reportedly telegraphed to North Bay by W. Vansickle, the agent at Azilda, only a minute after it had occurred, but it took rescuers at the scene, including passengers from train No. 2, several hours to free some of those trapped in the wreckage. As is often the case after a big accident there was confusion over the identities and number of the victims. An unidentified body was supposed to be that of a Mr E. May until he arrived in Winnipeg very much alive and unhurt. Another fatality was supposed to be a Nelson Schalz of New Hamburg, Ontario, and his name was not immediately removed from the list of dead despite a telegram from New Hamburg that he was still living. The final number of dead was probably 12 but possibly 14.

After the accident Engineer Thurlow and Conductor Kirby of the harvester's special found that the angle cock at the rear of the baggage car of the express was closed, thus shutting off the air and explaining why the brakes on the vehicles behind the baggage car had failed to apply. Hanging loose beside the angle cock were several links of a heavy stay chain which could, it was theorised, have struck and turned off the angle cock. Messenger Clerk Alguire, who was riding in the baggage car, spoke of the vehicle

[943] *The Canadian Statesman*, Bowmanville, Ontario, Canada, 19th September 1906

vibrating and swinging from side to side as the train came fast down the hill from Cartier, adding weight to the theory that it was the loose chain which had struck the angle cock and caused the train to run away.[944]

The inquest on the victims of the disaster ended on 19th September with a verdict by the coroner's jury that, "Thomas Puddicombe and others came to their deaths near the east switch at Azilda on the C.P.R. by train No. 1 going west and train No. 2 going east colliding at that point; that collision was due to a defective air brake on No. 2 train, the defect being a closed angle cock on the rear end of the mail car. We find that the train crew took the usual precaution in the handling of the train." [945]

An accident in Ecuador on the evening of Boxing Day, Wednesday 26th December 1906, was described rather patronisingly in a newspaper report as having been caused by 'simple natives having fooled with a train's brakes.' The railway at Mocha had only been opened the week before by the Guayaquil and Quito Railway Company and the arrival of a train at the station there was treated by the local inhabitants as a matter of great interest. A group of them closely inspected a passenger car which had been detached from the rear of the train. Unfortunately, one of the party, without understanding what he was doing, released the brakes, whereupon the car ran back down an incline for about a mile before it left the rails on a curve and overturned, killing or injuring all of its occupants. Seven lives were lost.[946]

Six miners met their deaths in an accident on Tuesday 21st May 1907 while they were returning to their homes in Dayton, Tennessee, United States, from the Dayton Coal and Iron Company mine. A yardman was conducting shunting operations without using an engine by moving two heavily laden cars with the aid of a heavy metal bar when the trucks got beyond his control and running down a 35 feet gradient (about 1 in 171) collided with the rear of the miner's train. About two hundred miners were riding in coke cars and several others were hurt. The men who lost their lives were named as Bert Best, Daly Green, Elijah Huff, Tom Jarmon, John Teels and James Towery.[947]

Confusing and mutually contradictory reports in different newspapers make it difficult the exact mechanism of an accident on Wednesday 22nd April 1908 on the Laquin Lumber Railroad at Whalen, Pennsylvania, United States. The train involved in the accident had set off just before 8.15 a.m., its engine pushing a single flat car on which 25 Italian labourers were riding, on their way to work. What it appears most likely happened is that the train encountered a runaway car rushing down the mountainside towards it. Engineer Stitler reversed his machine in an attempt to minimise the force of the impact and having shouted to the labourers to 'jump for their lives', did so himself and sustained only slight bruises. A brakeman also jumped and was more badly hurt while Fireman Roy Lyons remained on the engine and escaped unhurt anyway. The labourers, too frightened

[944] *The Winnipeg Tribune*, Winnipeg, Manitoba, Canada, 15th September 1906
[945] *Manitoba Free Press*, Winnipeg, Manitoba, Canada, 21st September 1906
[946] *Fairbanks Evening News*, Fairbanks, Alaska, United States, 27th December 1906
[947] *Daily Crescent-News*, Defiance, Ohio, United States, 22nd May 1907

to jump, huddled together on the car and waited for the inevitable crash. The runaway car struck the train and the reversed engine and both cars then careered down the bank at high speed until they reached a set of derailing (catch) points which threw them into an earth bank. The engine itself was little damaged but four of the labourers were killed outright and 10 others were badly hurt. The injured were put on a special train for Ralston. Another labourer died shortly after the train left Laquin and it would seem that in all seven or possibly nine labourers lost their lives, the others succumbing after reaching Williamsport Hospital.[948 949 950]

A Union Pacific Railroad work train, consisting of just engine No. 1508 and caboose, in charge of Conductor John Murphy, with Engineer Joseph Clinton and Fireman Percy Hanson, was occupied gravelling the line on the Borie cut off, west of Cheyenne, Wyoming, United States, on Tuesday 10th November 1908. By 8 p.m. the work had been completed for the day, the train had returned to the main line and was just about to set off to return to Cheyenne for the night when it was run into from behind by extra freight train No. 223 train, of which the crew of the work train knew nothing.

Freight train extra No. 223 was in charge of Conductor McCormick with Engineer J. C. Schley and Fireman Mons Christensen. The events leading up to the accident began about a mile west of Granite Cañon, where the extreme cold and frosty rails caused the train to begin to get out of control, the air brakes refusing to hold it. The train of an engine and 30 cars was scheduled to stop at Granite Cañon but instead it rushed through there with Engineer Schley whistling for the hand brakes to be applied. The Granite Cañon operator immediately telegraphed his headquarters at Cheyenne that there was a runaway coming towards them. The yards at Cheyenne were cleared so that the runaway train could have a clear run through them and onwards to Archer Hill, eight miles beyond, where it was anticipated that the adverse gradient would serve to halt the runaway. All would have gone fine had not the work train backed off the Carr-Borie cut-off line at Borie, 11 miles short of Cheyenne, right in the path of the runaway. The runaway was travelling at reportedly 70 miles an hour when it collided with the caboose of the work train, crushing it as if it were paper and piling the engine and 31 cars of the runaway on top of the work train. The whole of the wreckage was piled into a space not more than 200 feet in length and in places 40 feet high.

In the caboose at the moment of impact were Conductor Murphy, Brakemen Clarence Stitt and C. W. Rodgers and three unnamed Japanese labourers. All were killed outright, as were Engineer Schley, Fireman Christensen and Brakeman J. D. Duncan of the freight extra. Seriously injured were Engineer Clinton and Fireman Hansen of the work train and Brakeman Ed Tracey of the freight train. Burning coals from the fireboxes of the wrecked locomotives set fire to the wreckage and despite the efforts of a squad of Cheyenne firemen with a chemical engine and a hundred railroad employees most of the waggons of the freight train were consumed. The line was completely blocked for 12 hours. A newspaper reporter noted that 'Coal, silver bullion, grain, groceries, cement and many other classes of freight litter the ground in the vicinity of the wreck. Within an hour of the

---

[948] *The Bradford Era*, Bradford, Pennsylvania, United States, 23rd April 1908
[949] *Indiana Evening Gazette*, Indiana, Pennsylvania, United States, 23rd April 1908
[950] *The Philadelphia Inquirer*, Philadelphia, Pennsylvania, United States, 23rd April 1908

accident the railroad company had guards on the ground to protect the freight that escaped destruction. Few of the persons that visited the scene realized that the heavy metal pigs scattered about were silver, the general impression being that they were lead.' [951]

Some mystery surrounds the number of Japanese labourers killed in the accident as another newspaper stated that an Attorney Donavan, who had been a passenger on train No. 3, which was held up at the scene of the wreck, was told at Borie that 25 or 30 Japanese had been killed and their bodies consumed by the fire.[952] However, a report of the inquest on the victims, concluded at Cheyenne on 18th November, spoke only of ten victims, probably the nine already detailed together with Fireman Hansen, who was said to have been fatally injured.

The inquest jury brought in a verdict that, "We find that said extra 223 got beyond the control of the engine and the train crews of the same between Buford and Granite Cañon and could not be stopped before running into said work train, causing the wreck in which said Schley, et al. met their death.

"Said work train came off of a branch road of the Union Pacific railroad known as the Borie cut-off and after switching some cars pulled in on the east bound track of the Union Pacific railroad and was just leaving for Cheyenne when struck by extra 223.

"We further find, in our best judgement, that said extra 223 got uncontrollable by there being some defect in the braking power.

"We also find that there was no operator at Borie at the time of this accident, which in our judgement had there been one, the life of said John Schley, et al, might have been saved, for he would have had time to warn the crews on the said work train of the approaching runaway train extra 223.

"We also find that the Union Pacific railroad company have been negligent for a year past and up to the time of the accident for not inspecting the air brakes of freight trains at the Laramie city yards before leaving for Cheyenne. The same being true for extra 223. This train left Laramie without being inspected by an air brake inspector at about 3.30 o'clock on the afternoon of 10th November 1908."

The Union Pacific conducted its own investigation. It disagreed with the jury's findings and blamed Engineer Schley for the accident. General Superintendent W. L. Park stated that, "As to the correctness of their findings, {that Engineer Schley was to blame for the accident} three of the oldest engineers , among them the chairman of the general board of adjustment, admitted that the train in question, if in their charge, would have been brought down the hill safely, and of this there is, of course, no doubt.

"A great many of the oldest and most experienced engineers and conductors were interrogated and in every instance corroborated the findings. Also the board of review, composed of the general officers, support this in every particular.

"Naturally in such an accident gossip is rife and numerous theories advanced. As to the equipment and its condition, the United States government have the matter of safety appliances under close surveillance, with inspectors constantly on the road.

---

[951] *The Cheyenne Daily Leader*, Cheyenne, Wyoming, United States, 12th November 1908

[952] *The Ogden Standard*, Ogden City, Utah, United States, 12th November 1908

"The Union Pacific railroad stands well at the head in this particular, having received but one or two fines since the inauguration of the safety appliance act some years ago.

"This particular train was a through time freight of exceptionally high class equipment and of much less than the ordinary tonnage; inspected in every detail at Ogden and Green River.

"Excessive piston travel, caused by wear or straining of brake shoes or leverage is the only factor entering into the question of Laramie inspection. Every railroad man knows that these conditions do not arise until days, or weeks, after such inspection is made, and would only apply to cars that were drifting around in service between Green River and Laramie, where thorough inspection is in vogue."[953]

The Pouch Tunnel accident in France on 15th December 1908, which involved runaway waggons and a collision in a tunnel, is described in the Tunnel Accidents chapter.

A very brief report in the London *Times* mentioned an accident near Dehra, India, on Sunday 23rd May 1909. Sixteen waggons ran down an incline and collided with a morning passenger train. The engine and 15 vehicles were smashed, and the line was completely blocked. Eight 'natives' and two railway servants were killed, and 14 people were injured.[954]

A runaway waggon on an interurban line caused a major disaster in the far west of Canada on Wednesday 10th November 1909. The accident occurred near Lakeview, just outside of Vancouver, British Columbia, on the British Columbia Electric Railway.

The events which led up to the accident began in the early morning, with the departure from the Carrall Street station at 5.10 a.m. a train consisting of electric locomotive No. 503 and a single Canadian Pacific Railroad flatcar loaded with timber bound for an iron foundry at Nanaimo Road.

The little train was in charge of Conductor Fred Wiggins with Engineer David McDonough and Brakeman William Burrows. When the train reached the foundry siding the train crew discovered that they could not immediately put their waggon into the siding because a boxcar was in the way. Instead the brakes were put down on the flatcar, its wheels were blocked with a piece of wood, the engine was uncoupled, went into the siding and was coupled up to the boxcar, the flatcar having been left standing on the main line. Engineer McDonough then brought the engine and attached boxcar out onto the main line and set back gently onto the flatcar. The vehicles were equipped with automatic couplings and so should have coupled up on their own as soon as they came together. This done, Conductor Wiggins released the brake on the flatcar, Brakeman Burrows removed the wooden scotch, Engineer McDonough was signalled that all was ready, and he began to pull the two vehicles forward to the siding switch preparatory to setting back into the siding with them.

[953] *The Cheyenne Daily Leader*, Cheyenne, Wyoming, United States, 19th November 1908
[954] *The Times*, London, England, 25th May 1909

The forward movement had only gone a short distance when Conductor Wiggins felt a jerk and saw that the coupling between the two cars had broken. Worse, the flatcar with its load was beginning to run back, the gradient of the main line at this point being severe, steeper than two percent (1 in 50). The conductor seized a block of wood and thrust it beneath the wheels of the car, but the momentum of the latter was already too great, and it just rode over the obstruction and kept going. Brakeman Burrows was aboard the flatcar and attempted to apply its brakes but impeded by the load of timber he could not apply them sufficiently hard to stop the car. After it had run a hundred feet or so at increasing speed he was forced to jump off to save his life, especially as further on the line ran over a trestle where to jump would have been fatal. Engineer McDonough followed the runaway down the line with his locomotive but without being able to do anything to help.

Meanwhile at 5.50 a.m. the first passenger car of the day set off from Carrall Street, carrying 24 people, mostly company employees en route to work in the car shops at New Vancouver, together with a handful of non-railway passengers. It was in charge of Conductor William Harris with Motorman George Thorburn. Believing that he had a clear line at least as far as the end of the double-track section at Central Park Motorman Thorburn was running at full speed, with dawn just breaking, as his car rounded the curve approaching Lakeview station, where the headlight of the car would have illuminated the side of the track rather than straight ahead. He must therefore have had little warning of the runaway car coming towards him. He did, apparently, just have time to shout to his passengers to jump for their lives before the collision, and one man was able to do so, and escaped. For the remainder his warning came too late. The cars met with terrific force and more or less stopped dead, whereupon the heavy timbers slid forward off the flat car and acting as a battering ram swept the passenger car from end to end.

Alarmed by the noise of the collision the inhabitants of Lakeview rushed to the scene and began the rescue work. 'Men tore wildly at the debris or strained at the huge timbers to release their fellow men, but it was more than half an hour before the work was done, one after another giving place to new helpers as they fell back exhausted. By that time all that remained of fourteen strong men lay side by side on boards in one of the station buildings, while nine sufferers lay in another. The women from the neighbouring houses brought stimulants and rough bandages and did all in their power to ease the pain of the injured, until the arrival of the surgeons who were telegraphed for from both Vancouver and New Westminster. Two Grandview cars were rushed to the scene with aid, and in one of these the remains of the dead were conveyed to Vancouver, the injured being hurried to the general hospital.' [955]

Those who were killed outright included Motorman Thorburn, and Conductor Harris later died of his injuries, making a 15th fatality.

The disaster was the subject of a lengthy inquest which concentrated on the question of whether the couplings could separate after being properly made. Experienced railwaymen such as Conductor F. J. Coughlin described instances of apparently perfect couplings failing from a variety of causes such as worn knuckles pulling out under extra strains or on curves and uneven track causing drawbars to slip by. Brakeman J. J. Tapscott gave further examples of couplings slipping. The difficulties and dangers to

[955] *The Daily News*, New Westminster, British Columbia, Canada, 11th November 1909

brakemen of operating the brakes on loaded lumber cars were highlighted. On Saturday 4th December, having watched a demonstration of cars being repeatedly coupled in a siding, the jury returned a verdict exonerating all the employees involved. Their verdict read: 'We find no evidence to show that there was any criminal negligence on the part of the crew in charge of the freight train, nor have we found that the crew have disobeyed any rules in connection with the performance of their duties on the morning of the accident.

'With regards to the leads from the main line to the Nanaimo Road siding, we would suggest that some provision be made to insure safer conditions that now exist at that point, say by connecting the two tracks by a crossover south of the siding, in order that cars can be placed inside instead of outside the siding, as at present.

'We also suggest that the method of moving and keeping track of trains by night be reduced to a more definite system.

'We would recommend that all Interurban lines be equipped as speedily as possible with equipment and rules such as are approved by the Railway Commission.

'We strongly recommend that the city council take immediate step for the establishment and maintenance of an up-to-date city morgue, wherein all post mortems, inquests and inquiries may be held, without the loss of time and inconvenience which has hitherto existed.' [956]

Exactly how the flatcar had come to run away was never established.

An accident in Mexico on 23rd June 1910 involved a troop train in charge of a Mexican conductor and an American engineer. The troops were on their way from Sonora state in the north of the country to the capital, Mexico City, and had begun their journey down the west coast of the Pacific Ocean on the gunboats *Tampico* and *Guerrera,* arriving at the port of Manzanilla on Saturday and Sunday, 18th and 19th June. From there they boarded the train of the Ferrocarril Nacional de México (Mexican National Railway) for either Mexico City or the state of Guadalajara (different newspaper reports state different destinations). The train apparently did not start until Wednesday 22nd, the accident occurring early the next day. It was in charge of Conductor Martinez with Engineer Watson and consisted of either 10 or 13 coaches (the same report gives both figures!). The passengers being carried were soldiers of the Third and Sixth Infantry regiments, together with their wives and children.

Disaster struck as the train was proceeding towards Plantanar. As the train was ascending a steep incline a coupling gave way and the four rear vehicles began to run back down the hill, the brakes being unable to hold them. After running six miles they left the rails on a curve (Paso Blanco Curve) near Zapotilie station, went over an embankment and were 'dashed to pieces'. There were estimated to have been 150 passengers riding in the runaway coaches. A good number of them jumped out when the coaches began to start back down the bank and so escaped but the number of victims was still high. Most of the casualties were in the coach immediately in front of the caboose, which chanced to be the one carrying the officers and army surgeons and their families, the ordinary soldiers and their baggage riding in the forward vehicles of the train. It was

[956] *The Daily News*, New Westminster, British Columbia, Canada, 6th December 1909

initially reported that 10 officers and soldiers, 21 women and eight children had lost their lives. Those killed included Lieutenant Colonel Octavio Castro of the Third regiment, two army paymasters and two lieutenants.

After the accident Engineer Watson backed the remainder of his train to the scene of the wreck and sent a messenger to the nearest telegraph station. Assistance was quickly sent from Zapotlan, Colima and Guadalajara; until this arrived the injured were treated by those of the army surgeons travelling on the train who were themselves unhurt. A special train brough 50 of those injured, many of them women and children, to Guadalajara, together with the bodies of some of the dead. The injured were taken to both the civilian and the military hospitals in Guadalajara. Engineer Watson and Conductor Martinez were arrested and lodged in jail at Zapotlan pending an investigation into the disaster.[957] [958]

On Friday 4th October 1912 the brakes of Compañia de Andaluces mixed train No. 30 failed on the final descent to the terminus at Alicante, Spain. The engine of the train collided with the buffers at the end of the line, demolished the wall behind them and crossing the concourse penetrated the station waiting room, which was crowded with passengers. The train had reached a high speed on the falling gradients. A telegram from Madrid on 6th October spoke of nine dead and no less than 122 injured, some of them seriously. Among those killed was a ticket clerk.[959] [960]

By a coincidence the 1910 troop train disaster in Mexico was followed rather less than three years later by a worse accident with similar features in Eastern Europe. This second catastrophe occurred in Bulgaria early on Monday 12th May 1913, at a time when the country was making preparations for what would later become known as the Second Balkan War, initially between Bulgaria and an alliance of Serbia and Greece with the latter two countries later being supported by Romania, Turkey and Montenegro. The accident involved two military trains which were following one another up an incline between Buk and Drama, Bulgaria. A coupling broke on the leading train and 25 of the 30 vehicles (either passenger carriages or possibly goods waggons) making up the train ran back down the incline to collide with the second military train of 28 vehicles. It was reported that 150 soldiers died and another 200 were injured.[961] [962]

Thirty miles inland from the western side of India the land rises steeply from the coastal plain to the Deccan Plateau, 2,000 feet above sea level, a natural barrier which had to be overcome in the building of the Great Indian Peninsula Railway eastwards from Bombay (now Mumbai). The railway company constructed a single trunk line east of Bombay as far as Kalyan, where it divided into two branches which surmounted the climb to the plateau on separate inclines. The northern branch ran up the Thull Ghat incline to Igatpuri and onwards to Calcutta (now Kolkata) and the plain of the River

[957] *The Cullman Democrat*, Cullman, Alabama, United States, 30th June 1910
[958] *Abilene Daily Reporter*, Abilene, Texas, United States, 26th June 1910
[959] *La Vanguardia*, Barcelona, Spain, 5th October 1912
[960] *The Dominion*, Wellington, New Zealand, 8th October 1912
[961] *The Times*, London, England, 12th May 1913
[962] *The Railway Gazette*, Vol 18 pg. 620

Ganges, while the southern branch via the Bhor Ghat incline to Poona (now Pune) and Madras (now Chennai). Both inclines had ruling gradients of 1 in 37 and each was the subject of runaway train accidents at different times, the Bhor Ghat incline back in 1869 and the Thull Ghat incline on Thursday 15th October 1914.

The 1914 accident began when the 2-8-4 tank engine banking a heavy goods train up the Thull Ghat incline began slipping, while a drawbar broke near the front of the train. The train then began to run back down the bank. To cope with such eventualities, catch points had been provided and the mishap should thus have ended with nothing worse than a pile-up of goods waggons, expensive perhaps in terms of material damage but nothing worse. Unfortunately, a pointsman held the catch points closed so that the runaway waggons, instead of being harmlessly derailed, continued down the main line until they collided with a following passenger train, killing eight people.[963]

The long civil war in Mexico, which lasted from 1910 until 1920, brought in its wake a number of railway disasters in the country. Many of which were the result of acts of sabotage and will thus be dealt with in that chapter. On Monday 18th January 1915, however, a major catastrophe in the country was the result of a runaway train. Normal communications with the outside world had been completely disrupted, at least in the area where fighting was in progress, and news of the accident did not reach the United States until the beginning of March, and then only in the form of a letter written on 14th February by Mrs John Howard, a woman missionary, to the American Board of Commissioners for Foreign Missions. In her letter she described how after the capture of Guadalajara on 18th January by General Carranza, the governor of the town (a new governor installed by General Carranza?) had ordered that a special train be arranged to bring the families of the triumphant soldiers from Colima to Guadalajara.

Mrs Howard wrote, "There were more than twenty cars, simply packed with humanity, the roofs covered with men and women and many slung under the cars in a most perilous position even for ordinary travel.

"At the top of the steepest grade, coming down, the engineer lost control, the cars rushed down the long incline, throwing human freight on both sides and finally plunging into an abyss.

"Nine hundred people were on the train and only six were unhurt. More than 600 were killed outright. Some of the Yaqui Indians committed suicide when told of the accident to their families."

The newspaper report added that Mrs Howard was well known to the board and her report was regarded as undoubtedly being correct.[964]

A grossly overloaded interurban car featured in a runaway accident in Canada on Wednesday 7th July 1915. The Niagara Falls Park and River Railway was an electric line which had opened in 1893 to convey pleasure seekers from Lake Ontario excursion steamers docked at Queenston, Ontario, up the Niagara escarpment to various picnic site,

---

963 *The Railway Magazine*, 1982, pgs. 9-10.
964 *The Washington Post*, Washington, D.C., United States, 7th March 1915

the Brock monument and Niagara Falls. Since 1901 the company had been owned by the International Railway.

On the day of the accident hundreds of picnickers had used the railway, including two large groups of Toronto Sunday school children from the Woodgreen Methodist and St John's Presbyterian churches. A change in the weather had a hand in the disaster. At 7.15 p.m. it suddenly began to rain, and many picnickers rushed to board a car to take them back to Queenston. The car in question was an open cross-seated trolley, designed to carry just over 80 passengers, but no attempt was made to limit the number boarding, and it set off down the incline carrying about 160 passengers, sitting, standing and even hanging on to the outside of the car. In charge of the car was Conductor George Carswell and driving it was Motorman Sidney Boyt.

Near the start of the descent a safety switch (catch points) had been installed to divert any car running away at this point into a dead end siding, but there were no further such points provided beyond. Motorman Boyt brought his car past the safety points without any problem but first the trolley pole separated from the overhead wire, preventing the motorman from reversing his car, and then the car's air brakes failed to work properly.

A newspaper reporter detailed the events leading to the wreck, 'The cause of the accident, according to Mr Arthur Lipman, was the trolley coming off the wire, the brakes not working and the motorman not having a supply of sand to spill on the rails. Dr Lipman said that the motorman of the car told him these facts. Dr Lipman was a passenger on the ill-fated car and received slight injuries to his head.

'The scene of the accident was at the last curve the trolley line takes before it reaches the station and wharf at Queenton and brings it within a few yards of the river, just where it breaks from the Niagara gorge. The current is swift, the water turbulent, with dangerous eddies near shore. The river bank on the Canadian side falls precipitately to a great depth.

'I was here that the car left the track. The end of the descent had been reached. In its course down from Brock's monument, curving away from the river, the railway hugs the shoulder of the heights for a considerable distance, then bends back sharply on itself and travels a distance of half a mile before it reaches the final curve at the river's brink. The grade is heavy all the way and motormen are supposed to make the descent slowly.

'It is stated that the motorman lost control of the car through the brakes failing to respond at a considerable distance up the heights, and as a result the car dashed down the track. The screams and cries of the women helped to unnerve the motorman and attracted the stupefied attention of those above and those awaiting the car on the quay below. There was no possibility to give aid.

'Two of the sharp curves the madly speeding cars negotiated in safety, and as it swept around them the spectators held their breath. On the third and last curve the car plunged from the rails. It completely jumped the tracks, and, wiping away a trolley pole as though it were tinder plunged over the embankment. It landed on its side and plunged some few feet along the ground, cutting down a tree as though with a knife.' [965]

The first on the scene after the accident were soldiers of the 19th Militia Regiment who had been on duty guarding the Queenston-Lewiston International Bridge against possible

---

[965] *Manitoba Free Press*, Winnipeg, Manitoba, Canada, 8th July 1915

German sabotage. News of the accident was also sent to a military camp at Niagara-on-the-Lake and motor cars took doctors and medical equipment to the scene of the accident. The relief operation was carried out speedily in pouring rain and was completed in an hour and a half. It was still raining hard when a steamer carrying the dead and injured reached Toronto just before 11 p.m. Every available ambulance supplemented by private motor cars was used but several trips were still necessary to take all the casualties to the hospital. Less seriously injured victims, some limping or wearing bandages, were helped by friends to street cars.

Eight or ten passengers were killed more or less outright and with the death of further victims from their injuries the final death toll was 15, with 50 or 60 more injured.

An extensive inquiry was made into the accident. At the inquest the coroner's jury heard that, 'A car crowded far past the danger point, tracks weed covered and slippery with rain, the lack of a sand "punch" and a conductor who made no attempt to put the trolley pole in place after it had jumped the wire, or to use the extra set of brakes under his care, were the chief causes of the street car tragedy of more than a fortnight ago.'

Motorman Boyt testified that, 'the tracks along the stretch where the car got beyond control were overgrown in parts with weeds so that when the brake and reverse were alternately applied the effect was simply to make the car skid; that he made no attempt to use the auxiliary hand brake and that though there was sand in the sand box, there was no punch to spread it with.'

Conductor Caswell, 'admitted that his car was overcrowded. He collected 157 fares, "and didn't get them all," yet his car had a seating capacity of only eighty-four. When the car began to get out of hand and the trolley pole jumped the wire. Caswell hung to the trolley rope, made no attempt to replace the pole and never thought of the rear brake, which, according to the rules of the company, he should have jumped to as soon as the car was in danger. "I was waiting for a signal from the motorman," was his only explanation.' [966]

Despite these statements by the conductor and motorman when the jury returned a verdict on 4th August it was that the accident had been due to the excessive speed of the car, caused by the inefficient working of the brakes on a wet and greasy rail; that the motorman had done all in his power to avert the accident, but also that there had been a lax observation of the rules on the part of the officials and employees of the company, particularly regarding the use of sand and the loading of trains.

Prior to the verdict and kept secret from the jury, lest it influence their verdict, the vice president of the International Railway, E. R. Dickson, had been arrested on a charge of criminal negligence in connection with the accident, then released on payment of $20,000 bail. [967]

Following a preliminary hearing on 1st September the charge against the vice president was dropped and that appears to have been the end of the matter.[968]

A major disaster in Hungary on Sunday 26th September 1915 involved a hospital train

---

[966] *The Brandon Weekly Sun*, Brandon, Manitoba, Canada, 29th July 1915
[967] *Manitoba Free Press*, Winnipeg, Manitoba, Canada, 5th August 1915
[968] *Wreck!* by Hugh A. Halliday, pgs. 136-41

and a train of oil tanker waggons. It happened at about 9 p.m. between Vidrány and Mezölaborcz. Hospital train No. 920a was in the section and the tank train, No. 924a, which was following, should have been held at Vidrány, but travelling at about 80 kilometres an hour it ran past two stop signals there and into the occupied section. The gradient on this stretch was 25‰ (1 in 40) and the train must have been running very fast when it collided with the rear of the hospital train. The whole of both trains was thrown off the rails, even the engine at the rear of the tank train, separated by 31 waggons from the front of it. Escaping petrol from the damaged tank waggons set the wreckage and made any attempt at rescue operations difficult. In all 36 people burnt to death, including the driver of the tank train, whose charred body was found in the cab of his locomotive. Perhaps surprisingly only four of those who died were passengers on the hospital train. The fact that the death toll was still so high was explained by the fact that there were a number of unauthorised passengers riding in the brake van of the tank train. Only two people who had been riding on that train, who had jumped off back up the line, escaped with their lives.[969]

Disaster befell a special train carrying convalescent soldiers and men on leave in southern France on Tuesday 19th October 1915 on the Chemins de fer de Paris â Lyon et â la Méditeranée line from Roanne to Lyon via Saint Étienne. From Roanne southwards there was a lengthy climb for about 25 kilometres to Balbigny. The train of 23 carriages, bringing 1,000 soldiers from the front, had left Roanne at 6 a.m. and was due at Saint Étienne at 7.30 a.m. It had passed through St-Cyr-de-Favières station, 10 kilometres from Roanne, and before the next station was stopped at the entrance to Vendranges tunnel, where unspecified work was in progress, perhaps the emergency replacement of a broken rail or suchlike. A number of the soldiers got out of the train at this point to stretch their legs. While the train was standing at the tunnel portal a coupling gave way and 10 carriages began to run back down the gradient. When the train first began to run back some of the soldiers who had remained on board had jumped out, but a large number did not.

Another train had meanwhile arrived at St-Cyr-de-Favières station from Roanne and to avoid the runaway coaches from colliding with this train they were diverted into a siding. They then smashed through the buffers at the end of the siding and six of the vehicles fell into a 30-foot-deep ravine.

After the accident assistance was immediately requested and soon arrived, while the prefect of Loire and the sub-prefect of Roanne also went to the scene. The injured were taken to Roanne or went on to Saint Étienne by a train leaving at 10.45 a.m. At least 17 soldiers died in the accident and about 40 were injured. A note in the Wikipedia account of the accident states that the real toll could have been higher but reports of the disaster were supressed due to wartime censorship, but news of the accident does not really seem to have been censored very much as it was mentioned in different newspapers the following day.[970] [971]

As it was wartime there were also unsubstantiated rumours that the accident had

[969] *Eisenbahnunfälle Band I* by Ludwig Stockert pg. 22 (in German)
[970] *Le Figaro*, Paris, France, 20th October 1915
[971] *Le Temps*, Paris, France, 21st October 1915

resulted from an act of sabotage. The accident was investigated by the public prosecutor and a court martial sentenced the driver and head conductor to respectively 16 and 18 months' suspended sentences for manslaughter.[972]

[972] Webpage – Il y a 100 ans, une catastrophe ferroviaire faisait dix-sept morts (in French)

CHAPTER 8

BRIDGE FAILURES

Some of the most serious railway accidents of the 19th century had resulted from the failure of bridges. Bridges may fail for a variety of reasons. The inherent design of the structure may be faulty, the materials used in its construction may be poor and inferior, the materials may deteriorate over time or its foundations may be washed by a stream or river which has become exceptionally swollen following abnormally heavy rainfall. With improvements in bridge engineering failures due to faulty design had greatly diminished by the start of the 20th century, but accidents, including major disasters, continued to occur from causes such as bridge foundations being washed away.

The first bridge accident of the 20th century was a fairly minor affair. At 7.30 a.m. on the morning of Sunday 13th January 1901 a westbound Northern Pacific Railroad was wrecked at the mouth of the Selah canyon, seven miles from South Yakima, Washington, United States. The accident happened at a 100-foot-long bridge across the Kittitas creek canyon. As the train was slowly crossing the bridge Engineer R. Baird felt the structure giving way beneath his engine. He threw his throttle wide open and got his engine clear of the bridge, but the tender fell 18 ft. All the coaches of the train except the last Pullman sleeping car were derailed. The dining car spanned the gap in the bridge, crushing the tourist car, and was partly covered by one of the sleeping cars. A rail entered the door of the day coach, passed through the centre of the car and exited through the roof, without touching any of the passengers. Four passengers, mail clerk P. McElwell and Porter G. W. Turner were injured but no one was killed. Riding alone on the train from Colfax were three little daughters of Harry Hulin, en route to join their father in Tacoma; Elsie, aged 15, Bertha, aged seven and Freda, aged five, who suffered a slight bump to her head. The children were delayed for 18 hours by the wreck but were cared for by fellow passengers and trainmen. Heavy rain the day before combined with melting snow had led to a large volume of water rushing down the canyon and this had undermined the supporting timbers of the bridge, causing it to fail.[973]

[973] *The Colfax Gazette*, Colfax, Washington, United States, 18th January 1901

More serious than the previous accident was one which occurred near Glendon, North Carolina, United States, on the Durham and Charlotte Railroad, on Thursday 4th April 1901. A material train fell through a trestle across a deep river, killing the engineer and three other men. The accident received but scant coverage in the press.[974]

Another United States accident occurred when a timber bridge on the then new Stonybrook cut-off line of the Eastern Railway of Minnesota, 300 feet long and only a year old, gave way under a train of 34 iron ore cars, each laden with 50 tons of ore, at 3 p.m. on Sunday 4th August 1901. Eight of the cars were actually on the structure when it collapsed. The engine and 16 cars ended up in the St Louis River. Head Brakeman Jerry O'Reilly was killed and Engineer T. Phillips, Fireman H. G. Odell and Assistant Trainmaster John Postal were injured. It was expected that the line would be out of use for two weeks. The reason for the failure of the bridge was not stated.[975]

A Texas and New Orleans Railroad passenger train due at Sabine, Texas, United States, at 11.45 a.m. on the morning of Friday 9th August 1901 was crossing a trestle across Keith Lake bayou, seven miles above Sabine Pass, when the structure gave way, precipitating the engine and all nine cars into the water. The engine turned three quarters over, but the engineer and fireman were able to get out, the fireman being badly scalded and suffering an injury to his back. The 17 passengers on the train also all survived, according to a newspaper report, with nothing worse than a few bruises and a good ducking, but those in the Jim Crow coach might easily have drowned, trapped in the submerged car, had it not been for the fortunate presence nearby of a tug. The *Ernest*, commanded by Captain Wood, chanced to be up the Port Arthur canal in the vicinity, fishing, when its men witnessed the accident. Making for the scene the crew of the tug were able to cut away the roof of the Jim Crow car with axes and release the negro passengers, the conductor and a newsboy when the water was up to their necks. The rear part of the car reserved for the white passengers did not go into the water and they were able to escape by climbing out of the windows.

It was expected that all the freight, express and mail would be able to be retrieved from the water and it was arranged that the *Ernest* would meet the train every day, acting as a ferry to carry the passengers and mail across the gap until the trestle could be repaired.

The cause of the failure was unusual, probably unique. Its timber piling had been attacked and honeycombed by teredo worms until it was fatally weakened. The teredo worm or naval shipworm lives in saltwater and it was supposed that the railroad officials had overlooked the fact that the water in the bayou was connected via the Port Arthur canal with the Gulf of Mexico, was subject to tidal ebbs and flows, and had been very salty the past two months, thus enabling the worm to colonise the area around the trestle and attack the piling.[976]

---

974 *The Marion Daily Star*, Marion, Ohio, United States, 5th April 1901
975 *The New York Times*, New York, New York, United States, 5th August 1901
976 *Shiner Gazette*, Shiner, Texas, United States, 14th August 1901

Another bridge accident in Texas, United States, at about 1 a.m. on Christmas Day, Thursday 25th December 1901, was caused by a fire. A northbound Missouri-Kansas-Texas Railway of Texas bound for St Louis was due to arrive in Dallas, Texas, at 11.20 p.m. on 24th December but was running well over an hour late. The train was in charge of Conductor C. C. Morgan, with Engineer Jenks Clark and Fireman William Kearney.

Engineer Clark applied the air brakes and reversed his engine as soon as he saw that the Five Mile Creek trestle was on fire, which he could not do before he had almost reached the structure because it was preceded by a sharp curve. Then he and his fireman had jumped, falling down a 15 ft embankment. Both were badly hurt, the engineer suffering a broken shoulder and leg and his fireman a fractured skull and severe bruising. The other trainman hurt was negro porter Joel Henry, who sustained possibly fatal internal injuries.

The engine succeeded in crossing the bridge, but the baggage car left the track, presumably because of a partial collapse of the burning trestle, and the baggage car then in turn derailed the engine, which tore up the track for some distance before overturning into a ditch. The remainder of the train came to a stand on the burning bridge and caught fire from it. The passengers were able to escape from the burning train by scrambling down the embankment and only two of them, Rose Lull and Mary White, both coloured, were much hurt. There was also time to save the contents of the mail, baggage and express cars and passengers and crew were able to push the rear Pullman car back out of the reach of the flames, but on the other hand many of the passengers had to abandon their luggage and personal effects to the flames. Two coaches, a combination baggage and mail car and a chair car could not be saved and were eventually consumed by the flames.[977]

Conductor Morgan walked the six miles into Dallas to raise the alarm and a relief train with surgeons and also carrying newspaper reporters was sent to the scene. This train eventually returned to Dallas at 11.30 a.m. with the injured trainmen and all the passengers.

Isadore Harris, of Waxahachie, was a passenger in the chair car, and had his hair partly burned off. He described how the train stopped with a terrible jerk near the bridge over Five Mile Creek, and added, "We had to jump about twelve feet. No one hesitated. There were six white women in the chair car. They acted exceedingly cool and self-possessed, and the male passengers aided in their rescue and that of several children, through windows and a door at one of the platforms. The men had presence of mind enough to form a volunteer force and push the sleeper, which had remained on the track, back out of danger and put the women and children in it."

The accident caused little delay to traffic, trains being worked in and out of Dallas over the Houston and Texas Central Railroad via Hutchins and Lancaster.

The cause of the fire remained a mystery. A train had passed over the bridge two hours before the one involved in the accident, and at that time there had been no sign of any fire.[978] There was no mention of any evidence that the blaze had been started

---

[977] *The Cheyenne Daily Leader*, Cheyenne, Wyoming, United States, 26th December 1901

[978] *The Quincy Daily Herald*, Quincy, Illinois, United States, 26th December 1901

deliberately. Possibly it had been started by a spark from the engine of the preceding train setting some combustible material alight.

The short-lived South Carolina and Georgia Extension Railroad, shortly to become part of the Southern Railway, narrowly escaped a major accident on Thursday 19th June 1902, thanks to the quick reactions of the engineer of the train involved and a considerable amount of luck. A mixed train was crossing a bridge over the Broad River near Shelby, North Carolina, United States, when Engineer Albert felt the structure giving way and opened the throttle of his engine wide. As a result, most of the train accelerated clear of the bridge, only one coach and three boxcars falling with it. The coach fell 50 ft to the edge of the river but was caught by a stone pier and saved from going into the water. One man named Leroy Griggs, of Shelby, lost his life and Conductor Edward Turner, Trainmaster Dock Phifer, a negro named Riley Whisenant and a dozen unnamed passengers were injured. The river was swollen by heavy rains and had clearly undermined the foundations of the bridge.[979] Other reports stated that it was a coach and two flat cars which went into the river. Trainmaster Phifer saved the life of the young son of Mr D. D. Wilkins, who was trapped under the debris. The trestle was about 800 feet long.[980]

By 1900 a main line railway had arrived at the town of Temuco, in Chile, and had been extended southwards, but there was pressure for a branch line to be built westwards from Temuco to the town of Nueva Imperial and onwards to the port of Nehuentue on the Pacific coast. The branch was duly opened as far as Nueva Imperial in May 1902 and its continuation for the 22 kilometres to Carahue was also well in progress. The main engineering work between Nueva Imperial and Carahue was a long bridge across the Cholchol River, just outside of Nueva Imperial, the bridge being designed by Engineer Don Juan de la Fuente and built using a mixture of local woods including olive, cinnamon, cypress and tepa. By the beginning of August 1902 the structure was complete, and a ceremonial opening of the line was arranged for Tuesday 5th August. A stage was erected near the end of the bridge and various dignitaries were invited to the opening ceremony, including the mayors of the provinces of Cautin and the commune of Nueva Imperial, the governor of the department of Imperial, the local stationmaster, José Salazar, a local judge and so on. The band of the Nueva Imperial Mutual Relief Society was also present, conducted by Nicolás Naranjo.

The train which was to make the inaugural run over the bridge consisted of a locomotive, two ballast waggons loaded with sandbags and a carriage occupied by the workmen and labourers who had built the bridge and line. In charge of the train was Conductor Don Clianor Barril with Driver Luis Fernando Muñoz and Fireman Don Guillermo Campos.[981]

Although the train was hardly very substantial, for some reason which is not explained in the only account of the accident which the author has seen, the railwaymen seem to have had considerable misgivings about taking their train over the bridge. Driver Muñoz

[979] *The Polk County News*, Columbus, North Carolina, United States, 26th June 1902
[980] *The Gastonia Gazette*, Gastonia, North Carolina, United States, 20th June 1902
[981] Webpage: Amigosdeltren.cl – La tragedia de Cholchol (in Spanish)

was said to have been nervous, and to have walked out onto the bridge, followed by his conductor and the bridge engineer. Having gone a few metres, they exchanged some sentences and the driver was seen making hand gestures in front of the engineer.

"The bridge will not withstand the excessive vibration of the train," Driver Muñoz told Engineer de la Fuente.

"It will resist much more weight than that of the train," the engineer replied.

"More weight no doubt," the driver insisted, "but not the vibration. It will collapse."

One of the mayors then came up and wanted to know what the problem was, but got no reply. The men went back towards the train, the engineer climbing up onto the front of the train together with his assistants (it is a little difficult to understand the Spanish text, but it appears certain that the locomotive propelled the carriage and waggons out onto the bridge, with the carriage leading). The driver asked that someone be stationed alongside the lever for uncoupling the engine from the train, so that this could be done at the first sign of trouble. The conductor told his driver to calm down, to which the driver replied that he was perfectly calm, just worried about what might happen. The conductor got up on to the engine.

The train then proceeded slowly onto the bridge with the engineer greeting the assembled dignitaries while the band played Beethoven's *Triumphal March*. The leading carriage passed the first bride pier. Suddenly a loud noise was heard; the bridge creaked, it trembled, and the noise became a loud rumble. The engine, stopped, reversed, it was detached from its train and jumped forward, just as the whole bridge collapsed. The reversal of the engine and its whistle of alarm caused the band to stop playing. The carriage and waggons, together with the engineer, his assistants and all the workmen disappeared into the river in a cloud of dust, debris and water which rose to a height of several metres. The dream was over.

The bodies of the engineer and 40 workmen were later recovered. It is not entirely clear from the Spanish account, but it seems likely that the driver was able to reverse his engine clear of the bridge in time and that he, the fireman and the conductor survived. What remains unclear is why the bridge collapsed, and why the driver was so sure that it would. Did he just not like the look of it? Had he, perhaps, some knowledge of bridge engineering so that he could see it did not look right?

The branch remained open as far as Nuevo Imperial and a new bridge, this time of steel construction and designed by Gustave Eiffel, was completed in 1909, the first train arriving in Carahue on 30th July of that year.[982]

The bridge disaster in Chile was closely followed by one in India. Involved in this disaster was a Madras – Bombay mail train, which was crossing a bridge near Magnapatnam in southern Bengal when the structure collapsed. By the time of the report the bodies of 50 victims had been recovered, 'including those of eight Europeans, four of whom were soldiers.' Only 25 of the passengers who had been riding on the train survived, these including all the first-class passengers. Many of the survivors were badly hurt and had to be taken to hospital, though some had miraculous escapes, including two

---

[982] Ibid.

Roman Catholic bishops, who were carried down the river for several miles.[983] 'A portion of the mails has yet to be recovered.' [984] The river was in flood and the whole of the train had fallen into the river except the rear brake van.[985]

A few days after the disaster it was reported that 62 bodies had been recovered.[986] Flooding due to the same heavy rains which had caused the bridge failure had destroyed 20 villages in the area and left 6,000 families homeless.

According to the official report on the night of the accident the area above the bridge experienced "a sudden and violent fall of rain such as had never been approached in magnitude during the existence of the railway." This report stated that 75 passengers and four railway servants died, and six passengers were injured. The disaster occurred on the Madras Railway at about 3.30 a.m. on 12th September 1902.[987]

The Pullman passengers on a Maricopa and Phoenix Railroad train had a lucky escape on the evening of Wednesday 29th October 1902. A bridge across the Salt River at Tempe, Arizona Territory, United States, collapsed as a mixed train was crossing it. The engineer felt the bridge giving way beneath him, jumped and escaped but Fireman Goddard was slightly injured and an unnamed Indian, possibly a trespasser, was killed. The passengers were saved because there were an unusually large number of freight cars at the front of the train. As well as the engine two freight cars and a loaded stock car dropped 20 feet to the river below while the Pullman car which came next in the train stopped suspended from the bridge and the train passengers escaped with a bad shaking up.

After the accident Superintendent Porter, with a corps of men, made up a train from Mesa and the passengers were brought to Maricopa in time to make their connections with Southern Pacific Railroad trains to Los Angeles and San Francisco.[988]

News having been received on Sunday 8th June 1903 of the washing away of seven hundred yards of a Seaboard Air Line Railroad trestle on the southern approach to the Broad River Bridge, near Carlisle, South Carolina, United States an engine with a pile driver and a crew of 10 was sent from the northern side during the afternoon to repair it. The damage to the structure was, however, more extensive than had been realised, and when the engine reached the middle of the bridge it collapsed, precipitating bridge, engine, pile driver and crew into the river. Three lives were said to have been lost.[989]

A bridge at km 55 known as 'De Entrambasaguas' ('The Waterway') over a stream

---

983 *The Telegraph*, Brisbane, Queensland, Australia, 16th September 1902
984 *The Times*, London, England, 15th September 1902
985 *The Mid Surrey Times and General Advertiser*, London, England, 19th September 1902
986 *The Brisbane Courier*, Brisbane, Queensland, Australia, 19th September 1902
987 Administration Report on the Railways in India for the year 1902
988 *Corona Courier*, Corona, California, United States, 8th November 1902
989 *Savannah Morning News*, Savannah, Georgia, United States, 9th June 1903

which was a tributary of the River Jiloca near its junction with the main river, collapsed beneath train No. 6, the Valencia – Calatayud mail train, between Calamocha and Luco de Jiloca, Spain, at 10.08 p.m. on Monday 22nd June 1903.[990] Fire broke out in the wreckage and consumed four vehicles of the train.

At the time a severe storm was causing damage at several points along the line of the Central del Aragon and the swollen stream doubtless caused the failure of the bridge. The latter was a small structure, only two metres long and 4½ metres high. It was reported that some of the mail had been saved, though much damaged. The engineer responsible for the line and the director of company left for the accident site. News of the accident caused great dismay in Valencia, from where most passengers had come. The first official report of 23rd June that there were two dead and 12 injured, five seriously and that the driver and fireman had gone into the river with their but were unhurt. On 24th June a relief train returned with seven injured and all the uninjured passengers.

It was reported that engine and leading two coaches had gone into river while the remainder of the train which had stayed on the track had caught fire. A severe storm had been in progress at time of the accident. Many passengers, who had reportedly had to swim to safety by the light of the burning train and the flashes of lightning, blamed a defective bridge for the accident. The engine boiler exploded at the moment of the collapse and the whole of the train not in the river had quickly caught fire. The dead were now identified three Piarists (members of a Roman Catholic educational order); two officers of the Correos encargados de la Ambulancia and a line engineer, while 14 injured had been extracted from the wreckage. There were said to have been 50 passengers on the train. The uninjured civil guards on the train had helped the passengers out through the windows. It was believed that a number of bodies had been swept away by the river.[991]

A further newspaper report on 27th June stated that eight people were still missing and added that it was likely to take two months to repair the track and the damaged bridge.[992]

The final number who lost their lives remains uncertain; a newspaper report two months after the accident apparently spoke of 20 dead.[993]

What remains at the beginning of 2020 almost certainly the worst railroad accident in American history occurred on Sunday evening, 7th August 1904, at Hogan's Gulch, a mile north of Eden station, Colorado. (The location of the wreck is also given as Steele Hollow, but this place was actually two miles further north.) The unlucky train involved was a Missouri Pacific south and eastbound express, train No. 11, running on Denver and Rio Grande Railroad tracks. Train No. 11, the *World's Fair Flyer*, as a newspaper referred to the train, was said to be the fastest train out of Denver by the Denver and Rio Grande route and it usually carried a heavy passenger list. The train would become No. 8 upon arrival at Pueblo. It was formed of a locomotive No. 1009, Missouri Pacific baggage, smoking, and chair cars, Pullman sleeping cars *Casimir* and *Wyuta* and Denver

[990] Federación Castellano Manchega de Amigos del Ferrocarril: Accidentes Ferroviarios en España (in Spanish)
[991] *Noticiero Granadino*, Granada, Spain, 24th June 1904
[992] Ibid, 27th June 1904
[993] Federación Castellano Manchega de Amigos del Ferrocarril: Accidentes Ferroviarios en España (in Spanish)

and Rio Grande dining car *Glenwood*. It was in charge of Conductor James Smith, with Engineer Henry Hinman and Fireman David Mayfield, and had left Denver at 5 p.m., with a two hours and five minutes run to Colorado City, and a run of an hour and ten minutes from there to Pueblo.

The location of the accident was a bridge over Hogan's Gulch, referred to by a newspaper as Dry Creek, about eight miles north of Pueblo, a place said to be one of the most dangerous points on the road and not far from Buttes, where two years earlier washouts had paralysed the service over the line for some time, though no accidents involving loss of life had occurred then. The bridge was 96 feet long and crossed an arroyo, a stream which was normally dry, and only flowed after sufficient rain. Such arroyos are prone to flash floods after thunderstorms. Hogan's Gulch itself was about 50 feet wide and 15 feet deep, with steep banks. Less than half a mile below the bridge the creek emptied into the Fountain River. Two hours before the accident there was no water in the creek.

Trains Nos. 3 and 15 had preceded train No. 11 over the bridge and reached Pueblo safely and the last train over the bridge before it had failed was train No. 4, which had passed over it in the opposite direction about three-quarters of an hour earlier. There had been heavy rains in the area, but no danger was anticipated at Hogan's Gulch, a recent inspection having shown the bridge to be in first-class condition. Just before the accident there were reports of heavy rain in the vicinity of Eden station, but there was no intimation of a cloudburst. No reports of any problem on the line were received at Pueblo until news of the wreck was received just before 9 p.m. A cloudburst must have occurred though up in the hills above the line, just before the train reached the bridge and a torrent of water had rushed down towards the structure, demolishing a wooden road bridge and carrying its remains towards the trestle. The impact of this debris together with the force of the torrent together caused the railroad trestle to collapse, plunging the engine and front part of the train, comprising the baggage, smoking and chair cars, into the water. The force of the current ripped the trucks (bogies) from the cars and thus lightened they were carried down into the Florence Creek before being beached downstream. Engineer Hinman lost his life in the wreck but Fireman Mayfield had a miraculous escape. Crawling from the cab of the wrecked locomotive he had searched vainly for his engineer, then staggered towards Eden station to raise the alarm. On the way he had met Trackman Henry Eiten, who had been sent up the line by his section foreman to check that all was well given the storm and had become alarmed when he had witnessed the accident to the train from the distance. Eiten had asked the fireman who he was and what he was doing. Told that he was the fireman, Eiten had said that he was coming to warn trains of a heavy water running under bridge No. 111 A. The fireman had replied, doubtless with considerable bitterness, "You're a hell of a flagman. The train is in the ditch and my pardner is dead." The two of them had then continued towards Eden station and had met Agent Frank Jones, coming from there to check all was well. The message to Pueblo from Eden advising them of the wreck reached there a few minutes before one from Pinon on the far side of the collapsed bridge (see below). Having sent the message Agent Jones had hitched a buggy and he and the fireman had driven to the wreck.[994] 'When he {Agent Jones} got there not a person could be found who had gone down in the wreck. The cries had been followed by a complete silence, and apparently the

[994] *Tragedy at Eden* by Dow Helmers, pgs. 45-6

passengers in all three coaches had gone down to death. It was reported that there were 100 passengers in the three cars that went down. Chief Despatcher Butler states that this is about a correct estimate of the number. It is probable that a complete list of the dead will never be made known. The conductor's reports were on his person and have not been recovered.'

Agent Jones had realised that there was nothing that he could do, and the fireman was in a bad state. He returned to Eden with the fireman and the latter, who had collapsed, was taken to Pueblo on the first of the relief trains and was admitted to hospital in a serious condition. Fireman Mayfield related later that as the engine had reached the trestle it had seemed to rear up and then fall over onto its right side, the engineer's side. He had jumped and escaped.

A newspaper report stated that, 'When Supt. Bowren, of the Pueblo division, was notified from Pinon station {north of the bridge, on the far side of the creek from Eden} by passengers who had escaped and who had walked through the blinding rain to the nearest telegraph station, he immediately sent a hurry call for all surgeons and nurses in the city to follow him on relief trains, and taking an engine and car, ran quickly to the scene of the wreck. One train followed another from the city in quick succession, and every available physician responded to the call accompanied by a large supply of medicines and a number of nurses. In the meantime, measures were taken for the equipment of a hospital train and that left Pueblo later in the night. In addition to the surgeons and nurses a carload of coffins were sent out.' [995]

The message from Pinon had been sent from the Atchison Topeka and Santa Fe depot, there being no night operator at the Denver and Rio Grande one, by Denver and Rio Grande Special Agent Richard Brunazzi, who had been riding in one of the Pullman cars. He had in fact only had to run part of the distance to Pinon when he was able to command a handcar, which enabled him to cover the distance considerably faster. His telegram (or one of them) read, "Our No. 11 into washout near M.P. 111. Engine and three coaches cannot be seen or heard of. Two, Three and Fifteen will have to be diverted. Protect all trains from south and send engine from north to take sleepers and diner. No one hurt on this part of train." [996]

Along with Engineer Hinman, Conductor Smith had also lost his life. Apart from the fireman 28 passengers and the six crew of the Pullman car were reported to have survived the disaster, all but three of them being from the rear vehicles of the train, which had not gone into the water. One of only three passengers to escape from the chair car was John Killen, of Pueblo, who according to one account was found unconscious lying on the top of one of the cars which had been washed down the river. Another account stated that he had crawled out of the wreckage a mile below the bridge and had been so badly hurt and exhausted that he was unable to talk coherently.[997] Another of the three was Henry Gilbert of Oklahoma, but his escape was marred by the loss of his wife. 'One of the saddest stories is that of Mr and Mrs Henry Gilbert of Oklahoma City, who had been married only two months and were on their honeymoon trip. Gilbert is one of the three men who marvellously escaped from the chair car. His wife, who was sitting beside him, was lost,

---

[995] *The Weekly Gazette*, Colorado Springs, Colorado, United States, 11th August 1904
[996] *Tragedy at Eden* by Dow Helmers, pgs. 42
[997] *Deseret Evening News*, Salt Lake City, Utah, United States, 8th August 1904

and although he is unscathed his loss has made him a nervous wreck. He is now in the Pueblo hospital. He has been wire chief of the Western Union Telegraph company at Oklahoma City.'

Two other passengers escaped because for different reasons they did not catch their intended train. 'Among the many narrow escapes from a vision of sudden death on the wrecked train Sunday night was that of a young man whose home is in Pueblo and who was the guest for the day of Mr James F. Burns of this city {Colorado Springs}. In the afternoon the young man had a telephone conversation with his mother in Pueblo, in the course of which she urged him to return on the 7 o'clock Missouri Pacific train. It was the understanding that he should do so, but he was urged by his hosts to remain for the Sunday evening supper, which was delayed. Accordingly, the party were just seating themselves at the table at the time when the ill-fated train was putting out from the depot. About 10 o'clock came a frantic telephone call from the nearly distracted mother. News had just reached Pueblo that the Midland Pacific train leaving Colorado Springs at 7 o'clock with awful loss of life. The mother was assured that her son had taken a later train than the one he intended and about 1 o'clock she telephoned again that he had arrived safe.

'Another fortunate experience was that of J.F. Powers of Colorado City, foreman of the water service on the Midland Railway. Yesterday he was wrathful over the perverse luck that kept him from a contemplated trip; today he is thankful over a providential escape perhaps from death. He had planned to take the 7 o'clock train to Pueblo but wanted first to cash a check. The check was for a sufficiently large amount to make it difficult to find anyone who had the cash to spare so Mr Powers went from one place to another and finally abandoned his trip in disgust. When he read yesterday's Gazette he felt disposed to have the check framed and keep it as a talisman which once saved his life. However, he believes that lightning doesn't strike place in the same place and left last night over the Missouri Pacific for St Louis.

'Passengers on the Pullmans were in the meantime attempting to assist the victims, but no trace could be found of the cars which had been precipitated into the stream. The water was so deep that the engine and tender were entirely immersed and the rain falling in such torrents that there was danger of part of the track being washed out.'

The accident was also witnessed by a Mr and Mrs Walter Hart, travelling in a horse and buggy along the parallel road, and after some time they reported the accident to Bragdon, Atchison Topeka and Santa Fe station, from where a message was telegraphed to Pueblo.[998]

Nothing much could be done in the darkness, but, 'Shortly after daybreak many hundreds of persons began swarming to the scene of the wreck and the greatest excitement prevailed in this city, where all kinds of rumours were circulated, and it was with difficulty any authentic news could be obtained. The banks of the river were lined as far down as the Fourth-street viaduct with people anxiously watching for bodies that might float past, and some were taken out several miles from the scene of the accident. The fire and police departments of this city placed men at work patrolling and dragging the river all the way from Pueblo to Eden. Their heroic efforts have availed but little owing to the swollen condition of the stream. Much better progress will come shortly as

---

[998] Ibid, pgs. 37-9

the river is falling, and when its normal condition is reached, it is not thought great difficulty will be had in recovering the bodies of the ill-fated passengers. On account of having no boats and prevalence of quicksands at the scene, the east side of the river could not be reached, and here, it is expected, many bodies will be found.

'The rescuers are suffering great hardships, many of them having been constantly at work since last night, without food or relief of any kind, the cold of last night nearly exhausting them.

'A force of nearly 300 well-organized men are now at work under the direction of able officers and rapid developments are promised.

'The list of missing also has been augmented every few minutes, adds to the horrible details of the story.

'At 11 o'clock the river is receding slowly, and bodies are being discovered on sand bars practically covered by the mud and sand washing over them. They are being brought to the city and placed in the morgues, where 23 so far have been taken, only six of which have been identified, although the bodies have been viewed by thousands of people.'

Another report written the day after the disaster stated, 'Latest estimates of the loss of life by the wrecking of Denver & Rio Grande train No. 11, (the Missouri Pacific flyer), south and east bound, at Dry Creek last night, put the total at 100 or more. It is believed there were more than 125 people on the ill-fated train, and only about two dozen survivors have been accounted for. With the breaking of day, the full horror of the scene which was concealed to a great degree by the mantle of night, became apparent. Wreckage is visible in all directions, dead bodies being visible here and there in the piles of debris from the cars, driftwood and mud. Many of the bodies were carried down Fountain creek by the wall of water, which had force enough to carry several coaches nearly four miles away from the point where they went through the bridge.' This statement as to the distance the coaches were carried is contradicted in the next paragraph and is clearly an exaggeration, for the account continues, 'The engine got almost over, but slipped back, and the baggage car, smoker and chair cars plunged into the torrent. The engine fell on its right side. The chair car was carried half a mile down Fountain creek. The baggage and smoker were washed down the stream and have not been found. The diner and sleeper did not go down.

'The bodies of two women and a girl, probably 15 years old, which were the first to be recovered, were found half a mile below the wreck. These two were among the occupants of the chair car. The bodies when found were covered with mud and have not been identified.

'Chief of Police Shoup of Pueblo and 25 police are on the ground with 200 people participating in the search for bodies. It is thought 125 bodies were lost. Undertaker Collier is at the scene and the bodies are being placed in boxes and brought to this city. The chair car was found a mile from the scene of the accident half filled with sand, under which a number of bodies were buried. The express car was found near the scene of the wreck with the safe open and the contents gone.' [999]

The inquest on the victims of the accident lasted 11 days. At its conclusion the jury pronounced a lengthy verdict, touching on all points of the wreck and blaming the

[999] *Deseret Evening News*, Salt Lake City, Utah, United States, 8th August 1904

accident on the negligence of the Denver and Rio Grande company. The chief points of the verdict, as summarised by a Colorado newspaper, were:

- The crew of train No. 11 was blameless.
- The railroad was to blame for not having constructed a bridge which would withstand the impact of the county bridge (the floodwater had first demolished a more flimsy waggon bridge 1,000 feet upstream from the railroad bridge, and it was the remains of this bridge which had been swept down and struck the piers of the railroad bridge, fatally weakening it).[1000]
- Bridge No. 110B was not of the best class of bridges used by railroad companies.
- The weakness of the bridge consisted of the inferior class to which it belonged.
- The lack of telegraphic communication between Eden and Buttes made it impossible for the dispatcher to send any word to the train after it had left Buttes.
- If a regular system of track walkers had been employed train No. 11 might have been flagged and the disaster avoided.
- Bridge No. 110B was under the charge of the section gang at Pinon, five miles away, instead of the one at Eden, one mile away.
- The bridge should have been constructed to carry all the water that the gulch would carry.
- A one span metal structure across the stream would probably have allowed the county bridge to pass without the railroad bridge sustaining any damage.[1001]

Robert Shaw in his book considers, in the present author's opinion correctly, that the verdict of the jury in blaming the railroad company for the disaster, was unreasonably harsh, and a case of 'being wise after the event'.[1002] The harshness of the verdict was perhaps a reflection of the magnitude of the disaster.

The total number of dead reported by the Federal Administration Bureau of Railroad Safety was 96, which did not include the name of Tony Fisher, who died a month after the accident. In addition to this a further 14 persons were reported as 'missing'. Assuming that these 14 were all on the train and that their bodies, buried in the sand and mud, were never recovered, the number of dead rises to 111. This would imply that in addition to Conductor Smith, Engineer Hinman, Brakemen J. F. Turner and Henry Hough and Baggageman Thomas Reese a total of 106 passengers had lost their lives.[1003]

Much less serious than the Eden disaster was an accident near Mitchell, Georgia, United States, at a location also given as Belle Springs, early on Wednesday 5th October

[1000] *A History of Railroad Accidents, Safety Precautions and Operating Practices* by Robert B. Shaw, pg. 304

[1001] *Semi-Weekly Iowa State Reporter*, Waterloo, Iowa, United States, 30th August 1904

[1002] *A History of Railroad Accidents, Safety Precautions and Operating Practices* by Robert B. Shaw, pg. 305

[1003] *Tragedy at Eden* by Dow Helmers, pgs. 140-143

1904. A passenger train on the Augusta Southern Railroad, in charge of Conductor C. D. Davis, with Engineer J. H. Wrenny and Fireman N. S. Nevilles, ran onto a burning trestle which gave way beneath it. Baggagemaster W. S. Shirley was fatally injured and died an hour later and the conductor, the engineer, the fireman, Mail Clerk T. R. Harper and four passengers were hurt. There was time to save all the baggage, express and mail but the tender, compartment car and two coaches were burnt, and the 60-foot trestle was entirely consumed by the fire, the origin of which was unknown. 'Traffic as resumed by transferring to other trains,' the newspaper report added.[1004]

The failure of another fire damaged bridge on Friday 24th March 1905 again resulted in a fatality, this time of a passenger. The bridge concerned was No. 5702, situated a mile and a half south of Arthur City, Texas, United States, on the St Louis-San Francisco Railroad. The approach to the bridge was at the foot of a down grade and around a bend so that the engineer of northbound passenger train No. 6, which was running two hours late, did not discover the danger until he was close upon it. He then opened the throttle wide in an attempt to get across the burning structure. When the engine struck the bridge the ties (sleepers) gave way, letting the weight down onto the stringers (the horizontal beams of the bridge). The engine and forward part of the train got across, but it was too much to hope for that all the train would do so, and a chair car and two sleeping cars were derailed and turned over onto their sides. The passengers were subsequently able to exit the cars by climbing out of the windows on the uppermost side. Miss Maggie Cooney, maid to Mrs Frank Hays, of St Louis, was fatally injured and died in hospital the same evening. She was clearly unlucky for otherwise only half a dozen passengers were slightly hurt.

An extra train was made up at Paris, Texas, after the accident and brought physicians to the scene. Southbound passenger train No. 5 was almost due when the accident occurred at 3 p.m. It was stopped in time, the passengers from it were transferred to the extra train and two coaches and two sleeping cars of it were worked back to St Louis as train No. 6. The line was cleared about midnight, when the remainder of train No. 5 continued its journey southwards and the 7.20 p.m. northbound train, which had been held up several hours, could finally proceed as well.

The bridge consisted of two bents (piers) and it was discovered that the ties on the north side of the east bent were burnt almost in two. It was theorised that the fire had been set deliberately but the only evidence for this was that the locomotives were fitted with spark arresters and were carefully examined before starting their journeys, ruling out in theory any possibility of a fire having been started by hot coals dropped from the ash pan.[1005]

The engine, tender, baggage, mail and smoking cars of Wisconsin Central Railroad passenger train No. 1 plunged from a high steel bridge into the Red Cedar River a mile west of Colfax, Wisconsin, United States, at 11 a.m. on Wednesday 7th June 1905. A travelling salesman for the Dearburn Feather Duster Company of Chicago, who was a passenger on the train, described how the accident had happened to a newspaper reporter. He said that the train crew had obviously had misgivings as to the safety of the bridge

---

[1004] *The Weekly Star*, Wilmington, North Carolina, United States, 7th October 1904
[1005] *The Galveston Daily News*, Galveston, Texas, United States, 25th March 1905

and that the train had been pulled up on its approaches, the conductor and engineer alighting and going forward on foot to try and determine if it was wise to cross it with the train. They made up their minds that it was safe, and the engineer and fireman returned to the engine cab. The conductor had intended to accompany the engine crew across the bridge on the locomotive, but seeing some hobos riding on the train he changed his mind and went back the train to put them off. Though he did not know it at that time the hoboes had probably saved his life.

The commercial traveller related, "I remained standing on the platform of one of the rear coaches, for I thought I would be where I could jump if the bridge should start to break down. The train started ahead. When the engine reached the middle of the bridge that part began to creak and then seemed to go straight down and the engine with the engineer and fireman on board disappeared in the flood beneath. The mail and baggage cars followed. One end of the baggage car was up out of the water and we could see the baggage man crawling out of that end of it and swim ashore. The smoker went in and remained standing on end. Beside the engineer and fireman, a lad stealing a ride was drowned. No one else was injured."

The bodies of Engineer George Phillips and Fireman William Severns were recovered some days later. Although the travelling salesman only spoke of one hobo there were unconfirmed reports that five tramps were missing, believed drowned. The baggageman had climbed through a window and a mail clerk crawled out through a ventilator, both having lucky escapes. The passengers also climbed to the top of the smoker and got out, badly bruised and shaken but none fatally hurt. The remainder of the train remained on the track and was to be returned to St Paul and diverted via the Chicago St Paul Minneapolis and Omaha. The fortunate conductor was named Hayes.

The centre pier of the bridge had failed after it was undermined by floodwater.[1006]

An accident occurred on the Ferrocarril Central Mexicano at Kilometre 317, between Irapuato and Silao, Mexico, early on the morning of Sunday 2nd July 1905. A bridge collapsed beneath a southbound passenger train and, according to one report the baggage car and a third-class coach fell into the river. Another report stated that the engine and mail and express cars also fell with the bridge. Seven passengers died and 33 were injured but the *Galveston Daily News* was able to reassure its readers that no Americans had suffered in the accident. The track was washed away in four places near Irapuato as well as for several miles south of Marfil.[1007]

A new branch of the Chicago and North Western Railway under construction in Wyoming, United States, was the scene of a bad accident on Sunday 26th March 1906. At 5 p.m. a work train was returning to Casper from the end of the branch at Wolton (renamed Hiland 1914). It consisted of an engine, a coach, two flat cars and a water car. The coach was filled with Austrian and Italian labourers. The spring thaw was in progress, all the streams were swollen with meltwater and in many places there was water on the track. As the engine of the train crossed a bridge near Natrona the structure, which

---

1006 *The Weekly Telegram*, Eau Claire, Wisconsin, United States, 9th June 1905
1007 *The Galveston Daily News*, Galveston, Texas, United States, 4th July 1905

had been undermined by the floodwater, swayed, cracked, began to settle and then fell with a roar. The engine got to the far side, but the coach fell with the bridge and was then crushed by the following flat cars and the heavy water car. A fire, started by the stove, then broke out in the part of the coach which was not under water and threatened to incinerate the dead and injured before they could be released, but the uninjured train crew and less badly injured passengers formed a bucket brigade and were able to extinguish the blaze.

A newspaper reporter wrote, 'Not one of the men in the passenger coach escaped unhurt, and the drop into the stream, the demolition of their car by the water car which fell upon it and the inrush of muddy water reduced the survivors to a state of panic. Those least hurt fought desperately to get out of the wreck and away from the rushing water, while the most severely injured and those pinned down by wreckage begged and pleaded with their more fortunate companions not to abandon them. Men battled madly with one another for the first opportunity to get through exits, and men pinned down clung to those who were free and implored their assistance.'

It was to be many hours before help arrived from outside. One of the trainmen ran to Natrona after the wreck and news of the accident was telegraphed to Casper. A relief train was immediately prepared and set off for the scene with two surgeons on board but got no further than a bridge over the North Platte river which had been likewise weakened by the high water and was, it was decided, not in such a condition as to bear the weight of the train. The relief train then returned to Casper and surgeons, officials and workmen set off instead on handcars for the 20-mile trek to Natrona. The North Platte bridge was safely crossed, and they had covered several miles when a heavy snowstorm blew up, the snow made the line impassable and the rescuers themselves were in need of relief. One of the handcars, manned by four strong men, then started off back to Casper and reached the place after a struggle. A second relief party then set off by road but did not reach the first party until well into the night. The combined parties then pressed on and finally reached Natrona at noon the following day, 19 hours after the accident. Meanwhile a bridge party with a pile driver had been despatched from Rapid City, South Dakota, the bridges west of Casper repaired, and a relief train could finally also be sent to the scene.

Before the rescue party and the relief train arrived those on the scene had worked through the night to free the trapped and injured, they were taken to Natrona and received such treatment as could be given in the section house and several ranch houses. The relief train with the injured returned to Casper during the afternoon.[1008] Ten men were killed and 17 injured, four of whom it was feared at the time of the report might also die.[1009]

The accidents due to the foundations of trestles and bridges being damaged by floodwater continued in the United States with one on Friday 24th August 1906 at Moccasin Gap, Virginia, on the Virginia and Southwestern Railroad. It was 2 a.m. and pitch dark when the accident happened and so Engineer Robert Burton had little if any warning of danger, especially as he was running at 30 or 40 miles an hour. The river was very high but previously it had risen much higher and the trestle had stood firm, so the

[1008] *Stromsburg Weekly News*, Stromsburg, Nebraska, United States, 5th April 1906
[1009] *The Enterprise*, Sheridan, Wyoming, United States, 30th March 1906

engineer did not anticipate any danger. The engine and eight cars plunged into the Moccasin River. Engineer Burton was seriously injured but somehow managed to keep his head above water and after floating a considerable distance downriver he succeeded with great effort in reaching the riverbank. Fireman J. A. Cleet and Brakeman R. E. Booher were pinioned beneath the wreckage of the train and drowned. Their bodies were recovered in the afternoon. The property loss was heavy.

The first reports of a bridge accident spoke of the loss of 100 or perhaps even 200 passengers when a Chicago Rock Island and Pacific Railroad passenger train went through a bridge over the Cimarron River near Dover, Oklahoma, United States, early on Tuesday 18th September 1906.[1010] [1011] A telephone message reported that 225 passengers had gone into the water and only six were safe. A private message stated that 47 bodies had been taken to Dover and placed in the morgue there. Communication was however lost after the receipt of this message and there was some doubt as to its accuracy.

The train consisted, according to one report, of a locomotive, combination baggage and mail car, smoker, day coach and Pullman sleeper car. Train No. 12 had left Fort Worth, Texas, at 9 p.m. on Monday 17th and was due in Chicago, Illinois, at 9.30 a.m. on Wednesday 19th. The engineer had very little warning when the train ran onto the bridge at about 5 a.m. The whole of the train except the rear sleeping car went into the river when the bridge collapsed. Another report stated that the engine, baggage and mail cars had gone into the river, the smoker partly so, and that two Pullman cars had remained on the track. The engineer, fireman, expressman and mail clerks had escaped on a pile of driftwood.

Clyde Glest (probably the same person as the one whose name was elsewhere given as Lloyd Zeist) was a passenger who escaped from one of the cars that went into the river. He told a newspaper reporter, "We were running very slowly and carefully when we could feel the track giving way and heard the engineer whistle for emergency brakes. I started for the floor with the porter, and just as we got outside of the car, I felt the track give entirely away, and fell into the river. The porter and I managed to escape to the top of the smoking car, and floated down the stream several hundred yards, when the waves and driftwood dislodged us, and we were cast out into the main current. In the meantime, I had taken off my clothes, and that aided me much in getting to shore. The porter went down a few minutes after being dislodged. I immediately started for the railroad station where I notified the agent."

Another report stated that many passengers, more dead than alive, had been taken to hotels at Kingfisher. Relief trains had been sent to the scene from Chikasha, Oklahoma City and Herington, Kansas.[1012]

In another accident the same day a heavy freight train went through a different bridge across the Cimarron River on the St Louis San Francisco line between O'Keene and Ames, Oklahoma. 'The engineer, fireman and head brakeman cannot be found.'

A further report from Kingfisher on the day after the accident stated that 20 people

---

[1010] *Semi-Weekly Waterloo Courier*, Waterloo, Iowa, United States, 21st September 1906
[1011] *The Evening Independent*, Massillon, Ohio, United States, 18th September 1906
[1012] *Reno Evening Gazette*, Reno, Nevada, United States, 18th September 1906

were still missing. Efforts were being made to locate the smoker, which was at a point 110 yards below the bridge. This report continued, 'When the train went off the bridge the baggage car, smoker and day coach broke loose and swung downstream. The baggage car sank for an instant. The smoker half floated, with one end above water, the couplings between it and the day coach broke, and, submerged the car was carried down the river for a distance of over 100 yards. None of the persons who escaped saw anyone leave the smoker, except four persons who stood on the roof of that car when it was struck by a portion of the wrecked bridge. They were knocked into the water and rescued finally several miles down the river clinging to debris.

'Piles of wreckage in the Cimarron river at points one and two miles below the scene of the wreck were examined by men in boats today. An effort was made to find the 20 persons who are believed to have been in the smoker and still missing. With long poles the debris, consisting of splintered boards, furniture from coaches and baggage was probed, but not a body was found.

'When the train took the bridge, which had been undermined, the piling and rails that held it in place snapped, and the train, excepting two Pullmans in the rear, was precipitated into 25 feet of water.

'It was reported here by telephone that bodies of five men had been taken out of the river 10 miles southeast of the scene of the wreck earlier today. Another telephone report from down the river said that a passenger who escaped from the wreck on the debris had been lodged in the middle of the river and that it was impossible to rescue him, owing to the swift current.

'Bodies of passengers who were buried in the sinking smoker will probably never be recovered. For a short time following the wreck a small portion of the smoker was visible. It sank slowly in quicksand. Although the water had fallen five feet today the smoker could not be touched with 10-foot poles. Waves heavy with sand lash continually over the spot where the smoker, believed to contain missing bodies, is thought to lie.' [1013]

On 27th September it was stated that eight people were confirmed dead, 20 or more were injured and the same number were still missing.[1014]

On 23rd October it was reported that 'In spite of protests from public officials and others, no effort has ever been made by the Rock Island to raise the wrecked sleeper, in which some fifteen or twenty people are supposed to have perished.' [1015]

The final death toll in the accident appears to have remained unclear.

At 7 a.m. on Saturday 19th January 1907 Atchison Topeka and Santa Fe Railroad train No. 30 was wrecked at a bridge over Kill Creek, Kansas, United States, a quarter mile from De Soto. The exact sequence of events which led to the wreck was not wholly clear; either the stone abutment on the west side of the bridge gave way, the engine plunged into the water and its boiler exploded, or else the engine suffered a boiler explosion which damaged the bridge and threw it into the water; but the former possibility, that the bridge failed first, seemed the more likely. Engineer Bartell and Brakeman Shaw were

[1013] *The Sun*, Baltimore, Maryland, United States, 20th September 1906
[1014] *The Warren Mail*, Warren, Pennsylvania, United States, 27th September 1906
[1015] *The Daily Ardmorite*, Ardmore, I.T, United States, 24th October 1906

killed but Fireman W. W. Dortch escaped by climbing onto a raft of wreckage, upon which he floated three miles to Loring before being rescued. He was suffering badly from exposure but not otherwise hurt.

Of the 14 cars of the train one each carrying lumber, miscellaneous merchandise and hogs, an empty bunk car, one car each carrying cement and household goods and a flat car loaded with rails were destroyed, they being piled up high upon the engine, which ended up out of sight in the creek. The bridge itself was completely destroyed.

A tramp named Logue or Hogue, from Lawrence, Kansas, had a remarkable escape. He had been riding in one of the empty box cars. After the accident he walked out of the pile of wreckage in the creek, apparently unhurt, and left on a Union Pacific Railroad train for Kansas City, presumably riding blind baggage again.

By the time of the newspaper report wrecking crews from Lawrence, Argentine and Topeka, the latter with a derrick (breakdown crane) were at the scene. The pile of wreckage was standing 25 feet high in the creek and had dammed it. The water had formed an entirely new channel for itself by cutting into the bank on either side of the wreckage. The creek had swollen to about 150 feet wide. It joined the Kansas River 200 yards downstream. It was estimated that it would take two days to rebuild the bridge and re-open the line. The 15 or 20 telegraph wires across the creek at the bridge were also broken and linesmen were busy repairing them to restore communication.

De Soto station was said to have seen much confusion after the accident, with three trains being held there and passengers transferred from one train to another and then back to the first. Finally, it was decided to divert trains over the Ottawa cut-off, necessitating a change of engines. A particular problem was the need of world heavyweight boxing champion Bob Fitzsimmons to get to Atchison for an engagement that evening. He and his manager made a special fuss about their having been held up and the importance of their not being further delayed; they were said to have been 'nearly distracted'. Finally, the company laid on a special train to take them to Topeka. Otherwise trains were diverted via the Union Pacific route.[1016]

The failure of a trestle across the Fleming creek three miles east of Flemingsburg, Kentucky, United States, on the narrow-gauge Cincinnati Flemingsburg and Southeastern Railroad on Friday 10th May 1907 did not, probably, come as any great surprise, for the location was stated to have been regarded as one of the most dangerous places on the road and it had been frequently predicted that it was only a matter of time before the structure gave way. As the train entered the trestle from the east the structure began to sway, but before Engineer Card Moore could do anything it had collapsed.

Riding on the train, on a business trip to Poplar Plains and Hillsboro, was the Flemingsburg Passenger and Freight Agent, Charles Foley, who described to a porter what happened next. 'He says that he was sitting on the left-hand side of the coach, next to the coloured compartment, and as the train moved out on to the trestle at a moderate rate of speed he was looking at the trestle, which is in a curve, and when the engine, freight car and coach were all on the trestle, the end of the coach being barely on the trestle, he saw one of the struts give way, when the whole structure collapsed, the whole

[1016] *Lawrence Daily World*, Lawrence, Kansas, United States, 19th January 1907 train being turned or careening to the right, piling up in a mass of wreckage at the bottom of the ravine, the coach standing on end at an angle of 45 degrees to the rear and about the same from left to right.' The trestle was 50 feet high.

Two passengers, Harry Thomas of Covington and a second person who was either a labourer or travelling man, according to different accounts, and who was later identified as Thomas Fowles, of Morehead, were killed outright. Engineer Moore, Fireman Clarence Browning, Conductor John Moore and all but one of the 20 passengers on the train were injured, several it was thought fatally. The only passenger to escape injury was the 16-month-old daughter of Harry Thomas, who had been sleeping in her mother's lap.

In a short time after the wreck three hundred people had arrived at the scene, the women acting as nurses. Physicians attended from Elizaville, Ewings, Mt Carmel and Poplar Plains. Ironically the train involved was reported to have been a funeral train, carrying the body of a brother of Mr Charles Rankin, who was one of the passengers possibly fatally injured, to Hillsboro for internment. The corpse was mangled almost beyond recognition in the accident.

The *Times-Democrat*, the leading local weekly newspaper, urged a rigid investigation into the cause of the accident, asserting that the trestle had been pronounced unfit for use several times. The collapse was reportedly due to rotten timbers.[1017]

The next accident to be considered was the first one of the 20th century to arise from a faulty design. Seventy-five lives were lost, but none of them were train passengers, for the bridge collapsed while it was still under construction. A train was involved, however, inasmuch as a locomotive and nine trucks loaded with iron were on the structure when it fell, so it has been thought worthy of inclusion here.

The accident in question occurred on Thursday 29th August 1907 and was the collapse of the Quebec Bridge across the St Lawrence River, from Sainte-Foy to Lévis, in Quebec province, Canada.

Construction of the bridge was begun, after years of proposals, in October 1900. At this point the St Lawrence River was two miles wide, with waters which were 180 feet deep at its middle. The river could run as fast as 8 or 9 miles an hour and had a tidal range of 18 feet. During the winter months it froze, and ice could pile up as high as 50 ft. Bridging it was, therefore, not an easy task. The final design chosen was a cantilever structure, as recommended by noted French engineer Gustave Eiffel. It was to be the longest cantilever bridge in the world at that time. The bridge was to have massive granite and concrete piers with a centre span 1,800 feet long, standing 150 ft above the river.[1018]

Construction of the bridge was by the Phoenix Bridge Company, who were given a deadline of 31st December 1908 to complete the structure. Should they fail to do so they were to be fined $5,000 a month until it was completed. As the work of construction progressed workers and supervisors noticed deflections in some of the cords, and such deflections continued to be reported. Some of the engineers were unconcerned about the reports, others were of the opinion that the deflections had been present when the parts

1017 *Public Daily Ledger*, Maysville, Kentucky, United States, 11th and 13th May 1907.
1018 Webpage – Failure Case Studies: The collapse of the Quebec Bridge, 1907

left the factory, though the manufacturer insisted they had not been. A young civil engineer, Norman McLure, believed that the deflections had occurred after the cords were installed, but his superior, consulting engineer Theodore Cooper, disagreed. The cords were placed under observation, however, and the defection of cord A9L was noted to have increased from ¾ inch to 2¼ inches in less than two weeks. The opposite cord, A9R, was bent in the same direction. There was growing concern, and an erection foreman halted the work on 27th August while McLure and Cooper, at the time in New York, discussed the problem. Then, after a reassurance from another civil engineer, Edward Hoare, on the following day, the erection foreman ordered work resumed, due to fears that otherwise too many workers might leave the job and work have to halt for the season. A telephone message from the Phoenixville, Pennsylvania office of the Phoenix Bridge Company confirmed that it was safe to resume work.

In New York Messrs Cooper and McLure, unaware that work on the bridge had restarted, wired Phoenixville at 1.15 p.m. on 29th August with instructions not to add any more weight to the bridge until the matter had been further considered. The message was ignored pending the arrival of the chief engineer. He arrived at about 3 p.m. but did not act pending a meeting with McLure. McLure reached Phoenixville at about 5.15 p.m. and after a brief discussion it was decided to leave the matter until the next morning. Fifteen minutes later, at 5.30 p.m. the bridge collapsed. Time had run out.[1019]

Half of the bridge, from the south shore to mid-river, crumpled up and fell into the water. Ninety men were at work on it at the time, and the whistle had just blown for them to cease work for the day when they heard a grinding sound. They turned to see what was happening, and the next moment the cry was raised, "The bridge is falling."

'The men made a rush shorewards, but the distance was too great for them to escape. The fallen section of the bridge dragged others after it, the slipping girders and cables booming like a crash of artillery. Terror led fleetness to the feet of the frightened workmen as they sped shorewards, but only a few of them reached safety before the last piece of the work of the south side was dragged into the river. Nearer shore the wreckage of the bridge did not go below the surface of the water, and the workmen who remained above were rescued and taken to the hospital at Lévis.

'The steamer *Glenmount* had just cleared the bridge when the first section went. The water thrown up by the debris came clean over the bridge of the steamer. The captain at once lowered boats. The small boats plied backwards and forwards over the sunken wreckage for half an hour, but there was no sign of life. The twisted iron and steel held its victims in a terrible deathgrip. A few floating tubes and the broken strands of the bridge toward the north shore were the only signs that anything unusual had happened. There was not a ripple on the smooth surface of the St Lawrence as it swept along toward the gulf.

'All the men drowned were employed by the Phoenixville bridge company and the sub-contractors of Montreal.

'At 10 o'clock tonight sixteen bodies had been picked up, and of the eight men in the Lévis hospital two are not expected to live throughout the night.

'At the moment of collapse a locomotive and nine loads of iron had gone on the

[1019] Ibid.

bridge. They went down with the wreck.' [1020]

In view of the magnitude of the catastrophe the Governor General of Canada appointed a Royal Commission to examine its causes. Its major findings were:

• The collapse resulted from the failure of the lower chords in the anchor arm. The failure of these chords was due to their defective design.

• We do not consider that the specifications for the work were satisfactory or sufficient, the unit stresses in particular being higher than any established by past practice. The specifications were accepted without protest by all interested.

• A grave error was made in assuming the dead load for the calculations at too low a value and not afterwards revising the assumption. This error was of sufficient magnitude to have required the condemnation of the bridge, even if the details of the lower chords had been of sufficient strength, because, if the bridge had been completed as designed, the actual stresses would have been considerably greater than those permitted by the specifications. This erroneous assumption was made by Mr Szlapka {Chief Designing Engineer for the Phoenix Bridge Company} and accepted by Mr Cooper, and tended to hasten the disaster.

• The loss of life on August 29, 1907, might have been prevented by the exercise of better judgement on the part of those in responsible charge of the work for the Quebec Bridge and Railway Company and for the Phoenix Bridge Company.

• The failure on the part of the Quebec Bridge and Railway Company to appoint an experienced bridge engineer to the position of chief engineer was a mistake. This resulted in a loose and inefficient supervision of all parts of the work on the part of the Quebec Bridge and Railway Company.

• The work done by the Phoenix Bridge Company in making the detail drawings and in planning and carrying out the erection, and by the Phoenix Iron Company in fabricating the material was good, and the steel used was of good quality. The serious defects were fundamental errors in design.

• The professional knowledge of the present day concerning the action of steel columns under load is not sufficient to enable engineers to economically design such structures as the Quebec Bridge. A bridge of the adopted span that will be unquestionably safe can be built, but in the present state of professional knowledge a considerably larger amount of metal would have to be used than might be required if our knowledge were more exact.

A recent analysis of the accident has concluded that 'At Quebec the greatest bridge in the world was under construction in 1907 under severe financial constraints, with inadequate funds provided for either engineering work or the bridge itself. These constraints had delayed engineering analysis and led to adoption of unconservative specifications. When the miscalculation of the dead load was identified, the measures taken to re-analyse the structure were not adequate. On this project, virtually every conflict between safety and economy was resolved in favour of economy. Most of the poor engineering decisions were made by the prominent consulting engineer, Theodore Cooper. Cooper's reluctance to travel to the site, based on his poor health, led to

---

[1020] *The Daily Colonist*, Victoria, B.C., Canada, 30th August 1907 confusion about responsibility and site supervision. When skilled ironworkers observed the growing deflections, indicating a gradual collapse of the structure, a confident site supervisor might have realized the gravity of the situation and halted construction. However, the engineers on the site lacked the confidence and the authority to contradict Cooper's judgements.' [1021]

A further disaster occurred in 1916 during the rebuilding of the bridge, costing the lives of 13 workmen; this time no train was involved. The bridge finally came into use in 1917.

On Monday 25th November 1907 a bridge over the Riuedecañas stream, between Cambrils and Hospítalet Spain, on the Valencia – Tarragona line of the Caminos de Hierro del Norte de España (Nord Railway), collapsed beneath a Valencia – Barcelona express train. The express which had left Tarragona at 10.51 with 86 passengers, meet with disaster at km 253. The bridge had stone piers and 12 steel spans and was approximately five metres high. After the engine, leading van, a third-class and a first-class carriage had passed over the span the structure gave way. The next first-class carriage fell into the stream and dragged the following carriage, also first-class, with it, the two ending up in a splintered pile of wood, iron and bodies. The following third-class and postal composite carriage ended up suspended from the bridge and the rear van remained on the track. The engine, off the track, ran 50 metres beyond the bridge, embedded in the soil along with the leading van. The third vehicle which cleared the bridge went down the embankment. The ruins of the carriages which had gone down into the river presented some horrific sights, including the body of a little girl of six or eight, and two middle-aged men, clearly well-off, whose bodies were so locked together that they could not be separated.

The bridge pier which failed had been reinforced with wooden beams, having been in a poor condition for some time. It was common knowledge that the bridge was unsafe; even local tramps who sheltered beneath the other arches of the bridge at night would not dare to take refuge under the portion of it which collapsed. At 3.45 p.m. a relief train arrived at the capital (Tarragona) with 16 injured victims. It was met at the station by little groups of people who showed by their expressions the anxiety which they were suffering, and members of the Civil Guard strove to maintain order. When the train pulled in Red Cross workers began transferring the injured from the carriages to ambulances, assisted by railwaymen and local officials. In addition to the passengers, Conductor Vincente Guillen, Sub-conductor (?) Francisco Gallego, Driver Chirivella, Fireman Reselló and Inspector Juan Oliver were injured. Those killed included Don Gustavo Trost, director of the Barcelona branch of the Crédit Lyonnais bank. The bodies of 19 victims had been recovered at the time of the first reports on the accident.

The unanimous opinion was that the accident due to poor state of the bridge.[1022]

Francesco Gallego described his experience of the accident. "The express in which I was a riding had left Barcelona just after 8 a.m. and consisted of a locomotive, a leading van in which I was riding, three first-class carriages, one third-class, a postal van and a rear van. Up to the point where the accident happened to journey was made without

[1021] Webpage – Failure Case Studies: The collapse of the Quebec Bridge, 1907
[1022] *La Vanguardia*, Barcelona, Spain, 27th November 1907

incident or delay. There were a good number of passengers. The train was travelling at about 60 kilometres an hour. When the accident happened, I was in the leading van. There was a severe jolt and I tried to find out what was happening. At that moment the engine, the van I was in and the following carriage were split from the rest of the train with such force that I was thrown from where I was sat completely out of the van and down to the stream below. I was knocked out. When I regained consciousness it all seemed like a dream and I was very confused. They wanted to take me to hospital in Tarragona, but I asked to go to Valencia so as to be with my family."

The engine involved in the accident, No. 1479, had been involved in another accident the previous 23rd November. The director general of the Nord company, the engineer and other officials visited the scene of the accident. It was stated that a speed restriction which had imposed over the bridge because of its condition had been lifted to avoid delays to trains. The number of dead had now increased by the next day to 27, with 21 injured.[1023]

A couple of days later Driver Chirivella, who had suffered severe burns and other injuries, became the 28th fatality.[1024]

Atchison Topeka and Santa Fe Railroad passenger train No. 3, the *California Limited*, was wrecked two and a half miles east of Hardy, Arizona Territory, United States, at 8.50 p.m. on Sunday 28th June 1908. A wooden bridge about 70 feet long and six feet deep across a small arroyo had caught fire and collapsed beneath the train. Due to potential track problems the train was not running at more than 30 miles an hour as it reached the bridge. According to some of the passengers Engineer Clarence Currin had given a short whistle immediately before the train reached the bridge, other passengers heard nothing.

As a result of the accident the engine was completely buried beneath its tender and the mail car. The mail car and the combination baggage and smoking car were demolished, and the dining car was badly broken up. Engineer Currin and Fireman Benjamin Thomas were killed outright, as was one passenger, Charles Partridge, of Redlands, California, who by misfortune had gone to the baggage car a short time earlier to feed his two dogs. The animals themselves were unhurt and the baggagemaster escaped with a few bruises. Two mail clerks were bruised and shaken and a conductor, two cooks and six waiters in the dining car were painfully but not fatally hurt. Finally, 40 or more passengers suffered cuts and bruises.

Immediately upon receipt of news of the accident at Winslow a relief train was made up there with medical supplies and Drs Brown, Sprankle and Hathaway. Wrecking trains with men and material were also despatched from Winslow and Gallup. The body of Mr Partridge was returned to Winslow together with the injured passengers.

Investigation showed that the bridge had been entirely consumed, leading to the suspicion that arson had been committed, and the matter was placed in the hands of the secret service department. Adding weight to the theory was the fact that the previous train was an eastbound one which had left Winslow about 1 p.m., and it was thought that if a spark from the train had set the bridge alight the fire would have been noticed by the

[1023] Ibid, 28th November 1907
[1024] Ibid, 1st December 1907

local section crew. A shoo-fly was built around the wrecked portion of the train. It was said that both Engineer Currin and Fireman Thomas had been considering giving up railroading, having saved up sufficient money to purchase farms.[1025]

Just over a month after the Hardy wreck a similar accident happened to eastbound passenger train No. 2, the North Coast Limited, of the Northern Pacific Railroad about five miles east of Allard, Montana, United States. On Saturday 10th August 1908, the train was travelling at about 50 miles an hour and had just rounded a curve when it ran onto a burned bridge over Glendive Creek. The locomotive went into the creek and toppled over, and the baggage, mail and smoking cars piled on top of it. Burning ties set fire to the train but the blaze was soon extinguished with water from the creek. The day coach also left the track and ended up tilted on end; its occupants were only slightly hurt.

Fireman B. A. Matthews and an unknown hobo were killed and Engineer C. D. Litch, Express messenger J. C. Wilson and 20 or 30 passengers were injured. The express messenger was badly scalded, and it was feared that he would die.

One of the passengers, Nate Kateley, of Glendive, described the accident. "The first we knew was a sudden stop and jerk, and then things began to crash, and everything seemed to be in the air, including the car seats and the people. Then quiet suddenly ensued. It seemed to me that the smoker turned bottom side up and then righted itself again, slewing to one side and lying across the track. We went outside, through the end door, and then we saw the cars on fire. We went in and helped the cripples out. The baggage car was also on fire but the baggage and express were removed and the flames were easily extinguished.'

A relief train was speedily sent from Glendive, taking all the physicians in the town to the scene, and it returned with the injured, the worst hurt being taken to the Grace hospital. A wrecking train was also sent to the scene, but it was expected to be at least 12 hours before the line could be brought back into use. Westbound train No. 3 was a few minutes late, or a double accident might have resulted.[1026]

On the evening of Thursday 19th August 1909, a Ballarat – Maryborough mixed passenger and goods train enjoyed what was described as a miraculous escape at McCallum's Creek, seven miles from Clunes, Victoria, Australia. The creek was crossed by a bridge of seven metal girder spans supported by six bluestone piers, each eight yards apart, and was 38 ft high. The gulch through which the creek flowed was 150 or 200 yards wide, the opening having been reduced by the throwing up of an embankment on the Talbot side of the creek. Swollen by excessive rainfall, the waters of the creek had accumulated on one side of the bridge and had caused the subsidence of one of the piers on the Talbot (Maryborough) side of the bridge.

The mixed train had left Ballarat for Maryborough just after 8 p.m., in charge of Guard Harry Walsh, with Driver J. Wright and Fireman McInerey. It consisted of an 'R' class locomotive, eight goods waggons, two passenger carriages and a guard's van. Driver Wright had been cautioned to go carefully as there were fears that the heavy rains

1025 *The Winslow Mail*, Winslow, A.T., United States, 4th July 1908

1026 *The Anaconda Standard*, Anaconda, Montana, United States, 11th August 1908 might have damaged the track, and the train approached the bridge slowly, but steam was then applied to take it up the climb on the Talbot side of the bridge. The night was very dark, with what were described as drifting showers of rain. It was fortunate that the momentum of the train had increased, for the extra weight of it caused the bridge to give way, but the momentum was enough to carry the engine to the other side of the gap, where it came to rest embedded in the embankment. The following goods waggons piled up in the creek and the leading carriage went over the precipice, coming to rest leaning down at an acute angle, its front end resting on the wreckage of the waggons below.

Guard Walsh had applied the Westinghouse brake as soon as he had realised that something was wrong. Driver Wright and Fireman McInerey were thrown from their locomotive into the creek during the bridge collapse but escaped with cuts, bruises and shock. One of the eight passengers in the second carriage suffered a broken arm when he was thrown about it but that was the limit of the casualties. There had been only one passenger in the first carriage, a Mrs Matthews, of Burnie, Tasmania, who was 'in a delicate condition' (pregnant).[1027] She was rescued from her rather precarious position by an inspector. The passengers in the second coach were able to scramble out of its rear end. Guard Walsh made his way back to Clunes to raise the alarm.[1028]

The recovery of the engine and rolling stock and repair of the bridge and track was a major exercise, not helped by further torrential rain, and almost a month later the work was still in progress.[1029]

In contrast to the Quebec Bridge disaster, which occurred on the line before it had even been opened, an accident near Whoop-Up, Alberta, Canada, on Monday 31st January 1910, happened on a line which had been taken out of use. The original Canadian Pacific Railway line between Lethbridge and Fort MacLeod had become a bottle-neck with only low speeds possible over it and passenger trains taking two hours to cover the 38½ miles between the two places, with freight trains taking three hours. Consequently, a new line was built, five miles shorter, with a viaduct across the Old Man River, and opened on 1st November 1909.[1030]

With the new line opened, work began on dismantling the structures on the old line, and it was a work train engaged in this task which met with the accident. Just west of Whoop-Up were two wooden bridges across coulees (deep ravines) about 90 feet deep separated by a narrow steep ridge. A work train was standing on the bridges, with the engine in its middle, and the 'Galician' (Eastern European?) workmen were busy tearing down the timbers of the west bridge. Several of the joists had been removed when without warning the structure collapsed, and train and workmen, all but one of whom had been on the west bridge at the time, fell into the ravine. The east bridge also collapsed at the same time. One man, named J. Robski, was killed outright, 14, including Foreman J. Smith and Engineer George Munro, were seriously injured, and Brakeman W. E. Smith and Roadmaster Charles. Bjork escaped with slight injuries.

'The sight that met the first arrivals on the relief train was a most appalling one. A

1027 *The Horsham Times*, Victoria, Australia, 24th August 1909
1028 *Australian Railway Disasters* by Kenn Pearce, pgs. 50-52
1029 *Geelong Advertiser*, Geelong, Victoria, Australia, 13th September 1909
1030 Webpage – Atlas of Alberta Railways: The Crow's Nest Line

heap of broken, twisted timbers filled both coulees, while here and there prostrate bodies were scattered about the ground, where small fires had been built beside them and some efforts made to make their suffering less intense. The pain that some of the men suffered must have been intense. The moans of the injured before the doctors were able to give them relief were pitiable. The work of removing the injured men to the train was one of the most extreme difficulty. A light fall of snow and the steep sides of the coulees made the ascent a difficult one to make at best. Most of them had to be taken up the side of one coulee and then down and across the steep slippery sides of the other. The few men who arrived on the scene at first were soon supplemented by many more, and the men were removed with all possible speed to the train and brought into the city, the train arriving here {Lethbridge} about six o'clock.'

General Superintendent A. Price of the Canadian Pacific went to the scene and declared to a newspaper that the collapse was purely accidental and was probably caused by the engineer applying the air brake, resulting in a sudden jar which had proved too much for the weakened bridge.[1031]

A second workman died of his injuries the day after the accident, a third man must have succumbed shortly afterwards and on 4th February Engineer Munro became the fourth fatality.[1032]

An inquest was quickly held. On its first day Roadmaster Bjork, in charge of the work of dismantling the bridges, admitted that he had no theoretical or practical knowledge of bridge construction and no knowledge of the strain placed upon a bridge when one of its abutments was removed. The bridge and building inspector had given him instructions as to how to proceed. Six piers at the west end of the bridge had been dismantled. The crew had just come back to work at 1 p.m. after the dinner break when the collapse occurred. He had told the engineer to go out onto the bridge very carefully. He had not examined the bridge before work began.

Fireman Fred Summers said that his engine had been standing still when the collapse had occurred without warning. There had been no sign of any weakness of the bridge. The engine weighed about 80 or 85 tons.[1033]

At the resumed inquest Bridge and Building Master Robert Sinclair stated that he had given Roadmaster Bjork general and detailed instructions as to how to proceed. He had not left any written instructions for Bjork but considered that with his experience Bjork should have been capable of doing the job. He had always found Bjork to be a reliable man, who obeyed his orders.

Conductor McKillop stated that he had been in charge of the work train. He said that as far as he had been aware the bridge had been intact when his train ran onto it. They had worked on the bridge for about four hours in the forenoon. The train had not been moved off the bridge when they went to dinner. The air hoist had been left on the bridge about the middle. When they returned the air hoist had been coupled up and they had moved towards the west end of the bridge.

Brakeman Chapman said that he had been riding on the first flat car of the train and

---

1031 *The Lethbridge Daily Herald*, Lethbridge, Alberta, Canada, 1st February 1910
1032 Ibid, 5th February 1910
1033 Ibid. 2nd February 1910

had been speaking to the fireman when the bridge gave way.

Roadmaster Bjork, recalled, stated that he had been in charge of two gangs, a bridge gang and an extra gang, but the bridge gang, under Hansen, had been withdrawn a week or so before the accident.

A Mr Seaman stated that he had been a roadmaster for several years. The bridge was what was known as a "bent" bridge. He had dismantled such bridges before. He did not think that the bridge would have gone down if the bolts had been left in the rails. The appearance of the rails showed that they had been removed. He did not think that the packing bolts had been in either at the time. Mr Sinclair's instructions he thought had been sufficient. He would not let an engine onto a bridge with one end disconnected and had never seen that done. He did not consider that Bjork had been a competent man for the job. In his opinion the bridge had collapsed because the bolts had been taken out of the rails and the engine had been taken onto the bridge when it was in that condition.

The jury returned a verdict that Robiski had come to his death by being crushed between two flat cars of a train on the 7.9 bridge and that the structure had failed while its demolition was under the supervision of Roadmaster Bjork, who had been totally incompetent to do the job. They recommended that a further investigation be made by the proper authorities.[1034]

Six months after the catastrophe near Green Mountain, Iowa, in March 1910 it was again a Chicago Rock Island and Pacific Railroad which was in trouble, on Friday 23rd September 1910, two miles east of Clayton, Kansas, United States, this time on its own line. Westbound passenger train No. 27, from Kansas City to Denver, was running at full speed when it ran onto a small bridge which had been washed away by a cloudburst. The engine and mail car went down into 20 feet of water and the smoker was almost telescoped by the following chair car, many of the passengers in these two cars being killed instantly. Others were carried into the raging water and it was many hours before their bodies could be recovered. The day coach and Pullman car were thrown from their chairs and berths by the shock, but these vehicles did not go down into the gap and the passengers in them were soon engaged in the rescue work or else hurried to Clayton to give notice of the accident. From Clayton news of the accident was telegraphed to divisional headquarters, and within half an hour a wreck train, physicians and nurses were hurrying to the scene from Norton, Colby, Goodland and Phillipsburg.

The train had been running almost on time and was proceeding along a straight stretch of track. The first warning of danger which Engineer Pickinbaugh must have had must have been when he found that he was running into water. Reversing his engine had not prevented the engine and two leading vehicles going into the gap, and the smoker tilting up and being smashed by the chair car, but by sticking to his post, putting his engine into backward gear and applying the brakes, rather than jumping, he was credited with saving the lives of a hundred or more passengers in the rear of the train. The three Pullman cars remained on the rails. The engineer was one of those who lost their lives.

By the following day the bodies of 16 victims had been recovered. In addition to Engineer Pickinbaugh Fireman William Mills, Conductor J. W. Usher and Baggageman

[1034] Ibid, 7th February 1910

A. V. Hauffman were killed. Mail clerk F. L. Olney and about a dozen passengers were injured, some scalded by steam which rose in clouds from the engine when it entered the water.[1035]

The 30-foot-high bridge which had failed crossed an arroyo which as noted previously is a normally dry stream bed which only fills and flows occasionally after heavy rain; flash flooding after thunderstorms is common in deserts or semi-desert areas where arroyos occur. In the present instance a tremendous amount of heavy rain had fallen in the area which fed the arroyo, transforming it into a raging torrent which carried the bridge away before the train arrived. The accident occurred just after 1 a.m., while the storm was still raging.[1036]

A fire-damaged trestle caused another serious accident in the United States on the evening of Sunday 28th May 1911. A heavy double-headed train, loaded with coal from the Clinchfield mines in Virginia, had just passed onto the Seaboard Air Line Railroad at Bostic, North Carolina, and was making its way eastwards to the coast, when only two miles into its journey it reached a trestle over Watkin's Creek at 5.15 p.m. As often happened the crew of the leading engine did not discover that the trestle was on fire until too late. The structure collapsed beneath the two engines and they plunged 40 feet to the ground below, 17 of the 29 steel cars of the train falling on top of them. Engineer R. M. Green on the pilot engine and Engineer J. M. Lindsey and Fireman Ray Dooley on the train engine were all killed outright. Fireman Early Lewis of the pilot engine, who had jumped at the last moment, and Brakeman Lon Neely, who had been riding on the tender of the train engine, were seriously and perhaps fatally injured, but Conductor Captain Frank Howell, riding in the caboose at the rear of the train, escaped unhurt. A wrecking train was despatched from Monroe with physicians. The cause of the fire was not stated.[1037]

On Thursday 23rd November 1911 France suffered a major bridge disaster near Montrevil-Bellay. The accident came only four years after another bridge disaster at Ponts-de-Cé (described in the next chapter) and hence caused something of a public outcry, though the accidents had quite different causes.

There had been bad weather in Western Europe for some days, with heavy rainfall. The rivers of the Loire basin were all in flood, and the River Thouet had risen 1.2 metres on the day preceding the accident to its highest ever recorded level. It was feared that the flooding would affect the safety of the railway and the bridge across the River Thouet near Montrevil-Bellay, which consisted of two metal spans, each 54 metres in length, supported by a central pier, with almost three kilometres of approach embankments. After an inspection of the line a decision was made to suspend freight traffic and to pilot the passenger trains over the affected area.

On the day of the accident omnibus train No. 405 left Angers for Poitiers at 5.02 a.m. It consisted of two class 030 tender engines, two vans, a first-class, two third-class and

[1035] *Lincoln Daily Star*, Lincoln, Nebraska, United States, 24th September 1910
[1036] *Saint Anne Record*, Saint Anne, Illinois, United States, 30th September 1910
[1037] *The Charlotte News*, Charlotte, North Carolina, United States, 29th May 1911

two second-class coaches and another two vans. Two of the vans were carrying livestock. At Vaudelnay-Rillé the train was joined by three track workers, an inspector acting as a pilotman and two fitters. The train then proceeded along the embankment at about six kilometres an hour but was going rather faster as it reached the bridge. It had almost crossed the structure when, at 7.15 a.m. the central pier collapsed, taking the adjoining spans and most of the train with it. The two front vans and the first-class coach were swept some hundreds of metres down the river, whilst the two engines and their tenders and the leading third-class coach were caught in the remains of the bridge spans and totally submerged. The following third-class coach was half submerged in the river as it hung off the end of the bridge.

The coupling behind this coach snapped and the two second-class coaches and the two rear vans remained on the line. Conductor Rabier was in the rear of the train and the only railwayman to emerge from the accident unscathed; he immediately set off to Vaudelnay-Rillé for assistance.

Thirteen passengers from the second third-class coach were able to climb up onto the portion which was out of the water but were stuck there. Rescuers on the bank threw a rope to the trapped passengers and one person was thus rescued, but when the attempt was repeated the rope broke and the second passenger, a local merchant, fell into the water and was drowned. The rescue attempt using ropes was then abandoned pending the arrival of soldiers from the 6th Engineering Regiment, who reached the spot three hours later and were able to rescue the remaining passengers using a boat. Other passengers, who had been swept down the river, managed to cling to overhanging trees and were eventually rescued by local residents. The only railwayman thus saved was Driver Bernard of the pilot engine, who after catching hold of a pig which was also being carried down the river was able to climb into a tree, exhausted and with a broken leg.

Uncertainty reigned for a considerable time over how many victims had died in the disaster, with estimates of 30 or even 40 dead. A week after the accident only eight bodies had, however, been recovered. Over the next few months the remains of further victims were discovered, some several kilometres downriver, and the final number of dead was put at 14, seven of them railwaymen.

Coincidentally, at the time that the disaster came to the attention of the French parliament. M. Jénouvrier, the member for Ille-et-Villane, had been denouncing the State railway's operation of the Ouest network and calling on M. Victor Augagnier, the Minister of Transport, for an inquiry. With news of this new disaster M. Jénouvrier announced that "bridges could not be swept away without warning" and that it was "the duty of the administration" to monitor them. In response a judicial investigation, headed by M. Marchesseau, Saumur investigating judge, was set up, with three experts: a public works contractor together with mining and roads and bridges engineers. A separate commission of inquiry was set up in parallel with three members chaired by the vice-resident of the general bridge council. No misgivings had previously been entertained as to the safety of the bridge, concern being focused instead on the approach embankment. Further protection to the foundations of the bridge had also been provided in August 1910.

The cause of the failure of the bridge was not entirely clear. Driver Bernard spoke of bridge slowly collapsing beneath his engine, while Ernest Préaubert, a local geologist, announced that the ground on which the bridge had been constructed was the worst in the world for building a bridge on. These statements lent weight to the theory that the bridge had failed due to the washing away of its foundations by the floodwaters combined with the extra weight of the train. An examination of the bridge by divers three days after the accident had found, on the other hand, that the bridge pier had broken not at its base but some metres higher up, casting doubts on the soundness of the original construction. A year later the Minister of Public Works, announcing the results of the investigation that he had ordered into the disaster, stated that the collapsed pile had not been built in accordance with the plan approved by his office, but that the builders had deliberately made a less than perfect job of the construction in order to save money. He did not give any details.

Whatever the causes of the weakness of the bridge an obvious contributing factor to the disaster was the running of a heavy double-headed train. The second engine was certainly not needed just to haul the train, it was attached solely to work it back to its home shed at Thouars. By coupling it to the passenger train a timetable path was saved over working it back separately as a light engine, but at the expense of fatally straining the bridge. Given that freight trains had been suspended because of their weight the decision to attach the second engine to the omnibus train seems very strange.

The accident led to criticisms in the (presumably right-wing elements) of the Press over the state running the railways and at the same time showed what a poor condition many of the railways had been in when they were taken over by the state. The Thouars viaduct on the same line was inspected and shown to be in good condition but was reinforced anyway, while M. Clavielle, the director of the State Railways, admitted that, for example, that there were 800 cracks in Eaplet viaduct, near Rouen, but that in view of the amount of repairs to be done elsewhere it would be necessary to continue using it, with due precautions.

The Montrevil-Bellay bridge was speedily rebuilt, the new structure being especially designed to cope with flood conditions and a swollen river. The need to urgently bring the line back into use was made especially pressing by the need for the running of trains over the route in connection with planned military manoeuvres in the following September.[1038] [1039]

At least seven people were thought to have lost their lives in the failure of a Chesapeake and Ohio Railway bridge over the Guyandotte River at Guyandotte, West Virginia, United States, on New Year's Day, 1st January 1913. The bridge was being double-tracked and pending the completion of the work a temporary structure was in use.

Just before 11 a.m., and shortly after a passenger train had crossed the bridge, a heavy westbound freight train had reached the centre of the structure when there was a crash and the bridge gave way. Engineer F. E. Weber went down with the bridge and was killed but his fireman and the brakeman had alighted from the train and walked over the bridge in front of it, so escaped. Not so lucky were the men who had been working on the bridge at the time. There had been 15 of them, 11 of whom went down with it, and only five of whom were rescued. The others were caught beneath the engine and cars and drowned.

1038 Wikipedia – Accident ferroviaire de Montrevil-Bellay (in French)
1039 *The Times*, London, England., 24th and 25th November 1911

The bridge was said to be a "hoodoo" bridge, this being the third fatal accident on it. Railroad officials blamed the collapse on the bridge piling having been undermined by the high water in the river. It was declared that they had been repeatedly warned of the danger since the river rose.

The ironwork of the bridge was said to be a total loss and it was estimated that it would take several weeks to repair the structure. In the meantime it was hoped to run passenger and freight trains over the Baltimore and Ohio line, while the financial loss was put at $500,000.[1040]

An accident on the Louisville and Nashville Railroad early on Wednesday 6th August 1913 was an indication in more ways than one of how times were changing. It happened at Noonday, Georgia, United States, six miles north of Marietta. Torrential rain which had fallen for three hours before the wreck happened had washed away the bridge before a fast freight train reached it at about 1 a.m. and in the darkness Engineer B. K. Farris had no chance of spotting the danger before his train ran into the gap. According to Conductor J. L. Brenenizer, who with a brakeman was riding in a caboose at the rear, the train was running at 15 to 25 miles an hour at the time. The locomotive, six gondola cars loaded with coal, three cars loaded with hay and two of the following three cars, which were carrying automobiles, plunged into the gap, the remainder of the train remaining on the track. The cars all piled up on top of the engine and Engineer Farris, Fireman T. S. C. Channell and Brakeman R. A. Merrill, who was also riding on the engine, were all killed. A fourth fatality was Section Hand Mike Jackson. The three men on the engine, the *Atlanta Constitution* noted, were white, while Jackson was coloured, a distinction which was clearly regarded as important at this date in the American South. The section hand's body, found half buried in a mud bank, was the only one which could be quickly recovered.

One indication of the changing times was of course that even at this relatively early date, before the outbreak of the First World War, part of the load of the freight train was automobiles, doubtless new ones straight out of the factory and very likely model T Fords. Another was that the almost new bridge which had failed had been constructed not of wood, stone or steel but of reinforced concrete, still a relatively new construction material at this date.

The cause of the accident would seem to have been fairly obvious, the washing away of the foundations of the bridge by a torrent of water – in the late afternoon the waggons remaining on the track were still under three feet of water – but the newspaper stated that an investigation was to be launched into the cause of the failure.[1041]

The Highland Railway suffered a bridge accident in Scotland on Thursday 18th June 1914, which though not that serious in relative terms, with only five deaths, was the worst in the history of the company. The accident happened on the company's newer direct line from Aviemore to Inverness, north of Carr Bridge station. North of Carr Bridge station the single line crossed the Dulnan River, continued on an embankment, passed through a

[1040] *The Sun*, Baltimore, Maryland, United States, 2nd January 1913
[1041] *The Atlanta Constitution*, Atlanta, Georgia, United States, 7th August 1913

cutting and then arrived at a bridge over the Baddengorm Burn, normally a small stream which fed into the Dulnan River about 1,200 yards below the railway bridge. The bridge over the burn consisted of a single 15-foot-span brick and stone arch with substantial abutments. It had been last inspected a fortnight before the accident. Chargeman Donald Campbell had walked over the bridge between 7 and 8 on the morning of the 18th, when there had been very little water in the burn. When a storm developed at about 3 p.m. he had been about a mile and a half further north and had examined several other bridges which sometimes caused trouble, while he had never known the Baddengorm Burn to rise very high. The last train to pass over the bridge before the accident was the 11.55 a.m. up goods from Inverness to Perth, which had passed at 1.37 p.m., some time before the storm had begun.

The 11.50 a.m. passenger train from Perth to Inverness was in charge of Guard William MacCulloch, with Driver Murdo Ross and Fireman Charles Ross, and consisted of an engine and tender, a Caledonian Railway horse box and brake composite carriage, and a composite carriage, brake third-class carriage, composite carriage and brake van, the last four all being Highland Railway vehicles.

Driver Ross told the Board of Trade inquiry into the accident that as his train had approached the Baddengorm Burn bridge he had seen nothing wrong with the track, though there had been a lot of water pouring out into the strath (valley) on the down side of the line. As his engine crossed the bridge, travelling at about 20 miles an hour, he felt it sway or tip to the left. He did not apply the brakes, but the train was soon brought up anyway by their automatic application as it became divided. As soon as his engine had come to a stand he had jumped off and leaving it in charge of his fireman had walked back to the bridge, where he had found that the leading three vehicles were on the track but the fourth vehicle, the brake-third, had gone down into the burn. At this point the bridge was still intact enough for him to cross it to the Perth side but as soon as he had done so it was carried away. The third vehicle, on the north side of the bridge, and the fifth vehicle, on the south side, were upright, and the passengers in them were able to climb out, but they soon sank down due to the embankments being washed away and the third vehicle tipped over onto its side. The unlucky fourth vehicle fell over onto its side at right-angles to the burn. Its top and one side were washed away, and the rush of water made it impossible to approach it. Four passengers somehow escaped from it, the other five were drowned, [1042] though at least one of the passengers attempted to rescue them, putting his own life in peril in doing so. 'An American gentleman named Crawford rushed forward to render assistance. Suddenly the bank gave way under his feet and he was hurled down into the water and swept from view. A good swimmer, he fought doggedly for his life, and at last succeeded in grasping a floating plank. He was dashed hither and thither, being bruised and cut on the jagged rocks. At last he was swept across a flooded field and succeeded in scrambling into the branches of a tree.'

The passengers who lost their lives were:

Duncan McPherson, agent for the Northern Steam Navigation Company, Inverness.

W. J. Morrison, an Inverness auctioneer.

Mrs Waitman, of Keswick.

[1042] Board of Trade report

Miss Guthrie, niece of Lord Guthrie.

An unidentified elderly gentleman, apparently from Devonshire.[1043]

There were fears that Sir Oliver Lodge, a world-renowned physicist, had been a victim as a trunk with his name was found among the wreckage, but in fact he was safe in London, the trunk belonging to one of his daughters who was a passenger elsewhere on the train.

Forester and farmer Robert McKinnon, whose farm bordered the burn a short distance away from the bridge, told the inquiry that he had never previously, in 35 years, experienced a storm of such severity as on that day, or water more than a third up the bridge abutment, or of the bridge ever having been blocked by debris.

The inspecting officer, Lt-Col E. Druitt, noted that above the bridge the burn ran in a narrow gorge, so that when a great volume of water came down the burn the width of the bridge was not sufficient to let it pass through without damming it up, whereas beyond the bridge the water could spread itself out over a wider area of low-lying ground. It was also possible that the bridge had been blocked by debris, several trees having been washed away. The area above the bridge had thus become a reservoir and the bridge itself a sluice through which water had poured with great velocity. The result was the tearing out of the granite pitching on the downstream side of the bridge, followed by the scouring away of the gravel beneath the pitching and the undermining of the abutments and wing walls on the bridge. When the train came onto the bridge the weight of the engine had caused the abutments to sink and the arch of the bridge broke as the fourth vehicle of the train was on it. After Driver Ross had walked back over the bridge the remainder of it collapsed and the floodwater subsequently washed away the embankment and widened the gap until it was 32 feet wide at the bottom and about 100 feet wide at the top. Almost all of the bridge and its foundations were washed away.

In conclusion the inspecting officer remarked that, 'The bridge itself was of very solid construction, but the possibility of such a sudden rush of such a large volume of water through it could never have occurred to those responsible for design and erection of the bridge in question, or no doubt foundations of a different character would have been provided…I am informed by the Engineer to the Highland Railway, that the new bridge, which will be 26 feet 6 inches in span, will have its abutments and wing walls formed of mass concrete, with a concrete invert not less than 2 feet deep extending right out to the ends of the wing walls. At each end of the invert a thick curtain wall will be put down to a depth of 10 feet. On the north side the foundations will be carried down to a depth of 7 feet up to the top of the invert, and on the south side the side the new bottom will be founded back into the original solid of the hill face, and the foundation will be 2 feet 6 inches to 3 feet below the top of the invert. In this way the bed of the burn from end to end of the wing walls will be a monolithic solid mass, and in necessary, at the bottom of the down-stream end of the invert, a concrete apron extending downstream from the curtain wall, dipping down steeply under the bed of the burn, will be put in.

'It will be seen from the evidence that the bridges on this section of the line are carefully watched and examined after heavy floods, and that this bridge had never previously shown the slightest sign of being undermined by the heaviest flood since its erection in 1893. I consider, therefore, that the collapse of the bridge was due to no want

---

[1043] *The Times*, London, England, 19th, 20th and 22nd June 1914 of attention of care on the part of those responsible for its maintenance, but to an event that could not have been foreseen and guarded against.' [1044]

An accident at Lizy-sur-Ourcq, France, on Wednesday 16th September 1914, was an indirect result of the First World War, which had broken out a month and a half earlier. In their opening campaign the Germans had advanced far into France, until they were within 15 kilometres of Paris on 5th September. In order to impede their advance bridges had been blown up, including one on the railway at Lizy-sur-Ourcq. The Great Retreat, as it was known, was finally stemmed between 6th and 12th September by an Allied counter attack known in English as the First Battle of the Marne and in French as Le Miracle de la Marne.

On the day of the accident a Red Cross train carrying wounded soldiers was by mistake routed onto the line crossing the destroyed bridge, and its engine and at least two coaches fell into the River Marne. A report in a French newspaper, repeated the next day in a British newspaper, assured its readers that most of the wounded in the affected carriages had, however, been rescued.[1045] [1046] The seriousness of the accident was, however, probably minimised for morale purposes for according to official railway archives there were in fact 40 dead.

[1044] Board of Trade report

[1045] *Le Petit Journal*, Paris, France, 17th September 1914.

[1046] *The Times*, London, England, 18th September 1914

CHAPTER 9

OPENING BRIDGE ACCIDENTS

Accidents involving opening bridges appear to have occurred in only a handful of countries, with the majority occurring in the United States. It seems that most countries managed to a large degree to avoid building such bridges, which from a railway point of view are a constant source of expense and worry. Shaw states that the early railroads on the eastern seaboard of the United States were forced by opponents of the new form of transport to build such bridges over even insignificant waterways where they were never really needed.[1047]

Brake failure was the alleged cause of a drawbridge accident to a fast freight train at Elizabeth, New Jersey, United States, on the Long Branch division of the Central Railroad of New Jersey on Tuesday 12th August 1902. Eight cars, including one loaded with dynamite, piled up on top of the engine which ended up in mud 20 feet below the surface. The protecting signals had been at danger but Engineer Beany stated that his brakes had failed to work and so he had been unable to stop his train.

Fireman Webster and Brakeman Patrick Mansfield lost their lives. Engineer Beany, who had jumped, suffered a broken leg and internal injuries.

Traffic north and south to Long Branch and North Grove was delayed five hours, trains having to be diverted over the Lehigh Valley Railroad to Roselle and thence via the Central of New Jersey main line. Passengers were reportedly ferried over the Elizabeth River in small boats.

The salvage of the wreck was carried out very carefully in view of the dynamite.

The first passenger train of the day from Leiden to Woerden in the Netherlands on the morning of Tuesday 23rd December 1902 was involved in an accident when train No. 291 fell through an open bridge over the Dubbele Wiericke between Bodegraven and

[1047] *A History of Railroad Accidents, Safety Precautions and Operating Practices* by Robert B. Shaw, pg. 187

Woerden. Head Conductor J. Priem was trapped in the wreckage of the partially wrecked baggage van and lost his life. His colleague, Conductor J. F. Hartman, was more or less badly hurt and the driver and a trainee driver were also injured. The eight passengers were unhurt.

A combination of circumstances had led to the accident. Firstly, the bridge tender had overslept. He explained later that his alarm clock had failed to go off. His duty should have begun at 3.12 a.m. when he had to check the line, the bridge and his signals. At 6.02 a.m., 20 minutes before the booked departure time of the first train from Bodegaven, he had to exchange the so-called bridge signal with Bodegaven. When the Bodegaven official attempted to contact him there was of course no reply, but the signal was also received by a bridge tender at another bridge nearer Woerden, who did reply, and the Bodegaven official, thinking that it was the Dubbele Wiericke bridge tender who was answering his call, assumed that the bridge was properly closed and advised his stationmaster that train No. 291 could proceed.

The greater part of the train crossed the seven-metre gap in the line, but coach AB 2157 stuck in the gap and the rear two vehicles stopped short of the opening. The engine separated from its tender, shot forward 20 metres and overturned. The tender turned upside down and the following baggage van and two coaches smashed into it and one another and were badly damaged. The two rear vehicles of the train were little damaged. The bridge itself also suffered considerably in the accident.

After the accident a wooden footbridge was laid across the waterway to enable passengers to make their way between trains on either side of the gap. A temporary bridge was brought into use after only four days and a new permanent bridge in July 1903.

The accident served to highlight a difference in the manner in which the failure of a bridge tender to report for duty was dealt with, depending upon whether the bridge was situated at a station or away from a station. In the former instance the conductor of the train was to take the bridgetender's place but in the case of bridges away from stations this was not the case and it emerged that several times when the Dubbele Wiericke bridge tender had previously failed to answer the bridge signal at the proper time, the stationmaster had nevertheless allowed a train to proceed. The accident could thus have happened even without the mistake by the telegraphist. By an order of 14th July 1904 the regulations were amended to require all bridges to be so manned before a train was allowed to cross them.[1048]

At 8.45 p.m. on Monday 8th August 1904 the locomotive and nine cars of eastbound Louisville Henderson and St Louis Railroad freight train No. 64 went through an open swing bridge at Spottsville, Kentucky, United States. Engineer Walter Riedel, Fireman William Lishen and an unknown number of tramps went down with the train into the Green River and were killed. Conductor D. C. Benton and Brakemen William Newton and Orchard Pullman were able to run to the rear of the train and escaped. By Wednesday the bodies of two tramps had been recovered, together with those of the engineer and fireman. Neither of these tramps had been identified but the man was stated to be an umbrella mender. (How was it known that he was an umbrella mender if he had not been

---

[1048] *Spoorwegongevallen in Nederland 1839-1993* by R.T. Jongerius, pgs. 82-4 (in Dutch)

identified?) The other tramp turned out to be a woman, dressed in men's clothing. She was described as a 'high-diver'. What a high-diver was is unclear over a century later, the only meaning given in a dictionary of American slang which fits at all is 'a pickpocket', but that does not seem all that appropriate. The umbrella mender was carrying some of the woman's hair in his pocket, so they were supposed to have been travelling together.

The bridge had been opened for the steamer *Crescent City*. Just before the boat reached the bridge its pilot saw that the train was not going to stop, and the steamer was halted just short of the bridge, the train then falling just in front of it. A searchlight on the steamer was turned on to search for any victims but without success.

How the accident had happened was unclear. The regulations dictated that all trains must come to a stand before crossing the bridge, being then signalled forward by the bridgetender when it was safe to do so. According to the bridgetender the train had not stopped at all, and his signal at danger, a red light, had not been cleared to white. Why Engineer Riedel had failed to pull up was unknown. The only theory which could be advanced was that there had been a failure of the air brakes.[1049] [1050]

Engineer Elijah Pate, Fireman C. B. Bates and Extra Engineer John Dobbins lost their lives when train No. 13 of the Central of Georgia Railway from Albany to Dothan and Hartford met with an accident at the Chattahoochee River Drawbridge, near Columbia, Alabama, on the evening of Monday 20th February 1905. The train, in charge of Conductor C. C. Wightman, had left Albany a little late, and with rails which were slippery due to rain it was found impossible to regain the time.

The drawbridge was protected by signals a mile distant, half a mile distant, thus three in all. Engineer Pate was known to be a careful engineer and had crossed the bridge many times before. According to a passenger on the train Engineer Pate slowed up and stopped in obedience to the signals, then gave two shorts blasts on his whistle, this being the normal signal given in response to a signal from the bridge tender to proceed and started the train forward again. The bridge tender, on the other hand, denied having given any such signal, and stated instead that he had signalled to the engineer to stop.

The draw was open and the next moment the engine went over the edge of it. As it fell the brake pipe broke, the brakes were applied, and the coaches remained on the line. The leading baggage, mail and express car went partly over the bridge and was held there at an angle of 45 degrees by the couplings. The locomotive alone plunged into the water. Mail Clerk Eugene Hightower, Baggagemaster Oscar Baldwin and Express Messenger W. Parker were all badly shaken up but not otherwise hurt and managed to scramble out of the baggage car unaided.

Train No. 14 had crossed the drawbridge just a short time before, meeting train No. 13 at Hilton, on the far side of the river. Engineer J. M. Stewart of that train said afterwards that he had seen the *Queen City* steamer in the river waiting for the bridge to be opened and that he had had a premonition that something might happen, to the extent that he had called to the crew of train No. 13, "Watch out for the drawbridge." He said that he had not been surprised when he had learnt of the accident.

[1049] *The Breckenridge News*, Cloverport, Kentucky, United States, 10th August

[1050] *Hopkinsville Kentuckian*, Hopkinsville, Kentucky, United States, 16th August 1905

The last act of Engineer Pate had been to sound the danger whistle as his engine fell from the bridge. Fire, caused either by an exploding oil lamp or an overturned stove in the baggage car, set the three coaches which remained on the bridge on fire and all the mail, baggage and express contents were consumed.

The *Queen City* had only just cleared the draw, which was only open 15 feet and would have been closed in another few seconds. The crew of the steamer witnessed the accident. Lifeboats were lowered, searchlights brought into use and a search made for survivors from the engine crew, but without success.

Engineer Dobbins had ridden in a first-class coach over the 60 miles from Albany to Hilton and had been in the cab only a few minutes when the accident happened.

It was impossible to account for Engineer Pate's mistake.[1051]

A drawbridge accident in early April 1905 was prevented by a derail. The location was the San Joaquin drawbridge, 14 miles south of Stockton, California, United States, on the Southern Pacific Railroad. At dawn on Saturday 8th April 1905 a southbound freight train of about 35 cars was approaching the drawbridge at a fair speed when according to Engineer White the semaphore signal protecting the bridge was thrown up in his face by the bridge tender, leaving him no time to stop his train before the engine and four cars left the track, 100 yards of which was torn up. The engineer and Fireman Cy Larue jumped and escaped. Spilt oil caught alight after the derailment, damaging the engine cab and a small trestle on the approach to the drawbridge.

Wrecking trains were sent from Tracy and Stockton, the blaze was extinguished, and the track cleared in the afternoon. Meanwhile, trains were diverted onto the Atchison Topeka and Santa Fe at Tracy. Had the train not been derailed by the derail it would have gone into the river from the open draw.[1052]

A negro excursion train on the Atlantic Coast Line Railroad from Kinston, North Carolina, to Norfolk, Virginia was within six miles of its destination when it fell through an open drawbridge at Bruce, Virginia, United States, on Thursday 17th August 1905. According to the regulations all trains must stop before crossing the bridge, but the train approached at a speed of 15 to 20 miles an hour and failed to stop. The engine and first two coaches of the train went off the drawbridge and the engine, leading coach and part of the second were submerged in 16 feet of water. It was difficult for some time to estimate the number of casualties and figures of 50 or even 100 dead were quoted, but two days later only 13 bodies had been recovered. At that time it was believed that the bodies of a score or so of other victims had floated away with the tide when the first coach was retrieved from the river or were buried with the second coach in the mud of the Elizabeth River. Possibly though, only 17 passengers lost their lives, together with Bridge Tender Collins Ferguson, who was knocked from the structure by the falling train.[1053] Engineer D. L. Reig and his coloured fireman, Alfred Cooper, survived. 'Among the victims the only white ones were Edward Joliffe, the Manager of the excursion, and

[1051] *Early County News*, Blakely, Georgia, United States, 23rd February 1905
[1052] *Oakland Tribune*, Oakland, California, United States, 8th April 1905
[1053] *Brainerd Daily Despatch*, Brainerd, Minnesota, United States, 18th August 1905

Edward Forbes, who assisted him, both of Greenville, N. C.' [1054] Forbes was Joliffe's nephew. About 100 passengers were injured, most only slightly. Wrecking trains with physicians were sent to the scene from Norfolk and Portsmouth.

On 19th it was reported that 'All day long operations of the rescuing and wrecking crews continued. When derricks brought the first coach up and out from its submerged position it hung at an angle and the strain loosened the ends of the car. As a result seats, cushions, etc., were washed out and down into the river, and with them the dead imprisoned, and it is claimed that a large number of the bodies have been washed away with the sweep of the tide.

'Thousands travelled to Bruce's station Friday. The work of the divers was impeded by high tide and a stiff gale. Chains were placed under the cars by divers, and wrecking engine tugged and pulled at the coaches until they were brought to the surface, when they were gradually brought to the beach. In a mass of debris lay the engine and tender, and these have not been brought out yet. Remnants of the draw, completely destroyed, tangle about them, and the divers must clear away timbers, twisted iron and tangled wreckage.

'The Atlantic Coast Line officials have given out no statement in the matter, but Cleveland Thomas, an eye-witness, says that the small "Danger" flag which was placed on the track to warn the approaching Kinston train, though formerly a red flag, had faded almost completely white from continuous use during the past five years. Thompson declares that Engineer Reig remained by his engine up to the very last and did not jump as first reported. He went down with the engine and was badly injured about the head and body. He is now in the Sarah Leigh hospital here {Norfolk} and it is said he will recover. J. J. Thompson, aged 60 years, a farmer and father of Cleveland Thompson, rushed to the scene of the wreck with an ax and leaping 20 feet to the top of the second coach cut a hole in the top of the car and rescued 35 people just before the car dropped into the water and became almost completely submerged.'

Apart from the badly faded danger flag the explanation for the disaster probably lay in the fact that Engineer Reig was making his first run over the line, having just come from East Radford, Virginia, and employment on the Norfolk and Western Railroad.[1055] He was unaware that he was approaching a drawbridge and did not see the small danger flag placed by the bridge tender. When he did see the stop sign 300 feet from the bridge it was too late; by then the brakes had already been applied by a porter who had opened the conductor's valve in one of the cars. He had, however, also failed to see a fixed warning signal half a mile from the drawbridge, or, in broad daylight, notice its open span at right-angles to the track, and Shaw in his book accuses him of sheer inattention.[1056] It seems strange though that he was given charge of a train to operate over a line on which he had never travelled before without having been given a pilot.

On Saturday evening, Saturday 29th September 1906, the drawbridge at Clarksville, Tennessee, United States, had been opened for the steamer *Buttorff*, in charge of Captain Tyner. At the same time northbound Louisville and Nashville Railroad passenger train

---

[1054] *The Landmark*, Statesville, North Carolina, United States, 22nd August 1905

[1055] *The Evening Democrat*, Fort Madison, Iowa, United States, 19th August 1905

[1056] *A History of Railroad Accidents, Safety Precautions and Operating Practices* by Robert B. Shaw, pg. 194.

No. 102 was approaching. The signals had been placed at danger to protect the bridge, but Engineer Frank Porter reportedly failed to notice them until the last moment and when he did do so the brakes would not work. As a result the locomotive, mail and baggage cars fell 40 feet into the Cumberland River, two passenger coaches remaining on the bridge. Six or seven men went down with the front of the train, five of them were rescued, three of them clinging to the top of the baggage car 100 yards from the bridge and a fourth from the top of the mail coach 200 yards from the bridge. Engineer Porter and Express Messenger Wood were drowned.[1057] [1058]

It was stated that had the steamer been three minutes earlier the train would have fallen on top of it, probably resulting in great loss of life.[1059]

Fourteen months after the Bruce, Virginia, disaster, on Sunday 28th October 1906, a considerably worse accident was to occur on the West Jersey and Seashore Railroad, a mile from Atlantic City, New Jersey, United States; the most serious accident of this type, in fact, to have occurred anywhere, with the exception of that at St Hilaire, Canada, back in 1864. The resort of Atlantic City was separated from the mainland by a deep tidal channel known as The Thoroughfare, and it was at the opening bridge across this channel that the disaster happened.

Conductor James Curtis, who had been in charge of the ill-fated train, told his story of what had happened. "Our train was the Atlantic City express, scheduled to leave Camden at 1 o'clock in the afternoon and reach Tennessee Avenue at 2.30 p.m. We were a little late in starting, and Scott, the motorman, kept hitting it up a little to catch up. The trip had been uneventful up to the time we jumped the bridge.

"There were seventy-seven passengers, according to my count of the tickets, in the first two coaches. I should say the passengers were about equally divided between men and women. In the last coach there were some women, but in the extreme end there were a lot if musicians, members of a band who were going to play a Sunday afternoon concert at one of the piers.

"What caused the accident I only know from hearsay. It all came so quickly that no explanation was possible. I was in the forward part of the middle coach when I heard a crashing noise in front and realised the train had jumped the track. Before I could realise it fully, however, the second car toppled, and the next instant our car splashed into the water.

"It would be impossible for me to tell what happened then. There were screams and curses from all parts of the car. The water did not seem to be coming in very fast, and had it not been for the intense darkness I might have thought that the car was fully half out of the water. The shock of the fall was not great, but most of the passengers had, I suppose been looking out of the window, and they all knew the awful danger.

"Almost before I could think what to do a big negro who had been in the car dashed himself against a window, smashing it open and letting in a torrent of water. I could not see what was going on in the other parts of the car, but believe that others were attacking

[1057] *The Bee*, Earlington, Kentucky, United States, 4th October 1906
[1058] *The New York Times*, New York, New York, United States, 30th September 1906
[1059] *The Paducah Evening Sun*, Paducah, Kentucky, United States, 1st October 1906

the windows, so I rushed at the window through which the negro had gone and fought my way through the torrent.

"I am a pretty big man and it was hard getting through the window. My breath was almost gone, but once through I gave one spring and swam with all my might, reaching the surface quickly. I should say the cars lay in about fourteen feet of water. The first two cars of the train had disappeared beneath the water, while a small portion of the rear of the third car showed above water.

"I will not talk any more about the accident. It is too dreadful. Just say that I smashed my way through a window and got out – I – I will not talk any more." [1060]

Another newspaper told of the escape of the brakeman of the train and his rescue of several passengers. 'The brakeman of the train, Roland Wood, as the train dashed off the tracks dashed to the rear of the train. As the conductor escaped, Wood threw open the door, and, climbing up on the abutment, hauled several of the passengers up without aid to a place of safety. Those in the front of the car were drowned in their seats, seven members of the Italian band known as the Royal Artillery Band meeting death there. A fourteen-year-old girl got out of the car by breaking a window. A workman, who had rushed from the tower house to the place where train went off, saw her swimming about in the water. Throwing a rope to her, he hauled her up on the bridge and cared for her in the tower house.'

Describing the bridge, the same newspaper wrote that it stood low in the water and was an ordinary wooden bridge with no protection at all. From bank to bank there were no uprights or even iron railings on its sides.

Regarding the bridge tender the paper wrote, 'Daniel E. Stewart, the aged tender of the Thoroughfare draw, was panic-stricken when the train plunged into the stream. He ran shouting towards the city end of the bridge and begged every one he met to hurry for help.

'"There was nothing the matter with the bridge," Stewart declared, "I opened the draw for a schooner about ten minutes before the 2 o'clock outbound train passed, and if there had been anything the matter with the track an accident must have happened then."' [1061]

The newspaper pointed out that the slower moving outbound train might, however, have been able to negotiate a faulty piece of track more easily than the faster moving inbound train.

Charles Kessler, the owner of the schooner *Sinbad*, which was the vessel which had passed through just before the accident, saw it happen, and quickly summoning a volunteer crew he had taken his vessel to the wrecked train. He managed to recover seven bodies which had floated to the surface but could not rescue any survivors. Meanwhile Chief C. W. Maxwell of the Atlantic City Fire Department was early on the scene with a dozen firemen. 'Already a score of men were on the bridge. Many had ropes which they had torn from the rigging of boats or snatched from boathouses. The bridge tender was well-nigh overcome. His face was blanched, but he worked in a frenzied way with the rest.

[1060] *The Philadelphia Inquirer*, Philadelphia, Pennsylvania, 27th October 1906
[1061] *The New York Times*, New York, New York, United States, 29th October 1906

'Men were swimming near the edge of the drawbridge a moment after the three cars plunged down. Their heads, bobbing up and down, could be seen by the bridgeman, who hastened towards the spot as soon as he could cross the ties. Two men sank as he gasped wide-eyed at the water. As he looked another man shot like a buoy from the water and struck out for the piling which supports the draw. It was James Curtis, the conductor, who had hurled his weight against a car window until the glass gave way and he shot through. He reached the piling, and the bride tender gave him a hand.

'The current is believed to have carried several of the swimming victims many feet away from the bridge. The terminal of the West Jersey & Seashore, the line on which the accident happened, is nearly a mile from the bridge. On the platform was a crowd of a hundred men and women waiting for the arrival of the train. They had friends and relatives on the train. This message flashed over the wires to Station Master Brennan's office, not twenty feet from where the train stood:

'"No. 1065 has gone over the bridge. Send help quick."

'The message came through from the signal station at the east end of the bridge. The dispatcher had seen the accident but did not lose his nerve. He knew the train crew well and realized that his friends had plunged probably to their death. As his message came off the wires at the terminal there was a scene of great activity. Concise, short orders, were given, and soon a wrecking crew was hurrying towards the bridge. Dozens of messages were flashed to Camden and to Philadelphia.

'General Manager Atterbury and several others hastened shoreward on a special train. A wrecking crane with crane attachment started but even before the railroad began the work of rescue a dozen physicians, several ambulances and many patrol waggons were rattling through Atlantic City's streets all heading for the bridge. On the shore of the thoroughfare they waited while police and firemen assisted the survivors across the perilous woodwork of the bridge.

'Most of the wounded had to be carried. The crews of the Inlet boats helped take the injured from the water. The groaning, bleeding men lay on the decks while the sailors offered restoratives. No one held back from the work of rescue. Some were weakened by the sight of blood, but they worked as best they could. The injured were taken to the Atlantic City Hospital, where doctors and nurses awaited them. Several trips were made by the ambulances before all the injured were removed. Soon after the accident a crowd of many hundreds gathered about the banks of the Thoroughfare and the east end of the bridge, pressing close to the little stairway from the ground to the elevated structure.

'Through the long lines of anxious faced men and women the injured were carried to the waiting patrols and ambulances. Now and then a groan of a bleeding sufferer drew a responsive murmur of sympathy from those nearby. Now and then a wildly excited man elbowed his way through the crowd, and, springing up the steps, rushed as close as he could get to where the last car protruded from the water. He then scanned the faces of the injured, trying to find some relative who had been on the train.

'It was not long before the crowd at the station here learned that their relatives and friends had been thrown from the bridge. A man wearing a fur coat sprang from an automobile in front of the station. "My God," he cried, "The Philadelphia train has gone into the river. I saw it go at the end of Missouri Avenue."

'The waiting crowd heard him. The men and women quickly crowded about the office of the ticket seller.

'"I am afraid it is true," he told them.

'Many women burst into tears. Men also wept. One woman fainted and was carried into the waiting room. With one impulse the men and women started for the bridge, in trolley cars and hacks. They joined the other crowd at the east end of the bridge. The wrecking crew arrived before 3 o'clock, but even before this the crews of the inland boats and other rescuers had dragged many dead and wounded from the cars.

'Boat hooks were used by the sailor men. They fished about in the rear car as far as they could reach. But the water was too deep for them to reach the first and second cars where most of the dead lay. Men with axes cut out the upper side of the last car as soon as helpers arrived. As they cut into the wood many bodies were exposed.

'Some were dragged out and placed beside the other dead. A baggage car was run out on the westbound rails and this was used as a temporary hospital and morgue. Into this James Curtis, conductor, was taken after he had stood shivering on the bridge until he could stand no longer. His leg was hurt, but he said he did not feel the pain.'

At 10.30 p.m. the Pennsylvania Railroad, which controlled the West Jersey and Seashore line, issued a statement that there were not more than 57 dead. All were passengers except for the motorman, Walter Scott. The statement said that the drawbridge had been properly closed and locked, the signals had been clear and that until the cars could be raised it was impossible to determine the cause of the accident.

Motorman Scott should not have been working the train anyway, but he had arranged to swop shifts with his colleague so that he could spend the first part of the day with his wife, a decision which it turned out cost him his life.

The electrified line had only been opened on 15th September. Three coach passenger trains made the journey between Camden and Atlantic City at hourly intervals between 7 a.m. and 11 p.m., taking 90 minutes for the 60-mile journey.[1062]

Asked early on, before the cars had been retrieved from the water, as to the possible cause of the wreck, General Manager Atterbury had stated that, "Perhaps a broken flange or some part of the car's gear gave way." [1063] Examination of the wreckage failed, however, to reveal any such defect.

At the inquest into the disaster, which was concluded barely a week after it had happened, the coroner's jury placed the blame upon Bridge Tender Stewart, who they decided had failed to ensure that one of the rails was properly in its socket, and that the projection of said rail had caused the accident.

The jury's verdict 'reached after considerable disagreement on many points' read, "The said Walter C. Scott came to his death on Sunday, 28 October, 1906, at or about 2.30 o'clock in the afternoon, by the derailment and plunging into the waters of Beach Thoroughfare of the electric train on the West Jersey and Seashore railroad that left Camden at 1 o'clock on said day; said derailment being caused by the north-west end of the south-west rail the draw of the bridge over Beach Thoroughfare, in or near Atlantic City, in said county, being out of its proper position, due, we believe, to the neglect or

[1062] Ibid.
[1063] *The Hickory Democrat*, Hickory, North Carolina, 1st November 1906

oversight of the man on the draw.

"We strongly recommend that the men employed to operate the drawbridge inspect each with the other ends of the draw before the signal is given for trains to pass the opening ends of the draw." [1064]

Stewart, on the other hand, swore that the rail had been in place, and that the derailment must have been due to some other cause. He was said to have been thin, attenuated and grey and to have aged considerably. He remained at home, surrounded by his sympathising wife and family. Most of the railroad men agreed with him, while experts who had examined the bridge and rail agreed that the accident could not have happened in the way the jury believed.

The bridge tender also had the sympathy of the coroner, '"We will arrest bridgetender Stewart when it is certain bail can be procured for him," said Coroner Gaskell upon his return from May's Landing tonight, where he had a long conference on the subject with District Attorney Abbott. "I intend to reserve the taking into custody of Stewart for myself," he remarked, as I don't want to see the old man go to jail. In fact, if the old man's friends cannot raise the bond that will be demanded to ensure his appearance before the grand jury I will take enough interest myself to secure his release. I am still of the opinion that the throwing of the blame entirely upon him is an outrage."

'The Coroner is voicing public indignation, which is growing over the Coroner's jury's action in allowing the railroad to escape censure in connection with the wreck. He is not satisfied with the conclusions of the jury, and neither is Prosecutor Abbott. These two officials concur in the belief that the railroad should have been held jointly responsible in the verdict. They objected to placing the blame alone upon the shoulders of Stewart, who they consider had a faulty mechanism to control.

'Public sympathy is turning towards the bridge tender, and when his arrest comes there will be a purse raised to aid in his defence, owing to the feeling among citizens that the railroad should have been made to share in the burden of responsibility for the catastrophe.

'Jurors were surprised when informed of the criticism of their finding. Senator Edward S. Lee, the foreman, reasserted his declaration that there was no other conclusion to be reached than the one contained in the verdict.

'The arrest of Stewart will bring before the grand jury the details of the wreck, and the grand jury may make further inquiries and also render an opinion as to the cause of the disaster. Something may be said by the jury over the employment of old men at bridges, which is reported to be the habit of railroads, and declare in favour of younger and more active men, whose eyesight and sight are acute.' [1065]

On 17th January 1907 the grand jury, sitting at May's Landing, announced that after several days' deliberation they could find no evidence to hold or even censure Daniel Stewart. On the other hand they found that the West Jersey and Seashore Railroad had been negligent in not seeing that the construction of the bridge was safe for the travelling public. They had examined the bridge and found it defective. There were no mechanical signals at either end of the bridge to show that the rails were connected in their proper

---

[1064] *Lebanon Semi-Weekly News*, Lebanon, Pennsylvania, United States, 8th November 1906
[1065] *The Philadelphia Inquirer*, Philadelphia, Pennsylvania, United States, 9th November 1906

place. One rail or all the rails might be out of their proper position and the white signal could be given to approaching trains in either direction that all was right. They further believed that the railroad company should be instructed to have defective rails repaired at the earliest possible date. The jury further recommended that the speed limit for trains passing over the drawbridge should not exceed eight miles an hour.

Upon learning that he had been exonerated the bridge tender told a group of his friends, "I thank God that He did not let me go down to the grave with the blame for that awful wreck hanging over me. It was wrong for the Coroner's jury to attempt to place the blame for that frightful thing on me. I did all that any man could have done under the circumstances to avoid it. It was fearful I had to stand and watch 58 men, women and children drown before my eyes. I can see them struggling in the water yet. I will always see them but thank God I am cleared of the blame at last." [1066]

The Interstate Commerce Commission also investigated the disaster and theorised that when the drawbridge had been closed and the supports removed from the rail ends for some unexplained reason these had failed to drop into position, remaining sufficiently out of alignment to derail a train travelling at a fairly high speed. Had the misalignment been only slight it might not have been noticed by Bridge Tender Stewart when he removed the red flag placed in the track when the draw was open but might still have been sufficient to derail the train. The draw was interlocked with the signals but there was no detection to check that the rails were properly in place. There was, however, no definite proof that this was how the accident happened and it remained unexplained.[1067]

Similar to the Atlantic City disaster, though much less serious, was one at Jacksonville. Florida, United States, on Christmas Eve, 24th December 1910. An Atlantic Coast Line Railroad passenger train came to a complete stand on the approach to the drawbridge and was then signalled forward by the bridgetender. All appeared in order but the inexperienced bridgetender, aged 22 and with only seven weeks' service, had failed to notice that for some reason one of the movable rails at the end of the bridge had failed to drop into place when it was last closed. The engine alone fell into the water, killing the engineer.[1068]

Thick fog was doubtless the main factor in a drawbridge accident at the Hoffman Swing Bridge, near South St Paul, Minnesota, United States, on the morning of Wednesday 16th October 1912. A train on the St Paul Bridge and Terminal Railroad was heading for the stock yards at about 7 a.m. It consisted of a locomotive, 18 cars loaded with cattle and a caboose. As it approached the bridge the latter was open for the steamboat *Hiawatha.* The train was running slowly but did not stop and the engine and six or eight of the following cars plunged from the opening into the Mississippi River.

The sound of the crashing engine and cars, the bellowing of cattle and the cries of men about the scene reportedly brought hundreds of people to the scene and a rescue operation

---

[1066] Ibid, 18th January 1907

[1067] *A History of Railroad Accidents, Safety Precautions and Operating Practices* by Robert B. Shaw, pgs. 195-6

[1068] Ibid, pg. 196

was at once started. Fireman Frank Webber and Brakeman John Garvin were injured but survived. Engineer Charles Kramer drowned; his body was recovered soon after the accident. One of the cars was left hanging over the edge of the bridge. The cattle in it were roped and dragged into the river. They swam ashore but a number of them were so badly hurt that they had to be put down.

Bridgetender Daniel Tibbet stated that he had not blown his whistle, the usual signal for the train to advance. The newspaper report stated that 'An investigation of the bridge tender's story will be made.' It seems clear that the bridgetender was not fault and that what happened was that in the fog Engineer Kramer had lost his bearings and reached the bridge sooner than he had expected to do so.

The cars were owned by the Great Northern and Chicago Milwaukee and Pacific companies. The financial loss was estimated at about £16,000.[1069]

On Tuesday 24th June 1913 the crew of a freight train on the Baltimore and Annapolis Short Line Railroad had a narrow escape when their train went through a drawbridge over the Severn River near Annapolis, Maryland, United States. Engineer Lewis Thomas lost control of his train on the long downgrade to the trestle, where the drawbridge had been opened for a vessel. The whole crew of five jumped, though the engineer and Fireman Andrew Smith left it very late to vacate their engine. Probably they hoped up to the last moment to avoid an accident, but the momentum of the train pushed it onto the draw, which was within two feet of closing. Lacking its proper supports, that end of the span collapsed, and the locomotive and seven cars fell into 40 ft of water, the caboose alone remaining on the rails.[1070]

An opening bridge accident occurred in Germany at midnight on Saturday 26th July 1913. The last passenger train of the day left Hilkenborg station and approached a swing bridge over the Ems River. The bridge was open and the signals at danger, but the driver of the train overran them, only realising his mistake when his train was about 100 metres from the bridge. The driver then made an emergency brake application and as the train was only going slowly he almost managed to pull up in time, but the two leading axles and one driving axle of the locomotive went off the end of the bridge and it tipped. Fortunately the coupling between the engine did not give way and the engine was held in position by the remainder of the train. The engine crew put out the engine fire with wet coal and the few passengers in the train were conducted back to the beginning of the bridge via the running boards of the carriages.

The bridge tender at the west end of the bridge had seen the train running onto the open bridge but was unable to close it immediately as a sailing ship was passing through. As soon as he could he began to close the bridge, and there was only a two-metre gap when the train went over the edge of it.

Recovery of the engine was difficult. A floating crane was chartered from the Kaiser's shipyard at Wilhelmshavn. It set off on 28th July, ran into heavy seas but reached the

[1069] *The Fargo Forum and Daily Republican*, Fargo, North Dakota, United States, 16th October 1912

[1070] *The Washington Post*, Washington, D.C., United States, 25th June 1913 accident site on the evening of the 29th July. At 11:00 on 31st July the engine was back on the rails and traffic could be resumed already that afternoon.

The train driver was awarded a fine for overrunning signals.[1071]

[1071] *Historische Eisenbahn-Katastrophen* by Bernhard Püschel, pgs. 96-7 (in German)

CHAPTER 10

TRAIN ACCIDENTS INVOLVING BRIDGES

Bridges could feature in train accidents in one of two ways. Either the bridge itself could collapse for a number of reasons such as faulty design, faulty construction or because its foundations had been damaged; or a train could fall from a bridge, the structure itself remaining intact. Bridge failures have been dealt with in a previous chapter; this chapter will concern itself solely with accidents where the bridge itself was not at fault. It may seem perverse to list such accidents separately instead of in the chapters dealing with primary causes of the accidents, the procedure adopted elsewhere in this book, but the fortuitous presence of a bridge has such a potential for transforming what might have been only a minor accident into a major disaster that it has been decided to treat such accidents separately, in the same way as accidents occurring in tunnels have been given a chapter of their own.

Early on the morning of Thursday 31st October 1901 a Chicago Great Western Railroad northbound doubled-headed extra freight train, hauled by engines No. 146 and 5, and in charge of Conductor James Druse, was wrecked at a wooden bridge over the Little Wapsie River near Dunkerton, Iowa, United States. A heavily laden car load of wheat, second from the caboose, broke down and left the track, taking with it the two cars ahead of it, the following car and the caboose. The five vehicles rolled down the embankment and the caboose ended up in the water. The bridge was totally wrecked and 1,000 feet of the track north of it was torn up. Riding in the caboose with the conductor were Mr J. W. Emery, a prominent stockman of State Center, and his uncle, Mr J. B. Veneman, of Des Moines. Mr Emery was struck on the head by a heavy iron hook and killed; his uncle and the conductor were slightly hurt.

Pending the repair of the track, bridge trains were diverted over the Burlington Cedar Rapids and Northern Railroad between Oelwin and Independence and the Illinois Central Railroad between Independence and Waterloo.[1072]

[1072] *The Cedar Rapids Republican*, Cedar Rapids, Iowa, United States, 1st November 1901

The mail train which had left Bilbao, Spain, at 9.10 a.m. on the morning of Saturday 26th June 1903 for Barcelona via Logroño and Zaragoza met with disaster between San Asensio and Cenicero, at a bridge over the Najerilla River. The train was a long one, with 16 passenger carriages and three mail and luggage vans, hauled by two engines. It reached the bridge at about 3 p.m. The pilot engine crossed the bridge safely, but the second engine left the rails and plunged from the bridge, taking the whole of the train with it.

Survivors reportedly gave harrowing descriptions of the accident, of the screams of the victims and of the distressing incidents which had followed the catastrophe. Many of the victims were so stunned and bewildered that they were unable to take any part in the rescue work. It was several hours before relief trains arrived from Miranda and Logroño with doctors, nurses and 100 soldiers. Over 140 passengers were rescued, and early estimates put the number of dead at between 90 and 110. As has frequently been the case these early estimates were considerably exaggerated; the correct figures appear to be 44 dead and 84 injured.[1073]

The passengers on the train had included a number of invalids going to the sulphur baths at Fitero and a party of wealthy Bilbao mine owners going to purchase mines at Teruel. The corpses of victims floating down the river after the accident included those of railwaymen, sisters of charity (nuns) and police officers who had been en route to Barcelona to deal with a strike there. The fireman of the train engine was scalded to death in the fall of his engine from the bridge.

A reported unsavoury incident in the aftermath of the accident was the attempted theft by a railway guard of a pocket book containing the equivalent of £80 sterling from a dead passenger. 'The bystanders when they learnt of this act were with difficulty restrained from lynching him.' [1074]

The work of clearing the wreckage away apparently only proceeded slowly, not helped by bickering between the military and the civil authorities.[1075]

The reason for the derailment was not immediately established. There was some suggestion that the train had been travelling at excessive speed, but this was denied by other witnesses.

Possibly on the same day as the bridge accident a train was derailed on a setoff points on the Plencia line, injuring three passengers, one seriously.[1076]

Northbound Southern Railway passenger train No. 15 went through Fishing Creek Trestle, three miles east of Yorkville, South Carolina, United States, at 11.30 a.m. on Thursday 3rd September 1903, and fell 50 feet to the stream below. Engineer Henry Brickman, Fireman Fred Rhyne, Mail Clerk C. J. Smith and three unknown negroes were killed and Conductor Ed Turner, Flagman Whisonant, Baggagemaster Dukes and about 20 passengers were injured, several of them possibly fatally. The train had consisted of an

---

[1073] Federación Castelleno Manchega de Amigos del Ferrocarril: Accidentes Ferroviarios en España (in Spanish)
[1074] *St James' Gazette*, London, England, 29th June 1903
[1075] *The Times*, London, England, 30th June 1903
[1076] *La Vanguardia*, Barcelona, Spain, 28th and 29th June and 1st July 1903

engine and three cars and was carrying about 40 passengers. The entire trestle gave way.

A passenger, Mr B. F. Willford, of Charlotte, who was prominent in the rescue work, told his story to a reporter. "The train was running at about its usual speed when it struck the bridge. Soon after passing out up on the trestle I felt the car give a lurch and heard the crashing in front as the engine and two front cars went down. I instantly grasped the back of the seat in front of me and succeeded in keeping my place. The car in which I was riding dropped off the end of the broken bridge and on to the wreckage of the cars already down. As soon as my car stopped I scrambled out through an opening in the side, into an opening made by the breaking of timbers, and began helping the other occupants of the car to places of safety. There were about 15 passengers in the first-class car, including three or four ladies, two girls and one boy. I helped one or two young men to their feet and urged them to aid me in getting the other passengers out, but they were apparently so confused that they did not lend a hand, but went out, and I got no help until people from the outside came to the rescue. The seat on which I had been sitting was torn to pieces, and the back of my coat was badly cut, and I received a blow on the head." [1077]

The general opinion immediately after the accident was that it had been due to the old and rotten wooden timbers of the trestle and the inquest jury brought in a verdict to that effect, blaming the Southern Railway for carelessness in permitting the structure to remain in use in such a condition.[1078]

Despite the general view and the verdict of the coroner's jury an investigation by the South Carolina State Railroad Commission resulted in a different conclusion. Colonel John Averill stated in mid-October that the failure of the trestle had probably resulted not from rotten timbers but from a derailment due to a broken wheel on the front truck of the engine.[1079]

The next bridge accident to be considered occurred at Danville, Virginia, United States, on Sunday 27th September 1903. Like the accident at Vaughan, Mississippi, featuring Engineer Casey Jones, rather more than three years earlier, the accident was later the subject of a ballad and became famous as 'The Wreck of Old 97'. It would seem that like the earlier accident it was due to the engineer running too fast in an attempt to make up lost time.

Southern Railway southbound train No. 97 had left Washington D. C. at 8 a.m. and was a fast mail train, 'carrying local and Northern mail for Southern points'. Behind the locomotive it consisted of two postal cars, an express car and a baggage car, and was in charge of Conductor J. Blair with Engineer Joseph A. "Steve" Broadey and Fireman A. J. Clapp. Just north of Danville station was the Stillhouse trestle, which was on a sharp curve at the end of a falling gradient. The trestle was 500 feet long with a maximum height of 75 feet. As the engine of the train negotiated the curve and had travelled 50 feet onto the trestle it left the track, while a portion of the trestle gave way under the weight of the engine. The engine fell through the opening, followed by the cars, and landed in the mud at the bottom of the creek. The engine was not greatly damaged but was partially buried by the wreckage of the cars, which were 'reduced to a great mass of broken wood

[1077] *The Landmark*, Statesville, North Carolina, United States, 8th September 1903

[1078] *The Marion Star*, Marion, South Carolina, United States, 9th September 1903

[1079] Webpage – Fishing Creek Train Wreck

and twisted steel and iron.' The accident occurred at 2.50 p.m.

Engineer Broadey, Fireman Clapp, Conductor Blair, Flagman Moody and Mail Clerks P. N. Augenbright, S. Chambers, D. T. Flory and J. L. Thomson, plus Mail Clerk Thomson's 12-year-old son were killed outright, and seven mail clerks were injured, one fatally.

Describing the immediate aftermath of the derailment a newspaper reporter wrote, 'When the cars struck the bottom of the ravine the mail bags from the mail cars were torn to pieces, and the letters and parcels flew in every direction. The wreckage immediately took fire, but the flames were extinguished before they gained any headway, by the fire department of Danville, Va., which had been summoned. No mail, it is thought, was destroyed. The contents of the express car were scattered about in great confusion and were practically destroyed.

'Several thousands of people quickly gathered and began the rescue work. The bodies of nine persons, who had apparently suffered instant death, were found. As these were beyond human aid, the rescuers turned their attention to the injured. They were quickly placed on stretchers and removed to the hospital at Danville. In the meantime Superintendent Coapman and a large force of men arrived at the scene of the wreck from Danville. The superintendent took immediate charge and directed the work of recovering the bodies. The bodies of the engineer and fireman were found beside the engine, showing that they had made no attempt to jump as the train went down. The bodies of the mail clerks were found under mail bags and other portions of the debris.

'Mail Clerk Spier is considered to be the most seriously injured, and no hope is entertained for his recovery. He sustained injuries to his spine and head. All the other clerks reported to have been hurt are expected to recover.

'As soon as the rescue work had been completed Superintendent Coapman and his force began to clear away the wreckage, which completely blocked traffic over that division of the system until this morning. The mail sacks and loose mail were gathered up and placed in a secure place. All the debris of the cars was cleared away, but it was impossible to pull the engine from the mud. While the clearing was in progress the trains were sent on by way of the Norfolk and Western and the R. F. and P. lines.'

A 16-year-old boy named Rutled Armisted, from Lynchburg, who had been in the express car, was supposed to be still buried in the wreckage. He reportedly regularly travelled between Lynchburg and Danville and back. His and Mail Clerk Spier's deaths brought the probable total number of fatalities in the accident to 11.

Two boys had been stood under the trestle when the accident occurred and one of them had clearly been severely traumatised by what he witnessed. 'He has flighty spells and imagines that he can see trains flying through the air. His parents are very much exercised, but it is hoped that the aberration is only temporary.'

Six crates of canaries in the express car were little damaged, and the birds reportedly regaled the rescuers with song from nearby trees after their accidental liberation.

Some of the people who drove out from Danville came perhaps more to gawp at the wreckage than to assist and ended up seeing more than they wanted to, for it was reported that ladies who had come out to the wreck had fainted at the sight of the bodies, some of which were badly crushed.

Southern Railway Vice-President Finley stated that the trestle had been little damaged and was back into use by 9.10 a.m. on 29th. According to him eye-witnesses reported that the train had been travelling at 30 or 35 miles an hour, while a report from Danville on 28th stated that the speed of the train as it approached the trestle had not been above ordinary and Engineer Broadey had reduced speed to about 30 miles an hour as per the regulations.[1080] However, other sources stated that the train had been going too fast. It had been running an hour late, and it was theorised that Engineer Broadey had been running fast to try and regain some time, unaware perhaps of the risk that he was taking.[1081] The exact speed of the train it was impossible to say but another account stated that it was between 50 and 60 miles an hour, [1082] which possibly was close to the truth. At any rate the speed was apparently too high to enable the train to safely negotiate the curve.

On Wednesday 30th September the coroner's jury at the inquest on the victims of the accident found that the excessive speed of the train had been the sole cause of the accident.[1083]

Excessive speed was possibly also the cause of another trestle accident on the Southern Railway exactly three weeks after the Danville accident, on Sunday 18th October 1903. Engineer Jim Nichols on southbound freight train No. 81 was running fast, anxious to reach Dallas, Georgia, United States, where the train had a meet with northbound vestibule passenger express No. 18. The train consisted of engine No. 345 and 25 cars and was supposedly travelling at 60 miles an hour as it came downhill to the Pumpkinvine trestle. Approaching the trestle he reduced speed somewhat, but the iron trestle vibrated violently and his engine swayed. He put on steam again to try and clear the structure, but the northern half of it collapsed, taking with it the engine tender and 13 cars loaded with corn, oats, cotton and apples. The trestle was 360 feet long, 77 feet high and had 11 spans, six of which went down. The engine itself remained on the track but Fireman Fagala went down with the tender and lost his life. The rear of the train remained on the track and Conductor Sorrels and the flagman were not hurt. According to one newspaper a much worse accident could have followed, for a passenger train, carrying 60 members of the *Johnny Comes Marching Home* Company, was behind the freight train. This train was stopped up the line by an unnamed negro brakeman, presumably the aforementioned flagman. The company made considerable collection of money to reward the hero negro but, according to the newspaper report, he had by then disappeared.[1084]

The failure of the trestle was variously attributed to the excessive speed of the train and to a rail having been removed by train wreckers.[1085]

Several American newspapers carried very brief accounts of an accident to a passenger train on the railway between Chinandega and El Vieja, in Nicaragua, on

[1080] *Times-Despatch*, Richmond, Virginia, United States, 29th September 1903
[1081] *Oxford Eagle*, Oxford, Mississippi. United States, 1st October 1903
[1082] *The Atlanta Constitution*, Atlanta, Georgia, United States, 28th September 1903
[1083] *Times-Despatch*, Richmond, Virginia, United States, 1st October 1903
[1084] *Savannah Morning News*, Savannah, Georgia, United States, 20th October 1903
[1085] *The Dallas New Era*, Dallas, Georgia, United States, 23rd October 1903

Tuesday 2nd February 1904. A train was reported to have been derailed and wrecked while passing over a high bridge and more than 20 first-class passengers killed, and many others injured.[1086] [1087] No more information about this accident has come to light.

A major disaster occurred in France on Sunday 4th August 1907 on the Angers – Poitiers line of the Chemins de fer de l'État. Passenger train No. 407 was due to leave Angers Saint-Laud station at 11.25 a.m. that Sunday morning, in charge of Conductor Pierre Grenet with Driver Bourreau and Fireman Leclair. There was heavy traffic and an extra coach had to be added to accommodate all the passengers. Perhaps for this reason the train was also slightly delayed, pulling out at 11.28. It consisted of engine No. 2517, five passenger carriages and two vans. Stops were made at Maître-École and Pyramide (Trélazé) stations and the train then approached the Saint Maurille bridge across the River Loire at Ponts-de-Cé. At Pyramide station the passengers, mostly slate workers from Trélazé, had had to get into the third-class carriage at the front of the train, the remaining carriages being already full.

Fifty metres short of the Saint Maurille bridge the train became derailed, the rails buckling beneath it, and as it passed onto the bridge the engine smashed against the deck of the bridge and fell from it 10 metres to the river below, followed by its tender, a van and the leading third-class carriage. Most fortunately the shock of the front of the train leaving the bridge snapped the coupling at the rear of the third-class carriage, the Westinghouse brakes came into action and the remainder of the train came to a stand on the bridge, with the next two carriages left more or less hanging from the structure. There was a panic among the passengers in the rear of the train as they scrambled and fought to get out of the train and off the bridge, while those in the coach which had fallen into the river struggled to reach the bank. Where the disaster occurred the river was four or five metres deep and many of the passengers could not swim, while those who could had to overcome a strong current to reach safety.

The accident happened a kilometre and a half from the nearest house and it was some time before rescuers reached the scene. When they did there were heartbreaking scenes on the riverbank, for example a man named Normand, from Trélazé, was running up and down searching for his wife and child, and was so crazed with grief than when the body of his child was brought to him that he failed to recognise her. M. Normand also lost his wife in the accident. Fireman Leclerc, who was a strong swimmer, was able to reach the bank despite a head injury, but was sobbing at the loss of his driver, who was not able to escape.[1088] Conductor Grenet also survived the accident but in addition to Driver Bourreau 26 passengers lost their lives, mostly through drowning, the dead including half a dozen children. Twenty others were injured.

Rumours that the accident was the result of an act of sabotage were soon discounted. It was quickly established that on a hot summer's day the rails had buckled on the approach to the bridge, derailing the train, and the impact of the derailed engine and vehicles had been too much for the bridge decking to withstand.[1089]

1086 *Savannah Morning News*, Savannah, Georgia, United States, 9th February 1904

1087 *The Standard*, Ogden, Utah, United States, 9th February 1904

1088 *Le Petit Parisien*, Paris, France, 5th August 1907

1089 *La Catastrophe des Ponts-de-Cé* by Fabrice Rabarin (in French)

A fortnight after the disaster M. Louis Barthou, the Minister of Public Works, issued a circular to all the railway companies directing them to carry out extra inspections of the track during periods of hot weather in order to prevent further derailments due to 'sun kinks'. Many other bridges were also inspected and renovated in the weeks following the accident.[1090]

By chance the Southern Railway suffered another trestle accident near Dallas, Georgia, United States, on Tuesday 7th January 1908, only a short distance from that on the Pumpkinvine trestle back in 1903. This time the accident happened on the Copper Mine Creek Trestle, between Dallas and Hiram. The accident involved the second section of a Chicago – *Florida Limited* express, said to be 'the finest Pullman train which ever pulled out of Cleveland, Ohio.'

Due to arrive in Atlanta shortly after noon, the train had left Chattanooga, Tennessee, in charge of Conductor Y. R. Norris and Engineer James Edwards. With the engineer on the locomotive were Fireman Moses Baldwin and Road Foreman of Engines D. H. Snapp. The passengers on the train were all Ohio people en route to vacations in Florida or Cuba, most of whom were in the dining car at the time of the accident. At about 2.40 p.m. the train was late and running fast. It had just passed the point where the Seaboard Air Line was carried overhead on a high trestle and was running on an embankment, in the middle of which was a wooden trestle about 10 feet high over Copper Mine Creek. As the train reached the trestle the large mogul engine left the track, it was generally supposed because the track had spread, and the jolt caused the trestle to fail. The engine overturned and ended up against an embankment and the seven coaches of the train either fell through the trestle or rolled over and over down the approach embankment. The entire train left the track but the fire and cooking range in the dining car fell out as it turned over and so no fire resulted in the wreckage. The passengers had to climb out of the windows of the overturned cars or were later cut free by rescuers.

As soon as possible word was sent to Hiram for help, every physician for 20 miles went to the wreck and relief trains were sent from Rome, Dalton and Atlanta (two). The Seaboard Air Line track passed close to the wreck and a train on that line stopped and took up about 50 passengers who were able to walk the half-mile between the two trains and carried them to Atlanta. The more badly hurt passengers were taken as soon as possible on a relief train to Hiram where they were treated.

The three men on the locomotive were all fatally scalded by escaping steam and all died within a few hours, while two passengers, Mrs Emma Hoover and Miss Florence Studebaker, both of Cleveland, Ohio, were said to be fatally injured.[1091]

A local newspaper complained that it was impossible to obtain a complete list of the injured because of a lack of courtesy on the part of the Master Mechanic of the road, whom the reporter had presumably upset, perhaps by an over eager pursuit of information for his paper.[1092]

[1090] *Le Petit Journal*, Paris, France, 17th August 1907

[1091] *The Sandersville Herald*, Sandersville, Georgia, United States, 15th January 1908

[1092] *The Dallas New Era*, Dallas, Georgia, United States, 9th January 1908

Whether a bridge accident in China on Monday 23rd November 1908 was caused by an initial derailment or due to a primary failure of the structure is not clear from the brief reports of the accident. The first reports seem to imply that the evening express on the Chinanfu and Tsing Tau railway was derailed on the bridge near Nunkow Yu, close to the frontier with the German colony of Tsing Tau. The train was running at high speed at the time and the engine and leading coaches fell from the bridge to the river below. All the Chinese passengers were said to have either drowned or been severely injured; 16 were reported to have perished. The bridge was badly damaged and likely to be out of use for some weeks.[1093]

A further report from Tsing Tau on 25th November stated that only nine people had died and about 20 were injured. The bridge was stated to have been only about three metres high and to have crossed one of the many local sandy river beds. There was an unconfirmed theory that the accident had been caused by one end of a girder slipping off a bridge abutment. The first of the two locomotives hauling the train apparently reached the far bank before the bridge collapsed and the second locomotive ran against the stone abutment. The three following vehicles piled up against the locomotive before the train came to a stand, leaving three vehicles, one containing cattle, still standing on the track. The front engine broke free from its train and ran ahead, undamaged. Immediately after the accident fire broke out, caused by broken oil lamps in the carriages and the fire from the second engine, and the bodies of some of the victims were so burnt as to be unrecognisable.

The report concluded by stating that, 'The passenger service to Tsinafu is being maintained by transferring the passengers across the river to a train on the other side. The freight traffic will naturally be very much deranged.' [1094]

A disaster occurred on one of the largest bridges of the Ferrocarril Central del Perú (Central Railway of Peru) at Chaupichaca, between Matucana and Casapalca, on 8th February 1909. Repair work was in progress on the bridge involving the renewal of some of its girders, and a work train was standing on the structure when it was struck by its engine, No. 33, which had run away. While the work was in progress the engine had had to go up the line a few miles to Tamboraque for water. It was a wet day and a small smelter was in operation at Tamboraque. Smoke from the smelter had combined with the rainfall to form a soapy solution on the rails which greatly reduced the available adhesion. In addition the line fell very steeply to the bridge at a gradient of no less than four percent (1 in 25). The result was that the engine got out of control on its return journey and when its brakes were applied it just slid on the slippery rails.

When the engine appeared out of a tunnel close to the bridge a crane was at work on the span and there was no time for anyone to get clear.[1095] The shock was sufficient to cause part of the bridge to collapse, taking with it into the ravine the train, the runaway engine and a number of men, some of them employees of the American Bridge Company and the others Jamaicans employed as labourers. All of those who fell with the bridge lost

[1093] *The Salt Lake Herald*, Salt Lake City, Utah, United States, 24th November 1908

[1094] *North China Herald and Supreme Court and Consular Gazette*, Shanghai, China, 5th December 1908

[1095] *Railways of the Andes* by Brian Fawcett

their lives; in total they numbered 14. Those who died included Mr Niblock, the deputy manager of the company, Conductor Antonio Viera, Driver Pedro Bastante, Fireman Eleodoro Mariscal and Brakesmen Eleazar Escudero and Daniel Alcázar.[1096] A party of American officers from the Pacific fleet had a narrow escape.

Traffic over the line was held up for six weeks, [1097] but the bridge was reopened with a new span in April 1909.

As in a similar accident in 1916 at Penistone, England, it was decided that retrieving the locomotive from where it lay would be too difficult and expensive to be worthwhile. Instead parts which could be useful were stripped off and removed and the remainder of the engine was abandoned to rust away. No. 33 had been considered something of a 'hoodoo' engine anyway and had apparently been renumbered from No. 37 after being involved in a series of derailments.[1098]

Canada suffered one of its worst railway disasters at about 1 p.m. on the afternoon of Friday 21st January 1910. Canadian Pacific Railway train No. 7 from Montreal, Quebec, to Minneapolis, Minnesota, United States, which had left Sudbury at 12 noon with about 100 passengers. It consisted of a locomotive, a mail and baggage car, an express car, a second-class coach, a colonist car, a first-class coach, a diner and Pullman car, in charge of Conductor Thomas Reynolds with Engineer Trelford and an unnamed fireman. The train passed Webbwood, Ontario, and shortly arrived at a bridge across the Spanish River. This was a high cantilever structure over a river about 250 feet wide and 30 deep, though as it was winter in Canada its water was frozen.

The train was travelling at about 45 miles an hour. The engine, baggage and mail car and express car were crossing the bridge normally when the rear wheels of the second-class coach left the rails and the colonist car was completely derailed and smashing into a bridge girder was split in two, 'as cleanly as with a knife'. The second-class car was dragged across the bridge, minus its rear wheels; its passengers unhurt, if shaken. The first half of the colonist car burst into flames, the second half, the first-class car and the diner fell into the river, though half of the diner remained out of the water. The Pullman car turned over onto its side but did not go into the river. The cars which went into the river smashed through the ice and vanished under the water. The passengers who were in the cars which went into the water were, in the words of a newspaper reporter, 'caught like rats in a trap' and many of them were drowned, but a few were saved thanks to the heroism of the train's conductor and others.

'In the dining car was witnessed one of the greatest struggles and most superhuman efforts that ever a man put forth. The first call for dinner had just been made and Conductor Tommy Reynolds, of North Bay, was one of the diners. With him was W. J. Bell, the lumberman of Sudbury, and David Brodie, police magistrate of the district of Sudbury, besides several others. In an instant all were submerged above their heads. Reynolds extorted his passengers to suspend themselves from the hat racks, chandeliers and other projections, and diving down in the ice-chilled waters, broke through the

---

[1096] *Historia del derecho civil peruano siglos XIX y XX* by Carlos Augusto Ramos Nuñez pg. 161 (in Spanish)
[1097] *Railways of Peru* by Rodrigue Levesque, pg. 29
[1098] *Railways of the Andes* by Brian Fawcett

window and emerged on the outside of the car.

'By the merest chance the car had lifted sufficiently to allow him to get up between the ice and the car. Mounting the roof, he reached in and pulled out little Alfonso Rousal, six years old, of the "Soo," Ont., and Police Magistrate Brodie, of Sudbury, a medium small man, through the ventilators. The rest were too big, no axe was available and with superhuman strength this brave man actually tore that section of the car to pieces and opened a space through which he rescued eight passengers single-handed, all that came alive out of the water, with the exception of Brakesman Morrison, who was on the rear of the first-class coach with an unknown, who also managed to come to the surface.

'The first to come out was Mr W.J. Bell, of Sudbury. He is resting at his home and 30 stitches were required to close up his gaping wounds, and he has received a very severe fracture to his chest. Police Magistrate Brodie has three broken ribs and his face and scalp are terribly cut. He is confined to the general hospital.

'Conductor Reynolds himself has bruises and cuts innumerable, but after being fixed up, insisted on returning to his family to North Bay.

'The chef and waiters in the diner were trapped in a heart-rending manner, tangled up in their little coop-like department. All were rescued but scalded and cut in a horrible manner. They are in a Sudbury hospital. The dining car was about the only one to escape without a scratch.'

First reports spoke of 75 dead and then of 63 dead and 37 injured.[1099]

A summary inquest held by Coroner Jones at Webbwood on 23rd January returned a verdict of 'death from causes unknown' but there were protests that this was not enough, and a more thorough inquest was to follow, held by Coroner Howie at Sudbury, as well as a separate investigation by the Canadian Railway Commission. By 24th January 34 bodies had been recovered and divers were continuing their search for more. Meanwhile the papers carried the stories of surviving passengers. Thus, 'R. Burrows, local manager of the Bell Telephone company, who was one of those rescued from the Pullman car which plunged into Spanish river on Friday, gives a graphic account of some of the incidents in connection with the accident.

'"When the Pullman went down the embankment," he said, "I was thrown into a corner and the next thing I remembered was that I was covered with debris, including glass from the windows which cut my face severely. I made my way from under it and proceeded down the car, where I met one of the pluckiest women I ever met. We heard a pounding on the stateroom door at the front of the car and endeavoured to open it. The door being blocked, we picked up a portion of one of the seats and broke it in. We found a lady and a little boy in the room. The lady said she had not been injured. We endeavoured to place her in safety, but she insisted upon helping others not so fortunate as she. For that reason she devoted her attention to the injured. We found out later that she had received a bad wound on her head.

'"I saw a man make his way from the colonist car, apparently unhurt. He endeavoured to make his way up the track but fell from the bridge and was drowned.

'"I remember seeing four commercial travellers sitting in the first-class car, engaged

[1099] *The Winnipeg Tribune*, Winnipeg, Manitoba, Canada, 22nd January 1910 in a game of cribbage. They all went down. About three hours after the accident I found one of the men at the top of the first-class car floating in the water. There was just room between the water and the top of the car for a body to float. I believed him dead, and pulled him through the ventilator, and he was finally brought back to life. He had been in the icy water for three hours.

'"After the fire in the second colonist car had subsided, I could see beneath the bridge, among the matter which had fallen through, a number of partly burned bodies and human bones."'

Further down the page the newspaper carried another story. 'William Dundas of Ottawa, mail clerk on the ill-fated train, tells the following graphic story:

'"We left Nairn Centre at 12.42, and nothing unusual was noticed until we reached the bridge at Spanish River. When about half a train's length from the bridge, I felt the train pulling in a very rugged manner behind, and I knew that a portion of the train behind me was off the track. For a distance the train pulled on, then I felt the air applied. I felt her pulling up and directly we were down on the other side of the bridge, I jumped.

'"Immediately the second-class car jumped the track it struck an upright of the bridge and telescoped. The first part remained on the bridge and burned while the other half simply leaped out into the stream, with the other two coaches behind it. One half of the broken car took its victims with it. Those in the first half were burned, while those in the second half were drowned.

'"A few were rescued from the burning car; the dead taken from the half which jumped into the river were recovered and numbered ten. The first four taken from the half which was burned were alive but are terribly injured, and I doubt if they are still living.

'"Directly I jumped from the train I looked about to render some assistance. There were only two women to be seen. The shrieks of the injured and dying could be heard, but I could not see anyone but the two women on our side of the river. There were only three of us to do anything, the remainder were on the other side of the bridge and could not get across. Later we were joined by two men who were engaged in bridge construction further up the river; we five worked and did all we could for five hours before assistance reached us.

'"The brakeman was thrown completely into the air and fell on the ice. No sooner had he struck than he jumped up, and scaling the bank came to the assistance of the others, though he was terribly bruised and shaken up after being thrown a distance of thirty feet to the ice.

'"The greatest thing I ever saw was the case of a young Norwegian of about 25 years old, when we extricated her she had three tremendous wounds across her head, all running parallel to her forehead. The top one was the worst and laid her skull open for an inch across. Her right leg was terribly torn and injured from the ankle to the knee, laying bare the bone.

'"In this condition she walked along the track and sat for five hours awaiting the arrival of the doctors. So great was the wound in her leg that the needle would not reach across, and I had to hold the parts together while the doctor sewed the leg up. During this operation she did not move a muscle or even wince. It was the greatest fear of endurance I ever witnessed.

'"Over the other side of the river I could see Bob Burroughs, district manager of the Bell Telephone company, who was travelling to the Soo, working like a Trojan. What was the worst of all was our helplessness when the two coaches leaped into the river. They went through the ice like bullets and did not break the ice more than a foot on each side of them."

'When the mail clerk left no bodies had been taken from either of the sunken coaches. All in these have perished except the dining car attendant and the brakeman. The Pullman car lies on its side a wreck. The priest from Blind River is dead.' [1100]

The Sudbury inquest, on the death of one passenger, as representative of all those who had died, concluded on Friday 4th February. The jury's verdict was that, "Jacob Rysynszyk came to his death as a result of the derailment of part of train No. 7, on Jan. 21st, 1910, at a point on the 'Soo' branch of the C.P.R., where the railway crosses the Spanish river, and on the easterly approach to the bridge.

"Said derailment being caused by the forward car of the first-class car leaving the track and plunging over the embankment, followed by the dining and sleeping car, also causing derailment of the second-class car. But this jury is unable to determine the cause for the said first-class car leaving the track.

"We would recommend that the following be thoroughly investigated by the Royal Commission for Canada:

(1) Are three sectionmen sufficient to keep an eight mile section in the winter time, in the rigorous climate of Northern Ontario?

(2) Should openings be placed in the roof of cars, capable of being opened either from inside or outside of the car, suitable for the escape of passengers from overturned cars?

(3) Should emergency tools be carried at convenient places outside of cars as well as inside?" [1101]

The report of the Royal Commission has not come to light.

The exact cause of the accident was never established. It was either a broken rail or a defect in the coach of the train such as a broken wheel, but it was impossible to establish which. In either case with the technology of the day the accident was basically unavoidable.

The final death toll was put at 43.

Like the Spanish River disaster an accident on the Atlantic Coast Line Railroad on Saturday 25th March 1911 near Ocilla, Georgia, United States, was due to an unavoidable cause. Train No. 95, the *Dixie Flyer*, from Chicago, Illinois, to Jacksonville, Florida, was crossing the Allapaha trestle at 7.20 a.m. when the axle of the driving wheel of engine No. 211 broke. The train had only been going at a moderate speed when the accident had happened without warning. The engine remained on the trestle but the baggage car behind it was telescoped, four hundred feet of the trestle collapsed and the first- and

---

[1100] *The Winnipeg Tribune*, Winnipeg, Manitoba, Canada, 24th January 1910

[1101] *The Times & Guide*, Weston, Ontario, 11th February 1910 second-class coaches and the first Pullman sleeping car fell into the river. Another report stated that the baggage car, two express cars, two day coaches and a Pullman had gone off the trestle. The four Pullman cars at the rear of the train remained on the track. Initially it had been reported that the whole train had gone into the river and that 40 had died but in fact there were only seven or possibly eight deaths. Those who definitely lost their lives were Conductor Charles Parnell, Fireman Lucian Ellis, Porter Whidden, Express Messenger Woodward and passengers O. F. Bomwart, of Henderson, Kentucky, W. Culpepper, of Tifton, Georgia and Mrs W. D. Fletcher, of Rowland, Illinois. Express Messenger and Baggagemaster Powell and seven passengers were injured.

Following the accident a hospital train left Waycross for the scene at 8.40 a.m., also a relief train from Tifton and company doctors on a handcar from Willachoochee, the nearest station on the Waycross side of the wreck.

A few weeks earlier engine No. 211 had made a record run with the same fireman and Engineer Harry Wyche. It was not stated if Engineer Wyche was driving the train on the day of the accident. The line was expected to be out of use for 36 hours.[1102]

A train of eight cars, one carrying 80 miners and the other seven loaded with cement, broke through the Rancagua Bridge, Chile, possibly after becoming derailed, and fell 150 into the ravine below. The accident happened near the Braden copper mine on Friday 24th February 1911. Eighteen of the miners were killed and the remainder injured, as was the American engineer of the train, named Albert Bragenton, who was from San Francisco.[1103]

A passenger train from Port Arthur to Grahamstown, Cape Colony, South Africa, due at Grahamstown at 10.20 a.m. on Saturday 22nd April 1911 at the Blaauwkrantz Bridge on the Kowie Railway. The third vehicle of the train became derailed 50 yards short of the bridge, which crossed a ravine with a maximum depth of 250 feet in one span. Reports that the bridge itself had collapsed proved incorrect, it remained intact, but the derailed part of the train, a van and four carriages, toppled down the embankment and over the side of the bridge. Initially the derailed vehicles had remained more or less in line but a check rail at the beginning of the bridge proved their undoing. The engine and leading two vehicles of the train remained on the track, the coupling at the front of the derailed third truck having snapped. The guard of the train jumped from it and likewise escaped but the passengers were not so lucky and at least 29, and possibly as many as 32, lost their lives.

Interviewed afterwards the engine driver said that he had no warning of what had happened until his engine, deprived of its load, had suddenly shot forward. He had then put on full steam and gone on to Grahamstown for help.

'Some of the dead and injured and a mass of wreckage were caught on a ledge about 120ft down. The removal of the debris from this point and from the crags was most difficult, but in six hours the doctors and nurses extricated the dead and injured, who

1102 *Weekly Times Enterprise and South Georgia Progress*, Thomasville, Georgia, United States, 30th March 1911

1103 *Washington Post*, Washington, D.C., United States, 25th February 1911 were hoisted up in sacking to the edge of the gorge. The faces of the dead were contorted, and the bodies horribly mutilated.

'A little girl had a miraculous escape. She was caught in one of the forks of the bridge and held by one foot over the ravine. The rescuers had to scale the girders to a height of 200ft., and brought he down safely, crying for her mother, but unhurt. A huge crowd of whites and natives awaited the arrival at Grahamstown of the trains bringing the dead and injured, and affecting scenes were witnessed.'

Most of the dead and injured were taken in the relief train to Grahamstown, though those living south of the bridge were taken to Kowie.[1104]

In the Union Assembly on Monday 24th April the Minister of Railways, Mr Sauer, announced that the number of dead was 29 and was expected to rise to 30. He stated that the railway belonged to a private company and that therefore no responsibility attached to the government in connection with the accident. A Commission of Inquiry would be appointed to investigate the accident, which would include an expert in railway matters. On the same day most of the victims of the accident were buried at Grahamstown, all businesses, banks and courts being closed. Meanwhile the bridge, being undamaged, had already reopened to traffic.[1105]

At the inquest into the accident on 26th August a policeman was questioned as to why the debris from the accident had been burnt on the Sunday morning, the burning wreckage enveloping the bridge in a blue haze. He stated that it had been burned by order of the manager of the railway company. Had the wreckage not been already examined by an investigator this action could well have destroyed any evidence as to a defective vehicle which might have caused the accident. A more obvious cause, however, which could not be, or at any rate was not, covered up, was the condition of the track, for, 'Railway officials testified to finding fifty bad sleepers between the place of the first and final derailments. The dogspikes were loose in the rotten sleepers, and the road was in very poor order. Two bad timbers were found on the bridge, one at the spot where the train went over. It was further stated that the line curved twice between the place of the first derailment and the bridge.' [1106]

At the conclusion of the inquest on the victims of the accident on 8th May the coroner's jury found that it was due to the derailment of the train on the curve approaching the bridge. They added a rider that the derailment had been due to the poor condition in which the permanent way and rolling stock were maintained and recommended that they be immediately examined and subject to annual examinations in the future.[1107]

Union veterans of the American Civil War and their families were the principal sufferers in an accident on the Lehigh Valley Railroad at Manchester, New York, United States, on Friday 25th August 1911. They had been attending an encampment of the Grand Army of the Republic at Rochester and had boarded the train at Rochester which

---

[1104] *The Standard*, London, England, 24th April 1911
[1105] *The Times*, London, England, 25th April 1911
[1106] *The Standard*, London, England, 27th April 1911
[1107] *The Times*, London, England, 9th May 1911

connected to train No. 4 (that involved in the accident) at Rochester Junction in order to reach their homes in Pennsylvania and New Jersey that night.[1108] To accommodate this additional traffic two extra coaches were added to train No. 4 at Rochester Junction and from there it consisted of locomotives Nos. 2476 and 1804, an express car, a mail car, a baggage car, four coaches, the Pullman sleeping car *Austin*, a dining car, two coaches (those added at Rochester Junction), the parlor car *Emelyn* and two coaches. It was running about 40 minutes late when it reached Manchester.

The train had reduced speed on the approach to Manchester yard as the signals had been at caution and was probably not travelling at more than 25 miles an hour as it passed through the station. Just beyond was a steel girder deck bridge about 178 feet long with two bents across a waterway known as the Canandaigua Outlet, leading from Canandaigua Lake. The bridge had substantial masonry abutments and stood about 40 feet above the water. As the enginemen began to increase speed after passing the station and close to the bridge the train became derailed. Two engines and leading five vehicles kept the track, one truck of Grand Trunk Railroad coach No. 2187, sixth in the train, left the track; the Pullman car *Austin* ended up leaning over, the dining car went down the far embankment, the tenth vehicle, Lehigh Valley coach No. 297, ended up on its side in the creek 40 feet below the bridge, and the eleventh coach, Lehigh Valley coach No. 294, ended up with one end in the creek and the other, far end, poking up above the level of the bridge. The parlor car *Emelyn* and the last two coaches remained on the track. Most of the fatalities occurred in the tenth and eleventh vehicles, the two Lehigh Valley coaches added at Rochester Junction for the G.A.R. veterans, which were of wooden construction.[1109]

After the accident, 'Residents of Manchester were also brought quickly by the shrieking of the engines and went to work with a will. There was delay in getting notice to the railroad officials at other points and so summoning help, because the cars in thrashing along the track had cut the telegraph and telephone wires. Telephones in the town were used, however, and it was not long before physicians and nurses from Canandaigua, Geneva and Rochester were on the way in relief trains, while wrecking crews were hastened here as fast as possible.

'There was much confusion in the early efforts at rescue. Many of the passengers who had relatives in the smashed cars ran about in hysterics and added to it. The volunteer workers had no leadership, but simply went at the wreckage as best they could, digging hard wherever shrieks or groans guided them to the injured and pushing away wreckage that held the bodies of dead. It was as a result of this that some of the rescuers removed means of identification from bodies and were unable afterward to tell from which ones they had come.

'It was necessary to chop through the sides and bottoms of the cars and the work of removing the victims went with painful slowness. Death had come swiftly to many a large number of the dead having had their skulls crushed in when they were thrown with terrific velocity against the car seats and projections. The mortality was high among the elderly passengers.

'The station was turned into a temporary hospital and the injured were attended by

---

[1108] *The New York Times*, New York, New York, United States, 26th August 1911
[1109] Interstate Commerce Commission report

local physicians until those who had come from other places were able to give relief. The dead were taken in such vehicles as could be had to the Morgue to Shortsville, two miles away, to await identification and the orders of the Coroner.

'When the special relief train arrived from Rochester about sixty of the injured were placed aboard and, attended by doctors and nurses, were rushed to that city, to be accommodated at its hospitals. Several children were taken to the Clifton Springs Sanitarium.

'The second car, after standing on its end, fell into the gulch this evening. Late tonight three persons were taken out of the vestibule. Two were dead and the other, a small boy, was still alive though severely injured.

'Several priests were aboard the train, and two nuns, who were killed. The priests were among the dying and administered the last rites of the church to them. Father Thomas J. Hendrick, of Ovid, Father William Harrington, of Ithaca, and Father James Griffin, of Waverly, had received a bad shaking up and suffered from some shock, but they did not spare themselves.

'The priests turned their attention at first to the Roman Catholics, but as Father Harrington was passing one woman she called to him, and when he asked if she was a Catholic she replied that she was not, and then asked for a prayer. The priest knelt and offered one. After that the priests administered to all who would accept.

'Sisters Hortensia and Veda of a Geneva convent, though suffering from shock and bruises, also went among the suffering and dying to render aid and comfort.'

A newspaper reporter managed to find two passengers who briefly related their experiences of the accident. 'James McDermott, a Grand Army veteran of Scranton, said as he left this city {Wilkesbarre, Pennsylvania}:

'"I was sitting in one of the forward cars when I felt the wheels of the car in which I was riding leave the rails and go running over the ties of the trestle we were crossing. In another second I was rolled out of my seat. Then the train came to a standstill. I looked out of the window and saw two of the cars back of me over the embankment. I left my car and went back, but I could render little assistance. Two men soon appeared with axes and they started to cut through one of the wrecked cars in which many veteran passengers were imprisoned. The sight of those who were injured, carrying the dead and injured from the cars, was sickening even to a hardened soldier."

'J. W. Morris, of New Brunswick, N. J., who was a delegate to the G.A.R. convention, also was interviewed:

'"I was in the War of the Rebellion and saw much bloodshed, but today's accident came so unexpectedly that I was unnerved. I was in one of the forward cars, the front trucks of which left the rails."' [1110]

With the death of one of the injured passengers some days after the accident the final toll was 28 passengers and one railroad employee dead, and 59 passengers and four employees injured.

The cause of the derailment was immediately obvious, the Lehigh Valley Company issuing a statement on the night of the accident that it had been due to a broken rail. The

---

[1110] *The New York Times*, New York, New York, United States, 26th August 1911 accident was also one of the first for which a report was issued by the Interstate Commerce Commission. After describing the train, the location and the circumstances of the accident the commission noted that the line was signalled by the Hall Automatic Block System, which meant that any interruption in the track circuits in the vicinity of the derailment, such as would be caused by a broken rail, would automatically place and maintain the semaphore signals at danger. As the Manchester towerman had been able to clear his signals for the approaching train, it was clear that the rail could not have been broken prior to its arrival. It was therefore very probable that it was the admission of steam to the cylinders of the two locomotives to get the train under full headway after clearing the station which had taken the load off the pilot wheels of the engines, resulting in an extra load on the driving wheels, which had put too much pressure on the defective rail, causing it to break.

The rail in question had been manufactured by the Bethlehem Steel Company by the open hearth process, being rolled on Christmas Eve 1909. It had been laid at the end of October 1910 and shortened in April 1911 in connection with the installation of a new frog in a crossover. After the accident it was found to be broken into many pieces. Seventeen pieces were found but not all of the rail. Inspection showed that it was what was known as a "piped rail", a defect caused by slag having got into it in the furnace. There were also transverse fissures in the head of the rail, defects which were not due to its manufacture but had developed after it was laid.

The commission noted that once the rail had been laid its defects could not have been detected even by careful inspection, but that they could and should have been discovered and the rail rejected when holes were cut for new fishplates after the rail was shortened in April 1911. As to the more general question of defective rails, it was extremely difficult to suggest any means of preventing future accidents in the light of the current knowledge, and the commission recommended that a most complete and searching examination should be made of the whole question.[1111]

The mixed train which left Tallarook for Mansfield at 9 a.m. on Monday 25th September 1911 managed to get no further than a bridge over Native Dog Creek at Harvey's Gully, Victoria, Australia. The train consisted of two passenger coaches and three trucks. As it reached the bridge it became derailed and a composite coach and at least one of the trucks fell between 20 and 30 feet from the bridge to the gully below. No one was killed but 16 of the 35 passengers on the train were more or less injured.

The accident happened at a somewhat remote location between Molesworth and Cheviot stations, so it was some time before assistance could be obtained, though help did not, perhaps, come as quickly as it might have done. In the meantime the uninjured passengers and train crew did what they could to help the casualties. The accident happened at 11.50 a.m. According to newspaper reports the Yea stationmaster wired for the casualty train to Seymour at 12.10 p.m. and it set out at 1.30 but did not reach the accident site, 20 miles away, until 3.40 p.m., while a second relief train was despatched from Melbourne Spencer Street, 90 miles away. Given the reported remoteness of the accident, the speed with which news of the accident reached Yea, several miles away, seems a little surprising, but presumably the engine of the train, which was not derailed,

[1111] Interstate Commerce Commission report

was hurried down the line to convey the news; the newspaper accounts do not say. The Seymour relief train came equipped with a makeshift telephone but when an attempt was made to set it up it could not be made to work.

Commissioner McClelland ordered a Board of Inquiry to be held into the accident in addition to the internal departmental inquiry.

Driver J. Redmond, who had been in charge of the engine of the wrecked train, was one of the most experienced drivers in the service and very careful. He stated that the restriction in speed at the curve where the accident had occurred was always carefully observed.[1112]

The report of the Board of Inquiry into the accident was published on 21st October. It did not have an easy job deciding how the accident had happened. The track was in good condition, as was the rolling stock, including the empty cattle truck which was supposed to be the first vehicle to have left the rails. There was no evidence of any obstruction on the line and the driver had, in the opinion of the board, been travelling within the permitted speed. The theory put forward by the board was that with the train running at close to its maximum permitted speed the empty cattle waggon would have been oscillating. At the point of derailment there was a change in the gradient from 1 in 87 to 1 in 40 and a momentary expansion of the couplings might have caused a jolt to middle and lower portions of the train sufficient to lift a wheel of the cattle truck from the rails and cause the derailment.[1113]

In a sort of parallel with the Manchester disaster the year before, it was a train carrying Confederate American Civil War veterans which was wrecked near Eastabuchie, Mississippi, United States, on Monday 6th May 1912.

The Vansandt Confederate Veterans Special was carrying passengers from points in Texas to a reunion at Macon, Georgia. It consisted of locomotive No. 293, a combination baggage and passenger coach, two passenger coaches, two tourist sleeping cars, a dining car and five sleeping cars and was running as train extra No. 293 with Conductor White, Engineer W. A. Woods and Fireman Cicero Jones. The accident occurred on the New Orleans and Northeastern Railroad.[1114] [1115]

Bridge No. 160, from which the train fell, was a wooden trestle 140 ft long over what was often a dry creek, but which had some water in it at the time of the accident due to recent heavy rainfall. At about 8.40 a.m. the train, which was running an hour and 20 minutes late, became derailed at a point 67 feet south of the trestle. The engine and tender and the combination baggage and passenger car passed over the bridge before overturning onto their sides about 180 feet further north, while the following two passenger coaches remained more or less upright. Unfortunately the next two vehicles, the tourist sleeping cars, crashed through the trestle, telescoped one another and were practically destroyed. Only the front truck of the sixth vehicle, the dining car, left the rails and the five rear sleeping cars kept the track. Engineer Woods, his negro fireman and

[1112] *The Mount Alexander Mail*, Castlemaine, Victoria, United States, 26th and 27th September 1911

[1113] *The Argus*, Melbourne, Victoria, Australia, 23rd October 1911

[1114] *The Daily Item*, Georgetown, South Carolina, United States, 7th May 1912

[1115] *The Waterloo Press*, Waterloo, Indiana, United States, 9th May 1912 another railroad employee who was riding on the engine were killed as the engine rolled down an embankment, as were six of the passengers in the two coaches which went into the water. The dead passengers included three women and two children; none of the veterans themselves lost their lives. Two Pullman porters, a dining car porter and 53 passengers were injured.[1116]

The accident did not receive much coverage in the press but was briefly mentioned and at least one newspaper reported on its aftermath. 'The nurses and doctors found the veterans aiding their comrades as best they could. Many of the most seriously injured old men went to work with remarkable courage. As soon as possible the sleepers were pulled into Hattiesburg with eight bodies and those most seriously hurt.

'Hattiesburg citizens, as soon as they heard of the disaster, made arrangements to care for the living and the dead. The United Daughters of the Confederacy, the Kings Daughters, the Masons, the Elks and other benevolent organisations sent representatives to the scene.' [1117]

Conductor White told the Interstate Commerce Commission inquiry that he had been riding in the second car when the train became derailed. Climbing out of a window he had gone forward to the engine, then had checked that the flagman was protecting the rear of his train. He had then looked at his watch and it was 8.42 a.m.; he put the time of the derailment at 8.40. Going forward again he had seen that Trainmaster Stollenwerck was going forward towards Eastabuchie and he had then assisted the passengers. He thought that the front trucks of his car had been the first to leave the rails and put the speed of his train at 30 miles an hour or perhaps a little more.

Flagman Ellison said that when the train had come to a sudden stop he had gone back to flag an approaching train. He put the speed of the train at 25 to 30 miles an hour and said that he had wondered why it was going so slowly.

Mr F. J. Lees, an eyewitness to the accident, also gave evidence, and confirmed that there had been nothing unusual about the speed of the train and that the engine and leading vehicles had passed over the point of the initial derailment safely but had been pulled off the line by the derailed vehicles behind them.

The commission found that the derailment had been caused by the soft and uneven condition of the track, which had been in bad condition for some time. It had been necessary to repair the track twice in the 10 days preceding the accident and the track was in the process of reconstruction at the time of the derailment.

The commission pointed out that the first three cars of the train, which had steel underframes, had been little damaged. The two wooden tourist cars in which the casualties had occurred had been one 26 and the other 32 years old and it stated, 'What happened to these two cars cannot but lend emphasis to the recommendation contained in the Commission's last report to Congress that the adoption of steel cars be required.' Likewise the bridge, though in good condition and well maintained, had not been strong enough to withstand the impact of the derailed vehicles. Had modern coaches or a stronger bridge been provided the number of casualties would doubtless have been reduced.

[1116] Interstate Commerce Commission report

[1117] *The Cedar Rapids Republican*, Cedar Rapids, Iowa, United States, 7th May 1912

Finally the Interstate Commerce Commission noted that the schedule for train extra No. 293 as per its train order had called for a speed only two minutes slower than that of the fastest train on the division. With the track in the condition in which it had been a slow board should have been erected to limit the speed of all trains, especially in view of the prolonged rains which had prevailed throughout the south, which could not have failed to weaken a piece of track already in soft condition.[1118]

A mixed train on the Lancaster and Chester Railroad with 79 passengers, many en route to a play-off baseball game in Chester, met with an accident at Hooper's Creek, South Carolina, United States. A freight car in the train was derailed on or approaching a trestle and the shock of the derailment caused the structure to collapse beneath the train. Five people died and the claims for damages almost bankrupted the company. According to an internet web page the accident occurred on Monday 30th June 1913, but it has not been possible to trace any newspaper reference to it around this date.[1119]

On Sunday 19th October 1913 the second section of northbound passenger train No. 4 on the Mobile and Ohio Railroad consisted of locomotive No. 353, a baggage car and three coaches, in charge of Conductor Pruitt with Engineer Beasley. It was carrying a detachment of troops from Fort Morgan, Alabama, to the Alabama-Mississippi Fair at Meridian, Mississippi. It left Mobile, Alabama, at 12.05 p.m., passed State Line, Mississippi, the last open telegraph office before the point of the accident, at 1.54 p.m. and at about 1.50 p.m. became derailed 300 feet south of bridge No. 67-B, about three miles short of Buckatunna, Mississippi, United States. The engine was not derailed and passed over the bridge, but its tender was, and its derailed wheels damaged the ties to the extent that they caused the bridge to collapse. The baggage car came to rest with its forward end jammed in the bank of the creek, badly damaged, and 10 feet of the first coach were telescoped by the second coach. All the coaches were very badly damaged, and 17 soldiers were killed, and 139 soldiers and six railroad employees were injured. Most of the fatalities occurred in the baggage car and first coach; only one passenger in the third coach was killed. In the baggage car the casualties were caused by the iron express safe and heavy baggage and equipment being thrown from one end of the car to the other.

The train was believed to have been travelling at about 55 miles an hour at the point of derailment.

The bridge was described as having been of substantial construction and would not have failed but for the derailment. There were no guard rails, but it was thought doubtful if their provision would have averted the accident as the derailed wheels had been two feet from the rails by the time that the bridge was reached. The bridge was of wooden construction and was 225 ft long with a maximum height of 24 ft. It was well constructed and maintained and had received frequent attention in the way of repairs and renewals. The line fell towards the bridge for four miles at an average gradient of one-half of 1% (1 in 200).

[1118] Interstate Commerce Commission report

[1119] Webpage – Lancaster & Chester Railway

Engineer Beasley told the Interstate Commerce Commission (ICC) investigators that this had been only his second trip driving a passenger train. He thought that he had left Mobile at 12.12 p.m. (other employees indicated that it had actually been 12.05) and that he had been running at about 35 miles an hour when the jumping of the tender had indicated that it was derailed. He had immediately reversed the engine and made an emergency brake application, by which time the engine had reached the bridge. He had shut off steam at milepost 66, a mile short of the point of derailment. On previous trips he had slowed down for the curve at this point but had not done so on the day of the accident. He had only made one brief brake application since passing State Line. He had timed the train as taking a minute and 35 seconds to cover the mile between mileposts 65 and 66 (giving a speed of 36.3 miles an hour). A slow order received at Mobile had meant reductions of speed at three different points, with about a minute lost at each. He did not know the maximum permitted speed for passenger trains but did not think that his train had exceeded 40 or 45 miles an hour anywhere beyond State Line.

The conductor and fireman of the train put its speed at the moment it had left the track at 30 to 35 miles an hour, the flagman thought 40 to 45 miles an hour and the general agent of the passenger department, who had been riding on the train, agreed with the estimate of the conductor and fireman.

One of the officers, Lieutenant Cunningham, told the inquiry that there had been considerable discussion among the men regarding the speed of the train, many thinking it excessive. After leaving Vinegar Bend the speed had seemed to increase and had done so decidedly in the few minutes before the derailment.

Private Thompson was one of six men riding on the rear platform of the last car of the train. He said that he had been timing the train with a friend, he calling out the mileposts while his friend held a watch. The last two miles before the derailment had been covered in 65 and 64 seconds, giving speeds of 55.4 and 56.3 miles an hour, and with the train's speed increasing he had remarked that the next mile would probably be covered in an even 60 seconds.

Private Walker corroborated the statement of Private Cunningham. He said that the speed of the train had increased beyond State Line and that the last mile had been covered in 56 seconds.

Roadmaster Mulvoy stated that he had examined the scene after the accident. All the broken bridge ties and timbers had been in good condition and he had found nothing defective. The track at the site of the accident had been considered a prize section at the last annual inspection. He considered the bridge to be perfectly safe at the speed at which passenger trains were allowed to run.

Bridge Supervisor Bennett said that the bridge had been found to be in good condition at inspections on 17th July and 24th September.

Section Foreman Brennan stated that he had last inspected the line two days before the accident, when he had gauged the track and found it to be all right. The joints had been tight and in good condition.

Superintendent Pigford stated that Engineer Beasley had passed a first-class examination on the rules for road service in May. The ICC noted, however, that the engineer did not know the maximum permitted speed for passenger trains, which according to Rule 91-D was 50 miles an hour for first-class trains.

The commission concluded that the track at the point of derailment had been well maintained and bridge No. 67-B had been of substantial construction. Its collapse was clearly the result and not the cause of the derailment. There was nothing in the condition of the engine or tender which might have caused the train to leave the track. Opinions as to the speed of the train varied, but taking into consideration the testimony of the soldiers travelling on the platform at the rear of the train, the fact that the train had been running on a falling gradient for the last four miles with only a short application of five pounds of air, and that the distance of nearly five miles from State Line had been covered in about five minutes, it was likely that the speed of the train had been not less than 55 miles an hour when it left the rails.

While the cause of the derailment could not be determined with certainty, it was believed that it had been due to excessive speed on a 3-degree curve which had a superelevation of only 3½ inches on the outside rail. The excessive speed and insufficient superelevation had probably caused the left wheels of the tender to lift from the rails due to its high centre of gravity and the surging of water in its tank. When the tender had rocked back the flanges of the wheels had come down on the ball of the rail, run in that position for a short distance and then dropped off on the outside.

Engineer Beasley had joined the railroad when he was 14, had been a fireman for three years and eight months and an engineer since May 1913. He had received 30 demerits for driving a freight train at excessive speed on 3rd August.

The commission noted that excessive speed had been a factor in many derailments. Its report included a table of 16 derailments caused wholly or partly by excessive speed since 1st July 1911. Of these three were due to the trains being operated in excess of the permitted speed whilst in 13 there had been no speed limit imposed. A comment from a previous report was quoted which stated that, 'in many place and on many roads in this country track is not properly constructed or sufficiently well maintained to provide for the safe operation of trains at high speed. It is believed that means should be taken to ascertain the limit of speed at which can be safely operated and to provide and to provide an adequate margin between this limit of safety and the highest speed permitted or attained.'

Eleven of the 16 derailments had been tender derailments. These formed 'one of the most perplexing problems with which operating and mechanical offers have to contend.' They resulted due to a number of causes and the real cause could often not be determined. On account of their short wheel base, high centre of gravity and the surging of water from side to side within their tanks they were subject to forces with which it was particularly difficult to deal, and which were aggravated by any deficiencies in the track. Such derailments had occurred more or less frequently on every large railroad in the country. They should be given the closet possible study with a view to definitely ascertaining their causes and eliminating them as far as possible.[1120]

A bad accident in the Dutch East Indies on Thursday 2nd April 1914 was caused by an animal on the line. A morning train collided with a buffalo on or near a bridge. The train was derailed, and the bridge then collapsed beneath it so that the engine and leading five coaches fell into a river near Tanjong-Priok, not far from Batavia on the island of Java.

[1120] Interstate Commerce Commission report

Twenty 'natives' lost their lives in the accident. The Europeans chanced to be riding in the (presumably first-class) coaches at the rear of the train and were unhurt.[1121] [1122] [1123]

[1121] *The Times*, London, England, c4th April 1914
[1122] *Railway Gazette* Vol 20, pg. 500
[1123] *The Portland Guardian*, Portland, Victoria, Australia, 4th April 1914

CHAPTER 11

TUNNEL ACCIDENTS

In his classic work on British railway accidents, *Red for Danger*, L. T. C. Rolt remarks on the acute trepidation felt by Victorian travellers when they were passing through a tunnel, and their oft repeated pleas to the railway authorities that all tunnels should be artificially illuminated, pleas which, probably for practical reasons if nothing else, fell upon deaf ears. He goes on to point out that these fears were by no means unjustified, especially in the period before the introduction of the absolute block system, when trains were merely spaced by the time interval system. If a train should chance to break down in the open there was a good chance that the crew of a following train might be able to see it in time to pull up, even if there was insufficient time for the guard of the first train to go back and protect its rear. In the smoke- and steam-filled darkness of a tunnel there was no such hope. Furthermore, should a serious collision occur in a tunnel the confined space would be likely to result in the wreckage being hideously compressed, and should a fire break out in a long tunnel provided with ventilating shafts, the latter could act as chimneys. In addition the rescue work would be hampered by the darkness and the difficulty of working in a confined space where the use of breakdown cranes was impossible, and the only resort was to drag the wreckage bodily out of the tunnel.

The single line through Kingwood Tunnel, West Virginia, United States, on the Baltimore and Ohio Railroad, was the scene of a collision early on the morning of Monday 11th February 1901. An eastbound freight train had been banked through the tunnel by engine No. 1610. The engine, operated by Engineer James Dulling Jr., with Fireman Dulling (his son?), had the right of way to return back through the tunnel at once. It set off back but midway through the tunnel, which was about three quarters of a mile long, it collided with engine No. 1866, operated by Engineer William Linn, and pulling another heavy freight train which was also being banked in the rear. The collision, at 4.15 a.m., was described as having been terrific, with 10 or 12 cars being crushed between the two engines. Despite this the two engineers and two firemen, only the fireman of the helper engine was hurt to any extent, and he was able to walk out of the tunnel. The other members of the engine crews helped clear the wreckage, and their escape from death was described as miraculous. The narrowness of the tunnel meant that it was not cleared of wreckage until after 2 p.m. The collision was blamed on a failure to exhibit the danger signal to the second freight at West End.[1124] [1125]

At 1.55 a.m. on the morning of Saturday 29th June 1901 the residents part of Baltimore, Maryland, United States, were awakened by a boom and a rumble which caused large numbers of them to peer out of their windows in their night attire. Those in the immediate neighbourhood of the occurrence found that a hole about 30 feet wide had appeared in the middle of East Hoffman Street. In the midst of the hole water was pouring from both ends of a water main and the size of the hole was increasing until it threatened the residences lining the street, causing their inhabitants to hastily evacuate carrying their most valuable possessions. In addition gas pipes had broken and there was soon an unbearable smell of escaping gas.

What the residents were unlikely to have realised at this point was that a train was also involved in the affair. Pennsylvania Railroad train second No. 78 had left the Union station at 1.48 a.m. for New York and was passing through Union Tunnel, on the Pennsylvania and Western Maryland Railroad Joint stretch of line. Engineer John Moxley, describing the accident, said, "I heard the crash, felt the jar, and knew that a sudden resistance had been encountered by the train. It was not a sudden shock, as when a train strikes a hard obstacle, but it suggested running into an overturned carload of broom corn clippings.

"I immediately reversed my lever and brought the train to a full stop.

"Our crew went back in the tunnel to find the trouble and saw the rear car covered with the earth and rocks. I shut off all steam and the fireman banked the fires and then we made our way out of the Broadway end of the tunnel." [1126]

The engineer, Fireman Stewart and the rest of the crew went to the Gay Street tower to report the accident. Train second No. 78 was the second section of the Florida express and consisted of an engine and tender, three express cars of perishable goods, three Pullman sleepers and a day coach.

Pennsylvania Railroad General Manager Hutchinson told a newspaper reporter that, "As the train was coming through the tunnel at the usual slow rate of speed for such places the engineer suddenly felt that it was running over something on the track, and before he could do anything to stop her the engine was thrown off the track by a pile of sand which had fallen through the roof of the tunnel. Just as the engineer and fireman were banking the fires a moment after the derailment there was a sudden downfall from above which completely covered the middle of one of the express cars and half covered the other two. The passenger coaches were not hit by the fall. One of the express messengers in getting out of his car to get away from the danger, and that was the only injury which resulted from the affair.

"What caused the tunnel to cave in is what we would like to know but will not be able to find out for some time yet. The general theory is that the water pipe broke and the great force of the stream thus liberated gradually weakened the roof of the tunnel."

[1124] *The Sun*, Baltimore, Maryland, United States, 12th February 1901
[1125] *The Evening News*, Cumberland, Maryland, United States, 12th February 1901
[1126] *The Sun*, Baltimore, Maryland, United States, 29th June 1901

In a similar vein Pennsylvania Railroad Assistant Engineer Ford stated, "I understand that there is a 10-inch main in the bed of the street over the tunnel and that it carries a pressure of 100 pounds to the inch. When that broke it cut away the earth in all directions and made a gravity increasing cavity which weakened the roof of the tunnel and caused the accident."

The railroad officials' theory as to the cause of the roof failure was hotly disputed by the city council. Thus, 'Chief Engineer Quick, of the Water Board, stated yesterday that the four-inch water pipe mentioned in connection with the accident to the Union Railway Company's tunnel could in no wise been the cause of it. The four-inch main was not laid deeper than three or four feet from the surface, and the distance from the main to the crown of the tunnel's arch was not less than 20 feet. If there had been a leak on the main of any importance it would have been most likely to show first on the surface. The failure of the tunnel was the cause of the broken pipe.' [1127]

The dispute as to the cause of the tunnel collapse was resolved on 3rd July when the Pennsylvania Railroad admitted responsibility through its attorney. This admission followed an examination by Building Inspector Preston, who stated that his investigation had shown conclusively that the first brick ring at the bottom of the arch of the tunnel, known as the "soffit", had disintegrated. In the rings above this the bricks had not adhered to the mortar and did not support the arch. Pieces of timber used in the construction had also decayed and were in bad condition. He was of the opinion that the tunnel had failed because of defects in its original construction. A further examination was necessary to determine just how far the defects extended.[1128]

It was expected that it would take at least a fortnight to reopen the tunnel. Meanwhile, Pennsylvania trains were diverted via the Baltimore and Ohio line out of Baltimore, regaining their own line at Bay View.

A bad mistake by a signalman led to a double collision in Bowling Tunnel, England, on the Lancashire and Yorkshire Railway, on Thursday 24th October 1901. Fortunately the accident did not have serious consequences.

Signalman Joshua Parkinson was in the last hour of his late turn duty in Bowling Junction cabin at the east end of the tunnel. At 9.08 p.m. he accepted the 6.10 p.m. goods train from Low Moor to Leeds from Low Moor No. 4 cabin at the tunnel's west end. In passing through the tunnel the train became divided, leaving 16 waggons and the brake van standing in the tunnel. The front of the train passed Bowling Junction cabin at 9.16 p.m. and Signalman Parkinson failed to notice that it had no tail lamp or side lamps. At the subsequent Board of Trade inquiry he could only excuse his carelessness by pleading that the cutting in which his box stood had been full of smoke and that he thought that he had seen the shadow of a tail lamp. The Inspecting Officer, Major E. Druitt, was not very impressed with this excuse.

Having thus cleared back the goods with half of it still in the tunnel Signalman Parkinson immediately accepted another goods train, the 9.05 p.m. Low Moor to Laisterdyke, from Low Moor No. 4 cabin, and Signalman James Towns in the latter cabin

---

[1127] *The Sun*, Baltimore, Maryland, United States, 1st July 1901
[1128] *The Sun*, Baltimore, Maryland, United States, July 1901

sent the train on at the same time. Meanwhile in the opposite direction the 9 p.m. Leeds to Manchester passenger train had been offered to Signalman Parkinson by Hall Lane box at 9.13. It passed Bowling Junction at 9.18, at about the same time as the Laisterdyke goods entered the tunnel from the Low Moor end, with the result that the Laisterdyke goods collided with the detached vehicles of the Low Moor to Leeds train, two waggons of that train were derailed, fouling the opposite running line, and almost at once they were run into by the passenger train.

Back in Bowling Junction cabin Signalman Parkinson had been advised at 9.26 by the man at Hall Lane that only part of the first goods train had arrived there and had replied, "He must have left the remainder in the tunnel." He then rang Signalman Towns at Low Moor and asked if anyone had arrived at that end of the tunnel and was told, "No." He must have then spent some very worrying minutes wondering how bad an accident he had caused before Signalman Towns rang him back to say that the passenger train had arrived there and its driver had said that they had struck something in the tunnel. A little later the fireman of the Laisterdyke goods arrived at his cabin to tell him of the collision, closely followed by the guard of the Leeds goods. Signalman Parkinson was lucky inasmuch as the casualties in the accident were limited to the shaking up of the aforementioned guard; unfortunately in his distressed state he then compounded his sins by failing to send the "Obstruction Danger" signal to the boxes on either side, an omission commented adversely on by the inspecting officer.

As might be expected Major Druitt placed the entire blame for the two collisions on Signalman Parkinson for failing to check that the Leeds goods train had passed complete with a tail lamp. Driver Francis Cody of the Leeds goods also came in for some slight criticism. Noticing that his train was incomplete about the Bowling Junction advanced starting signal he had first stopped it and asked his fireman to go back and check that the last waggon was fit to travel before proceeding to Hall Lane box. The inspecting officer thought that Driver Cody might have exercised better judgement had he proceeded at once to Hall Road box, when he might have got there in time to prevent the first collision, but conceded that the driver had acted as he thought best.

Driver James Burchall of the Laisterdyke goods had had no chance of avoiding the collision with his view obscured in the smoke-filled tunnel.

It was impossible to say what had caused the disconnection of the thirteenth waggon of the Leeds train, a private owner waggon belonging to the Liversedge Coal Company.[1129]

The Liverpool Overhead Railway was a suburban line in England which ran from the centre of Liverpool northwards for seven miles to suburban Seaforth. When originally opened as a six-mile line it had run throughout on a viaduct, but it was later extended at each end, the southern extension passing through a tunnel about 800 yards to a terminus at Dingle. It was in this tunnel section that a serious fire occurred on the evening of Monday 23rd December 1901. The terminus station was situated within a widened portion of the tunnel and consisted of an island platform. Electric trains worked the line.

At 5.58 p.m. on 23rd December the train which was due to arrive at 5.52 approached

[1129] Board of Trade report

but stopped 80 yards short of the platform because its rear motor had failed, due, Mr Alex Trotter of the Board of Trade suggested, to a breaking down of the insulation of a cable connected to the motor which had given rise to an electric arc which had caused a rush of current opening a circuit breaker and stopping the train. Motorman Robert Ashbee had then made several attempts to restart the train but without success, and each time the current was turned on a vivid flash was produced which ended up setting fire to the woodwork of the coach. A gale was blowing from the west, causing the fire to spread from coach to coach until the whole train was enveloped after about 12 minutes. At first the 29 passengers on the train were urged by the guard and motorman to keep their seats as there was no danger. The motorman and Guard Charles Maloney having made futile attempts to control the blaze to passengers had to leave the train and the majority of them left the station and made their way home with no ill effects. A few, however, lingered to watch the fire and the efforts made to extinguish it.

Station Foreman Thomas Rendell and Car Cleaner Joseph O'Brien were called to assist and about 10 minutes later Station Foreman Rendell sent two telephone messages to the generating station for the current to be cut off, one via the National Telephone Company network and the other via the railway network. The current was duly cut off but too late to affect the fire and all it did was to plunge the station into darkness. The fire also gave rise to large volumes of hot and toxic smoke which soon filled the station.

The last people to escape from the station were Signalman William Owen, Booking Boy George Gough and a passenger, Mr George Stewart, all of whom lost consciousness at some point in doing so. It was impossible to be certain what happened to the victims who lost their lives in the accident, but it seemed that Station Foreman Rendell had guided two passengers, James Beadon and Forbes Bingham, to an air shaft where they had hoped to find safety. Instead the shaft had become a flue up which the smoke and fumes were drawn and all three died there of suffocation. Car Cleaner O'Brien was guided to the foot of a staircase by the boy Gough but although the boy escaped the car cleaner did not, perhaps because he lost his way in the smoke and darkness, and he crawled back to the platform and succumbed there, while Motorman Ashbee and Guard Maloney stayed too long at the train and became the fifth and sixth fatalities.[1130]

The inquest on the accident resulted in a verdict of "Death due to accidental suffocation", but the jury considered that a great oversight had been made in storing sleepers in a dead end siding near the end of the down platform.[1131]

Lt-Col H. A. Yorke, who conducted the Board of Trade inquiry jointly with Mr Trotter, noted the remarkable way in which the fire had spread from the incoming train first to the sleepers, then to the signal box and then to a spare train standing at the far end of the station, but stated that he did not think that the company could fairly be blamed for storing some sleepers in the corner of a wide tunnel. He noted, though, that no timber of any sort would be stored in the tunnel in the future.

Concluding the report, Lt-Col Yorke stated that the cause of the fire had been, as detailed by Mr Trotter, a defect in one of the motors of the train, but that no serious danger would have arisen had Motorman Ashbee acted with a moderate degree of prudence, and disconnected the defective motor, instead of making repeated attempts to

---

[1130] Ibid.

[1131] *The Times*, London, England, 9th January 1902

get it to work. Motorman Ashbee, Guard Maloney and Station Foreman Rendell had also, the inspecting officer wrote, displayed a lamentable lack of judgement in not hurrying all the passengers out of the fire and not permitting any of them to loiter. Furthermore although the railwaymen deserved credit for attempting to prevent the fire spreading had they realised sooner that the train was doomed they too might have escaped. The eventual cutting off of the current had done no good and by plunging the station into darkness had increased the difficulty and danger of the situation.

The accident showed the advisability of removing as far as possible all woodwork in the vicinity of electric equipment on trains, of using non-flammable and smokeless insulating material, and of using as little woodwork as possible in stations on underground railways, with platforms of stone or concrete and buildings such as signal boxes of brick or iron. Mr Trotter made similar recommendations in an appendix to the report.[1132]

After the inquest Booking Boy Gough was presented with the Fire Brigades' Association silver medal for his help in rescuing passengers.[1133]

The inexplicable failure of an engineer resulted in a disastrous rear collision in Park Avenue Tunnel, New York, United States, at 8.20 a.m. on the morning of Wednesday 8th January 1902. The tunnel, on the approach to the New York Central and Hudson River Railroad Grand Central terminus, had come into use in 1876 when the lines were depressed to eliminate a number of level crossings. It was a sort of half-tunnel, of the four tracks through it only to the two outside tracks were covered over, the two centre tracks being open. Smoke and steam tended to linger in the tunnel and there had been several previous collisions in it, especially the one on 20th February 1891 dealt with in the previous volume in this series.

On the morning of the accident a New York New Haven and Hartford Railroad accommodation train from Danbury, Connecticut, had arrived at Mott Haven Junction, east of the tunnel, at 8.07 a.m. and was there switched to the most easterly track (No. 1). This train was due in Grand Central terminus at 8.17 a.m. Its last two cars were reserved for New Rochelle passengers and there were about 60 of them in its rearmost car.

In front of Danbury train on track No. 2 was a White Plains local train, due at the terminus at 8.14 and which passed Mott Haven Junction at 8.03½ with Engineer John Wisker and Fireman E. W. Fyler on its locomotive. This train, however, was stopped at 110th Street to give precedence to a Croton local train and at 96th Street the Danbury train was crossed over to No. 2 track in front of it in accordance with the normal operating policy of trains using the two centre tracks where possible. The White Haven train was thus held there until the Danbury train had cleared 86th Street. The Danbury train, meanwhile, progressed as far as 56th Street but was stopped by signals there waiting for a platform to become available at Grand Central and Signalman Charles Flynn in the tower at 59th Street kept his signals at danger against the following White Haven train. White was apparently still used as a "Clear" indication on the New York Central at this date, with green as a caution signal, so his home signal at 58th Street was showing red and its associated distant signal at 63rd Street was green.

1132 Board of Trade report
1133 *The Times*, London, England, 9th January 1902

As an additional precaution Flagman Fred Barnum of the Danbury train had gone back 200 yards with a red lantern and a torpedo to protect his train, laying the torpedo on the rail between 57th and 58th Street and positioning himself there.

According to the initial newspaper report on the accident the White Plains train had approached at about 20 miles an hour. The semi-automatic signal at 77th Street was white but that at 63rd street was, as already stated, at caution. Engineer Wisker failing to react by shutting off steam and applying the brakes Fireman Fyler had called out, "Green, green." It seemed to the fireman that the engineer heard him but he had taken no action. Shortly the home signal at 58th Street came into view, at "Danger". Engineer Wisker had still done nothing and Fireman Fyler had shouted, "Red, red." Meanwhile Flagman Barnum had been waving his red lantern at the White Plains train, it had exploded the torpedo he had laid and a second torpedo at the home signal, placed there automatically when the signal was red, and a 12-inch gong also sounded at the signal. When the White Haven train passed him the Flagman Barnum had smashed his lantern against the engine cab and as the train passed the home signal Fireman Fyler had jumped back into the tender. Finally, when his engine was only about a hundred yards from the Danbury train Engineer Wisker had seen its rear end, reversed his engine and threw on the air brakes. By then it had been too late. The engine of the White Haven train ploughed halfway through the rear car of the Danbury train, the locomotive itself being very little damaged. Engineer Wisker had stuck to his post and remained on his engine until he was arrested on the order of the coroner.

A New York newspaper described the scene. 'That a terrible tragedy had occurred was evident immediately. Above the hissing of escaping steam could be heard shrieks and groans of pain. There was a din of breaking glass and crackling woodwork. The boiler of the engine was inside the wrecked car. Dead and wounded were on each side and in front of the boiler, filled as it was with boiling water and steam.

'The force of the engine's impact had telescoped the rear car on to the car ahead. The passengers who were sitting in the forward portion of the car, therefore, were pushed backward. Those sitting in the rear were plunged upward and forward. Only a space of some nine feet remained between the headlight of the engine and the platform of the second car. In this small space and in the area above and on the side of the intensely hot boiler were jammed the bodies of the sixty passengers who had occupied that car. Men and women were piled in inextricable confusion. Dead bodies in several instances pinned to the floor wounded women. The dead were terribly mangled.

'The damage done to those in the rear train and in the forward cars of the Danbury train was largely due to fright. A few were bruised or cut, but many fainted or suffered severe nervous shocks. As soon as could be the men who maintained presence of mind opened the doors of the other cars and enabled the passengers to alight. The tunnel was so filled with fumes, however, that little could be seen one way or another.'

Brakeman Frank Manent of the White Plains train went forward immediately after the collision and stationed himself on No. 3 track to protect the obstruction from any outbound train.

The noise of the collision was heard by a policeman in Park Avenue, above the tunnel, and the fire alarm was sounded. Acting Battalion Chief Freel and his men were soon on the scene and scaling ladders were used to access the train from the road above. It was found necessary to cut open the roof of the cars to get to the trapped passengers, who were in danger of scalding to death from steam from the engine boiler. The dead and injured were thus removed from the train, the bodies of those killed being laid on cushions removed from the train which had been laid on track No. 3. The injured were removed to various hospitals, some of which were almost overwhelmed by the number of casualties, especially the Presbyterian Hospital. Trains into and out of Grand Central station were delayed up until 4 p.m., when the lines were cleared; the passengers from many incoming trains were let off at Mott Haven and Harlem stations.[1134]

Seventeen passengers in all were killed outright or died later and a large number of others were injured.

Engineer Wisker made a statement after the accident. "I did the best I could under the circumstances. The tunnel was choked with smoke and steam. I was practically running in the dark. I could not see anything. I did not see the green light or the red light and I did not hear any torpedoes. I heard no shouts or warnings. Because I was running in the dark, and not because of any signals, I shut off steam. Next I applied the service brakes, and still being in the dark, I put on the emergency brakes, and then the crash came.

"An outgoing train had passed a few seconds before and the smoke and steam from this train united with my engine to blot out the light. I was on the alert all the time. I had a premonition that something was wrong before the crash came. I was looking ahead anxiously and trying to penetrate the cloud of smoke and steam for the signals, but I could see nothing. It is true that I was a little behind time and was anxious to get in as soon as possible."

Tests made after the accident by the New York State Railroad Commission confirmed the engineer's story, as a newspaper reported on 11th January. 'It was demonstrated again and again during the test that whenever a train passed and dense cloud of smoke and steam was discharged, which completely obscured the signals and made it impossible to see objects more than eight or ten feet away. The obscurity lasted at times thirty seconds and at others as much as sixty seconds. How an engineer riding on a locomotive travelling at high speed could be expected to see a signal that might be hidden from view for half a minute or a minute was a puzzle to the investigators who made yesterday's test.'

It also seemed likely that at least the torpedo placed automatically on the rail when the home signal was at "Danger" did not explode. 'Another important development was the fact that the automatic torpedo-system which supplements the light signals is defective. At the signal station at Fifty-ninth street, where the railroad officials declared a torpedo exploded to warn Engineer Whisker, six separate attempts were made yesterday to explode the torpedo by running engines over it, and the torpedo failed to explode every time. Expert mechanics and signal men were called by the embarrassed railroad officials, and they tried to put the mechanism into working order; fresh torpedoes were used and two kinds of locomotives employed, a heavy one and a light one, but the torpedo did not explode. The striking apparatus - it is called an anvil - was wedged up, but even then the torpedo did not explode.' [1135]

On 24th January the coroner's jury at the inquest on the victims of the accident found

---

[1134] *The New York Times*, New York, New York, United States, 9th January 1902
[1135] *The World*, New York, New York, United States, 11th January 1902

the New York Central Company responsible for the accident, absolving Engineer Wisker from blame. The jury found that the four short block sections through the tunnel, with signals which were very difficult to see, were a constant source of danger, with engineers having to rely on their sense of location, and condemned the railroad company for having done nothing to alleviate a danger which had been present for 10 years. The same day the State Railroad Commissioners came to the same conclusion.[1136]

Various solutions were proposed. Electrification of the tunnel was ruled out because of the difficulty of providing enough electric power for a long train, the inadaptability of the third-rail electric system with the maze of tracks at the Grand Central terminus and the amount of time which would be occupied in switching from electric to steam locomotion at Mott Haven Junction. Another suggested solution was increasing the ventilation within the tunnel so that the air could be entirely renewed within two minutes.[1137] The Railroad Commissions proposed the operation of the tunnel as a single-block section rather than four and suggested that this was entirely feasible, but the railroad company strongly disagreed. *The World* newspaper stressed that there needed to be 'SOME relief plan at once'.

Despite the verdict of the jury and the opinion of the Railroad Commissioners Engineer Wisker was indicted for manslaughter by the Grand Jury in February.[1138]

It was over a year before the case came to trial but on 24th April 1903 a jury found Engineer Wisker "Not Guilty" of manslaughter.[1139]

Robert Shaw is of the opinion that Engineer Wisker was culpably negligent in overrunning the signals, missing the flagman's lantern and the exploding torpedoes [1140] (if they did explode) and it is clearly the case that countless other engineers over the years must have managed to cope with the poor visibility without having an accident. Also the rules laid down that a signal which could not be seen where one should have been shown must be treated as a "Danger" signal. But clearly the railroad company also bore a great deal of responsibility for not trying harder to improve matters.

Electrification of the lines through the tunnel was finally completed in February 1907.

The South Wales Mineral Railway consisted of 11 miles of single track from Glyncorrwg to Incline Top, from where a wire-worked incline descended at 1 in 10 to Briton Ferry. There were crossing places at Ton Mawr and Cymmer, 3½ and 7¾ miles respectively from Incline Top with a signal box at Ton Mawr, with telephonic communication with Incline Top, while the office at Cymmer had telegraphic communication with both Incline Top and Glyncorrwg. Only three trains a day normally worked over the line, crossing one another at Cymmer unless special instructions were issued by the traffic manager at Briton Ferry or Glyncorrwg. The line was not equipped for passenger traffic; between Cymmer and Ton Mawr it passed through the 1,109 yard Gyluchy Tunnel.

1136 Ibid, 25th January 1902

1137 Ibid, 11th January 1902

1138 *New York Times*, New York, New York, United States, 2nd March 1902

1139 Ibid, 25th April 1903

1140 *A History of Railroad Accidents, Safety Precautions and Operating Practices* by Robert B. Shaw, pg. 239

Although passengers were in theory not provided for Glyncorrwg was an isolated spot, and the custom for 40 years had been for passengers to ride over the line either in empty trucks or the guard's van so as to enable them to get to and from the market at Neath, close to Briton Ferry. Thus, on Saturday 16th August 1902 there were seven passengers riding in the brake van of the 4.30 p.m. train from Glyncorrwg. As they were not being charged for making the journey the Board of Trade considered that they were not passengers in the ordinary sense of the word, though this seems rather a question of splitting hairs.

As already stated the Glyncorrwg train should, according to the normal method of working, have crossed the opposing train, which had left Incline Top at 4 p.m., at Cymmer. A fortnight earlier, however, Mr Tallamy, a cashier for the colliery company, had arranged for the two trains to cross at Ton Mawr, and hence had been able to get home earlier. On 16th August cashier John Perrott decided that he wished to do likewise, and according to his version of the conversation he had asked Traffic and Telegraph Clerk George Stephens if it was "all right" to take the train through to Ton Mawr. Stephens had said that it was and he (Perrott) had assumed that Stephens would make the necessary arrangements.

Mr Stephens agreed that cashier Perrott had told him that he wanted to cross the up train at Ton Mawr, but according to his version he had arranged for Perrott to make the arrangements at Cymmer, as cashier Tallamy had done a fortnight earlier. There was no third person present, so it was impossible to say which version was correct, though just before the train started from Glyncorrwg its driver, Rowland Hughes, asked Stephens if it was all right to run through to Ton Mawr and was told that it was, no mention being made of any arrangements at Cymmer.

After the 4.30 p.m. train had departed Mr Stephens wired Cymmer to advise them of its departure but received no reply. He then wired Incline Top to be told that the 4 p.m. up train had also departed. He then asked Incline Top to telephone Ton Mawr to stop the up train there, but the Ton Mawr signalman had left duty. With no one to contact at either Cymmer or Ton Mawr there was thus no way of stopping either train. They met in the middle of Gyluchy Tunnel. Driver Hughes saw the up train entering the tunnel and was able to bring the down one to a stand, but the collision was still a violent one. Both drivers, both firemen, cashier Stephens and the seven passengers were all injured, some of them seriously and two of the passengers fatally.

Major E. Druitt was again the Board of Trade inspecting officer. He decided that none of the crews of either train was to blame. Regardless of what was said between Stephens and Perrott it had been Stephens' duty to ascertain from Incline Top that the up train knew that it had to wait at Ton Mawr before he let the down train leave Glyncorrwg.

The inspecting officer did not blame Revell, the man at Cymmer, for going off duty, as he was only doing the job for a fortnight a year as holiday cover and was only doing what was always done on Saturday afternoons after the coal traffic had ceased. Somewhat illogically he considered Signalman Barnes at Ton Mawr much to blame for doing exactly the same thing, and going off duty at 4 p.m. after being advised by the Port Talbot company that their traffic had ceased.

Since the accident the Ton Mawr signalman and the clerk at Cymmer had received definite instructions that they must stay on duty until after the passing of the last train. As regards the crossing of trains instructions had been issued that no train must cross another other than at Cymmer without obtaining a special running permit from the officer in charge, or pass the point named on the ticket without obtaining a fresh ticket.

Major Druitt also considered that no train should be allowed to leave Glyncorrwg, Cymmer, Ton Mawr or Incline Top without definite information having been received that the line was clear, for which a telegraph line would need to be installed between Cymmer and Ton Mawr. Finally, he recommended that the company should have only one traffic manager rather than two.[1141]

The Dingle tunnel fire of 1901 was followed by a much more serious accident with similar features in Paris, France, on Monday 10th August 1903. Like the Liverpool Overhead Railway the Paris Metro had only been open a few years at the time of the catastrophe, and also like the English railway it used electric traction. A third similarity was that line 2 North of the Chemin de fer Métropolitan de Paris, that on which the accident occurred, ran partly on a viaduct, but at Combat station it descended into a tunnel, which continued to the terminus at Nation station.

At 6.53 p.m. on 10th August train No. 43 was in charge of Conductor Cavayé with Motorman Chauvin. The train consisted of a motor car at each end between which were four second-class and two first-class unpowered trailers. When it stopped at Barbès station, at the start of the elevated section of line, Motorman Chauvin noticed smoke coming from the floor of the front motor car. The train passengers were asked to get off the train, the shoes were lifted from the third rail, a fire extinguisher was employed and at least one bucket of water was directed at the seat of the fire. These measures appeared to extinguish the blaze, but the disabled train was now blocking the line and there were no sidings available before the line re-entered the tunnel. The fatal decision was then made to return the train to Nation station under its own power, the train crew being unaware that the fire had started not just because the motors had overheated, but due to a short circuit. The now empty train left Barbès station at 7.05 p.m. in a cloud of black smoke and ran through La Chapelle and Aubervillers (now Stalingrad) stations without stopping. As soon as the train had restarted the fire on car M202 had reignited and was now fiercer than before. Upon reaching Allemagne (now Jaurès) station the train therefore stopped again and a fire extinguisher was procured. The train then entered the tunnel and proceeded to Combat (now Colonel Fabien) where the train stopped again. Here wooden paddles were used to lift the shoes from the rails. The shoes themselves then caught fire and the current had to be turned off for five minutes.

At this point it was decided that the train could not be taken to Nation under its own power and the decision was made to propel it there using train No. 52, the four-car set which was following. This train had picked up the passengers left at Barbès by train No. 43 and was now at Allemagne. The passengers on train No. 52 were then all told to get off at Allemagne and train No. 52, now empty, proceeded to Combat and was coupled to the rear of train No. 43. The combined train left Combat station at 7.32 p.m., powered by a single motor car and with the short circuit on car M202 still feeding the fire. Just beyond Combat, at Belleville, there was a siding where the disabled train could have been left but for disputed reasons it was not shunted there. According to the company this was

[1141] Board of Trade report

because Motorman Chauvin refused to allow his train to be left there, but the motorman alleged that it was because the points leading into the siding would not work.

The combined train of 12 cars thus continued towards Nation, eight stations further on. At Couronnes, the next station, Stationmaster Didier, alarmed by the sight of the train on fire, tried to stop it, but Motorman Fleuret of train No. 52 shouted that he thought that they could make it to Nation. He was wrong. Shortly after leaving Couronnes there was a violent short circuit, a series of explosions and a rapid worsening of the fire. The train stopped at the entrance to Ménilmontant station, the platforms of which were crowded with passengers, and its crew were forced to abandon it. Soon the whole train was on fire and thick smoke from it spread to the only exit from the platforms, forcing some of the passengers to flee along the tunnel towards the next station at Père Lachaise.

The problems with train No. 43 had not led to a suspension of traffic and train No. 52 was followed by train No. 48, another four-car set, which had picked up the passengers from trains Nos. 43 and 52 at Allemagne station. By the time that this train reached Couronnes station smoke was drifting back through the tunnel from Ménilmontant, while the train could not in any case proceed because the signal remained at danger with train No. 43 stopped in front of it. After train No. 48 had been waiting at Couronnes a few minutes Conductor Jouffroy of train No. 52 and one of his colleagues reached the station, having walked back through the tunnel, and they announced that train Nos 43/52 were burning fiercely. As if to reinforce the point thick black smoke had followed the two men through the tunnel and began to flood Couronnes station.

Realising the danger of the situation the Couronnes stationmaster and the conductor of train No. 48 tried to get the passengers from that train to leave the station. A few did so but many did not. A good number of them had already been ejected from one train at Barbès and then another at Allemagne and naturally were less than happy to repeat the exercise a third time. They remained on the train, waiting for it to resume its journey, or demanding refunds of their tickets, and there were angry confrontations with the station staff. Meanwhile at Ménilmontant the fire, powered by the short-circuit, continued to burn. The electric current had been cut at the sub-station there but as the line was not divided into electrically isolated sections power continued to be fed to the train from other sub-stations.

At about 8 p.m. the fire damaged the circuit controlling the station lighting at Couronnes and all the lighting failed. Panic then set in amongst the passengers still on the train or on the station platforms as they attempted to escape. Candles, lanterns and matches proved useless in the thick smoke or went out due to lack of oxygen. There was only one exit from the station and some managed to find it and get out but others, unfamiliar with the station or disorientated in the darkness, made their way to the north end of the station, where there was no way out other than via the tunnel to Belleville. A few made it out through the tunnel, but many died congregated at the end of the station platform.

When the fire brigade arrived they were unable to enter the station due to the intense heat, some of them suffering burns when they attempted to go down the stairs to the platforms. All that they could do was to wait for the fire to subside. Those who escaped but were suffering from smoke inhalation or burns were treated by a chemist on the Boulevard de Belleville before being taken away to hospital. M. Louis Lépine, the Paris Prefect of Police, went to the scene, as did M. Fulgence Bienvenüe, the chief engineer of the system. The firefighters were finally able to access the platforms in the early hours of the following morning when the fire had died down.

In all, 84 people died in the catastrophe, seven at Ménilmontant, 75 at Couronnes and two in the tunnel towards Belleville. Many more were injured. Trains Nos. 43 and 52 were completely burnt out and Ménilmontant and Couronnes stations were badly damaged. At the time that it happened the disaster was the worst to have happened on any urban transport system.

Within just over a week of the fire a whole series of corrective measures was ordered to prevent any repetition of the tragedy:

Immediately

- Responsible managers to be assigned to each section of line, ready to take charge in case of any accident
- Motormen must isolate short-circuited motors from the power supply
- Lighted signs to be provided at station exits
- Temporary firefighting stations to be provided pending the installation of hydrants.
- Stations to have unobstructed exits

Within 15 days

- Electrical components to be fully isolated
- Flammable materials to be removed, especially from motorman's cabs
- Fire hydrants to be installed
- A second protected lighting power supply to be installed

By 1st November

- Each line to be divided into electrically isolated sections
- Station exits to be widened [1142]

A commission of inquiry into the accident published its report in December 1903. This highlighted two problems in particular. One, relating to the fact that the same electrical circuit supplied the power to the trains and the station lighting, had already been recognised in the recommendations made a week after the accident, as noted above. The other concerned the vulnerability of the all-wooden train sets to fire, with cables and electrical equipment placed beneath the wooden floors of the cars and highly flammable gutta-percha used as insulation. The trains were forthwith redesigned, first with all-metal motorman's cabs, and then from 1906 with metal entirely replacing wood in their bodies.

The eight-car sets were also soon reduced to seven-car sets, with the two motor cars at the front. Within a few years these were replaced by multiple-unit trains which used much lower currents, thus greatly reducing the risk of a fire due to a failure of the electric motors.

The railway company attempted to place all the blame for the accident upon its

[1142] Wikipedia – Paris Métro train fire

employees, especially the motorman of train No. 43, whom it said had disregarded the instructions as to what to do in the event of a motor failure. Conductor Cavayé and Motorman Chauvin of train No. 43, Conductor Jouffroy of train No. 52 and M. Renaud, the Combat stationmaster, subsequently stood trial and were all found guilty with extenuating circumstances, bearing in mind their hasty recruitment and lack of experience. Conductor Jouffroy and Stationmaster Renaud were given one month's suspended prison sentences and fined 2,000 francs; Motorman Chauvin was fined 3,000 francs and Conductor Cavayé 2,000 francs. The Chemin de fer Métropolitan de Paris was found to be financially liable; it had already paid out 2,000,000 francs in compensation to the victims.[1143]

Missouri Pacific Railroad through freight train No. 174 left Jefferson City, Missouri, for St Louis at 6.50 a.m. on the morning of Monday 7th March 1904, followed by local freight train No. 102, also for St Louis, 20 minutes later. Train No. 174 had only covered four miles when it became divided in Moreau tunnel, Missouri, United States. It would seem that it was only belatedly that the three railroad employees in the caboose of train No. 174 realised what had happened. Brakeman Charlie Weyant was apparently just about to leave the caboose to go back and flag the train when it was run into by train No. 102. The brakeman was killed outright, as were Conductor Charles Shoemaker and Brakeman Frank Nevis, whose bodies were almost entirely consumed in the fierce fire which broke out after the collision.

Train No. 102 was a light one, consisting of only five cars and a caboose, and was therefore probably making considerably better time than the train in front, so the interval of 20 minutes between them at the start may well have diminished considerably by the tunnel. No one on the colliding train was hurt and it was backed out of the tunnel.

The fire made the work of rescue very difficult. Two fire engines were brought up on a train from St Louis but the only water available as from the Osage River, some distance east of the tunnel, necessitating the engines having to make repeated trips between the tunnel and the river, and little headway was made in tackling the fire until the evening. In the end the fire was practically left to burn itself out. Matters were further complicated after the fire had finally been extinguished by a part of the tunnel roof collapsing when an attempt was made to remove the wreckage.

Conductor Sam Jones was in charge of train No. 102, with Engineer Christine. They had made desperate efforts to free Conductor Shoemaker, who had apparently been writing at his desk at the front of the caboose and was pinioned there in the collision. The conductor's watch was found to have stopped at 7.45 and Brakeman Weyant's at 7.42; the collision was estimated to have occurred at 7.17. [1144] A boxcar of lumber in the first train had caused the fire.[1145]

There was a possibility that a fourth man, brakeman named Parks, had been on the first train and his body had been consumed in the fire, as he had disappeared after the

[1143] Wikipedia – Accident de metro du 10 août à Paris (in French)
[1144] *Sedalia Evening Sentinel*, Sedalia, Missouri, United States, 8th March 1904
[1145] *Sedalia Weekly Sentinel*, Sedalia, Missouri, United States, 11th March 1904

accident.[1146]

Early on Saturday 15th April 1905 a passenger train collided with a freight train in Ramo Tunnel, Italy, near Genoa. Twelve passengers died and 32 were seriously injured. Many of the passengers on the train were soldiers of the 54 Regiment. A special train carrying doctors and medical equipment was sent to the scene.[1147]

Early in the morning of Sunday 23rd July 1905 a passenger train from Holzminden to Paderborn collided with rubble from the collapsed roof of a tunnel at Altenbeken, Germany. Thirteen people were injured.[1148]

An accident in Spain on Saturday 10th August 1907 was fairly serious but could easily have been a lot worse. A late-running freight train from San Sebastian had travelled as far as the tunnel adjoining Otzaurte station when it was run into from behind by the following Sud express. As a result of the collision 11 waggons of the freight train caught fire and one of them, which was loaded with explosives, then blew up. Perhaps surprisingly no one was killed but Driver Izaguirre of the express, together with Conductor Lucio Atarma and Brakeman Villaba, who were presumably on the freight train, were all more or less badly hurt, especially the driver, who suffered several broken ribs. Five other people were slightly hurt.

The express engine was badly damaged, as was the tunnel, and work was immediately begun to shore it up for fear that its roof would collapse. The fire also damaged telegraph poles, interfering with communications.

A relief train was despatched from San Sebastian as soon as news of the accident was received and the injured were taken back to there.

By chance a passenger on the express was the Undersecretary to the Ministry of the Interior, the Count del Moral de Ciatrava, and he immediately sent off a telegram to the ministry giving an account of the accident.

Passengers were forced to transfer between trains, being forced to make a considerable detour.[1149]

In the late afternoon of Tuesday 15th December 1908 freight train No. 2320, consisting of an engine 42 waggons and two brake vans, including waggons-foudres (trucks with large barrels containing wine) and coal trucks, with a total weight of 630 tons, left Brive for Limoges, on the main line of the Chemin de fer de Paris à Orléans. It was due to leave Brive at 3.37 p.m., but was late. From Brive (100 metres above sea level) the line climbed for about 40 kilometres to La Porcherie (450 metres above sea level) at a ruling gradient of about 1 in 100. Beyond Allassac, in the Vézère gorge, between Estivaux and Vigeois stations, the rails were wet and despite the sanding of the rails, engine No. 5078

1146 *Sedalia Sunday Sentinel*, Sedalia, Missouri, United States, 13th March 1904
1147 *Trenton Times*, Trenton, New Jersey, United States, 15th April 1905
1148 *Eisenbahn-Katastrophen in Deutschland* by Hans Joachim Ritzau pg. 90 (in German)
1149 *La Vanguardia*, Barcelona, Spain, 10th August 1907

first slipped and finally stopped. During an attempt to restart the train a jerk caused the coupling between the 6th and 7th wagons to break. The 38 rear vehicles, lacking automatic brakes, began to run backwards, their two brakemen being unable to hold them. The Estivaux stationmaster, seeing the waggons coming back down the slope, telegraphed to his colleague at Allassac in the hope that he would be able to contact him in time to stop omnibus train 742, due to leave Brive 5:43 p.m. and arrive at Limoges at 8:30 p.m., which was following. His message came just too late; the passenger train had left for Estivaux three minutes before it was received at 6.15 p.m.

A collision was now inevitable, and it would take place shortly afterwards at about 6.30 or just before, about four kilometres north of Allassac, and by an unfortunate chance within the 157-metre-long tunnel near Pouch, situated immediately beyond a bridge over the Vézère River. As the locomotive and the first three cars of the passenger train entered the tunnel, travelling at about 55 kilometres an hour they were struck by the 38 runaway vehicles. Given the speed of the two trains, the impact probably took place at around 90 km/h. The front of the engine was lifted and wedged against the tunnel roof by the coal cars, and its tender and most of the coaches it was drawing crashed into it. Only the rear van and the last two carriages of the passenger train, which remained outside the tunnel, would be relatively spared.

The site being inaccessible by road, the first rescuers, who came on foot from Allassac and Estivaux stations, took a long time to reach it, and, devoid of heavy equipment, proved to be powerless in the face of the magnitude of the disaster. Furthermore, flames from the locomotive's firebox escaped and fed by the alcohol spilt from several smashed wagons-foudres, set fire to all of the wreckage, and the tunnel was soon transformed into a furnace. Thus, having reached the engine, the rescuers had to retreat before the rapid progression of the fire and to helplessly witness the dreadful agony of Fireman Lefort, who was burnt alive beneath the tender holding him prisoner. The fire, fuelled both by the mainly wooden waggons, and by their load, also prevented the intervention of the relief train from Brive.

When it was finally possible to access the interior of the tunnel to clear the wreckage, the state of the charred human remains found among the debris made it difficult to count and identify the victims in order to assess the magnitude of the disaster. Thus, while the number of injured could be fixed with certainty at 28, that of the dead was less certain. At first estimated at 14, it was increased to 26 when it was found that out of a squad of 15 track workers had boarded the train at Brive for Vigeois, only three of whom could be found. Among the dead were three who died later in hospital; Driver Pistre and two of the passengers.

The line was blocked for about two weeks, because after the wreckage have been cleared, it was found necessary to partially rebuild the tunnel roof, part of which had become detached under the effect of the heat. Until its reopening, traffic was diverted to the west by Périgueux, or on a substantially parallel route, via Objat, Saint-Yrieix-la-Perche and Nexon.[1150]

The disaster was a tragic illustration of two serious shortcomings of the railways of the time as regards safety: the fragility of the couplings and the shortcomings of the braking system. Thus, the 38 freight cars that had run away were not equipped with air

1150 *Le Petit Journal*, Paris, France, 17th December 1908

brakes, and only two brake vans were provided. The two brakemen had remained at their posts to the end and were among those killed. On Friday, December 18th, in the Chamber of Deputies Léon Betoulle, Socialist deputy of Haute-Vienne, challenged Louis Barthou, Minister of Public Works on these questions, emphasising in particular the need for the State to check compliance by railway companies with the requirements on the strength of couplings, and their proportionality with the traction force of the engines. François Labrousse, senator from Corrèze, made a similar intervention in the Senate.[1151] Another defence against runaways, the provision of catch points on steep inclines, seems not to have been considered.

The emotion aroused by the story, widely published in the newspapers, of the horrible death of the fireman Lefort, begging for his sufferings to be ended, was also the source of a debate on a theme hitherto little discussed in France, at least in the popular press, that of euthanasia. A newspaper reporter was told by Dr Bosrçdon, one of the first to arrive on the scene, that the latter had been so tormented by the ordeal of the fireman that had he had a revolver he would have put the man out of his agony.[1152]

The Chemins de fer de Paris â Lyon et â la Méditeranée station at Olliergues, France, in the mountains not too far from Clermont Ferrand was situated between a tunnel and a viaduct on the one side from Giroux and another tunnel on the far side towards Vertolaye. This latter tunnel was the site of a serious collision on the morning of Monday 4th December 1911. At a time variously stated in different newspaper reports as just before 8 a.m. and just before 9 a.m. the Olliergues stationmaster was advised of the approach along the single line of a light engine from Vertolaye. A short time later passenger train No. 2903 from Saint Germain des Fossés, running considerably late, arrived at his station from Giroux. It was at this point that the fatal mistake occurred. The station work being completed, the stationmaster, in a moment of absentmindedness, signalled the passenger train away, completely forgetting for a moment that the engine from Vertolaye was still approaching. The passenger train had barely covered a couple of hundred metres from the station and was just entering the tunnel when it met the light engine, which was reportedly moving at some speed. Both locomotives were badly damaged in the collision but the carriages of the passenger train fortunately withstood the shock better.

Driver Charles Demai and Fireman Coutourier on the light engine were both killed and Driver Frédéric Fournier on the passenger train was so badly hurt that his life was despaired of. Fireman Henri Comte of the passenger train escaped with bruises, Conductor Vincent Brière suffered a slight head injury and four passengers were also slightly hurt.[1153] The passengers were well enough to be able to continue their journeys on another train brought to the far end of the tunnel.

The brother of Driver Demai worked in the central telegraph office at Clermont Ferrand and by an unfortunate chance was on duty that morning so that he took the telegram reporting the accident and the death of his brother. The Olliergues stationmaster, whose name was not given in the newspaper reports, was a respected employee with a good record. He was unable to explain his mistake and was greatly

---

1151 Wikipedia – Accident ferroviaire du tunnel de Pouch (in French)
1152 *Le Figaro*, Paris, France, 17th December 1908
1153 *Le Petit Journal*, Paris, France, 5th December 1905

upset by what he had done.[1154]

At 2.45 p.m. on the afternoon of Tuesday 6th August 1912 a serious collision occurred near Lozanne station, France, on the Chemins de fer de Paris â Lyon et â la Méditeranée. On the approach to Lozanne station was the junction of the lines from Bourbonnais and Paray le-Montai to Lyon and from Roanne to Lyon. Omnibus train 2855 from Paray le-Montai to Lyon usually preceded express train No. 847 from Vichy to Lyon through the junction. On the afternoon in question the omnibus train was running late and the signalman at the junction therefore decided to give priority to the express instead and set the carré arrêt (stop signal) against the omnibus train. As an additional precaution detonators were placed on the rails to alert the driver of a train running past the signal at danger. Unfortunately, in this instance, the detonators, instead of preventing an accident, appear to have caused one which might otherwise have been averted. Driver Ravaud of the omnibus train somehow failed to see the signal set against him but heard the explosion of the detonators and brought his train to a stand as quickly as possible, but not before it had run through the junction and into the tunnel between the junction and Lozanne station. Had the explosion of the detonators not caused him to pull up he could probably have continued to Lozanne station in safety. At the moment that the omnibus train had run through the signals the Vichy express was approaching, travelling at 90 kilometres an hour with clear signals. Its driver, M. Jacoton, seeing the omnibus train shooting through the junction in front of him applied his brakes in emergency and was able to reduce its speed somewhat but not to avoid running into the rear of the omnibus train in the tunnel.

The rearmost van and three carriages of the omnibus train were telescoped and badly damaged and the wreckage caught fire. The Lozanne station staff immediately began a rescue operation and were soon being assisted by local firemen but it took a considerable time to release the dead and injured passengers from the dark and smoke filled tunnel. Twenty injured were finally removed from the train together with four bodies, and were despatched on a rescue train to Lyon, which was reached at 5.30. Ambulances were waiting for the train at Lyon Perrache station from where the injured were taken to different hospitals.

Three of those who were killed outright were M. Joseph Grand, of Lyon; Mme Leyseene, of Sathonay and Muyard, of Saint Cyr le Chatou. The fourth person could not be immediately identified. A fifth passenger, Mme Couturier, died of her injuries shortly after being released and it seems likely that two more passengers succumbed later.

None of the passengers on the express, who included M. Jean Richepin, of the Académie Français, was hurt.[1155] [1156]

A bullfight in the Spanish town of Irún on Sunday 13th July 1913 led to a head-on collision between two electric trains on the line from Hendaye in a tunnel near Irún, Spain. Several coaches were overturned in the collision and some of the passengers were

1154 *Le Petit Parisien*, Paris, France, 5th December 1905
1155 *Le Petit Parisien*, Paris, France, 7th August 1912
1156 *The Times*, London, England, 8th August 1912

trapped beneath them. The collision occurred at 4.30 p.m. Heavy rain impeded the rescue operation. Five people were killed outright and another 27 injured, one of whom died later. Three of the dead were soon identified Félix Ibarrondo, Celestina Medinaviatia, a baker from Renteria, and Señorita Guílarte. Señor Ibarrondo was identified as a company employee, possibly he was the motorman of one of the trains.[1157] [1158] [1159]

A mistake by an experienced signalman in a busy signal box led to a collision at Liverpool, England, on Wednesday 15th October 1913. On the afternoon of that day the 2.30 p.m. express left the Cheshire Lines Committee station at Liverpool Central on time. It consisted of Great Central Railway engine No. 1016, five Cheshire Lines coaches for Manchester, a Great Eastern Railway coach for Harwich and two Great Central coaches for Grimsby with Head Guard Edward Lloyd, Rear Guard C. E. Bishop, in charge of the Great Central coaches, Driver Edward Howson and Fireman James Jones.

The train had only travelled 400 yards and was 200 yards within St James Tunnel outside of the station when Driver Howson noticed that it was slowing up and the vacuum pressure had dropped, and decided that someone must have pulled the communication cord. Not wishing to stop in the tunnel he put the large ejector on to release the brakes and continued to St James station, where he stopped with two coaches still completely and a third coach partly within the tunnel. Driver Howson spoke to the St James signalman, Walter Pipprell, and to Guard Lloyd, who told him that the communication cord had not been pulled on any of the front coaches and asked him to draw the rear of the train out the tunnel. As Driver Howson was about to do so his train was run into from behind. Five passengers and Guard Bishop were killed outright and 57 passengers on the train were injured, one fatally. Six passengers on the colliding train were also injured.

Signalman Pipprell at St James station told the Board of Trade inquiry into the accident that the first train had arrived at 2.35 p.m., and as soon as Driver Howson had explained the problem he had told the Liverpool Central signalman what had happened. In reply the latter had told him that the 2.35 p.m. passenger train to Stockport was on its way to him, and the collision occurred at, he estimated, 2.37½ p.m. Signalman Pipprell stated that he had replaced the up distant signal as the Manchester train emerged from the tunnel, and the electrical repeater had shown that it had gone to danger. He had replaced the up home signal as well when the train was halfway along the platform.

Signalman William Thompson at Liverpool Central explained at the inquiry that as soon as the Manchester train had departed he had been busy with some shunting operations at the station end of the box, away from the block instruments. He saw Booking Boy Edward Shorleson at the block instruments at the Manchester end of the box and heard a bell ringing but could not make out what was being rung because of the noise of the shunting train. Without troubling to look at the block instrument or to ask the booking boy he had jumped to the conclusion that St James must have cleared back the Manchester train and without further ado had pulled off the starting and advanced starting signals for the Stockport train. Seeing the train leaving the booking boy had asked,

---

[1157] *La Vanguardia*, Barcelona, Spain, 14th, 15th and 16th July 1913
[1158] *Neuchatel Feuille Davis de Neuchatel*, Neuchatel, Switzerland, 14th July 1913
[1159] *San Antonio Light*, San Antonio, Texas, United States, 14th July 1913

"Where's the Derby going?"

"Right away, you've sent him on, haven't you?"

"No," was the reply.

"What were you doing at the instrument? I saw you sending it."

"I wasn't sending it, I was refusing the down local," Shorleson had replied.

Signalman Thompson said that he had then given St James one beat on the bell, told the booking boy to ask the St James signalman if he had cleared back the Manchester train and had himself run to put back the advanced starting signal in the hope of stopping the Stockport train. Asked what the St James signalman had said, Booking Boy Shorleson had replied that Signalman Pepperel had told him, "I have not cleared yet, he is standing here."

"Tell him to put back his signals at once, if he hasn't already done so, and that the Stockport train is coming." Checking the indicator of the track circuit which extended 55 yards beyond the advanced starting signal Signalman Thompson said he had seen it change from red to white, indicating that the Stockport train was on its way to St James.

The Stockport train had only three bogie coaches and a six-wheeled guard's van. Driver Stephen Millward stated that he had seen the Liverpool Central starting signal off, and the advanced starter off from about 20 yards away, with the St James distant beneath it at danger. He had reduced speed. He did not see the St James home signal at all and was relying on his fireman to observe it; his train was travelling at about eight miles an hour when the collision occurred without warning.

Fireman James Barker of the 2.35 p.m. train stated that he had seen the advanced starting signal off with the distant signal on but did not see the St James home signal at all.

Guard William Diggles of the Stockport train said that his train had left on time. He had not noticed any of the signals. He put the time of the collision at 2.37 exactly.

The inspecting officer, Lt-Col E. Druitt, considered that the Stockport train had been past the Liverpool Central advanced starting signal before Signalman Thompson had had chance to replace it. Tests showed that with the tunnel in the smoky condition that it would have been on the day of the accident it was unlikely that Fireman Barker would have been able to see the St James home signal before the collision. The timings given by Guard Diggles would have implied an average speed of 22½ miles an hour between Central and St James stations, but the inspecting officer considered Driver Millward's estimate of the speed at the moment of the collision (Fireman Barker had put it at 10 miles an hour) to be more accurate. The degree of damage caused to the rear coaches of the Manchester train was explained by the fact that it had been standing with its brakes on.

Signalman Thompson and his two colleagues, Signalmen Thomas Bryden and Charles Arkinstall had, stated the inspecting officer, all claimed that at certain times the work of the box was too much for one man and they had been obliged to let the booking boy work the block instruments in order to avoid delay. It seemed likely that because at certain busy times an assistant signalman was employed that the signalmen at other times had seen no harm in letting the booking boy work the box instead. While conceding the fact that the accident could just as easily have happened had an assistant signalman rather than Booking Boy Shorleson been operating the block instruments, Lt-Col Druitt stated that booking boys should never be permitted to work the block instruments until they were fully qualified. What he would have made of railway enthusiasts working boxes, as has happened in various boxes from time to time, one hates to think.

Given the number of lever movements made at certain times of the day the inspecting officer considered that it was worthy of consideration whether a second qualified signalman should not be employed in the Central box at busy periods of the day, even though the signalmen had never applied for assistance (doubtless because they just got the booking boy to help instead).

It was hard to understand, the inspecting officer went on, how a signalman of such long experience and exemplary record as Thompson could have failed to carry out the most elementary rule of block working and cleared his signals for the Stockport train without first checking that it had been accepted by St James box. The most likely explanation was that it was so unusual for the 2.35 train to be stopped at St James that he had acted mechanically and without thinking when he had cleared the signals.

A contributory factor to the collision had been the difficulty in observing the St James home signal due to the smoke filling the tunnel. The 990 yard bore had only one proper ventilating shaft to permit the escape of smoke. The inspecting officer therefore recommended that the track circuit which at present ended beyond the advanced starting signal be extended right through the tunnel to St James station, locking the advanced starting signal at Central at danger until the last vehicle of the previous train had reached St James station.

There was no direct evidence as to the communication cord having been pulled in one of the rear coaches of the Manchester train, but it seemed certain that this must have been the case.[1160]

On Sunday 14th December 1913, unnoticed by anyone living nearby, the south portal of Harras (Zschopau) Tunnel on the Chemnitz – Roßwein line in Germany collapsed. Some hours later double-headed Chemnitz – Roßwein passenger train No. 1414, which was running late, ploughed into the mass of rubble blocking the exit from the tunnel, the pilot engine being forced to the roof of tunnel. One source states that four passengers in the leading coach of the train lost their lives, while seven were seriously and 27 slightly injured.[1161] Another source states that there were 10 dead and about 60 seriously injured.

The general director of the Saxon state railway company submitted that the accident had resulted from an 'Act of God', and that the railway company was therefore not at fault and not liable for damages, but the courts eventually ruled that as the collapse of the tunnel had occurred some hours earlier it was in fact just an ordinary railway accident for which the company were liable.[1162]

Cochem or Kaiser-Wilhelm Tunnel, on the strategically important main line between Koblenz and Trier, was at 4,203 metres for many years the longest in Germany. It carried a heavy traffic. Like many tunnels it was poorly ventilated and its constant dampness

---

[1160] Board of Trade report.

[1161] *Eisenbahnunfälle Bind I*, by Ludwig Stockert pg. 90 (in German)

[1162] *Eisenbahn-Katastrophen in Deutschland* by Hans Joachim Ritzau pg. 119 (in German)

meant that the rails of the track within it quickly rusted and frequently broke.

On the morning of Saturday 27th December 1913 Rottenführer (Ganger) Krämer discovered two broken rails in the tunnel. He fixed emergency bands onto them and intended to change the rails over the night of 27th/28th. He did not bother to tell his supervisor of his intention. As a rule there was a two-hour gap in the traffic after the passing of train D 124. At 21:30 Krämer telephoned Ediger-Eller station at the west end of the tunnel to find out what trains were running and was told that train D 124 was being followed by freight train 6480. Fahrdienstleiter (Signalman) Kaster told him, "For the moment you had best not change the rail. Anyway ask Cochem, I'm not responsible for the section through the tunnel to them."

Krämer did so and spoke to Fahrdienstleiterpraktikant (Trainee Signalman) Heinz at Cochem, "Is there anything following train 124?"

Heinz: "No."

Krämer: "Can the line be blocked after the express has passed?"

Heinz: "Yes." But Heinz had freight train 6480 at his station, where it was to be overtaken by the passenger train.

Krämer went to his gang. They waited for train D 124 to pass and then Krämer sent Rottenarbeiter (Platelayer) Schleifenbaum back with three detonators to protect the obstruction.

Meanwhile at Cochem Heinz suddenly realised that freight train 6480 was still waiting to proceed. He tried to contact Krämer via the tunnel telephone but in vain. Krämer was already in the tunnel with his gang and did not hear the call. Kaster at Ediger-Eller did hear the call though and asked Heinz, "What's up?"

Heinz explained, "I gave Krämer leave to begin working on the track as soon as D 124 had passed but I have train 6480 waiting here. He will have to postpone changing the rail for a moment but I can't get him to answer."

Kaster replied, "I told Krämer that he must not change the rail until after train 6480 had gone, so you can let the train go."

What Kaster did not mention to Heinz was that he told Krämer to contact Heinz about blocking the line. Heinz now believed that everyone understood what was happening, and he let train 6480 proceed.

Platelayer Schleifenbaum told the investigation afterwards that he did not know where he had laid the detonators. He alleged that when he heard the goods train coming he had put the shots down 450 metres in the rear of the broken rail. According to the regulations they should have been 700 metres back. In the smoke-filled tunnel the driver of the freight train could not see any other signals and was unable to pull up his train before it reached the break in the track. It was derailed and fouled the opposite running line just as goods train 6577 was approaching. In the subsequent collision with the wreckage the driver and two guards of train 6577 were killed and Ganger Krämer was seriously injured.

Signalman Kaster, Ganger Krämer and Trainee Supervisor Klinge, who had been teaching Heinz at Cochem, subsequently stood trial at Koblenz. Kaster was acquitted,

Krämer was sentenced to two weeks' and Klinge to two months' imprisonment.[1163]

A collision in Cumbre Tunnel, Mexico, on 4th February 1914 was the result of one of the many serious acts of sabotage carried out in Mexico during this period and is dealt with in the Sabotage chapter.

A gross breach of the regulations was blamed for a minor collision in No. 1 Tunnel, between Cowan and Boronia South signal box, New South Wales, Australia, on Thursday 21st January 1915. A light engine which had banked a goods train to Cowan and was returning to Hawkesbury entered the tunnel while it was still occupied by a stock train on its way from Sydney to Newcastle. Guard Johnson of the stock train was thrown about the brake van and stunned; he was lucky to escape more serious injury. The driver and fireman of the colliding train were shaken up. The brake van and a sheep van were wrecked, and the tunnel was blocked. It was cleared at 8.43 a.m., by which time the Moree mail had been held up at Gosford for three hours and the Glen Innes mail at Wyong for half an hour. The collision occurred at about 1.50 a.m.[1164] [1165]

How the accident had happened became clear from evidence given at the appeal against dismissal of Signalman Alexander Hudson, fourth-class night officer at Cowan.

Driver Albert Swanberg told the inquiry that as train No. 67 was passing through the tunnel he had noticed that it was slowing up. He had concluded that too many brakes had been pinned down just north of Cowan and had got off his engine and released them. Just then there was a very heavy bump. Stopping his engine he had gone back through the tunnel, met the driver of the colliding engine and then found his guard in the smashed up brake van.

Driver Francis Faulkner of the banking engine said that the distant signal at Cowan had been on, the home and starting signals off and the advanced starter was cleared as he had approached it.

Signalman Thomas Wylie at Kurangal stated that he had got line clear for train No. 61 from Cowan at 12.26 and the arrival signal for it from Cowan at 12.55. For train No. 67 he had received line clear at 1.06 and the arrival signal at 1.31. The engine had arrived from North Hornsby at 1.27, he had received line clear for it from Cowan at 1.31, sent it on at the same time and received the arrival signal for it at 1.45. He had had to repeat the bell signals to Cowan for train No. 67 but not for train No. 61, or the engine.

Signalman T. Brown at Boronia South stated that he had given line clear for train No. 61 at 12.42. At 1.30, in response to an inquiry from Hawkesbury, he had rung Cowan to ask what time train No. 61 had passed and had been told 12.51. At 1.45 he had accepted a light engine from Cowan, and at 2 a.m. Cowan had telephoned him and told him that the engine was chasing train No. 67 through the tunnel.

Traffic Inspector Robert Reeves stated that Hudson had told him that he had no recollection of dealing with either of the goods trains or the engine.

1163 *Eisenbahn-Unfälle in Europa* by Erich Preuß pgs. 149-50 (in German)

1164 *The Sun*, Sydney, New South Wales, Australia, 21st January 1915

1165 *The Sydney Morning Herald*, Sydney, New South Wales, Australia, 22nd January 1915

Signalman Hudson pleaded that he had felt very ill that night and had applied for a block porter to relieve him, to be told that relief would come on train No. 38. He attributed his sleepiness to his illness and his need to take chlorodyne (a popular sedative at the time – a mixture of an alcoholic solution of opium, cannabis and chloroform).[1166] He had been suffering from anxiety because his wife was in a delicate state (pregnant) and a nervous condition. He considered that there had been contributory negligence on the part of Signalman Brown at Boronia for not informing him at once when Brown knew that train No. 67 was in the section as well as the engine. It is not all that clear from the only newspaper report what this means, but the inquiry from Hawkesbury must have related to train No. 67, so when Brown had confirmed that train No. 61 was the one which had passed Cowan at 12.51, which he must have cleared back to Cowan at about 1.05, he should presumably, according to Hudson's argument, have wondered where train No. 67 was.

Finally, Signalman Hudson pleaded that he had received a terrible shock when he had been told, at first, that Guard Johnson could not be found, and that it would be a lasting lesson to him.

The Board was not impressed, and his appeal against dismissal was turned down.[1167]

Chesapeake and Ohio Railway eastbound passenger train No. 214 met with an accident in the short Pinnacle Tunnel, near Altman West Virginia, United States, on the Coal River branch, on Tuesday 17th August 1915. The train consisted of an engine, an express car, a mail and baggage car, a smoking car and two day coaches, in charge of Conductor Halstead with Engineer Wentz. It had left St Albans, West Virginia, at 8.10 a.m. for Sovereign, West Virginia, left MacCorkle, the last open telegraph station, at 8.43, and was travelling at about 20 miles an hour when it became derailed in Pinnacle Tunnel, just over three and a quarter miles further on, at 9.53 a.m.

Pinnacle Tunnel was only 307 feet long and the engine, express and mail and baggage cars came to a stand outside of its east end with the rear of the express car and the mail car derailed. In the tunnel the smoking car and front of the following day coach were derailed, and the smoking car struck and displaced the upright timbers and brought down the top timbers lagging, rocks and earth from the roof of the tunnel. The roof of the smoking car was crushed and the four dead and nine injured passengers from the accident were all from this car.[1168]

A passenger in the train told a newspaper reporter, "The train was steaming along at its regular rate of speed eastward. Suddenly there was jumping and jarring, and the passengers knew that the train had left the track. The engine, however, had left the tunnel, dragging the train after it.

"When the baggage car left the track it jumped over the rails and swerved to one side, tearing out about 60 feet of timbering which supported the roof. The timbering was heavy, about fourteen inches by 12 inches, and this was ripped out, leaving no support for the roof.

[1166] Wikipedia – Chlorodyne
[1167] *The Co-operator*, Sydney, New South Wales, Australia, 18th March 1915
[1168] Interstate Commerce Commission report

"With the support gone the refuse of stone and dirt which had gathered from the natural roof and was only supported by the artificial roof crashed down on the smoking car and carrying with it the timbering of the car and tin and glass. In this wreckage were caught those who were hurt and killed.

"One of the dead passengers must have stuck his head out of the window when the smoking car left the track for one of the pillars torn down was splattered with blood and matted hair, showing that the man was not killed by the fall, but by being struck by one of the props.[1169]

Engineer Wentz told the Interstate Commerce Commission inquiry that he had felt his engine give a little jerk as it proceeded through the tunnel and thinking that was a broken rail had applied the air brakes. Looking back he had seen that the baggage car was derailed and when his train had stopped he had gone back to find what was wrong. He had then cut off the engine and proceeded to Altman to notify the dispatcher. Returning to the tunnel he had found the broken rail. He had noticed no problem on his westbound trip through the tunnel at 5.58 a.m.

Fireman Turner stated that he had also felt the jerk. He put the speed of the train at 18 miles an hour.

Conductor Halstead said that he had been collecting fares at the time of the accident, of which he had had no warning before the application of the brakes. Examining the track after the derailment he found the broken rail.

Section Foreman Walker stated that he had examined the track four days before the accident, when it had appeared to be in a good condition. No regular track walker was employed, but occasionally in bad weather a sectionman was used. The track was inspected about twice a week.

Track Supervisor Owan stated that he had found an old crack four or five feet in length under the broken rail. It could not have been detected by ordinary inspection. The rail had been second-hand when laid. He did not think that a regular track walker was necessary.

Divisional Engineer Botts stated that the rail had broken 95 feet inside the tunnel. The breaks were all new apart from a piped head. The tunnel roof had broken in 45 feet beyond the point of derailment.

Trainmaster McLaughlin believed that the rail had broken under the rear wheels of the locomotive, but the front trucks of the baggage car had passed over the break before the rail became displaced.

The Interstate Commerce Commission concluded that the derailment had been caused by a broken rail due to an internal defect known as piping arising from a manufacturing fault. The fault could not be discovered by ordinary inspection.

The crushed smoking car had been sound and of good construction for the branch line service on which it was used.[1170]

[1169] *The Raleigh Herald*, Beckley, West Virginia, United States, 20th August 1915

[1170] Interstate Commerce Commission report

CHAPTER 12

LEVEL CROSSING COLLISIONS

Wherever roads and railways intersect they must do so either by means of a bridge or on the level. Bridges have the advantage of allowing road and rail traffic to operate entirely without interference from one another, but they are much more expensive to provide, especially in flat countryside, due to the need to provide either extensive earthworks for approach embankments or to excavate cuttings, as well as the cost of providing the span itself. Where underbridges are provided there is often also the need to build retaining walls and there can be problems with drainage.[1171] The provision of level crossings, especially if, as in many parts of the world, they are unstaffed, is the much cheaper alternative, but where they exist, they introduce an additional source of danger to both road and rail users. Robert B. Shaw in his outstanding book on American railway accidents, *A History of Railroad Accidents, Safety Precautions and Operating Practices*, writes that he has excluded ordinary grade (i.e. level) crossing collisions from consideration on the grounds that they are not proper train accidents, inasmuch trains usually suffer much less harm from such accidents than the road vehicle involved. This book, on the other hand, will attempt to include the most serious level crossing accidents and any others that have features of interest, even if none of the casualties in them, if any, occurred to train users.

The visibility at Hamilton Avenue Crossing, Oak Park, Michigan, United States, was somewhat restricted by adjoining factory buildings, and this circumstance and carelessness on the part of the road user were the main causes of an accident there on the morning of Friday 14th June 1901. Major George Buckingham drove onto the crossing without stopping to check if it was safe to do so with a double carriage (one drawn by two horses). A Pere Marquette Railroad passenger train was approaching at high speed but Engineer Wiggins said afterwards that he had whistled for the crossing. The first that had known of the accident was when his engine had struck the carriage. Major Buckingham, his daughter Abbie, Mrs Thomas Applegate and Mrs William Humphrey, the two last-named from Adrian, were all killed. The badly mangled bodies of the Adrian

---

[1171] *Backtrack*, Volume 24, pg. 88.

women were found on the locomotive; they had been guests of the major during a state General Army of the Republic (Union Civil War veterans) encampment. The pilot of the engine was broken.[1172]

At 11.45 p.m. on Saturday 17th August 1901 a crowded Long Island Railroad train returning from Manhattan Beach to the ferry at Bay Ridge collided with a trolley car of the 15th Street Line at the level crossing with Gravesend Avenue and Avenue D at Kensington Junction, New York, United States. The train was running at almost full speed. It struck the trolley about its centre, smashing it to pieces. The motorman of the trolley car and a policeman riding in its cab were killed outright and 25 passengers were injured, some very seriously; initially it was reported that three more had died soon after the accident.[1173]

Edward Rock, in charge of the tower guarding the crossing, was arrested after the accident on a charge of negligence, though he insisted that the signals had been set at danger against the trolley car and had been disregarded by its motorman.[1174] By Monday 19th it was clear that in fact Towerman Rock was not to blame and that the dead motorman, John Sirinek, in charge of Brooklyn Rapid Transit car 1082, was the only person at fault. Brooklyn Rapid Transit President J. L. Greatsinger stated that, "The responsibility for the accident rests for the dead motorman of the car. All the facts point to him as the author of the disaster. William Rock, in the signal tower, saw the trolley and train approaching from opposite directions. The junction at Gravesend Avenue and Avenue D is V shaped. The train had the right of way, as all trains have, and the motorman should have stopped and waited for the train to pass.

"As it was, even if the train had not been struck by the trolley car, the car would have come off the track, as the switch was locked for the train to pass. Mr Meneely, Mr Wheatley and myself visited the Seney hospital early Sunday morning and saw to it that the injured passengers were given the best of attention."[1175]

A week after the accident President Greatsinger announced that his company planned to install derailing switches at signal stations as an additional safeguard, as recommended by the State Railroad Commissioners.[1176]

The very same evening as the New York accident another collision between a train and a trolley car occurred in Chicago, Illinois. A trolley car of the 47th Street line was struck by a Pennsylvania Railroad train at the level crossing at 47th Street and Stewart Avenue. This accident was more serious than the New York one; again the motorman, Willis Bowman, lost his life, but four trolley car passengers were also killed. Nine other trolley car passengers and its conductor, V. J. Fulton, were injured. According to newspaper reports although the crossing was in any case protected by gates the custom was for the car to stop short of it, the conductor to alight and go forward onto the track and from his position there signal to the motorman whether he could proceed or whether

[1172] *South Haven Sentinel*, South Haven, Michigan, 22nd June 1901
[1173] *The Brooklyn Daily Eagle*, New York, New York, United States, 18th August 1901
[1174] *The New York Times*, New York, New York, United States, 18th August 1901
[1175] *The Brooklyn Daily Eagle*, New York, New York, United States, 19th August 1901
[1176] *The New York Times*, New York, New York, United States, 27th August 1901

he must wait. On the night in question the trolley car had paused momentarily but before Conductor Fulton had time to alight it had lurched forward, smashed through the wooden gates and struck the side of a passenger train at the junction of its second and third coaches. Motorman Bowman lived for a short time after the accident but was not in a state to give any explanation for the collision.

Flagman Frederick Wallrock, stationed at the crossing, said afterwards, "I don't know what was the matter with the motorman. I saw the train coming, lowered the gates, which were in good condition, and had a light swinging from them. I saw the car coming also, but as the gates were down, I never dreamed of its coming right on. It didn't seem to me that it stopped at all. There was a smashing sound as it came through the gates almost at full speed. I yelled at the motorman, but it was all over in a second."

Towerman Albert Hass stated that he could see Motorman Bowman tugging at the brakes and heard him shout that he could not stop the car, implying a failure of the trolley car brakes as a cause of the collision.

Engineer Gates, Fireman E. J. Eckhard and Conductor S. McMahon were all detained by the police until they were satisfied that they were not responsible for the accident. The train was eventually allowed to proceed to Fort Wayne, its destination.[1177]

The internal combustion engine was only invented in 1884 and the first motor car followed in 1885 or 1886. [1178] In previous volumes of this series it has been noted that the first level crossing collision involving a motor vehicle apparently did not occur until 1st May 1904, in France.[1179] This, it would seem, is incorrect; there was in fact an accident two and a half years earlier, on 30th October 1901.

On the Wednesday morning in question a party of six automobilists, as the newspaper called them, were touring part of Long Island on a sightseeing trip in the company of Mr H. B. Fullerton, special passenger agent of the Long Island Railroad. The party included M. Henri Fournier, 'the celebrated French automobilist'. Despite the fact that the party included someone who worked for the railroad, and would, it would seem, have been therefore well aware of the dangers of level crossings, the yellow car entered a crossing near Westbury station, New York, United States, without any thought of danger. M. Fournier, who was driving, failed to notice a light engine, on its way from Mineola to Hicksville to pick up a train. The engine struck the car, wrecking it, and all six of its occupants were injured, Mr Fullerton, it was feared, fatally, though the next day he was rallying. Ambulances took the injured to the Nassau County Hospital in Mineola.[1180]

M. Fournier told a newspaper reporter afterwards, "We were running at a moderate rate of speed of about seven miles an hour, and on approaching the railroad I heard no bell or saw no flagman. On coming from behind the hotel, or rather from the front, I first caught a glimpse of the engine. I had not time to cross in front of it and out only hope was to stop and turn about. I put on the brakes and turned the machine to the right but in spite of it all the engine struck us, and I found myself hurled through the air. The engine

---

[1177] *The Fort Wayne Journal-Gazette*, Fort Wayne, Indiana, United States, 18th August 1901

[1178] Wikipedia – History of the Internal Combustion Engine

[1179] Wikipedia – Liste des accidents ferroviaires en France au XX siècle (in French)

[1180] *The Brooklyn Daily Eagle*, New York, New York, United States, 30th October 1901 gave no warning whistle and we were all unaware of its presence until it struck us."

The second report on the accident identified the car as a 10 horsepower Mors with a Tonneau body.

As regards the crossing, it was stated to be one of the most dangerous on the main line of the Long Island Railroad, and it was noted that only a few days previously it had been the site of a near collision with automobiles driven by W. K. Vanderbilt Jr and Foxhall Keene.

Ironically only the previous month the railroad company had applied to the State Railroad Commissioners for permission to remove the level crossing and replace it with a bridge carrying the railroad over the depressed roadway, the railroad company to pay half of the cost. This idea was vigorously opposed by local citizens on the grounds of the considerable expense to the taxpayer, the need to re-site stores and properties and so on, and the commissioners therefore refused to permit the railroad company to do away with the crossing.[1181]

Presumably because of its novelty the accident was very thoroughly covered by the local press.

Teenage boys enjoying themselves in the snow by riding on a bobsled were the victims of a tragic accident near South Bethlehem, Pennsylvania, United States on the evening of Thursday 15th January 1903. The street on which the boys were coasting downhill crossed the Philadelphia and Reading Railroad at the bottom of the slope. The sled carrying the youths, who were friends aged between 13 and 15, reached the level crossing at the foot of the hill simultaneously with the train. Five boys, four of whom were named Dennis McGinley, John Coyle and Owen and Patrick Cahill, were killed instantly and a sixth, John Heuber, was thought to be fatally injured. The name of the other victim was not known to the press.

The engineer of the train did not even realise that there had been an accident until he stopped at South Bethlehem station, a mile further on, where blood and bones were discovered on the pilot of the locomotive.[1182]

Newark, New Jersey, United States, was the scene of another tragic level crossing accident on the morning of Thursday 19th February 1903. Every weekday morning between 8 and 9 a.m. during school term time the North Jersey Street Railway company ran a special trolley car for the accommodation of high school pupils. On the morning of the accident the car was, as usual, crowded with boys and girls, many of whom had transferred from other trolleys, and the front platform of the car was full of boy passengers. The route of the trolley car took it over the Clifton Avenue Crossing in Newark with the Delaware Lackawanna and Western Railroad, 'long known as a dangerous spot', according to a newspaper reporter.

As the trolley car approached the crossing at 8.40 a.m., also nearing it was the Lackawanna's *Passaic Express*, which had started from Summit, New Jersey, at 8.05

[1181] Ibid, 31st October 1901

[1182] *The Evening Independent*, Massillon, Ohio, 16th January 1903 a.m. Motorman Peter Brady saw the train and tried to stop the car, but it slid upon the icy rails, crashed through the nearside gate, and came to a halt with its front platform projecting across the railroad line. A moment or so later the engine of the express struck the trolley car, pushed it aside and the car then overturned. The force of the collision threw a number of the children beneath the wheels of the train. Eight children were killed outright and about the same number injured. Engineer Oscar Barclift, driving the express, suffered cuts and bruises and the motorman of the trolley car was badly hurt. A ninth child died the next day.

'Mounted Policeman Stucky of the fourth precinct was an eyewitness to the horror. He saw the train approaching and said that the trolley car was about four feet over the track when it was struck. There were about twelve children on the front platform grouped around the motorman. It is not known whether they hampered the motorman's movements so that he was unable to apply his brakes to stop the car.

'Passengers on the train heard the screams of the children two blocks away and before the train struck the car. As showing the force of the collision, the front part of the trolley car was picked up three blocks down the road between the two tracks.

'When the train reached the station in Newark the engineer and fireman were arrested by Special Policeman Thomas McDonald, employed by the railway police, and sent to the police headquarters. The engineer was so dazed that it was necessary to pull him off the engine.

'The conductor of the trolley car, a man named George Gould, said after the accident, "It is the custom for trolley conductors to leave their cars as this crossing is approached and, as the car stops, to run ahead and see if the track is clear. I should have done that this morning, had I not seen that the gates were lowered. Naturally I concluded that the car would stop, and the first thing I knew of any danger was when we went crashing through the gates."'

The mayor of Newark, Henry Dorremus, was soon on the scene, and with tears streaming down his face and shaking with emotion he told a reporter, "I was completely unnerved. I cannot tell who is to blame, but whoever is will be made to suffer."

One of the surviving pupils, Elizabeth Mowbray, was also interviewed by a reporter. "We always caught that car in going to the high school in the morning, and as usual, we all met in the car. Of course, among a lot of high school children, there is also considerable merriment. We were as merry as ever and were laughing when we approached the school. Before coming to the school it was necessary to cross the railroad tracks.

"Just before reaching the tracks we felt a jar and heard the conductor say something about the motorman breaking through the gates. I learned later that the gates were down, and that the motorman had lost control of the car and that it had broken through the gates and was on the railroad tracks.

"We were still laughing and joking when there was a shrill whistle, and the next instant was the terrific crash. How I escaped with my life I have not the least idea. The shock was so great that for a time, I do not know how long, I was unconscious. When I became conscious, I was sensible of severe pains about my back. And then, horrors! There beside me in the snow, was the severed foot of a girl I had just been talking with. I knew it was hers because of a fragment of the dress which stuck to the shoe. About me, as I looked around, I saw the bodies of several of my girlfriends. It was horrible! I was fainting when they carried me away."[1183]

The jury at the inquest on Ernestina Miller, one of the victims of the accident, returned a verdict on 7th March which placed most of the blame upon the trolley car company for allowing the rails at the crossing to be covered with over two inches of ice and snow, for allowing an overloaded car to be run, for permitting eight or possibly ten passengers to ride on the front platform of the car, where they would likely have made it difficult for the motorman to properly do his job, and for not installing a derail at the crossing. The street railway company alleged that the Lackawanna company would not allow the installation of a derail, but the jury pointed out that there was one already in use at the Bloomfield Avenue Crossing, Montclair, on the Lackawanna line.

The railroad company were found to be at fault for running trains at a higher speed than permitted by city ordinances and the Board of Street and Water Commissioners and the Police Department were blamed for permitting violations of such ordinances. The jury recommended the abolition of all grade crossings within the city limits.[1184]

To be as stubborn as a mule is a well-known expression and two mules which were stubborn, or perhaps merely interested in self-preservation, were the cause of an accident at Redcomb Junction, Ohio, United States, eight mile east of Cincinnati, on Friday 10th July 1903. The animals were pulling a waggon carrying nine people. As the party was going over a level crossing at Redcomb Junction a Pennsylvania Railroad passenger train was approaching. The engineer of the train, seeing the waggon on the crossing, whistled a warning, whereby the mules stopped and stubbornly refused to budge. The animals themselves were over the crossing but the waggon was squarely upon it and was smashed. Robert Copenhagen, two of his children, the child of a neighbour and a man named William Poole were killed outright. Mrs Copenhagen and William Booth were supposedly fatally hurt. The youngest Copenhagen child, two years old, and a man named Charles Johnson were painfully hurt but thought likely to recover.[1185]

Wednesday 17th August 1904 was a race day in Chicago. Illinois, United States. In the late afternoon a 48th Avenue electric trolley car and two trailers of the Consolidated Traction System, 'packed to the running boards with racegoers' was struck by an eastbound Chicago Great Western Railroad express on Colorado Avenue level crossing in Austin, a Chicago suburb. The trolley cars approached the crossing at some speed just as the express, travelling according to different accounts at 20 or 40 miles an hour, came around the corner. Motorman Michael Ryan reportedly applied the brakes with full force, but the brakes failed to work, and the trolley car ran onto the crossing as the train was passing and struck it between the engine and tender. The motor car was smashed to pieces and the leading trailer knocked over, smashed and dragged down the track for 100 feet. The couplings with the second trailer broke and it escaped damage. Mrs Frances Rautman, William Irving, Mrs Jeremiah Shuckrow and a fourth unidentified woman were killed outright, and Motorman Ryan was reportedly fatally injured, sustaining a fractured

---

[1183] *Evening Express*, Lock Haven, Pennsylvania, United States, 20th February 1903
[1184] *The New York Times*, New York, New York, United States, 8th March 1903
[1185] *Adams County Union Republican*, Corning, Iowa, United States, 15th July 1903

skull. Twenty-three other trolley passengers were seriously injured.[1186]

Another, contradictory account, stated that the trolley had come to a stand short of the crossing, the conductor had gone ahead but had failed to see the approaching train and had signalled the trolley forward. After the accident two trolley car conductors, W. H. Condon and Steve Moran, were arrested together with Flagman Michael P. Meyer. There were no gates or barriers at the crossing, but it was protected by a flagman. There were eight tracks, comprising two main tracks and six side tracks. Different witnesses disagreed as to whether or not the flagman had signalled the trolley to stop.[1187]

Less than three weeks after the Austin accident another trolley car accident happened in St Louis, Missouri, United States. At 1.30 p.m. on Saturday 3rd September 1904 a Wabash St Louis and Pacific Railroad shuttle train was on its way from the World's Fair grounds to the Union station when it collided with Maplewood trolley car No. 314 on the Sarah Street Crossing. The trolley car was smashed and 50 or more of its passengers were scattered 'bleeding, wounded, dying and dead' along the line for 100 yards. The pilot of the Wabash locomotive was crushed in. The motorman and conductor of the trolley car were unhurt but seven of its passengers were killed outright and about 20 more injured.

Wabash Engineer A. W. Burbank, Crossing Watchman Fred Uetz, Flagman James Wease and trolley conductor Pat Sheehan were all arrested pending the coroner's inquest. Motorman Theodore Cook was not arrested, 'because of his indisposition', but a policeman accompanied him to his home and remained there to guard him.

As usual, statements as to how the accident had happened were conflicting. Engineer Burbank stated that he had seen the car stopped on the track, realised it was impossible to stop, had reversed his engine, applied the brakes and prepared to jump but had not done so.

Fireman Jerry Wilson said that he had jumped after his engineer had motioned for him to do so.

Motorman Cook did not speak to reporters but was said to have alleged that one of the fuses of his car had blown out as it was crossing the track and by the time that he had dealt with the problem his car had been run into.

Bystanders stated that the trolley had stopped on the crossing and the conductor had not got off it.

Flagman Uetz said that he had seen the train coming and had rung the signal bell. The trolley car had stopped short of the crossing, then gone forward. The conductor had not left the car but had been on it when it was struck. The crossing safety gate (barrier) at the crossing had not been lowered because of orders from the railroad company not to do so because there was no barrier on the south side of the crossing.

A newspaper report of the accident wrote, 'The scene of the wreck, at Sarah street and the tracks, is one that precludes the possibility that the train could not have been seen by the crew of the trolley car. The right of way of the road west of Sarah street is flat and level. There are three tracks. The train was approaching on the extreme southern one and

[1186] *Austin Daily Herald*, Austin, Minnesota, United States, 18th August 1904
[1187] *The Daily Journal*, Moline, Illinois, United States, 18th August 1904

the trolley car stopped twelve feet north of the north track. A view unobstructed by the flagman's shanty, which is at the edge of the track, could have been had of the train, which the flagman saw at a semaphore pole about seventy-five yards away.

'The street car approached within ten feet of the first track. It stopped, and if the conductor did not get off, although he said he did, contrary to the statements of others who saw the accident, the motorman could not have failed to see the train. The flagman saw it and warned all passers of its approach by his bell. He also noticed that the street car stopped momentarily, and then started.

'H. H. Culp, a patient at the hospital, who suffered a wound in the stomach and head, which subsequently caused his death, said the conductor got off the car and signalled the motorman to pass on, saying, "Come on, we can make it." After that Culp knew nothing. He said the collision lifted him from his seat, and when he recovered consciousness he was being carried to the ambulance. {Although he had been able to make a statement after the accident Culp died on the operating table that evening, probably from internal haemorrhage.} Miss Alice Atkins, another victim, had a similar statement to make. The passengers who stood on the rear platform escaped without serious injury, and none of them could be found.

'Sheehan, the conductor of the train, was arrested and taken to the Eighth District holdover. He declined to make a statement. He answered questions in mono-syllables to the effect that he had attended to his duty and that he did not see why the car stopped on the track, if it did stop. To the statement that it was his duty to explain his share of the responsibility for the accident and that it was due the public he answered between puffs of his pipe:

'"That don't cut no ice. I haven't got nothing to say."

'In contrast, the engineer of the train, A.W. Burbank, talked freely. He said that he saw the street car stop on the track, he could not stop the train, but tried to do so. He estimated the speed of his train at about twelve miles an hour. Persons who saw his train pass the crossing, Ed Shannon, of No. 4340 Clayton avenue was one of them, said the train was running at the rate of thirty miles an hour. Burbank admitted that his train was on time and that the schedule for the five-mile run from the Fair grounds to the Union Station was based on a running time of thirteen minute – a speed of about twenty-two miles an hour.'

H. L. Magee, Wabash general superintendent, stated that, "It appears to me that there can be no doubt that this wreck was caused by the carelessness of the crew of the trolley car. From all that I have ascertained, the motorman moved his car on our tracks before stopping, and the conductor did not cross the tracks as required by law.

"The accident is, of course, very much to be regretted, but in no way can blame be attached to the shuttle train service. This train was running on its schedule time and in accordance with its orders.

"The crew of this train, as well as the men employed on all shuttle trains, are experienced operators and the most reliable men in our employ. They have been impressed to regard the safety of surface traffic and pedestrians above everything, and to exercise all possible care that no accidents shall occur through running at a high rate of speed within the city's limits." [1188]

It has not been possible to trace a result of the inquest but it would seem that the conductor and motorman of the trolley were clearly at fault.

On the afternoon of Monday 1st February 1905 members of the Ladies Aid Society of the Universalist Church attended a gathering at Waldwins farmhouse, near Arkport, New York, United States. They set off back to Hornellsville in the early evening, at a time when it was just getting dark, riding in three horse-drawn sleighs, and their route took them over a level crossing, known as Stephen's Crossing, with the Pittsburg Shawmut and Northern Railroad. There was an uninterrupted view from the crossing along the railway line in both directions, but in the twilight and with snow falling visibility must have been considerably restricted.

As the sleighs approached the level crossing a train was seen to be approaching. The driver of the leading sleigh urged his horses on and made it over the crossing safely. The women in this sleigh attempted to warn the following sleighs of the danger and according to some accounts succeeded in doing so, and the driver of the second sleigh tried to pull up but could not check it quickly enough and when it did finally stop the sleigh was directly across the track.

The train was No. 9, a passenger working due in Bolivar at 8.40 p.m., with Conductor John McLaughlin, Engineer William Johnson and Fireman S.L. Smith. Fireman Smith told a newspaper reporter what happened. "We left Hornellsville on regular time, 7.15. It was snowing and the wind was howling, making it difficult to see ahead. We were running about 25 miles an hour. As the crossing was approaching the whistle was blown and bell rung. The first we knew the engine crashed into something. The air was put on at once and the jar was so great that the doors in the cars flew open. The train came to a standstill in about four lengths. The engineer and I jumped off and ran to the pilot where we saw two women lying across the pilot, one dead, one alive. I said, 'Isn't this horrible?' We supposed that two people had been riding in a sleigh and that there were no others hurt. We backed the train to the crossing and there on the left side was all the rest of the dead and injured. I think the sleigh was hit in the center, and it appeared that all except the three persons on the front seat were sitting on the bottom of the sleigh. The sleigh drove onto the track from the right-hand side. It was a diagonal crossing, not straight across. It was just 7.20 when the sleigh was hit."

Conductor McLaughlin said, "The first I knew of the accident was the jar from the air brakes. I was sure something had gone wrong. The train crew ran out and we found two women lying on the pilot. We backed up at once and found the others lying in the snow, slaughtered like sheep. It was an awful sight. We placed the dead and wounded in the train and did all we could for the wounded. On the way to Hornellsville I was trying to quiet a woman who was injured terribly, supporting her in my arms. 'Am I going to die?' she asked. I replied, you fainted, and you will be all right soon. 'My God, what will become of my children?' she moaned. In a minute or two more she was dead in my arms. The passengers and train crew did everything possible for the dead and dying and as long as any of us live we will never be able to efface the horrible sights from our memories."

[1188] *The St Louis Republic*, St Louis, Missouri, United States 4th September 1904

The driver of the second sleight, Elisha Quick, was badly hurt, but made a statement. "I was driving along at a sharp canter when I saw the flash of the headlight and heard the ladies in the front sleigh call out not to attempt to cross. The women in my sleigh saw the train at the same time, and realizing their peril, began to scream and shriek for help. Their cries and the glare of the light so frightened the horses that I was unable to stop them, and they sprang upon the track. They succeeded in clearing it, but just as the sleigh was half way across, the engine struck us with a terrible crash, and I remember nothing more until I found myself being picked up out of the snow. I could not hold my horses."

One of the surviving women from the second sleigh said, however, that she and her friends had seen the headlight of the train and had begged their driver to pull up and not to try and cross in front of it, but he had refused to heed their warning and had whipped up his horses to try and beat the train.

Of the 13 women in the sleigh seven were killed outright, three soon died of their injuries and the other three were seriously hurt.[1189] [1190]

The jury at the inquest on the victims of the collision found that "said Elisha Quick, driver of the said team, had ample opportunity to observe the approaching train, in time to have averted the accident. That in view of all the circumstances connected with the case, carefully and impartially considered, it is the opinion of this jury that said Elisha Quick was guilty of either carelessness or very poor judgement, or both, and is thereby more or less responsible for the terrible calamity, but we do not believe the facts warrant a criminal prosecution. We do not find that any blame can be attached to the Pittsburg Shawmut and Northern Railway Company in connection with the accident." [1191]

An American street car featured in another bad accident late on Sunday 20th August 1905, not in an obvious large city such as New York, Chicago or St Louis but in Butte, Montana. A few minutes after 11p.m. a so-called 'jumbo' car, No. 13, was returning from the amusement park at Columbia Gardens, well filled with passengers. It was proceeding along Utah Avenue when it came to the level crossing with the Butte Anaconda and Pacific Railroad. According to the conductor of the car, Fred Hinkle, it was brought to a halt at the crossing, he gave the usual two bells for the motorman, Alfred Jackson, to go on, and the car started over the crossing. The car had barely entered the crossing when Jackson clearly saw the rear of a train coming towards him, for he tried to put the car in reverse. In his panic, however, he failed to do so, and the car continued forward until a third of it was either over the crossing or on it. Then it was struck by the train. The street car was overturned and a heavy freight car fell onto it. Some of the street car passengers, seeing what was about to happen, had time to jump clear and escape at the last moment.

A number of other lucky passengers were thrown off the car as the impact occurred. One of them was a street car employee, Adolph Heilbronner, who realising the seriousness of what had happened ran to the nearest telephone and called out all the doctors and ambulances in town, together with the fire department and the police. The passengers who had been thrown off and were only slightly hurt made their way off homewards or else joined the crowd which was rapidly gathering around the site. Many

---

[1189] *Bolivar Breeze*, Bolivar, New York, United States, 2nd February 1905
[1190] *The Cazenovia Republican*, Cazenovia, New York, United States, 9th February 1905
[1191] *Bolivar Breeze*, Bolivar, New York, United States, 23rd February 1905

other passengers were trapped beneath the overturned car. Those at the rear of it were not trapped too badly and once the car had been lifted a little they could be got out. Those in the middle of the car were more difficult to release, some of them held down by seats, but were released with some difficulty. Those at the front of the car, which was pinned down by the freight car, were in the worst position.

'To give an idea of the situation it is only necessary to say that for a third of its length the big jumbo car was directly under the freight car. It lay with its bottom up and front wheels in the air, just clear of the edge of the freight. Under this, where it would be hardly supposed there would be found anything alive, were heard cries. One was the cry of a girl calling to her mother to "take me out." The car door farther back was not crushed so flat, being held partly by some of the woodwork of the upper part of the car and seats. Pinned between the roof of the overturned car and the inverted floor were several injured passengers.

'As soon as the firemen arrived, railroad men and citizens generally began to gather up means of raising the weight upon the imprisoned people. Towards the rear and centre of the jumbo this was not such a hard task. When it came to getting to those under the front part and the freight as well it was a different matter. Jackscrews had to be used to raise the end of the freight car, and heavy timbers to shove it up as the men worked in order to prevent it toppling over and causing another horror.

'Under the flattened end of the jumbo car and the freight car were seven persons. The rescuers worked for nearly two hours before the last one was out. As the work went on rumours spread that swelled the number under there to a dozen or more, and no one who saw the condition of the wreck expected to see any of them taken out alive. Several were taken out alive, however, though some will die later. The last victim was taken out about 20 minutes to 1 o'clock this morning. It was a girl, and she was hurried off to one of the hospitals.'

At the time of the initial report eight trolley car passengers were listed as dead and about 25 passengers, Conductor Hinkle and Motorman Jackson were reported injured.[1192]

By the following day it was reported that no further deaths were expected. Pending the coroner's inquest the entire crew of the freight train had been placed under arrest – Conductor C. W. Hoagland, Engineer John Logan, Fireman Fred Sanford, Head Brakemen W. P. Dalton and Second Brakeman J. J. Cruse. Conductor Hinkle and Motorman Jackson of the trolley car, both badly hurt, were not arrested. As usual there was a conflict over who was responsible for the accident. At least one of the passengers allegedly stated that the trolley car had not stopped before entering the crossing, but street car manager Wharton stated, "I am confident the motorman, who is an old and experienced man, made the stop there. Many people tell me he did. I don't believe the freight had a man on the end car. In fact, I have reason to believe the man was back two cars. The train came out of a narrow street. There was a freight car on each side of the street, so our motorman, in a brightly lighted street, was unable to see a dark object coming out from between two other dark objects." [1193]

The first witness to be called at the inquest was Engineer Logan, but as his engine had been at the far end of the train he could say little of use regarding how the accident

[1192] *The Anaconda Standard*, Anaconda, Montana, 21st August 1905
[1193] Ibid, 22nd August 1905

had happened.

Conductor Hoagland stated that the collision had occurred at 11.10 p.m. He had been standing on top of the last boxcar to check that the line was clear. The speed of trains over the crossing was restricted by the rules to eight miles an hour. They had the right of way over road traffic. When he had seen that the motorman was not going to stop he had hollered at him, but he had still kept going. The conductor said that he had then signalled "Stop" to his brakeman with his hand lamp and the latter had passed the signal back to the engineer. The train was moving at three or four miles an hour but because of the delay in transmitting the signal to the engineer he believed that it would have taken the train 60 feet to stop and the crossing was only 30 feet away. He said that about a third of the street car had been on the crossing when it stopped, and he saw the motorman trying to back up.

Head Brakeman Dalton said that he had been standing on the fourth car and had transmitted the stop signal from the conductor to the engineer, who could not himself see the conductor. He had felt the jolt as the air brakes were applied and knew that his signal had been seen.

Fireman Stanfield said that the engine bell had been ringing continuously. He did not think that there had ever been a watchman at the crossing.

Several passengers then gave evidence, with conflicting accounts as to whether they had seen the train conductor riding on the freight car, whether the street car had stopped before entering the crossing and so on. Young Conductor Hinkle (19) also gave evidence, but, the newspaper reporter wrote, 'The witnesses' explanation as to his own accident at the time of the disaster made it clear that he knew comparatively little or nothing as to the conditions surrounding the tragedy.' Following the conductor more passengers gave evidence and the inquest was then adjourned.[1194]

Two more days of evidence from passengers and others resulted in more conflicting testimony.[1195] Then on the fourth day of the inquest Motorman Jackson was called. He testified that on arriving at the crossing he had stopped the car briefly three feet from the track and had looked up and down the line but seen nothing. He had then given the customary foot-bell signal that the track ahead was clear and in reply had received the two-bell signal to go ahead. Just as his car had got onto the track he had noticed the freight car and as he tried to reverse the collision had occurred. There had been no lights at the crossing. He had seen the conductor aboard the box car. The evidence of other witnesses followed that of the motorman and the inquest was again adjourned.[1196]

After hearing more evidence the coroner's jury finally reached a verdict on Monday 28th. They found that "the collision was the result of carelessness on the part of the Butte Street Railway Company.

"We find that the street car was in charge of Conductor Frederick R. Hinkle, a youth 19 years of age, and the trolleys and signals were in charge of Master Bucher, a boy 13 years of age.

"We find that the street car did not come to a full stop before going on the railroad tracks.

[1194] Ibid, 23rd August 1905
[1195] Ibid, 24th and 25th August 1905
[1196] Ibid, 26th August 1905

"We find that a street lamp was burning at or about the intersection of Iron street and Utah avenue.

"We find that the conductor of the railroad train was upon the box car which collided with the street car and transmitted stop signal to the brakeman, who in turn transmitted stop signal to the engineer, and that the train was running at a speed of about four miles an hour.

"We find that there is a city ordinance requiring railroad companies to provide gates or flagmen at the following railroad crossings within the city of Butte; Utah avenue, Maryland avenue, Wyoming street, Main street, Montana street, and that there was neither flagman nor gates at the crossing of Utah and Iron street on the night of August 20, 1905.

"We find that on Aug 6, 1902, the petition of the Butte, Anaconda and Pacific Railroad company to substitute automatic electric arm gongs instead of gates or flagmen at the crossings on Montana street and Wyoming street was granted by the city council of Butte.

"We find that the city council has been negligent in enforcing the ordinances governing railroad and street railway crossings." [1197]

As might be expected the verdict of the jury did not satisfy everyone, Mayor MacGinniss reading a statement at a meeting of the city council a couple of days later in which he condemned the verdict as unjust and stating that no blame should be attached to the council, but that the blame lay with the railroad company.[1198]

There was a lucky escape from a much more serious accident on the Spring Avenue Crossing, St Louis, Missouri, United States, at 9.45 a.m. on Friday 3rd November 1905. A Missouri Pacific Railroad passenger train in charge of Conductor Merrifield, was, as a newspaper described it, 'speeding along at a clip of 25 miles an hour', when it collided with a mule and waggon on the crossing. The mule was killed outright, and the waggon was demolished. Edward Krauth, in charge of the waggon, was thrown 30 feet but not hurt.

What might have made the accident much more serious was the fact that the waggon was loaded with 3,000 pounds of gunpowder and dynamite. But for the fact that the explosives were packed into 40 thick wooden boxes the whole lot could have exploded, wrecking the train.

Krauth stated afterwards that he would not have attempted to cross had the gate on the north side of the crossing not been open, or if he had been warned by Gatekeeper Michael Kelly not to cross. Kelly for his part insisted that he *had* warned Krauth of the danger, and that the accident would not have happened had not hind wheel of the waggon not become embedded in a pool of mud which almost covered the south rail of the track.[1199]

Very early on the morning of Sunday 1st September 1907 a motor car containing the

[1197] Ibid, 29th August 1905

[1198] Ibid, 31st August 1905

[1199] *Sedalia Sunday Democrat*, Sedalia, Missouri, United States, 5th November 1905

Marquis Pallavicini, Signor Malvano (a lawyer), a mechanic and a chauffeur was on its way to attend a motor racing meeting at Brescia, Italy. As the car arrived at a level crossing near Magenta station at 04:30 the chauffeur, perhaps not expecting to encounter any trains at that hour of a Sunday morning, failed to notice that the crossing barriers were lowered. He failed to stop the car and it crashed through the crossing barrier just as a freight train arrived. There was a terrible collision and the petrol tank of the car exploded. Somehow the mechanic escaped unhurt and the chauffeur suffered only a slight head injury. The other two occupants of the motor were less fortunate. Signor Malvano died on the spot and the Marquis survived long enough to be taken to the Magenta hospital but succumbed to his injuries soon afterwards.[1200]

A trolley car taking 25 employees to the pin factory at Oakville was struck by a New York New Haven and Hartford Railroad extra freight train on Main Street Crossing, Waterbury, Connecticut, United States, at 6.30 a.m. on Friday 29th November 1907. The car was struck with great violence in the middle. Five employees – Annie Corcorry, Robert French, Walter Hayes, Jane Kelly and Sarah Ryan – were killed, and five more employees were seriously injured. The Oakville Pin Company was shorthanded, so the trolley car was less than half filled, it was usually carrying about 70 people.

It was alleged that the wrecked trolley had been closely following another one over the crossing, the first vehicle getting across safely. The gates were normally lowered upon the approach of a train but had remained up and it was said that the flagman had waved the second car across. The rules of the trolley car company also required the conductor to cross the tracks on foot and check that all was clear before the trolley entered the crossing but this, it was said, was not done. The engineer, it was stated, had no way of avoiding the collision, though he had applied the brakes.

After the accident many of the injured were taken to nearby stores, while those unable to walk were laid alongside the track before ambulances and carriages took them to their homes. The bodies of those killed were cared for by relatives.

Conductor John Dillon and Flagman John Flavin were arrested, but the motorman of the trolley car, Charles Larnad, was not held.[1201]

At the close of the inquest on Walter Hayes, one of the victims of the accident, Deputy Coroner Walter Makepeace, who held the inquiry without a jury, found that Conductor Dillon and Flagman Flavin were both criminally responsible for the accident. The coroner, having noted that the engine of the train was or should have been plainly visible to the conductor, stated, "The gateman, however, failed to lower the gates until the train was about on the car. Conductor Dillon continued to give the motorman the signal to advance up to the moment the car was struck. When Engineer Beauchamp saw the car he blew his whistle, and made every effort to stop his train, but could not, and the car was struck, and several persons killed, and others injured."

Conductor Makepeace found that Motorman Larnad had used proper care both prior to and after he saw the engine approaching, and that Engineer Leo Beauchamp had also used due care and complied with the rule of the company, and that the company itself had

---

1200 *The Standard*, London, England, 2nd September 1907

1201 *The Sun*, Baltimore, Maryland, United States, 30th November 1907 complied with all the orders of the state railroad commissioners.[1202]

Two days after the accident another car, containing 10 people and in charge of Conductor Rogers, had a narrow escape from being hit by the Winstead express on another Waterbury crossing with Bank Street. The car smashed one of the gates as it was being lowered. Motorman John Lynch and Gateman Daniel Campbell each blamed the other's carelessness for the near miss.[1203]

A Saturday night basketball game ended in tragedy early on Sunday morning, 23rd February 1908, at West Nyack, New York, United States. The game had been between teams from West Nyack and nearby Spring Valley. After the game, and a dance which finished at midnight, 21 young persons from Spring Valley, basketball players and supporters, had had supper before setting off homewards at 12.30 a.m. in two horse-drawn waggons, with 9 people in one and 12 in the second. The first waggon was being driven by Nelson May, aged 18.

What happened when the first waggon reached a level crossing with the West Shore Railroad at 12.40 a.m. was disputed. Gateman Konrad Kaufman alleged that the gates had been lowered five minutes before its arrival for the passage of a New York Ontario and Western Railroad express (that company had trackage rights over the West Shore) and that May had forced his way through them. This was disputed by friends of the dead and injured who asserted that the gate had been open, and Coroner Van Orden said he was not satisfied with a deposition made by the gateman. The latter declared that at the moment of the collision he had been in his shanty 'fixing the fires', and claimed that he had known nothing of the accident until 10 minutes afterwards.

At any rate the horses crossed the first of the two tracks and were on the second when the waggon was struck by the express, which was three hours late and travelling at 60 miles an hour. The waggon was struck almost in the middle and although Engineer Tierney knew that he had struck something it took him half a mile to stop. Five of the occupants of the waggon were carried along on the pilot of the engine. One of the five, Miss Edith Bird, although somewhat injured and presumably much shocked, was remarkably calm, and referring to her fellow passengers had said, "I guess most of them are dead." After she had been carried to one of the coaches of the express she was able to give the names of all her companions.

A brakeman went hastily through the coaches of the express. He was unable to find a physician but did manage to locate a nurse, who set to work treating the injured. As soon as was possible the express set off for Weehauken and Engineer Charles Tierney made what railroad men said was the fastest run ever to get there. Ambulances were waiting at Weehawken to rake the injured to hospital. Meanwhile a pair of foam-specked horses came running through the streets of Spring Valley, dragging the remains of the shattered waggon and bringing the first news of the accident.

They stopped at the livery stables of George Young, whence they had set out earlier in the evening. They were unhurt, through in their panic after the train had struck the waggon they had apparently run through one of the crossing gates.

[1202] *Journal Courier*, New Haven, Connecticut, United States, 7th December 1907
[1203] Ibid, 2nd December 1907

By the next morning it was known that six people had died – Nelson May, Miss Jeanette Palmer, who died on the train en route to Weehauken; George Reith, Misses Bertha and Edith Singer, who were sisters; and George Slinn, who died in the ambulance. Three others were injured – Miss Edith Bird, Henry Deiterlein and Warren Palmer. Harry Deiterlein died Tuesday morning.[1204][1205]

Engineer Tierney was the first witness at the inquest on Nelson May, taken as representative of the victims of the accident. The engineer stated that he had left Middletown at 11.5 p.m., three hours and 28 minutes late. He said that he had been running at 45 miles an hour and had whistled when 1,400 or 1.500 feet north of the crossing. He had seen nothing on the crossing until he was within 20 feet of it, when he thought that he saw a horse. He had not noticed if the gates were up or down.

The other trainmen could add nothing to the engineer's testimony.[1206]

On the second day of the inquest William Scheck Jnr. Testified that he had found the gates broken when he arrived at the crossing and this was confirmed by Gateman Kaufman.[1207]

The resumed inquest heard testimony from Conductor Thomas Smith and Baggagemaster L. V. Cannon of the express that after the accident Miss Bird had told them that the driver of the waggon, Nelson May, had been drinking and was "full". The inquest was adjourned again for testimony from Miss Bird.[1208]

The evidence of the last two witnesses, Miss Bird and Mr Warren Palmer, the surviving waggon passengers, was taken at their homes a few days later. In the end it proved of little use as neither could speak of the events leading up to the collision and Coroner Van Orden, despite producing a voluminous report running to several hundred pages upon the accident, was unable to come to any conclusion as to how it had occurred, and he said that the question would have to be answered by lawyers in damage suits. In the absence of concrete evidence as to the liability of either he placed no blame upon either the railroad company or the waggon driver.[1209]

Early on the morning of Sunday 29th February 1909 a party of five set off on a pleasure trip from St Louis, Missouri, United States, in an automobile, described as a four-cylinder Cadillac. Their tour took them to nearby Creve Coeur. One of the two women in the party had a Kodak (camera) and having snapped some views of the St Louis side of the lake expressed a wish to photograph its opposite side. The driver of the car, Mr Theodore Witte Sr., a buyer for the Witte Hardware Company, therefore set off down Creve Coeur Lake Road. This road crossed the Chicago Rock Island and Pacific Railroad line about two and a half miles from Vigus station. The crossing was described as being a dangerous one, the road crossing the line at an oblique angle with high brush obscuring the view on either side. There was no flagman and there were no gates.

[1204] *The Sun*, Baltimore, Maryland, United States, 24th February 1908
[1205] *The Times*, Haverstraw, New York, United States, 29th February 1908
[1206] *Middletown Daily Argus*, Middletown, New York, United States, 5th March 1908
[1207] Ibid, 6th March 1908
[1208] Ibid, 11th March 1908
[1209] *Orange County Times-Press*, Middletown, New York, United States, 27th March 1908

The car reached the crossing at the same moment as Rock Island train No. 682, heading for St Louis and travelling at 35 miles an hour. The vegetation prevented Mr Witte from seeing the train and the train crew from seeing the motor. As the train got to the crossing the front wheels of the automobile had just crossed the second rail and the pilot of the locomotive struck it squarely in the centre. Sitting in the front seat of the car alongside of Theodore Witte was Mr Frederick Witte, a departmental manager for the Witte company and probably Theodore's brother. The two men were thrown 15 feet into a ditch and both were fatally hurt. Riding in the back of the car were Mrs Carl Klinge, Theodore White's sister-in-law, his three-year-old son, Theodore White Jr, and Miss Halcyon Campbell, a friend of Mrs Theodore White. All three were killed instantly. The body of the toddler was found in a culvert 150 yards from the crossing, while the engine carried the remains of the car a quarter-mile down the track, the bodies of the two women being carried along on the engine pilot.[1210]

At the inquest on the victims the engineer and fireman testified that the engineer whistle had been sounded and its bell was ringing as they approached the crossing. They did not see the car until the last moment.[1211]

The victims of a level crossing accident at Haverstraw, New York, United States, on Thursday 16th June 1910, were not returning home after an evening out or out on a pleasure trip, nor even going to or from their work but rather were mourners at a funeral, together with their pastor and the driver of the vehicle. The accident happened to the leading coach of a funeral procession at the foot of Graveyard Hill. As the coach came up to a level crossing with the West Shore Railroad a light engine was also approaching it, on its way southwards to pick up a train at Weehawken.

What exactly happened was unclear. An electric bell rang in the gateman's shanty as he was eating his dinner and he automatically pulled a lever which lowered both heavy crossing barriers. As they came down they trapped the funeral coach on the crossing after it had got halfway across. Young Beebe Williams, the son of a local livery owner and the driver of the coach, thrashed his horses in a vain attempt to get them to force their way through the gate but without success and a moment later the engine struck the coach so that, according to eye-witnesses, it literally blew up. The engine stopped 150 yards down the line. The body of Beebe Williams lay 30 yards down the track and those of Mr and Mrs E. V. Siefred, of New City, New York, were 20 yards further on. The Rev A. Ramoth, pastor of the New City Methodist Episcopal Church, was also killed outright while Mrs Perry Kessler suffered a fractured skull, two broken legs and other injuries and died shortly afterwards. They victims had been returning from the burial of Mrs Siefred's father, which had left her a small fortune.[1212]

It has not been possible to trace any report of an inquest or investigation into the accident but in August it was reported that the West Shore Railroad had awarded $15,000 to the widow of the Rev Ramoth, while the claims of the relatives of the Siefreds and of William Beebe were outstanding. Whether or not this implied that the gateman had been at fault was not stated, though it would appear from the newspaper description of the

1210 *Quincy Daily Herald*, Quincy, Illinois, United States, 30th August 1909

1211 *The Chillicothe Constitution*, Chillicothe, Missouri, United States, 30th August 1909

1212 *The Corrector*, Sag Harbor, New York, United States, 18th June 1910 accident that he was, and that preoccupied with eating his meal he had lowered the crossing barriers without checking that the crossing was clear of vehicles, thus trapping the funeral coach on the line.[1213]

Four prominent residents of Baltimore, Maryland, met their deaths along with their chauffeur in a level crossing accident at Mill Lane Crossing, near Cape May, New Jersey, United States on Tuesday 9th August 1910. The victims were Mr Frederick Feldner, a well-known lawyer and company director; his wife, Mrs Mollie Feldner; their daughter, Mrs Doris Mergenthaler; their son-in-law, Mr Fritz Mergenthaler, 'inventor of the linotype machine'; and Milton Jones, the chauffeur. The party had left Atlantic City, where they had been staying the previous few days, that morning for Cape May, where they had planned to stay until Saturday.

The crossing was an open one, unprotected by gates or a flagman, and the road crossed the tracks of both the Atlantic City (a Philadelphia and Reading subsidiary) and West Jersey and Seashore (a Pennsylvania subsidiary) railroads. On either side of the road, parallel with the railroad tracks, a high field of corn obscured the view. The large Lozier automobile was reportedly being driven at considerable speed and overtook several other cars on the way to the crossing. No one in the vehicle apparently anticipated danger and the driver did not reduce speed as he reached the crossing despite the restricted view. Miss May McNeill, who lived near the crossing, saw both the car and a Pennsylvania Railroad express approaching the crossing and waved to the car driver to warn him but he took no notice of her. When the train and car met on the crossing the train was marginally in front, so that the car ran into the side of the locomotive. The train was reportedly travelling at 60 miles an hour and the car was smashed and overturned.

Engineer Joseph Wheaton stopped the train as soon as he realised what had happened. Several physicians chanced to be on it, and they went back to the wreck of the car and the victims. Each body in turn was checked for any sign of a heart beat but all were dead. The engineer was said to be prostrated by grief at the accident. Conductor George Taylor said that the whistle had been blown for the crossing as usual and that the collision was unavoidable.[1214]

The inquest on the victims of the accident was held the following day.

Conductor Taylor stated that train No. 13 had been running to time. His engineer had shut off steam and whistled a quarter of a mile before the crossing. He (the conductor) knew nothing of the crash before it happened.

Engineer Wheaton stated that he had whistled at all whistling posts and at Prayer Point, above the crossing. He continued, "I think the automobile hit the engine. I didn't see the automobile on the track but caught a glimpse of it twice its length away as it came down the road. I did not sound my whistle when I saw it, for I did not have time."

Fireman Evan Brown testified that he was sitting on the box ringing the bell when he heard a crash. He saw nothing of the car. The proper signals were given.

Baggagemaster Eldredge Doughty of train No. 13 was also positive that the proper

---

[1213] *Orange County Times-Press*, Middletown, New York, United States, 12th August 1910
[1214] *The Sun*, Baltimore, Maryland, United States, 10th August 1910

signals had been given.

Train passengers and lineside residents also confirmed that the train had whistled for the crossing.

Miss McNeill related how she had attempted to warn the driver of the car of the danger and, in contradiction to the newspaper account the day before, stated that the automobile had not been going fast at all.

Also called were Engineer Lewis Corson and Conductor Somers Lloyd of a Reading train due at Cape May at 5.50 p.m., who testified that their train had been running six minutes late that afternoon and there had been no racing between the two trains which might have resulted in reckless speed.

Several witnesses spoke of the crossing being a dangerous one and there were calls to make it and other crossings safer. Thus, a Baltimore newspaper reported that, 'The accident has aroused a determination on the part of business men and large real estate dealers here who are interested in promoting safety in motoring to urge the establishment of safeguards at all railroad crossings. Manager Doyle, who has entertained numerous large automobile parties from Philadelphia and other points demanded flagmen for June 11, when a run of 135 machines came to Cape May, but the Reading and Pennsylvania both failed to comply with his demand.

'It is now probable that the county authorities will take a hand. Residents in the immediate neighbourhood report many narrow escapes and say that the railroad companies ought to install automatic gates or at least a system of signal bells for the protection of the general public.'

As regards the speed of the car, the newspaper wrote, 'It has been established beyond a doubt, irrespective of the corroborative evidence of residents along the road, that the Feldner automobile was going at a reasonable pace. A machinist at the hotel garage examined the car before it had been touched by anyone and found that the speed was in the first notch, which indicates between 8 and 10 miles an hour.

Since the wreck has been brought in many have touched it, and the fact that both spark and throttle are advanced to high speed gave rise to a story of scorching. But evidence of eyewitnesses corroborates the report of the machinist that the automobile was on low speed when it was struck.

'It became positively known that late today that Fritz Morgenthaler himself was driving the machine, as several chauffeurs said they saw him at the wheel. Jones was sitting beside him. Morgenthaler was known as a careful driver. Why no member of the party heard the engine whistle, for there is no doubt that it did whistle several times, is a mystery that, it is believed, will never be solved.'

After deliberating for two hours the inquest jury returned a verdict of 'Accidental death', with a recommendation that safety appliances be placed at the crossing.[1215]

A Michigan United Railways street car was struck by westbound Michigan Central Railroad express train No. 31 on East Main Street Crossing, Kalamazoo, Michigan, United States, at 10.07 p.m. on the evening of Saturday 12th November 1910. Five street

---

[1215] Ibid, 11th August 1910

car passengers and Motorman Ward Abbott were killed more or less outright and 26 were injured, four it was thought at the time fatally.

At the time of the accident Conductor Vern Van Horn of the street car was standing on the crossing between the two Michigan Central tracks and had signalled to the motorman that it was safe to proceed. The conductor claimed that he had neither seen nor heard the train until it was close upon him, when it was too late to stop the car. The train hit the front of the car, smashing it to pieces, and the bodies of living and dead passengers were carried a block on the pilot of the engine before the train could be stopped. To add to the horror the passengers who remained in the car received burns and shocks from the electric current, which made all the metal parts of the damaged car 'live'.[1216] The conductor had waited for a freight train to clear the crossing on the other track but had then failed to notice the approaching express. The street car was crowded with passengers who had attended a concert given by Giuseppe Campanari, a noted baritone, at the Y.M.C.A.[1217]

By the end of the inquest upon one of the passengers on 17th November, one of those injured had died, bringing the number of people killed to seven. The inquest jury blamed the railroad company, the city authorities and the state railroad commission for the accident but exonerated the street car conductor. Their verdict read:

"Said Theodore Moske, while a passenger on the Michigan United Railway city line car No. 242, met his death by being struck by a Michigan Central railroad express train, No. 31, on East Main street crossing, on or about 10:07 on November 12,1910, and we believe said accident was caused through the negligence of the Michigan Central railroad company by not giving proper protection to the public over such a dangerous crossing, and exceeding the speed limit at the East Main street crossing of the city ordinance and state law and by obstructing the view of the main lines by side tracks upon both of their main lines so near to said crossing.

"We feel the city should be censured for not enforcing the ordinance regarding the speed limit of trains inside the city limits and the proper protection of grade crossings; also that the state railroad commission should be censured for allowing such an important crossing in such a dangerous manner, also that the conductor of the Michigan United Railway company's car used all due care and diligence in protecting his car and the passengers thereon in the performance of his duties." [1218]

On Christmas Day, Sunday 25th December 1910, a mistake by crossing keeper led to a bad accident in France. The express from Orleans to Le Mans which usually passed Châteaudun at 11 38 a.m. was two kilometres from the latter place when it collided with a car at a level crossing, the gates of which had been recklessly left open by the crossing keeper. The driver tried to stop the train but could not avoid the collision. The vehicle was thrown a great distance by the train, which was running at 75 kilometres an hour.

Six of the nine people in the car, Monsieur and Madame Barthélémy Bouilion – 56 and 47 years old respectively, Monsieur and Madame Pardessus Barthélémy, a six-year-

[1216] *The Warren Republican*, Williamsport, Indiana, United States, 18th November 1910
[1217] *The Daily Tribune*, South Haven, Michigan, United States, 14th November 1910
[1218] Ibid, 18th November 1910

old girl and a six-month-old baby were killed. The other three, Mr. Belianger, who was the driver of the car, a six-year-old son of Monsieur and Madame Pardessus Barthélémy, and a six-month-old baby girl, who was being nursed by one of the couples, were injured. The little girl, in particular, was very badly hurt.

The victims all lived at Ouzer-le-Doyen (Loir-et-Cher), and had been returning from a baptism at Donnemain.

Mr. Mouchât, the Chateaudun examining magistrate, immediately opened an investigation and took statements from the driver and fireman of the express and from Madame Barbault, a widow who was the crossing keeper. Madame Barbault was then charged with reckless homicide but in view of her large family of five children she was provisionally set free.[1219]

On Sunday 2nd July 1911 a picnic was in progress at Goldberg's Farms on Trotting Course Lane, Long Island, New York, not far from Ozone Park station on the Long Island Railroad branch to Rockaway Beach. It was decided that some beer was needed for refreshment and a party was sent to get some, bringing it via the small Jockers Hotel nearby. Thus, at 4.30 p.m. a party of two men, two women and a boy was on its way back from the hotel to the picnic with a keg of beer, riding in a heavy spring waggon being driven by Miss Mary Sessler, housekeeper at the hotel. Trotting Course Lane descended a steep hill, at the foot of which the lane crossed the Rockaway Beach line. As the waggon approached the crossing Flagman Louis Ammenworth stepped slowly out into the road to stop it, as an electric express from Rockaway Beach had been signalled and was close at hand.

Ammenworth waved his flag to warn those on the waggon to stop but the horse pulling the waggon seemed to suddenly become unmanageable and dashed down the hill onto the line. He said that he had signalled the waggon to stop in plenty of time had the driver been able to control the horse. When it became clear that the waggon was not going to stop he had turned and waved his red flag at the advancing train. Motorman Harry Lefferth saw the signal, shut off power and applied the brake in emergency, but the train was 100 feet beyond the crossing before it could be pulled up.

The waggon was smashed in the collision and all five of its passengers were killed. They were Miss Ida Cohen and her brother Jacob, Miss Leah Goldberg, Jacob Simon and Elmer Weinberg, aged 10. Miss Sessler suffered a fractured skull and it was feared that she would die; the horse was also killed. Motorman Lefferth was badly cut and bruised, but later took the train on to Manhattan. Conductor Richard Creed, who had been standing on the platform of the train beside the motorman, was struck by flying wreckage, knocked from the platform into a ditch and his left arm broken.

News of the accident soon reached the Goldberg Farm and soon many picnickers hurried to the scene. Help was called for and ambulances soon arrived from Jamaica. Miss Sessler and Conductor Creed were hurried off to hospital and the ambulance men were kept busy reviving women picnickers who had fainted.

Charles Jocker, the owner of the hotel, was said to have been heartbroken at the accident. He had rented the horse and waggon to the picnickers to carry the beer, the

[1219] *Le Figaro*, Paris, France, 26th December 1910

intention being that once it had been dropped off Miss Sessler would return with the empty waggon. He explained that picnics were held at the farm every Sunday. No liquor was on sale there, so when the picnickers became thirsty some of them volunteered to go and buy some. He did not know where they had bought the beer, but they had arrived at his hotel rolling the keg along the road. There they had asked if they could hire a rig.

It was not clear why the horse had apparently run away, as there had been nothing to frighten it, and thought possible that in fact rather than doing so it had just been overpowered by the weight of the heavy waggon and its load on the steep hill.

Ironically, the day before the accident the crossing had been one of several mentioned in a Public Service Commission order requiring the Long Island company to substitute gates for flagmen at dangerous crossings. Witnesses to the accident, considered, however, that it would have needed a strong gate to stop the horse and waggon involved in the collision.[1220]

At 4.15 p.m. on Wednesday 19th July 1911 a chauffeur-driven car containing a party of four prominent women from Saratoga Springs was returning there from a trip to the Berkshire Hills and Williamstown, Massachusetts. As the car reached a level crossing at Raceville, New York, United States, seven miles north of Granville, the second section of Delaware and Hudson Railroad passenger train No. 138 was also nearing the crossing. Seeing the train the chauffeur, Charles Ostrander, changed to a higher gear to beat the train over the crossing. In doing so he 'lost the spark', in the words of the newspaper report, and the car stalled on the crossing. Before he could restart the car, or its occupants get out, the train had struck it. Mrs Charles Grant and Mrs Charles Wells were killed outright; Mrs F.M. Robinson died on the train before it reached Granville and Mrs George Yates expired while being carried to the office of Dr Davies there.[1221] Chauffeur Ostrander suffered two broken legs, a broken arm and internal injuries and died later as well.[1222]

In what might be described as a typical level crossing tragedy a large horse-drawn waggon described as a hayrack and carrying 30 young people home from a Saturday evening dance was struck by a train on South Commercial Street Crossing, on the outskirts of Neenah, Wisconsin, United States, on Sunday 24th September 1911. The riders were returning from the home of a Peter Hanson on the Oshkosh road and were said to have been singing as the hayrick pulled up the slight incline to a level crossing with the Chicago and North Western Railway. As the waggon passed over the crossing its rear half was struck by a speeding northbound passenger train. Twenty-three of the passengers were scattered 200 feet along the track, dead, dying or badly hurt. Those seated at front of the hayrack escaped almost without a scratch. Thirteen of the occupants were killed more or less instantly and ten seriously injured.

The aforementioned Peter Hanson was the person driving the hayrick. He escaped unhurt. He said that he had seen the headlight of the North Western train, had

[1220] *The New York Times*, New York, New York, United States, 3rd July 1911

[1221] *The Berkshire Evening Eagle*, Pittsfield, Massachusetts, United States, 20th July 1911

[1222] *The North Adams Transit*, North Adams, Massachusetts, United States, 21st July 1911 overestimated how far away it was and had urged the horses pulling the hayrick forward. None of the other surviving passengers had realised the danger that they were in.

Mangled bodies were strewn along the track and six were taken off the pilot of the engine. Dismembered parts lay about and three of the dead were so badly mangled that they could only be identified by the clothes that they were wearing. Two of those killed were thrown against a flagman's shanty so violently as to overturn it. It was expected that four of the injured would die. Almost all the victims were employed at the Neenah paper mill.

Anton Brezinski, one of the uninjured survivors, spoke to a newspaper reporter. "I was standing up in front of the waggon near the driver holding a lantern to light the road for him, and we were both on the lookout for an approaching train. There is a huge advertising billboard running up to the track and this completely hid the train from our sight. The board is sixty feet long and about twenty feet high, painted a dark grey, giving it the appearance of the sky in the background, and deceived us in thinking that the track was clear.

"There was no warning of any kind, no whistle or bell, and the first intimation I had was when I saw the headlight appear from behind the billboard and less than thirty feet from us. We were then directly on the track and it was too late to avoid the accident. Further back in the waggon the others were talking and laughing among themselves and could not have seen the train anyway. When the crash came it literally swept the waggon from under me and I landed squarely on my feet."

It was reported that both Engineer Frank Spooner, who had been in charge of the engine of the train, and Peter Hanson had become nearly insane after the accident. The engineer had tried to take the train to the end of its run, but by Appleton his mental state had deteriorated so much that he had to be relieved and taken to a secluded spot to recover. Peter Hanson was also placed in the care of physicians.

No blame, it was stated, was attached to either Engineer Spooner or Conductor Keers, in charge of the train. The crossing was stated to have been a dangerous one.[1223]

At the inquest on 3rd September on the victims of the accident, who it was by then reported numbered 14, Peter Hanson was described as a pathetic figure. He cried several times as questions put to him brought back memories of the tragedy. He admitted that he had drunk three beers at the dance. Three times, he said, he had looked up the track before passing the billboard, 100 feet from the crossing, but had seen nothing and had heard no whistle. He had not looked again until directly on the crossing, when the train had still seemed a safe distance away.

Tony (Anton) Brezinski testified that there had been no noise in the hayrick (so much for the reporter's earlier story that the passengers had been singing). He had been fixing a blanket around the head of Mary Schwarzbauer when he had looked up and yelled, "Here comes a train." Hanson had whipped up the horses, but too late. He said that he had bought beer for the party, but no one had been drunk.

Mention was also made of a bill which had been introduced into the state legislature which would have stipulated that railroad companies should place alarm bells at all crossings within the corporate limits of cities and villages. The railroad commission had

---

1223 *The Wauwatosa News*, Wauwatosa, Wisconsin, United States, 29th September 1911 asked that the bill be referred indefinitely or until the commission had determined whether or not it was a matter which came within its jurisdiction. The inquest was then adjourned.[1224]

At the resumed inquest Chicago and North Western General Manager W. D. Cantilion admitted that in order to satisfy the demands of the travelling public the company must violate city and village ordinances and state laws which attempted to regulate the speed of trains. The reason of course was that the ordinances were unrealistic. If trains were obliged to reduce speed to say 10 or 15 miles an hour through every city and village every train journey of any length would be a very tedious and time-consuming affair. It was also noted at the inquest that the state railroad commission had as its goal the elimination of grade (level) crossings in Wisconsin. Over a century later there are still a lot left, and as elsewhere in the world much of the elimination of crossings has been due to the closure of the lines on which they were found.

After half an hour's deliberation the inquest jury found that the (now 15) deaths had resulted from the carelessness of driver Peter Hanson and would not have occurred had the crossing been properly guarded.[1225]

On the morning of Sunday 14th January 1912 five women servants of Mr Alexander Biddle and Mrs Harriet Massey wished to attend morning mass at Saint Dominic Roman Catholic church in Torresdale, Pennsylvania, United States. The church was two miles from the houses where the servants lived, the morning was dark, the temperature stood at zero and the roads were covered with snow. However, Mr Brunner owned a livery stable so to avoid a long unpleasant walk through the snow the servants arranged with Mr Brunner to have his driver, Charles Davidson, convey them to church in a horse-drawn 'bus (contemporary spelling of bus, showing that it was a contraction of omnibus). They set off just before 6 a.m.

A mile from the church and half a mile from Torresdale station the road the party were taking crossed the Pennsylvania Railroad main line at Linden Avenue Crossing. As they neared the crossing a freight train rumbled past. There should have been a flagman protecting the crossing but the regular flagman, Jonathan Randall, had been given two days' leave. An Italian named Sebastian Taglione, who lived several miles away at Bristol, was supposed to be taking his place but was not there either.

With no one at the crossing to warn him Davidson drove onto the crossing behind the freight train right into the path of a Pittsburgh to New York flyer which should have left the Broad Street station in Philadelphia at 4.26 a.m. Delayed by the bad weather, it did not pull out of there until 6.13 a.m. and was running fast in an effort to regain some time. The train was travelling at 60 miles an hour when it struck the 'bus, killing the horse and smashing it 'like an eggshell'. Davidson was thrown 50 feet and the remains of the 'bus and its passengers were scattered along the track for half a mile. The driver and all five of the servants were killed instantly. The names of the servants were Rose Gallagher, 18; Agnes Garrity, 22; Bridge Malloy, 42; Mary O'Connor, 19 and Mary Roddey, 20. It seemed likely that none of those with the 'bus had seen the train at all, while the engineer of the train only saw the 'bus at the last moment.

[1224] *The La Crosse Tribune*, La Crosse, Wisconsin, United States, 3rd October 1911
[1225] Ibid, 10th October 1911

A brakeman ran a quarter-mile to a house to telephone for assistance. Police waggons were sent from Tacony and Frankford police stations and the remains of the bodies were collected up and taken to the undertaker. Over an hour after the accident the relief flagman finally appeared, sneaking away again it was said when he saw the police and realised what had happened.[1226]

At the inquest it developed that Divisional Supervisor, W. F. Rench, had granted Randall leave the day before. The crossing was supposed to be manned between 6 a.m. and 8 a.m. Rench said that he was aware that Taglione could not get to the crossing before 6.40 a.m. and that Randall should have stood in for him until he arrived. The coroner's jury, however, exonerated both Randall and Taglione and Coroner Knight took exception to the supervisor's efforts to blame Randall for the accident.[1227]

'No one left to tell the story' was one of the headings of a newspaper story relating to an accident at Owens Crossing, near Elletsville, Indiana, United States, early on the morning of Tuesday 19th March 1912. The day before the accident the five men who were the victims of the accident had all been working in a glass factory at Spencer, Indiana, when the machinery had broken down, meaning that they could do no more work that day. With their afternoon suddenly free they decided to hire a surrey (a four-wheeled carriage) and two horses and take a trip to the nearby town of Bloomington. This was done and they spent the afternoon and evening there, walking about the town and some of them drinking, though it was said that one of them, possibly the driver of the surrey, had had nothing to drink.

About midnight they set off back to Spencer. It was raining so the curtains of the carriage were closed. They reached Owens Crossing, about a mile south of Elletsville, at the same moment as a northbound Chicago Indianapolis and Louisville Railroad, (Monon) express. Whether the driver of the surrey was one of those who had been drinking, or whether in the darkness and rain he simply failed to notice the approaching train, it was of course impossible to say, but the surrey entered the crossing in front of the train, was demolished and strewn along the track. The railroad was on a curve on the approaches to the crossing and the engineer only glimpsed the vehicle at the last moment. The train ran on to Elletsville before it could be stopped. The body of one of the men was found on the pilot of the engine and thinking that there were probably others the enginemen awakened local residents to help them search for more victims. After three hours all five bodies had been found. The men were identified by papers found on them as Hall Brown, of Spencer; Arthur Farmer, of Summitville; Jesse Rodges (Hodges?) of Greenfield; Martin Wardin, of New Albany and Emmitt Williams of Kansas City. None of the men had much money upon them having presumably spent it all in Bloomington enjoying themselves. One of the men was married, the others were single. Neither horse was killed, though one was seriously injured.[1228]

On Sunday 30th June 1912 the Tourné skat club (skat is a three-player card game, very popular in Germany) arranged an excursion to Deutsch Lissa, travelling in several

---

[1226] *The Philadelphia Inquirer*, Philadelphia, Pennsylvania, United States 15th January 1912
[1227] Ibid, 20th January 1912
[1228] *The Bedford Daily Mail*, Bedford, Indiana, United States, 19th March 1912

charabancs of the Breslau Omnibus Company. They started their homeward journey to Breslau (now Wroclaw) at 23:00. Near the village of Pilsnitz, Germany, the high road they were taking crossed the double-track Breslau – Stettin main line at Schrankenposten 4, where the level crossing was protected by lifting barriers. As the charabancs reached the crossing the barriers were raised. The first vehicle crossed safely, but the second was still on the track when it was struck by Glogau – Breslau passenger train No. 647. The charabanc was totally destroyed in the collision and 10 of its occupants were killed outright and 15 more or less seriously injured. Four of the victims were carried 250 metres on the locomotive. Only two people on the charabanc escaped unhurt: a woman sat inside and the chairman of the club, who was standing on the running board at the rear of the vehicle and doubtless had time to jump clear when he saw the train. A doctor who chanced to be a passenger on the train was able to give immediate first aid.

At night the busy crossing was manned by a relief crossing keeper. Fritz Standke had come on duty at 18:00. On the way to work he had stopped off at an inn for a glass of beer but by 23:40 he was completely sober. Express train No. 92 passed the crossing at 22:45, and he had then opened the barriers and sat at a table reading a newspaper. He said later that he had felt drowsy and had thus not reacted to the warning signal indicating the approach of train P 647, which was only a minute late at nearby Herrnprotsch. A minute or so later he woke up and looking at his watch saw that the train was due, whereupon he had jumped up and rushed outside, too late to close the barriers or stop the train. It was a clear moonlit night, but the double-headed train was travelling at 75 kilometres an hour and its drivers could do nothing to avoid the accident.

Regarded by his superiors and willing and reliable, the relief crossing keeper tried to kill himself immediately after the accident but was prevented and arrested. Following a trial at Breslau he was sentenced to two years' imprisonment and costs. He could not afford to pay damages, being totally destitute. He was also dismissed by the railway company. He was released from prison following a Royal pardon in July 1913.[1229]

The carelessness of a car driver led to at least six deaths on a level crossing at Alexis, Michigan, United States, on Sunday 28th July 1912. Those killed were farmer Rudolph Cousino, who was probably the car driver; his six- and twelve-year-old sons and four-year-old daughter; his sister-in-law, Mrs Louise Cousino and Eva Navarre. Mrs Mamie Cousino, Rudolph's wife, and Lafayette Cousino, his brother, were probably fatally injured and two-year-old Opal Cousino, daughter of Louise and Lafayette, was slightly hurt. Visibility at the unguarded crossing was obscured by a cornfield and the car driver had driven onto the track without stopping to check that it was safe to do so. The car was hit in its centre by a Lake Shore and Michigan Southern Railroad passenger train and the victims were thrown 100 feet. The train was stopped as quickly as possible and the victims were placed in the baggage car of the train. Train passengers who had assisted in the rescue of the victims were overcome by what they saw.[1230]

In 1912 Rebecca Street level crossing in the Pittsburgh suburb of Wilkinsburg, Pennsylvania, United States, bore the alternative name of 'Dead Man's Crossing'. It was

[1229] *Eisenbahn-Unfälle in Europa* by Erich Preuß pg. 37 (in German)

[1230] *The Waterloo Press*, Waterloo, Indiana, United States, 1st August 1912 said that no less than 55 people had lost their lives at the crossing in the previous 10 years. On Thursday 3rd October 1912 the crossing again lived up to its name when five people lost their lives in a collision involving a car and a Pennsylvania Railroad express.

A large number of people witnessed the accident for the crossing was located at Wilkinsburg station. The semi-centennial of the town was being celebrated and thousands stood opposite a reviewing stand watching a parade. John E. Beck of Wilkinsburg, superintendent of the Westinghouse Air Brake Company. Had offered to take the Rev W. L. Nicholson, pastor of the First Presbyterian Church, Mrs Nicholson and their four-year-old son, Charles, for a ride. Mrs Nicholson got into the automobile but then changed her mind and got out again. Mr Beck started off with the pastor and young Charles, reached the level crossing and was signalled to stop. He attempted to do so but was unable to pull up before the front of his vehicle was straddling the tracks. Seeing the train approaching at 50 miles an hour he tried to back up, whereupon his car stalled. The car and its occupants were hurled into the air, twisted and fell onto the station platform, striking Mrs Mary Diamond and killing her instantly, though a baby in her arms was unhurt. Mrs Terese Polonino, Mrs Diamond's mother, who had three more of Mrs Diamond's children with her, was crushed by the car against the wall of the station and also killed. The three children were little hurt. All of those in the car lost their lives and an Italian track worker had to go to hospital after he was struck by flying debris.[1231]

Initial reports blamed a failure to lower the crossing gates for a bad collision between a Lake Shore and Michigan Southern Railroad coal train and a street car on Center Street Crossing, Ashtabula, Ohio, United States, on Monday 16th December 1912. The accident occurred at 8.30 p.m. Conductor Freeman Mullen of the street car, who was not hurt, went on over the crossing ahead of the streetcar and evidently not seeing the coal train signalled to his motorman to come on. As the car reached the tracks Motorman James McCutcheon saw the advancing train when it was too close to do anything. With a tremendous crash the locomotive, pulling 20 steel hopper cars loaded with coal, struck the street car and in doing so was derailed. The heavy cars by their momentum forced the remains of the street car and the derailed engine 50 feet along the track, demolishing fences and telephone and trolley poles. The coal cars were derailed in turn, one of them crashed through F. H. Dorman's grocery store, spilling coal within. Another tore away the side of Mrs Frances Culver's house and eight feet of the Richard's brothers brick warehouse was demolished. The engine crew had time to jump and escaped injury.

A rescue operation was quickly organised, with all the ambulances and doctors in the city attending. Dangling live wires, brought down by the collision, made the rescue work hazardous. Of the 14 street car passengers exactly half lost their lives. Those killed were Mrs Frank Bartlett, Mrs W. H. Cook, Mrs John Espey, Mrs George Kitson, Mrs Laura Leaphart and Mr and Mrs David Stowe. All were from Ashtabula except Mrs Espey from Willoughby. Most of the remaining passengers and Motorman McCutcheon were injured and the latter could give no account of the accident. Mrs Rose Thompson of Ashtabula, who had been stood at the crossing waiting to cross was also caught in the wreckage and became the eighth fatality.[1232]

[1231] *Xenia Daily Gazette*, Xenia, Ohio, United States, 4th October 1912

[1232] *The Marion Weekly Star*, Marion, Ohio, United States, 21st December 1912

Motorman McCutcheon was arrested after the accident and held pending the inquest, but the conductor was not. The inquest was opened the day after the collision when Towerman James Hoolihan was asked the obvious question – why he had not lowered the gates? In reply he stated that he had attempted to do so but had been prevented by the street car, which had shot over the crossing. He said that he had sounded the gong and shouted to the crew of the street car, "For God's sake, what's the matter with you, can't you see the train coming?"

Motorman McCutcheon testified that, "Conductor Mullen went ahead. The gates were not down. He gave me the signal to come ahead. Then, after I had gotten under way, he saw the train coming and called for me to stop, but it was too late. I saw I couldn't stop the car without stopping it on the tracks on which the train was coming. The only thing to do was to go ahead and try and get the car across. This I tried to do. The gates not being down, there was nothing to warn me until Mullen called for me to stop." [1233]

Conductor Mullen stated that he had signalled his motorman to stop when he saw the danger, but his signal had gone unheeded.

The train crew testified that their train had not exceeded 15 miles an hour, other witnesses put it at nearly 40 miles an hour.[1234]

The inquest was adjourned at this point. When it was resumed and concluded Coroner Harry Austin blamed the accident on the failure of Towerman Hoolihan to lower the crossing gates in time and on the engine crew exceeding the local speed limit. There was no mention of Towerman Hoolihan being charged with manslaughter.[1235]

Shortly before 11 p.m. on Friday 3rd January 1913 Chicago and North Western Railway passenger train No. 26 struck a buggy in which seven people were riding three quarters of a mile west of Denison, Iowa, United States. Mr and Mrs John Reimers, their five-month-old son, Mr Carl Evers and his daughter Minnie, aged 14 were all killed outright, while Mr Evers' six-year-old son was fatally injured and died in hospital. Mr John Reimers' son, also aged six, was injured. The party had been going home after spending the evening trading in Denison.

On the same day two people were killed and a third was probably fatally injured in a level crossing collision with a car at Dunscombe, Iowa.[1236]

During a storm on Wednesday 22nd October 1913 strong winds blew an empty coal waggon out of a goods yard at Waldenburg-Altwasser and along the line towards Neu Weißstein, Germany, until it collided with an electric tram on a level crossing. Fourteen people on the tram were killed.[1237] [1238]

[1233] *The Fort Wayne News*, Fort Wayne, Indiana, United States, 17th December 1912
[1234] *The Evening Herald*, Norwalk, Ohio, United States, 18th December 1912
[1235] *Xenia Daily Gazette*, Xenia, Ohio, United States, 31st December 1912
[1236] *The Muscatine Journal*, Muscatine, Iowa, United States, 4th January 1913
[1237] *Eisenbahnunfälle Band I* by Ludwig Stockert pg. 64 (in German)
[1238] *Eisenbahn-Katastrophen in Deutschland* by Hans Joachim Ritzau pg.129 (in German)

Four United States soldiers and their civilian chauffeur died when the car they were riding in was struck by a Galveston Houston and Henderson Railroad passenger train on a level crossing at Texas City Junction, Texas, late on Sunday 23rd November 1913. The victims were Privates Frudenthal Hill, John Livingstone and John Parkinson, Orderly Frederick Proctor and Wyly Sloman, the chauffeur.

Another soldier named Henry Smith was seriously injured. The soldiers were stationed at a camp at Texas City. Why the men were riding around in a chauffeur-driven car, unless it was a taxi, was not stated in the brief reports of the accident.[1239]

Lamar Avenue Crossing, on the outskirts of Memphis, Tennessee, United States, was the scene of a collision between a Union Railroad of Memphis freight train and a street car on the evening of Sunday 4th January 1914. Five street car passengers lost their lives and two more were seriously injured. Three of those who lost their lives were T. J. Caveney, of Cleveland, Ohio; Thomas Norrill, of Memphis; and T. J. Vall, of Jackson, Mississippi. The other two fatalities were not immediately identified. Those injured were Mr T. S. Ladd and Miss Lilian Troy.[1240]

The freight train was reportedly operating on a line which officials had ordered removed several days earlier and it was said that there had been no lights at the rear of the freight train and there had been no flagman at the crossing. In view of the failure of the railroad to comply with the order, and the accident Mayor E. H. Crump of Memphis and Commissioner of Public Utilities Thomas Dies took unilateral action and the tracks were ripped up by their employees, guarded by the police, the day after the accident. Later, though, they had to be replaced following an application by the railroad company to the Chancery Court and the issuing of a restraining order.[1241] The order enjoined the city officials from interfering with traffic on the railroad pending the hearing of arguments on 9th January regarding the application by the city for an injunction enabling the city to compel railroads to abolish level crossings and substitute subways.

The matter was a little more complex than was explained in most newspaper reports for at the site of the level crossing accident the Union Railroad of Memphis tracks ran parallel with those of the Nashville Chattanooga and St Louis Railroad. Both sets of tracks over the crossing had been torn up, an action which officials of the latter company protested was unwarranted in view of the fact that their road had not been a party to the accident. Major Crump argued, however, that the city had been within its rights, and that when the Nashville Chattanooga and St Louis company had been given permission to construct a line across the street the right had been reserved to annul the concession at any time.[1242]

Tram line 3 of the Süddeutschen Eisenbahngesellschaft crossed the Oberhausen – Wanne railway line at Schrankenposten 149 between the two aforementioned German towns. As in at least some locations in the United States the regulations of the tramway

[1239] *Stevens Point Daily Journal*, Stevens Point, Wisconsin, United States, 24th November 1913
[1240] *The Commercial Tribune*, Cincinnati, Ohio, United States, 5th January 1914
[1241] *The Indianapolis Star*, Indianapolis, Indiana, United States, 6th January 1914
[1242] *The Galveston Daily News*, Galveston, Texas, United States, 6th January 1914

company laid down that the tramway employees should stop and check that the crossing was safe to negotiate before crossing the line. As the crossing was in any case provided with barriers Motorman Niemeyer did not bother to make these checks.

The morning of January 16th 1914 was foggy. A workman named Kraus arrived at the crossing to repair the lamp. He joined Crossing Keeper Klein in his room and the two of them must have been making a fair amount of noise, as the crossing keeper heard neither the warning signal that a train was approaching, nor the bell signal (Vorwecksignal) given from Poste 147 to Poste 148 and Poste 149 when the train passed the first crossing. He should also have realised from the working timetable that the train was due, as the express was only a minute late, and have lowered the barriers. Because of the fog visibility was down to 300 metres. The barriers remained up as the express approached the crossing, the tram driver did not bother to check that the line was clear, and a collision ensued, killing four people on the tram, one of them Motorman Niemeyer himself, and injuring a number of others.[1243]

On Monday 26th January 1914 passenger train P 221 collided with a waggon-load of bricks on the level crossing at Schrankenposten 23, Germany, located on the Mülheim (Rhein) – Düsseldorf main line. The casualties in the accident were limited to one railwayman, who was injured by flying bricks.

The line was a very busy one, with over 250 trains daily passing over it.

At first the accident appeared to have been simply the fault of the Relief Crossing Keeper, Nikolas Merten, and he was arrested and brought to trial. In court he asserted that he had not been given sufficient training for his job. He had been doing the job for 32 days but had received no formal training or instruction and had not been examined on his duties. There also an error in the working timetable, which showed train 221 as coming from Kuppersteg instead of going to Kuppersteg. The regular crossing keeper had got used to the error and said nothing about it to Merten. For the inexperienced relief keeper it was a trap, and it did not occur to him that a train with an odd number must be going *to* Kuppersteg. The Bahnmeister (supervisor) who had failed to detect the error in the working timetable was disciplined, while Merten was acquitted in court, but no longer employed as a crossing keeper.[1244]

Four people riding in an automobile were killed outright and the other two in the vehicle were fatally injured in a level crossing collision at Austin, Indiana, United States, four miles north of Scottsburg at 3 p.m. on Thursday 10th September 1914. The Belch family had left Dayton, Kentucky, several days earlier to visit relatives in Austin. Mr Ellsworth Belch was driving the car and managed somehow to stall its engine just as it got onto the crossing. At that moment southbound Pennsylvania Railroad passenger train No. 36, with Conductor Wilbur Robertson and Engineer Thomas Wren. Seeing the car Engineer Wren made desperate efforts to stop the train but the car was demolished. Ellsworth Belch and his wife, their daughter Amelia, aged 20, and Leona Trulock, aged 12, daughter of Marshall Trulock, were killed outright. Mrs Mary Belch, one of the

---

[1243] *Eisenbahn-Unfälle in Europa* by Erich Preuß pg. 38 (in German)
[1244] Ibid, pgs. 34-5

Austin relatives who was Ellsworth Belch's aunt, and Beulah Robins, a 14-year-old Austin orphan, were placed on the train and hurried to Seymour but died before it reached there.[1245] [1246]

Eleven people reportedly lost their lives and more than 15 were injured on the evening of Thursday 17th September 1914 in a train/streetcar collision a quarter-mile from Binghampton, Tennessee, United States, just outside of Memphis and not very far from the scene of the Lamar Avenue Crossing accident the previous January, described above. An eyewitness to the accident stated that two different freight trains had been switching at the crossing, one no doubt a Union Railroad of Memphis train and the other belonging to the Illinois Central Railroad, which had trackage rights over the Nashville Chattanooga and St Louis. The Raleigh Springs street car from Memphis, consisting of a motor car towing a trailer, waited for the Union Railroad train to clear the crossing and its conductor then wrongly assumed that the way was clear, signalled his motorman to proceed and the street car was driven onto the line in front of the Illinois Central train. The motor car cleared the crossing, but the trailer was caught by the train, thrown a hundred yards and went down an embankment. According to the conductor there had been 75 passengers in the trailer at the beginning of the journey, 40 of whom had just alighted at Binghampton. The conductor and a few of the passengers who saw what was about to happen were able to jump off the trailer in time.

Within an hour nine bodies had been retrieved from the wreckage. The victims were entirely locals. All the ambulances in the city were called to the scene and a fire engine was also stated to have been busily employed clearing the wreckage of 20 or so smashed and overturned freight cars, though one would have thought that the railroad would have employed their own wrecking crane for that job.[1247] [1248]

The reaction of Mayor Crump, if he was still in office, to another street car crossing accident does not seem to have been recorded.

On Tuesday 6th October 1914 a car containing a party of prominent middle-aged men was returning to Murcia, Spain, from a bullfight which they had attended at Alicante. As the touring car in which they were riding came to a level crossing near Almoradí its driver, Don Ramón Servet, was clearly going too fast and or was not driving with due care and attention, for he smashed through the chain erected across the road to stop traffic. All might still have ended without tragedy had he been able to break through the far chain as well but instead the car came to a stand on the crossing and was immediately run into by a mail train, which reportedly carried it almost two kilometres down the line.

Other motorists arriving at the crossing found the bodies of Don Servet and his four companions. All of them were dead except Señor Pascual Sandoval, who was immediately taken to Almoradí. It was found that he was not as badly hurt as had been feared and that he could probably be removed to Murcia the next day.

1245 *The Bedford Daily Democrat*, Bedford, Indiana, United States, 11th September 1914
1246 *The Plain Dealer*, North Vernon, Indiana, United States, 17th September 1914
1247 *The Evening Post*, Marysville, Ohio, United States, 18th September 1914
1248 *Lebanon Daily Herald*, Lebanon, Indiana, United States, 18th September 1914

Don Servet, 55, was the owner of licensed garages in Almoradí and Murcia and the father-in-law of Señor Sandoval, the latter being the only child if the Marquises of Peñacerrada. The other men in the car had been Don Diego Fontes, 45, an aristocrat and businessman who owned a hotel and two theatres: Señor Laso, a provincial timber inspector, and Don Eustasio Viviente, 53, a highly regarded doctor and a member of the same fraternal organisation as Don Servet. All of those killed had children.[1249] [1250]

A misunderstanding of a hand signal by a student motorman led to an accident resulting in the death of 15 people and injury to about 40 at 5.29 p.m. on the evening of Wednesday 14th April 1915. The location of the tragedy was the West Jefferson Avenue Crossing in Detroit, Michigan, United States. A westbound street car of the Detroit United Railways, crowded with workpeople on their way home, was struck by a backing Detroit Toledo and Ironton Railroad freight train of 25 cars. The trainee motorman, John Westover, aged 34, and his instructor, Motorman Richard Vallade, were arrested a few hours later on charges of manslaughter.

As the car had reached the crossing it had stopped, and the conductor had alighted and run ahead to check that the way was clear. Seeing the train close at hand he had reportedly signalled to Motorman Westover to wait. Instead, it was stated, the latter, thinking that he was being called forward, had applied power and the heavy car had come rapidly down a slope and onto the crossing. The string of freight cars, pushed by a switch engine, struck the middle of the car and pushed it 100 feet down the track. Some of the dead and injured passengers fell out of the car as it was being pushed along, others were crushed in a splintered mass of steel and wood.

It took several hours to remove the bodies of all the dead from the wreckage. They were laid out in a row along the street pending the arrival of the coroner and later removed to undertakers.

Ambulances and private cars were used to take the injured to hospital. The wrecked street car struck a grocery store and the building of Delray station but no one in either place was hurt.[1251]

Most of the victims were immigrant women who had been working at the Scottes Dillon tobacco factory. A local newspaper reporter described some of the heartrending scenes at the morgue. 'Mrs Sophia Jotout, 34 years old, of No. 83 Delray ave. is one of those who went to work, Wednesday morning, in the tobacco factory. She did not come home last night.

'Her husband, Dan, came to the county morgue, before daylight, and was led down the aisle of the death room. Sheet After sheet was pulled back. Time after time, the husband winced, only to turn his head away and say, "No." Finally he turned about and rushed from the room hysterical. "She no there." He cried.

'Jotout was sent home and told to return later. Clerk De Laurier is certain this man's wife is one of the unidentified victims, as other women in the car, who worked with her, are positive that they saw her in the mass of humanity that was caught in the wreck.'

[1249] *ABC*, Madrid, Spain, 8th October 1914

[1250] *La Vanguardia*, Barcelona, Spain, 9th October 1914

[1251] *Defiance Crescent-News*, Defiance, Ohio, United States, 15th April 1915

Coroner Burgess stated, "Someone ought to go to jail for this. That crossing is one of the safest in the city,. There is a clear view of the track, coming or going. Negligence alone can be the cause of the accident, and I propose to do all I can to fix the blame." [1252]

Motorman Westover gave evidence at the beginning of the inquest on Gail George, one of the victims of the disaster. His testimony was confused and contradictory on some points, such as whether he had applied power to bring the car onto the crossing or whether his instructor had. He testified that he had never before seen a flagman at the crossing and was not depending on a signal from one. He said that he had received a book of instructions and had glanced through it but not read it through and he did not know if it contained anything regarding conduct at level crossings. Continuing, he asserted that he had had his arm on the controller, holding the car in position, when Vallade had reached over his arm for the reverse lever and in doing so had caused his arm to operate the controller, moving the car forward.[1253]

On the second day of the inquest Motorman Westover was subjected to a fierce cross-examination by Prosecuting Attorney Jasnowski. He again denied having read his rule book or being aware of anything in it regarding conduct at level crossings. The control box of the wrecked car was also brought into court and he demonstrated with it how he said the accident had happened.

Richard Vallade followed Westover into the witness box and was questioned regarding a previous minor crossing collision in which he had been involved, which had not been his fault. He then testified that although he had a rule book he had never been examined on it. The inquest was again adjourned.[1254]

The next day the coroner's jury boarded a Detroit United street car together with Westover and Vallande, who assumed the positions they had occupied on the night of the accident. Vallande, who the newspaper reported described as 'a careful witness', demonstrated that it would have been impossible for him to have reached over Westover's arm and pull the controller, as the latter had said. Although he might have touched the trainee's arm, he said, he did not know how the power had been turned on. He admitted that he had allowed Westover to violate the company's rule by stopping within 50 feet of the crossing. According to him the car had "drifted" towards the crossing and he said he believed Westover had applied power. He said that previously he had told Westover of the 50 feet rule, but the latter had replied that other motormen stopped closer. Having received a signal from the conductor, Vallande testified, Westover had moved forward, stopped three feet from the crossing to let a woman passenger off, then he (Vallande) had seen the train coming and had told Westover, whom he supposed had also seen the train, not to start. Instead Westover had released the air and the car had drifted towards the crossing, when he had shouted, "Stop the car! Reverse her!" He had reached for the reverse and in the mix-up the power was thrown on and they were struck. He might have touched the student's arm.[1255]

At the resumed inquest Motorman Vallande blamed the lack of a red flag from the flagman at the crossing for the collision, stated that Conductor George Asam had

---

[1252] *Detroit Times*, Detroit, Michigan, United States, 15th April 1915
[1253] Ibid, 20th April 1915
[1254] Ibid, 21st April 1915
[1255] Ibid, 22nd April 1915

signalled the car to stop and admitted that under the company rules the responsibility for safety rested equally on the conductor and the motorman.

Conductor Asam testified that the flagman had come up to him as after he (Asam) had crossed the track. The flagman only had a green flag, which was under his arm. The car had stopped, let off several passengers, moved forward and stopped close to the crossing to let off an elderly woman. When the car had restarted a second time with the train approaching the conductor said that he had run towards the car, throwing up his hands and hollering. He did not see Westover restart the car.[1256]

The statements by Westover, Vallanda and Asam that the street car had made a second stop close to the crossing were disputed the following day by Attorney Jasnowski, who pointed out that this second stop had not been mentioned in their original statements. Conductor Asam's statement that the flagman had only been carrying a green flag was also disputed.

Towerman Fred Sutton, the flagman, then gave evidence. He testified that he had derailers, two for the railroad line and one for the street car, but they were out of order. He had had green and red flags. As the freight train had approached he had given its engineer a "highball" to proceed with his green flag. The street car had come to a stop about 10 feet from the crossing and the conductor had alighted, followed by an elderly woman. When Conductor Asam had come over the crossing he had indicated that a train was approaching and he was pretty certain that Asam had said, "I see it." The car had restarted, stopping over the third rail, and he (Sutton) was positive that he had heard simultaneous shouts of 'Go ahead,' and 'Back up.' Then the train had struck the car.[1257]

State Railroad Commissioner Charles Cunningham stated on Saturday morning, 24th April, that he had been assured by the general manager of the railroad company that gates would be put up at the crossing within 10 days and full interlocking installed within 30 days. At the inquest that day Towerman Sutton continued his evidence, stating that cars usually stopped within five or six feet of the crossing. He did not think either Westover or Vallande saw the red flag he was waving. He saw the conductor raising his arm to stop the car when it restarted, but Vallande was still looking at the woman who had alighted. After the accident Vallande had blamed the conductor. Sutton said he had also tried to stop the car with his red flag. Sutton's testimony was said to have agreed almost word for word with his original statement.

Engineer Clarence Augustus testified that he had been pushing a train consisting of two cars of pig iron and 23 box cars of soda ash. He received his signals from the switchman on a box car third from the engine, who got them from the conductor who got them from another switchman at the front of the train. His train had been moving at six miles an hour. He saw the near switchman giving the emergency "wash-out" signal by vigorously flapping his arms, had immediately applied the brakes, the engine wheels had skidded, and he had put it into reverse. It took about four car lengths – 160 feet – to stop. Air (brakes) was never used during switching because the air would have to be "bled" every time a car was switched.[1258]

The inquest then had to be delayed a couple of days because of the death of a relative

---

[1256] Ibid, 23rd April 1915
[1257] Ibid, 24th April 1915
[1258] Ibid, 26th April 1915

of a juror. When it again resumed Conductor William Taylor and Brakeman Bert Johnson of the freight train testified that the accident would not have happened had the derailers not been out of order as they would have stopped their train. They had not been working for two years. Taylor and Johnson corroborated Sutton's statement that Westover and Vallande had been watching the woman who had alighted and not looking towards their train.[1259]

The following day State Railroad Commissioner Cunningham stated at the inquest that a hearing of the commission had discovered on 6th January 1914 that the derail had been out of order for a year and the Detroit Toledo and Ironton was ordered to put it in order by 1st May and immediately provide a flagman. In the summer of 1914 he had made two inspections of the crossing and had spoken to President James Kurn of the railroad company about the unprotected state of 10 crossings. Nothing more had been done. He believed that the railroad company had not improved the crossings due to a lack of money. Automatic bells at the crossing were out of the question because they would ring almost continuously, and people would ignore them.[1260]

The penultimate day of the inquest was occupied largely with arguments between the railroad and street car companies as to who was responsible for the lack of a derail at the crossing.

Wilfred McCann, assistant general manager of the Detroit United Railway, testified as to various notifications made to the railroad company that the derail was out of order. Leo Butzel, attorney for the Detroit Toledo and Ironton, countered that at one time the railroad company had been in the hands of the receiver, and that afterwards there had been a dispute between it and the street car company over the question of costs. [1261]

At the conclusion of the inquest the coroner's jury found Motorman Vallande, Conductor Asam and Towerman Sutton all guilty of criminal negligence. No mention was made of Motorman Westover, presumably because he was still under training. The jury also severely censured the state railroad commission for not enforcing the law regarding the putting in order of the derailer. Coroner Burgess expressed the opinion that it was a just and fair verdict.[1262]

On the evening of Wednesday 16th June 1915 the nine-year-old son of master roofer Kahn was leading an ox-drawn waggon over the level crossing at Kell station, Germany. His father had been called up for war service, his mother was walking 70 metres behind the waggon. The crossing was protected by barriers, manned by Under Assistant Gielen, who was both signalman and supervisor.

Goods train 7954 had been reported as passing Reinsfeld, 7.2 kilometres distant, at 20:09 and its booked maximum speed was 30 kilometres an hour. Gielen calculated that it should arrive at Kell at 20:29. He stood at the crossing controls, ready to lower the barriers as soon as the waggon had cleared the line. Suddenly, at 20:19, the train arrived and struck the waggon. Fortunately the boy saw it in time to jump clear, but the waggon

[1259] Ibid, 28th April 1915
[1260] Ibid, 29th April 1915
[1261] Ibid, 30th April 1915
[1262] Ibid, 1st May 1915

was destroyed.

On the day of the accident the goods train had consisted of just an engine and one luggage van. Furthermore its driver had grossly exceeded the authorised speed, travelling at 60 kilometres an hour instead of 30, and thus covering the distance in half the time he should have done. When he was 110 metres from the crossing he saw the raised crossing barrier and the waggon but could not pull up in time. He was taken to court at Trier, charged no doubt with the equivalent of dangerous driving, and sentenced to a month's imprisonment and costs.[1263]

In the United States if the 4th July falls on a Sunday then the Independence Day celebrations are held on Monday 5th July instead. Such was the case in 1915. The heavy traffic in connection with the celebrations meant that the *Queen of the Valley* express of the Philadelphia and Reading Railroad left Reading, Pennsylvania, 15 minutes late on the evening of that day and it made a stop at Hershey, Pennsylvania, to take up passengers who had attended the celebrations there and were returning to Pittsburgh. Meanwhile George Cassel (another newspaper gives the surname as Castle) and his family had been attending the celebrations in nearby Hummelstown and as the express left Hershey were returning home in two mule-drawn carriages. Their homeward trip had barely begun when they reached Railroad Street Crossing in the middle of Hummelstown. The mules pulling the first team had just reached the westbound railroad track when the headlights of the express, which was travelling at a reported 40 miles an hour, came into sight about 100 yards away. The driver attempted to hurry the mules across, but the animals would not respond quickly enough, and the train struck the carriage, smashing it to pieces. The passengers in the carriage were hurled in all directions and the mules carried on the front of the engine until the train stopped, when they fell off, dead.

Engineer Swavely of the express knew that his engine had struck something, but not what. It took half a mile for the express to stop and it was then backed to the crossing and a search was made, with the assistance of Hummelstown people who had either seen or heard the crash, when the bodies of the victims were found. George Cassel, Mrs Harrison Cassel, his daughter-in-law; Russel Cassel, his grandson; John Cassel and Elizabeth Cassel, probably his children, were all killed outright. Mrs George Cassel, the other occupant of the carriage, was fatally injured and died in Harrisburg hospital the following morning. Harry Cassel, riding in the second carriage, saw his wife and son killed in front of him.[1264]

The crossing at Hummelstown was said to be one of the most dangerous on the line, being situated on a slope and having no warning lights. During the daytime it was guarded by a watchman, but he finished duty at 7 p.m. None of the newspaper reports on the accident mention what time the accident occurred, but the Cassels had not left the celebrations until after a fireworks display so the accident probably happened at 10 or 11 p.m. An immediate investigation into the accident resulted in John Dohoney of the Public Service Commission recommending that the Philadelphia and Reading provide a night watchman as well. The railroad company agreed, and a night watchman was provided from the following evening.

[1263] *Eisenbahn-Unfälle in Europa* by Erich Preuß pgs. 35-6 (in German)
[1264] *Lock Haven Express*, Lock Haven, Pennsylvania, United States, 6th July 1915

The *Harrisburg Patriot* was not impressed that, to use a well-known idiom, the stable door had only been shut after the horse had bolted. It commented that, "There is only one thing better than the decision of the Public Service Commission to have the Reading station a night watchman at the Hummelstown grade crossing where six persons were killed Independence day. That is the decision should have been made long ago. It is difficult to imagine a more heart-breaking tragedy than that, coming as it did on the homestretch of a pleasant observation of a nation's holiday. Somebody erred, of course, and the prospective investigations will be futile unless this point is determined. These enquiries ought to develop why on such an occasion when crowds thronged the streets and highways some protection was not afforded at grade crossings." [1265]

The deafness of the car driver was one of the factors in an accident near Whitehall, New York, United States, on the afternoon of Sunday 5th September 1915. Mr Nelson Norton, aged 67, was driving to his home in Whitehall in a touring car with Mrs Norton, 52; his daughter, May, 20; his son, John; his daughter-in-law, Mrs John Norton, 30 and his granddaughter, 3. The crossing where the accident occurred was unprotected and visibility was impaired by a heavy growth of brush. Mr Norton's car reached the crossing at the same time as a Delaware and Hudson Railroad excursion train which was said to have been travelling at 50 miles an hour. The engineer of the train whistled as he approached the crossing and the whistle was heard by the other occupants of the car but not by Mr Nelson. When his passengers succeeded in attracting Mr Norton's attention to the train he attempted to turn the car into a ditch, but it swerved back onto the road directly into the path of the locomotive and was cut in two. The bodies of the victims and the remains of the automobile were strewn along the track for 600 feet. Five of the occupants of the car were killed outright, John Norton suffered a broken leg and internal injuries, but it was expected that he would recover.[1266]

Sunday 24th October 1915 saw several level crossing accidents in the United States, including one which cost seven and very likely eight lives at Fraser, Michigan. There were two level crossings with the Grand Trunk Western Railroad main line at Fraser, situated 100 yards south of the station and 200 yards north of it, the latter being where the accident happened. A car driven by farmer William Stoldt and containing, as well as him, his wife Rachel; his sister-in-law, Miss Minnie Engel; and his children, Pearl, 17; Esther, 15; Hazel, 12; Mabel, 10 and Martha, 6, was struck by northbound passenger train No. 18. Visibility at the crossing was somewhat restricted by a house and a grove of trees. Witnesses stated that the train had whistled for the crossings. It was theorised that possibly Mr Stoldt had wrongly assumed that the train would stop at the station. He was the only survivor of the collision and suffered multiple fractures. Spoken to after the accident when he briefly regained consciousness he had no recollection of what had happened. He begged to know what had become of his family, but it was thought best not to tell him that they were all dead. His own recovery from his injuries was considered doubtful.[1267]

[1265] *The Hershey Press*, Hershey, Pennsylvania, United States, 8th July 1915
[1266] *The Philadelphia Inquirer*, Philadelphia, Pennsylvania, United States, 6th September 1915
[1267] *The Fort Wayne Sentinel*, Fort Wayne, Indiana, United States, 25th October 1915

Heavy fog was blamed for a level crossing accident near Berthold, North Dakota, United States, on Friday 17th December 1915. A Great Northern Railroad train struck a bus carrying six schoolboys, all less than 16 years old, home from their lessons; presumably, given the date, it was a horse-drawn vehicle. John Ahern, the driver of the bus, brothers Leonard and Earl Miller, brothers Severt and Anton Hagen and Harry Arnold were all killed. Harry's brother Leonard was probably fatally hurt.[1268]

[1268] *Salt Lake Tribune*, Salt Lake City, Utah, United States, 18th December 1915.

CHAPTER 13

RAIL/RAIL CROSSING COLLISIONS

At locations where railways and roads intersected the idea of having them cross on the level had obvious advantages in terms of economics over the cost of building an over or underbridge, especially if the level crossing was unmanned and the construction of a bridge would require the building of substantial approach embankments or possibly approach cuttings leading to an underbridge. With road users being limited in the early days to pedestrians, cyclists and horse-drawn vehicles the dangers to a train from a collision on such a level crossing were also likely to be minimal. For railways crossing other railways on the level, though, the situation was very different. A collision with another train on a crossing was likely to be far more serious than a collision with, say, a horse and cart, and furthermore a train driver approaching a level crossing and seeing another train nearing or on it could not pull up in anything like the distance that a road user could. It is no surprise therefore that in most countries, where the railway system was either built directly by the state or under state supervision, that rail/rail crossings were not provided. It was only in countries where the railways were built by competing private companies without state interference that such crossings existed, most especially in the United States, where lines were often built as cheaply as possible and the economy of not needing to construct a bridge was doubtless an important consideration. Thus, the United States had many rail/rail level crossings, Great Britain at one time had about 20 or so and its colonies, such as Ireland and Canada, had a handful. One or two did also exist elsewhere, such as at Istravoz (later Hilal), Turkey, where the Oriental Railway crossed the Smyrna – Cassaba Railway.[1269] As has been pointed out the American companies also seemed to have a perverse proclivity for building rail/rail crossings where they could easily have been avoided, perhaps because in the early days train speeds were low, trains infrequent and night operation unusual so that the dangers were not fully appreciated.[1270] By 1901 every one of the 15 or so remaining rail/rail crossings in Great Britain was fully signalled and there were hardly any further collisions on them. In the United States, on the other hand, where many crossings remained with little or no signal protection,

---

[1269] Wikipedia – Hilal railway

[1270] *A History of Railroad Accidents, Safety Precautions and Operating Practices* by Robert B. Shaw pg. 99

collisions on them continued to occur regularly.

The first rail/rail collision of the 20th century occurred on Monday 4th February 1901 at South Adrian, Michigan, United States. The contestants in the collision were Wabash Railroad passenger train No. 1, the *Continental Limited*, which was double-headed, and a Detroit and Lima Northern Railroad train. The Wabash train had the right of way; the brakes of the Lima Northern train allegedly failed. No one was much hurt, the engine crews having jumped, but there was considerable material damage. The Lima Northern engine was much heavier than the two Wabash engines. The Wabash pilot engine, No. 329, was so badly damaged in the collision that it had to be loaded onto a freight car to be taken to Fort Wayne, Indiana, for repairs. The train engine, No. 656, had its right side damaged, but was able to make its own way to Fort Wayne to be patched up. In doing the damage to the two Wabash engines the Lima Northern locomotive was itself 'utterly demolished'. Both lines were blocked, and the Wabash trains had to make a 50-mile detour.[1271] The *Continental Limited* finally arrived in Decatur, Illinois, eight hours late.

None of the passengers on train No. 1 was injured but 'Monkey Burrows, the express messenger on the Continental, stood paralysed watching the two trains come together, and when the collision came he was slammed to the floor much as he is accustomed to unload trunks. Beyond a bad shaking up he was uninjured.' [1272]

There were two more rail/rail collisions before the end of February. On 20th of the month there was a minor accident at Lima, Ohio, involving Erie and Lake Erie and Western trains. Then on 22nd a slightly more serious one at Hanley Falls, Minnesota. This time Great Northern and Minneapolis and St Louis Railroad passenger trains, both westbound on their respective routes, were involved. The St Louis train was timetabled to pass the crossing first but was five minutes late. It was going over the crossing when the Great Northern train, apparently under full steam, struck it between the fourth and last carriages. The coaches were derailed, the track was ripped up and the colliding engine was wrecked. Standing on the platforms of the two coaches were a brakeman and a newsboy, who both jumped, the brakeman straining his shoulder in doing so and the newsboy breaking his ankle. Among the passengers W. J. Wolford of Dawson suffered a fractured skull and it was feared he might die; Mrs Nash suffered slight scalp injuries and C. H. Leaman of Minneapolis lost three teeth and had cuts to his face. A wrecking train was sent to scene, carrying Dr Penhall to care for the injured. The accident happened at 2.50 p.m.[1273] [1274]

On the afternoon of Saturday 13th July 1901 the Kansas City and Northern Connecting Railroad passenger train which was due to leave Pattonsburg, Missouri, for Kansas City at 2.20 p.m. started a little late. The train reached Weatherby, Missouri, where there was a crossing with the St Joseph branch of the Chicago Rock Island and Pacific Railroad, at

[1271] *The Fort Wayne Journal-Gazette*, Fort Wayne, Indiana, United States, 6th February 1901
[1272] *The Daily Review*, Decatur, Illinois, United States, 5th February 1901
[1273] *The Redwood Gazette*, Redwood Falls, Minnesota, United States, 27th February 1901
[1274] *The Winona Republican and Herald*, Winona, Minnesota, United States, 22nd February 1901

3.15 p.m. At that moment a Rock Island extra wheat train was approaching the crossing. According to different reports all trains were required by rule to whistle and/or stop short of the crossing, but the Rock Island train did neither and ran through the crossing at 25 miles an hour as the Quincy train was on it. The engine of the freight train cut through the passenger train at the junction of the baggage car and the day coach, throwing both into a ditch and being itself overturned. About 25 passengers, Superintendent Willis Stoneburner, Conductor Harvey, Brakeman Dallou and Mail Clerk Skilling were injured. At first it was thought that no one had been killed but half an hour after the accident farmer Elisha Price, 80 years old, was found buried in the wreckage. He was still alive when found but died shortly afterwards. He lived within sight of the accident which claimed his life. The engine crew of the Rock Island train jumped and escaped. It was supposed that the Rock Island engineer had assumed that the late-running Connecting Railroad train had already gone. Depending on where they resided the injured were either taken back to Pattonsburg or on to Kansas City.[1275 1276 1277]

In typical American fashion in the distance of about 100 miles the Kansas City and Northern Connecting Railroad crossed in succession on the level the Wabash (at Kirksville), the Iowa and St Louis (at Novinger), the Chicago Burlington and Quincy (at Milan), the Chicago Milwaukee and St Paul (at Galt), the Chicago Rock Island and Pacific (at Trenton), the Wabash again (at Pattonsburg), the Chicago Rock Island and Pacific again (at Weatherby) and finally the Chicago Burlington and Quincy again (at Osborn).[1278]

At Bluffton, Ohio, United States, at about 10 a.m. on Monday 18th August 1902, it was a Lake Erie and Western Railroad passenger train which was running late. Its engineer reduced speed for the crossing with the Northern Ohio Railroad but even so allegedly tried to get his train over the crossing in front of one of that company's freight trains. It was a serious error of judgement for the engine of the passenger train had barely cleared the intersection when the Northern Ohio train, reportedly travelling at speed as its engineer tried to beat the passenger train, crashed into it. Both engines, the baggage car of the passenger train and a number of box cars of the freight train were derailed and the Erie engine overturned. The passenger cars remained on the track. Fireman Charles Peltner of the Erie train was badly hurt and Brakeman A. Lawrence, Bridgeman W. H. Kepner and passenger B. J. Considine, of Portland, Indiana, were also hurt. It was considered miraculous that no one had lost their life. A crowd of several thousand soon collected to view the wreck. Workmen cleared the lines within a short time.[1279]

There was a lucky escape in a collision at Butler, Pennsylvania, United States, on Saturday 4th October 1902. Freight trains belonging to the Buffalo Rochester and Pittsburgh and Western Pennsylvania Railroads were involved. The Western Pennsylvania train, in charge of Conductor Frank Dixon, was a heavy one and the rails

---

[1275] *The Quincy Evening Herald*, Quincy, Illinois, United States, 15th July 1901
[1276] *Plattsburg Leader*, Plattsburg, Missouri, United States, 19th July 1901
[1277] *The Quincy Weekly Whig*, Quincy, Illinois, United States, 18th July 1901
[1278] Railroad Atlas of North America: Prairies East and Ozarks
[1279] *Defiance Crescent-News*, Defiance, Ohio, United States, 19th August 1902

were wet and slippery. Engineer Gilbert Kennedy was unable to hold the train as it approached the crossing with the Buffalo Rochester and Pittsburgh crossing, and it cut clean through the middle of a train of that company which was on the crossing. Following the collision Engine No. 1396 went down a 10-foot embankment and overturned but Engineer Kennedy and Fireman Frank Gerrard, who had decided not to jump, escaped unhurt, as did the other members of both trains. Flagman Robert Thompson at the crossing was slightly cut about the face by flying wreckage and his signal house (probably just a hut) was carried down the embankment with the engine. What was especially fortunate was that eight freight cars were wrecked and one of those worst damaged was loaded with dynamite, which did not, however, explode. Engineer Kennedy had been back at work less than two months after having been injured in a wreck back in June at Sharpsburg.[1280]

Derails (trap points) where installed could prevent a rail/rail crossing collision. Thus on Tuesday 17th February 1903 at Piqua, Ohio, United States, a Pittsburgh Cincinnati Chicago and St Louis Railroad (Panhandle) train and a southbound Cincinnati Hamilton and Dayton Railroad (C.H. & D.) freight train were both approaching the crossing of the two roads shortly after 10 p.m. Priority was given to the Panhandle train and the derail was open against the C.H. & D. one. As at Butler a few months earlier Engineer Smith of the latter train was unable to control it 'on account of the rails being slippery' and the engine, tender and one car went into the ditch and the locomotive overturned, pinioning the engineer beneath it. He had jumped but slipped as he hit the ground and was caught by the falling engine. A portion of the cab of the locomotive had to be cut away to release him and he suffered a broken leg, but after treatment at the Piqua Hotel he could be taken home. Removal of the tender and car allowed the track to be cleared; the engine was left in place pending the arrival of the wreck train from Lima.[1281]

On Monday 21st September 1903 Atlantic Coast Line Railroad and Atlantic and Danville Railroad freight trains collided on the crossing of the two lines at Boone, Virginia, United States. Both trains and one of the engines were wrecked and the damaged engine set fire to the wreckage. Brakeman Isaac West was burnt to death and Engineer Richard Carrington was seriously injured. 'The accident probably was caused by a misunderstanding of signals.' [1282]

Brighton Park Crossing, Illinois, United States, on the southern outskirts of Chicago, was the scene of a bad collision on Tuesday 9th August 1904. Baltimore and Ohio Railroad local passenger train No. 17 from Garrett, Indiana, to Chicago, had stopped at the crossing as per the rules, and the semaphore signal having been cleared for it was then proceeding over the crossing on the Baltimore and Ohio Chicago Terminal track. Meanwhile a Chicago and Erie Railroad freight train of 27 cars was being backed westwards from the stockyards, its path taking it over the crossing. Brakeman Charles Arnold, riding a car at what had become the front of the reversing train, maintained that

[1280] *The Butler Citizen*, Butler, Pennsylvania, United States, 9th October 1902
[1281] *The Piqua Daily Call*, Piqua, Ohio, United States, 18th February 1903
[1282] *Savannah Morning News*, Savannah, Georgia, 22nd September 1903

he saw the semaphore signal cleared for his train to proceed, and he signalled to his engineer at the other end of the train to do so. The next instant he saw the target go to danger and immediately signalled for his engineer to stop. Approaching the crossing on the Chicago and Erie line there was a curve, with a clump of trees on its inside. Either Engineer George Bodley was unable to see the brakeman waving his arms or he was not looking out properly, for rather than pulling up he opened the throttle and the freight waggons were reportedly moving at 30 miles an hour when they reached the crossing. In the collision the first, empty, freight car was tossed aside and the second smashed into one of the coaches of the passenger train. By chance the worst damage to the passenger train was done at the point where the family of J. W. Schwarz was seated. Mr Schwartz, a Baltimore and Ohio freight conductor from Garrett, was seriously but it was thought not fatally hurt. Mrs Schwartz and their children, Earl, aged nine; and twins Chester and Lester, aged four, were all killed. Eight other passengers and Brakeman C. W. Jones of the passenger train were injured.

Following the accident no less than eight railroad employees were arrested pending the inquest. They were Flagman Matthew Clarissy, Conductor Charles Allen, Engineer J. W. McLaughlin and Fireman H. McPheeter, of the passenger train; and Conductor P. F. Waller, Engineer Bodley, Fireman John Landgraf and Brakeman Harry Moore of the freight train, but apparently not Brakeman Arnold.[1283]

No report as to the findings of the inquest has been traced.

Beaumont, Texas, United States, was the scene of a fairly minor collision on the evening of Thursday 17th August 1905. A Gulf and Interstate Railroad train was pulling into the station at about 9.30 p.m. across the tracks of the Southern Pacific Railroad when it was struck by a train of flat cars belonging to the latter company. A flat car struck the Gulf and Interstate train between its engine and tender, knocking them into a ditch, and they were badly wrecked. Engineer Dan Collins had two bones in his right leg broken and Fireman H. Stephenson was badly bruised. The Gulf and Interstate claimed that Engineer Collins had had the right of way.[1284]

What was described as 'a peculiar accident' happened at Jerseyville, Illinois, United States, on the morning of Sunday 3rd September 1905. Northbound Chicago and Alton Railroad passenger train No. 204 and southbound Chicago Peoria and St Louis Railroad (the Bluff Line) passenger train No. 3, the Springfield and St Louis accommodation, were involved. In charge of the Alton train was Conductor Shumin with Engineer Strunk, while the Bluff Line train was in charge of Conductor Jones with Engineer Eberle. The two engines met exactly on the crossing and both were thrown into the ditch. The Bluff Line engine ended up alongside the Alton track, the Alton train went down into the street and almost into the residence of Fred Byers. Engineer Eberle and Fireman Parker both jumped, the fireman collided with a telegraph pole. The crew of the Alton engine did not jump; both were badly bruised. Both engines were 'dismantled' and the front end of the baggage car was derailed and damaged. None of the passengers was injured.

[1283] *The Fort Wayne Weekly Sentinel*, Fort Wayne, Indiana, United States, 10th August 1904

[1284] *The Galveston Daily News*, Galveston, Texas, United States, 18th August 1905

The crew of each train blamed those of the other for the collision. Those of the accommodation claimed that they had had the right of way, having brought their train to a stop and whistled. Engineer Strunk of the Alton train claimed that had given the necessary signals and had not heard the whistle of the other train or seen it approach.[1285]

At 10.30 p.m. on Thursday 19th April 1906 southbound Pittsburgh Cincinnati Chicago and St Louis Railroad (Panhandle) train No. 80 collided with a Chicago Indianapolis and Louisville Railroad (Monon) freight train on the crossing at Lacrosse (now spelt La Crosse), Indiana, United States. Just before the impact Fireman J. W. Wilson and Brakeman C. E. Snyder jumped, the brakeman suffering two broken legs and the fireman a slightly sprained ankle. Engineer J. M. Carney remained on the engine and was unhurt. Four car lengths back from the car of the Monon train which was hit was a car loaded with dynamite, so it was another narrow escape, especially as the cars which were struck also caught fire after the collision. The 12.40 a.m. southbound passenger train from Chicago brought the injured men to Logansport, where they arrived at 1.30 a.m.[1286]

Who was to blame for the collision was not clear, though it was speculated that the fact that the Panhandle train was late and probably trying to make up time might have been significant.[1287]

At 10:45 a.m. on the morning of Sunday 8th July 1906 the engine of a Chicago and Erie Railroad freight train was run into by a passenger train on the Delphos and Ironton branch of the Cincinnati Hamilton and Dayton Railroad at the crossing of the two roads at Spencerville, Ohio, United States. The Dayton engine struck the rank of the Erie engine and then overturned into a ditch, burying Engineer Harry Clayton beneath it. There hot water and steam poured upon him and he was scalded to death in a short time. According to one report an unknown woman passenger, the fireman and the brakeman were hurt.[1288]

At the time of the newspaper reports blame for the accident had not been fixed but an official for the Dayton company apparently told a reporter that 'the target tender had given signals to the engineers of both trains to come ahead.' [1289] It seems likely that the reporter had got his facts wrong; quite why would the target tender seek to deliberately engineer a collision?

At about midnight on Monday 13th/Tuesday 14th August 1906, Louisville and Nashville and Southern Railway freight trains contested the right of way at the Georgia Carolina and Northern crossing at the extreme western edge of Atlanta, Georgia, United States. 'Both trains were under good headway when the impact came.' The Louisville and Nashville train got to the crossing first and was halfway across when it was run into by the Southern train, the car at the point of impact being cut in two and smashed to

1285 *Decatur Herald*, Decatur, Decatur, United States, 5th September 1905
1286 *The Logansport Pharos*, Logansport, Indiana, United States, 20th April 1906
1287 *The Logansport Journal*, Logansport, Indiana, United States, 20th April 1906
1288 *Hamilton Democrat*, Hamilton, Ohio, United States, 9th July 1906
1289 *Lima Times-Democrat*, Lima, Ohio, United States, 9th July 1906

pieces. The engineer and fireman of the Southern train jumped, escaping with cuts and bruises, while a brakeman named Scrutchin riding on the roof of the car immediately in front of that which was struck was thrown to the ground and knocked unconscious. He was taken home and given medical treatment; the extent of his injuries was not immediately known. A Seaboard Air Line Railroad train which had set off from Atlanta at midnight had to return whence it had come and take a different route because the line was blocked by the accident. There was a flagman stationed at the crossing and it was not at once apparent how the trains had collided.[1290]

Only a few hours after the Atlanta collision on 14th August 1906 northbound Missouri-Kansas Texas-Railroad train No. 6 from Texas to St Louis was struck on the crossing with the St Louis-San Francisco Railroad at Fort Scott, Kansas, United States, by a switch engine of the latter company. The passenger train was six hours late. A chair car and a Pullman sleeper car were overturned. The chair car went over completely but a telegraph pole stopped the fall of the sleeper car. Twenty-five passengers were injured, mostly only slightly, and all were able to continue their journeys, now even more delayed, after treatment by local physicians.[1291] [1292]

A St Louis-San Francisco Railroad passenger train due in Birmingham, Alabama at 5.45 a.m. on the morning of Saturday 12th October 1907 was run into by a Louisville and Nashville Railroad freight train on the crossing at Pratt City, Alabama, United States, six miles west of Birmingham, at 6 a.m. The freight train struck squarely the Pullman car *Leeds*, third from the end of the train. Every passenger in the car was injured and Pullman Conductor J. C. Smith was killed instantly. It was believed that all of the eight passengers who were hurt would recover.

Both trains had approached the crossing at low speed. Engineer Donohue of the freight train stated that he had seen the passenger train in time but had been unable to pull up because all of the cars of his train had not been equipped with air brakes.[1293]

The apparent misunderstanding by the motorman of an interurban passenger car led to a collision at West Toledo, Ohio, United States, on Saturday 15th February 1908. The accident occurred at the crossing of the Toledo and Western Railroad electric interurban and the Michigan Central Railroad, though the other train involved in the accident is referred to in all the reports as belonging to the Cleveland Cincinnati Chicago and St Louis (Big Four) Railroad, which presumably had trackage rights over the Michigan Central.

Eyewitnesses to the accident stated that shortly after 8 p.m. the interurban car had arrived at the crossing and its conductor had alighted to check that the crossing was clear. Seeing the Big Four passenger train coming the conductor had signalled to his motorman, James Meyers, to wait, but the latter had either disregarded, or more likely

---

[1290] *The Atlanta Constitution*, Atlanta, Georgia, United States, 14th August 1906
[1291] *San Antonio Daily Light,* San Antonio, Texas, United States, 14th August 1906
[1292] *The Lincoln Daily Star*, Lincoln, Nebraska, United States, 14th August 1906
[1293] *The Commercial Tribune*, Cincinnati, Ohio, United States, 13th October 1907

misunderstood, his signal, and applied power. The passenger train was late and was heading for Detroit at almost a mile a minute. The interurban car was caught in the middle, lifted high in the air and thrown 150 feet. Motorman Meyers and seven passengers were killed, and a dozen injured, some possibly fatally. The passengers who lost their lives were Mrs William Fisher and Ray Fisher (her brother?) of West Toledo; Clarence Beech of Sylvania, Mrs John Neipp and her children Lincum, aged eight and Theodore, aged six, of Toledo; and Hiram Parker, a farmer of Sylvania. All the Toledo ambulances were rushed to the scene and the injured were taken to St Vincent's Hospital while the bodies of the dead were carried to the West Toledo station of the Lake Shore and Michigan Southern Railroad nearby.[1294] [1295] [1296]

Less than a week after the West Toledo accident, on Friday 21st February 1908, there was a crossing at the crossing of the Baltimore and Ohio and Lake Shore and Michigan Southern Railroads at Elyria Junction, Ohio, United States. The crossing here was protected by a derail but this failed to prevent the accident. Lake Shore passenger train No. 50 and a Baltimore and Ohio freight train approached the crossing at the same time. The passenger train was given priority and the derail was open against the freight train. The latter must have been travelling at considerable speed for its engine left the rails at the derail but continued forward and struck the passenger train anyway. The colliding engine then overturned, and Baltimore and Ohio engineer James Vigals was crushed beneath it and killed. His fireman, Charles Richard, was badly cut and Engineer John Neihaus and Fireman Gus Herbert of the Lake Shore train, a mail clerk on the Baltimore and Ohio train and two passengers were also injured. The baggage car, mail car and one passenger car of the Lake Shore train were derailed.[1297]

At least seven people died and over 20 were injured in a sidelong collision at Oakland, California, United States, on Saturday 4th July 1908. Unlike many such rail/rail crossings in the United States this one was fully signalled from an adjoining tower (No. 3) and was protected by derails (trap points) but these were not enough to prevent a collision. Both lines over the crossing were owned by the Southern Pacific Railroad.

The trains involved were one from Santa Cruz in charge of Conductor Fred Corey, with Engineer Everett Barry and Fireman Herbert Drake, and an Alameda local train in charge of Conductor George Cosgrove with Engineer William Mark and Fireman C. T. Ferris. All the evidence suggested that it was the Santa Cruz train which was at fault. Towerman Carlton McCurdy had decided to give the Alameda train precedence over the crossing and had cleared his signals for it, with the result that the signals were necessarily against the Santa Cruz train and the derail was open. The derail did its job in forcing the train from the rails, but as at Elyria Junction in February it was travelling so fast that it simply ploughed along on the ties over the crossing. Just as it was on the crossing it was struck by the Alameda train, which had been running under clear signals. The smoker car of the Santa Cruz train chanced to be on the crossing at the moment the Alameda train

---

[1294] *Sandusky Daily Register*, Sandusky, Ohio, United States, 16th February 1908
[1295] *Logansport Daily Tribune*, Logansport, Indiana, 16th February 1908
[1296] *The Goshen Democrat*, Goshen, Indiana, United States, 18th February 1908
[1297] *The Van Wert Daily Bulletin*, Van Wert, Ohio, United States, 22nd February 1908

reached it and it was in this car that the dead and injured were found.

A number of passengers told their stories to local newspaper reporters. A. C. Payne, of 1211 Castro Street Oakland, was riding in the car of the Santa Cruz train immediately behind the smoker and was looking out of the window when he was suddenly thrown across the car on top of a woman passenger. He said, "I jumped from the rear door, and as soon as I saw I could not be injured I joined the rescue forces. I pulled an old woman and child out who were more frightened than hurt. The little girl, I think, was cut, and so was the woman. There was a small panic inside the car with people pushing and shoving and pushing to get out of their death prison. This car was not as bad off as the smoker and I told people to take care, that they could not be hurt. The crash was over, I told them."

Morris Walter, of 803, Broadway, Oakland, said, "I was riding in the smoker in the sear next the aisle. Suddenly the conductor threw open the door and cried, 'keep cool, keep your seats, there is an accident ahead.' Then came the crash. My friend, with whom I was sitting, Leslie Albee, was killed instantly. People to the right and to the left of me were torn and mangled. All that saved me was the fact that I was riding next the aisle. All was confusion and screams. It seemed at first as though all on the train were killed." [1298]

At the inquest on the victims of the accident Engineer Barry admitted approaching the crossing at a speed of 40 miles an hour, despite a city ordinance limiting the speed of trains to 22 miles an hour, stating in his defence that there was no board to indicate that he was within the city limits. He was running fast because his train was late, and he was anxious to reach the Mole in time to enable passengers to catch the ferry across the bay to San Francisco. He said that he had seen the signal protecting the crossing and the open derail at the same moment when his train was 150 yards from them, and that he could have stopped but for oil on the track.

The jury returned a verdict that the deaths of the accident victims had been caused "by the engineer of the westbound Southern Pacific Company train No. 57 not being properly piloted and disregarding the danger signals at said point and running across Webster street over the tracks of the northbound Webster street local train and his train being run into by said local train.

"We further believe that said company should establish speed limit signs within the corporate limits of the city of Oakland and that when engineers are transferred from one division to another some method of determining their competency should be adopted before being placed in charge of a locomotive." [1299]

At 7.45 a.m. on the morning of Friday 17th December 1909 a Macon and Birmingham Railroad train was unloading its passengers at Harris, Georgia, United States. The station at Harris adjoined a crossing with the Central of Georgia Railway line and the Macon train was stopped with part of it over the crossing. Before the station work could be completed the train was struck by a northbound Central of Georgia train, which demolished the baggage car of the Macon train and derailed a day coach. Postal Clerk W. H. Newman, who was in the baggage car, was seriously injured and died on the Sunday afternoon, while two men who were stood on the ground near the track, J. L. Williams

[1298] *Oakland Tribune*, Oakland, California, United States, 5th July 1908
[1299] Ibid, 7th July 1908

and W. A. McHargue, were crushed by the overturned car and also killed. One report mentioned a fourth death, that of coloured Macon employee Jack Holt of La Grange. Several railroad employees were injured; Joint Agent at Harris City H. C. Rawle (also reported as killed), Macon Baggagemaster Phillips and Porter Will Griggs and Fireman Jim Coleman of the Central, who sprained his back in jumping from the engine. A passenger, Miss Tidwell of Harris City, was caught under the baggage car and slightly hurt. Engineer Rogers of the Central train stated that he had seen the Macon train in plenty of time to have stopped but claimed that his brakes had failed. His locomotive, Atlanta and West Point Railroad engine No. 16, was sufficiently unscathed by the collision to be able to work the train on the Atlanta. There was slight damage to a portion of the depot and to the track.[1300 1301]

What was described as a spectacular wreck occurred at 6.50 p.m. on Friday 19th January 1912 at Mitchell, Indiana, United States, at the crossing of the Chicago Indianapolis and Louisville (Monon) and Baltimore and Ohio Railroads. Monon freight train No. 75 was pulling slowly over the crossing when it was struck by Baltimore and Ohio freight train No. 88 running reportedly at 40 miles an hour. Engineer C. R. Trueblood of the Baltimore and Ohio train stated that he had had no occasion to use his brakes after passing Loogootee, about 25 miles to the west, until cresting the brow of a hill just short of Mitchell. When he applied them there they had failed to work and as the train had descended the hill its speed had increased. Unable to hold his train Engineer Truebood and his fireman had jumped; the latter was slightly hurt. The Baltimore and Ohio engine cut clean through the Monon train but in doing so was derailed, ran up the platform of the Baltimore and Ohio station and stopped just short of the telegraph office. Three cars behind the engine and four cars of the Monon train were derailed. Two cars of potatoes, one of flour and one of beer were destroyed and a car of machinery was partly destroyed. Had there been anyone on the platform of the Baltimore and Ohio station they would have been lucky to have escaped but fortunately it was deserted at the time as no passenger trains were due.[1302 1303]

A collision at the crossing at Hazlehurst, Georgia, United States, on Tuesday 7th January 1913, cost the life of Engineer George Little of Southern Railway through passenger train No. 5. His train was in collision with a Georgia and Florida Railroad freight train.[1304] A newspaper stated the next day that the two railroad companies declined to give out any information regarding the wreck and it was not yet clear who was responsible for it.[1305]

A severe rainstorm was blamed for the collision between Baltimore and Ohio Railroad passenger train No. 15, the westbound Wheeling and Chicago express, and an interurban

---

[1300] *The Lagrange Reporter*, Lagrange, Georgia, United States, 24th December 1909
[1301] *Americus Times Recorder*, Americus, Georgia, United States, 23rd December 1909
[1302] *Seymour Daily Republican*, Seymour, Indiana, United States, 20th January 1912
[1303] *The Bedford Daily Mail*, Bedford, Indiana, United States, 20th January 1912
[1304] *The Waycross Journal*, Waycross, Georgia, United States, 10th January 1913
[1305] *The Atlanta Constitution*, Atlanta, Georgia, United States, 9th January 1913

car of the New Midland Power and Traction Company west of Cambridge, Ohio, United States, at 7.30 a.m. on the morning of Sunday 13th January 1913. In accordance with the normal procedure the conductor of the interurban had alighted from his car after it had stopped short of the crossing and walked forward to check that all was clear. Looking down the track through the rain the conductor saw the headlight of the approaching express but mistook it, he said, for a flash of lightning, and he signalled his car forward. As it got to the crossing it was struck by the express running at 40 miles an hour. The lightly constructed car was demolished by the engine of the train and its wreckage and the bodies of the passengers scattered along the railroad right of way. The engine of the train was derailed but no one on it was hurt. None of the 30 passengers on the interurban had had sufficient warning to jump and four of them were killed, one almost certainly fatally injured and 15 others hurt. Fred Rainey, of Cambridge, clerk in a hardware store; and Dora Fairchild, a tin mill worker, were killed outright. William Carter, of Littleton; and Mr McNeale, of Columbus, died after reaching hospital and a fifth passenger, Sheldon Spiles, of Byesville, was so badly hurt that his death was expected hourly. Help was summoned from Cambridge as soon as possible and every physician and nurse in the city went to the scene, together with ambulances.[1306] [1307]

Bad weather, in this case fog, was also blamed for a collision at Pembroke, North Carolina, United States, at 7.40 a.m. on the morning of Friday 30th January 1914. Atlantic Coast Line Railroad passenger train No. 78 was standing on the crossing with the Seaboard Air Line Railroad when it was struck by westbound Seaboard passenger train No. 19. The colliding engine hit the other train at the junction of two coaches, knocking both off the track. One of the derailed cars in turn struck and slightly damaged the station buildings. Four passengers were hurt, none seriously. Engineer C. A. Schafer of the Seaboard train was the person most seriously hurt, spraining his ankle when he jumped from his engine. He explained afterwards that in the fog and darkness he had missed the signals and had only seen the Atlantic Coast train when he was close upon it, when he had applied the brakes and jumped. It took four hours to clear the wreckage.[1308]

A serious crossing accident on Friday 17th July 1914 originated in a cause which we have not come across before. It involved a freight train of the Virginian Railway and an electric train of the Virginia Railway and Power Company. It has been a little difficult to decide which chapter to include this accident in, as the Virginia Railway and Power Company route had many of the characteristics of a street car line, but contemporary newspaper reports refer to it as an interurban line, so it has been included here rather than in the level crossings chapter.

The accident occurred at Fairmount Park, Virginia, United States. At this point the Virginian Railway had a single track while the Railway and Power Company line was double track. Neither railway used block signals, the former being operated on the train order system and the latter under street car rules. The crossing of the two was, however, protected by a pair of gates which normally stood across the track of the steam railway

[1306] *The Titusville Herald*, Titusville, Pennsylvania, United States, 14th July 1913
[1307] *Defiance Crescent-News*, Defiance, Ohio, United States,14th July 1913
[1308] *The Robesonian*, Lumberton, North Carolina, United States, 2nd February 1914

but when open for a train on that line stood right across the electric tracks. The gates were provided with two red lights to give an additional warning after dark. All trains, steam and electric, were required to come to a full stop before proceeding over the crossing.

Westbound extra freight train No. 469 of the Virginia Railway consisted of locomotive No. 469, three loaded cars, 68 empty cars and a caboose. It was in charge of Conductor Knox and Engineer Ivey and had left the Sewalls Point terminus at 12.05 a.m. Westbound Railway and Power Company train No. 83 consisted of motor car No. 3205 and trailer car No. 2239. It had departed from Willoughby Pier at 11.55 p.m. on 16th July, five minutes late, and was in charge of Conductors Franklin and Whitley and Motorman W. J. Atkinson.[1309]

At 12.30 a.m. the freight train was two thirds of the way over the crossing when it was run into by the electric train, which was travelling at an estimated speed of between 30 and 35 miles an hour when it collided with the 46th car of the freight train. Motor car No. 3205 was described as having been totally destroyed in the collision while the trailer car was somewhat damaged. Seven interurban passengers were killed outright or fatally injured and 68 passengers, Motorman Atkinson and the two conductors of the electric train were injured. Three of the cars of the freight train were derailed but little damaged.

The cause of the collision was simple enough. Motorman Atkinson testified at the subsequent Interstate Commerce Commission inquiry that he had shut off power at Simpson's Crossing, about 550 feet short of the crossing, and had then fallen asleep. He said that after what would now be referred to as a microsleep he had come to when his train was two car lengths from the crossing.

It further developed that during the two weeks prior to the accident Motorman Atkinson's hours of duty had been from 6.15 a.m. to 1.35 p.m. and again from 6.35 p.m. until 12.50 a.m. On the early shift he would reach home, four miles away, at 2.30 p.m. and get three hours' sleep before his late shift. At the end of the late shift there was no car to take him home so to avoid the walk home he would try and sleep on a seat in the car barn.

Mr A. T. Cunn, Vice President and General Superintendent of the company, stated that he was familiar with the motorman's working hours and mentioned another accident at the crossing due to a motorman failing to pull up short of it.

A passenger, Mr W. K. Lopez, related that after passing Simpson's Crossing he and other passengers had noticed the red light on the gates and had called out to the motorman, but without effect.

The commission came to the obvious conclusion that the accident had been caused by the electric train failing to stop short of the crossing. Motorman Atkinson was an experienced employee with a good record and the only reason that could be given for his failure was his falling asleep. In the 24 hours preceding the accident he had been on duty 13 hours and 45 minutes and had not had opportunity for adequate rest between spells of duty. The officers of the Virginia Railway and Power Company were aware of this but had done nothing about it, making them censurable for dereliction of duty to the public. Steps should be taken to rearrange the working hours of employees so as to allow them to obtain adequate rest.

[1309] *The Kokomo Daily Tribune*, Kokomo, Indiana, United States, 17th July 1914

The commission noted that traffic over the crossing was exceptionally heavy, with 164 electric trains daily on weekdays rising to 400 on Sundays and 500 on public holidays. There were 14 daily trains on the Virginian Railway. The commission in the final sentence of its report concluded that, 'A proper regard for the safety of the public would seem to require that where traffic of such density exists the crossing should be protected either by separation of grades, an interlocking plant, or by derails which would require trains on the electric road to come to a full stop and send a man ahead to the crossing and line up the rails before it would be possible for the electric train to pass.' [1310]

On Wednesday 21st October 1914 the penultimate coach of a Chicago Rock Island and Pacific Railroad suburban train was struck by the engine of a Chicago and Eastern Illinois Railroad special train at the crossing of the two roads at Auburn Park, Illinois, United States. The coach was struck at an angle of 45 degrees and 24 persons were seriously injured, almost everyone in the coach being hurt. Most of the injured passengers were from Morgan Park and Blue Island. The crossing was described as a temporary one as the Rock Island tracks were being elevated. No other details of the accident appeared in the newspaper reports and it was not considered bad enough to warrant an Interstate Commerce Commission investigation.[1311]

Another collision at a crossing of a very busy interurban line and a steam railroad occurred on 7th May 1915, this time in Los Angeles, California, United States, The lines involved were the double track of the Pacific Electric Railway and a single track freight branch of the Atchison Topeka and Santa Fe Railroad.

On the day of the accident westbound Santa Fe extra freight train No. 480 had left Pasadena, 9½ miles away, at 12.40 p.m. It consisted of locomotive No. 480, running tender first, eight empty cars and a caboose and was in charge of Conductor McClellan and Engineer Lorin Wells. It was diverted from the main line onto the freight branch just short of the crossing to make way for a passenger train. The Pacific Electric train, westbound from South Pasadena, consisted of electric car No. 422 in charge of Conductor Carrier with Motorman Baker. It had left Echandia Junction, its last stop, on time at 1 p.m.

The collision occurred at 1.05 p.m., engine No. 480 striking the electric car near its centre, pushing it down the line 25 feet and crushing it. The rear trucks of the engine were derailed. Five passengers were killed, and 40 passengers and one employee were injured.

Engineer Wells told the Interstate Commerce Commission investigators that his train had been travelling at 10 miles an hour when, 200 feet from the crossing, he had seen an eastbound electric car approaching the crossing. Fearing that it would not get across in time he had made a service brake application, releasing the brakes again as that car cleared the crossing. He was aware of the westbound car in the distance but was concentrating on the eastbound one and did not expect it to try and cross ahead of his train. Even when the westbound car got closer he expected it to run right up to the

---

[1310] Interstate Commerce Commission report

[1311] *Englewood Times*, Englewood, Illinois, United States, 23rd October 1914 crossing and stop, as cars often did. Then the speed of the car seemed to increase, he realised that it was not going to stop and made an emergency application of the brakes. At the moment of the collision he put the speed of his train at 8 miles an hour and that of the electric car at 20 miles an hour. He did not see the crossing flagman. The bell on his engine was ringing continuously.

Fireman LeFevre of the freight train likewise stated that he had not seen any flagman at the crossing and Conductor McLellan agreed that he had not either.

Motorman Baker stated that the speed of the Pacific Electric car had been 8 miles an hour as it neared the crossing. He said that he had seen the freight train about 200 feet north of the crossing but had thought that it was stationary. He had then looked for the flagman and seeing him had given one clang of his gong. The flagman was sitting at his shanty and gave him an "all right" signal with a green flag, which he had acknowledged with two clangs of the gong. He had not again noticed the freight train until he was almost upon the crossing, when he had seen the engine close at hand and realising that there was likely to be a collision he had applied full power.

Crossing Flagman Ostrow testified that just prior to the collision he had been sitting at his shelter, where he had an unobstructed view of both tracks, talking to Flagman House of the main line crossing. He had noticed the eastbound car and gave it the proceed signal with his green flag. A minute and a half or two minutes later he saw the freight train and a few moments later the westbound Pacific Electric car. He put the speed of both the freight train and the westbound electric car at about 15 miles an hour. He said it was customary for him to give signals to the cars while seated in front of his shelter.

Motorman Walters of the eastbound car stated that he had sounded his gong and received a proceed signal from the flagman. He saw the engine of the freight train but was not sure if it was moving. The westbound car had passed him about the middle of the bridge over the Los Angeles River just beyond the crossing.

Mr W. H. Rice, an experienced railroad man, witnessed the accident. He saw the flagman signalling with his green flag but was unable to distinguish which of the two cars he was giving the flag to. He estimated that there was an interval of half a minute between the east and westbound cars passing over the crossing and put the speed of both the train and westbound car at 8 or 10 miles an hour.

The Interstate Commerce Commission concluded that the accident had been caused by Flagman Ostrow giving a proceed signal to the eastbound car in such a way that it could be misinterpreted as applying to the westbound car. It pointed out that the flagman's sole task was to protect the crossing and by his presence make it unnecessary for every electric car to come to a stand and its conductor having to check that it was safe to cross. The flagman had an unobstructed view of the crossing and yet saw neither the steam train nor the westbound car until they were practically on the crossing. There could be no excuse for such dereliction of duty. A contributory cause of the accident was the failure of Motorman Baker to approach the crossing with his train under control and at a speed such that he was able to stop within half of the distance that the track was seen to be clear, as per Rules 64 and 184.

Having thus so soundly condemned Flagman Ostrow the commission's report went on to reveal that he had formerly been employed as a conductor and had been given the job of flagman after an accident in which he had lost both of his legs. He thus had difficulty in walking about and was not as active and agile as would be ideal for someone employed at such a busy crossing.

Traffic over the crossing was exceptionally heavy, more so even than that at Fairmount Park, with over 1,000 electric cars daily combined with numerous freight and switching movements on the Santa Fe line. As in its report on the Fairmount Park accident the commission considered that a proper regard for public safety at such a busy crossing required either the grade separation of the lines, an interlocking plant, derails or gates to guard against a misinterpretation of signals.[1312]

Cattle were the only casualties in a crossing collision at Hackstaff, California, United States, at 9.30 p.m. on Thursday 25th November 1915. A Nevada-California-Oregon Railroad stock train was standing over the crossing with the Western Pacific Railroad line while some loaded cars were detached for transfer to the other line when a Western Pacific train approached the crossing at high speed, 'contrary to the railroad and California law', which required it to stop short of the crossing. The colliding train reportedly did not even whistle. Surprisingly only one animal was killed in the car which was directly struck despite its being 'smashed to kindling', but two other cars were also wrecked and in all seven head of cattle were killed outright and two more crippled. The roofs of the two damaged cars had to be removed to release the animals inside them. None of the train crews was hurt. Although the only newspaper report on the accident does not say so, it seems probable that the engineer of the Western Pacific train had lost control of it.[1313]

[1312] Interstate Commerce Commission report

[1313] *Nevada State Journal*, Reno, Nevada, United States, 27th November 1915

CHAPTER 14

FIRES

Fires often broke out on trains following derailments or collisions. By the beginning of the 20th century the gradual elimination of coal stoves for train heating removed one potential hazard but the substitution of gas for oil for train lighting proved to be another danger, and a number of accidents were made much more severe by fires started by leaking gas. The introduction of electricity as a means of traction proved another new source of potential train fires, as shown by the Dingle and Paris Metro accidents covered in the Tunnel Accidents chapter. This chapter will, however, confine itself to what might be termed primary fires, that is accidents where there was no precipitating derailment or collision but merely a fire.

The first primary train fire of the 20th century was a minor affair. As the down Scotch express which had left London Euston at 5.15 a.m. on Friday 22nd February 1901 was approaching Lockerbie station on the Caledonian Railway in Scotland, one of the vans in the centre of it was seen to be on fire. The blaze was quickly extinguished, and little damage was done to the van itself. Considerably more damage was done by the water used to put the fire out, which ruined a large quantity of London newspapers. A mail bag also caught fire, but its contents were saved.[1314]

On Saturday 8th June 1901 six waggons of a London and North Western Railway goods train caught fire between Wolverton and Roade, England. The train was carrying new sleepers which had recently been saturated with creosote as a preservative. Consequently they burnt very fiercely, and all traffic had to be stopped for an hour and a half, delaying several important expresses. The engine and part of the train were detached and with little water available in the vicinity to fight the fire it had to be left to burn itself out. Five of the waggons were wholly destroyed and the sixth severely damaged.[1315]

[1314] *The Times*, London, England, 23rd February 1901
[1315] *The Times*, London, England, 10th June 1901

The Caledonian Railway was in trouble again on Thursday 20th June 1901. A sleeping car on an up West Coast express which had left Aberdeen at 8 p.m. caught fire at Marykirk, Scotland. Its passengers were transferred to other carriages in the train, the fire causing great excitement. The sleeping car and a corridor carriage were burnt out.

'Two ladies travelling to South Africa lost almost all their effects. One of them was going there to get married, and all her wedding presents were destroyed. The other lady lost £40 in cash and a ticket to South Africa and had to return to Aberdeen.'

The cause of the fire was not stated.[1316]

As the South Eastern and Chatham Railway Ostend boat train which had left London at 11 a.m. was passing through Canterbury East station, on Tuesday 15th October 1901 the signalman noticed that one of the carriages towards the rear of the train was enveloped in flames. According to the newspaper report he telegraphed Kearnsey, England, which seems a little strange as there were intermediate stations at Bekesbourne, Adisham and Shepherds Well where the train could presumably have been stopped. Possibly it was the Shepherds Well signalman rather than the Canterbury one who saw the fire. At any rate the train was eventually stopped when it was found that one of the axle-boxes of a carriage belonging to the Holborn portion of the train was ablaze and that the fire had spread to the lower part of a carriage full of passengers. These were transferred to other carriages, the coach was detached, and the train set off again after a delay of 25 minutes.[1317]

The Dingle Tunnel fire of 23rd December 1901 has been covered in the Tunnel Accidents chapter.

On Friday 4th July 1902 a motor car of the Metropolitan Elevated Railway with 200 passengers on board caught fire as it was passing through Augusta Street station, Chicago, Illinois, United States. A panic ensued and several passengers were only prevented from jumping from the moving train by the intervention of the guards. When the train came to a stand there was a mad scramble to get off in which some of the passengers were crushed, two women being seriously injured. The fire could not be got under control before it had destroyed the motor car and damaged the adjoining one. The passengers were compelled to make their way back along the track for 300 feet to reach the nearest station.[1318]

In accordance with the regular working of the down London and North Western Railway newspaper train which had left London Euston at 5.15 a.m. on Friday 18th July 1902 three vans containing daily and weekly newspapers for Birmingham and the South Staffordshire districts were detached from it at Rugby, to be attached there to the ordinary train. When the latter arrived at Stechford, England, smoke was seen to be issuing from

---

1316 *The Times*, London, England, 21st June 1901
1317 *The Times*, London, England, 16th October 1901
1318 *The Brooklyn Daily Eagle*, New York, New York, United States, 5th July 1902

one of the vans, and examination showed that a large portion of its contents were on fire. As soon as the doors of the van were opened the fire flared up. The van was uncoupled from the remainder of the train and shunted to a position beneath a water column, by which the fire was put out. There had been three tons of papers made up into bundle in the van, most of which were seriously damaged by the combination of fire and water. Only a small number of London newspapers were thus available in Birmingham and the surrounding area until more could be sent down later in the day. The fire was supposed by the railway officials to have been due to spontaneous combustion, though something such as a carelessly discarded cigarette end would surely be more likely.[1319]

The 11 p.m. New York New Haven and Hartford Railroad passenger train from New York for Boston arrived at New Haven, Connecticut, United States at 1 a.m. on Thursday 14th May 1903 and was due to depart for Worcester and Boston at 1.10. As soon as the train arrived the pipe supplying gas was connected to the rear end of Pullman sleeping car No. 2032. A moment later there was a slight explosion and the car was filled with flames. The car quickly filled with smoke and its 20 passengers had some difficulty in getting out. A rush was made for the front door, with the passengers being unable to see through the smoke and in the rush a woman was knocked down. An elderly man whose berth was close to where the fire had started was somewhat affected by the smoke but seemed to recover once he had got out into the fresh air.

Every window of the car was broken to let out the smoke. Firemen had been called and were soon able to put out the blaze, after which they searched for missing valuables.

There had been no time for the passengers to collect their possessions and after they had reached the platform they were concerned about valuables which they had left behind. Thus, 'B. D. Kaplan, of 227, East 68 street, New York, who escaped with hardly any clothing on and was black from the smoke, claimed that he had about $30,000 in money and securities in the car. A hunt was made, and a wallet was found. It was wet but it contained four orders for $5,000 each and other securities belonging to Mr Kaplan. Mr Kaplan stated that he had recovered all his valuables except some money which had been burned up with his clothes. He could not tell exactly how much.

'A. Freschl, of Chicago, said that he had lost about $100 in money and a watch and pin. He also said he had lost a business memorandum for which he would not have taken $1,000.

'Many other passengers claimed to have lost more or less, though it is thought that when a thorough search of the car is made most of the stuff, except the clothes, will be found. The handbags of most of the passengers were soaked, but their contents seemed to be all right.

'The passengers escaped to the depot with very scant clothing on, and some were only in their night garments. The railroad officials, as soon as they could, did all they could to make them comfortable, providing some of the men with overalls and shoes.'

The passengers either continued their journeys on other early morning trains or booked into hotels.

[1319] *The Times*, London, England, 19th July 1902

It was believed that in some way the connection had been broken when the pipe had been attached and the escaping gas had been ignited by the brakeman's lantern.[1320]

A passenger coach on an eastbound Lake Shore and Michigan Southern Railroad fast mail train caught fire between Elyria and Olmstead Falls, Ohio, United States, on Monday 20th July 1903. The fire reportedly became sufficiently threatening for the women and children among the passengers to alight from the train once it had been brought to a halt – the male passengers were obviously made of stronger stuff! A bucket brigade of trainmen extinguished the fire.[1321]

The disastrous fire which occurred on the Paris Metro on 10th August 1903 happened in a tunnel and is described in the Tunnel Accidents chapter.

In 1903 the British mails for India and Australia were taken through France by a special train which left Calais Harbour station (Calais Gare de Ville) at 3.25 a.m. every Saturday morning. Should the number of mail vans be too great to include in one train some of them were attached to the Peninsular and Oriental (P & O) Company's special train which preceded the mail train. This was the case on Saturday 29th August. The mail that day consisted of 2,340 bags and two mail vans had to be attached to the P & O train. Between Amiens and Creil on the Chemins de fer du Nord the first compartment of one of the two vans, in which the postal official was riding, caught fire. When the train stopped at Creil it was discovered that 35 of the 69 sacks in the compartment had been partly burned. The sacks had contained printed matter which from the charred remains appeared to have been destined for Bombay. The cause of the fire was unknown.[1322]

A fire in a Russian mail train between Moscow and St Petersburg on Tuesday 10th November 1903 proved to be a very expensive affair; 437 foreign parcels and 11 sacks of foreign mail were destroyed. The material value of the destroyed items was estimated as $3,500,000. The cause of the fire was again stated to have been spontaneous combustion.[1323]

Soon after the 9.10 p.m. goods train for Melbourne had left Toorak, Victoria, Australia, on the evening of Tuesday 26th April 1904, it was seen that some waggons loaded with straw were on fire. Three straw-laden waggons had been placed next to the engine, followed by a waggon of wood which also caught fire. When the train reached Broadford station the waggons were shunted into a siding and the fire was extinguished, but by this time the waggons were little more than iron skeletons. Although the straw waggons were covered with tarpaulins it was supposed that a spark from the engine might

---

[1320] *The Daily Morning* Journal Courier, New Haven, Connecticut, United States, 14th May 1903
[1321] *Defiance Crescent-News*, Defiance, Ohio, United States, 20th July 1903
[1322] *The Times*, London, England, c2nd September 1903
[1323] *Crawford County Democrat*, English, Indiana, United States, 19th November 1903

have caused the fire.[1324]

Sparks from the chimneys of steam engines were of course a common cause of fires despite the use of tarpaulins to protect vulnerable loads and the use of spark arresters.

A Midland Railway special train carrying *The Darling of the Gods* theatrical company, their luggage and scenery from Buxton to Yarmouth, caught fire outside of Stamford Tunnel, England, on the night of Sunday 7th August 1904. A luggage van containing dresses, musical instruments and other property was consumed. The loss was estimated at several hundred pounds.[1325]

A mixed train from Port Pirie to Petersburg, South Australia, on Thursday 11th January 1906 consisted of some passenger carriages and 20 waggons, three of which were loaded with saltpetre, which caught fire sooner after the train had set off. The burning load soon formed a lava-like slag which escaped from the waggons onto the track and set fire to the sleepers. Water poured onto the burning saltpetre hardened, forming a mass on the track which was difficult to remove. After two and a half hours the train returned to Port Pirie with the woodwork of the saltpetre waggons burnt to cinders and 'looking as if they had been through the crater of a volcano.' [1326] There was no mention of anyone being hurt.

The members of Mr J. C. Williamson's Opera Company had what a newspaper reporter described as 'a thrilling experience', on the Tasmanian Government Railway on the morning of Sunday 25th February 1906, though they would probably have used other words to describe the event. The opera company had just concluded an extremely successful season at Launceston and at 10 a.m. on Sunday set off on a specially chartered train across Tasmania to the city of Hobert in the south, where they were to continue their season at the Theatre Royal on Monday evening.

The train had covered about seven miles, and the company were just settling down to enjoy the scenery, when first one person and then others smelt burning. While they were discussing what to do one of the company, at considerable personal risk, made his way along the carriage footboards to find what was wrong and discovered that a property truck was on fire. Even at this relatively late date it would seem that the carriages of the Tasmanian Government Railways were not provided with communication cords, for the passengers had to resort to shouting and waving from the windows to attract the attention of the train driver, while some of the women passengers were reportedly panicking, fearful that the fire might spread to the passenger carriages. The driver's attention was soon captured, and the train was stopped. The burning truck and that next to it, which was loaded with scenery and was the leading vehicle of the train, were uncoupled and drawn 100 yards down the line, where railwaymen and male passengers set to work to bring the fire under control and save as much of the contents as possible.

The train stopped near Breadalbane station and from there Mr Harry Bracy, Mr

[1324] *The Evening News*, Sydney, New South Wales, Australia, 2nd May 1904

[1325] *The Times*, London, England, 9th August 1904

[1326] *1291 Morning Post*, Cairns, Queensland, Australia, 12th January 1906

William's representative, was able to telephone to Launceston for assistance. The chief clerk at Launceston, Mr C. Harrison, set off with another engine, a pair of trucks and a gang of men and the contents of the burning waggons were transferred, while the main train continued to Hobert after an hour or so's delay. No one was sure when the train arrived at Hobart what part of the contents of the burning trucks had been saved and there was an anxious wait until the salvaged contents of the trucks arrived at 9 p.m. It then transpired that some of the private property of members of the company had been lost, together with some theatrical wardrobes and the scenery of *Princess Ida* and *The Mikado*. It was noted that a fortunate aspect of the affair was that even so the company's season would not be interfered with, for they were due to perform *The Gondoliers* at the Theatre Royal, the scenery and wardrobe for which had already been sent down the previous week. The management would thus have plenty of time to replace the lost scenery and properties.

The fire was supposed to have been caused by a spark from the engine.[1327]

Another fire broke out in an express passenger train of the Tasmanian Government Railway going in the opposite direction, from Hobart to Launceston, on Tuesday 20th November 1906. A non-smoking first-class compartment of the train began to fill with smoke and a young lady passenger became so alarmed that she attempted to jump from the moving train. Fortunately she was restrained by another male passenger who then had to get out onto the footboard of the carriage to alert the driver. The train was stopped between Tunbridge and Antill Ponds stations and the fire extinguished. Again it had apparently originated from a spark from the engine which had lodged in the sash of a carriage window. Passengers stated that there had been a big blaze as well as much smoke, though the actual damage done to the vehicle was slight.[1328]

The unintentional actions of a passenger were the probable cause of a fire on a Pullman sleeping car which resulted in a fatality near Loveland, Iowa, United States, while it was possible that the actions of the passenger who lost his life led to his death in the accident. The train involved early on Sunday 6th January 1907 was Chicago and North Western Railway No. 8, the *Los Angeles Limited*, on its way eastwards to Chicago, Illinois. The victim who lost his life was Louis De Bario, of Washington, D. C., a clerk to the House Committee on Irrigation, whose family came from Laramie, Wyoming. He was occupying the Pullman car *Redfield*, marshalled third from the rear of the train.

Shortly after midnight the train had left Council Bluffs, Iowa, for its next stop at Missouri Valley, 22 miles to the east. Near Loveland, two miles short of Missouri Valley, the *Redfield* car was found to be on fire. It was filled with smoke and its passengers were groping about in the darkness. The train was stopped at Loveland, the passengers from the *Redfield,* minus their baggage and most of them in their nightclothes, were moved into the rear two sleeping cars, which were detached and left at Loveland, and the front of the train was run as fast as possible to Missouri Valley, where there were water tanks. During the journey the fire, which had previously been smouldering with mostly smoke, suddenly became a fierce blaze. Once the water tanks had been reached the fire was

1327 *The Mercury*, Hobart, Tasmania, Australia, 26th February 1906
1328 Ibid, 21st November 1906

quickly put out and the trainmen entered the Pullman to retrieve the possessions of its passengers. It was then that they found the found the badly burnt body of Mr De Bario lying in an aisle beside his suitcase. He had evidently been overcome by smoke as he tried to make his escape.

There were two possibilities as to how Mr De Bario had become a victim of the fire. One was that he had been overlooked when the other passengers had been hurriedly evacuated at Loveland and the other, perhaps more likely possibility, was that he had most unwisely returned to the burning car to retrieve his suitcase.

An engine was sent to Loveland to collect the two sleeping cars left there and bring them to Missouri Valley, where the passengers from the *Redfield* were given clothing, most of them having lost all their possessions in the burning car. Three other passengers were slightly hurt; Mr O. H. Johnson, of Salt Lake City, Utah, suffered slight burns to his head and neck; Mr Henry Cabrow, also of Salt Lake City, cut his wrist and Mrs B. Higgins, of Carbondale, Pennsylvania, suffered shock.[1329]

The origin of the fire was not known for certain but the berth in which it had begun had reportedly been occupied by a woman who had been using an alcohol lamp to heat milk for a nursing bottle for her baby.[1330]

Trains carrying the equipment of opera and theatrical companies seem to have been rather disposed to catch fire at this period, for on Monday 28th October 1907 another was in trouble on the North British Railway in Scotland. Near East Fortune a train carrying several theatrical companies who were en route from Scotland to various English towns was seen to be on fire by its driver. He decided it was best to continue a few miles further to a location with a water column (this must have been Dunbar). On reaching the water column copious quantities of water were poured onto the burning vehicle, but the efforts of the railwaymen and *The Beautiful Friend* theatrical company failed to prevent the total destruction of the van, together with the whole of its scenery and baggage. The train was delayed two hours. The company had been travelling from Dundee to Fencehouses, County Durham.[1331]

There were no casualties in a fire on a Lancashire and Yorkshire Railway electric train on Thursday 30th April 1908 and the accident escaped the attention of the press. It was, however, considered serious enough to warrant a full Board of Trade investigation and report.

The fire, which was stated to have been a very severe one, occurred just after 1 p.m.; the train was fortunately carrying few passengers and there was no panic. Shortly after leaving Freshfield station the motorman received a long continuous ring on the bell, the signal to make an emergency stop, which signal was by chance given not by the guard, at that point unaware that there anything wrong, but by the fire itself involving the bell wires. The inspecting officer, Mr A. P. Trotter, noted that this was merely a fortunate circumstance and the fire might equally as likely have destroyed the bell wires preventing

[1329] *Burlington Evening Gazette*, Burlington, Iowa, United States, 7th January 1907
[1330] *Santa Fe New Mexican*, Santa Fe, New Mexico, United States, 7th January 1907
[1331] *The Times*, London, England, 29th October 1907

the guard signalling to the motorman.

A short circuit, the origin of which was not clear, caused severe arcing which lasted for two or three minutes. This primary fire, which was severe enough to melt several pounds of iron, then caused a secondary fire in the intermediate carriage where the short circuit occurred. The secondary fire probably originated beneath the floor of the carriage, then spread to the vestibule, where it burnt fiercely, and into the interior of the wooden carriage. Seat cushions caught fire, much woodwork was burnt, and the hot air scorched the varnish and even the door at the far end of the carriage. Mr Trotter stressed the value of end doors and free communication between the carriages, noting that probably no passenger could have survived in that carriage.

Mr Trotter considered that more important than the origin of the short circuit was the persistence of the electric current after it had been cut off. The circuit breakers had opened at Formby and Freshfield, the points between which the short circuit had occurred, at 1.08 p.m. The attendants had closed them but two minutes later they had opened again. The inspecting officer remarked that had the attendants guessed that there was something wrong when the circuits had opened automatically the first time the fire probably would not have spread and been easily controlled. But circuit breakers frequently opened and the practice on the line was to replace them at once on the assumption that they had come out because of a momentary overload. Meanwhile the motorman of the train had applied a short-circuiting bar across the rails to prevent any further supply of current. The fire was then extinguished with chemical extincteurs and the train resumed its journey after a delay of no more than 20 minutes.

Mr Trotter recommended that in the construction of new rolling stock and the repairs to existing stock care should be taken to protect cables from water not only under ordinary conditions but when there was a high wind or driving snow. He also recommended the arrangement of the layout of feeders, circuit breakers and section switches to limit the amount of current distributed to each section of line, and that the sections should be, where possible, of such a size that only one train was likely to start on them.[1332] There was no mention of substituting steel for wooden carriages, though all-steel carriages had come into use on the Paris Metro two years earlier.

The last Metropolitan Railway passenger train of the day from Baker Street was running into Swiss Cottage station, England, at about 1 a.m. on Tuesday 6th October 1908 when an explosion was heard, and flames were seen to be issuing from the luggage compartment behind the motorman's cab. So rapidly did the fire spread through the leading coach of the train that the motorman was forced to jump out before the train had come to a halt. Of the 20 passengers on the train three or four were seated in the leading coach but all were able to get out in time. The coach itself was burnt out and the footbridge beneath which the train had come to a stand and its adjoining stairs also caught fire. St John's Wood Fire Station was but a few yards away and firemen were quickly on the scene. Water was obtained from a hydrant in the street and after about half an hour the fire was put out. The train was then towed away to Neasden depot for repair. The cause of the fire was not definitely ascertained but it was believed to have originated

[1332] Board of Trade report

in the explosion of a naphtha lamp in the luggage compartment.[1333]

In mid-October 1908 the town of Metz, Michigan, United States, was destroyed by a forest fire. When all hope of saving the little town from the fire had gone its inhabitants fled on a relief train via the Detroit and Mackinac Railroad, but three miles from Metz the train met with disaster. During the afternoon of Thursday 15th October 1908 the 200 people in the small town telegraphed north and south asking for relief, knowing that the railroad was their only means of escape. At the time that the message was received in Hawks, the next station north of Metz, the regular southbound freight train was there. A special train of three empty box cars and two coal gondolas was therefore made up and rushed down the line to Metz, and on arrival there the men, women and children were invited to board it with such household goods as they could carry. The train then proceeded southwards, but at Konieczny Crossing, two or three miles from Metz, Engineer John Kinville saw piles of burning cedar ties on either side of the track. Opening the throttle, he tried to hurry past the danger spot, but the heat had distorted the rails and his engine was derailed. The train immediately caught fire from the burning ties and the occupants of the box cars had to flee to save themselves, though the trainmen made a vain attempt to re-rail the engine.[1334]

One of the survivors of the accident, Wesley Irwin, a Bay City travelling man, described the accident. "The train left Metz about 6 o'clock Thursday evening and ran into the fire three miles south of that place. A burning pile of cedar had warped the rails and into the trap the train leaped, carrying its human cargo to destruction. The five box cars immediately took fire and it was with difficulty the 200 people in them escaped.

"In a deep-bodied steel coal car were located the women and children and as the heat increased they were unable to escape. Men were forced to seek places of safety and could give no aid. It was terrible to witness, and I hope I will never be compelled to go through such agony again. It made one's heart sink as the pitiful cries of the women and children reached the ears of the fleeing men, calling pleadingly not to be left along in that seething furnace. The women clung to their little ones and begged for aid. Agonizing cries rent the air and caused the blood in one's body to run cold.

"A party of seven of us took to the woods and were forced to jump over burning logs and run through fire to make our escape. It was the only chance to take, and we thought it preferable to remaining with the women and children and meeting the same fate. For five miles we ran through smoke and fire, which seemed every moment to be taking our strength away. When within half a mile of Posen my eyesight ran out and I could only depend upon the railroad track to guide me through. Only two of the original party reached Posen that night. The others straggled along one at a time." [1335]

Fourteen women and children died in the coal car, as did Fireman Arthur Leem and Brakeman William Lee. Another account stated that Brakeman William Barrett (a different person from Brakeman Lee?) had attempted to take refuge in the water tank of the engine tender but had boiled to death. According to one report Engineer Kinville also died at the scene but another account stated that *Conductor* Kinville and Engineer Foster

---

[1333] *The Times*, London, England, 7th October 1908

[1334] *New York Times*, New York, New York, United States, 23rd October 1908

[1335] *The Pointer*, Riverdale, Illinois, United States, 17th October 1908

(Conductor Foster and Engineer Kinville?) had escaped by crawling along the tracks on their hands and knees.[1336] [1337] [1338]

One of the vans of an express train running through the Orange River Colony was discovered to be on fire near Brandfort at 10 p.m. on the evening of Wednesday 6th October 1909. The three postal officials riding in the van raised the alarm but by the time that the fire could be dealt with it had already taken a good hold and it was only extinguished after the van had been detached and run to a water tank at Brandfort station. The mails for Pretoria, Lorenzo Marques, the Transvaal country districts, and some portions of the Rand were totally destroyed, together with three or four bags containing business letters for Johannesburg. The remainder of the Johannesburg mail was being carried in another van and was not affected. Mails for Army headquarters in Pretoria which had left Southampton on the mail steamer on 18th September were also lost. All the registered letters in the van were saved.

Initially the fire was blamed on one of the usual culprits – a spark from the engine. Later this theory was discounted, a searching inquiry having established that the fire had originated in the explosion of some chemical matter being sent in a postal packet to Pretoria, 'contrary to the regulations and to the repeated warnings of the authorities.' The wholesale destruction of mail was likened to the burning of Army mails in June 1900 by Christaan de Wet during the Second Boer War.[1339]

Very brief reports in British and American newspapers gave notice of a bad train fire near Vosnessenska, Russia, on the evening of Tuesday 21st February 1911. A kerosene oil lamp in a third-class carriage of a passenger train exploded and set the carriage on fire. Seven passengers died before they could escape, and 32 others were more or less badly burnt. The injured were taken to Perm.[1340] The entire train was reportedly consumed by the fire.[1341]

On the morning of Friday 28th July 1911 a cattle van attached to a London and North Western Railway passenger train caught fire near Blaby station, England, just south of Leicester. The van contained a heifer named *Aldershaw Duchess*, belonging to a Captain Harrison, of Lichfield, which had won first prize at the Leicester show. The animal was burnt to death, doubtless to the great displeasure of the captain, who was presumably a passenger on the train, and at the cost of a considerable sum in damages to the railway company.[1342]

As a passenger train from Melbourne to Bairnsdale was running between Fernbank

---

[1336] *Trenton Evening Times*, Trenton, New Jersey, United States, 16th October 1908
[1337] *New York Times*, New York, New York, United States, 23rd October 1908
[1338] *Fort Wayne Sentinel*, Fort Wayne, Indiana, United States, 16th October 1908
[1339] *The Times*, London, England, 8th and 9th October 1909
[1340] *The Times*, London, England, 23rd February 1911
[1341] *Mitchell Daily Republican*, Mitchell, South Dakota, United States, 22nd February 1911
[1342] *The Times*, London, England, 29th July 1911

and Lindenow, Victoria, Australia, on Sunday 3rd February 1912, and was about two miles from Fernbank, a passenger observed that the canopy of canvas and tarpaulin between the first and second-class coaches at the front of the train was on fire. He first attempted to beat out the flames with his hat, but failed. Fortunately passenger trains in Victoria were by this date equipped with communication cords, unlike their counterparts in Tasmania a few years earlier. The passenger pulled the alarm chain and the train was quickly pulled up. The 35 passengers in the two leading coaches quickly evacuated them, alarmed by the crackling sound of burning timber. All got out safely but left behind portmanteaux and parcels which were consumed by the flames. The train crew and some of the passengers then attempted to fight the fire. The guard's van following the first-class carriage was uncoupled and pushed back out of the way but the two 70 ft corridor coaches, which were only a few months old, were burnt out.

The cost of the damage was estimated at £3,500. The cause of the fire was undetermined.[1343] [1344]

In a repetition of the accident at Blaby the previous year three prize cattle, Devons, part of the herd of Lord Portman, of Bryanston, Blandford Forum, Dorset, were either suffocated or burnt to death in a fire involving a cattle truck on the Somerset and Dorset Joint Railway on Tuesday 28th May 1912. The cattle were being taken to Blandford from the Bath and West of England Show at Bath. The fire was observed shortly after the train had started from Bath and fortunately before it had entered the long Combe Down Tunnel. A cowman and a boy jumped from the train before it could be pulled up, the cowman sustaining a rather bad cut about his head. He was able to go home, however, after treatment at Bath hospital. The animals lost were *Bryanston Good Will*, which won first prize for Devon heifers in milk; two-year-old *Hestercombe Dasher*, which gained second prize for Devon bulls; and *Expert,* recommended in the same class. A spark from the engine chimney was believed to have ignited straw in the cattle truck.[1345]

On about Monday 10th March 1913 on passenger was killed outright, four were fatally injured and another 25 were less seriously injured in a fire in a railway carriage near Rudna Poosejowski, Russia (now in the Ukraine). The fire had been caused by the accidental dropping of a bottle of methylated spirits.[1346]

The catalogue of mail train fires continued with one near Doble, Victoria, Australia on Friday 10th October 1913. The driver and fireman of a train from Stawell to Ballarat spotted the fire and the train was stopped two miles from Doble. Meanwhile the guard had thrown the Melbourne mail bags out through the window, saving them from the flames by this rather desperate measure. Once the train had stopped the burning van was detached but the fire spread so quickly that there was no time to save the Ballarat or Geelong mails, the travellers' samples or other merchandise in the van. By the time that

---

[1343] The Sydney Mail, Sydney, New South Wales, Australia, 7th February 1912
[1344] *Northern Star*, Lismore, New South Wales, Australia, 17th February 1912
[1345] *The Times*, London, England, 29th May 1912
[1346] *Stevens Point Daily Journal*, Stevens Point, Wisconsin, United States, 11th March 1913

the fire had been put out all that remained of the van was its iron skeleton.[1347]

Wartime censorship did not prevent a brief mention in the British press of a fire on a morning Highland Railway mail train near Ballinluig, Scotland, on Friday 2nd October 1914. Sixty bags of mail and 100 packages destined for the Royal Navy Grand Fleet at Scapa Flow were destroyed. In addition the guard of the train was seriously injured by the explosion of a gas cylinder.[1348]

Sparks from the engine chimney and probably a poorly covered load of cotton resulted in a fire in two waggons of a goods train as it approached Rathen station, Saxony, Germany, on Monday 10th May 1915. No one was hurt.[1349]

1347 *The Daily Mercury*, Maitland, New South Wales, Australia, 13th October 1913
1348 *The Times*, London, England, 3rd October 1914
1349 *Eisenbahnunfälle in Sachsen* by Claus Schlegel pg. 117 (in German)

CHAPTER 15

CHEMICAL EXPLOSIONS

Railway accidents in the early 20th century involving explosions were of two types; those which might be defined as chemical explosion and locomotive boiler explosions. Accidental chemical explosions, such as those due to mishaps with blasting powder or munitions, were rare, whereas locomotive boiler explosions were more common. This chapter will deal with chemical explosions.

A gas cylinder exploded in one of the cars of eastbound Atchison Topeka and Santa Fe Railroad passenger train No. 8 on the evening of Monday 16th December 1901 near Garden City, Kansas, United States. In the resulting fire the smoker, chair car and one sleeper car were burnt out. According to most accounts none of the passengers was hurt but Conductor John O'Day and Porter Robert Pennington were badly cut about the head and face.[1350] One newspaper spoke of several passengers being slightly hurt and of a trainman, presumably Porter Pennington, having been killed.[1351] The gas involved in the explosion was acetylene, used briefly by the Santa Fe and Texas Midland companies for train lighting. It was not a success, the gas being dangerous, giving off a nasty smell and being corrosive.[1352] A newspaper commented a few days after the accident that, 'The Santa Fe has had three explosions of acetylene gas in a month. Except for this fault it has given good satisfaction.' {!} [1353]

A piece of rough switching (shunting) led to a disaster on the evening of Monday 12th May 1902. The scene of the disaster was the Sheraden yards of the Pittsburgh Cincinnati Chicago and St Louis (Panhandle) Railway near Pittsburg, Pennsylvania, United States.

The Sheraden yards had over 30 tracks at the time of accident these contained several hundred cars. A train consisting of 10 tank cars of refined petroleum and naphtha was

[1350] *The Norfolk Weekly News Journal*, Norfolk, Nebraska, United States, 20th December 1901
[1351] *Sedalia Evening Democrat*, Sedalia, Missouri, United States, 18th December 1901
[1352] *The American Passenger Railroad Car Part 2* by John H. White
[1353] *The Quincy Journal*, Quincy, Illinois, United States, 21st December 1901

being made up for the west. Five cars, two containing refined petroleum and two naphtha, were switched with excessive force against some standing cars in the course of making up the train, and one of the cars of naphtha was damaged. A stream of inflammable liquid immediately began to pour out onto the track. Seeing that had happened the trainmen tried to pull the damaged cars away. It would shortly be getting dark, and the lampman had already made his rounds and lit the switching ground signals. As the cars were being pulled back the leaking naphtha fell onto the little flame of a switch signal and an explosion followed almost at once.[1354] This, however, was only the start.

'The first explosion sent showers of burning naphtha over the freight station, close at hand, and also enveloped a number of carloads of coke and lumber that were close by, and in a moment all were blazing. Knowing that there was imminent danger of another explosion, there was no attempt to save the freight station or the cars, for the time, and all sought safety in flight. This fact saved many lives, for a 5 o'clock, fifteen minutes after the first explosion, the two cars of refined petroleum that had been damaged by the bursting of the tank of naphtha, and were leaking, blew up with a terrific report. No one was injured by this, however.

'On the south side of the yards, cut out of the hillside, is a road on which are the tracks of the Sheraden branch of the Pittsburg Railway company. Above these tracks is a rounded hill bare of trees. Across the cut, on the other side of the yards, a hill full 200 feet high, rises sheer above the roundhouse.

'Many persons, among them many children returning from school, gathered on these hills to watch the fire. Suddenly a third explosion occurred about a mile away.

'Between the time when the first tank of naphtha was damaged in switching and sprung a leak, and the moment when it exploded, enough of the dangerous liquid had escaped to work its way into the sewer that empties into the Ohio river near the mouth of Cork's run. Burning oil, too, found its way into the sewer, and as soon as it reached the open air at the mouth an explosion followed. The force of the explosion blew in one side of Seymour's hotel, on the hillside above the bridge, where five people were injured. All of the windows in the neighbourhood were broken by the concussion, and nine other persons were injured by falling glass. The traction bridge across the run at this point was almost entirely demolished and a car that had passed just a moment before was all but dashed to pieces.

'More people came in on the cars from Pittsburg and were added to the number of spectators. Railroaders were at work, railroad officials were near, also newspaper men and others.

'At 6.30 o'clock two more naphtha tanks in the center of the train exploded.

'There was an awful roar, loud enough to be heard in the town of Pittsburg, five miles away, and a sheet of solid flame shot up from the wreckage and enveloped both hillsides even to their tops.

'There was a moment's lull, as though every living thing in the vicinity had been annihilated, and then the cries, the screaming for help, and the blanks in the crowd that told of the explosion's dread results.

[1354] *The Bradford Era*, Bradford, Pennsylvania, United States, 13th May 1902

'On the hillside to the north of the yard several houses were burning, the trees were afire, the scant vegetation was blackened and eaten away, and masses of burning liquid from the tanks below could be seen here and there where they had fallen.

'On the tracks themselves was a fleeing crowd of men dazed by the shock of the explosion, hurrying to escape from the deluge of fire that had threatened to overwhelm them. But these were safe.

'The charred bundles of clothing, almost shapeless, that lay where death or worse than death had caught them, were past all help save where some movement or a piercing cry told that life was still left in the blackened form.

'On the hillside to the south, where the crowds, almost three-quarters of them children, had gathered on the traction company's tracks to watch the fire, there was a line of the bodies of those who had been caught in the flames, but were still alive. They cried piteously for help, those who were too badly hurt to run away cried in agony and asked that they be put out of their misery. Fleeing in all directions except towards the burning wreckage were those who had been more fortunate, but who had still suffered in the accident.'

By the late evening there were 22 known dead and at least 200 others injured, 75 per cent of whom it was feared would not recover.[1355] Those who died included Brakemen Albert McKean and John Swan.

A terrific explosion of a car loaded with dynamite occurred at the western end of the Pennsylvania Railroad yards at Crestline, Ohio, United States at 8.03 p.m. on Sunday 1st November 1903. By great luck no one was close to the site of the explosion and no one was killed, though Yardmaster W. H. Courter and Clerk Carl Geisinger, who were in the yard office half a mile away, were badly hurt. Operator J. A. Redmond was also injured by a piece of flying debris.

A newspaper reporter described the results of the explosion. 'The direct damage to the railroad company was extensive. The car of dynamite stood in about the centre of the yards, with eight tracks on the side of it, filled with cars. The explosion blew the cars next to it into small pieces and distributed them over the adjoining pasture fields. The cars further away were splintered and turned over. The debris formed, as it were, a huge funnel, with the hole torn in the ground as the small end. The ground seemed to have been parted and pushed and the sidetracks described graceful curves around it. The heavy rails were bent, some of them almost at right-angles. The hole around which the debris was distributed was 100 feet in diameter and 55 feet deep.

'Some of the heavy steel cars were torn to pieces, others were bent and twisted, turned completely over and stood on end. The wooden box cars were blown to splinters which scattered all over the place.

'Several big oil tank cars were wrenched apart and the heavy sheets, top plate, twisted into fantastic shapes and thrown a quarter of a mile away. Several strings of box cars on the other side of a small divide were blown into fragments and bales of hay with which they were filled, were broken open. The debris caught fire immediately after the explosion and burned fiercely. The loose hay on the other side of the divide caught and

[1355] *The Evening Independent*, Mansfield, Ohio, United States, 13th May 1902 the hundreds of bales fed the fire for hours. Cars were pulled away from the fire, but a large number of them were so badly disabled that they were allowed to burn, furnishing light to the men clearing up the wreckage. E. T. Whiter, of Pittsburg, superintendent of that division, estimated that between 150 and 300 cars were totally destroyed. All of these were loaded with general merchandise.' [1356]

As for the town of Crestline itself, a mile or so away, 'When the concussion came every building in the city seemed to be lifted from its foundations, while walls fell, windows crashed and roofs shook, threatening to fall and crush the occupants. Churches filled with worshippers were thrown into a screaming mass of humanity and those who were enjoying the comforts of their firesides were hurled to the floor and it seemed as if the world were at an end.

'In the city the state of affairs almost beggars description. Almost every window is broken, and the business district presents a horrible sight. Thousands of dollars was caused to business districts alone. The Union depot and both banks suffered a great deal, but the churches are probably the worst damaged. Services were being held in all of them when the explosion occurred, and in each case the people thought it was their particular church which had collapsed.

'Women screamed and fainted, and it took the might efforts of the men to prevent a stampede and cause death to hundreds. Every possible conveyance was pressed into service to convey the stricken women to their homes.

'There has never been a case in the history of the country when such widespread ruin and disaster has been wrought by a single explosion. At first railroad men could not be induced to go near the wreck, fearing a second explosion, and this gave the fire a better start.'

The sound of the explosion was clearly heard in the towns of Mansfield, Bucyrus and Galion and even at Wooster, 54 miles away, buildings were shaken, windows rattled, and people thought that there had been an earthquake.[1357] Property loss was estimated as between £$280,000 and $300,000.

All the evidence as to how the explosion had happened was destroyed in the blast, but Superintendent Whiter stated that, "The men were just making up a freight train to go west. The switching crew dropped a cut of 10 cars down onto the remainder of the train. The car containing the dynamite was about the middle of 40 cars. The jar might have set off a can of the stuff or caused a can to topple over and explode, exploding the others. About two minutes elapsed after the car was coupled on before the explosion occurred. Just who is at fault I am not prepared to say. A thorough examination will be made." [1358]

At 7.45 a.m. on the morning of Friday 21st September 1906 a carload of dynamite standing on the track near the Louisville and Nashville and Southern Railroad union depot at Jellico, Kentucky, United States exploded. The car, belonging to the Pennsylvania Railroad, had contained about 20,000 pounds of high explosives packed into 450 boxes and consigned to the Rand Powder Company at Clearfield, Tennessee.

1356 *The Mansfield News*, Mansfield, Ohio, United States, 3rd November 1903
1357 *The Newark Advocate*, Newark, Ohio, United States, 2nd November 1903
1358 *The Mansfield News*, Mansfield, Ohio, United States, 3rd November 1903

The town of Jellico straddles the state line between Tennessee and Kentucky and the station stood at the end-on junction of the Southern Railroad line to the south and the main line and Packard branch to the north. The exact site of the explosion was in Kentucky, on the Louisville and Nashville line.

A large part of the business district of Jellico was devastated as well as many private residences on the Kentucky side of the town. It was said that the blast had left one seventh of the population of the two towns homeless. The number killed in the explosion was initially estimated at 12, with scores of others injured and an estimated $500,000 of damage.

As to the cause of the explosion there were two conflicting theories. One was that three men had been shooting at a target placed on the car and that a bullet had set off the dynamite. There were reports, or rumours, that a man had been seen firing at the car with a rifle.[1359] The alternative explanation was that while the car had been standing on the siding a car of pig iron had been roughly shunted against it, as very likely had happened at Crestline, and that the impact had caused the explosion.[1360]

The initial death toll of twelve was reduced the next day to seven, with five possibly fatally injured.[1361] At the coroner's inquest, however, there as testimony that about 15 unknown men had been standing near the car when it had exploded, all of whom had supposedly lost their lives, while a local country woman was searching for five children whom she had given permission to go to Jellico to see a circus, and who had disappeared. The naked body of a girl aged about two was found in a field a mile from the scene of the explosion and there was evidence that a family of four had also been victims. It was concluded that the number of dead was at least 22, with five still missing.

The coroner's jury clearly did not accept the story of the men using the explosives van for target practice, a theory which would be convenient for the railroad companies as it would free them from blame. Instead they passed a verdict that the explosion had resulted from the criminal negligence of the Louisville and Nashville or Southern Railroads or both or their employees.

Meanwhile Mayors Baird and Stewart of the two Jellicos issued an appeal for help, which read in part, "The gravity of the situation in Jellico, Ky., is more appalling than we at first anticipated. We at first decided to decline the offers of outside assistance that have come to us from all parts of the country, but now we declare that contributions of money or other material assistance will be gratefully received.

"Many of the people in humbler circumstances have lost practically everything. Some will never be able to re-establish themselves in business or otherwise if outside aid is not extended to them.

"Contributions may be sent to D. D. Scott, recorder of the city of Jellico, Tenn., or Frank Snyder, clerk of the city of Jellico, Ky." [1362]

Slightly less than four months after the Jellico explosion, on 19th January 1907, there

[1359] *The Landmark*, Statesville, North Carolina, United States, 25th September 1906
[1360] *The Owingsville Outlook*, Owingsville, Kentucky, United States, 27th September 1906
[1361] *Eau Claire Sunday Leader*, Eau Claire, Wisconsin, United States, 23rd September 1906
[1362] *The Sun*, Baltimore, Maryland, United States, 25th September 1906

was another disastrous one, this time involving a passenger train. Cleveland Cincinnati Chicago and St Louis (Big Four) Railroad accommodation passenger train No. 3 from Indianapolis, Indiana to Mattoon, Illinois, consisting of a locomotive, a combination baggage and mail car, a smoking car and a day coach, was passing through Sandford, Indiana, United States, at 8.50 p.m. on a dark and wet Saturday evening when there was a tremendous explosion as it passed the powder car of a freight train standing on an adjoining track. The blast threw the entire train from the track, its well-filled coaches were demolished, and the engine was hurled 50 feet. Engineer Edward Welch and Fireman Harry Jarred were thrown into the air and landed in a muddy field 100 feet from their locomotive. The fireman was badly hurt, the engineer only slightly so. Fire broke out in the wrecked coaches and there was no efficient water supply to fight it, while bitter cold added to the ordeal of the survivors. The whole train quickly became a mass of flames, and rescuers could only reach passengers who had been thrown out of it by the blast or had been able to crawl out of the wreckage. Two men who were still alive when the relief train from Terre Haute reached the scene could not be freed or moved and were cremated.[1363]

Apart from the train not a home in Sandford escaped damage from the blast. Windows were shattered, dishes and furniture broken and in some cases doors torn from their hinges. Nevertheless the residents of the village did all they could to help the injured by the light of the burning coaches, opening their homes and reportedly working heroically all night. Telegraphic communication was destroyed by the explosion and the first news of the disaster received in Terre Haute, nine miles away, was brought by someone who had walked there, though the shock of the blast had been felt there and beyond. Relief trains with physicians and supplies were sent to the scene from both Terre Haute and Paris, Illinois. Seven of the injured were brought back to hospital in Terre Haute and 22 others were taken to Paris. By a couple of days after the accident the number known to have died in it had reached 23.

Mr C. E. Marrs, who lived nearby and whose house was turned into a temporary hospital, said, “I was standing in my yard when I suddenly heard a terrible sound. It seemed to rend the air. All at once what looked like balls of fire went whirling into the air high above the trees.

“I went over to where I heard screaming. There was a terrible scene. I never saw anything like it. People were running around the place wringing their hands like crazy persons. The wreck was already in a blaze. I could hear screams coming from the wreck.

“When I found that they were taking people who had been hurt to the houses nearby I went to help. I came into my own house and saw a man die as I entered the door.” [1364]

Section Hand Charles Picker, who was also involved in the rescue, added, “It seems that God chose the worst night possible for the accident. It was so cold and windy and dark. It is tragic to think of the suffering.” [1365]

Eight cars of the freight train were also burnt in the wreck; the crew of the freight train managed to pull the others away. The car of powder had been en route from Concord Crossing, Massachusetts to East Alton, Illinois, the Big Four having received it from the

[1363] *The Lake County Times*, Hammond, Indiana, United States, 21st January 1907
[1364] *The Warren County Democrat*, Monmouth, Illinois, United States, 24th September 1907
[1365] *The Washington Pose*, Washington, D.C., United States, 21st January 1907

Grand Trunk Railroad. Whether the car actually contained gunpowder or dynamite which was being shipped at powder rates was queried. It was reported that it was a common deception for dynamite to be forwarded as gunpowder as it meant a saving of four-fifths of the expense to the shipper.[1366]

The freight had been put into the siding at Sandford to allow train No. 3 to overtake it.

As at Jellico there were various theories as to the cause of the explosion. The crew of the freight train suggested that the concussion of the passenger train running by had caused it. A few people, with perhaps the Jellico accident in mind, suggested that a tramp or a drunkard had fired a shot at the car. A third theory was that gas had entered the car from a leaking oil pipe nearby and that a spark from the engine had set it off, but the car was sealed, and this theory had few contenders. It was also noted that "black powder", that is gunpowder, would require a naked flame to set it off. On the other hand the ground was not torn up as one would have expected from a dynamite explosion and it was soon possible to reopen the line to traffic.[1367]

At the inquest at Paris, Illinois, on the victims of the disaster the coroner's jury returned a verdict that the accident had involved an explosive of greater power than gunpowder and had probably been caused by the jar when the passenger train made an emergency brake application, which almost all the witnesses agreed it had done just prior to the explosion.[1368]

An unusual accident occurred at Hannover Hbf, Germany, on Friday 7th August 1907, when the air reservoir of engine No. 502 exploded, killing a passenger waiting on the platform.[1369]

On Friday evening, 9th August 1907, a car containing 100 packages of dynamite arrived at Essex, Ontario, Canada, on the Canada Southern Railway, and was placed on a siding to be taken down the branch to Amherstburg the next morning. The explosive was to be employed to blast a lime kiln crossing in the Detroit River there. The branch train on Saturday morning, 10th August, was in charge of Conductor Thomas Barry with Engineer David Cottrell, Fireman Jas Madigan and Brakemen Leo Conlon and Joseph McNarry. The dynamite car had been picked up but before the branch train could start it had to wait for the mail train from St Thomas. Meanwhile Brakeman McNarry had discovered that drops of nitro-glycerine (the actual explosive component of dynamite) were leaking from the cars onto the rails. Conductor Barry drew the attention of Station Agent A. O. Stimmers to the matter and the two of them, the brakemen and others went into the car to readjust the boxes and try to stop the leak. It was a very hot day.

Half an hour later, without warning, the dynamite car exploded. At that moment Brakeman McNarry was on the ground as the train was being set back onto the passenger coach, waiting to tell the engineer when to stop, while Brakeman Conlon was riding on the rear of the dynamite car and Station Agent Stimmers was standing nearby, watching

---

1366 *The Lake County Times*, Hammond, Indiana, United States, 21st January 1907
1367 *The Weekly Sentinel*, Fort Wayne, Indiana, United States, 23rd January 1907
1368 *Cook County Herald*, Arlington Heights, Illinois, United States, 8th February 1907
1369 *Katastrophen der deutschen Bahnen Teil II* by Hans Joachim Rizau pg. 31 (in German)

to see if the nitro-glycerine was still leaking. The force of the explosion reduced everything within a radius of 300 yards to ruins. The engine, tender and cars of the train were reduced to scrap metal and a hole was formed 20 feet deep and 50 feet across. The new stone depot was torn to pieces, the freight station and nearby cottages destroyed, Greene's elevator levelled and Laing, Ritchie & Co's planning mill collapsed. Brakeman Conlon was blown to pieces and the charred body of Brakeman McNarry was found under some engine coal. A Dr Bryan of Essex was named as a third fatality, killed by the shock. Conductor Barry, Engineer Cottrell, Fireman Madigan, Station Agent Stimmers and Mrs Mary Morton, of Ypsilanti, Michigan, a passenger, were all seriously injured. Practically every building in the town of 1,500 was damaged and scores of residents were injured by falling walls or cut by flying glass. The shock of the blast was felt 16 miles away in Windsor, where it was thought that an earthquake had occurred, and people rushed out into the streets. When news of the accident was received special trains with doctors and nurses were sent to Essex from Windsor, Ontario and Detroit, Michigan.

The financial loss was estimated at about $200,000.[1370] [1371] [1372]

The inquest upon the victims of the explosion opened at Windsor on 14th August. George Delong, an explosives expert, testified that when dynamite was not sufficiently cured nitro-glycerine sometimes readily leaked from the absorbent and could then be readily exploded by friction. Friction from iron pipes placed in the same car could have precipitated the explosion. The inquest was adjourned until the 23rd.[1373] [1374]

At the resumed inquest Frederick Julien, another explosives expert from the Pluto Powder Company brought in a box of dynamite and brought out a high dynamite cartridge for the jury to examine. This caused a panic; none of the jurors could be persuaded to examine the cartridge and many fled the room and did not return until the dynamite had been restored to its case and removed from the courtroom.

Julien testified that his company despatched most of its products over the Pennsylvania Railroad, which had a paid inspector who carefully examined all of the dynamite before it was placed in the cars.

The inquest jury returned a verdict holding the Michigan Central Railroad Company (the controlling company of the Canada Southern) responsible for the accident, presumably because the company did not employ an inspector to examine the dynamite sent over its system like the Pennsylvania.[1375]

The bravery and self-sacrifice of a Mexican engineer named Jesús Garcia saved the lives of possibly several hundred people at the expense of his own. A narrow-gauge train ran from the town of Nacozari for eight kilometres to the Pilares copper mines. Ore from the mines was brought down by train to concentrators in the town, from where the concentrate could be despatched further via a standard gauge line. In the opposite direction the railway was used to bring supplies to the mines. At about 2 p.m. on the

[1370] *The Winnipeg Tribune*, Winnipeg, Manitoba, Canada, 12th August 1907
[1371] *The New York Times*, New York, New York, United States, 11th August 1907
[1372] *The Weekly Sentinel*, Fort Wayne, Indiana, United States, 14th August 1907
[1373] *The Daily Canadian*, Nelson, British Columbia, Canada, 15th August 1907
[1374] *The Daily Colonist*, Victoria, British Columbia, 15th August 1907
[1375] *Manitoba Free Press*, Winnipeg, Ontario, Canada, 24th August 1907

afternoon of Thursday 7th November 1907 Engineer Garcia had already brought two trains of ore down to Nacozari and was returning to the mines with a train which included two open waggons loaded with dynamite, detonators and fuses and several waggons of baled hay. Strictly against company regulations, which decreed that waggons loaded with dynamite must be placed at the very rear of the train, the two waggons of explosives were marshalled next to the engine. Who was responsible for placing them there was not stated, but possibly it was related to the fact that the train had no conductor that day as he had been taken ill. As the train set off sparks from the chimney of the wood-burning locomotive lodged in one of the waggons loaded with explosives. Other railwaymen riding with the train tried desperately to extinguish the fire, without success, and a brakeman shouted to the engineer that of one of the cars was on fire.

By the time that Engineer Garcia had turned, climbed up onto the tender and seen that the fire was well under way, the train was at the main street of the town. He realised that if the load exploded there, the detonation would probably set off in turn the company's main dynamite store in the town, with catastrophic results. There was also the fact that the line climbed steeply from the town towards the mines, so that if he just abandoned the engine at that point it might run short of steam down the line and the train roll back into the town before it exploded. He decided he must stick to his post, knowing that the only way that he could save the town from destruction was to get the train well away from it before the fire set the explosives off. He shouted, "Salta por tus vidas!" (Jump for your lives) to the other trainmen and they did so and fled, calling out warnings as they did to people near the train. Then he opened the throttle wide. The only other person still on the train was a boy named John Chisholm, aged 13, who had been riding on the last car. It was supposed that the noise made by the train had prevented him from hearing the warning shouts, or perhaps he was just reluctant to jump off a moving train.

Some distance outside of town, just beyond a secondary loading area known as Camp 6 the line surmounted a ridge and then descended and Engineer Garcia must have hoped to get his train past this spot and out into open country, where he could risk jumping from the engine and leave the train to its fate without it being likely to roll back into town. As he passed Camp 6, whistling continuously to warn people of the danger, the labourers there, unaware of course of the danger, stood by the track to watch the train pass, doubtless wondering why it was going so fast and whistling so much. Then, at 2.20 p.m. the explosion came. The shock of the blast threw people in Nacozari to the ground and smashed almost every window in the town. The noise of the explosion was heard at least 10 kilometres away. According to contemporary newspaper reports, of the train, the track, Engineer Garcia and the boy Chisholm there was no trace.[1376] It would seem, however, that this was an exaggeration for the sake of making a better story, for at any rate the locomotive, No. 2, survived and was later sold to an American company.[1377]

The train had covered about six kilometres from the town when it exploded. About 12 labourers from Camp 6 lost their lives, possibly also a second railwayman who had remained on the train.[1378]

Jesús Garcia subsequently became a national hero. Two years after the accident a

---

1376 *The New York Times*, New York, New York, United States, 9th November 1907
1377 Webpage – Did You Know? The Hero of Nacozari
1378 *Santa Fe New Mexican*, Santa Fe, New Mexico, United States, 9th November 1907

monument to him was erected in Nacozari, sadly by then his fiancée, Maria de Jesús Soqui, had died of a broken heart. He was posthumously awarded the American Cross of Honor and the town was renamed Nacozari de Garcia.[1379]

Early on the morning of Tuesday 3rd March 1908 an eastbound Cleveland Cincinnati Chicago and St Louis (Big Four) Railroad train stopped at Litchfield, Illinois, United States. There it was examined, and no hot axle boxes were discovered. It set off again and had got up speed when 10 minutes later, at 1.30 a.m., two cars loaded with powder in its centre exploded.[1380] Cars were thrown into the air by the force of the blast and landed 50 feet away. A dozen cars were wrecked, and their remains scattered over an area of 20 acres. The wreckage caught fire and involved the whole train, which was loaded with cotton, lumber and general merchandise. The Litchfield fire department attended but about fifteen cars and their contents were consumed.[1381]

None of the train crew was hurt and unlike some of the previous explosions mentioned this one occurred away from any populated area so no lineside residents were involved. Riding some cars away from the powder cars, however, were two tramps, Paul Dougherty, of Paxton, Illinois; and Lloyd Gaw, of San Antonio., Texas. Both were thrown into the air and badly hurt; they were taken to St Francis hospital in Litchfield. Paxton had received an honourable discharge from the United States Navy and was, perhaps, on his way home.[1382] [1383]

The rails at the scene were twisted and tangled for a thousand feet and the explosion made a large hole in the ground. Houses were shaken and the noise of the blast was plainly heard in towns 20 miles distant.

The wreckage was still burning in the late afternoon and until then the heat was so intense that no one could approach very close. The line was blocked all day, but a large force of men set to work once the fire had been brought under control and it was hoped to re-open the line the next day.[1384]

Thirty passengers on a Kansas City Southern Railroad train were injured, several it was believed fatally, when a powder mill exploded at Holmes Park, Missouri, United States, on 11th November 1908, as the train was passing. The train was two blocks from the mill when the blast occurred, shattering coach windows and showering the passengers with broken glass. The train was not derailed and took the injured to Kansas City hospitals. One mill employee was killed, and several buildings were destroyed.[1385]

On Tuesday 16th December 1913 an explosion of benzine occurred in the coach of a passenger train near Biadoling, a location not far from Cracow and at this date in Galicia

[1379] Webpage – Did You Know? The Hero of Nacozari

[1380] *The Daily Reflector*, Jeffersonville, Indiana, United States, 4th March 1908

[1381] *The Daily Free Press*, Carbondale, Illinois, United States, 4th March 1908

[1382] *The Daily Reflector*, Norwalk, Ohio, United States, 4th March 1908

[1383] *The Weekly Sentinel*, Fort Wayne, Indiana, United States, 4th March 1908

[1384] *Centralia Evening Sentinel*, Centralia, Illinois, United States 4th March 1908

[1385] *The La Crosse Tribune*, La Crosse, Wisconsin, United States, 12th November 1908

(Austrian Poland). The coach was quickly set on fire and the workmen riding in it pulled the communication cord. As soon as the train had stopped they jumped out to escape the flames. Unfortunately an express was passing at speed on the other running line at the moment that they jumped, and eight of them were struck by it and killed.[1386]

Various breaches of the rules regarding to the carriage of explosives by rail led to an explosion at Jetalsar station, India, on the metre-gauge Gondal-Porbandar Railway on 14th February 1914. The transport of 10 casks of gunpowder involved two transhipments. The fact that the casks were carrying explosives was overlooked and they were placed in an open waggon along with other items of general merchandise. A spark from the engine apparently caused a fire which led to the explosion. Three railwaymen were killed, and the rolling stock was badly damaged.[1387]

Yet another accident involving mail being carried by rail occurred on Thursday 12th November 1914 at Rydal, New South Wales, Australia. A departmental motorcycle (one owned by the railway department?) stowed in the brake van of the Forbes mail train suddenly exploded while the train was stopped at the platform of Rydal station. The van caught fire and was rapidly enveloped in flames. The fire, probably fuelled by the petrol from a full motorcycle tank, burnt fiercely and in about five minutes the van and its contents, including the mails, were destroyed. Guard Small, who was in charge of the train, was scorched about the hands and face but escaped more serious injury. The accident caused considerable alarm and upset among the passengers, many of whom had also lost all their luggage. The train did not reach Sydney until 7 a.m., an hour late, because of the mishap. The van involved was not a postal one but an ordinary railway van. Why the motorcycle should take it upon itself to explode was not explained.[1388]

Early on the morning of Thursday 2nd December 1915 a mysterious accident occurred to westbound Chicago Milwaukee and St Paul Railroad train No. 17, in charge of Conductor Wescott with Engineer Greene. As the train was running at speed near Paralta, Iowa, United States, an explosion occurred in its express car of sufficient magnitude to blow off the side door of the vehicle. Worse the interior of the car was almost immediately a mass of flames. All that Express Messengers C. H. Bronson and Forest Howe could do was to make their way through the opening and hang on grimly to the outside hand rails until the train crew realised what had happened and stopped the train, because the position of fire prevented them from accessing either the door to adjoining car or the air valve by which they could have stopped the train.[1389] Shortly after the explosion the train passed through Paralta station, where there was a signal tower, but by chance this was situated on the north side of the line, the opposite side to that with the opening, and the towerman noticed nothing. The two men had almost given up hope of their plight being discovered and were considering dropping to the ground when the train finally stopped, Conductor Wescott having smelt the smoke and realised that something

---

[1386] *The Bathurst Times*, Bathurst, New South Wales, Australia, 19th December 1913
[1387] Railways in India: Administration Report for the year 1913-14
[1388] *Evening News*, Sydney, New South Wales, Australia, 12th November 1914
[1389] *New Castle News*, New Castle, Pennsylvania, United States, 3rd December 1915

was wrong. By then the two men were badly burned about the face and hands. They were given first aid by the company physician at Marion and subsequently admitted to St Luke's hospital in Cedar Rapids. It was feared that Express Messenger Howe might lose an arm it was so badly burnt.

The remainder of the train was separated from the burning express car and the trainmen and passengers saved what they could of the contents of the car. Most of the express matter was bound through from Chicago, Illinois, to Omaha, Nebraska.[1390] $30,000 worth of currency in a supposedly fireproof safe was found to be badly singed and $1,000 worth of silver had melted and formed a congealed mass on the floor of the car. The goods piled towards the front of the express car were those which least suffered from the fire, they included 'a huge pile of silks, sweaters, general wearing apparel and nearly a bushel basket full of watches of all kinds and descriptions.' [1391] The cost of damage to the car and its contents was estimated at $75,000.

Conductor Wescott told a newspaper reporter that in 30 years as a passenger conductor he had never had a fire on a train before. The first that he had known of anything being wrong had been when he and his brakeman had smelt smoke shortly after leaving Paralta. They had hurried to the baggage car where they could put their heads out of the side doors and had looked out on the north side without seeing anything wrong. Crossing to the other side they had seen smoke issuing from the express car and had pulled the emergency cord and the train stopped. The express messengers said that they had been hanging on outside for about seven miles.[1392]

As to the condition of the two express messengers the day after the accident a newspaper reporter wrote, 'The two injured men were reported today to still be in much misery from their burns, their hands being in particularly bad condition. The wives of the men arrived in Cedar Rapids late Thursday, having hastened here immediately on hearing of the accident. Mrs Howe was the first to arrive and when she was ushered into the room at St Luke's hospital where the two men were, it is s said that she was unable to determine which of the two was her husband until she identified him by his voice. Both men had their heads swathed in bandages and the fact that they were about the same size and build led to the inability of Mrs Howe to make identification of her husband.' [1393]

After the accident there were wild rumours of the remains of high explosives and pieces of 'infernal machines' (bombs) having been found but an investigation determined that a much more likely cause was the escape of acid from an item of cargo which had not been properly marked.[1394] The cargo also included a quantity of moving picture film, and the celluloid film used at the time was notoriously unstable, caught fire very easily, burnt rapidly giving off toxic gases and was liable to explode in a confined space.

On Saturday 4th December 1915 Private Henry Holland of the Royal Army Medical Corps was travelling from Wigan to Liverpool, England, with a fellow soldier. He had with him the nose cap of an unexploded German shell which his brother had brought

[1390] *The Perry Daily Chief*, Perry, Iowa, United States, 2nd December 1915
[1391] *The Evening Gazette*, Cedar Rapids, Iowa, United States, 3rd December 1915
[1392] *The Oxford Mirror*, Oxford Junction, Iowa, United States, 9th December 1915
[1393] *The Evening Gazette*, Cedar Rapids, Iowa, United States, 3rd December 1915
[1394] Ibid.

back from France as a souvenir. While the other man was examining the shell it went off. Private Holland was wounded in the abdomen and died in hospital from peritonitis, the other man lost his right hand.[1395]

At the inquest the jury returned a verdict of 'Accidental death' and the coroner warned the public about the careless handling of war souvenirs.[1396]

[1395] *The Times*, London, England, 6th and 10th December 1915
[1396] *The Standard*, London, England, 11th December 1915

CHAPTER 16

BOILER EXPLOSIONS

The railways of the early 20th century were powered almost entirely using steam engines, and locomotive boiler explosions were therefore much more common than chemical explosions. Although the explosion of a locomotive boiler could often prove fatal to the engine crew it was rare for train passengers to suffer and probably because of this information regarding boiler explosions in countries outside of Britain and the United States has been hard to obtain.

Boiler explosions could occur in three different ways. Firstly, the boiler barrel itself could fail under normal working steam pressure due to inherent weakness or corrosion. Secondly, the safety valves provided to release steam to avoid an excessive pressure, could fail, either because they had been deliberately interfered with to obtain a higher pressure to get the engine to work harder or because they had been incorrectly assembled. Finally, the engine firebox could collapse, usually due to the water having been allowed to get too low so that the firebox crown sheet overheated and weakened.[1397]

A special Lancashire and Yorkshire Railway coal train from Glasshoughton to Goole was running between Knottingley and Sudforth Lane, England, at a speed of about 20 miles an hour on the evening of Tuesday 11th March 1901 when the boiler of the engine, No. 676, exploded, killing the driver and fireman instantaneously. The boiler of the engine was blown off its frame and ended up in a field 50 yards away. Twenty-four waggons of the train were derailed, 10 being thrown down an embankment.

The train was in charge of Guard Albert Atkinson, with Driver Marshall and Fireman G. H. Coggan, the two men who lost their lives. Guard Atkinson told the subsequent Board of Trade inquiry that leaving Glasshoughton his train had consisted of engine No. 676, 52 loaded coal waggons and his van. As his train was approaching Sudforth Lane he had heard a loud report and his first thoughts were that the engine had fallen into the adjoining canal. His train had come to a sudden stop. He had got down from his van, lit his handlamp and gone forward towards the engine. Finding his path blocked by derailed

1397 *Locomotive Boiler Explosions* by Christian Hewison pg. 11

waggons he had had to go down the bank into the field to get around them. When he had reached the engine he had found that its boiler had gone. He had shouted to the driver and getting no reply had gone on to Sudforth Lane signal box to tell the signalman that the lines were blocked and to wire to Wakefield for the breakdown train. Guard Atkinson added that before leaving Glasshoughton he had asked the driver if he intended to stop at Knottingley for water, but his driver had said that they had enough water to get to Whitley Bridge.

The Board of Trade inquiry into the accident was conducted by Major E. Druitt, assisted by Mr Thomas Carlton of the Consultative Branch of the Board of Trade's Marine Department. Mr H. A. Hoy, chief mechanical engineer of the Lancashire and Yorkshire company, had claimed at the inquest on the driver and fireman that the explosion had resulted from Driver Marshall letting the water in the boiler get too low, and at the Board of Trade inquiry introduced the possibility of an explosive cartridge having got mixed up with the coal.[1398]

Mr T. Hurry-Riches, locomotive superintendent of the Taff Vale Railway, in a long statement, supported the low water theory, but a recent author, C. H. Hewison, has described most of Mr Riches' evidence as nonsense.[1399]

Mr Carlton came to a different conclusion as to the cause of the explosion. He found that a group of stays on the left-hand side of the firebox had been the ones which had first failed. The stays were manufactured from an alloy, called bronze, of Mr Hoy's own invention, consisting of a mixture of 62 pounds of copper, 38 pounds of zinc and six ounces of iron. The alloy was claimed to be stronger than copper and perhaps more elastic and giving but in fact proved to possess very little elasticity, which was why the stays had cracked, while the strength of the stays depended mainly on that of the screw threads in the weaker copper of the firebox.

The stays had leaked into the firebox, a little at first and then more. Boilersmiths had attempted to stop the leaking by hammering the stay heads excessively, probably causing some of the cracks. The circulation of the water had also been reduced by the water spaces only being 2½ wide in their lower parts, causing the stays to overheat so that they lost almost 50% of their tensile strength. This combination of defects had led to the explosion.[1400]

In the conclusion to the report Major Druitt stated that he considered it due to the memory of Driver Marshall and Fireman Coggan to point out that there was no evidence that the explosion had been due to any want of attention or care on their part. It was satisfactory to note that all the 'bronze' stays were being replaced with copper ones. It was unsatisfactory that leaking boilers were not always repaired before the engines were sent back into service. A report should be sent to headquarters in the case of engines such as No. 676 whose boilers were reported leaking day after day, so that the boilers could be carefully examined. Mr Carlton's remarks as to the design of large boilers and the very low pressure tests to which they were subjected should receive the greatest possible attention of the company. Finally Major Druitt was not happy about the fact that the

[1398] Board of Trade report
[1399] *Locomotive Boiler Explosions* by C.H. Hewison, pgs. 110-12
[1400] Ibid, pg.111

engine had been running tender first.[1401]

The boiler of engine No. 16 of the Huntingdon and Broad Top Railroad exploded on Friday 10th May 1901 at Mount Dallas, near Huntingdon, Pennsylvania, United States. Conductor Charles Hollingshead, Engineer A. S. Berkstresser, Fireman Stanton Edwards and Brakeman Clinton Richey were killed instantly, the bodies of the conductor and engineer were thrown 400 yards across the river. The entire firebox and cab of the engine were carried away.[1402] The only trainman to survive the explosion was Flagman William Garner, who had been at the rear of the train. The engine had been hauling the second section of a set of empty coal cars and was standing just behind the first section at the time of the explosion. According to the engineer's time-piece the explosion had occurred at 6.35 a.m. There were rumours after the accident that the engine had been in an unsound condition, but nothing seems to have been substantiated.[1403]

The boiler of the engine of a passenger train exploded as it was about to leave the station at Lezama, Spain, on the Ferrocarril de Bilbao a Lezama on Friday 4th October 1901. The locomotive was destroyed, parts of its boiler being carried eight metres into the air to land in the courtyard of the local administrative headquarters. The naked body of the fireman, Antonio López, was found beneath the remains of the engine and the unnamed driver and Stationmaster Claudio Sanmarti were seriously injured. The signalman was found buried under the debris with the driver and fireman but seems to have escaped serious injury. Many other people were less badly hurt, and the carriages of the train and surrounding buildings were damaged by the blast, the wall of a house being demolished. Help was soon on hand and the worst injured were taken to hospital. A priest administered the Last Rites to the fireman, who was clearly beyond aid.[1404] [1405]

The following day it was reported that only the stationmaster, who was possibly fatally hurt; the driver, who had lost an eye and was suffering from burns, and three other victims remained in hospital.[1406]

A locomotive on the Ferrocarril Nacionales de México (Mexican National Railway) exploded at Sabinas, Mexico, on Monday 14th October 1901. The engineer, the fireman and a little girl, the daughter of the pumper at Aure (presumably a place near Sabinas) lost their lives.[1407]

Wabash Railroad engine No. 710 suffered a boiler explosion two miles north of Boody, Illinois, United States, at about 9.20 p.m. on Thursday 24th October 1901. Fireman Thomas Evers and Head Brakeman Turner Holland lost their lives and Engineer

---

1401 Board of Trade report
1402 *The Locomotive*, June 1901
1403 *Bedford Gazette*, Bedford, Pennsylvania, United States, 17th May 1901
1404 *La Vanguardia*, Barcelona, Spain, 5th October 1901
1405 *The Times*, London, England, 7th October 1901
1406 *La Vanguardia*, Barcelona, Spain, 6th October 1901
1407 *The Locomotive*, November 1901

F. M. Donnelley was injured. Interviewed after the accident, he said that the explosion had happened without any warning as the coal train he had been drawing had just turned the crest of Boody hill and was plugging along at about 25 miles an hour. The fireman had been sick, and the head brakeman had been helping him; he (the engineer) had been firing between Morrisonville and Taylorville. He had noticed that the two steam gauges showed exactly the same pressure. When the explosion came he could sense that he was going somewhere but had no idea what had happened. Finding himself pinned down he had shouted for help but none came. He had then been able to loosen his clothing where it was caught in several places and had finally been able to extricate himself. Several minutes later he saw the rear brakeman (who had gone back to Boody for help) and went to meet him. A newspaper report explained that engines of the class involved in the explosion had their cabs built over the ends of the boilers to give what was known as a three-quarter deck cab, which had a narrow space on each side for the train crew. The fireman and head brakeman had been on one side and the engineer on the other. When the boiler had exploded the locomotive had been lifted from the rails and the 1,850 tons of train behind it had pushed it off the track and overturned it, crushing the fireman and head brakeman beneath. The engineer on the other side had escaped.[1408]

A fourth person injured was a coloured tramp named George Ellis or Anthony, who suffered mainly from shock. He also reportedly had a leg blown off, but this was not counted as an injury as it was a wooden one.[1409]

The boiler of an engine exploded at Deutschlandsberg station, Austria, 'on the Graz Koeflach Railway,' on Sunday 2nd February 1902. The driver, fireman, guard and manager of the goods station were all killed. The explosion was attributed, rightly or wrongly, to the men on the engine having allowed the water in the boiler to get too low.[1410]

On Monday 3rd March 1902 the boiler of Southern Pacific Railroad Mogul locomotive No. 1701 exploded near Callender, California, United States, killing Engineer Richard Duggan and Brakeman J. C. Henshaw and seriously injuring Fireman W. Postlewaite. The engine was wrecked. It was only pulling a caboose. The sound of the explosion was heard over a wide area.

Conductor S. C. Jones said, "Just as the engine and caboose reached Deuman hill, there was a violent trembling of the locomotive, twice repeated, and the explosion followed a second later." An informed source remarked that the conductor's statement implied that some of the engine stays had first given way, precipitating the failure of the boiler.[1411]

Just after 5 a.m. on the morning of Tuesday 27th May 1902 Atlantic Coast Line Railroad engine No. 60 was hauling local freight train No. 91 and was within a hundred yards of the Hull-street level crossing in Manchester, Virginia, United States, when it

---

[1408] *The Daily Review*, Decatur, Illinois, United States, 25th October 1901
[1409] *The Locomotive*, November 1901
[1410] *The Standard*, London, England, 4th February 1902
[1411] *The Locomotive*, April 1902

blew up, causing considerable damage to the town.

'Windows in adjoining houses were shattered, the ends of two frame structures, which form a part of the planning mills opposite the road at the point of the accident, were ripped away at the point of the accident, and solid small stones are sticking in the woodwork of houses in the vicinity. Two freight cars were demolished, and the track bed distorted out of all semblance to itself. Pieces of iron were thrown for hundreds of yards, one of these tearing a hole in a fence 200 yards away large enough for a man to pass through. Other small pieces were sent flying in all directions, and the marvel of it all is that fewer lives were lost.'

An eyewitness stated, "I was standing there at the tree waiting for the train to pass when the explosion came. I was looking at the engine when it happened. First there came, apparently without any warning, a most terrific explosion, the loudest I have ever heard, sounding more like thunder than anything else, and immediately I saw a lot of things and they might have been iron and they might have been men, fly into the air. The engine stood straight up, just like a man would, and then sank down, falling off the track, where it is now.

"The freight car immediately behind the engine was thrown over to one side in front of the engine, and the car immediately behind this was torn from its trucks and stood up for a moment, then falling back. The iron pieces fell in a shower within a radius of some 200 yards, and the whole ends of two buildings were torn off. When I had seen this much I decided I had all I wanted and got back behind the tree."

Engineer Robert Gwathney and Trainman James Winston were killed outright, and Fireman John Taylor died from his injuries the same evening. Trainmen Stephen Basser and William Savage were seriously injured.

The initial newspaper report on the accident blamed it on the age and condition of the locomotive, stating, 'So far as experts have been able to learn, the explosion was due entirely to the age and weakness of the engine and its boiler. It was stated authoritatively yesterday that the engine was on its last run, its danger having been realized by the superintendent of motive power of the road and was to have been retired in a day or so. Now it lies a jumbled mass of iron, half overturned in a ditch near the track at the point of the explosion, and those victims whose lives are not claimed are in the Retreat for the Sick.

'The engine which blew up has been in actual service for many years. Bolts and rivets were rusted through the accumulation of the water and the weakness of the locomotive was shown by the fact that in the explosion the boiler gave way in no particular part, but gave way as a whole, there being several vents where the steam under normal pressure ripped it to pieces. An expert engineer, who looked over the wreck yesterday, declared that the engine should have been in the junk pile instead of on the road.' [1412]

At the inquest, which was quickly held, different experts gave differing opinions. This 'Mr George Paul, machinist and engineer, testified on oath that he had seen the wrecked engine and that he did not consider from what he saw that the boiler was safe. "I am positive," he said, "that there were twenty-five stay bolts that were broken before the explosion."'

[1412] *The Times*, Richmond, Virginia, United States, 28th May 1902

Similarly, 'Mr W. H. Dickerson, an engineer of twenty-five or thirty years' experience, said he found twenty-seven broken bolts, which he believed had been broken before the explosion. One of the hangers of the crown street was broken, and it was an old break. The broken bolts were in a space of eight or nine inches, and he said that he thought that the ones which were broken already gave way first.'

On the other hand, 'Mr R.H. Simmonds, a machinist, said he had examined the engine and found only four only four old broken bolts.'

The inquest jury returned a verdict that they had found from the evidence that the explosion had been due to a defective boiler.[1413]

The Liberty Bell is an iconic symbol of American independence. Commissioned in 1752, it became famous after the publication of a short story claiming (falsely) that it had been rung on 4th July 1776. It is kept at Philadelphia, Pennsylvania, but between 1885 and 1915 it was sent around the country to various expositions and patriotic gatherings.[1414] At the beginning of June 1902, the bell was in Charleston, South Carolina, and a party of prominent dignitaries was on its way from the Quaker City to Charleston to collect it. They got as far as Aquia, Virginia, on the Richmond Fredericksburg and Potomac Railroad when, at 10.15 p.m. on Friday 6th June 1902, the engine of their train suffered a boiler explosion and was partially derailed, the dining car and two Pullman sleeping cars leaving the track. The other two sleeping cars which formed the remainder of the special were not derailed. None of the passengers was injured but Engineer John Pumphrey and Fireman C. G. Saunders were badly hurt. One hundred yards of track was torn up. The passengers were taken back to Washington D.C. by the engine of a following freight train in the remaining two sleeping cars, to continue their journey to Charleston by an alternative route.[1415]

Baltimore and Ohio Railroad locomotive No. 1724 had just come out of the shops near the end of 1902 and was supposed to be in first-class condition but on only its second trip since its overhaul its boiler exploded on Sunday 2nd November 1902 between Halethorpe and St Denis, Maryland, United States, just a few miles out of Baltimore. Engineer E. W. Biggs, Fireman O. W. Hunt and Head Brakeman C. O. Stalling were all killed. The engine crown sheet had failed. The boiler was blown clear of the running gear and the wheels of the engine did not leave the track.[1416]

By an unlucky chance on Tuesday 20th January 1903 the explosion of the boiler engine No. 8069 hauling an eastbound freight train of the Pittsburgh Cincinnati Chicago and St Louis (Panhandle) Railroad near Bowerstown, Ohio, United States, occurred just as it was passing a westbound freight train on the other track. The boiler was thrown into the air and came down onto the second car of the westbound train. Twenty-six cars plunged down an embankment. Fireman J. G. McCurdy and Brakeman Frank Clemens of

---

[1413] Ibid, 29th May 1902
[1414] Wikipedia – Liberty Bell
[1415] *The Times*, Richmond, Virginia, United States, 7th June 1902
[1416] *The Locomotive*, December 1902

the eastbound train and Engineer J. D. Ward and Brakeman Petrie of the westbound train were all killed. Engineer Daniel O'Connell of the eastbound train was fatally injured. Three hundred yards of both tracks were torn up. The crown-sheet of the engine of the eastbound train had failed.[1417]

Engineer Robert McAuliffe was so badly hurt in a boiler explosion on the Grand Trunk Railway at Stirling, Ontario, Canada, on Sunday 5th July 1903, that he died from his injuries later in the day. Fireman Porter was also injured, but less seriously.[1418]

On Saturday 22nd August 1903 an excursion train was run by the Northern Pacific Railroad to carry a party of 500 members of The Elks (a fraternal organisation whose full name is The Benevolent and Protective Order of Elks) from Portland, Oregon, to a clam bake (a social gathering at which seafood is cooked) at Olympia, Washington. At 10.35 a.m. the train met with an accident a mile and a half west of Chehalis, Washington, United States. It was running at about 40 miles an hour when its passengers heard a sharp report and realised that it had become derailed. In another 200 feet the train came to a halt. All the coaches except the last one were derailed and the first three went down a 30-foot embankment.

The engine involved was No. 237, reportedly one of the oldest passenger engines on the line. It was in charge of Engineer Robert Green and Fireman Leo Doskey, both of whom were injured. One of the train passengers, Charles Farmeline, a confectionary dealer and Elk, was killed outright, as was a man named Sam Condon, whose name was first given as Charles Gales, and who was described as a tramp, though at the time of the accident he was working his way by passing coal to the fireman. In addition to the engine crew Conductor J. A. Walker, Head Brakeman Borse and about 40 passengers were injured, one of whom, Mrs C. B. Brown, was very seriously hurt and died later.

A passenger described the accident. 'Thurston Daniels, son of ex-Lieutenant Governor Thurston Daniels of Vancouver, was in the first car and among the seriously hurt. He said: "It all happened in a minute's time. We were thrown down among the seats, and we heard a crashing of timbers after we came to a stop. The car turned over and pitched down the embankment on the west side of the track, and the last thing of which I was conscious before they pulled me out and placed me upon the grass on the bank was the screaming and groaning of the people around me."'

An injured passenger, Mr A. L. Jenkins, brought news of the accident to Chehalis at about 11 a.m. A special train consisting of an engine and flat cars was despatched from Chehalis with Dr Kennicott of that place. There were plenty of doctors on hand for Doctors Chalmers, Cornelius, Hickey and Littlefield of Portland had been passengers on the train and Doctors Coleman and Dow of Chehalis, Dumon of Centralia, Mowdell, Redpath and Swan of Olympia and Allen of Tacoma all lent assistance. The injured were loaded onto the flat cars and quickly taken to Chehalis, where they were accommodated in Dr Kennicott's new hospital building.

There was no doubt as to what had happened. 'The accident was without doubt caused

---

1417 Ibid, February 1903

1418 Ibid, August 1903

by the explosion of the boiler. It was witnessed by J. C. Bower, a rancher living about two hundred yards from the track. He states that there was a tremendous puff of smoke and steam, accompanied by a deafening explosion, and then the engine toppled over. As it lay on its side near the track, the full effect of the explosion on the interior of the machine was shown. The crown sheet was torn loose from both ends and from the forward end and hung over the front of the fire box. From appearances, when it was blown loose it carried everything under it down onto the track and derailed the engine. Not finding a vent to admit of its escaping quick enough, the steam shot out through the boiler tubes and varied away the boiler-head casing, pieces of it lying some distance ahead of the engine.' [1419]

The explosion of the boiler of the engine of a goods train derailed it near Aerseele, Belgium, on or about Monday 28th December 1903. The driver, fireman and conductor of the train were critically injured.[1420]

A spectacular boiler explosion occurred at Paris St Lazare station on the Chemins de Fer de l'Ouest on Monday 4th July 1904. Engine No. 626 had arrived with a train from Nantes at 8 a.m. and was to work another train out at 3 p.m. It was therefore standing in a siding under steam when its boiler suddenly exploded at 10.45 a.m. Pieces of the boiler weighing over a tonne were blown into the air and considerable damage was done to the station and surrounding streets, but despite the violence of the explosion no one was killed and there was only one case of serious injury, the arms and legs of Driver Lecomte, who had been stood near his engine, being peppered with ballast. Fifteen people were slightly hurt, amongst them the passengers on an incoming train and the residents of an adjoining street. The cause of the explosion was not stated.[1421]

The boiler of the locomotive of a Grand Trunk Railway westbound freight train exploded about a mile west of Vaudreuil, Quebec, Canada, on Saturday 17th December 1904. Engineer Thorpe and Brakeman Benoit were badly scalded.[1422]

In a repetition of the January 1903 accident at Bowerstown on Saturday 4th February 1905 two trains were involved in a boiler explosion on the New York Central Railroad at Whitesboro, New York, United States. The boiler of the engine of westbound passenger train No. 23 from Boston and New York exploded just as eastbound train No. 30, the Buffalo Express, was passing on the other track. The force of the blast derailed all 13 carriages of the eastbound train 'as if they were a string of toy coaches'. The forward coaches were hurled for a distance of from 10 to 40 feet into an adjoining field while the rear Pullman sleeping cars toppled into a ditch. None of the vehicles of the westbound train left the track. Engineer John Allen and Fireman John Brennan were killed, the former living a few minutes after the accident. About 25 passengers on the eastbound

---

[1419] *The Chehalis Bee-Nugget*, Chehalis, Washington, United States, 28th August 1903
[1420] *Journal de Genève*, Geneva, Switzerland, 29th December 1903
[1421] *La Presse*, Paris, France, 5th July 1904
[1422] *The Locomotive*, January 1905

train were more or less hurt, most of them from the sleeping car which had been directly opposite the locomotive of the westbound train when it blew up.

The accident occurred in the early hours of the morning and many local residents were awoken by the roar. There was thick snow on the ground and the locals came to the assistance of the injured on sleighs, by which several of the injured were taken to the Whitesboro Sanatorium. It was also bitterly cold, with a temperature 18 degrees below zero Fahrenheit, making the rescue operation painful and difficult. It took an hour to release the passengers from the sleeping car opposite the wrecked engine, the rescuers working with benumbed hands and feet and some being only partly clothed as they had hurried from their homes. None of the passengers was reportedly dangerously hurt.

Among those riding on the Buffalo Express had been Supreme Court Justice Hooker, who was badly bruised and suffered apparent internal injuries, and Fred J. Titus, professional cyclist and former husband of comic opera star Edna May. He was not much hurt but lost a diamond pin in the wreckage.

A relief train was sent from Utica, three miles to the east, and a special train forwarded the passengers of the westbound train to Buffalo.[1423] [1424] [1425]

In contrast to the Whitesboro accident most of the victims of a boiler explosion on the Pennsylvania Railroad the following month were track labourers who chanced to be working opposite the engine of a westbound freight train when it exploded at Bolivar, Pennsylvania, United States, on the morning of Wednesday 15th March 1905. Engineer John Riblett had let the water in his boiler get too low; he was blown high in the air and over an embankment and was killed instantly. The unnamed fireman appears not to have been injured, possibly he was up on the tender out of harm's way. Less fortunate were the labourers, a number of whom were severely or fatally scalded. The worst hurt, Washington Wambaugh, died in hospital the same evening; Emanuel Hull succumbed the next morning and another man, Frank Gil, died on the Saturday evening. Brakeman Theodore Bennett and six other members of the track gang were more or less hurt. The boiler was hurled into the air and fell 90 feet in front of the wrecked locomotive. A train standing on the opposite track shielded people standing at the station from injury.[1426]

The explosion of an engine boiler on the Ferrocarril Central Mexicano (Mexican Central Railway) on Wednesday 29th March 1905 four miles south of Ciudad Juárez, Mexico, led to the instant death of Fireman Samuel Saunders, whose body was blown 200 feet, and more or less serious injury to Engineer Martin Campbell and Brakeman Pedro Santano. The crown sheet of engine No. 851 had failed.[1427] [1428]

On the morning of Monday 4th December 1905 a passenger train left Maitland at 5.38

[1423] *Brainerd Daily Despatch*, Brainerd, Minnesota, United States, 4th February 1905
[1424] *The Philadelphia Inquirer*, Philadelphia, Pennsylvania, United States, 5th February 1905
[1425] *The Sun*, Baltimore, Maryland, United States, 5th February 1905
[1426] *The Indiana Progress*, Indiana, Pennsylvania, United States, 22nd March 1905
[1427] *The Locomotive*, April 1905
[1428] *Mitchell Daily Republican*, Mitchell, South Dakota, 31st March 1905

a.m. for Sydney, in charge of Guard Charles Janssen, with Driver James Pead and Fireman William Pearce. When the train was between Metford and Thornton, New South Wales, Australia, the boiler of its engine, No. 163, failed, and there was a great rush of steam into the engine cab on the driver's side. Driver Pead was severely scalded all over his body and Fireman Pearce from the hips down, but he managed somehow despite his injuries to make his way along the top of the locomotive to its front where he was able to open the valve to apply the Westinghouse brakes and stop the train, which came to a halt near the Thornton distant signal.

Having discovered what had happened Guard Janssen put the injured driver and fireman into the brake van and extinguished the engine fire with water and dirt. The Singleton goods train chanced to be at Thornton and its engine took the passenger train on to Newcastle, 15 miles further on, where the driver and fireman were admitted to hospital.[1429]

Driver Pead died of his injuries the following day.[1430]

At the adjourned inquest on Driver Pead on 20th December Richard Bryant, steam shed inspector at Hamilton, testified that he had examined the engine after the accident and had found that a plate on the outer shell of the firebox on the right-hand side had fractured over a distance of about 3ft 9½. The fracture would allow steam and water to escape into the engine cab. The flaw was an internal one which could not have been detected before the accident. The safety valves had been in good order. His opinion was that the flaw had already been present when the boiler was first constructed.

Leading Boilermaker James Rigby of Hamilton corroborated Mr Bryant's evidence. A number of other witnesses were examined and gave similar evidence, all agreeing that there had been a flaw, probably dating from the construction of the boiler and which could not have been detected, that the safety valves had been working properly and that there was no evidence of low water. The inquest was then adjourned to hear the evidence of Fireman Pearce, still in hospital recovering from his injuries.[1431]

The resumed inquest heard evidence from Mr Samuel Barraclough, lecturer in mechanical engineering, who agreed that the presence of a long-standing flaw in the boiler was the most likely explanation for the accident, but added that this was merely a surmise, and the whole of the upper plate would have to be cut out to provide a definite answer. He thought it most unlikely that such a flaw could be detected during a routine workshop examination, hidden as the crack was along the joint. The inquest was again adjourned.[1432]

Fireman Pearce gave evidence on the final day of the inquest, at the end of January 1906, as did Guard Janssen. They were followed by Mr Barraclough, who testified that he had examined the boiler at the university and had found further cracks, hidden by the lap, when it was taken to pieces. He did not think that a flaw had been present when the boiler was originally constructed by Messrs Beyer Peacock. The problem could not have

---

[1429] *Newcastle Morning Herald and Miner's Advocate*, Newcastle, New South Wales, Australia, 5th December 1905

[1430] *Singleton Argus*, Singleton, New South Wales, Australia, 7th December 1905

[1431] *Newcastle Morning Herald and Miner's Advocate*, Newcastle, New South Wales, Australia, 21st December 1905

[1432] Ibid, 6th January 1906

been discovered at a workshop examination.

Mr Barraclough was followed by other witnesses and there was a discussion of the superiority of butt-joints as opposed to lap-joints, the boiler of engine No. 163 having had the latter.

At the conclusion of the inquest the Coroner, Mr Hibble, returned a verdict of 'Accidental death'. He recommended that in future all boilers should be constructed with butt-joints and that special examinations should be made in the workshops of all boilers with lap-joints. He also stated that he was not convinced that the 25 per cent margin used in boiler tests, which Mr Barraclough had stated was ample, was actually sufficient. Finally he had warm words for the bravery of Fireman Pearce in making his way along the boiler of a moving engine despite being badly scalded.[1433]

The Lancashire and Yorkshire Railway was fated to suffer another boiler explosion on Monday 9th April 1906, though this time with less serious consequences than the Knottingley one five years earlier. The 9.30 p.m. passenger train from Stockport to Colne was passing through The Oaks station, England, when the accident occurred.

Guard William Lowthian, who was in charge of the train, was the first witness to give evidence at the Board of Trade inquiry. He stated that his train had just passed The Oaks station when he heard someone shouting from the lineside to put on the brakes. He had done so, thinking that a passenger must have fallen from the train, but then found that it had been the fireman, who had told him that the engine fire had blown out all at once, practically throwing him and the driver off the engine. None of the passengers was hurt.

Driver William Lambert said that it had been dark at the time of the accident (10.21 p.m.) but the lights in the cab were good, the gauge glass was clean, and he could see that there was water in the boiler. His engine had been working fine. The first sign of trouble was as they were passing The Oaks station, when he heard a hissing noise in the firebox. Three or four hundred yards further on what he described as a flush of hot water drove him from the cab and he retreated to the side of the locomotive, clinging onto the hand rail. There was no time to close the regulator.

Fireman G. Engledow agreed that the engine had been working perfectly normally up until immediately before the accident. The first intimation of a problem had been at the home signals approaching The Oaks. Thinking that it was the injector blowing back he had shut off, but the hissing noise had continued, then there was a slight report and he and the driver had been driven from the footplate, though unlike his driver he did not get any water on him. He described himself as 'being a bit black and blue', which considering that he had apparently jumped from the footplate while the train was running at between 20 and 30 miles an hour was hardly surprising.

The shed foreman at Colne, where the engine, No. 869, was washed out on 4th April, stated that he had looked inside the boiler after it had been serviced and had seen nothing wrong. It had been clean, and there had been no sign of bulging or overheating.

Mr James Yates, the boiler inspector, whose sole occupation, Mr Carlton pointed out, was to travel around the company's lines examining boilers, testified that as he had seen

1433 Ibid, 31st January 1906

no signs of bulging or overheating he had not bothered to look at the firebox top at all, being satisfied that it was perfectly clean.

Like the 1901 Knottingley accident the Board of Trade inquiry was conducted by Lt-Col E. Druitt (as he had now become) with the help of Mr Thomas Carlton of the Marine Department, or rather, as in 1901, Mr Carlton, as the expert on boilers, did the investigation and his remarks were approved by Lt-Col Druitt. The accident to the boiler, Mr Carlton determined, had involved the bulging of the crown of the firebox due to the failure of 60 stays, 57 of which had entirely broken away, leaving a hole of upwards of 50 square inches through which the steam and heated water of the boiler had escaped. The overheating was due, in his opinion, to an accumulation of loose scale on the firebox crown, which had been inadvertently or carelessly left when the boiler had been last cleaned and washed out at Colne five days before the explosion.

Although, Mr Carlton said, the driver and fireman had maintained that there was plenty of water in the boiler, explosions due to low water were frequent, so he had carefully investigated this possibility. Having done so he had found no evidence that there had been insufficient water, but a considerable quantity of scale, weighing about 72 lbs.

In his conclusion to the report which Mr Carlton had made Lt-Col Druitt simply commented that the explosion had been due to the fact that the periodical washing out of boilers at Colne had not been attended to with sufficient care.[1434] C. H. Hewison has commented that either the Colne shed foreman was not doing his job properly or else the staff were under too much pressure to get too much done in too short a time.[1435]

The cause of a boiler explosion on the Pennsylvania Railroad at 5.30 a.m. on Friday 27th April 1906 was unclear. The accident involved the engine of an eastbound freight at Ducklow Tower, Steelton, Pennsylvania, United States. Engineer J. F. Good was killed and Fireman C. H. Lefever and Brakeman J. J. Wallower were reportedly fatally injured. The body of the engineer was found half submerged in the Susquehanna River near the tower. The injured men were taken to Harrisburg Hospital.

The train had left Harrisburg only a few minutes before the explosion and it was said that it was impossible for the water to have got too low in that short distance. Instead the theory was put forward that a stick of dynamite which had been accidentally dropped onto the track had been struck by the engine. No evidence was put forward, however, to support this hypothesis.[1436]

Almost exactly a month after the Steelton accident, on Saturday 26th May 1906, a locomotive of the Continental Coal Company exploded at Tulsa, Indian Territory, United States. Again the explosion was blamed on dynamite, this time not a stick which had accidentally fallen onto the track but dynamite which had been deliberately placed there to wreck the train by the owners of property adjoining the line who were opposed to its operation. The locomotive was reportedly blown into fragments, which given that the

---

[1434] Board of Trade report

[1435] *Locomotive Boiler Explosions* by Christian Hewison pg. 113

[1436] *The Philadelphia Inquirer*, Philadelphia Pennsylvania, United States, 28th April 1906 engine crew were not hurt must have been a considerable exaggeration.

Convinced that this had been an act of sabotage, the authorities had gone to the length of arresting a man named F. W. Cassner, cashier of the Tulsa First National Bank, who owned property along the line, for participation in the affair, and other arrests were expected to follow. There was said to be great excitement in the town, more trouble was expected, and people were arming themselves.[1437]

The Locomotive magazine, however, was very doubtful of the sabotage theory and thought it much more likely that it was an ordinary boiler explosion. It wrote, 'We have known of many cases in which boiler explosions have been attributed without adequate reason to high explosive like dynamite, by persons who, being unfamiliar with the fearful destructiveness of steam, felt impelled to adopt the dynamite hypothesis to account for the observed violence of the explosion.' [1438]

At 9.30 p.m. on Sunday 11th November 1906 the southbound *Sunset Limited* express of the Southern Pacific Railroad was passing through the station at Sargents, California, United States, when the boiler of its engine, No. 3027, exploded. The bodies of Engineer Sam Gillespie and Superintendent Joseph Goodfellow were picked up a short distance from the wrecked locomotive. Fireman James Bindon was thrown 60 feet and died a few hours later. None of the train passengers was badly hurt. The explosion demolished the station buildings and the boiler, weighing it was said 100 tons, was thrown forward 300 feet along the track where it was then struck by the boilerless remains of the locomotive, which pushed it along another 20 feet through the gravel and ties before the train came to a stop. The tender, mail car, chair car and two tourist sleepers were derailed and some of the passengers in them were slightly injured. The rear six cars of the train remained on the track undamaged and none of their occupants was hurt.[1439]

The Southern Pacific Railroad managed to have two boiler explosions in California, United States, within a few hours on 9th and 10th April 1907. The first explosion occurred to local freight train No. 242, in charge of Conductor Harry Gates. The train was climbing into the Tehachapi Mountains and close to Tunnel No. 12 when the explosion occurred at 11.10 p.m. on Tuesday 9th April. There were two helper engines assisting the train and it was the rear helper engine, No. 2618, which exploded, killing Engineer Oliver Machion and Fireman W. Vaughan. Their bodies were found, some time after the accident, in a small canyon 160 feet from the engine. Brakeman H. B. Jones, riding in the cupola of the caboose next to the helper engine, was thrown through the air and was unconscious for several hours. Conductor Gates, sitting at his desk in the caboose, was thrown to the floor but not hurt. He said that had the explosion happened inside the tunnel the structure would have been demolished, the enginemen at the front of the train killed and the line blocked for a long time. It was also a wonder that the oil which formed the load of the train had not been set alight by the explosion. Drs Kellog and West and Nurse Day went to the scene of the accident on a special train and attended to the injured brakeman. At Kern station Brakeman Jones was met by his anxious wife and went home

1437 *Sioux County Herald*, Orange City, Iowa, United States, 30th May 1906
1438 *The Locomotive*, June 1906
1439 *The Daily Californian*, Bakersfield, California, United States, 12th November 1906

some hours later.

Four hours after the tunnel No. 12 accident, just after 3 a.m. on Wednesday 10th April, engine No. 2703 was being taken from the roundhouse at Mojave to be attached to a train which was to leave for Kern at 4 a.m. Acting Hostler H. D. Earnest and Helper Shea were close to the locomotive when its crown-sheet dropped, the boiler exploded, and the two men were killed instantly. Parts of the engine were blown 150 feet and fell blocking the main line. No one else was hurt.[1440]

At the inquests upon the two accidents the juries brought in verdicts of 'Accidental death', without attaching blame to anyone. In the case of engine No. 2618 Master Mechanic J. Schelaberger testified that the explosion had been due to low water in the boiler exposing the crown-sheet. Chief Boiler Maker J. Blaine testified that the engine had been in perfectly good condition when it had left the shops on Tuesday after a thorough test. In the case of engine No. 2703 it was stated that it had arrived at Los Angeles a few hours earlier and had not been inspected.[1441]

At 10.30 p.m. on the evening of Tuesday 30th July 1907 Illinois Central Railroad northbound extra freight No. 804, a banana special bound for Chicago, was crossing southbound extra freight train No. 696 at Sitka, Tennessee, United States, when the boiler of engine No. 804 exploded as the engines were alongside one another. Engineer Sam Maloney, Fireman Robert Henderson and Head Brakeman William Guy, all of the northbound train, were killed outright, as were five tramps, names unknown, who had been riding on the southbound train. On the southbound train Engineer A. Beesley, Fireman J. J. Dennis, Conductor Wright, Head Brakeman Ben Boaz and Flagman Emmerson were all seriously injured and scalded. The conductor and flagman were blown about 50 feet. At least 16 cars were destroyed, and the track was torn up for about a hundred yards. A car load of iron sewer pipes, each weighing 1,500 lbs, was scattered over 500 yards. Fire broke out immediately after the accident and with no water available to extinguish it had to be left to burn itself out. Special trains were run to the scene from Milan and Jackson with doctors and nurses. While the line was blocked trains were diverted over the Louisville and Nashville and Mobile and Ohio tracks.[1442] [1443] [1444] [1445]

The Hull and Barnsley Railway did not suffer any major disasters during its relatively brief existence, but it did manage to have one or two accidents, including a fatal boiler explosion on Wednesday 25th September 1907. The accident occurred at about 3 a.m. in the sidings at Wath, England, the terminus of a single-line branch from Wrangbrook Junction. Tank engine No. 109 was in charge of Driver Brook with Fireman Alfred Heffer. They had come on duty at Cudworth at 2.50 p.m. the previous day and had made various short trips between Wath and Wrangbrook Junction. The branch usually closed at 1 a.m. but was open later than usual due to heavy coal traffic. They were about to make

[1440] Ibid, 10th April 1907
[1441] Ibid. 12th April 1907
[1442] *Hamilton Democrat*, Hamilton, Ohio, United States, 31st July 1907
[1443] *Centralia Evening Sentinel*, Centralia, Illinois, United States, 1st August 1907
[1444] *The Paducah Evening Sun*, Paducah., Kentucky, United States, 1st August 1907
[1445] *The Bolivar County Democrat*, Bolivar, Mississippi, United States, 3rd August 1907

the final trip of the day with a train to Wrangbrook, where the train would be left, and they would return to Cudworth with the light engine and book off duty. Fireman Heffer had gone to the signal box for the single line staff when the engine blew up, he was unhurt but Driver Brook, who had been on the footplate, was fatally injured and died in a few hours.

The Board of Trade investigation was as usual conducted by Mr Thomas Carlton, his report being certified by Lt-Col P.G. von Donop. Mr Carlton found that the boiler had exploded due to the failure of a group of stays, about 30 in number. He wrote in his report, 'The clenched heads of these stays were completely wasted away, and this part of the fire-box was in consequence dependent for support on the screwed parts of the stays in the stay holes, but owing to the repeated hammering and caulking of the ends to make them steam tight the ends had been seriously damaged and the stays had become too short, the ends being below the fire surface of the plate. In this condition they were unable to support the plate, and the latter was forced over the ends in the form of a bulge. Once the bulge started the surrounding part of the plate appears to have slipped easily and rapidly over the adjacent stays, many of which also were without proper heads, the scalding steam and water escaping through the stay holes into the firebox and thence to the atmosphere. When the bulge had extended the full length of the side of the firebox to the back plate and tube plate these crumpled in and the bulged side appears to have begun to tear away at the two upper corners simultaneously, and after completely tearing along the top, it was driven downwards, hinging along a line level with the top edge of the upper row of rivets attaching the bottom part of the side to the foundation ring, and it appears to have held on at this part until the plate itself had burnt through an angle exceeding 180 degrees from its original position. The plate was then blown to the left-hand side of the boiler, its flight in that direction being due to the bottom edge remaining attached to the foundation ring till the last. The failure of the side stays described above occurred with extreme rapidity, the whole operation lasting probably less than a second.'

Mr Carlton went on to detail the examinations and repairs made to the engine in the ten months prior to the accident, of which there were a number. It appeared that by no means all the problems reported had been corrected, instead Mr Edward Laws, the locomotive foreman at Cudworth, had reduced the boiler pressure slightly, from 170 lbs to 160 lbs. Boilersmith Ben Tillotson had also repeatedly informed him that the engine needed repairs, but he had decided to keep it working until another engine, No. 106, which was then itself being overhauled, was available to replace it.

Concluding the Board of Trade report, Lt-Col von Donop stated that the accident had been mainly due to an error of judgement on the part of Foreman Laws in keeping the engine in service too long instead of withdrawing it for repairs.[1446]

The boiler of a goods train exploded at Herlisheim, Alsace, Germany, on Wednesday 28th October 1908, obstructing the opposite running line in front of Basle – Köln express train D 161. Fortunately the driver of the express noticed the explosion taking place, made an emergency brake application and was able to reduce the speed of his train to the point where only a minor collision took place with the wreckage.[1447]

[1446] Board of Trade report

[1447] *Eisenbahn-Karastrophen in Deutschland* by Hans Joachim Ritzau pg. 60 (in German)

Lt-Col E. Druitt was back in the seat as the Board of Trade inspector in charge of the investigation of a boiler explosion at Cardiff Docks, on the Rhymney Railway in South Wales, on Wednesday 21st April 1909, Mr Thomas Carlton again providing the technical expertise.

Engine No. 97 had been at work on 17th April and when it was returned to the shed Driver Edward Davis noted in the report book that several repairs were necessary. The repairs were made, Fitter Daniel Rees noting that they had been completed. The boiler was washed out on the day before the accident and inspected by the foreman boilermaker. The engine fire was then lit, and Lighter-up Robert Brown came on duty at 6.30 p.m. on the 20th April, when no steam pressure was registering. At about 8 p.m. the pressure was 40 lbs and just before 9 p.m. he saw it was up to 60 lbs. Another employee named Ralph Brown noted that the pressure was about 110 lbs at 11 or 11.15.

Driver David Lewis took charge at 11.30 p.m. He found that the steam gauge was registering 200 lbs per square inch, the maximum it would register, and 50 lbs above the pressure at which the safety valves should open, but they had not done so. He tried to move the safety valve lever without effect. Deciding that the pressure gauge must be faulty, he mentioned it to Fitter John Stuart so that it might be replaced by another one. Accompanied by the driver the fitter went to Shed Foreman John Williams and explained what was wrong and asked what he should do. Without speaking to the driver or seeing the engine the foreman told the fitter that it was all right and the engine could go out.

Driver Lewis duly took the engine out at 11.47, picked up three waggons and a brake van and proceeded to Gaol Lane signal box, a mile and a quarter away and up a stiff gradient. The water in the glass fell somewhat so the driver opened the injector as usual to force water into the boiler, but neither injector would work. The waggons and van were left at Gaol Lane and the engine was run back down the bank to the shed. Seeing the engine returning, and realising that something must be wrong, Fireman Edwin Chambers, who often worked with Driver Lewis, left another engine and went across to find what was wrong. He saw that the steam gauge was at its maximum, that there were two inches of water in the gauge glasses and no steam was blowing off.

The engine was shunted to get it over a pit where the fire could be dropped, and efforts were made to get the injectors to work but they would not. Shunter Tom Parry got onto the footplate. One of the water gauge glasses broke, shortly followed by the other. Shunter Parry was helping Fireman H. Hawkes drop the fire when the boiler exploded. Fireman Hawkes, Shunter Parry and Cleaner W. Murphy, who had been 10 yards away from the engine, were all killed, Driver Lewis was badly hurt and Lighter-up Brown and Fire-dropper J. Williams less seriously so.

Mr Carlton determined that the explosion had arisen from a cause which we have not previously encountered. He noted that it was apparent that the boiler had burst at an exceedingly high pressure, sufficient to shear, break or tear all of the all of the upper pins in the sling stays connecting the firebox crown girders to the boiler shell. The broken stay pins were in such a position as that they could not have been affected by any overheating, and there were no signs of overheating anyway.

The evidence was, Mr Carlton continued, very clear as to what had happened. Observations indicated that the steam pressure as shown on the gauge had gradually risen until it was up to 200 lbs, the highest point on the dial. The safety valves were designed to open at 145 lbs but had not blown off at all. Thus either the steam gauge was faulty, or the safety valves were inoperative. The steam gauge had not been touched but the safety valves had had to be removed by a fitter to carry out one of the repairs requested on 17th April and had not been seen to work since. The injectors had failed to work at Gaol Lane because they were not designed to operate at such a high boiler pressure. A simple mistake by Fitter Rees in replacing eyepiece washers in the wrong place had led to the safety valves being unable to open at all, whatever the pressure. When Rees had finished his task he had told Foreman Mechanic James Baxter that he was done, but the latter had not checked the safety valves himself, nor had he told Night Foreman John Williams that the safety valves had been dismantled, nor had he left instructions that the operation of the safety valves should be checked following their reassembly.

Rhymney Locomotive Superintendent C. T. Hurry Riches suggested that an equally probable cause of the explosion was that the boilers had been short of water, that the injectors had suddenly been got to work and the inrush of water onto an overheated firebox had caused a sudden generation of steam. Mr Carlton rejected this theory, which he described as a very old one, out of hand, pointing out that not one out of 1,800 boiler explosions of all kinds reported under the Boiler Explosions Act had been from this cause. The coroner's jury, however, chose it over the wrongly assembled safety valve scenario, and passed a verdict that, "The deceased men were killed by the explosion of the boiler of an engine belonging to the Rhymney Railway Company, and that the cause of the explosion was that the man left on the engine after the driver had gone from it, got the injector to work, and that the flow of cold water onto the already heated plates suddenly generated a large amount of steam causing the explosion. They considered it purely accidental and that no one was culpably to blame."

As might be expected Lt-Col Druitt was not impressed by the jury's verdict and stated that he disagreed with it. He was sympathetic to the actions of Night Foreman Williams, a very old servant of the company who had died not long after the accident, saying that Start's report had been very vague and he (Williams) had not been aware that the safety valves had been dismantled and reassembled.

As regards Driver Lewis it had never occurred to him that the safety valves were inoperative. Having mentioned the subject to Fitter Stuart and been told that it was all right to take the engine out he had considered it his duty to do so. Driver Lewis had also been severely criticised at the inquest for not dropping the fire at once at Gaol Lane when he found the injectors would not but the inspecting officer considered that Driver Lewis had acted to the best of his judgement as there had been plenty of water in the boiler and the fire could be much more easily dropped at the engine pit only just over a mile distant.[1448]

Eleven employees of the Ferrocarril Nacional de México (Mexican National Railway) lost their lives and two more were probably fatally injured in a boiler explosion at Monterey, Mexico, early on Saturday 16th October 1910. The locomotive involved, No. 2758, was said to have been an antiquated machine which had been in the roundhouse at Monterey for some time undergoing repairs and was considered unfit for service. On the

[1448] Board of Trade report

morning of the accident it had been fired up for three quarters of an hour, but its steam gauge was still only registering 25 lbs pressure when it should have been registering almost 200. The engine had been run 200 yards out of the roundhouse when Engineer G. S. Monaghan noted that there must be something wrong with the steam gauge, as clouds of steam were escaping from the safety valve and the cylinders of the locomotive were responding to the throttle. With the gauge malfunctioning, though, he dared not proceed to pick up his train to take it to Victoria, Texas. At this point an unknown person sent a man to the deck of the deck of the boiler to inspect the safety valve, and this man made the fatal mistake of screwing down the safety valve. Predictably the engine boiler exploded and the cab, a portion of the water tank and the firebox were blown 100 feet, together with the bodies of the engineer and men who had come from the roundhouse to assist in the repair.

Those who lost their lives in the explosion were Engineer Monaghan, Fireman Juan Perez, Night Foreman Manuel Fernandez, Helpers Agapita Martinez, Porfirio Padron and Francisco Palacois, Machinist Helper Jesus Gonzales Montez, Mechanics Pedro Garza and Santago Cardenas and two other men whose names were not available. Inspector Manuel Hernandez and Catarino Perez were thought to be fatally injured.[1449]

At 6.30 a.m. on Saturday 4th March 1911 a so-called donkey engine employed at the sawmill of the Portland Railway Light and Power Company exploded a mile and a half from Estacada, Oregon, United States. At the moment of the explosion a number of men belonging to a logging crew were gathered around the engine warming their hands before they started work. Foreman James Pittman and four of the workmen were killed outright and a sixth man was fatally hurt and died in hospital a few hours later. The fireman of the engine, Thalmer Negbise, had put some wood into the engine between 5 and 6 and then returned to his boarding house for his breakfast. He was just coming back to the engine when it blew up and escaped with slight injuries. Where the engineer was, was not stated. The force of the explosion was demonstrated by the fact that young fir trees fully 12 inches in diameter were broken off near their roots by the explosion as if they had been hit by a tornado.

After viewing the scene of the accident and taking what testimony there was available Coroner Fox returned a verdict of 'Accidental death' due to too much water in the boiler and a faulty safety valve. The verdict was presumably based on assertations that the safety valve had only been registering 40 or 50 pounds before the explosion, though no one remained who could definitely state what it had been registering immediately before the blast. The cause of the explosion was not really known.[1450]

Seaboard Air Line Railroad freight train No. 19 left Johnston Street station, Raleigh on the evening of Friday 11th August 1911 for Hamlet in charge of Conductor Walter Finch with Engineer Steven Stogner and Fireman John Maxey. It had covered about five miles and was near Fetner, North Carolina, United States, when the crown sheet of engine No. 904 failed, sending masses of scalding steam into the cab. The conductor, who was riding on the engine, was the most seriously injured of the three men and died

[1449] *Galveston Daily News*, Galveston, Texas, United States, 17th April 1910
[1450] *Morning Enterprise*, Oregon City, Oregon, United States, 5th March 1911

later in hospital. The engineer and fireman were less seriously scalded.[1451] [1452]

Passenger train No. 49 of the New York Central and Hudson River Railroad, the *Boston and Buffalo Special*, was running five hours late when the boiler of its engine exploded near Oneida, New York, United States, in the early hours of Friday 26th January 1912. Engineer Alfred Bretzer was killed outright and Fireman John Kearns so badly injured that it was thought he could not live. The locomotive, tender, express and baggage cars left the track, but the seven following sleeping cars were not derailed. Some of the passengers were thrown from their berths but none was apparently hurt.

Myron Pebbles, a witness to the accident, spoke to a local newspaper reporter. "It was 6.25 when the accident happened. I was on my way to the barn, which is across the track, to do the morning milking. I saw the train as it approached the crossing and hurried to get over the crossing ahead of the train. I got by all right and was about 150 feet when the accident came. I felt the shock and heard the detonation. I turned around a once. A great mass of flame and fire was rushing through the air and over my head. I ran at once to the side of the locomotive. I first came upon the body of the fireman. He was {line missing in newspaper}. The engineer was all black, but he did not appear to have been injured. He was motionless, though.

"I noticed one thing, that his left hand was uncovered. It was pure white, and in marked contrast to the rest of hm. His clothing was blown from the lower part of his body. The fireman was conscious. When I reached him he was groaning. His arms had been torn out and looked as though they had been stretched out to twice their normal length. He was covered with blood. I ran down to the signal tower about a quarter of a mile east of the explosion point. I told the operator to hold up everything both ways.

"Then I went back to the wreck. James Kelley, who lives near here, joined me and we picked up the fireman and put him in the buffet car. The railroad men refused to allow us to touch the engineer until the coroner came.

"The operator at the tower notified Canastota and Oneida of the accident. C. Fred Campbell sent an ambulance from Oneida and John McMahon sent his ambulance from Canastota. Campbell's ambulance got stuck in a snow bank at the cross road leading to the wreck and I had to get a team to pull him out. Mr McMahon's ambulance took the engineer's body away a little after 7 o'clock."

The Oneida wrecking train had already been made up to answer another call and was at the scene within 30 minutes.

As to how the explosion had been caused, the newspaper wrote, 'The crown sheet of the boiler, it as seen by those who examined it, was twisted and torn. Men who have had experience of explosions who examined the wreck, said that the crown sheet had been red hot or even white hot. This head extended to a distance of some nine inches below the crown sheet. That this is so, they claimed, is shown by the discolorations indicating the extreme heat. Old railroad men say that such a heat indicated that the boiler must have been without water for half a mile before the accident.

[1451] *The Charlotte News*, Charlotte, North Carolina, United States, 12th August 1911
[1452] *The Wilson Times*, Wilson, North Carolina, United States, 15th August 1911

'The theory was advanced that the pipe leading from the tank on the tender to the boiler became frozen in the extreme temperature and that while the fireman believed that the injector was forcing water into the boiler, it was not accomplishing this, owing to the frozen pipe. This must have been the case, old railroad men stated to The Herald, for at least half a mile.

'Water may have been finally forced into the boiler and upon the white hot crown sheets with the resulting explosion. Another theory advanced was that the water did not get into the boiler, but that the heat generated a gas that continued to form until the explosion came.'

A worse accident was very narrowly averted, as eastbound passenger train No. 10 was passing at the moment of the explosion, and the three rearmost cars of each train were still opposite one another.[1453]

A disastrous explosion occurred in the shop yards of the Southern Pacific Railroad at San Antonio, Texas, United States, at 8.55 a.m. on the morning of Monday 18th March 1912. Engine No. 704 of the Galveston Harrisburg and San Antonio Railroad was being got ready for service when it exploded without warning and with great violence. Buildings, structures and people in the immediate vicinity of the engine were blown to pieces and damage and devastation as caused throughout the neighbourhood with roofs collapsing, trees uprooted, and windows blasted out. The roundhouse was blown to atoms and everything within a radius of 100 feet of the engine disappeared. An oil tender which had stood close to the locomotive was blown 150 feet and gallons of escaping oil immediately caught alight, the fire adding to the destruction. The boiler of the engine was hurled a block away and landed between two houses in Duval Street.

The United States Army at Fort Sam Houston immediately dispatched military ambulances to the scene together with a detachment of cavalry to maintain order, assisting the San Antonio police and fire department. The injured were taken to hospital and the dead to the morgue.

A public inquiry was opened the day after the accident but was superseded a couple of days later by an official inquiry under the auspices of the Interstate Commerce Commission. There were suggestions that the explosion had been the result of sabotage, especially in view of the prolonged labour dispute, and a safety valve was apparently found screwed down after the accident, but there seems to have been little real evidence to support this theory, and none to support the use of dynamite, as some had suggested. Mr J. A. Starling, Labour Statistics Commissioner, told a newspaper, "Speaking from forty years of experience in this line of work, I am convinced beyond doubt that the occurrence was caused by tremendous boiler pressure. The engine was one of the biggest in the country, and there was not a piece of boiler as big as that {a small bookcase in his office} to be found after the explosion. Of course there is no telling what the pressure was, as the steam gauge was wrecked, or whether there was the proper supply of water, but there is no question that boiler pressure was responsible for this terrible accident."

Another possibility was human error. The engine had been damaged in an accident at Seguin the previous December and had just been repaired. When it was first fired up on

[1453] *The Syracuse Herald*, Syracuse, New York, United States, 26th January 1912 the morning of the accident there had been a problem, the firing had been stopped and the pressure allowed to drop. The engine had then been fired up a second time. There was also the possibility, likelihood perhaps, that the new scab workers had lacked the knowledge and experience to do a proper job, and the explosion had resulted from their incompetence or negligence.

The third possibility to be considered was that the boiler itself had been defective; possibly there had been unnoticed damage caused in the Seguin accident.[1454 1455]

The Interstate Commerce Commission report appeared towards the end of May. It was severely critical of the Southern Pacific Company, whom it blamed for the explosion. In part it read, "The evidence shows that the law and the rules governing the inspection of locomotive boilers were disregarded by the railway company's inspector and the officials in charge of such work at this point in the matter of making and properly certifying to the reports required by law. A report stating that the safety valves had been set had been sworn on March 16, 1912, by the railway company's inspector and the roundhouse foreman who signed it as the officer in charge of such work. The evidence shows conclusively that the safety valves had not been set at that time, and as a matter of fact it was while this work was being done, on March 18, that the explosion occurred. The evidence also shows that the inspector failed to witness the testing of the steam gauge and that the injectors had not been tested at the time that this report was made out, notwithstanding these facts he certified under oath that this work had been done.

"The opinion was expressed at the investigation that nitrogylcerin or some other high explosive was used, but nothing was found to support such an assertion.

"The question has been raised by those who suggested this theory as to whether excessive steam pressure could cause such complete destruction of a boiler. The most violent explosions on record have been caused by excessive steam pressure. The destructive effects of steam boiler explosions is {sic} not caused by steam alone, which is contained in the steam space at the instant the initial rupture occurs but is due to the instant quantity of steam which is instantly generated from the water contained in the boiler.

"The railroad company was at fault in requiring or permitting inspections and reports in the matter which was not in accordance with the law, and in allowing such important and responsible work as setting safety valves to be performed by an employee of whose experience and judgement the testimony shows they knew practically nothing, and in keeping a boiler in service for which the factor of safety, as shown by test, was below the recognized standard." [1456]

The final number of dead was never established. It was soon stated that 26 men had been killed and 40 injured in the yard itself but how many non-railroad employees were killed or injured was not stated. There was also the difficulty that many of the men working in the shops were new there, many of them scabs who were replacing long-term employees sacked following a labour dispute and working under assumed names. It would seem likely that as many as 40 died and 50 or more injured. The Interstate Commerce Commission report, however, put the number of dead at 30.

[1454] Webpage – The Great Locomotive Explosion by Farrell I. Tucker

[1455] *The San Antonio Light*, San Antonio, Texas, United States, 18th, 19th and 20th March 1912

[1456] Ibid, 20th May 1912

On Monday morning, 29th April 1912, a South Eastern and Chatham Railway train of empty coaches left Tonbridge at 5.53 a.m. for Tunbridge Wells, England, to form the 6.50 a.m. passenger train in the opposite direction between these two places. It was in charge of Guard Frederick Blunden, with Driver Daniel Kadwell and Fireman Henry Pilcher. All went well until the train was within a mile and a half of its destination, when the train's brakes were applied, and it came to a sudden stand despite the signals being off. Going up to the engine Guard Blunden found the driver and fireman lying on the ground alongside the engine, the former apparently dead (he was in fact only stunned). The enginemen had either been blown or had jumped from the footplate following an escape of steam and boiling water due to the collapse of the roof of the firebox. The accident was the result of a Board of Trade inquiry, conducted under Major J. W. Pringle by Mr Thomas Carlton of the Marine Department.

Mr Carlton found that the roof of the firebox of engine No. 216 had collapsed over almost its whole area, 44 of the iron roof stays being broken and the nuts drawn completely off 23 other stays. The steam had escaped violently through the 68 holes thus formed. He concluded that the explosion had undoubtedly been caused by the roof of the firebox having become overheated, weakening the stay ends and nuts to such a degree that a large number of them had failed to withstand the ordinary working load of the boiler.

When Fireman Pilcher had come on duty at 4.30 a.m. he had tested the water gauges, found the boiler to be full of water and the steam pressure at 70 or 80 pounds. The locomotive had left the yard at 5.15 a.m., picked up the coaches, done a bit of shunting and arrived at the down platform. The fireman said that there was 170 lbs of pressure when the train set off. He had used the injector twice on the journey to feed water into the boiler on the climb to Tunbridge Wells (this was confirmed by the guard). Driver Kadwell could not help the inquiry as although he had recovered from his injuries he had no memory of the accident.

Mr Carlton said that he could not accept Fireman Pilcher's statement that the water gauge glass had been three-quarters full just before the explosion was not consistent with the condition in which the firebox had been found afterwards. It would have been singular if both gauge glasses had failed to work at the same moment and he could only assume that the fireman had been in error and had mistaken empty glasses for full ones. The engine had been ascending a gradient of 1 in 100 tender first so the water would have been lower at the firebox end than at the smokebox end. It was probable that the firebox top had been bare of water for at least 10 minutes before the explosion took place.

Major Pringle stated that he fully concurred with the opinions expressed by Mr Carlton.[1457]

The boiler of engine No. 1771, working goods train No. 5817 from Angers to Nantes, on the Chemin de Fer de Paris à Orléans, exploded near the station at Pointe, France, 14 kilometres from Angers, on the evening of Sunday 23rd June 1912. The driver and probably the fireman were killed, though the body of the latter had not been found at the

[1457] Board of Trade report

time of the newspaper report. At least part of the train was derailed and the line being blocked passengers were being carried by road past the accident site.[1458]

Passengers arriving in Laredo, Texas, at the end of June 1912 told of a disastrous boiler explosion at Saltillo, Mexico, on Friday 28th June. To substantiate her story one woman took a Kodak (camera) with her when she visited the scene of the accident and her photographs showed the wrecked locomotive and several badly mangled bodies. It was stated that engine No. 549 had been standing in front of Saltillo station on the Ferrocarril Central Mexicano (Mexican Central Railway) 'carrying too much steam', and the boiler had exploded as a result. There were a number of people at the station at the time, both railwaymen and passengers waiting for trains. The blast wrecked buildings nearby and damaged or shook others further away. Pieces of the engine were scattered for blocks around. Initially it was reported that 12 people had died and about 10 others were badly injured.[1459] Another report put the number of dead at 16.[1460]

The boiler of engine No. 4601 of the Chemins de Fer de Paris â Lyon et â la Méditeranée, which was pulling a goods train from Bourg-en-Bresse to Djion on Thursday 6th March 1913, exploded near Coligny, France, as a Rome-Paris express was passing on the opposite line. Fireman Simplet, aged 31, was killed outright and Driver Phillipe died of his injuries in hospital at Bourg-en-Bresse. The boiler of the engine was hurled forward 20 metres, the underframe of the engine derailed and the 60 waggons of the train forced backwards. Debris from the explosion slightly damaged the passing express but no one on it was hurt.[1461]

When the steamer *Fiona* arrived in Sydney, New South Wales, Australia, on 9th July 1913 from Fiji, it brought news of a boiler explosion at the Labasa Sugar Mill on the island. The explosion had occurred in the engine shed and it carried engine No. 4 up through the roof of the shed. It came down again onto the laboratory which adjoined the shed. There was then a second explosion and the remains of the engine came to rest on a nearby coal heap. Two Indian coolies were killed outright, two more were seriously injured and a European mechanic was reportedly badly scalded, despite which he was supposedly not badly hurt. The explosion presumably occurred at an unknown date in about June 1913.[1462]

The afternoon mixed train for Wolseley and Adelaide was about to about to leave the station at Mount Gambier, South Australia at 2.15 p.m. on the afternoon of Monday 2nd February 1914 when its engine exploded with a terrific roar. Fragments of the boiler and other engine parts were blown a considerable distance, the dome of the engine, weighing 4 or 5 cwt, was thrown 200 feet, while buildings near the station were damaged. Driver E.

1458 *Le Temps*, Paris, France, 25th June 1912
1459 *The Liberty Vindicator*, Liberty, Texas, United States, 5th July 1912
1460 *The Bryan Daily Eagle*, Bryan, Texas, 29th June 1912
1461 *Le Figaro*, Paris, France, 7th March 1913
1462 *The Daily Telegraph*, Sydney, New South Wales, Australia, 11th July 1913

E. O'Brien and Fireman William Taylor had a fortunate escape. Driver O'Brien escaped with a strained back while his fireman suffered a bad cut to his forehead needing six or seven stitches. Both also suffered from shock. They were treated and taken to their homes.

The engine which had exploded, No. 17 of the WX class, was about 37 years old but had been rebuilt in 1905. It was believed in the initial report on the accident that a defect of the boiler had caused the explosion.[1463] A departmental inquiry was no doubt made into the accident, but its results do not seem to have been made public.

Less than two months after the Mount Gambier accident South Australia suffered another locomotive boiler explosion. On Saturday 28th March 1914 a coal train consisting of two engines, 24 waggons and a composite brake van left Quorn at 4.05 p.m. that day for Port Augusta, in charge of Guard T. Yates with Driver Edward Rogers and Fireman J. Smith on the pilot engine and Driver W. Marston and Fireman A. G. Smith on the train engine. At milepost 257, two miles from Port Augusta, at a location known as Cudmore Hill, the boiler of the train engine, Y class No. 103, exploded. The engine overturned, its tender shot past and the following waggons piled up behind. Driver Marston was thrown clear and unhurt, but Fireman A. G. Smith was buried in the wreckage. The fireman was taken to hospital unconscious but recovered consciousness later. Guard Yates walked to Port Augustus to report the accident.

After the accident a deviation line was laid to enable train services to be resumed before the wreckage could be cleared and the 7.10 a.m. train from Adelaide arrived at Port Augustus early on Sunday morning, about eight hours late. Its passengers had arrived there some hours earlier, having been transferred between trains on each side of the accident site. The cost of the damage to the engine and tender and 20 waggons was estimated at £6,000.

It was discovered that there was a fracture 5 ft in length directly over the firebox. The fracture was clean along the first line of rivets and it was believed that the explosion had been caused by a defect in the plate at the centre of the fracture.[1464]

1463 *Border Watch*, Mount Gambier, South Australia, Australia, 4th February 1914
1464 *The Advertiser*, Adelaide, South Australia, 30th March 1914

CHAPTER 17

VANDALISM

'Accidents' due to vandalism and sabotage have always been a problem for the railways They are obviously very closely related, inasmuch as both represent deliberate attempts to cause trains to crash, as opposed to accidents caused through non-deliberate carelessness, negligence or purely accidentally. Here they will, however, be treated separately, as they are slightly different. This separation is perhaps a little artificial, and if the perpetrator of a given act is not caught or his or her motives not sufficiently obvious it may be difficult to decide which it was, but it has nevertheless been felt that the distinction to be worth making.

Vandalism is usually, though not invariably, carried out by either children or young adults and its results are most commonly, though not always, less serious than those following from acts of sabotage. The motives for train accidents resulting from vandalism are usually either simple curiosity, in the case of younger children – 'I wanted to see what would happen' – or in the case of older children and teenagers a reflection of general anti-social behaviour; so that such individuals who attempt to derail trains also carry out other acts of 'mindless vandalism', and perhaps also commit burglaries or carry out arson attacks.

Sabotage on the other hand is usually undertaken by adults, who employ methods more likely to cause death and injury, so that while a vandal might for example just place an object upon the track or jam a stone into a set of points, a saboteur would be more likely to employ methods such as blowing up a section of track using explosives.

A typical act of vandalism, as opposed to sabotage, occurred near Marshall, Missouri, United States, on the night of Wednesday 27th February 1901. The perpetrators were four boys who lived at a poor-farm (a residence where very poor people, termed paupers, were supported at public expense). They placed fishplates and small pieces of iron on the nearby line of the Chicago and Alton Railroad. A passenger train hit the obstructions. It was not derailed but a hole was made in the floor of a car and the air brakes were severed. The boys were caught and three of them (probably the older ones) were lodged in jail. They said that they had done it 'just for fun', a classic motive, therefore, for an act of vandalism.[1465]

Another deliberate attempt to wreck a train in Missouri a year later perhaps came closer to sabotage than vandalism. George Busch, aged 16, and George Young, aged 20, 'the sons of respectable parents', travelled from Trenton, Missouri, United States, to the next town at Princeton, 24 miles away, on the morning of Saturday, 29th March 1902. How they made the journey in the morning was not stated but very likely they stole a ride on a Chicago Rock Island and Pacific Railroad freight train, for in the evening they tried to come back by the same method but were put off by the conductor. They thus had a long, long, walk home. Five miles short of Trenton, by now doubtless utterly fed up with their marathon trek, they placed several heavy ties across the track. A passenger train was shortly due but by good luck a late-running freight train, 'pulled by one of the 1,400 ton class engines,' arrived first. The weight and speed of the train were enough to push the obstructions off the track without any harm being done. At the point where the obstruction was placed the line ran on a high embankment and a derailment, especially of the passenger train, could have been disastrous.

Word was telegraphed ahead, the two were arrested and one of them, which was not stated, confessed to the act. The motive for the act was clearly revenge, while dime novels (cheap melodramatic adventure novels popular between about 1860 and 1915) were also believed to have played a part.[1466]

More serious than the previous two incidents was the derailment of Seaboard Air Line Railroad train No. 34 at Evergreen, Florida, United States, on Wednesday 4th March 1903. Half an hour before the derailment eight-year-old Birdie Royals, 'a white boy', the paper informed its readers, told Postmaster A. Herring, who lived close by where the accident occurred, of his intention to open a switch and wreck the train. Probably not surprisingly the postmaster did not believe the boy, but the latter did open the switch and the wreck of the train followed.[1467] George Hopkins, variously described as a cook and a porter, was fatally injured and died the next day, and Engineer Edwin Pearce was also fatally injured and died the day afterwards.[1468] Fireman Mack Scott, Conductor W. H. Dowling, Roadmaster J. B. Newton, Waiters Sam Freeman and Edward Thomas and four passengers, including Mrs Daniel Butterfield of New York and Ernest Gitting of Baltimore, were all more or less hurt.

The train had been travelling at about 50 miles an hour when the derailment occurred at 2.30 p.m. The engine had overturned into a ditch when it ran onto the points, trapping the engineer, fireman and roadmaster beneath it. Several men lifted the engine and released them. Evergreen was merely a siding without telegraph communications, and Conductor Dowling immediately despatched someone to Yulee, four miles away, to raise the alarm. There the operator wired Jacksonville and a wrecking train was quickly made up and sent to the scene. Meanwhile the injured had been placed in the Pullman observation car and made as comfortable as possible by Dr Morris of New York, who had

---

1465 *Sedalia Evening Sentinel*, Sedalia, Missouri, United States, 28th February 1901
1466 *The Quincy Daily Herald*, Quincy, Illinois, United States, 31st March 1902
1467 *Savannah Morning News*, Savannah, Georgia, United States, 6th March 1903
1468 Ibid, 9th March 1903

been a passenger on the train.[1469]

Colonel Pellatt, another passenger on the train, said that if the accident had happened 20 minutes earlier many passengers would have been killed or injured, as the dining car, which like the locomotive had overturned, had been filled with passengers. The Colonel added that there had been a number of negroes living within 40 feet of the wreck but none of them had come near the train or offered any assistance but had merely stood in their doorways and looked on. The equipment of the train was reported to be badly damaged but not beyond repair.

Birdie Royals was placed under arrest but left in the custody of his mother. The officials were investigating whether any malicious persons had induced him to tamper with the switch [1470] but the boy himself simply said that he had "wanted to see what would happen." [1471]

On Friday 29th May 1903 passenger train P 983 was derailed by a sleeper which had been placed across the line at Drebkau, Germany, on the Cottbus – Großenhain line. One person was killed and 13 were injured. The perpetrator was apprehended and confessed to having committed the offence while drunk.[1472]

The 8.30 p.m. passenger train from Grays on the London Tilbury and Southend Railway ran into an obstruction near West Thurrock Junction, England, on the evening of Monday 12th February 1906. A search of the line revealed an iron bar, four feet long and about two inches thick, which bore wheelmarks showing that it had been struck by a train. There was a possibility that the affair was related to a reduction in railway staff, over 100 navies having been discharged during the previous fortnight, but it seemed more likely that it had been placed on the line by boys who had taken the bar from adjoining chalk quarries, where similar bars were used by workmen in breaking off chalk from the cliffs.[1473]

Southern Railway passenger train No. 38, 'the finest train operated by the company', which had left Atlanta at 12.40 p.m. on Sunday 23rd August 1908 for Washington, D.C., was wrecked a mile south of Buford, Georgia, United States, by an iron bar placed on the track. The locomotive was overturned, and Engineer B. F. Dewberry and Fireman Mayson Watkins were scalded to death. A number of the passengers were thrown about the train and shaken up, but none was seriously hurt. Almost 200 feet of track was torn up and three coaches were so badly damaged that they could not be moved.

'Engineer Dewberry was removed to a hotel in Suwanee and upon the arrival of the wrecking train was transferred in the private car of P. L. McManus, assistant general manager. On this car Mrs Dewberry had been rushed to the scene of the accident. The engineer died eight minutes later. His wife called his name and spoke to him endearingly.

[1469] Ibid, 5th March 1903
[1470] Ibid, 6th March 1903
[1471] Ibid, 6th March 1903
[1472] *Karastrophen der deutschen Bahnen Teil II* by Hans Joachim Ritzau pg. 26 (in German)
[1473] *The Times*, London, England, 14th February 1906

The horribly injured man attempted to answer, but his reply was unintelligible. In a few moments he was free from suffering. Dr T. H. Hancock had come up from Atlanta on the wrecking train, but the patient was beyond human aid.' [1474]

Shortly after the derailment Louis Coxey or Cooksie, aged 11 or 12, according to different reports, was arrested. He admitted having placed the obstruction on the line, and 'in the artlessness of his childhood he recounted how he had been influenced by two older youths, whose names he would not furnish, to place the bolt and await the wreck which was sure to follow.' [1475] [1476]

An eastbound passenger train, probably on the Chicago Burlington and Quincy Railroad but possibly on the Chicago Rock Island and Pacific (newspaper accounts disagree, and the lines run parallel) had a narrow escape on Monday 17th May 1909. The train struck and somehow picked up a tie at the east end of the yard at Knoxville, Iowa, United States and carried it almost four miles before the tie caught in a switch at Flagler, derailing the locomotive and leading mail car. The speed of the train had been reduced and not much damage was done, but the line was blocked for some time.

The tie was one which had been removed by section men some time before. How it had got onto the track was at first a mystery but a couple of days later two boys, Clyde Nicholas and Thomas Biggs, both aged 12, were arrested by railway detectives and Nicholas confessed that they had placed the tie on the track to see how far the train would knock it. The newspaper predicted that the two offenders would be sent to the reform school.[1477]

Another report stated that the boys had originally set out to catch gophers (burrowing rodents) but that sport having proved too tame they had decided to wreck a train as it was more exciting. Going to the yards they had tugged a big heavy tie onto the tracks, then hidden nearby and awaited events. Later, hearing of the wreck, they had gone to Flagler and had their quest for excitement partly quenched.[1478]

Minneapolis St Paul and Sault Ste Marie (Soo) Railroad passenger train No. 17, northbound from Chicago to Duluth, was wrecked on an open switch half a mile south of Fremont, Wisconsin, United States, at about 1.16 a.m. on Tuesday 5th September 1911. Engineer August Kuester, a man believed to be Fred Miller of Neenah and a third unidentified man were killed outright, Fred Fortune, of Detroit, Michigan was fatally hurt, while Mrs Margaret Garver of Duluth, a passenger, was less badly hurt. Miller, Fortune and the unidentified man were tramps stealing a ride.

Fireman Earl Heinsden had jumped and escaped with bad bruises. It was rumoured that he had seen the red light of the switch points and called to his engineer that the points were open, but the engineer had replied that the red light was that of a freight train drawing up in the siding. According to the Interstate Commerce Commission the fireman

---

1474 *The Atlanta Constitution*, Atlanta, Georgia, United States, 24th August 1908
1475 *The Weekly Banner*, Athens, Georgia, United States, 28th August 1908
1476 *The Fayetteville News*, Fayetteville, Georgia, United States, 4th September 1908
1477 *The Pella Chronicle*, Pella, Iowa, United States, 20th May 1909
1478 *The Daily Telegraph*, Atlantic, Iowa, United States, 21st May 1909

said that by the time that he had seen the open switch they were too close to it for him to warn his engineer. The engine and tender were a mass of wreckage and the trucks (bogies) were torn from several of the coaches, which ploughed deeply into the sandy soil, the baggage and express cars, two coaches and one sleeping car being derailed. The passengers in the three rear sleeping cars were severely jolted and some cut by broken glass. Railroad men said that the results of the accident would have been much more serious had the coaches been of wooden instead of steel construction.

The train involved was known as the "Velvet" or Duluth Flyer and was one of the fastest on the road.[1479] [1480] Forty minutes earlier train No. 3, the Chicago – St Paul flyer, had passed over the switch safely.

It emerged that the switch had been opened by 15-year-old Leonard Eske, the son of a farmer living about a mile from Fremont. The motive was revenge. Some time earlier the boy had asked the section foreman to let him have a ride on a hand car, but the foreman had refused. Eske had then conceived the idea of derailing a train with the belief that the section foreman would get the blame. At about 8 p.m. the evening before he had broken the lock on the points leading to the south passing track, but he alleged that he had not actually taken the lever out of the socket. It was conjectured that this was why train No. 3 had passed safely, but the vibration caused as it did so had caused the unlocked facing points to open sufficiently to derail the following train.[1481]

It can be seen that this derailment was one of those which came perhaps closer to sabotage than vandalism, but as Eske was only 15 it has been classified as vandalism.

On the other hand clearly the derailment of a Georgia Railroad freight train near Decatur, Georgia, United States, on about Thursday 18th March 1915 was vandalism rather than sabotage. Marc Ingram, aged eight, and Roy Wilson, aged ten, confessed to having put an iron bolt on the track, "Just because we wanted to see a train wreck. We've seen one in the movies, but we wanted to see a real one." No one was hurt. The youngsters had made a similar attempt several weeks earlier.[1482]

[1479] *The Daily Northwestern*, Oshkosh, Wisconsin, United States, 5th September 1911
[1480] Interstate Commerce Commission report
[1481] *The Times-Enterprise*, Albert Lea, Minnesota, United States, 13th September 1911
[1482] *Daily Times Enterprise*, Thomasville, Georgia, United States, 19th March 1915

CHAPTER 18

SABOTAGE

Four distinct motives for sabotaging trains may be distinguished (after Robert B. Shaw):

(a) As an aid to intended robbery, either of the train passengers or to steal, for example, bullion being carried by the train.

(b) To satisfy a personal grudge against the railway company, such as by an ex-employee who feels that he has been wrongly dismissed.

(c) To hinder the operation of the railway, as for example during a strike.

(d) As an act of guerrilla warfare or terrorism, perhaps against an occupying enemy power. Sabotage against the railways was a feature of most of most conflicts of the early 20th century, beginning with the (Second) Boer War.

The first 20th-century act of Boer sabotage occurred on Wednesday 23rd January 1901. A train carrying Lord Kitchener and a body of troops was proceeding towards Middelburg, preceded by an armoured train. The armoured train was derailed by dynamite near Balmoral, Transvaal, and the troop train coming along behind it a number of Boer presented themselves and opened fire upon it. The troops returned fire and after a heavy exchange the Boers retreated with casualties.[1483]

The Boer attacks on trains were frequent in the next couple of months. The very next day, Thursday 24th January 1901, an engine was derailed at an unstated location in the Orange River Colony. The Boers had laid 10 mines, only three of which had exploded. No one was hurt.[1484]

On Thursday 7th February 1901 the Boers derailed a train near Vlaklaagte,

---

[1483] *The Times*, London, England, 26th January 1901
[1484] Ibid, 29th January 1901

Transvaal.[1485]

A mine laid under the railway between Uitkyk and Middelburg, Cape Colony, blew up under a platelayers trolley on Sunday 10th February 1901. The explosion killed a ganger and two Kaffirs (an Afrikaans word for blacks, used without comment in *The Times* in 1901 but now regarded as racist).[1486]

On Wednesday 13th February 1901 a pilot train, consisting of an engine propelling three trucks, was crossing a culvert midway between Vaal and Greylingstad, Transvaal, when a mine exploded beneath the waggons. A party of Boers hidden in some kraals a quarter of a mile away then opened fire, several shots hitting the engine. An armoured train which had been following the pilot train retaliated with its "Pom-Pom" gun and men. More Boers appeared, bringing their number up to 300. The 4.7 gun at Greylingstad also opened fire. The Boers uncoupled the pilot engine and sent it back at full speed towards the armoured train, but the latter withdrew, and a rising gradient stopped the runaway engine. The British suffered no casualties, the Boers several; two trucks were wrecked.[1487]

Monday February 18th 1901 saw two Boer attacks. In the morning a train was derailed between Vereeniging and Johannesburg, Transvaal. The attackers were driven off before they could obtain much from the train. The British lost one man killed, and one slightly wounded.[1488] In the evening a northbound train was derailed at Taaibosch, Cape Colony, near De Aar. The engine and one coach were overturned. The fireman was slightly wounded, and the guard was missing. According to a report to the London *Times* for some reason the Boers also committed an atrocity. Some natives who had been riding in two waggons were ordered to get out and stand in a ballast hole. They were then fired upon by the Boers. Two were killed, a number injured, and the rest escaped into the darkness.[1489] Another report added that an escorting soldier had been killed and the train was looted.[1490] A month later it was stated that five lives had been lost in the attack; it is not clear if this included the two Africans shot by the Boers at the ballast pit. In retaliation Boers Jan Petrus Niaber, Sarel Ninaber and Jan Andries Nieuwoudt were sentenced to death for treason and murder by a military court and shot. Two other men – Mermanus Nieuwondt and Frederick Nieuwondt – were sentenced to five years' penal servitude.[1491]

Tuesday 19th February 1901 saw the Boers blowing up a supply train at Klip River, Transvaal, south of Johannesburg. They had drawn up carts and waggons ready to

[1485] Ibid, 11th February 1901
[1486] Ibid, 12th February 1901
[1487] Ibid, 15th February 1901
[1488] Ibid, 19th February 1901
[1489] Ibid, 20th February 1901
[1490] Ibid, 22nd February 1901
[1491] Ibid, 21st March 1901

remove the contents of the wrecked train but were frustrated by the arrival of armoured trucks from Lord Kitchener's train, which had been following the supply train. The Boers were driven off with slight British casualties, but the driver of the supply train was scalded to death when the boiler of its engine burst.[1492]

Another attempt to obtain supplies by sabotaging a train was made by the Boers on Saturday 16th March 1901 near Vlaklaagte, Transvaal. Two hundred and fifty of them under Commandant Buys had brought waggons and Cape carts to the lineside to be loaded with any supplies they might be able capture. Unfortunately for them the pilot train which preceded the mail train exploded the dynamite charge they had laid; its front truck being derailed but the remainder of the train passing over the opening. Troops from Vlaklaagte arrived and the Boers were forced to retire empty-handed.[1493] The Boers were desperate to obtain supplies by this stage of the war because the British had adopted a scorched earth policy, destroying crops, burning homesteads and farms and confining both Boer civilians and Africans in concentration camps.[1494] Two days later 150 Boers with a 'galloping Maxim' attempted to derail a train between Kaffirspruit and Vlakaagte by fastening sleepers to the metals but the attempt was foiled.[1495]

More successful was an attack by Commandant Buys and 400 men on a supply train three miles north of Vlakaagte, Transvaal, on Friday 22nd March 1901. A charge exploded under the middle of the train. The derailed trucks were defended by the escort force until superior numbers overwhelmed them and the Boers escaped with two waggon loads and several Cape carts full of mixed provisions. Three British were wounded, and several Boers were seen to fall. The engine and leading five trucks got away to Vaal station.[1496]

On Thursday 28th March 1901 a train was attacked between Balmoral and Wilge River, Transvaal, on the Delagoa line. The engine, four trucks and a bogie carriage were overturned or derailed, and the line was blocked for 30 hours. There were no casualties. 'It is satisfactory to know that more mounted troops are to be sent to guard this line, on which accidents have been frequent of late.' [1497]

Two Boer attacks on trains were carried out on Friday 29th March 1901. Another train on the Delagoa line was blown up at Wonderfontein, Transvaal, between Middelburg and Belfast. Two hundred Boers then rode up but were driven off by the arrival of a strong patrol of the 5th Lancers and an armoured train, leaving six dead and one wounded. In the second attack a train of empties was wrecked near Olifantsfontein, Transvaal. There were

---

1492 Ibid, 21st February 1901
1493 Ibid, 18th March 1901
1494 Wikipedia – Second Boer War
1495 *The Times*, 20th March 1901
1496 Ibid, 25th March 1901
1497 Ibid, 30th March 1901

no casualties.[1498]

Early on the morning of Tuesday 2nd April 1901 the Boers attacked a train a mile north of Mount Prospect, Natal, a charge of dynamite throwing the engine and trucks off the line. They then fired on the train and meeting no opposition came down from the neighbouring hills and set fire to it, retreating on the arrival of General Campbell's troops. The same day it was announced that the night service of trains north of Naauwpoort in Cape Colony had had to be again suspended due to a recurrence of attacks on trains.[1499]

In what was presumably a result of Boer sabotage, though not specifically stated as such in the report, a train was thrown off the rails near Smaldeel, Cape Colony, on Monday 8th April 1901. Two trucks with horses were smashed, one man was killed and five were injured.[1500]

By May 1901 the Boers had effectively lost the war. The British had offered peace terms on various occasions, notably in March 1901, but the "Bitter-enders" among the Boer commandos vowed to reject them and fight on, [1501] and isolated attacks on trains continued. On Saturday 4th May 1901 a train was thrown off the line below Mortimer, Cape Colony. Three trucks were burned. The line was cleared under the protection of an armoured train and traffic resumed in the evening.[1502]

A supply train was derailed near Godwaan, Transvaal, on Monday 20th May 1901 by a party of Boers led by Hinton, 'an ex-Guardsman and Johannesburg policeman'. The train driver was killed, and the fireman wounded. The Boers were disturbed before they could take possession of any supplies.[1503] On about the same date General Knox had set off to Ventersdorp Road in an armoured train to make an inspection. Night had fallen before his return and during the journey the front truck ran over some dynamite near Kroonstad, Orange River Colony, and was blown up. An officer had both legs broken and a sergeant was severely injured. Boers then fired on the train from only 70 yards away but men of the 4th Battalion of the Argyll and Sutherland Highlanders who were accompanying the train returned fire with a Maxim gun.[1504]

On Wednesday 26th June 1901 200 Boers under Commandant Viljoen attacked the two blockhouses protecting the railway at Brugspruit, Transvaal. One blockhouse fell but the other resisted until the arrival of an armoured train. A mine partly derailed the latter and the Boers, under the impression that it had been wrecked, rushed towards it, but were

[1498] Ibid, 1st April 1901
[1499] Ibid, 4th April 1901
[1500] Ibid, 13th April 1901
[1501] Wikipedia – The Second Boer War
[1502] *The Times*, London, England, 6th May 1901
[1503] Ibid, 24th May 1901
[1504] Ibid, 30th May 1901

mown down when within 10 yards of it by a fusillade of case shot and fled with many dead and wounded.[1505]

What was by a slight margin the most serious case of Boer sabotage during the whole conflict in terms of casualties was also the first attack of its kind on the Northern line. On Thursday 4th July 1901 a train was wrecked five miles north of Naboomspruit, Transvaal. Lieutenant Best and seven men of the Gordon Highlanders, one man of the Wiltshire Regiment, the driver, fireman and guard and four 'natives' were killed, one man of the Army Service Corps was missing, and 10 men of the Gordon Highlanders were wounded.[1506]

Another train was attacked and derailed near Naboomspruit, Transvaal, on Saturday 10th August 1901. A Boer prisoner was killed, three refugees wounded, and two soldiers slightly wounded. 'An armoured train coming up engaged the enemy, killing eight and wounding two.' [1507]

On Friday 30th August 1901 a forage train was blown up and burned by the Boers near Malasene, Transvaal, on the Delagoa Bay line, and a second train which went to its assistance was also derailed. There were no casualties.[1508]

A mixed train carrying civilian passengers and supplies was derailed by Boers on the morning of Saturday 31st August 1901 near Hammanskraal, Transvaal. A mine was exploded beneath the armoured truck in the centre of the train as the latter was going through a deep cutting. A party of about 100 Boers, 'believed to be under the command of Jack Hinton', were lying in wait and opened a fierce fire upon the train as soon as the mine went off. The officers made their way as quickly as possible to the armoured guard's van but Lieutenant-Colonel Vandeleur, D.S.O., of the Irish Guards, the senior officer, was slow to do so, stopping to warn the women and child passengers to lie down flat. As he was then making his way between the carriages he was shot dead. Of the armed escort of 45 men nine others were killed and 17 wounded, most of them belonging to the West Riding regiment. The Boers then boarded the train, took money, clothing and some bread from the passengers and some of the escort's Lee Metford rifles, but little ammunition.

According to the report of the *Times* correspondent in Pretoria the Boers also carried out another atrocity. Jacoba Page, the maid of two German refugees returning to Pietersburg with their children, reportedly left her compartment to appeal to the Boers not to shoot into the carriage and was shot dead for her trouble. This story was, however, apparently not true, Miss Page appearing in the casualty list a week later as having been

1505 Ibid, 8th July 1901
1506 Ibid, 6th and 8th July 1901
1507 Ibid, 13th August 1901
1508 Ibid, 6th September 1901

'slightly wounded'.[1509]

Prompted by the death of a prominent officer, the alleged murder of a maid (did he himself believe it? He presumably was not on the train) or just the apparently endless train wrecking, the *Times* correspondent in Pretoria wrote on 7th September 1901 that the only way to deter the Boers from wrecking trains was to have prominent burghers (Afrikaans citizens) of towns along the lines ride on the trains as hostages, as the Prussians had done during the Franco-Prussian war to deter French francs-tireurs.[1510] His suggestion led to considerable correspondence in *The Times*, some prominent people, including for example the writer Sir Arthur Conan Doyle, wholly supporting the idea, others roundly condemning it. Predictably the Dutch *Handelsblad* protested that it was a new violation of the rules of war, but British writers were also strongly opposed to the idea, pointing out that many countries, including Great Britain, had protested when the Germans had done the same thing in 1870.[1511] In the end the measure was not re-adopted (Lord Roberts had apparently tried it in June 1900 before almost immediately abandoning it) and there were no further attacks on trains anyway. The war ended with the defeat of the Boers and the signing of the Treaty of Vereeniging on 31st May 1902.

Westbound Great Northern Railroad passenger train No. 3, running five hours late and travelling at about 25 mile an hour, was wrecked a mile west of Essex, Montana, United States, at about 2 a.m. on the Monday 16th December 1901. The engine, baggage, mail and smoking cars passed the point of derailment safely, but the day coach, tourist, diner, buffet and Pullman sleeper cars were derailed and more or less wrecked, the day coach overturning. It also caught fire, but the blaze was extinguished by the passengers.[1512] One passenger, Otto Erickson, a Swede, was killed outright, three passengers were thought to be fatally hurt and another 14 were bruised and scratched.

The derailment was attributed to the parting of the rails. Assistant Superintendent H. A. Kennedy stated, "There is not the slightest doubt but that the wreck of No. 3 was the deliberate work of parties unknown but believed to be discharged employees."[1513]

In June 1902 employees of the Erie Railroad company were out on strike. Non-union men employed in the various collieries of the company were housed in one building and taken to their places of work every morning on a special train. When the train rounded a curve at Hughestown, Pennsylvania, United States, on Thursday 5th June 1902 it ran into a pile of rocks on the tracks by the strikers and the engine and car were derailed. No one was hurt.[1514]

The European portion of the Ottoman Empire known as Macedonia was struggling for independence at the beginning of the 20th century, with different armed groups fighting

---

[1509] Ibid, 2nd, 3rd, 6th and 24th September 1901
[1510] Ibid, 9th September 1901
[1511] Ibid, 10th, 12th, 13th, 20th, 21st and 24th September and 1st October 1901
[1512] *The Dakota Huronite*, Huron, South Dakota, United States, 19th December 1901
[1513] *Adams County News*, Ritzville, Washington, 18th December 1901
[1514] *The Evening Journal*, Wilmington, Delaware, 5th June 1902

the Turks and one another for either a wholly independent Macedonia or for the area to be under Bulgarian, Greek or Serbian control.[1515] It was against this background that a train was derailed by a charge of dynamite on the line from Salonika (now Thessaloniki, Greece) to Dedeagatch (now Alexandropoulis, Greece) just two kilometres from Salonika on Tuesday 28th April 1903. The engine was destroyed but the remainder of the train was little damaged. No one was hurt.[1516]

There was more sabotage in Macedonia on Wednesday 26th August 1903. An ordinary train from Sofia to Constantinople was blown up by dynamite at 11 p.m. near Kuleli Burgas station, 35 kilometres south of Adrianople. Six people were killed, three of them railwaymen and three foreign travellers. The management of the Chemins de fer Orientaux had received a letter from Saratoff warning it to suspend operations, which it had ignored.

An exasperated correspondent to the London *Times* wrote, 'This outrage can only be described as an act of inconceivable folly, calculated to diminish, if not to destroy, European sympathy with the Macedonian cause. As in the case of the Salonika outrages, the victims are Europeans, and the crime is committed just at a moment when foreign sympathy, especially in France, seemed aroused on behalf of the suffering Macedonians. By restoring to the methods of Anarchists the chiefs will not only incur the reprobation of the civilized world but will also strengthen the position of those who desire that Turkey should be allowed a free hand in supressing the insurrection.'

Those killed were listed as the train conductor, another conductor, a waiter in the restaurant car, two women and a child.[1517]

The *Autonomye*, a revolutionary newspaper, announced on Saturday 5th September 1903 that a Turkish troop train had been blown up near Nevagna, Macedonia, between Ürsküb (now Skopje) and Velos, the day before and many soldiers had been killed. However, an Associated Press correspondent who investigated the report stated that there was no truth in it.[1518]

On about Tuesday 7th June 1904 there was a dynamite explosion in a Turkish train between Gjevgjelia and Gemendje, Macedonia. No further details of the attack were given but the *Times* correspondent in Vienna remarked that 'the object of dynamite outrages such as that just perpetuated is precisely to revive uneasiness and insecurity. The *komitajis* {groups of rebels associated with the Bulgarian Macedonian-Adrianople Revolutionary Committees and the Supreme Macedonian-Adrianople Revolutionary Committee}[1519] are unlikely to undertake operations on any large scale this year, but very little effort is requisite to maintain disquietude and discontent in view of the really miserable plight of large districts of Macedonia.[1520]

[1515] Wikipedia – Macedonia (region)
[1516] *The Times*, London, England, 30th April and 1st May 1903
[1517] Ibid, 28th August 1903
[1518] *The Ogden Standard*, Ogden City, Utah, United States, 5th September 1903
[1519] Wikipedia – Komitadji
[1520] *The Times*, London, England, 9th June 1904

Far away from Macedonia, war had broken out between Russia and Japan in February 1904 and at an unknown date, probably in June 1904, a Russian train on the Chinese Eastern Railway was derailed south of Harbin, Manchuria, by a band of Chunchuses (referred to in the newspaper report as Manchurian bandits also known as "Red Beards"). Thirty-three Russians were killed and 58 injured.[1521] There was a possibility that the Chunchuses had been commanded by Japanese officers but it would appear that both the Russians and the Japanese had made use of the Chunchuses at different times.

The Macedonian sabotage continued with two separate attacks on Sunday 10th July 1904. At a junction near Dedeagatch (now Alexandropolis, Greece) a bridge was dynamited as a goods train was crossing over it. Eight waggons were wrecked, and a brakeman was killed. The other attack, on a Chemins de fer Orientaux train between Amatovo and Karosoula, was unsuccessful.[1522]

Seaboard Air Line Railroad passenger train No. 41 was wrecked shortly after 1 a.m. on Friday 9th September 1904 near Catawba Junction, South Carolina, United States, apparently as the result of an act of sabotage. A second accident followed the first. Train No. 41 consisted of a locomotive, an express car, mail car, two day coaches and a Pullman sleeper car, in charge of Conductor Richard West with Engineer Gaston Mears and Fireman Ed Roberts.

As the train was crossing a trestle over a lagoon a few hundred yards from a bridge over the Catawba River he felt the trestle giving way beneath him. Opening the throttle he managed to get his engine clear of the trestle itself, but it had left the rails and plunged down a 25 ft embankment beyond, turning over three times in the process. Neither the engineer nor his fireman jumped, Engineer Mears escaped with bruises, but his fireman was killed. The engineer told a reporter, "I did not feel the pain then so much, my greatest suffering coming from the knowledge that back behind me in the mud and water of the lagoon was a train of 40 people and that on behind this was an engine, almost running wild, with its freedom from weight, the knowledge that there was no way on earth to stop this engine and that in a few minutes it would plunge into the dark hole below, and into the mass of human life buried in the coaches." Engineer Mears finally managed to release himself and struggled to a farmhouse, from where he telephoned to Rock Hill for help. He added that he had been in several wrecks before and this was the third fireman to have been hurled to his death from his side.[1523]

Only one passenger died in the initial accident, a Mrs Black, who suffered a broken neck, while her husband, sat beside her, was injured, but was hailed for his actions after the accident.., 'The wreck developed at least one hero. Mr Black, whose wife was killed beside him, was the last man to leave the coaches, and what he did the very first thing was to walk, with all his bruised condition, to Catawba Junction, two miles away, get into the telegraph office and send to Munroe for help. He is a telegraph operator in the service

---

[1521] *The Register*, Adelaide, South Australia, 2nd July 1904

[1522] Ibid, 12th July 1904

[1523] *The Weekly Star*, Wilmington, North Carolina, United States, 16th September 1904 of the Norfolk and Western railroad and was going to Atlanta with his wife to testify there in a railroad suit. He hardly realizes the blow he has sustained and sat all day on the bank nearby with mournful eyes and muttering to himself, "My God, what an ordeal for a man to bear."' [1524]

Another account wrote that, 'Yet he {Mr Black} got out of the wreck, and with the section master and a section hand, who had arrived on the scene, supporting him on either side, hobbled two miles to Catawba Junction, broke into the telegraph office, and informed the officials of the wreck. "How I got off the message I do not know. I only know that somehow I made them understand that there had been a wreck," he said.'

As Engineer Mears had realised with the entire crew of the passenger train disabled by the accident there was no one to go back down the line to protect the obstruction and freight train No. 19, which was actually only an engine and caboose, had left Monroe 15 minutes after train No. 41, with Conductor Chapman and Engineer E. Y. Barksdale. With no warning of danger train No. 19 plunged from the damaged trestle onto the wreckage of the passenger train, killing Engineer Barksdale, while his unnamed fireman escaped almost unhurt. The fourth fatality in the accident was a woman later identified as Mrs Thomas McManus. Mrs McManus had not been very much hurt in the initial accident and had been struggling to get out. Another male passenger had given her some matches to light her way in the darkness, but she had then given them to Mr Black, who had been trying to find what had happened to his wife. Left in the darkness she was unable to escape before the engine of train No. 19 fell upon the carriage that she was in, killing her.[1525]

The likelihood that the initial derailment had resulted from a deliberate act was soon discovered, 'Geo S. Fitzwater, chief detective of the Seaboard Air Line, arrived on the scene early in the morning and took charge of the situation until organized help arrived. He said this afternoon that he had found some spikes and bolts and two angle bars which had been removed from the track with claw bars, and he said that he was confident that dirty work had been done. Mr Fitzwater sent a telegram to Superintendent E. Berkley at Atlanta, telling him that he had discovered evidence of the malice of this work and expressed his belief that someone had disconnected the joints in the lower half of the bridge. The stretch of track in that vicinity is a clear, straight, and apparently well-built one. The timbers of the bridge appear to have been wrenched from their fastenings and those that are splintered are of sound wood. There was nothing to suggest that the foundations of the structure was undermined by a washout, although the ground was somewhat marshy yesterday.' [1526]

Again in the United States, the *Diamond Special* passenger train of the Illinois Central Railroad was wrecked at Barclay, Illinois, 15 miles north of Springfield, at 3.35 a.m. on the morning of Wednesday 21st September 1904. The train had been running 45 miles an hour when it was thrown onto the top works of the Barclay Coal & Mining Co's mine, all but the rear sleeper car leaving the track. Engineer Walter Hayes and Fireman George Concannon were injured, the fireman fatally so. None of the passengers was badly hurt.

[1524] *The Macon Telegraph*, Macon, Georgia, United States, 10th September 1904

[1525] *The Weekly Star*, Wilmington, North Carolina, United States, 16th September 1904

[1526] *The Macon Telegraph*, Macon, Georgia, United States, 10th September 1904

The derailment was clearly the result of sabotage; the switch lock had been battered and broken, the switch opened to divert the train from the main line and the switch lamp extinguished.[1527] It could not immediately be determined who had committed the sabotage, or why.[1528]

On Sunday 1st January 1905 it was the turn of the Russians to derail a Japanese train. Patrols under Lieutenant Levshanovsky and Sub-Lieutenant Grunitcheff damaged an embankment and uprooted a number of telegraph poles 10 kilometres north of Hai-Cheng, Manchuria, on the Chinese Eastern Railway, A Japanese engine from Liau-yang was thrown off the line.

A report in the *Morgen Post* in Berlin picked up by several American newspapers gave vague details of several trains being derailed on the Trans Caucasus Railway being wrecked by Armenian revolutionaries near Baku, Russia, the trains being hurled over a precipice and all their occupants killed. The attacks happened on about Monday 20th February 1905.[1529] This would have been one of the opening events of the 1905 Russian Revolution, which had begun on 22nd January.

At 2.30 a.m. on Sunday morning, 14th May 1905, Atchison Topeka and Santa Fe Railroad passenger train No. 17 was ditched by train wreckers a mile east of Emporia, Kansas, United States. Conductor E. A. Taylor, Fireman F. A. Grover and four passengers were injured, two of the passengers possibly fatally so. The spikes and fishplates of two rails on the inside of the curve had been removed. The locomotive passed the loose rail safely, but the mail car left the track and was dragged a hundred yards along the embankment. The two express and baggage cars, smoker and two day coaches which were following went into the ditch, the end of a baggage car rearing up high enough to touch the telegraph poles. The two Pullman sleeping cars at the rear of the train remained on the track.

After an hour a relief train arrived from Emporia and the injured were taken there, while a hundred workmen laboured to repair the track.

This was the fourth attempt in four months to wreck passenger trains at the same spot. The previous attempts had consisted of ties being piled up onto the track which had just been pushed away. Three men were seen near a section house on Saturday night. On Sunday morning a track wrench and a claw-bar, clearly the tools used to loosen the track, were found to be missing, it was later discovered that they had been thrown into a pond. There was no clue as to the identity of the culprits.[1530]

On Sunday 24th September 1905 four minor Chinese officials were killed, and 20 other persons were injured at Peking station, China, when a bomb exploded in a private railway

[1527] *Centralia Daily Sentinel*, Centralia, Illinois, United States, 22nd September 1904
[1528] *The Weekly Telegraph*, Bloomington, Illinois, United States, 23rd September 1904
[1529] *The Seattle Star*, Seattle, Washington, United States, 22nd February 1905
[1530] *The Cuba Review*, Cuba, Missouri, United States, 18th May 1905

car containing the members of one of four missions which had been ordered abroad to study foreign political methods. One report on the incident implied that the bomber himself had also been blown to pieces. No possible motive for the bombing was stated.[1531]

The 1905 Russian Revolution continued and evolved and on 27th October the October Manifesto was presented to Tsar Nicholas II, demanding basic civil rights, the right to form political parties, moves towards universal suffrage and so on. The Tsar reluctantly agreed to the demands on 30th October and the measure was welcomed by liberals but denounced by radical socialists and revolutionaries who called for an armed uprising to overthrow the Empire. There was also a backlash from right-wing elements opposed to any concessions.[1532]

On Monday 30th October the *Times* correspondent in St Petersburg wrote, 'The situation in Riga is growing worse. Policemen have been shot, in collisions in the streets, and 300 persons who were attending a meeting had an encounter with a force of Dragoons, several being injured.

'At Tiflis {now Tbilisi, Georgia} the shops and all the industrial establishments and State institutions are closed. Firing and bomb-throwing are common in the streets, the strikers firing on droshkies {low four-wheeled open carriages}. All the Caucasian railways have ceased working, as well as the Transcaspian lines. A military train at Tiflis has been derailed. Several persons were killed and injured, and eight carriages were smashed.' [1533]

Note: The dates quoted above are all quoted as per the Gregorian calendar, which was already in use in Great Britain. Up until 1918 Russia continued to use the Julian calendar, the dates of which at this time were 13 days earlier, thus for example 30th October 1905 by the Gregorian calendar was 17th October by the Julian calendar.

The wrecking of a military train at Kutais, Russia, on Saturday 4th November 1905, is referred to by the newspaper correspondent as a collision so possibly the train ran into an obstruction placed across the track. Nine soldiers were killed and 18 injured, including a colonel. After the train had stopped the revolutionaries opened fire on it with rifles and the soldiers replied. Two revolutionaries were killed and eight arrested. 'The battalion then proceeded to the Ozurgeti district, where sanguinary encounters have taken place between the people and the troops. The battalion has now returned to Kutais.' [1534]

A freight train on the Warsaw – Brest line manned by soldiers of the railway battalion was derailed by strikers on Thursday 9th November 1905 near Novominsk, Russian Poland (now Mińsk Mazowiecki, Poland). The guard was killed, and several soldiers were injured.[1535]

[1531] *The Times*, London, England, 24th and 25th September 1905
[1532] Wikipedia – 1905 Russian Revolution
[1533] *The Times*, London, England, 31st October 1905
[1534] Ibid, 6th November 1905
[1535] *The Lincoln Daily Star*, Lincoln, Nebraska, United States, 9th November 1905

A train carrying a detachment of sappers to maintain communications between Livonia and Courland in the Baltic part of the Russian Empire (now areas of Latvia and Estonia) in the event of a strike of railwaymen was itself derailed by 'insurgents' on Thursday 14th December 1905 near Stockmansof, Livonia. The attackers then fired on the train, killing or wounding 25 men, and the remainder of the detachment retreated.[1536] Another report stated that half of the troops had surrendered to the rebels and the other half were still holding out after four days, but their position was desperate.[1537]

A train derailed near Gatchina, Russia, on Saturday 16th December 1905, presumably by sabotage, was reportedly carrying $10,000,000 in gold.[1538]

The Russian Revolution proper ended with the crushing of the Moscow uprising on 18th December but sporadic assassinations and attacks by revolutionaries, including attacks on railways, continued.[1539] Thus the locomotive of the incoming Riviera express was blown up at Vilna, Russia (now Vilnius, capital of Lithuania), on the evening of Monday 25th December 1905.[1540]

On Thursday 28th December 1905 the *Times* correspondent in St Petersburg reported that insurgents had wrecked four trains on the Baltic line between Narva and Reval, Russia (now places in Estonia). Two were goods trains, one was an ordinary passenger train and the fourth was carrying nearly a whole regiment of cavalry. There were many casualties.[1541]

On Sunday 31st December 1905 the *Times* correspondent in St Petersburg reported that, 'The *Novoe Vremya* announces that the whole Siberian Railway is out of hand and that Krasnoyarsk is in the hands of the revolutionaries.

'The partial failure of the railway strike has led to reprisals on the part of the strikers. It is reported that the bridge over the Volga at Syzran has been destroyed together with a passing train.' [1542]

Revolutionaries blew up a railway bridge as a passenger train was passing near the station at Jastozomb, Russian Poland, on the Dombrovo line of the Vistula Railway on Wednesday 3rd January 1906. The engine and luggage van (another report says four carriages) were wrecked, the fireman killed and the driver and four passengers dangerously injured. Simultaneously 30 revolutionaries attacked Jastozomb station,

---

[1536] *The Times*, London, England, 22nd December 1905
[1537] *The Commercial Tribune*, Cincinnati, Ohio, United States, 19th December 1905
[1538] Ibid.
[1539] Wikipedia – 1905 Russian Revolution
[1540] *The Times*, London, England, 27th December 1905
[1541] Ibid, 29th December 1905
[1542] Ibid, 1st January 1906

bound the employees, stole 800 roubles, and destroyed secret mobilisation documents and the furniture of the telegraph office. They then released the employees and made their escape.[1543 1544]

New York Central and Hudson River Railroad eastbound passenger train No. 34 was wrecked in the early hours of Saturday 26th May 1906 near Hoffman's, New York, United States, seven miles east of Amsterdam. It had been running at high speed at the time. The train consisted of an engine, seven express cars, two horse cars and three empty drawing room coaches. The locomotive and two express cars left the track and piled up, blocking the line for several hours. An express agent was cut and bruised, and nine horses were killed.

The derailment was due to the fishplates having been tampered with. About 100 local Italian section hands had struck for higher wages on Thursday and marched seven miles along the tracks, continuing their demonstrations on Friday when several of them were arrested for trespassing and fined.[1545]

Russian revolutionaries or their sympathisers were apparently behind the derailment of the Nord express from Paris at Charleroi, Belgium, on Sunday 29th July 1906. A piece of track had been removed, the engine and tender left the rails and the driver and fireman were killed. The coaches of the train were not derailed, and the passengers were unhurt. The attack was apparently aimed at the Grand Duke Vladimir who it was supposed was bound for St Petersburg on the train, but railway officials stated that he had not been on it anyway.[1546]

On Tuesday 3rd September 1907 Insterburg – Berlin express D 6 was derailed at Rehfelde, Germany, on the Küstrin – Berlin main line, rail fastenings having been loosened. One person died and nine were injured. Despite a large reward the perpetrator was never caught.[1547]

A gang of 50 men, described as robbers but possibly revolutionaries, destroyed a portion of the Transcaucasian Railway between Bejuk Kjassik and Karjasy stations, Russia, on the evening of Friday 3rd January 1908. Their object was to wreck and loot the mail train, but their plan was frustrated by the fact that the latter was preceded by a military train with an escort. The military train was thus derailed instead, and seven soldiers were wounded. The others fired at the saboteurs who returned fire and then vanished into the darkness. The mail train, which was said to have been carrying 2,000,000 roubles (£200,000), was held up at Karjasy overnight. In the morning an infernal machine (bomb), rifle cartridges and surgical bandages were found on a bridge

---

1543 Ibid, 4th January 1906

1544 *The Argus*, Melbourne, Victoria, Australia, 6th January 1906

1545 *The Sun*, Baltimore, Maryland, United States, 27th May 1906

1546 *The Iola Daily Register*, Iola, Kansas, United States, 30th July 1906

1547 *Karastrophen der deutschen Bahnen Teil II* by Hans Joachim Ritzau) pg. 31 (in German not far from the scene of the attack.[1548]

There was a terrorist outrage, as it was described, at Sokoloff, Russian Poland, 'a station on the Vistula railway', on Friday 10th January 1908. A bomb was exploded beneath the postal van of a train. Two soldiers escorting the train were killed and the engine driver, two other railwaymen, a postal official, a workman and five soldiers were injured. The terrorists then took money and parcels from the wrecked van and made their escape, firing their revolvers at the escort. The value of their booty was not known.[1549]

A major case of sabotage occurred on the Chinese Eastern Railway near Nikolskoya, Manchuria, on or about Sunday 30th August 1908. A passenger train was 'maliciously wrecked' and many persons were killed or injured, the latter including a General Oserovski.[1550]

An attempt was made on the morning of Saturday 28th August 1909 to wreck the East Indian loop mail train at Jamalpur, India, but the saboteurs were frustrated by the fact that it was preceded by a slow workman's train. Sixteen waggons of the latter were thrown off the line. 'The loss of life is unknown.' [1551]

The failure on the morning of Saturday 11th June 1910 of a bridge near Luleh Burgas station, Turkey, beneath the "Conventional" Chemins de fer Orientaux train was attributed to sabotage. A mail van and some of the passenger carriages fell into the water. A conductor was killed, and three persons were seriously and six slightly hurt. The intended victim was thought to have been Mahmud Shevket Pasha {a Turkish general and future Prime Minister} who had originally intended to travel by rail from Salonika but in the end took a steamer instead. Four bolts had been removed from the track 80 yards short of the bridge, derailing the train, the impact of the derailed vehicles on the bridge presumably causing the structure to collapse.[1552]

India suffered another 'railway outrage' on Wednesday 5th April 1911 when a passenger train on the East Indian Railway was derailed due to sabotage near Dinapur, a rail having been removed. An Indian was killed and three Europeans, cantonment magistrate Major Kirkwood, a Mrs Kirkwood and a Miss Walling, of Dinapur, were seriously injured. The East Indian Railway offered a reward of 1,000 rupees for information leading to the discovery of the offender, while it was reported that, 'The police have raided the residence of Anada Mohan Roy, a reported wealthy zemindar, or landowner, and son of the late Mohini Mohan Roy, a well-known pleader. They found five revolvers and a quantity of dynamite, and, according to the "Indian Daily News,"

[1548] *The Times*, London, England, 6th January 1908
[1549] Ibid, 11th January 1908
[1550] Ibid, 3rd September 1908
[1551] Ibid, 30th August 1909
[1552] Ibid, 13th June 1910

much picric acid and a complete scientific plant for the manufacture of currency notes were also discovered. Roy and another man have been arrested. The search of the house, which is a large, rambling structure continues, and all the approaches to the building are guarded by armed police.

'The police believe that they have discovered a nest of conspirators. Anada Mohan Roy is highly educated. He is a Master of Arts and passed first in chemistry at Calcutta University.'

A pleader was a lawyer who pleaded in court on behalf of his client.

An attempt had been made to wreck the mail train a month previously.[1553]

A major armed struggle, known as the Mexican Revolution, began in Mexico at the end of 1910 and lasted until 1920. The conflict began with a rebellion against the rule of the conservative President Porfirio Diaz, who was forced to resign in May 1911. Landowner Francisco Madero was elected to power but was opposed by both conservatives who saw him as too liberal and revolutionaries who saw him as too moderate. In February 1913 Madero was forced to resign and then assassinated and a counter-revolutionary regime under General Victoriano Huerta came to power. Huerta was forced out in July 1914 and a confused civil war began. In May 1915 Constitutionalist Venustiano Carranza took power, having defeated the forces of former Constitutionalist Pancho Villa and forced revolutionary leader Emiliano Zapata to return to guerrilla warfare. Agents of Carranza assassinated Zapata in April 1919 and Carranza himself was assassinated in May 1920.[1554] The conflict was characterised by a whole series of acts of railway sabotage, many reportedly with heavy casualties.

The first railway related incident of the Mexican conflict, though there was no accident or sabotage as such, appears to have occurred at Los Cajones on Wednesday 26th April 1911. A train bound for Mexico City was stopped by 400 rebels under the command of General Prudencia Figueroa. Thirty soldiers on the train were ordered to surrender but their officer, fresh from military training school, refused. In the ensuing fire fight the second lieutenant and 28 of the soldiers were killed and the remaining two men were captured.[1555] Then on the evening of Saturday 29th April the "weekend train" for Cuernavaca on the Ferrocarril Central Mexicano was within a few miles of its destination when rebels signalled for it to stop at El Parque. Mindful of what had happened at Los Cajones two days earlier the engineer instead put on steam. As the train passed them the rebels fired through the windows of the chair car and a short way down the line the engine left the rails, the rebels having torn up the track. The engineer was buried beneath his locomotive and three passengers and a negro porter were killed, while several other passengers were injured, either in the derailment or from the rebels having fired at the train. Among those killed was Dr Olssen-Seffer, a tropical agriculturalist employed by the government in the hope of developing the rubber industry.

The rebels ran down to the now stationary train and ordered its passengers to alight. The passengers did so and stood some distance from the wreckage while the rebels

---

1553 *The Standard*, London, England, 6th and 7th April 1911

1554 Wikipedia – Mexican Revolution

1555 *The Arizona Republic*, Phoenix, Arizona, United States, 28th April 1911 chatted among themselves. A few minutes later the rebels' reason for ordering the passengers from the train became apparent when a heavy freight train came around the corner. The passengers had not been permitted to warn its engineer and it collided with the wreckage of the first train. Its fireman was killed, and the engineer and a coal passer were badly scalded by escaping steam. It was thought that the rebels had expected troops to be riding on the passenger train. The line was, of course, completely blocked.[1556] The passengers were forced to walk into Cuernavaca. Workmen later cleared the line guarded by soldiers. The following day, Sunday 1st May, a southbound train from Laredo, Texas, to Mexico City, was stopped by rebels at Santa Maria. Rifles and ammunition were taken from it and it was then allowed to proceed.[1557]

Westbound passenger train No. 9 from New York to Chicago over the Canada Southern Railway was due in Ridgetown, Ontario, Canada at 3.32 a.m. on Wednesday 31st May 1911 but was running a few minutes late. It was in charge of Conductor George Marr with Engineer Stephen Quinlan and Fireman Russell Oakes. The train reportedly came to a complete stand at the distant signal, which was at "caution", and was only travelling at about 16 miles an hour at the home signal, which was at "proceed". The train was thus well under control when it was suddenly derailed, engine, tender and four baggage cars plunging into a ditch, crushing the engineer and fireman beneath them. None of the eight passenger coaches was derailed and none of the passengers was much hurt.

Superintendent Shearer of the Michigan Central Railroad, which controlled the Canada Southern, stated that it was clearly a case of sabotage. "The fishplates were taken off both ends of the rail and spikes pulled out and the rail removed and placed in the centre of the tracks, and, as far as we could see, its position had not been moved by the wrecking of the train," he said. Other evidence of sabotage was that a section house had been broken into and a crowbar was found beside the track. Crown Attorney Smith and Inspector Rogers, both of Chatham, Ontario, were informed of the affair and Michigan Central Detective Heenan was conducting an investigation, but there was no immediate clue as to who had been responsible.[1558] [1559]

There was a narrow escape from a possible disaster at about 10 p.m. on Thursday 29th June 1911 when an express from Le Havre to Paris on the Chemins de Fer de l'État was derailed near Pont de l'Arche, France. By good fortune no one was injured in the derailment and a train which passed in the opposite direction at the same moment avoided striking the wreckage by a margin of a few centimetres. An act of sabotage was clearly involved for a section of the left rail had been 'unpinned', the tools which had been employed being found close at hand. Also left lying around were copies of what were described as 'incendiary newspapers' such as *Le Libertaire* (an Anarchist publication), *La Guerre Sociale* (another Anarchist paper) and *La Bataille Syndicaliste.* Minor acts of sabotage affecting the railways, mostly involving the cutting of telegraph wires, were

---

1556 *The Galveston Daily News*, Galveston, Texas, United States, 1st May 1911
1557 Ibid, 3rd May 1911
1558 *The Lindsay Post*, Lindsay, Ontario, Canada, 2nd June 1911
1559 *The Winnipeg Tribune*, Winnipeg, Manitoba, Canada, 31st May 1911

reportedly almost a daily occurrence. The sabotage was probably connected with the failure of a general railway strike towards the end of the previous year, as a result of which some strikers had been imprisoned and about 2,000 railwaymen had been dismissed, though there had been Parliamentary agitation for their reinstatement.[1560] The minor acts of sabotage continued and acts were accompanied by placards indicating that it would continue until the dismissed strikers had been reinstated.[1561] On 9th July trains were derailed, near Orloy-Villers, at Chaville-Vélizy and Angouleme and the next day a train from Clermont to Paris ran into a mason's board 15ft long placed on the rails at a level crossing.[1562]

On 16th July the *Times* correspondent in Paris wrote that, 'The saboteurs probably know that special efforts are at last being made to arrest them. A service of motor-cars has been organized in order to patrol the lines, when possible, by night and without lights.' Quite how this would work the correspondent did not explain. The next day a train on the Chemins de fer de l'État collided with a block of concrete which had been placed on the line between Paris St Lazare and Les Invalides stations.[1563]

The expediency or otherwise of carrying out acts of sabotage was discussed with heated arguments for and against at the 22nd annual congress of railwaymen in Paris on 2nd August.[1564] Odd acts of sabotage continued at least until October.[1565]

Chicago Great Western Railroad passenger train No. 5 ran into an open switch and was derailed on the evening of Thursday 27th July 1911 at Edgeware Junction, Kansas, United States. Engineer W. Smith was killed and Fireman Hugo Smith was probably fatally injured. Both were from Des Moines, Iowa. Baggagemaster A. A. Gehrke was badly bruised and several passengers were injured, one seriously. The locomotive and baggage car overturned. Officials declared that train wreckers had opened the switch after prying off a lock.[1566] [1567]

There was a national railway strike in Great Britain in August 1911 in which there were a number of incidents involving strikers attempting to stop trains from being run, but only one act of sabotage directly involving a train seems to have been recorded. On Saturday 19th August 1911 a steel fishplate was wedged into a set of catch points near Gannow Junction, England, on the Lancashire and Yorkshire Railway. 'As the train approached the obstruction slipped out of position and the impact broke the points.' [1568]

On the evening of Sunday 8th October 1911 an attempt was made near Saidpur, India,

[1560] *The Times*, London, England, 1st and 3rd July 1911
[1561] Ibid, 6th July 1911
[1562] Ibid, 11th and 12th July 1911
[1563] Ibid, 17th and 18th July 1911
[1564] Ibid, 3rd August 1911
[1565] Ibid, 31st August and 10th and 17th October 1911
[1566] *The Muscatine Journal*, Muscatine, Iowa, United States, 28th July 1911
[1567] *The Joplin Daily Globe*, Joplin, Missouri, United States, 28th July 1911
[1568] *The Times*, London, England, 21st August 1911

to wreck the Darjeeling mail train, crowded with European passengers, including several high police officials, who were returning to Calcutta after their holidays. Eighteen feet of rail had been removed but a goods train was in front of the mail and was derailed instead, seriously injuring the engine crew. 'The police theory is that the outrage is the work of Anarchists, who have made at various time 30 attempts to derail trains on the Eastern Bengal Railway.' [1569]

Railway sabotage in Macedonia resumed at the beginning of December 1911. The various newspaper reports are difficult to decipher, names having changed, locations being hard to find on a modern map and different names apparently being given for the location of one attack. It would seem, however, that two attacks were carried out in Macedonia on Sunday 3rd December 1911, together with a third in Turkey, probably at the same time. In the most serious attack, in Macedonia, a bomb exploded under the train from Salonika (now Thessaloniki) to Constantinople (now Istanbul) near Akindjali. This accident was initially described as simply the explosion of a bomb under the guard's van with no other damage done but it would appear from a second report on an explosion near Poroy, which apparently relates to the same incident, that it was in fact more serious, three railway servants being killed. Another 'infernal machine', weighing 45lbs, which had failed to explode, was found in the wreckage. A second, less serious explosion, took place near Zelenikovo, Macedonia, on the line between Salonika and Ürsküb (now Skopje) and there was a third incident with three bombs exploding under a train near Keuprulu (now Köprülü) Turkey. Another bomb wrecked the goods depot at Keuprulu. The attacks were believed to have been the work of a Bulgarian revolutionary organisation which had recently crossed the frontier.[1570]

In China the Xinhai or Chinese Revolution had begun in 1911. This was actually a combination of many revolts and uprisings which culminated on 12th February 1912 in the abdication of the last Qing emperor, ending 2,000 years of imperial rule and the establishment of a republic.[1571] Towards the end of the revolution there were two major acts of railway sabotage.

On Tuesday 30th January 1912 a dynamite bomb exploded under a train carrying Imperialist troops from Honan to Siaokan (now Xiaogan), China. A following train then crashed into the disabled train, five coaches were telescoped, and many soldiers killed.

In what appears to have been a separate attack, though the reports are confused, a bomb was thrown at the Viceroy of Hukuang, who was riding in a third train. It struck the wrong car, killing several soldiers.[1572] [1573]

The other major act of sabotage in connection with the Xinhai Revolution occurred a few days later on Saturday 3rd February 1912 on the South Manchuria Railway. The pier

---

[1569] Ibid, 11th October 1911

[1570] Ibid, 5th, 6th and 7th December 1911

[1571] Wikipedia – Xinhai Revolution

[1572] *Warwick Examiner and Times*, Warwick, Queensland, Australia, 3rd February 1912

[1573] *The Daily Post*, Hobart, Tasmania, Australia, 1st February 1912 of a railway bridge near Shanhai-Kwan, Manchuria, was destroyed by 'Revolutionists' using dynamite. When a train from Mukden (now Shenyang) reached the bridge it fell into the stream, 'and many of the coaches were smashed to atoms.' Nine Chinese passengers were killed and 21 injured, but all the foreigners on the train escaped injury, perhaps because they were riding in a first-class coach at the rear of the train which did not leave the bridge. The attack was possibly an act of revenge as Revolutionist sympathisers in Mukden had been subject to barbaric attacks, some being dismembered and decapitated in the streets in front of their families.[1574]

Thirty-seven Federal soldiers died, and a score were wounded when a troop train from Mexico City bound for the south to relieve a point besieged by Zapatista rebel forces crashed through a bridge in Guerrero state, Mexico on Thursday 15th February 1912. (The newspaper report speaks of an Ixpla bridge over the Masusa River, neither of which the author has been able to identify, possibly it was the Ixtapa bridge. Another account gives the location as the Animas River Bridge.) The rebels had dynamited both the Ixpla (Ixtapa?) and Naranjo bridges, both being wrecked, and the train running onto the former a few minutes after the charges had been placed.[1575]

Also in Mexico there was fighting between rebels under General Campa and Federal troops around Escalón between Friday 22nd and Sunday 24th March 1912. On the Sunday the Federal troops entered Escalón, the rebel forces having fallen back to the town of Coralitos, a few miles to the south. Two thousand Federal troops followed the retreating rebels in three freight trains, preceded by an armoured train of three cars.

General Campa obtained a light engine and managed to persuade an engineer to drive it towards the advancing armoured train, the engine having been loaded with 40 boxes of dynamite. The engineer set the engine running down a grade, opened the throttle wide and then jumped off. The engineer of the armoured train saw the danger of the advancing engine but could do nothing about it. As a result of the collision the engine and leading car of the armoured train were telescoped, the other two cars were derailed, and the dynamite then exploded. Sixty federal troops were killed. The troops in the following three freight trains then detrained and prepared to defend themselves.[1576]

Mystery surrounds the circumstances of an alleged act of sabotage to a Federal troop train near Bachimba Pass, Mexico, on Monday 1st July 1912, mentioned by Robert Shaw in his *History of Railway Accidents Safety Precautions and Operating Practices.* [1577] According to a newspaper report in the early hours of 1st July the rebel camp heard a tremendous explosion just north of Consuelo station, in the federal camp a few miles away, accompanied by cries of distress which lasted half an hour. It was supposed that a Federal troop train had set off mines which the rebels had laid several days earlier and that many lives had been lost. The same afternoon Colonel Castulo Herrera announced in

---

[1574] Ibid, 6th February 1912

[1575] *The Daily Review*, Decatur, Illinois, United States, 15th February 1912

[1576] *The New York Times*, New York, New York, United States, 25th March 1912

[1577] *A History of Railroad Accidents Safety Precautions and Operating Practices* by Robert B. Shaw, pg. 408

El Paso, Texas, that he had received a message from Chihuahua giving details of the blowing up of a Federal train by electrically operated mines and the killing of all the troops in 20 coaches.[1578] However, Señor E.C. Llorente, the Mexican consul in El Paso, stated that he had been in communication with General Huertla's headquarters, who denied that any such attack had occurred, stating that the general was aware of the location of every rebel mine and that the detonation was probably the explosion by the vanguard (pilot engine?) of one or more mines.[1579] Where the truth lies it is impossible to say.

Three weeks after the alleged disaster to a troop train a passenger train from Mexico City to Cuernavaca was attacked by Zapatista rebels on Saturday 20th July 1912. The train consisted of an engine, a first-class and three second-class coaches and was carrying about 75 passengers plus an escort of 50 soldiers. A little beyond Parres station, at kilometre 67, the engine ran over a mine and overturned. As the train came to a stand about 500 Zapatistas opened fire on the second-class coach in which the soldiers were riding. The soldiers left the coach and fought the rebels from outside. Forty-three of the soldiers were killed, five badly wounded and two escaped. The rebels also fired at the passengers, one of the first of whom to be killed was the 10-year-old daughter of Captain Rosende Nunez, in charge of the escort. Having killed most of the soldiers and a number of the passengers the rebels looted the express and baggage cars, robbed both dead and living passengers and placed the bodies of both dead and wounded in the cars, which they set alight with oil from the engine.

A priest tried to intervene to prevent the killings, walking into the middle of the fray to seek out Delao, the leader, to whom he said, "Genoveno, what are you doing? Would you have the blood of innocent people on your soul?"

In all 41 passengers were killed, bringing the total number of dead to 84. One of the survivors remarked that most of the rebels were young, boys only just out of their teens.[1580]

A report from Salonika spoke of an explosion of dynamite on the railway at 6.30 a.m. on Sunday 21st July 1912 while a train was passing. The engine was thrown into the air and the tender blown to pieces, but no casualties were reported. The location was a place named Adjazlar, presumably in Macedonia, 'on the Turkish railway'.[1581]

Zapatista rebels in Mexico carried out another savage attack on a train on the afternoon of Sunday 11th August 1912. The engine of a passenger train was derailed by a loosened rail at a point 110 miles south of Mexico City and just north of Ticuman. The rebels then opened fire on the train with rifles before swarming down from the hillside to board the train. Their leader made no effort to restrain them and all the survivors of the escorting troops from the 14th Battalion and the passengers were killed, despite some

---

[1578] *Chicago Examiner*, Chicago, Illinois, United States, 2nd July 1912
[1579] *The San Antonio Light*, San Antonio, Texas, United States, 2nd July 1912
[1580] *The Liberty Vindicator*, Liberty, Texas, United States, 26th July 1912
[1581] *The Times*, London, England, 23rd July 1912

pleading for mercy, the looting being postponed until all the killing had been done. The only survivors were a few who had crawled away unnoticed from the train, and some of the train crew. As in the previous attack the rebels then set the train alight. The passengers were described as mostly belonging to the farming element and lower classes. Thirty-six soldiers and over 20 passengers were killed.[1582]

The passenger train from Constantinople (now Istanbul) arrived five hours late in Salonika (now Thessaloniki) on Wednesday 25th September 1912 after the explosion of a bomb beneath it 44 kilometres from Salonika, Macedonia. The track and engine were damaged but not the carriages and no one was hurt.[1583]

The First Balkan War broke out on 8th October 1912, a conflict between the Ottoman Empire (Turkey) and an alliance of Bulgaria, Servia, Montenegro and Greece. It would end in May 1913 with the defeat of Turkey, the loss by that country of the majority of its Ottoman Empire in Europe, the formation of the new state of Albania and the division of Macedonia between Bulgaria and Greece.

Early on in the conflict, on Tuesday 22nd October 1912, a train carrying 700 Turkish soldiers ran into some sheep on the line and was wrecked, 350 of the troops supposedly being killed. The accident happened at Seidler, probably in Macedonia, though it has not been possible to trace a place of that name. After the accident local peasants attempted to aid the survivors but many of them were massacred by the troops who accused them of having caused the disaster. This seems highly unlikely but offers an excuse to include the accident here in the sabotage chapter, together with the other confirmed acts of sabotage in Macedonia.[1584]

Zapatista rebels attacked a train at an unstated point between Mexico City and Ozumba, Mexico, 45 miles south of the capital, early on Monday 3rd February 1913. They derailed the engine in a cutting, then fired on the train from close range, concentrating on the troop car and killing or wounding all of the 20 soldiers escorting it. Four of the male passengers were also killed and the women passengers were carried off. 'A relief train took the wounded to Amecameaca.' [1585]

Seven German and American pay roll money guards on the Ferrocarril Nacional de México were killed and robbed by a band of 50 rebels near Vanegas, Mexico, near Vera Cruz, on about Friday 14th February 1913. Other railwaymen on the train were just robbed. Later 20 cars of a freight train were derailed and burned. All train services in the state of San Luis Potosi were subsequently suspended because of the danger of further attacks.[1586]

[1582] *The Nocona News*, Nocona, Texas, United States, 16th August 1912
[1583] *The Times*, London, England, 27th September 1912
[1584] *The Morning Bulletin*, Rockhampton, Queensland, Australia, 29th October 1912
[1585] *San Antonio Light*, San Antonio, Texas, United States, 3rd February 1913
[1586] *Urbana Courier-Herald*, Urbana, Ohio, United States, 15th February 1913

Very brief reports in the American press spoke of the sabotage of a Federal troop train by rebels above Guaymas, Mexico on about Monday 9th June 1913. The 'revolutionists' had mined the track and fixed a fuse so that the train would set off the charge. It was estimated that at least 200 soldiers were killed, and many others wounded.[1587]

A northbound train on the Ferrocarril Nacional de México was dynamited by Constitutionalists (the third faction in the Mexican Revolution along with the Zapatistas under Emiliano Zapata and those led by Pancho Villa) at Gomez Farias, near Vanegas, Mexico, on Sunday 21st September 1913. The locomotive passed safely over the dynamite charge, but the baggage car and two coaches were destroyed, the passengers killed being in the second-class coach. The telegraph lines north and south of the scene had been cut. The survivors were sent to Saltillo and troops were sent from San Luis Potosi to round up the Constitutionalists, whom it was said numbered about 1,000. It was stated officially that the number who had been killed in the attack, both soldiers and civilians, was 35.[1588] The train had started from Mexico City on Saturday evening and had turned back at Vanegas because bridges further north had been blown up.[1589]

It was not just Mexico which suffered serious acts of sabotage during this period for on Monday 29th September 1913 a train was derailed by brigands near Sosyka station, Russia, on the Vladikavkas Railway. Six carriages were smashed, 40 people were killed and over 100 injured. Some of the rails had been torn up. The object had been to rob the train.[1590]

A troop train was approaching San Salvador, Mexico, on the Ferrocarril Nacional de México, in Zacatecas state on Monday 27th October 1913 when rebel forces set off 'a tremendous charge of dynamite' under a trestle, with the result that 'the engine was torn into small bits and several cars with their human freight were splintered.' 140 Federal soldiers were reportedly killed and at least another 100 so badly hurt that they would be incapacitated for service for some time. The train was one of three, all heavily laden, and the other two trains were following so closely behind that they were showered with wreckage.[1591]

A passenger train on the Moscow – Kazan railway ran off the rails on the night of Wednesday 5th November 1913 between Nijni-Novgorod and Pensa, Russia. Fourteen people died and 15 others were seriously hurt. There was a suspicion that the accident had been maliciously caused.[1592]

[1587] *Goshen Weekly News Times*, Goshen, Indiana, United States, 13th June 1913
[1588] *The San Antonio Light*, San Antonio, Texas, United States, 23rd September 1913
[1589] *The Clayton Tribune*, Clayton, Georgia, United States 26th September 1913
[1590] *The Times*, London, England, 1st October 1913
[1591] *The Kokomo Tribune*, Kokomo, Indiana, United States, 28th October 1913
[1592] *The Times*, London, England, 6th November 1913

Unaware of the presence of northern Constitutionalist rebels so far south, a train carrying a body of government troops was derailed when it ran over a dynamite mine near Charcos, Mexico, on Tuesday 23rd December 1913. The train was practically destroyed and most of the soldiers who had escaped death in the initial derailment were killed afterwards by the rebels.[1593]

In January 1914 what was now the unified country of South Africa was in the throes of a serious railway strike, brought about mainly by the decision of the Railway Administration to retrench labour. The strike was short-lived and not very successful. It saw two derailments of trains, the first such events in southern Africa since the end of the Second Boer War. Both happened on Friday 9th January 1914; a goods train was derailed at Braamfontein and the front wheels of the pilot engine of the Cape mail were blown off by dynamite between Denver and George Goch. No one was hurt in either incident. A third attempt to dynamite the line near Witpoortje was discovered a few minutes before the passing of the Rhodesian mail.[1594]

An act of sabotage in Mexico on the night of Wednesday 4th February 1914 shares with the 1879 Tay Bridge disaster the unusual feature of being an accident to a passenger train which, like many shipwrecks, had no survivors. It happened in Cumbre Tunnel, in Chihuahua state, on the Ferrocarril de Nor-Oeste de México. Troops under General Máximo Castillo had pushed seven burning stock cars into the north end of the tunnel, where they set the tunnel timbers alight. Castillo's men then departed and a couple of hours later a passenger train entered the tunnel from the south end and collided with the burning cars in the tunnel. Engineer J. McCutcheon and his fireman were probably killed in the initial collision and some of the other occupants of the train may also have been. The remainder were probably suffocated by the smoke in the confines of the tunnel before they could escape. Rear Brakeman Juan Fernandez almost managed to escape; his body was found 200 feet from the tunnel entrance, in a sitting position with a handkerchief tied around his nose and mouth in a vain attempt to keep out the suffocating fumes. Other passengers probably tried to escape in the same manner but did not get as far.

The identified dead mentioned in the newspaper report included Head Conductor J. W. Weber, Conductor Thomas Kelley, Roadmaster J. Morris. Express Agent C. Marders, M. J. Gilmartin, superintendent of the company at Chihuahua; Henry Schoefield, superintendent at Juarez, and American passengers Mr K. L. Hatfield, an El Paso businessman; Mrs Lee Caruth and her five children, the family of a Madera engineer; and Mr L. E. Williams, an employee of the Madera Lumber Company. Of the approximately 44 passengers on the train about 20 had been Americans.

Robert Shaw, writing his book at a time before large numbers of newspaper archives were available online, states that there were no references to the accident in the year in which it happened, but in fact it was well covered by at least one Texas newspaper, which describes how a search of the tunnel was made four days afterwards when the fire had

---

1593 *San Antonio Light*, San Antonio, Texas, United States, 24th December 1913

1594 *The Times*, London, England, 10th January 1914

finally abated. 'Nothing but charred bones and buttons were found by the rescuing party which, with the aid of oxygen helmets and pulmotors, penetrated the Cumbre tunnel from the south portal as far as the locomotive and first two cars of the passenger train which entered the burning cavern... Entrance from the north portal will not be possible until tomorrow night at the earliest and probably not until the next day, as the heavy timbering is still burning. The rescuers who went in at the south end today found their way greatly impeded by not only the iron framework of the burned cars but by great masses of earth and rock which had fallen from the roof of the tunnel. Rock is still falling, making exploration dangerous.

'The wreckage is covered by from five to ten feet of earth and the only hope of finding any bodies is that they may have been covered by earth before having been destroyed by the flames, which is not probable. In the whole distance traversed today not a particle of woodwork of all the burned cars was found.' [1595]

The Americans blamed Castillo for the deaths of their subjects and when he moved to New Mexico he was arrested and imprisoned. Nothing could be proved against him and he had to be released.[1596]

A Ferrocarril Nacional de México passenger train from Tampico to San Luis Potosi was dynamited by Pancho Villa's rebels between Espinaso and Las Canoas stations, Mexico, on Thursday 12th February 1914. The train was carrying passengers and the usual military escort. When it was near Las Canoas the rebels exploded dynamite mines which had been laid beneath the track by means of an electric battery. 'Almost every car was shattered, burying the occupants in the debris. The wreck either ignited by the explosion or was fired by the rebels, who swarmed down upon it from the mountains surrounding the scene of the explosion.' [1597]

Fifty of the military escort were reportedly killed together with scores of passenger, including reportedly Americans and other foreigners. Of the train crew only the engineer and one brakeman escaped. Most of the passengers at the rear of the train escaped into the brush, some making their way to nearby stations.

A newspaper report noted that the attack showed the effrontery of the rebels as the Federals carefully guarded the Tampico line, which was used for the shipment of petroleum, and that the mines had been planted under the noses of the large Federal garrison at Cardenas, not far away. After the train had been destroyed by fire the rebels tore up a mile of track. A military train was despatched from Cardenas to the scene.[1598]

A troop train bound for Jalapa was blown up by Zapatista rebels near Lima station, Mexico, on the Ferrocarril Interoceánico de México on Saturday 21st February 1914. One report stated that over 100 persons had been killed but that the surviving soldiers had rallied after the explosion and set off in pursuit of the rebels, who had fled into the

---

1595 *The Galveston Daily News*, Galveston, Texas, United States, 9th February 1914
1596 Wikipedia – Máximo Castillo
1597 *Santa Ana Register*, Santa Ana, California, United States, 13th February 1914
1598 *The Star and Sentinel*, Gettysburg, Pennsylvania, United States, 21st February 1914

mountains.[1599] Other reports stated that all on board, including 55 officers and men and the train crew, had died. All the accounts agreed that the English engineer of the train had been among the dead. A following passenger train was fired upon by the rebels but escaped by backing rapidly.[1600]

Portugal was affected by a strike of railwaymen in February 1914 and there was considerable destruction of the infrastructure by strikers, with tracks being torn up, a bridge on the main line between Lisbon and Madrid being blown up and so on, according to passengers who arrived in Paris from Lisbon after a journey of 102 hours instead of the usual 32. The rails in a tunnel near Kabregas were torn up and a freight train was derailed at the mouth of the tunnel on 23rd February 1914.[1601]

Alabama Great Southern Railroad northbound passenger train No. 2 from New Orleans, Louisiana, to Cincinnati, Ohio, was derailed near Livingston, Alabama, United States, on Friday 18th September 1914. Eight passengers, a mail clerk and its engineer died in the accident and 28 passengers, five mail clerks, four Pullman porters and three railroad employees were injured.

The train consisted of a locomotive, mail and baggage cars with steel underframes, a wooden coach, a coach with a steel underframe, an all-steel, a wooden, an all-steel and three wooden Pullman sleeping cars and was in charge of Conductor Stone with Engineer William Jones and Fireman W. Stewart. The train had left York, Alabama, 9¼ miles from Livingston, at 2.54 a.m., 24 minutes late and was derailed on a switch 1½ miles south of Livingston at 3.7 a.m. while at running at about 50 miles an hour. Its engine was then diverted into a spur track, collided with a car of slag and overturned on its right-hand side against a mound of slag and railroad ties at the end of the spur. The tender was detached and thrown across the main track, the mail car skidded past the tender and the baggage and express car was thrown against the engine. The wooden coach was telescoped and destroyed, and the steel underframe coach had one side torn away. The Pullman cars escaped serious damage; it was their weight and momentum which had caused such damage to the front of the train.

The switch at which the derailment had occurred was provided with an eight-day lamp which was filled and cleaned each Saturday. The line was provided with automatic signals, and if the switch points were open more than ¼ to ⅜ of an inch, or a car was standing on the spur track foul of the main line, the associated track circuit would place and maintain automatic signal No. 262.2, placed 7,143 feet south of the switch, to "Danger".

The car of slag had been placed on the spur track half an hour before the accident by the crew of southbound freight train extra No. 202 because of a 'hot-box'. At that the switch had been in good condition. Extra No. 202 had been followed by extra No. 203, both of which had gone into the siding at Hixon to cross train No. 2. Automatic signal No. 262.2 was situated just north of Hixon and the crews of both trains had seen it change

---

[1599] *Lima Daily News*, Lima, Ohio, United States, 23rd February 1914

[1600] *Medicine Hat News*, Medicine Hat, Alberta, Canada, 23rd February 1914

[1601] *The Helena Independent*, Helena, Montana, United States, 27th February 1914 from "Clear" to "Danger", back to "Clear", back to "Danger" and back to "Clear" before train No. 2 had passed. It had then gone to "Danger" normally as the train had entered the block section.

Conductor Stone told the Interstate Commerce Commission (ICC) inquiry that it had been about 5 a.m. before he had had a chance to look at the switch after attending to the dead and wounded. The switch had been set for the spur track and its lock appeared to have been forced open. The lamp was then missing, having been removed subsequent to the accident by the Livingston agent. The agent had told him later that there had been no oil in the lamp because it had been poured out onto the ground.

Fireman Stewart stated that he had just finished firing when the derailment occurred. He was not sure whether or not his engineer had applied the brakes. He had not seen signal No. 262.2 himself but Engineer Jones had called it clear.

Section Foreman Gibson said that he had examined the switch at about 5.30 a.m. The points had been set for the spur track. There was no sign of the conducting rods having been struck by any derailed vehicle which could have forced them open. Earlier in the evening the switch lamp had been burning and in good order. There was always oil left in it when it was refilled every Saturday.

Engineer Jones survived for some time after the accident before dying of his injuries. A doctor who treated him at the scene said that the engineer had told him that he had had a clear block but had seen that the switch was open when he had neared it and had reversed his engine, applied the brakes and opened his sanders. He (the engineer) had also seen someone running away from the switch and had remarked that the perpetrator must have known his business and waited until the train had entered the block. The doctor added that Engineer Jones had been perfectly rational when he spoke and knew what he was saying.

Fireman Sims had been at York when Engineer Jones had reached there en route to hospital in Meridien. He said that he had asked the engineer how the accident had happened, and Jones had told him that he had seen a man throw the switch for the spur track in front of him. He had then made every effort to stop but had had no time.

Another witness spoke of finding sand on the track over the whole distance between the switch and the overhead bridge from which the switch first became visible. No other train had passed at that time which could have put down sand.

The ICC reached the fairly obvious conclusion that the accident had been caused by an open switch, and that the damaged lock and spilt lamp oil, together with the testimony of Engineer Jones, showed that it had been opened deliberately, apparently with malicious intent. The changing indications of signal No. 262.2 witnessed by the crews of the freight trains at Hixon were likely caused by the person tampering with the switch, who being familiar with the signalling had not finally opened it until the train was past the protecting signal.

It seemed likely, the report continued, that three of those passengers who were killed had been in the all-wooden car and five in the car with a steel underframe. While none of the steel underframe cars had been completely destroyed in the accident like the wooden car, it seemed questionable whether cars with steel underframes afforded a materially greater degree of safety than all-wooden coaches, though the underframe would prevent the car buckling or breaking in two, and in this respect they were an improvement on wooden cars. The small amount of damage sustained by the fifth, all-steel, Pullman car strongly supported the view that the greatest degree of protection to passengers could be afforded only by all-steel cars.[1602]

Bloodhounds were sent to the scene to try and pick up the trail of the saboteur but could do little because too many people treading around had contaminated the area.[1603]

The neutrality of Belgium had been guaranteed by the 1839 Treaty of London, which was signed by Prussia, but the outflanking of France proposed in the Schlieffen Plan required a German invasion of Belgium, and this immediately followed the outbreak of the war. Fears of sabotage by Belgian francs-tireurs, similar to that which they had experienced in France during the Franco-Prussian War of 1870-1, caused widespread repression from the occupying Germans, including the execution of about 6,000 Belgian civilians in the opening months of the war, but there was in fact almost no sabotage carried out.[1604] The only act of railway sabotage in Belgium committed out in the early months of the war appears to have been one at Bilzen, near Hasselt, on 25th September 1914. Here it was regular soldiers rather than civilians who destroyed the track, resulting in the derailment and destruction of a German military train. In retaliation the Germans burned the village.[1605]

Southbound train No. 135 of the St Louis-San Francisco Railroad, en route from Kansas City, Missouri to Fort Scott, Kansas, was derailed at 8.15 p.m. on Sunday 3rd January 1915 at Olathe, Kansas, United States. The train was in charge of Conductor Reynolds with Engineer Cheney and Fireman Weller and consisted of locomotive No. 1202, 24 loaded box cars and a caboose. It had stopped for water and was on its way again when it was derailed on an open switch 2,550 feet beyond the water crane. The engine went down an embankment, crushing to death the engineer and Brakeman Morrell, who had been riding on the engine. The leading car of the train, loaded with stock and household goods, was completely demolished and the caretaker of the goods and a 17-year-old trespasser (tramp) hiding in the car were also killed. After the accident the car caught fire, probably from an overturned stove.

Fireman Weller told the Interstate Commerce Commission (ICC) inquiry into the accident that he had seen that the points were wrongly set and had called a warning to his engineer. He was positive that the switch lamp had been out.

Conductor Reynolds, Flagman Benedict and Olathe Station Agent Sloan agreed that the switch lamp had been extinguished, the switch thrown for the wye track, and the switch lock and pin lying on the ground near the switch stand. The lock could not be opened without a switch key except by being forced open and there was no indication that it was damaged.

Olathe Operator Barrel stated that a southbound freight train had passed at 7.30 p.m. and there had been no train or train crew in the yard between that time and the passing of

1602 Interstate Commerce Commission report

1603 *The Galveston Daily News*, Galveston, Texas, United States, 19th September 1914

1604 Wikipedia – Belgium in World War 1 and The Rape of Belgium

1605 *The Helena Independent*, Helena, Montana, United States, 27th September 1914 train No. 135 Section Foreman Pappas said that he had lit the switch lamp in the afternoon, and it had been burning at 3 p.m. After the accident he had found that there was still oil in it, had re-lit it and it had stayed alight. He also mentioned that just before the accident three men had come to the box car where he lived and had asked to be admitted to get warm (newspaper reports said they had asked for food – possibly they wanted both) but he had told them that he was just going to bed and had refused them. A farmer who lived nearby had also seen four men hanging about the switch just before the accident.

The ICC concluded that the switch had been opened maliciously.[1606 1607]

A train between Vera Cruz and Mexico City on the Ferrocarril Nacional de México was wrecked by an explosion near Apizaco, Mexico on Wednesday 14th July 1915. Zapatista rebels had dynamited the train, which was loaded with Government officials. Thirty-five persons died and 40 were injured.[1608]

'The followers of Zapata declare they will obstruct all trains. After evacuating the City of Mexico General Zapata announced he would continue his campaign against General Carranza. He has established a powder and cartridge factory near Cuernavaca and it is expected that he will surround the capital and harass the Carranza government by blowing up other trains and also bridges between the City of Mexico and Vera Cruz.' [1609]

The First World War involved the colonies of British East Africa and the Mombasa – Voi section of the Uganda Railway was a frequent target of German attacks. The line was damaged on 4th and 6th July at Bura and near Maktau and at 11 a.m. on Sunday 22nd August 1915 a team under Lt Koch bombed the line at Maktau again. This time an engine was blown up and five waggons of supplies were derailed by a delayed action contact mine.[1610] On Sunday 5th September 1915 a train was derailed at Mile 237, on Thursday 9th September a train was wrecked at Mile 161 and on Thursday 7th October the Germans blew up the line at mile 249. In this last attack the engine of a train was overturned and most of the remainder of it was derailed but no one was hurt.[1611] Two Indian contractors who had assisted the Germans in this attack were tried and executed in January 1916.[1612] Another train was derailed at an unstated location on Tuesday 19th October, this time there were two fatalities.

A train on the Southern Pacific Railroad of Mexico travelling from Campo Verde to Torres was derailed by Yaqui Indians near Torres in Sonora province, Mexico, on Friday 24th September 1915. According to a report on the outrage received by radio in San Diego, California from Hermosillo via Guaymas, the Indians, who were about 60 in

---

[1606] Interstate Commerce Commission report
[1607] *The Morning Echo*, Bakersfield, California, United States, 5th January 1915
[1608] *Waxahachie Daily Light*, Waxahachie, Texas, United States, 15th July 1915
[1609] *The Galveston Daily News*, Galveston, Texas, United States, 19th July 1915
[1610] Facebook page – Nairobi Railway Museum
[1611] *The Times*, London, England, c3rd November 1915
[1612] Ibid, 15th January 1916

number, were deserters from one of the Mexican factional armies. Following the derailment the passengers, who were mostly women and children, were placed in a car of hay which was then set alight. Only about 20 of the passengers seem to have escaped, the other 60 burning to death.[1613]

Brief reports received in the United States mentioned the derailment by sabotage of a passenger train on Tuesday 5th October 1915 near Maltrata, Mexico. There were said to have been 50 killed and a score or so injured.

'No foreigners were on the train. Maltrata is in the vicinity where the Zapatistas have been operating and it is believed that the train was wrecked by some of these bands.' [1614]

Late on the night of Monday 18th October 1915 a southbound passenger train on the St Louis Brownsville and Mexico Railway was derailed near Olmito, Texas, United States, seven miles north of Brownsville. The train had been derailed by a gang of 20 Mexicans who claimed to be followers of Luis de la Rosa, 'leader of the so-called Texas revolution'. They had removed some spikes from a rail and lying concealed in brush had jerked the rail from under the nose of the engine with a wire, derailing the engine, baggage, mail and express cars. The passengers were robbed and the bandits fled, it was believed back to Mexico. Engineer H. H. Kendall was killed, crushed beneath his engine, the bandits shot and killed Corporal McBee, of the Third U.S. Cavalry and after the robbers had decamped an American shot dead an unidentified young Mexican passenger who had supposedly aided the bandits. In addition the bandits shot and probably fatally injured Dr E. F. McCain, Brownsville deputy state health officer.

As soon as news of the attack was received in Olmito a sheriff's posse left in automobiles and two companies of infantry in a special train. A trestle half a mile from the wreck was set on fire by the attackers, hindering the progress of the troops.[1615]

About a week after the accident two men named Cheno Flores and Luciano Chapa were arrested and Flores admitted to having taken part in the attack. He said that an "army" of 500 Mexicans had been raiding Texas for several months, operating in companies of 40 to 60 each under the command of Luis de la Rosa and Acera Pesano. He stated that their main object was to secure independence for Mexicans in Texas, robbery and looting being of secondary importance.[1616]

1613 *The Daily Advocate*, Victoria, Texas, 27th September 1915

1614 *The Syracuse Herald*, Syracuse, New York, United States, 7th October 1915

1615 *The Evening Review*, East Liverpool, Ohio, United States, 19th October 1915

1616 *Altoona Mirror*, Altoona, Pennsylvania, United States, 26th October 1915

CHAPTER 19

MISCELLANEOUS ACCIDENTS

The vast majority of railway accidents fall into one of a small number of categories such as derailments, collisions, bridge failures and so on. A few, though, do not easily fall into any obvious category, hence this short chapter on miscellaneous accidents. Some of the accidents listed in this chapter were fairly minor ones which have been included because of their unusual features.

Trains are obviously fairly massive objects, but even so the power of the wind can still sometimes be enough to overturn them, occasionally with serious consequences. No. 13 up mixed train left Rampore Haut (also spelt Rampur-hat), India, on the East Indian Railway, at 2.58 p.m. on the afternoon of Sunday 29th June 1902. It encountered a storm near the down distant signal of that station and after travelling about a mile 13 of its 15 vehicle were blown over by a cyclone. Thirteen passengers lost their lives and another 15, 'including a European lady, who was slightly hurt', were injured. The line was blocked for seven hours. Despite the damage done to the train as it was rounding a curve there was no damage to the surrounding countryside.[1617] [1618]

Coincidentally there were two more instances passengers train being blown over by the wind in the following months. On Sunday 24th August 1902, this time on a narrow-gauge railway between Athens and Kipissia, Greece. Narrow-gauge railways, because of their lighter engines and rolling stock, were of course more liable to such accidents. Again the nine-carriage train was rounding a curve when it was struck by a violent gust of wind and fell down the side of an embankment. There were no deaths, but 35 people were injured, five of them seriously.[1619]

[1617] Administration Report on the Railways in India for the Calendar Year 1902
[1618] *The Times*, London, England, 3rd July 1902
[1619] Ibid, 25th August 1902

Only six days later, on Saturday 30th August 1902, a westbound Chicago and North Western Railway passenger train which had left Winona at 1.50 p.m. was struck by a tornado at Meriden, Minnesota, United States, at 5 p.m. Engineer Mitchell was apparently unaware of the tornado before it struck but many of the passengers on the train said afterwards that they had seen the funnel-shaped cloud approaching. The engine escaped being struck by a few feet but the coaches of the train were blown down a steep embankment. Passengers Petersen Delmer, aged five; Miss Ethel Richardson of New Ulm, and a woman who was unidentified at the time of the report were killed, and 34 others were injured, two it was supposed fatally. Among the injured was Senator T. V. Knatvold, of Albert Lea, who suffered two broken ribs. A following westbound freight train discovered the wreck and returned to nearby Owatonna for assistance.

Conductor W. H. Kenzie, in charge of the train, sustained hip and back injuries and a badly cut forehead but said to the relief party when they arrived, "Never mind me boys, I'll be alright soon, attend to those who are in need of your services first."

Describing the accident Brakeman J. J. Reke said, "It was the most terrible mix-up I have ever seen. The cries and groans of the injured and dying were simply sickening. We had no warning of the storm whatever, and although it was getting very dark, we could see no cloud that looked as fierce as the one that struck our train. I was closing the transoms of the car when the train fell over, and was up on the arms of the seats in the forward car. Conductor Kenzie was also up on the arms of the car at the time lighting the gas. Bob Miller, the baggageman, was in the car assisting us, and it was a lucky thing that he was, for his car was blown to splinters and he would have been killed.

"Our only warning was the crashing of glass, and before we had time to think the cars were on their sides in the ditch. The two rear coaches fell squarely on their sides, and although the train was going at about a forty mile gait, I do not think the cars moved about two feet after leaving the track. The baggage car fell corner ways and was smashed into kindling wood. The other two cars fell straight over on their sides. I made my way to the rear door of our car and with an axe broke open the front door of the rear car where all the women and children were imprisoned. As the train went over it broke from the engine which ran down the track for some distance before it could be stopped. 'Dad' Mitchell was the engineer, and as soon as possible he reversed the locomotive and backed up to where the cars were." [1620]

To complete the series of wind related accidents on the evening of Sunday 28th September 1902 during a storm a goods train was blown off the rails between Vasto and San Salvo stations in Italy. Six waggons were wrecked and others more or less damaged. One of the men in charge of the train was killed and another was injured.[1621]

At first glance an accident on the Missouri Pacific Railroad at Cole Junction (River Junction), Missouri, United States, on Sunday 21st June 1903, was no different from many others. A westbound freight train, No. 17, had left Jefferson City and was taking the River Route at the junction when it was struck in the middle by eastbound extra

---

[1620] *The Winona Republican and Herald*, Winona, Minnesota, United States. 1st September 1902
[1621] *The Times*, London, England, 30th September 1902

freight train No. 76, which had approached the junction on the main line and failed to stop. A number of cars were piled up in a heap, killing horses and cattle. What made the accident unusual, and also more serious, was the fact that part of the cargo of one of the trains was a consignment of muriatic (hydrochloric) acid, which escaped as a result of the collision. Four men died in the accident and their bodies were so disfigured by the acid that they were unrecognisable.

The men who lost their lives were Fireman Mike Duvall, Brakeman William Atkinson and two unidentified tramps. A fifth man, Michael Finnegan, was probably fatally injured. Whether he was a railroad employee or a tramp was not stated but as, like the deceased fireman and brakeman, he was from Jefferson City he was probably a railroad worker, perhaps another brakeman.[1622]

When the Russo-Japanese War broke out on 8th February 1904 the Trans-Siberian Railway became of vital importance to Russia for the transport of its troops and supplies to the front. The railway was, however, still incomplete, the difficult section around Lake Baikal, the Circum-Baikal Railway, still being in progress. Trains had been carried across the lake on a special ice-breaker ferry, the *Baikal*, but in February 1904 the ice was too thick even for the ice breaker and as a temporary expedient tracks were laid across the ice.[1623] Several hundred locomotives and several thousand cars were taken over this railway, being hauled by draft animals. On at least one occasion, on about 20th February 1904, a locomotive being thus drawn by horses proved too heavy for the ice and fell through it to the depths of the lake.[1624] No attempt was made to run loaded trains over the ice, despite a report in a Vienna newspaper, the *Narodni Listy*, claiming that a train carrying three regiments of soldiers had fallen through the ice.[1625] Instead the soldiers negotiated the 38 kilometres across the lake on sledges, of which 1,000 were in use.[1626]

The famous San Francisco earthquake of Wednesday 18th April 1906 resulted in the derailment of a North Shore Railroad train at Point Reyes, California, United States, without it would seem any casualties. The derailed train became the subject of a much published photograph.[1627]

The passengers on at least one other train had a lucky escape, as one of them related to a newspaper reporter, "'I was on Southern Pacific train twenty miles from San Francisco when the shock came," said Frederick Droege of the Licking Rolling Mills of Covington, Ky. "A number of passengers were tumbled out of their berths. The conductor stopped the train and looked to see what he had hit. Nobody could find out what it was. Someone suggested an earthquake, and we ridiculed the idea.

"'We went on a few miles and, looking out of the windows, saw the chimneys of the houses had fallen over. Further on two great fissures appeared on each side of the track where the earth had opened. A little further we saw the Santa Fe track, next to ours, had

---

[1622] *Warrenton Banner*, Warrenton, Missouri, United States, 26th June 1903
[1623] Wikipedia – Circum-Baikal Railway
[1624] *The Times*, London, England, 29th February 1904
[1625] *The Brisbane Courier*, Brisbane, Queensland, Australia, 22nd February 1904
[1626] *The Times*, London, England, 2nd March 1904
[1627] *Earthquake Storms: An Unauthorized Biography of the San Andreas Fault* by John Dvorak

sunk out of sight where there was a deep fill. Then we realized it was an earthquake all right.'" [1628]

Slip coaches, which could be detached from the rear of trains to serve intermediate stations without the main train needing to stop, were an innovation which was quite popular in Great Britain at one time, though they never caught on anywhere else in the world. As noted in the previous volume in this series they featured in accidents from time to time. One such mishap occurred on the Great Eastern Railway at Marks Tey, England, on Saturday 29th December 1906.

The train involved was the 5.30 p.m. down train from London Liverpool Street to Ipswich. The main portion of the train, consisting of the engine and 12 coaches, was brought to a stand in Marks Tey because the signals were at danger, and it was then run into by the four detached slip coaches. Thirty-four passengers, divided between the two portions of the train, were slightly hurt.

Stationmaster Arthur Green told the Board of Trade inquiry into the accident that it had become foggy to the degree that he had decided that it would be unsafe to have the coaches slipped and he had instructed the signalman to maintain his signals at danger to stop the train. He had, unfortunately, not made the decision to stop the train until after it had left Chelmsford, and too late, therefore, to advise its train crew. Signalman William Nunn in Marks Tey Junction box therefore kept his signals at danger until he was advised too late by Foreman Porter that the slip coaches had been detached, whereupon he had belatedly cleared them.

Driver Harry Pizzy of the 5.30 train said that he had seen the Marks Tey Junction down distant signal at danger but being afraid that the slip coaches might have been detached he had not stopped at the down home signal, even though it had been at danger, but had kept going past not only that signal but the down starter as well before finally stopping.

Guard George Bonner was in charge of the slip coaches. He said that he had been specially looking out for the Marks Tey Junction distant signal but due to the fog and steam from the engine could not see either of its two lights. The speed of the train was not noticeably reduced, though, after the engine had passed the distant signal and he had assumed, therefore, that it must have been clear and had slipped the coaches as per usual.

Lt-Col P. G. von Donop found Driver Pizzy partially responsible for the accident by his failure to apply the brakes at the distant and this misleading Guard Bonner into thinking that the signal was clear, not to mention the fact that he had overrun two stop signals, though the inspecting officer does not seem to have criticised him for this.

As for Guard Bonner, when he realised that he had missed seeing the distant signal he should either have applied the automatic brake, bringing the whole train to a halt, or else have allowed the slip portion to be taken forward to the next station. He was also criticised for breaking the rules in relying on the Westinghouse brake to bring the slip carriages to a stop at the correct place and for failing to see the tail lamp at the rear of the main portion of the train, and Lt-Col von Donop found that Guard Bonner must carry the main responsibility for the accident. The inspecting officer refrained from making any

---

[1628] *The New York Times*, New York, New York, United States, 23rd April 1906 comments regarding the inherent dangers of slip coach working, or perhaps he did not feel that it was dangerous.

A major catastrophe with unusual features befell two Great Northern Railroad trains at Wellington, Washington, United States, on Tuesday 1st March 1910. Exceptionally heavy snowfalls had completely blocked the main line of the company on the descent from the 2.6 mile long Cascade Tunnel through the Rocky Mountains. At 6 a.m. on Wednesday 23rd February the westbound Spokane Express stopped at the east portal of the tunnel, and the line being blocked it remained there until the evening of Friday 25th, when it passed through the tunnel and as far as Wellington station, a mile beyond the west portal. Also on Friday it was joined at Washington by the Transcontinental mail train, which carried no passengers but was reportedly carrying 16 or 18 mail clerks.

The two trains remained stranded at Wellington over the weekend and into the following week. There was enough food at Wellington for a week but it was nevertheless a long, tedious wait. There were 12 women and six children among the possibly 70 passengers on the train. Attempts were made to keep spirits up by playing cards and telling stories to pass the time. On Sunday or Monday five or nine men (the accounts of the accident are very confused as to details) decided that they had had enough and they set off on a long and very hazardous journey to Scenic Hot Springs, which they eventually all reached. Such a journey was not practicable for the women, children or less fit passengers of course and they remained at Wellington, most of them sleeping in the train coaches though some stayed at the railway bunkhouse or in nearby cottages. About 16 track labourers and presumably also the 16 or 18 mail clerks were also sleeping on the train.

Disaster struck at 4.20 a.m. on 1st March when a mile-long avalanche swept everything at Wellington station from the tracks and into a narrow gorge 150 feet deep, burying it all under 70 feet of snow, earth, rocks and uprooted trees. A local newspaper wrote, 'Survivors of the avalanche which swept two passenger trains into the canyon at Wellington on the Great Northern who reached Scenic Hot Springs give vivid descriptions of the horror. When the snow slipped from the mountainside and swept the passenger trains, four huge electric motors, three locomotives and rotary snow plow, the water tank, part of the coal shed and the motor shed over the brink they say it was like riding in a storm-tossed ship.

'There was an electric storm raging at the time of the avalanche. Lightning flashes were vivid and a tearing wind was howling down the canyon. Suddenly there was a dull roar, and the sleeping men and women felt the passenger coaches lifted and borne along. When the coaches reached the steep declivity they were rolled nearly 1,000 feet and buried under 40 feet of snow.

'Chas Andrews, an engineer whose home is as Leavenworth, escaped from the doomed coaches 15 minutes before the avalanche descended because, he declared, he had received an occult warning. Andrews was awakened from a sound sleep at 4 o'clock in the morning by a dream voice, which told him to arise. He lay in his bunk for a few minutes, unable to go back to sleep or to understand the strange feeling of impending disaster. Moved by a power which appeared to Andrews to come from a source other than the physical, he arose and dressed. He walked to the bunkhouse, where other men were sleeping, and sat down alone. Fifteen minutes after he had left the train there was the might sound of the swiftly moving snow and crash of timber. He ran to the door and watched the white death move majestically across the track. Moved by the horror of what he had seen, and with the strange dream voice fresh in his memory, Andrews gave thanks more reverently than he had ever done before.

'Ira Clary and H. Purcell, conductors in charge of rotary trains, were among those who were swept into the canyon with the wrecked coaches. Purcell is a veteran of the Spanish-American war and he and Clary are close friends. They slept in the third-class car on the fast mail, No. 27.

'"I thought the end of the world had come," said Clary today. "The car in which I was sleeping appeared to be picked up and tossed about like a feather. Then it began to roll over and I was thrown from the top to the bottom and from the bottom to the top five or six times. "It seemed to me that I had been thrown about for 20 minutes before there was a crash that sounded above the roar of the sliding snow. The mail car struck a big tree which stood in our path and popped it open like an egg shell. When I realized anything I felt a sensation of suffocation and I found that I was buried six feet under the snow. I dug my way out and in a short time I was above the wreck.

'"I heard Purcell calling and I recognized his voice. I asked him, 'Is that Purcell?' and he replied, 'It sure is.' By the flashes of lightning I saw his hand sticking out of the snow and it took me 20 minutes to get him out. I was barefooted in the snow and was dressed only in shirt and trousers. When I gave the last pull on Purcell his boots fell on top of him. Purcell put on the boots and I put on his socks.

'"It was raining dismally and the trees all about us were cracking and screeching. We could hear people calling for help and we did what we could. We got Charlie Smart out unhurt and did what we could to assist others who were pinned down by wreckage or pinned in the snow. A fireman whom we know as Curly was fast near one of the locomotives. The steam from the engine had melted a big cavity beneath the snow crust and when Purcell went to help Curly he fell into this steaming cauldron. We thought that Purcell would be boiled alive." [1629]

By the evening of the following day the bodies of eight passengers and nine railroad employees had been recovered and another 38 passengers and 29 railroad employees were still missing, giving a possible total number of dead of 84 according to this estimate. A woman passenger was however extricated alive after being trapped under the snow in a Pullman coach for 12 hours.[1630]

By the next day there were reported to be 86 passengers, postal employees and trainmen either confirmed dead or missing and in addition 20 or 30 labourers were missing, so that at a conservative estimate there were 100 dead. There had in fact only been nine postal clerks on the mail train, not 18 or 20 as earlier stated, only one of whom, Alfred Hensell, of Spokane, survived. All those killed had been from the north-west and of those injured only one, the Rev Bishop Winget, of Chicago, was from the east.

By 12th March 72 bodies had been recovered and there was talk of building a 17-mile tunnel, which would be the longest in the world, from Chiwaukum to Skykomish, as the

[1629] *The Wenatchee Daily World*, Wenatchee, Washington, United States, 4th March 1910
[1630] *The Sun*, Baltimore, Maryland, United States, 3rd March 1910

only certain means of avoiding a repetition of the disaster.[1631] It was also stated that several of the passengers had asked Superintendent O'Neill of the railroad company to have the train moved back into the tunnel, where it would be entirely safe from avalanches, but he had replied that the train was perfectly safe at Wellington station.[1632]

Wellington station was renamed Tye, after the adjoining river, in October 1910 and a new, 7.8-mile-long tunnel between Berne and Scenic Hot Springs opened early in 1929.[1633]

The passenger train which left Trieste for Parenzo at 3 p.m. on Thursday 31st March 1910 was another victim of the wind, being blown into a ditch near Muggia, Austria (now in Italy). Three people lost their lives, three were seriously injured and a dozen were slightly hurt.[1634]

The Mann's Creek Railroad was a narrow gauge (3 ft) line which ran from the Chesapeake and Ohio Railroad main line at Sewell, West Virginia, United States, through rugged terrain to Clifftop and Landisburg. Owned by the Babcock Coal and Coke Company it mainly transported coal and lumber, but also carried passengers.[1635] On the evening of Tuesday 20th May 1913 it had a serious accident. There had been heavy rain that day. As the afternoon train from Sewell to Clifftop, consisting of an engine, three freight cars and a caboose containing a handful of passengers was proceeding slowly along the line at about 5 p.m. and had travelled about four miles from Sewell its engineer found the line blocked by a landslide. Seeing that it would take some time to clear the track the engineer decided to return to Sewell but had barely begun to do so when a second landslide blocked the line behind the train. All but one of the passengers then decided that they would have to make a night of it stuck on the train. The exception was a young man or boy named J. Muckles, of Charleston, West Virginia, who decided to walk back to Sewell.

A local newspaper described what happened next, 'Suddenly a thundrous roar broke out from above and an avalanche of earth descended upon the train, the ill-fated passenger coach being right in its path. It rolled the coach over and buried it under twenty feet of earth. The young survivor reached Sewell with the news about an hour later. A large force of men were rushed to the scene and put to work digging out the buried coach, which had been carried about six hundred feet down the mountainside. The rescue party could at first hear no sounds of life from the buried passengers, but when they reached the coach, three were found to be still breathing. There injuries were not so serious but that all will recover.'

The previously quoted newspaper stated that four passengers had lost their lives. They were; James Walker of Hinton, a travelling salesman for the Woolston-Mohler Grocery Company; Robert Young, of Landisburg, a log scaler; Alvin Gwinn, of Kentz Store,

---

[1631] *The Wenatchee Daily World*, Wenatchee, Washington, United States, 12th March 1910
[1632] *Estherville Democrat*, Estherville, Iowa, United States, 9th March 1910
[1633] Wikipedia – Cascade Tunnel
[1634] *Feuille d'Avis de Neuchatel*, Neucharel, Switzerland, 1st April 1910
[1635] Wikipedia – Mann's Creek Railroad

another log scaler, and three-year-old Robert Sellers.[1636] Other newspapers spoke of two other fatalities; a Mrs Amick of Landisburg and a young girl.[1637] Two other passengers, Mrs William Sellers and A. L. Knuckols, travelling salesman for the Charleston Hardware Company, were reported as being injured, as was Brakeman Charles Walters. The fate of the conductor, engineer and fireman of the train was not stated.

Rather surprisingly, in view of the multiple fatalities, the Interstate Commerce Commission did not bother to investigate the accident.

A passenger train on the narrow gauge line between Batavia and Buitenzorg, Java, Dutch East Indies, was derailed by an earthquake shortly after it had left Massing station on Thursday 5th June 1913. Only the engine and leading luggage van left the rails and no one was hurt.[1638]

At 2.06 p.m. on Wednesday 10th March 1915 the mail train from Madrid to Vigo met with an accident between Frieira and Figueira stations, Spain. The accident happened at kilometre 47, three kilometres from Frieira. As the train was passing through a cutting a large mass of rock and soil fell onto it, The mass spared the engine but crushed its tender and the following coaches, two third-class coaches being demolished. By the time of the second report 19 bodies had been recovered and there were 25 injured, of whom 11 were admitted to hospital and four were said to be in a serious condition. The dead included company director Don Felix Angoleti and Alfedro López Torregrosa, a tenor. Some of the injured went on to Vigo, others remained at Frieira. The mail van was not damaged.[1639]

The pioneer flight of the Wright brothers only happened in 1903 and the First World War was the first conflict in which aircraft were used to attack an enemy. The very first instance of an aircraft being used to attack a train apparently occurred on Saturday 27th March 1915, close to the front line between the Allied and German positions, presumably in France. An Allied aircraft dropped bombs on a German armoured train.[1640]

It was probably another three months before an Allied aircraft was again used to attack a train and this time the location was Przeworsk, Galicia. A Russian aircraft 'of the Ilya Mourometz type' bombed five long trains in the station there on Sunday 27th June 1915. A bomb struck one train and clouds of smoke were seen to rise from it. 'It would appear from German reports that the train carried artillery ammunition, and that, besides the killing of a number of men, 30,000 rounds of artillery ammunition were destroyed.'[1641]

[1636] *The Raleigh Register*, Beckley, West Virginia, United States, 22nd May 1913
[1637] *New Albany Tribune*, New Albany, Indiana, United States, 22nd May 1913
[1638] *Eisenbahnunfälle in Sachsen* by Claus Schlegel, pg. 97 (in German)
[1639] *La Vanguardia*, Barcelona, Spain, 12th March 1915
[1640] *The Times*, London, England, 31st March 1915
[1641] Ibid, 6th July 1915

As the months went on there were repeated air attacks on trains, mostly on the Western Front but also on the Easter Front and of Austrian trains by Italian aircraft. All the reported attacks seem to have been by Allied aircraft on Axis trains, but this is probably because the information has been gleaned from the British press and Axis attacks on Allied trains were not reported.

INDEX

TABLE OF MAJOR RAILWAY ACCIDENTS 1901-1915

Date	Location	Railway Company	Description	Passenger Fatalities
7 Feb 1901	Near Greenville, Pennsylvania, USA	Erie	Derailment, broken rail?	5
15 Feb 1901	Near Sydenham, New South Wales, Australia		Derailment excessive speed with unstable engine, slight track defects?	8
21 Feb 1901	Ruslin's Siding, New Jersey, USA	Pennsylvania	Head-on collision, single line, fire	13
17 Apr 1901	Saratoff, Russia		Runaway waggons, collision	13
28 Apr 1901	Near Lofa, China		Derailment, obstruction on line? Culvert collapsed under train?	19
7 Jun 1901	Daspoort, Transvaal		Rear collision	9
15 Jun 1901	Chailland, France		Runaway train, derailment	7
26 Jun 1901	Cass, Indiana, USA	Wabash	Derailment, culvert washed away	16
4 Jul 1901	'North of Nabroomspruit,' Transvaal		Derailment, sabotage	14
10 Jul 1901	Between Marshall and Norton, Missouri, USA	Chicago and Alton	Head-on collision, single line, fire	19
8 Sep 1901	Jamestown, North Dakota, USA	Northern Pacific	Runaway train? Collision	7
18 Sep 1901	Near Avon, Massachusetts, USA	New York New Haven and Hartford	Train diverted onto wrong line, collision	6

20 Sep 1901	Palota, Romania		Runaway train, collision, fire	32
27 Nov 1901	Near Seneca, Michigan, USA	Wabash	Head-on collision, single line, fire. Boiler explosion	?
14 Dec 1901	Between Perryville and Irene, Illinois, USA	Illinois Central	Head-on collision, single line, explosion, fire	6
20 Dec 1901	Keimberg, Germany		Collision with horse, rear collision in fog	12
21 Dec 1901	Melegnano, Italy		Collision	6
8 Jan 1902	Park Avenue Tunnel, New York, USA	New York Central and Hudson River	Rear collision in tunnel, fire	17
7 Mar 1902	Maxon, Texas, USA	Galveston Harrisburg & San Antonio	Derailment, broken rail, fire	10
30 Mar 1902	Between Barberton and Kaapmuiden, Transvaal		Runaway train, derailment, boiler explosion	47
12 Apr 1902	Near Machavie, Transvaal		Head-on collision, single line	16
29 Apr 1902	Cama, Missouri, USA	Atchison Topeka and Santa Fe	Derailment, track defects? Defective points? Excessive speed?	5
5 May 1902	Daspoort Cutting, Transvaal		Derailment, excessive speed?	15
6 May 1902	Near Moyenville, France		Derailment, track subsidence	8
29 Jun 1902	Near Rampore Haut, India	East Indian	Train blown over by the wind	13
3 Jul 1902	Near Gatchina, Russia		Rear collision?	10
4 Jul 1902	Gloversville, New York, USA	Mountain Lake	Runaway train, collision	12
7 Jul 1902	Near Guarda, Portugal		Derailment	6

31 Jul 1902	Khatauli, India	North Western	Derailment, engine defect, excessive speed	28
5 Aug 1902	Cholchol River Bridge, Chile		Trestle collapsed under train	40
24 Aug 1902	Bloomfontein, Orange River Colony		Collision with light engine	5
1 Sep 1902	Near Berry, Alabama, USA	Southern	Derailment, track defects?	30
12 Sep 1902	Near Mangapatnan, India	Madras	Bridge washed away; train fell into river	75
25 Sep 1902	Malden, Illinois, USA	Chicago Burlington and Quincy	Rear collision	5
27 Sep 1902	Arleux, France	Chemins de fer du Nord	Train diverted onto wrong line, derailment	24
4 Oct 1902	Bloemfontein, Orange River Colony		Rear collision	6
20 Dec 1902	Byron, California, USA	Southern Pacific	Rear collision	27
26 Dec 1902	Wanstead, Ontario, Canada	Grand Trunk	Head-on collision, single line	26
20 Jan 1903	Happy Hollow, Washington, USA	Great Northern	Rear collision	12
27 Jan 1903	Graceland, New Jersey, USA	Central of New Jersey	Double collision, fire	22
28 Jan 1903	Near Esmond, Arizona, USA	Southern Pacific	Head-on collision, single line, fire, runaway vehicle, collision	31
23 Feb 1903	Near Berea, Ohio, USA	Cleveland Cincinnati Chicago and St Louis	Head-on collision	6
20 Apr 1903	Red House, New York, USA	Erie	Sidelong collision, fire	7

7 May 1903	Near Dexter, Ontario, Canada	Canadian Pacific	Derailment, broken wheel, fire	12
4 Jun 1903	Near Stilwell, Kansas, USA	Missouri Pacific	Head-on collision, single line	8
13 Jun 1903	Sukheke, India	North Western State	Sidelong collision	11
19 Jun 1903	Near Raymond, Iowa, USA	Illinois Central	Head-on collision, single line	6
27 Jun 1903	Najerilla River Bridge, Spain		Derailment, train fell from bridge	43
7 Jul 1903	Rockfish, Virginia, USA	Southern	Train diverted onto wrong line, head-on collision	17
15 Jul 1903	Waterloo, England	Lancashire and Yorkshire	Derailment, engine defect	6
24 Jul 1903	Buchholz-Königstaße, Germany		Derailment, points moved under train	5
27 Jul 1903	Glasgow St Enoch, Scotland	Glasgow and South Western	Buffer stop collision	16
1 Aug 1903	Between Sucha and Kalvaria, Austria		Runaway waggons, collision	30
7 Aug 1903	Durand, Michigan, USA	Grand Trunk	Rear collision	22
10 Aug 1903	Between Couronnes and Ménilmontant, France	Chemin de fer Métropolitan de Paris	Fire in tunnel	84
27 Aug 1903	Between Pasian and Codroipo, Italy		Head-on collision, single line, fire	16
3 Sep 1903	Fishing Creek Trestle, Near Yorkville, South Carolina, USA	Southern	Derailment, broken wheel, trestle collapsed under train	5
27 Sep 1903	Danville, Virginia, USA	Southern	Derailment, excessive speed? Train fell from trestle	7
17 Oct 1903	Washington Crossing, New Jersey, USA	Pennsylvania	Rear collision in fog	17

19 Oct 1903	Between Cloudcroft and Russia, New Mexico, USA	Alamogordo & Sacramento Mountain	Runaway train, derailment	8
1 Oct 1903	Indianapolis, Indiana, USA	Cleveland Cincinnati Chicago and St Louis	Sidelong collision	16
14 Nov 1903	Kentwood, Louisiana, USA	Illinois Central	Rear collision, fire	32
21 Nov 1903	Near Palézieux, Switzerland		Collision with derailed engine	6
21 Dec 1903	Godfrey, Kansas, USA	St Louis-San Francisco	Derailment on points, fire	6
23 Dec 1903	Laurel Run, Pennsylvania, USA	Baltimore and Ohio	Derailment, obstruction on line (faulty loading), fire	63
26 Dec 1903	Near East Paris, Michigan, USA	Pere Marquette	Head-on collision, single line	18
27 Dec 1903	Between Sini and Gomharria, India	Bengal-Nagpur	Head-on collision, single line	12
6 Jan 1904	Willard, Kansas, USA	Chicago Rock Island and Pacific	Head-on collision, single line	18
2 Feb 1904	Between Chinandega and El Vieja, Nicaragua		Derailment, train fell from bridge?	20
9 Feb 1904	Near Sand Point, Ontario, Canada	Canadian Pacific	Head-on collision, single line	8
c8 Apr 1904	Aley, Lebanon	Chemin de fer de Damas Hamah et prolongements	Runaway train, derailment	8
18 Apr 1904	Near Zacatecas, Mexico	Ferrocarril Central Mexicano	Derailment, broken rail	8
30 Apr 1904	Wickes Siding, near Kimmswick, Missouri, USA	St Louis Iron Mountain and Southern	Derailment on points	9
7 Jun 1904	Between Gjevgjelia and Gemendje, Macedonia		Bomb exploded on train	?

22 Jun 1904	Between Luco de Jiloca and Calamocha, Spain		Bridge washed away, train fell into river, fire	14
June 1904	Near Harbin, China		Derailment, sabotage	33
3 Jul 1904	Litchfield, Illinois, USA	Wabash	Train diverted onto wrong line, collision, fire, sabotage?	23
10 Jul 1904	Midvale, New Jersey, USA	New York and Greenwood Lake	Rear collision	16
13 Jul 1904	Glenwood, Illinois, USA	Chicago and Eastern Illinois	Collision	18
7 Aug 1904	Hogan's Gulch, near Eden, Colorado, USA	Denver and Rio Grande	Bridge collapsed under train	106
31 Aug 1904	Near Richmond, Quebec, Canada	Grand Trunk	Head-on collision, single line	10
1 Sep 1904	Sintaluta, North West Territories, Canada	Canadian Pacific	Train diverted onto wrong line, head-on collision	5
4 Sep 1904	Machhpara,India	Eastern Bengal	Head-on collision	5
20 Sep 1904	Near Ferrara, Italy		Runaway engine, collision, fire	5
24 Sep 1904	Near Hodges, Tennessee, USA	Southern	Head-on collision, single line, boiler explosion, fire	62
10 Oct 1904	Near Warrensburg, Missouri, USA	Missouri Pacific	Head-on collision, single line	29
24 Oct 1904	Sadorus, Illinois, USA	Wabash	Collision between portions of divided train	5
11 Nov 1904	Azusa, Wyoming, USA	Oregon Short Line	Head-on collision, single line, fire	6
23 Dec 1904	La Chapelle, France	Chemins de fer du Nord	Rear collision in fog	13

19 Jan 1905	Cudworth, England	Midland	Double collision in fog, fire	5
c20 Feb 1905	Near Baku, Russia		Derailment, sabotage	?
3 Mar 1905	Clifton, Pennsylvania, USA	Pittsburgh Fort Wayne and Chicago	Rear collision, fire	5
6 Apr 1905	Near No. 9 Siding, Russia		Head-on collision, single line	8
15 April 1905	Ramo, Italy		Collision in tunnel	10
11 May 1905	Lochiel, Pennsylvania, USA	Pennsylvania	Derailment, traction shock, collision, fire, explosions, boiler explosion	20
7 Jun 1905	Red Cedar Bridge, near Colfax, Wisconsin, USA	Wisconsin Central	Bridge collapsed under train	5
17 Jun 1905	Near Patapsco, Maryland, USA	Western Maryland	Head-on collision, single line	18
21 Jun 1905	Mentor, Ohio, USA	Lake Shore and Michigan Southern	Derailment on points, boiler explosion, fire	15
2 Jul 1905	Between Irapuato and Silao, Mexico	Ferrocarril Central Mexicano	Bridge collapsed under train	7
27 Jul 1905	Hall Road, England	Lancashire and Yorkshire	Train diverted onto wrong line, collision, fire	21
7 Aug 1905	Between Spremberg and Schleife, Germany		Head-on collision, single line	15
13 Aug 1905	Kishman, Ohio, USA	New York Chicago and St Louis	Head-on collision, single line	11
17 Aug 1905	Bruce, Virginia, USA	Atlantic Coast Line	Train fell from open drawbridge	14
24 Aug 1905	Youreta, Siberia, Russia		Derailment?	22

1 Sep 1905	Witham, England	Great Eastern	Derailment, track defects, fire	10
9 Sep 1905	Near Stabley's Switch, Pennsylvania, USA	York and Dallastown	Collision	5
11 Sep 1905	53rd Street, New York, USA	Interborough Rapid Transit	Derailment, excessive speed, part of train fell from elevated line into street, fire	13
25 Sep 1905	Paoli, Pennsylvania, USA	Pennsylvania	Train diverted onto wrong line, collision	6
6 Oct 1905	Near Rostoff-on-the-Don, Russia		Derailment	27
30 Oct 1905	Near Sheffield, Missouri, USA	Atchison Topeka and Santa Fe	Derailment, track defects	10
4 Nov 1905	Kulais, Russia		Collision with obstruction on line? Sabotage	9
26 Nov 1905	Baker Bridge, Massachusetts, USA	Boston and Maine	Rear collision in fog, fire	15
6 Dec 1905	Between Ashay and Wilkins, Wyoming, USA	Union Pacific	Head-on collision, single line, fire	8
27 Dec 1905	Between Narva and Reval, Russia		Derailment, sabotage	?
29 Dec 1905	Syzran, Russia		Derailment, sabotage	?
31 Dec 1905	Lebedin, Russia		Collision	30
31 Dec 1905	Between Znamenska and Trepovka, Russia		Collision	20
16 Mar 1906	Near Abode, Colorado, USA	Denver and Rio Grande	Head-on collision, single line, in blizzard, fire	36
25 Mar 1906	Near Natrona, Wyoming, USA	Chicago and North Western	Bridge collapsed under train, fire	10
4 May 1906	Near Clover Creek Junction, Pennsylvania, USA	Pennsylvania	Head-on collision, single line	9

28 May 1906	Louisville, Kentucky, USA	Louisville and Nashville	Derailment, broken flange	9
16 Jun 1906	Pogranitchnaia, Manchuria	Chinese Eastern	Derailment	6
17 Jun 1906	Near Patty, Tennessee, USA	Louisville and Nashville	Collision with hand car	6
1 Jul 1906	Salisbury, England	London and South Western	Derailment, excessive speed	24
22 Jul 1906	Between Hamlet and Rockingham, North Carolina, USA	Seaboard Air Line	Head-on collision, single line	19
23 Jul 1906	Near Camden, Washington, USA	Great Northern	Derailment, track distorted by the heat? Track defects?	7
12 Sep 1906	Azilda, Ontario, Canada	Canadian Pacific	Runaway train, head-on collision	12
18 Sep 1906	Cimmarron Bridge, near Dover, Oklahoma, USA	Chicago Rock Island and Pacific	Bridge collapsed; train fell into river	20
19 Sep 1906	Grantham, England	Great Northern	Derailment, excessive speed, fire	12
4 Oct 1907	Lansingburg, New York, USA	Boston and Maine	Rear collision	7
6 Oct 1906	Near Piacenza, Italy		Derailment, cause unknown	5
7 Oct 1906	Near Marianovka, Russia		Collision	7
14 Oct 1906	Epernon, France	Chemins de fer de L'Ouest	Sidelong collision	12
27 Oct 1906	Near Hatton, Ceylon		Derailment, track subsidence	13
28 Oct 1906	Near Atlantic City, New Jersey, USA	West Jersey and Seashore	Derailment, cause unknown, train fell from drawbridge	56
12 Nov 1906	Near Woodville, Indiana, USA	Baltimore and Ohio	Head-on collision, single line, in blizzard, fire	70

29 Nov 1906	Near Lawyers, Virginia, USA	Southern	Rear collision, fire	7
26 Dec 1906	Mocha, Ecuador		Runaway coach, derailment	7
28 Dec 1906	Elliot Junction, Scotland	Dundee and Arbroath Joint	Rear collision in blizzard	21
30 Dec 1906	Terra Cotta, DC, USA	Baltimore and Ohio	Rear collision in fog	43
2 Jan 1907	Between Volland and Altavista, Kansas, USA	Chicago and Rock Island	Head-on collision, single line, fire	31
19 Jan 1907	Sandford, Indiana, USA	Cleveland Cincinnati Chicago and St Louis	Explosion, fire	25
19 Jan 1907	Near Fowler, Indiana, USA	Cleveland Cincinnati Chicago and St Louis	Head-on collision, single line in fog, fire	6
16 Feb 1907	205th Street, New York, USA	New York Central and Hudson River	Derailment, excessive speed, track defects?	23
12 Mar 1907	Alkmaar, Transvaal	Central South African	Derailment, culvert washed away	10
17 Mar 1907	Turnshiche, Manchuria		Collision	17
28 Mar 1907	Colton, California, USA	Southern Pacific	Derailment on points	23
10 Apr 1907	Near Pardee, Ontario, Canada	Canadian Pacific	Derailment, broken rail, fire	15
15 Apr 1907	Near Bartlett, North Dakota, USA	Great Northern	Derailment, sabotage? High speed over poor track?	5
20 Apr 1907	Between Dowbromino and Sowkino, Russia		Derailment	5

11 May 1907	Honda, California, USA	Southern Pacific	Derailment on points, engine defect? fire	33
21 May 1907	Dayton, Tennessee, USA		Runaway waggons, collision	6
25 May 1907	Near Sibola, Portugal		Derailment	7
28 Jun 1907	Butler, Manitoba, Canada	Canadian Pacific	Head-on collision, single line?	13
14 Jul 1907	Near Johnson City, Tennessee, USA	Southern	Collision, runaway engine, derailment	8
20 Jul 1907	Near Salem, Michigan, USA	Pere Marquette	Head-on collision, single line, fire	30
30 Jul 1907	Sitka, Tennessee, USA	Illinois Central	Boiler explosion, fire	5
4 Aug 1907	Near Les Ponts de Cé, France	Chemins de fer de L'État	Derailment, track distorted by the heat, bridge collapsed under train	25
7 Aug 1907	Between Tremessen and Talsee, Germany		Derailment, inadequate protection of PW work	8
24 Aug 1907	Near Coutras, France	Chemins de fer de Paris à Orléans	Collision with shunting goods	10
30 Aug 1907	Between Mattoon and Charleston, Illinois, USA	Central Illinois Traction	Head-on collision, single line	18
2 Sep 1907	Near Kanawha Falls, West Virginia, USA	Chesapeake and Ohio	Derailment, track defects?	5
3 Sep 1907	Near Caledon, Ontario, Canada	Canadian Pacific	Derailment, excessive speed	7
6 Sep 1907	Norris, Iowa, USA	Chicago Rock Island and Pacific	Derailment, excessive speed?	14
15 Sep 1907	Canaan, New Hampshire, USA	Boston and Maine	Head-on collision, single line in fog.	26

19 Sep 1907	Encarnación de Diaz, Mexico	Ferrocarril Central Mexicano	Head-on collision, single line, driver intoxicated	67
28 Sep 1907	Bellaire, Ohio, USA	Baltimore and Ohio	Train diverted onto wrong line, head-on collision	5
1 Oct 1907	Near Seoul, Korea		Derailment	?
10 Oct 1907	Near Budapest, Hungary		Collision, fire	7
15 Oct 1907	Shrewsbury, England	Great Western and London and North Western Joint	Derailment, excessive speed	11
22 Oct 1907	Near Rabshtyn, Russia	Vistula	Collision	18
24 Oct 1907	Kot Lakhpat, India	North Western	Collision	11
29 Oct 1907	Jagadhri and Kalanour, India	North Western	Head-on collision, single line?	9
31 Oct 1907	Oberzissen, Germany	Kempenich Light	Derailment, excessive speed	5
25 Nov 1907	Between Hospitalet and Cambrils, Spain		Bridge collapsed under train	15
17 Dec 1907	Between Block Hut at Mile 230-6¼ and Dadur Block Hut	East Indian	Head-on collision, single line?	8
25 Dec 1907	Between Ludhiana and Ladhowal. India	North Western	Head-on collision, single line	15
20 Jan 1908	Acqua Bella, Italy		Double collision in fog	9
15 Feb 1908	West Toledo, Ohio, USA	Michigan Central/ Toledo and Western	Side collision on rail/rail level crossing	6
6 Mar 1908	Between Tschergotur and Kurandy, Russia		Derailment	12
20 Apr 1908	Sunshine, Victoria, Australia	Victoria State	Rear collision, fire	45
22 Apr 1908	Whalen, Pennsylvania, USA	Laquin Lumber	Runaway train, derailment	9

24 Apr 1908	Gargantua Siding, near Maltrata, Mexico		Collision	28
28 Apr 1908	Denton, Michigan, USA	Detroit United	Head-on collision, single line	8
6 May 1908	Between Ghaziabad and Dajna, India	Oudh and Rohilkund	Head-on collision, single line, fire	75
16 May 1908	Dustin, Ontario, Canada	Grand Trunk	Derailment	9
21 May 1908	Contich, Belgium	Belgian State	Train diverted onto wrong line, collision	40
23 May 1908	Near Penza, Russia		Derailment	12
5 Jun 1908	Camp Parole, Maryland, USA	Washington Baltimore and Annapolis	Head-on collision, single line	8
26 Jun 1908	Itola, India	Bombay Baroda and Central India	Rear collision?	24
1 Jul 1908	Near Knob Noster, Missouri, USA	Missouri Pacific	Head-on collision, single line, in fog, fire	5
4 Jul 1908	Oakland, California, USA	Southern Pacific	Side collision on rail/rail level crossing	7
9 Aug 1908	Groß Tarup, Germany		Head-on collision, single line	9
30 Aug 1908	Near Nikolskoya, Manchuria	Chinese Eastern	Derailment, sabotage	?
25 Sep 1908	Young's Point, Montana, USA	Northern Pacific	Head-on collision, single line	20
26 Feb 1908	Gleisdreieck, Berlin, Germany		Sidelong collision, train fell from elevated line	18
1 Oct 1908	Sugar Ridge, Ohio, USA	Toledo and Ohio Central	Rear collision	6
15 Oct 1908	Kenieczny Crossing, Near Metz, Michigan, USA	Detroit and Mackinaw	Derailment, culvert destroyed by forest fire, fire	14

8 Nov 1908	Grisolles, France		Derailment, broken rail	10
11 Nov 1908	Little Woods, Louisiana, USA	New Orleans and North Eastern	Rear collision, fire	8
23 Nov 1908	Near Nunkow Yu, China		Derailment? Part of train fell from bridge, fire	9
3 Dec 1908	Between Mustafabad and Barara, India	North Western State	Head-on collision, single line	15
15 Dec 1908	Pouch Tunnel, France	Chemin de fer de Paris à Orléans	Runaway waggons, collision in tunnel, fire	22
4 Jan 1909	Nizza di Sicilia, Sicily, Italy		Rear collision	5
15 Jan 1909	Near Dotsero, Colorado, USA	Denver and Rio Grande	Head-on collision, single line, fire	26
22 Feb 1909	Delmar, Delaware, USA	Pennsylvania	Rear collision? Sidelong collision? Fire	6
24 Feb 1909	Riobamba, Ecuador		Derailment, track defects	23
24 Mar 1909	Between Mustafabad and Barara, India	Burma	Derailment, excessive speed, fire	7
23 May 1909	Near Dehra, India		Runaway waggons, collision	8
19 Jun 1909	Dune Park, Indiana, USA	Chicago Lake Shore and South Bend	Head-on collision, single line	8
19 June 1909	Near Ennore, India	Madras and Mahratta	Derailment track distorted by the heat	15
4 Jul 1909	Near Beudet, Haiti		Collision	10
24 Jul 1909	Hull's Point, Near Orrick, Missouri, USA	Wabash	Derailment, track damaged by floodwater	6

31 Jul 1909	La Crosse. Idaho, USA	Spokane and Inland Empire	Head-on collision, single line	15
14 Aug 1909	Near Husted, Colorado, USA	Denver and Rio Grande	Head-on collision, single line	11
28 Aug 1909	Jamalpur, India		Derailment, sabotage	9
28 Sep 1909	12th Street, Chicago, Illinois, USA	Pittsburgh Cincinnati Chicago and St Louis	Rear collision	6
8 Oct 1909	Between Kilmer and North Topeka, Kansas, USA	Atchison Topeka and Santa Fe	Head-on collision, single line	15
10 Nov 1909	Near Lakeview, British Columbia, Canada	British Columbia Electric	Runaway waggon, collision	13
15 Dec 1909	Reedy Fork, North Carolina, USA	Southern	Derailment, broken rail	11
21 Jan 1910	Spanish River, Ontario, Canada	Canadian Pacific	Derailment, broken wheel? Broken rail? Train fell from bridge, fire	43
29 Jan 1910	Stoat's Nest, England	London Brighton and South Coast	Derailment, defective vehicle	5
1 Mar 1910	Wellington, Washington, USA	Great Northern	Train struck by avalanche	53
21 Mar 1910	Between Green Mountain and Gladbrook, Iowa, USA	Chicago Rock Island and Pacific	Derailment, excessive speed with heavy train over poor track, faulty train makeup	49
30 Mar 1910	Mühlheim am Rhein, Germany		Rear collision	22
18 Jun 1910	Villepreux-les-Clayes, France	Chemins de fer de l'État	Rear collision, fire	23

23 Jun 1910	Near Zapotilie, Mexico	Ferrocarril Nacional de México	Runaway coaches, derailment	39
4 Jul 1910	West Middletown, Ohio, USA	Cincinnati Hamilton and Dayton	Head-on collision, single line	22
18 Jul 1910	Richmond, Victoria, Australia	Victorian	Rear collision in fog	9
8 Aug 1910	Near Ignacio, California, USA	Northwestern Pacific	Head-on collision, single line	12
14 Aug 1910	Saujon, France	Chemins de fer de l'État	Collision with shunting goods	37
24 Aug 1910	Near Duffield, Michigan, USA	Grand Trunk	Rear collision, fire	9
10 Sep 1910	Bernay, France	Chemins de fer de l'État	Derailment, high speed with unstable engine over poor track	10
20 Sep 1910	Rottenmann, Austria		Head-on collision	7
21 Sep 1910	Near Kingsland, Indiana, USA	Fort Wayne and Wabash	Head-on collision, single line	41
23 Sep 1910	Between Clayton and Dellvale, Kansas, USA	Chicago Rock Island and Pacific	Bridge washed away; train fell into stream	13
24 Sep 1910	Near Tipton, Indiana, USA	Indiana Union Traction	Head-on collision, single line	5
4 Oct 1910	Near Staunton, Illinois, USA	Illinois Traction	Head-on collision, single line	37
5 Dec 1910	Willesden, England	London and North Western	Rear collision	5
24 Dec 1910	Between Hawes Junction and Ais Gill, England	Midland	Rear collision, fire	12

4 Jan 1911	Near Cheney, Washington, USA	Northern Pacific	Rear collision	5
5 Jan 1911	Gaikas Loop, Cape Colony, South Africa		Derailment, cause unknown	18
13 Jan 1911	Batavia, New York, USA	New York Central and Hudson River	Rear collision in fog, fire	6
23 Jan 1911	Coke Ovens, Pontypridd, Wales	Taff Vale	Rear collision	11
14 Feb 1911	Courville, France	Chemins de fer de l'État	Double collision, fire	13
21 Feb 1911	Near Vossnessenka, Russia		Fire	7
24 Feb 1911	Rancagua Bridge, Chile		Derailment, train fell from bridge	18
22 Apr 1911	Blaauwkrantz Bridge, Cape Colony, South Africa	Kowie	Derailment, track defects, train fell from bridge	32
29 Apr 1911	Martin's Creek, New Jersey, USA	Pennsylvania	Derailment, inadequate protection of PW work, fire	9
4 May 1911	Between Lakaminia and Lakho, India	Bengal and North Western	Derailment, defective vehicle	8
29 May 1911	Near McCook, Nebraska, USA	Chicago Burlington and Quincy	Head-on collision, single line	14
10 Jul 1911	Near Dike, Oregon, USA	Oregon Trunk	Derailment, excessive speed, track defects, track distorted by the heat?	7
11 Jul 1911	Near Bridgeport, Connecticut, USA	New York New Haven and Hartford	Derailment, excessive speed through crossover, fire	12

17 Jul 1911	Müllheim, Baden, Germany		Derailment, excessive speed, driver intoxicated	14
27 Jul 1911	Hamlet, North Carolina, USA	Seaboard Air Line	Head-on collision, single line	10
28 Jul 1911	Grindstone, Maine, USA	Bangor and Aroostock	Head-on collision, single line, fire	5
25 Aug 1911	Manchester, New York, USA	Lehigh Valley	Derailment, broken rail, part of train fell from bridge	28
7 Oct 1911	Near Constantinople, Turkey		Derailment, sabotage	?
8 Oct 1910	Near Lingo, Missouri, USA	Chicago Burlington and Quincy	Derailment, defective vehicle	6
15 Oct 1911	Near Fort Crook, Nebraska, USA	Missouri Pacific	Head-on collision, single line	7
9 Oct 1911	Near Piraly, Brazil		Derailment, collision	6
23 Nov 1911	Near Montrevil-Bellay, France	Chemins de fer de l'État	Bridge collapsed under train	7
18 Dec 1911	Near Odessa, Minnesota, USA	Chicago Milwaukee and St Paul	Rear collision, fire	8
6 Jan 1912	Pont des Coquettiers, France	Chemins de fer de L'Est	Rear collision in fog, fire	10
30 Jan 1912	Between Siaokan and Honan, China		Derailment, sabotage, rear collision	?
3 Feb 1912	Near Shanhai-Kwam, China		Bridge dynamited under train	9
15 Feb 1912	Near Warrior Ridge, Pennsylvania, USA	Pennsylvania	Derailment, engine defect	6

15 Feb 1912	Animas River Bridge, Mexico		Bridge destroyed by sabotage; train fell into river	37
24 Mar 1912	Near Escalón, Mexico		Runaway engine, collision, explosion, sabotage	60
6 May 1912	Eastabuchie, Mississippi, USA	New Orleans and Northeastern	Derailment, track defects, trestle collapsed under train	6
18 May 1912	Near Paris Nord, France	Chemins de fer du Nord	Derailment, defective points, collision	12
16 Jun 1912	Malmslätt, Sweden		Head-on collision, fire	22
1 Jul 1912	Near Consuelo, Mexico		Derailment, sabotage?	?
4 Jul 1912	East Corning, New York, USA	Delaware Lackawanna and Western	Rear collision in fog	39
5 Jul 1912	Ligonier, Pennsylvania, USA	Ligonier Valley	Head-on collision, single line	22
14 Jul 1912	Western Springs, Illinois, USA	Chicago Burlington and Quincy	Rear collision in fog	11
20 Jul 1912	Near Parres, Mexico		Derailment, fire, sabotage and arson	?
31 Jul 1912	Near Lauromuller, Brazil	Estrada de Ferro Central do Brasil	Rear collision? Head-on collision? Fire	10
6 Aug 1912	Near Lozanne, France	Chemins de fer de Paris à Lyon et à la Méditeranée	Rear collision in tunnel, fire	7
17 Sep 1912	Ditton Junction, England	London and North Western	Derailment, excessive speed through crossover, fire	13
18 Sep 1912	Near Torreón, Mexico	Ferrocarril Central Mexicano	Collision with mules, derailment	11

3 Oct 1912	Westport, Connecticut, USA	New York New Haven and Hartford	Derailment, excessive speed through crossover, fire	5
4 Oct 1912	Alicante, Spain		Runaway train, buffer stop collision	8
22 Oct 1912	Seidler, Macedonia		Collision with sheep, derailment, sabotage?	350
30 Oct 1912	Jannowitzbrücke, Germany		Rear collision	5
11 Nov 1912	Near Montz, Louisiana, USA	Yazoo and Mississippi Valley	Rear collision, fire	15
13 Nov 1912	Irvington, Indiana, USA	Cincinnati Hamilton and Dayton	Train diverted onto wrong line, head-on collision, fire	11
3 Dec 1912	Near Dresden, Ohio, USA	Pennsylvania	Rear collision	11
15 Dec 1912	Guardia Mangano, Italy		Train diverted onto wrong line, head-on collision	13
10 Mar 1913	Near Rudna Poosejowski, Russia		Fire	5
7 Apr 1913	Near Ogulin, Croatia		Head-on collision, single line?	28
13 Apr 1913	Near East Greenfield, Quebec, Canada	Central Vermont	Derailment, track defects?	6
30 Apr 1913	Borgaon, India	Great Indian Peninsula	Rear collision	33
12 May 1913	Buk, Bulgaria		Runaway waggons, collision	150
20 May 1913	Between Sewell and Clifftop, West Virginia, USA	Mann's Creek	Train struck by landslide	6

23 May 1913	Between Poroi and Andjista, Bulgaria		Head-on collision, single line?	?
9 Jun 1913	Near Guaymas, Mexico		Derailment, sabotage	200
12 Jun 1913	Stamford, Connecticut, USA	New York New Haven and Hartford	Rear collision	6
c19 Jun 1913	Near Jenikeui, Bulgaria		Collision	55
19 Jun 1913	Near Vallejo, California, USA	San Francisco Napa and Calistoga	Head-on collision, single line	13
22 Jun 1913	Between Delmar Junction and Riggs, Iowa, USA	Chicago Milwaukee and St Paul	Collision	7
25 Jun 1913	McKellar, Ontario, Canada	Canadian Pacific	Derailment, excessive speed over track under repair?	8
28 Jun 1913	Near Panchura, India	East Indian	Derailment, track washed away	23
30 Jun 1913	Hooper's Creek, South Carolina, USA	Lancaster and Chester	Derailment, part of train fell from trestle	5
12 Jul 1913	Vineyard, California, USA	Pacific Electric	Rear collision	14
13 Jul 1913	Near Irún, Spain		Head-on collision, single line? in tunnel	6
13 Jul 1913	Near Cambridge, Ohio, USA	Baltimore and Ohio/New Midland Power and Traction	Side collision on rail/rail level crossing	5
26 Jul 1913	Near Bramminge, Denmark		Derailment, track distorted by the heat? Excessive speed?	15
c28 Aug 1913	Hokkaido, Japan		Derailment, track washed away	7

2 Sep 1913	Near Ais Gill, England	Midland	Rear collision, fire	16
2 Sep 1913	Near North Haven, Connecticut, USA	New York New Haven and Hartford	Rear collision	21
8 Sep 1913	Usküb, Serbia	Chemins de fer Orientaux	Collision	8
21 Sep 1913	Gomez Farias, Mexico		Derailment, sabotage	35
29 Sep 1913	Near Sosyka, Russia	Vladikavkas	Derailment, sabotage	40
6 Oct 1913	Dvinsk, Russia		Collision with light engine	14
15 Oct 1913	St James Tunnel, Liverpool, England	Cheshire Lines Committee	Rear collision in tunnel	6
19 Oct 1913	Buckatunna, Mississippi, USA	Mobile and Ohio	Derailment, excessive speed? Bridge collapsed under train	17
25 Oct 1913	Near Garyville, Louisiana, USA		Derailment	6
27 Oct 1913	Near San Salvador, Mexico	Ferrocarril Nacional Mexicano	Derailment, sabotage	140
29 Oct 1913	Warsaw, Prussian Poland		Train diverted onto wrong line? In fog, collision	7
3 Nov 1913	Near São Paulo, Brazil	Companhia Mogiana de Estradas de Ferro	Collision	50
4 Nov 1913	Mélun, France	Chemins de fer de Paris à Lyon et à la Méditeranée	Sidelong collision, fire	41
5 Nov 1913	Between Nijni-Novgorod and Pensa, Russia		Derailment, sabotage?	14
13 Nov 1913	Near Clayton, Alabama, USA	Central of Georgia	Derailment, broken rail	9

29 Nov 1913	Ceccano, Italy		Train diverted onto wrong line, collision	5
5 Dec 1913	Costesti, Romania		Collision	100
14 Dec 1913	Haras Tunnel, Germany		Collision with debris which had fallen from tunnel roof	10
15 Dec 1913	Near Darno, Austria		Collision	30
16 Dec 1913	Near Biadoling, Galicia		Explosion, fire	8
23 Dec 1913	Near Charcos, Mexico		Derailment, sabotage	?
24 Dec 1913	Oranjekanal, The Netherlands		Derailment, track defects?	5
2 Jan 1914	Woippy, Near Metz, Germany		Buffer stop collision	7
4 Feb 1914	Cumbre Tunnel, Mexico	Ferrocarril de Nor-Oeste de México	Collision with burning waggons in tunnel, sabotage, debris from tunnel roof fell onto train	50
12 Feb 1914	Las Canoas, Mexico	Ferrocarril Nacional de México	Derailment, sabotage, fire	?
21 Feb 1914	Near Lima, Mexico	Ferrocarril Interoceánico de México	Derailment, sabotage	?
c24 Feb 1914	Near Lake Magadi, Kenya		Derailment?	18
13 Mar 1914	Exeter, New South Wales, Australia		Head-on collision	14
2 Apr 1914	Near Tanjong-Priok, Dutch East Indies		Collision with buffalo, derailment, bridge collapsed under train	20

18 Jun 1914	Near Carr Bridge, Scotland	Highland	Bridge collapsed under train	5
17 Jul 1914	Fairmount Park, Virginia, USA	Virginian/Virginia Railway and Power	Side collision on rail/rail level crossing	7
20 Jul 1914	Empalot, France	Chemins de fer du Midi	Sidelong collision	6
5 Aug 1914	Near Tipton Ford, Missouri, USA	Kansas City Southern	Head-on collision, single line, fire	38
9 Aug 1914	Between Independentza and Piscu, Romania		Derailment, excessive speed, track defects?	6
23 Aug 1914	Saint-Laurent, France	Chemins de fer de L'Est	Collision	79
10 Sep 1914	Hex River Pass, South Africa		Derailment, track washed away?	9
15 Sep 1914	Near Lebanon, Missouri, USA	St Louis-San Francisco	Derailment, track washed away	26
16 Sep 1914	Lizy-sur-Ourcq, France		Train fell into river from damaged bridge	40
18 Sep 1914	Livingston, Alabama, USA	Alabama Great Southern	Derailment, sabotage	9
25 Sep 1914	Bilzen, Belgium		Derailment, sabotage	?
c18 Oct 1914	Between Marquise and Wacquinghen, France	Chemins de fer du Nord	Rear collision?	31
3 Nov 1914	Miechów, Russian Poland		Rear collision, fire, explosion	60
20 Nov 1914	Sambhu, India	North Western State	Rear collision	8
22 Nov 1914	Schönhauser Damm, Germany		Train diverted onto wrong line, rear collision	5
27 Nov 1914	Near Tula, Mexico		Collision	32

c28 Nov 1914	Near Liége-Guillemins, Belgium		Collision	?
5 Dec 1914	Riardo, Italy		Rear collision	7
9 Dec 1914	Between Gemmenich, Belgium and Aachen, Germany		Derailment	52
c20 Dec 1914	Kalish, Russian Poland?		Collision	400
1 Jan 1915	Ilford, England	Great Eastern	Sidelong collision	10
3 Jan 1915	Near Antwerp, Belgium		Sidelong collision	8
c5 Jan 1915	Near Batilly, Lorraine, Germany		Sidelong collision	19
11 Jan 1915	Between Nieborow and Lowitsch, Russian Poland		Head-on collision, single line	19
18 Jan 1915	Between Colima and Guadalajara, Mexico		Runaway train, derailment	600
3 Feb 1915	Konitz, Germany		Collision	6
10 Mar 1915	Between Frieira and Figueira, Spain		Train struck by rockfall	19
23 Mar 1915	Dirschau, Germany		Collision	6
20 Apr 1915	Between Mónyo and Kölnök, Hungary		Derailment, track subsidence	5
7 May 1915	Los Angeles, California, USA	Atchison Topeka and Santa Fe /Pacific Electric	Side collision on rail/rail level crossing	5
22 May 1915	Quintinshill, Scotland	Caledonian	Double collision, fire	223
21 Jun 1915	Montemorelos, Mexico		Derailment, excessive speed?	300
7 Jul 1915	Queenston, Ontario, Canada	International	Runaway train, derailment	15

14 Jul 1915	Apizaco, Mexico	Ferrocarril Nacional de México	Derailment, sabotage	25
12 Aug 1915	Orient, Ohio, USA	Baltimore and Ohio Southwestern	Rear collision	7
14 Aug 1915	Near Weedon, England	London and North Western	Derailment, engine defect	10
17 Aug 1915	Dietikon, Switzerland		Sidelong collision	5
22 Aug 1915	Between Eschenau and Fürth, Germany		Derailment, broken rail	5
20 Sep 1915	Plüschow, Germany		Collision	5
24 Sep 1915	Near Torres, Mexico	Southern Pacific of Mexico	Derailment, sabotage, fire, arson	60
26 Sep 1915	Between Vidrány and Mezölaborcz, Hungary		Runaway train, collision, fire	30
5 Oct 1915	Maltrata, Mexico		Derailment, sabotage	50
7 Oct 1915	Bischdorf, Germany		Rear collision	14
16 Oct 1915	Randolph, Kansas, USA	Union Pacific	Derailment, track washed away	11
19 Oct 1915	St-Cyr-de-Favières, France	Chemins de fer de Paris â Lyon et â la Méditeranée	Runaway coaches, buffer stop collision	17
22 Nov 1915	Near Muscogee Junction, Georgia, USA	Central of Georgia	Head-on collision, single line	7
17 Dec 1915	St Bedes Junction, England	North Eastern	Double collision, fire	14
23 Dec 1915	Rockledge, Tennessee, USA	Nashville Chattanooga and St Louis	Head-on collision, single line	10
28 Dec 1915	Bentschen, Near Posen, Germany		Derailment, excessive speed	19
28 Dec 1915	San Lazaro di Savena, Italy		Rear collision in fog	20

ABOUT THE AUTHOR

The author has been interested in railways for many years, ever since he was invited up to a signal box at the age of 14 back in 1966. Not long afterwards he 'cabbed' the engine of a goods train that was shunted out of the way of a faster train and its driver showed him a copy of *Red for Danger*, the late L. T. C. Rolt's book about British railway accidents. This was the first railway book that he ever bought, several months' saved up pocket money going towards a paperback copy. Later he came across a copy of *A History of Railroad Accidents, Safety Precautions and Operating Practices* by Robert B. Shaw, dealing with American railroad accidents, and his interest in the subject grew.

When the local signal boxes were all done away with in a modernisation scheme in the mid-1980s he began a more detailed study of world railway accidents. There was no internet back then, of course, and many hours were spent in the local reference library slogging through bound volumes of *The Times*, writing out longhand in large notebooks details of every accident found. Later it was possible to view the paper on microfilm and print pages, which made things a little easier, and with the internet the amount of information available vastly increased.

Some years ago, the author contributed to *Railway Disasters of the World* by P. W. B. Semmens (which deals only with 20th-century accidents) and more recently the idea came of writing his own book or books on world accidents, starting at the beginning. This book is the third in the series, with previous volumes covering the years 1811-1880 and 1881-1900. Further volumes will hopefully follow.

Printed in Dunstable, United Kingdom

91222497R00406